T0205270

Lecture Notes in Computer Science **12563**

More information about this subseries at http://www.springer.com/series/7411

Vladimir M. Vishnevskiy ·
Konstantin E. Samouylov ·
Dmitry V. Kozyrev (Eds.)

Distributed Computer and Communication Networks

23rd International Conference, DCCN 2020
Moscow, Russia, September 14–18, 2020
Revised Selected Papers

Springer

Editors
Vladimir M. Vishnevskiy
Institute of Control Sciences of the Russian Academy of Sciences
Moscow, Russia

Konstantin E. Samouylov
RUDN University
Moscow, Russia

Dmitry V. Kozyrev
Institute of Control Sciences of the Russian Academy of Sciences
Moscow, Russia

RUDN University
Moscow, Russia

ISSN 0302-9743 ISSN 1611-3349 (electronic)
Lecture Notes in Computer Science
ISBN 978-3-030-66470-1 ISBN 978-3-030-66471-8 (eBook)
https://doi.org/10.1007/978-3-030-66471-8

LNCS Sublibrary: SL5 – Computer Communication Networks and Telecommunications

This Springer imprint is published by the registered company Springer Nature Switzerland AG
The registered company address is: Gewerbestrasse 11, 6330 Cham, Switzerland

Preface

This volume contains a collection of revised selected full-text papers presented at the 23rd International Conference on Distributed Computer and Communication Networks (DCCN 2020), held in Moscow, Russia, September 14–18, 2020.

The conference is a continuation of traditional international conferences of the DCCN series, which took place in Sofia, Bulgaria (1995, 2005, 2006, 2008, 2009, 2014); Tel Aviv, Israel (1996, 1997, 1999, 2001); and Moscow, Russia (1998, 2000, 2003, 2007, 2010, 2011, 2013, 2015, 2016, 2017, 2018, 2019) in the last 23 years. The main idea of the conference is to provide a platform and forum for researchers and developers from academia and industry from various countries working in the area of theory and applications of distributed computer and communication networks, mathematical modeling, methods of control and optimization of distributed systems, by offering them a unique opportunity to share their views, as well as discuss the perspective developments, and pursue collaboration in this area. The content of this volume is related to the following subjects:

1. Communication networks, algorithms, and protocols
2. Wireless and mobile networks
3. Computer and telecommunication networks control and management
4. Performance analysis, QoS/QoE evaluation, and network efficiency
5. Analytical modeling and simulation of communication systems
6. Evolution of wireless networks toward 5G
7. Internet of Things and Fog Computing
8. Machine learning, big data, and artificial intelligence
9. Probabilistic and statistical models in information systems
10. Queuing theory and reliability theory applications
11. Quantum Information and quantum communication
12. High-altitude telecommunications platforms

The DCCN 2020 conference gathered 167 submissions from authors from 25 different countries. From these, 140 high-quality papers in English were accepted and presented during the conference. The current volume contains 55 extended papers which were recommended by session chairs and selected by the Program Committee for the Springer post-proceedings.

All the papers selected for the post-proceedings volume are given in the form presented by the authors. These papers are of interest to everyone working in the field of computer and communication networks.

We thank all the authors for their interest in DCCN, the members of the Program Committee for their contributions, and the reviewers for their peer-reviewing efforts.

September 2020

Vladimir M. Vishnevskiy
Konstantin E. Samouylov
Dmitry V. Kozyrev

Organization

DCCN 2020 was jointly organized by the Russian Academy of Sciences (RAS), the V.A. Trapeznikov Institute of Control Sciences of RAS (ICS RAS), the Peoples' Friendship University of Russia (RUDN University), the National Research Tomsk State University, and the Institute of Information and Communication Technologies of Bulgarian Academy of Sciences (IICT BAS).

International Program Committee

V. M. Vishnevskiy (Chair)	ICS RAS, Russia
K. E. Samouylov (Co-chair)	RUDN University, Russia
S. M. Abramov	Program Systems Institute of RAS, Russia
S. D. Andreev	Tampere University of Technology, Finland
A. M. Andronov	Riga Technical University, Latvia
N. Balakrishnan	McMaster University, Canada
A. S. Bugaev	Moscow Institute of Physics and Technology, Russia
S. R. Chakravarthy	Kettering University, USA
T. Czachorski	Institute of Computer Science of Polish Academy of Sciences, Poland
D. Deng	National Changhua University of Education, Taiwan, China
A. N. Dudin	Belarusian State University, Belarus
A. V. Dvorkovich	Moscow Institute of Physics and Technology, Russia
Yu. V. Gaidamaka	RUDN University, Russia
P. Gaj	Silesian University of Technology, Poland
D. Gracc	York University, UK
Yu. V. Gulyaev	Kotelnikov Institute of Radio-engineering and Electronics of RAS, Russia
J. Hosek	Brno University of Technology, Czech Republic
V. C. Joshua	CMS College, India
H. Karatza	Aristotle University of Thessaloniki, Greece
N. Kolev	University of São Paulo, Brazil
J. Kolodziej	NASK, Poland
G. Kotsis	Johannes Kepler University Linz, Austria
A. E. Koucheryavy	Bonch-Bruevich Saint-Petersburg State University of Telecommunications, Russia
Ye. A. Koucheryavy	Tampere University of Technology, Finland
T. Kozlova Madsen	Aalborg University, Denmark
U. Krieger	University of Bamberg, Germany
A. Krishnamoorthy	Cochin University of Science and Technology, India
N. A. Kuznetsov	Moscow Institute of Physics and Technology, Russia

L. Lakatos	Budapest University, Hungary
E. Levner	Holon Institute of Technology, Israel
S. D. Margenov	Institute of Information and Communication Technologies of Bulgarian Academy of Sciences, Bulgaria
N. Markovich	ICS RAS, Russia
A. Melikov	Institute of Cybernetics of the Azerbaijan National Academy of Sciences, Azerbaijan
G. K. Miscoi	Academy of Sciences of Moldova, Moldova
E. V. Morozov	Institute of Applied Mathematical Research of the Karelian Research Centre RAS, Russia
V. A. Naumov	Service Innovation Research Institute (PIKE), Finland
A. A. Nazarov	Tomsk State University, Russia
I. V. Nikiforov	Université de Technologie de Troyes, France
P. Nikitin	University of Washington, USA
S. A. Nikitov	Institute of Radio-engineering and Electronics of RAS, Russia
D. A. Novikov	ICS RAS, Russia
M. Pagano	University of Pisa, Italy
E. Petersons	Riga Technical University, Latvia
V. V. Rykov	Gubkin Russian State University of Oil and Gas, Russia
L. A. Sevastianov	RUDN University, Russia
M. A. Sneps-Sneppe	Ventspils University College, Latvia
P. Stanchev	Kettering University, USA
S. N. Stepanov	Moscow Technical University of Communication and Informatics, Russia
S. P. Suschenko	Tomsk State University, Russia
J. Sztrik	University of Debrecen, Hungary
H. Tijms	Vrije Universiteit Amsterdam, The Netherlands
S. N. Vasiliev	ICS RAS, Russia
M. Xie	City University of Hong Kong, Hong Kong, China
A. Zaslavsky	Deakin University, Australia
Yu. P. Zaychenko	Kyiv Polytechnic Institute, Ukraine

Organizing Committee

V. M. Vishnevskiy (Chair)	ICS RAS, Russia
K. E. Samouylov (Vice Chair)	RUDN University, Russia
D. V. Kozyrev	ICS RAS and RUDN University, Russia
A. A. Larionov	ICS RAS, Russia
S. N. Kupriyakhina	ICS RAS, Russia
S. P. Moiseeva	Tomsk State University, Russia
T. Atanasova	IICT BAS, Bulgaria
I. A. Kochetkova	RUDN University, Russia

Organizers and Partners

Organizers

Russian Academy of Sciences, Russia
RUDN University, Russia
V.A. Trapeznikov Institute of Control Sciences of RAS, Russia
National Research Tomsk State University, Russia
Institute of Information and Communication Technologies of Bulgarian Academy of Sciences, Bulgaria
Research and Development Company "Information and Networking Technologies", Russia

Support

Information support is provided by the Russian Academy of Sciences. The conference has been organized with the support of the "RUDN University Program 5-100."

Contents

Analytical Modeling of Distributed Systems

Distributed Systems Applications

Computer and Communication Networks

Power Domain NOMA Without SIC in Downlink CSS-Based LoRa Networks

Angesom Ataklity Tesfay[1(✉)], Eric Pierre Simon[1], Ido Nevat[2], and Laurent Clavier[1,3]

[1] University of Lille, CNRS, UMR 8520 - IEMN, 59000 Lille, France
{angesom.tesfay,eric.simon,laurent.clavier}@univ-lille.fr
[2] TUMCREATE, University Town 138602, Singapore
ido.nevat@tum-create.edu.sg
[3] IMT Lille Douai, Douai, France

Abstract. Low Power Wide Area Network, such as LoRa is one of the main building blocks of the Internet of Things. One of the main issues is to scale up the number of devices and one strong limitation comes from the downlink communication. In fact, the access point is constrained by the duty cycle, therefore it cannot address a large number of devices. We propose a superposition scheme to transmit multiple packets to multiple devices in the same frequency, time slot, and spreading factor. This scheme is applied to the specific physical layer proposed by LoRa, based on the chirp spread spectrum. Our proposal includes the power allocation scheme and the decoding technique that are very specific to this physical layer and show a significant performance improvement, increasing the number of devices that can be connected at least by ten compared to the classical LoRa-like system.

Keywords: IoT · LoRa · Power allocation · Scalability · CSS modulation

1 Introduction

Nowadays there is a rapid growth of the Internet of Things (IoT) network and more than 75 billion devices is expected to be connected to the network by 2025 [1]. Most of the IoT network requirements are related to operating in low power, low data rates, and wide-area connectivity [2,3]. Low Power Wide Area Networks (LPWAN) technologies, such as LoRa, provide a solution to these requirements. However, the huge challenge is to face the scale change in the number of communicating devices.

In this chapter we focus on the LoRa downlink. So far, this link is used to send few acknowledgments, not even necessarily for each packet. This link is

Supported by IRCICA USR CNRS 3380, COST ACTION CA15104 - IRACON and the French National Agency for Research (ANR) under grant ANR-16-CE25-0001 - ARBURST.

V. M. Vishnevskiy et al. (Eds.): DCCN 2020, LNCS 12563, pp. 3–13, 2020.
https://doi.org/10.1007/978-3-030-66471-8_1

very important to transmit not only feedback but also to transmit data to the devices and this will certainly require some update in the software as well as capability of the access point to transmit data to the devices. However, several limitations should be solved first: the complexity at the receiver has to be limited and it has to respect the duty-cycle, which significantly limits the scalability of the network.

Authors in [4] have shown that the duty-cycle limits not only the scalability but also the reliability of the network. In [5,6] the downlink feedback frames are shown to highly impact the network performance. However, no solution to remedy this problem is proposed.

To overcome this downlink limitation, we propose to simultaneously transmit multiple frames to multiple end-devices on a single channel (same frequency, same time slot and same spreading factor). As a consequence, we significantly enhance the scalability of the LoRa network. Our idea benefits from the chirp spread spectrum modulation and implements a joint multi-user detection. Multi-user access is, on a single channel, provided by the power domain NOMA (Non Orthogonal Multiple Access) scheme. Our contributions are

1. to propose a superposition transmission scheme for a chirp spread modulation in the downlink,
2. to design a complete multi-user receiving scheme, and
3. to propose a Power Allocation (PA) scheme to minimize the error probabilities at the receivers.

This paper is organized as follows: the description of LoRa technology, especially the coding at of the information, is provided in Sect. 2; in Sect. 3 the proposed scheme is presented and the decoding strategy is detailed. Section 4 analyzes the simulated results and Sect. 5 concludes the paper.

2 LoRa Information Encoding

2.1 LoRa Chirp Spread Spectrum Modulation

LoRa defines a physical Layer based on Chirp Spread Spectrum (CSS) modulation. This modulation is defined by its spreading factor (SF), ranging from 7 to 12. For a given bandwidth (B), it provides a trade-off between rate and communication range [7]. One symbol consists in a linear frequency change over the symbol duration T_s, where $T_s = 2^{\mathrm{SF}}/B$.

Let us consider the symbol transmitted by the ith user during the qth frame starting at $qT_s - T_s/2$, where $q \in \{0, Q-1\}$, with Q the number of symbols transmitted in a packet. The information carried by this symbol is represented by $m_q^{(i)} \in \{0, 2^{\mathrm{SF}} - 1\}$.

Information is carried by the position of a shift in the up-chirp. Indeed, the corresponding modulated chirp is obtained by left-shifting the raw chirp by a duration $\delta_q^{(i)} = m_q^{(i)}/B$ as illustrated in Fig. 1.

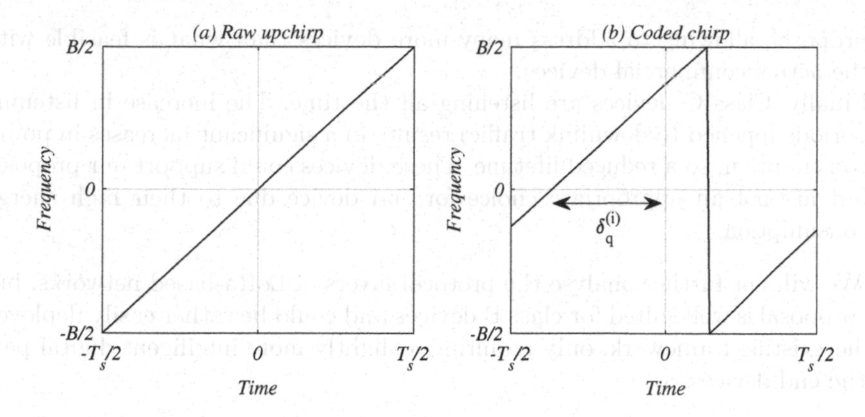

Fig. 1. Coding of the information from the (a) Raw up-chirp (b) Coded chirp associated with $m_q^{(i)}$.

The expression of the resulting coded chirp associated with the qth symbol of user i is then given by:

$$x_q^{(i)}(t) = \begin{cases} \exp\left(2j\pi\left(\frac{B}{2T_s}t^2 + \frac{m_q^{(i)}}{T_s}t\right)\right), & t \in \left[-\frac{T_s}{2}, \frac{T_s}{2} - \delta_q^{(i)}\right], \\ \exp\left(2j\pi\left(\frac{B}{2T_s}t^2 + \left(\frac{m_q^{(i)}}{T_s} - B\right)t\right)\right), & t \in \left[\frac{T_s}{2} - \delta_q^{(i)}, \frac{T_s}{2}\right]. \end{cases} \quad (1)$$

2.2 LoRaWAN - MAC Layer

If LoRa defines the PHY layer previously explained, LoRaWAN is an open standard which defines an adapted MAC protocol [8]. LoRa operates in the license-free industrial, scientific and medical (ISM) radio band. In ISM bands, when no listen-before-talk is used, duty-cycle is generally the main restriction of the networks for instance, 1% in EU 868 - 868.6 MHz. This duty-cycle significantly impacts the downlink and the gateway is extremely affected. When multiple uplink frames are received, the gateway cannot send downlink frames to all transmitters, which limits the capacity of downlink transmission and, as a consequence, impacts the scalability of LoRaWAN networks.

In the standard, three types of nodes (Class A, B and C) with different specifications are defined [9].

- Class A devices are the lowest energy consuming nodes and the cheapest ones. Two short downlink frames are scheduled after a transmission to send an acknowledgement, the end device listening in this second frame only when nothing is received in the first one. These devices are not well suited for our scheme because they do not share a common downlink frame.
- For class B, a beacon is sent by the gateway to synchronize the devices and open a common frame for downlink traffic. This frame could be used for our

proposal, allowing to address many more devices than what is feasible with the actual commercial devices.

- Finally, Class C devices are listening all the time. The increase in listening periods (opened to downlink traffic) results in a significant increases in power consumption, so a reduced lifetime. These devices could support our proposal but are not an appropriate choice for end device due to their high energy consumption.

We will not further analyse the protocol layers of LoRa-based networks, but our proposal is well suited for class B devices and could be rather easily deployed in the existing framework, only requiring a slightly more intelligent digital part on the end devices.

2.3 Link Budget

The link budget of a LoRa communication is given by $174 + 10\log_{10}(B) + \mathrm{NF} = -114$ dBm, where NF is the noise figure. At room temperature, it is equal to 6 dB. is $174 + 10\log_{10}(B) + \mathrm{NF} = -114$ dBm, The term 174 is the thermal noise in a bandwidth of 1 Hz and only depends on the receiver temperature.

3 Proposed Multi-user Scheme

In this paper, we propose to simultaneously transmit N packets in order to reduce the impact of the duty cycle. Those packets are transmitted in the same frame, using the same spreading factor and the same frequency band. This is well suited for Class B devices using the synchronized frame.

The basic idea of our proposed scheme is to simply add N packets from N different users at the gateway and to transmit the superposed signals. Different powers are allocated to each packet in order to differentiate them at the receiver, which is a power domain non orthogonal multiple access strategy. It is to be noted that a common preamble can be used at the beginning of the packet so that synchronization and channel estimation will be easier at the receiver side. This preamble needs to include information about the number of users and their ordering in terms of allocated power.

3.1 System Model

Let us consider a cell of radius R. The gateway is located at the center. Devices are distributed according to a uniform distribution within the cell (homogeneous Poisson point process). The distance between the gateway and a given device i is $d^{(i)}$. We consider a narrow band block fading propagation channel, i.e. ir is modeled as a single coefficient, constant during the whole packet transmission. It results in a channel gain given by $h^{(i)} = {d^{(i)}}^{-\eta/2} \cdot \chi^{(i)}$, where ${d^{(i)}}^{-\eta/2}$ is the path loss and χ_i the Rayleigh multi-path fading. The parameter η is the path loss exponent and will be chosen as 3.5 in the rest of our paper.

Our objective is to decode symbol q of a desired user that will be denoted by j. The M samples at times $t = nT$, $T = 1/B$ and $n \in [-\frac{M}{2}, \frac{M}{2} - 1]$ and $M = 2^{\mathrm{SF}}$, are given by:

$$r_q^{(j)}[n] = h^{(j)} \sum_{i=1}^{\mathrm{N}} \sqrt{p^{(i)}}\, x_q^{(i)}[n] + w_q^{(j)}[n]. \tag{2}$$

In (2), $p^{(i)}$ indicates the power allocated to user i and $w[n] \sim \mathcal{N}_{\mathbb{C}}(0, \sigma_n^2)$ is the complex Gaussian thermal noise.

In order to apply the NOMA scheme, a different power is allocated to each user. As we will describe in the following section, the information is carried after processing the received signal by the position of a peak in the Fourier domain whose amplitude is related to the allocated power. To keep a good signal to noise ratio at each receiver, more power is allocated to user with a weaker channel gains. However, if two users transmit the same information, a collision occurs, meaning that the two expected peaks fall at the same position. In order to be able to resolve such collisions, we propose a PA scheme where the addition of colliding peaks cannot result in the same amplitude as another peak (or addition of peaks).

To do so, we order users from the weakest channel gain ($i = 1$) to the strongest one ($i = \mathrm{N}$) using the channels estimated from a previous uplink. The power allocated to user i is then:

$$p^{(i)} = \frac{2^{i-1}}{\sum_{i=1}^{\mathrm{N}} 2^{i-1}} p_t, \tag{3}$$

where p_t indicates the total power transmitted by the access point.

3.2 Receiver

In a first step we use an adapted filter to the synchronization sequence in order to detect the preamble. The maximum output value of this filter gives the arrival time of the packet and allows to estimate the channel between the access point and the considered device j. Knowing $h^{(j)}$ allows us to compute the expected power of user j.

The decoding process can then be decomposed in the following steps: the received signal $r_q^{(j)}[n]$ is first multiplied by the complex conjugate of the up-chirp, $x^*[n]$ (called down-chirp, this operation will be called de-chirping in the sequel). Only considering the qth symbol, the signal obtained after de-chirping can be written as:

$$\begin{aligned} y^{(j)}[n] &= r_q^{(j)}[n] x^*[n], \\ &= h^{(j)} \sum_{i=1}^{N_u} \sqrt{p^{(i)}} e^{j2\pi \frac{m_q^{(i)}}{M} n} \mathbb{1}_{[-\frac{M}{2}, \frac{M}{2}-1]}(n) + w[n]. \end{aligned} \tag{4}$$

The second step is to apply a Fast Fourier transform (FFT) to $y^{(j)}[n]$ and to multiply by the conjugate of the channel estimate in order to correct the phase shift. The resulting signal is:

$$
\begin{aligned}
Y^{(j)}[k] &= \mathrm{Re}\left\{ \sum_{n=-\frac{M}{2}}^{\frac{M}{2}-1} \left(h^{(j)*} y^{(j)}[n]\right) e^{-2\imath\pi\frac{nk}{M}} \right\} \\
&= |h^{(j)}|^2 \sum_{i=1}^{\mathrm{N}} \sqrt{p^{(i)}}\, \delta[k - m_q^{(i)}] + W^{(j)}[k],
\end{aligned}
\tag{5}
$$

where $W^{(j)}[k] \sim \mathcal{N}_{\mathbb{C}}(0, |h^{(j)}|^2\sigma_n^2/2)$ is the FFT of the noise, $\delta[.]$ is Kronecker delta function, $\delta[n] = 1$ for $n = 0$ and $\delta[n] = 0$ for $n \neq 0$.

The classic idea of detection is to search for the peak that has the expected amplitude of the user of interest. However, when collision occurs this approach is not efficient. In this context, collision means two or more users transmits the same information, i.e., the same symbol at the same time, so that their peaks add. Therefore, we are rather interested in a multi-user detection scheme and for this reason first we derive the expression of the maximum likelihood detection in the frequency domain. We can write the received vector after FFT as $\mathbf{Y}^{(j)} = |h^{(j)}|^2\mathbf{X}^{(j)} + \mathbf{W}^{(j)}$, which counts M components given by (5) with $X^{(j)}[k] = \sum_{i=1}^{N} \sqrt{p^{(i)}}\delta[k - m_q^{(i)}]$. We observe that given $h^{(j)}$ and the transmitted symbols (so $X^{(j)}[k]$) the term $|h^{(j)}|^2 X^{(j)}[k] + W^{(j)}[k]$ is a Gaussian random variable with mean $|h^{(j)}|^2 X^{(j)}[k]$, variance σ_n^2, and $\mathbf{m}_q = \{m_q^{(1)}, \ldots, m_q^{(\mathrm{N})}\}$, the vector of information.

The optimal detector in the maximum likelihood sense maximizes the function

$$
\begin{aligned}
\Lambda &= \log \mathbb{P}\left(\mathbf{Y}^{(j)} | h^{(j)}, \mathbf{m}_q\right), \\
&= \sum_{k=0}^{M-1} \log \mathbb{P}\left(|h^{(j)}|^2 X^{(j)}[k] + W^{(j)}[k] \,\middle|\, h^{(j)}, \mathbf{m}_q\right).
\end{aligned}
\tag{6}
$$

Because the noise is Gaussian, it is quite straightforward to show that maximizing the likelihood function Λ is equivalent to minimizing the euclidean distance between the transmitted signal $\mathbf{X}^{(j)}$ and the received one $\mathbf{Z}^{(j)}$:

$$
\begin{aligned}
\hat{\mathbf{m}}_q &= \underset{m_q}{\mathrm{argmax}}\, \Lambda, \\
&= \underset{m_q}{\mathrm{argmin}} \left(M \log\left(\frac{1}{\sqrt{\pi |h^{(j)}|^2 \sigma_n^2}}\right) - \frac{\|\mathbf{Y}^{(j)} - |h^{(j)}|^2 \mathbf{X}^{(j)}\|^2}{|h^{(j)}|^2 \sigma_n^2} \right).
\end{aligned}
\tag{7}
$$

The difficulty is that each element of m_q, $m_q^{(i)}$, can take any value between 0 and $M - 1$. With N users it makes M^{N} possible combinations which rapidly becomes impossible to implement.

We consequently adopt another strategy to keep the multiuser detector but to take benefit from the fact that information should be found where peaks are. Indeed this can significantly reduce the number of combinations that have to be tested to solve (7).

To do so, we proceed in two steps:

- We fix a threshold and detect the peak above this threshold. The objective is to only get the peak of the desired user and the larger ones. If j denotes the desired user, this means that the threshold has to be less than $|h^{(j)}|^2\sqrt{p^{(j)}}$ but more than $|h^{(j)}|^2\sqrt{p^{(j+1)}}$ (we have ordered the users such that $p^{(1)} > p^{(2)} > ... > p^{(N)}$). In this work we set the threshold as

$$\gamma = \frac{|h^{(j)}|^2\left(\sqrt{p^{(j+1)}} + \sqrt{p^{(j+1)}}\right)}{2}. \tag{8}$$

  If user j is the weakest user we take $p^{(j+1)} = 0$ in (8).
- The decision rule depends on the number of detected peaks.
  - If j peaks are detected, it means that we have exactly the same number of peaks as the number we were expecting. In such a case we select the jth peak as the one corresponding to the desired user and we directly have the decoded symbol.
  - If the number of detected peaks is larger than j, it means that weaker peaks have collided and the resulting peak exceeds the threshold (this can also be due to noise). In that case we choose the peak with the closest amplitude to the expected one.
  - If the number of detected peaks is less than j, it means that strong peaks have collided, either including the expected one in a collision or not. if l is the number of detected peaks, $j-l$ gives the number of collisions. In such a case we will scan all possible combinations of $j-l$ collisions between the j strongest peaks. We then evaluate the distance in (7) between the reconstructed vector from the scanned combination and the received one, limiting this calculation to the $j-l$ detected peak positions (not a sum over all samples of the received vector). We select the combination giving the smallest distance and this gives us the decoded symbol.

Our approach starts to be complex when the number of expected peak (j in the described method) is large and the number of collisions ($j-l$) is also important. For instance with $j=5$ and $l=1$, the number of combinations is 10. When $j=10$ and $l=2$, the number of combinations is 750. The number can become huge, especially when l increases, for instance for $i=12$ and $l=4$, the number of combinations is 159027. Such cases can be prohibitive for an end device and adapted decoding procedure to reduce the search space have to be found.

4 Simulation Results

The performance of our proposal are evaluated through intensive Monte Carlo simulations. The simulated environment is described in 3.1. It is a disc with a maximum range $R = 10$ km. We randomly draw the different users in this environment. However, depending on the fading, the total channel gain can be too low to ensure that the user is able to connect to the network with the chosen SF. In such a case, i.e. when the received power is lower than the receiver sensitivity R_s, where, $R_s = -121.5, -124, -127, -129$ dBm for SF = 7 up to 10 respectively, $p_t = 14$ dBm, and $B = 250$ kHz, users cannot connect to the access point. They are discarded and drawn again (both the position and the Rayleigh fading). The detection of the packet and the channel estimation is performed using the preamble which is common to all users and transmitted with the full power so that it does not generate any errors and the channel estimation is accurate.

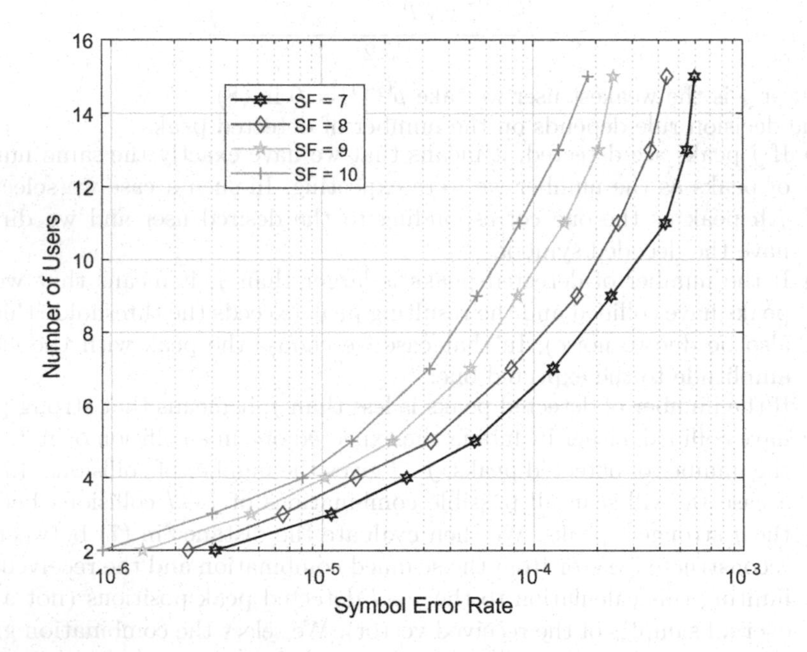

Fig. 2. SER for different N users, SF, and Noise levels, when B = 250 kHz.

Figure 2 presents the average Symbol Error Rate (SER) for various number of users N and SF. The results clearly show that the proposed scheme significantly improves the possible number of connected devices. The classical receiver can only support one user at time when the proposed scheme can go up to more than fifteen, keeping the SER below 10^{-3}. With channel coding, it is expected that the number of users is at least increased by an order of magnitude, whatever the SF considered.

Figure 3 shows the performance of decoding a single user with a different signal-to-noise ratio (SNR) when there are 9 other users transmitting at the same time. Once again, this figure illustrates the excellent performance of our proposal and that the proposed receiver outperforms the classical approach.

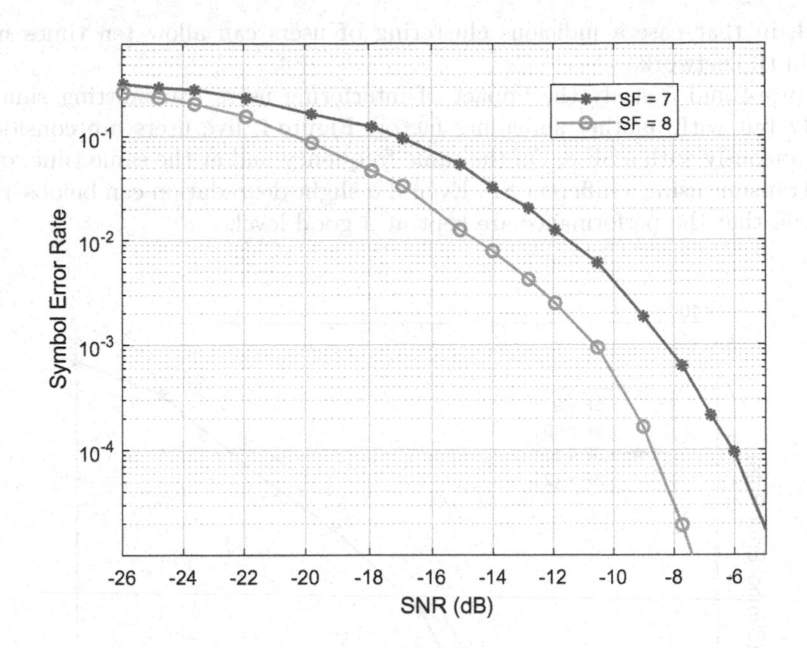

Fig. 3. SER vs SNR of one user in the presence of 9 other users, when SF = 7.

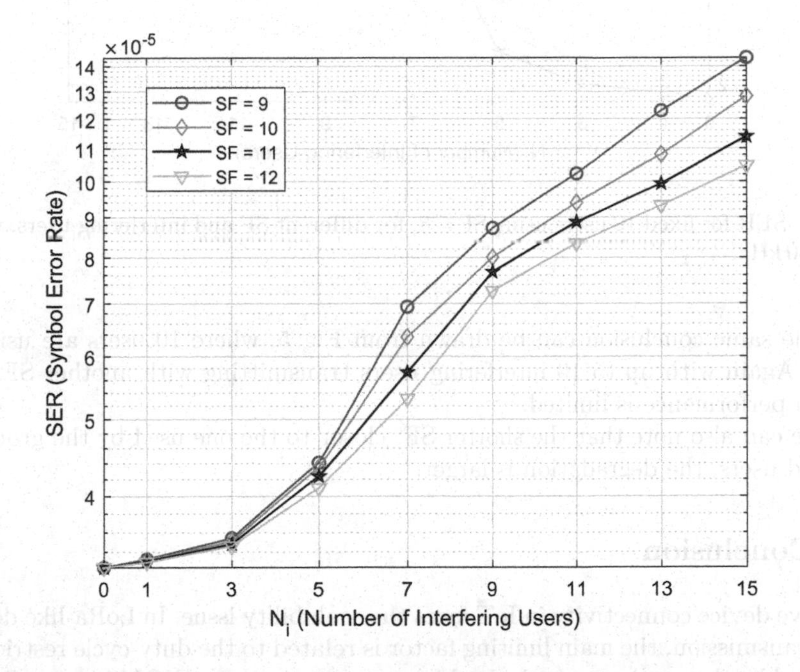

Fig. 4. SER for fixed 5 users using SF = 8, for different SF and interfering users, when B = 250 kHz.

Indeed, in that case a judicious clustering of users can allow ten times more users in the network.

Figure 4 and 5 study the impact of interfering users transmitting simultaneously but with another spreading factor. Figure 4, five users are considered simultaneously with a SF 8. On the same frequency and at the same time, other users transmit using a different SF. Even if a slight degradation can be observed, it is seen that the performance are kept at a good level.

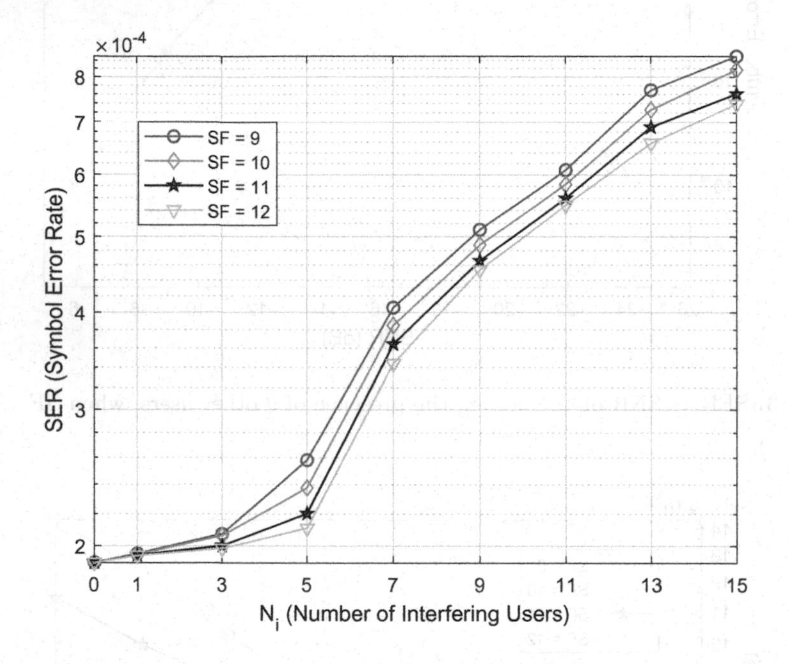

Fig. 5. SER for fixed 10 users using SF = 8, for different SF and interfering users, when B = 250 kHz.

The same conclusion can be drawn from Fig. 5, where 10 users are using a SF 8. Again with up to 10 interfering users transmitting with another SF, the loss in performance is limited.

We can also note that the shorter SF, closer to the one used by the group of desired users, the degradation is larger.

5 Conclusion

Massive device connectivity in IoT faces the scalability issue. In LoRa-like downlink transmission, the main limiting factor is related to the duty-cycle restriction imposed by the regulatory body. In this paper, we propose a NOMA transmission scheme and a multi-user detection to improve the performance of the downlink in LoRa-like networks in terms of scalability. The proposed system includes a power allocation scheme and decoding algorithm which are very specific to this physical layer. The results show a significant performance improvement, increasing the number of devices that can be addressed at least by ten.

References

1. Knud Lasse, L.: State of the IoT 2018: Number of IoT devices (2018). Accessed 22 May 2020, https://iot-analytics.com/state-of-the-iot-update-q1-q2-2018/
2. Raza, U., Kulkarni, P., Sooriyabandara, M.: Low power wide area networks: an overview. IEEE Commun. Surv. Tutor. **19**(2), 855–873 (2017)
3. Ikpehai, A., et al.: Low-power wide area network technologies for internet-of-things: a comparative review. IEEE Internet Things J. **6**(2), 2225–2240 (2019)
4. Adelantado, F., Vilajosana, X., Tuset-Peiro, P., Martinez, B., Melia-Segui, J., Watteyne, T.: Understanding the limits of lorawan. IEEE Commun. Mag. **55**(9), 34–40 (2017). https://doi.org/10.1109/MCOM.2017.1600613
5. Centenaro, M., Vangelista, L., Kohno, R.: On the impact of downlink feedback on lora performance. In: 2017 IEEE 28th PIMRC, pp. 1–6 (2017)
6. Mikhaylov, K., Petäjäjärvi, J., Pouttu, A.: Effect of downlink traffic on performance of LoRaWAN LPWA networks: empirical study. In: IEEE 29th PIMRC, pp. 1–6 (2018)
7. Semtech, AN1200.22: LoRa Modulation Basics, Technical report, Semtech (2015)
8. Patent, Low power long range transmitter (2014). https://patents.google.com/patent/EP2763321A1
9. T. M. W. 1.0, What is LoRaWAN, Technical report, LoRa Alliance (2015). https://lora-alliance.org/resource-hub/what-lorawanr

Simulation-Based Analysis of Mobility Models for Wireless UAV-to-X Networks

Vladislav Prosvirov[1(✉)], Viktoriia Khalina[1], Ekaterina Lisovskaya[1], Yuliya Gaidamaka[1,2], Jiri Pokorny[3], Jiri Hosek[3], and Konstantin Samouylov[1,2,3]

[1] Peoples' Friendship University of Russia (RUDN University), 6 Miklukho-Maklaya St, Moscow 117198, Russian Federation
gnarwhal18@gmail.com, viktoriya.khalina@gmail.com, {lisovskaya-eyu, gaydamaka-yuv,samuylov-ke}@rudn.ru

[2] Federal Research Center "Computer Science and Control" of the Russian Academy of Sciences, Vavilov St. 44-2, Moscow 119333, Russian Federation

[3] Brno University of Technology, Technicka 12, 616 00 Brno, Czech Republic
jiri.pokorny@vutbr.cz, hosek@feec.vutbr.cz

Abstract. Recently, the use of air base stations located on unmanned aerial vehicles (UAVs) has attracted great attention. Static deployment of a sufficient number of UAVs allows uniform wireless coverage in the demanded areas, where the existing cellular infrastructure has white spots or insufficient capacity. However, UAVs mobility may be required for applications, where UAVs are used to provide communications for mobile groups of users (e.g., massive sport or community events like marathon or music festival) or for patrolling tasks with relaxed requirements for data transmission delays (for example, when collecting information from a large number of mMTC sensors). In such tasks, the movement of UAVs can significantly increase the efficiency of the system, since in this case the coverage of the area can be provided by a smaller number of UAVs following the dynamics of ground users. Nowadays, more and more often the question arises about the mobile communications availability in a remote area, for example, during public events or search operations. The lack of on-demand connectivity with sufficient quality in such areas is unacceptable in modern conditions. Therefore, the study of the behavior of a dynamic UAV network is necessary for decision-making operation in such scenarios. The main contribution of this work is making the user mobility model more human-alike according to the real scenarios. The paper considers two models of UAVs movement, the effectiveness of which is estimated from the point of view of the coverage probability and average fade duration of the signal.

The publication has been prepared with the support of the RUDN University Program "5–100" (recipient K. Samouylov). The reported study was funded by RFBR, project numbers 18-00-01555 (18-00-01685) (recipient Yu. Gaidamaka) and 20-07-01064 (recipient Yu. Gaidamaka). This article is based upon the support of international mobility project MeMoV, No. CZ.02.2.69/0.0/0.0/16 027/00083710 funded by European Union, Ministry of Education, Youth and Sports, Czech Republic, and Brno, University of Technology.

V. M. Vishnevskiy et al. (Eds.): DCCN 2020, LNCS 12563, pp. 14–27, 2020.
https://doi.org/10.1007/978-3-030-66471-8_2

Keywords: UAV · User · Coverage probability · Mobility models · Wireless on-demand connectivity

1 Introduction

Currently, unmanned aerial vehicles (UAVs) are easily deployable air devices that can be used as base stations and repeaters to provide additional network coverage at various public events, in emergency situations, as well as to collect data and environmental monitoring [5,16,28]. UAVs can be classified into devices whose working height is measured in tens of kilometers and on devices operating at heights of several tens of meters, as well as on drones with a fixed or a rotating wing [22]. This article focuses on devices operating at low altitudes and with a rotating wing, as they have the ability to "hang" over the covered area.

1.1 Motivation

Thanks to the latest advances in technology, communications, embedded systems, sensor and storage technologies, as well as low cost and ease of deployment, UAVs have become one of the most attractive candidates for supporting and complementing existing and future wireless networks in various applications [3,19]. These applications begin with data collection and environmental monitoring and end up as repeaters and aerial base stations to increase network coverage and throughput on demand [23].

The use of air base stations deployed on unmanned aerial vehicles attracts much attention. Static placement of a sufficient number of UAVs allows uniform coverage of the communication area, which is important to ensure stable communication in areas remote from deployed cellular networks. However, UAV mobility may be in demand for applications where UAVs are used to provide communications for mobile groups of users (for example, video shooting a marathon or a concert). In such tasks, the movement of UAVs can significantly increase the efficiency of the system, since in this case the coverage of the area can be provided by a smaller number of UAVs.

The purpose of the work is to use the means of simulation to assess the coverage characteristics of mobile users traveling randomly one at a time or in groups for two territory patrol models with a different number of UAVs.

1.2 Related Works

There are various areas of research in the use of UAVs as a tool for boosting wireless networks, such as performance analysis of networks with UAVs [6–8,14, 17,21,27], the optimal location of UAVs [2,11,13,24,25], UAV mobility [10,15, 26].

In [14], the authors analyze the performance of data transmission from UAVs on a downlink, considering the type of communication device-to-device (D2D). Using stochastic geometry, the authors of [6] and [27] analyzed the characteristics of a network of static UAVs deployed at a constant height in a circular region, modeled using binomial and Poisson point processes, respectively. In [8], the coverage efficiency of a network with several UAVs was analyzed using a wireless transit connection. The probability of blocking in a network of several UAVs with non-orthogonal multiple access was analyzed in [21]. An analysis of the probability of covering a relative one fixed target user using Laplace transforms was performed in [7] and [22]. In [17], a system using machine learning was analyzed. The simulation area is a sphere filled with many smaller spheres with UAVs.

When calculating the optimal UAV location, the goal is often to provide an appropriate level of quality of service for various parameters. In this regard, the authors in [13] optimize the distance between two UAVs in cases of presence and absence of interference to obtain a certain probability of coverage in this area. In [11], based on the assumption that there is always direct visibility between devices, sequential algorithms were presented to obtain the minimum number of UAVs and their placement to cover a certain area. The use of various genetic algorithms to find the optimal location for UAVs is also now popular. So, in [2] and [25], the authors used the particle swarm algorithm and the bee colony algorithm, respectively. The latter also compares these two algorithms. Various optimization methods for determining the optimal UAV location were analyzed in [24].

UAVs, unlike conventional base stations, have high mobility, so many works are devoted to this aspect. In [26], the authors investigated the problem of energy efficiency on a certain circular trajectory and the general trajectory of a single UAV. The calculation of the optimal UAV trajectory for data collection was considered in [15], where the authors proposed to increase the speed of data transfer through the Internet of things devices. In [10], the trajectory of the motion of one UAV was studied, which complements the work of the ground-based base station using a circular motion around it. In [22], a model of the random movement of several UAVs in a cylindrical region is presented; in a similar work [7], a model of circular and spiral patrolling of a region is presented. Many of the above works mainly analyze the likelihood of blockages and various indicators of energy efficiency, for example, in [26]. However, most do not consider coverage probability as a primary metric. Authors who nevertheless focus on the probability of coverage in their works, for example, [22] and [7], analyze for one static user, not taking into account that in real conditions there can be many users and they can move to space. In this paper, we take the model of random movement in space and a change in height [22], as well as the model of circular patrol [7]. The analysis of the probability of user coverage is given, which is defined as the proportion of users for whom the signal power at the user's receiver from the UAV closest to the receiver exceeds a predetermined threshold necessary for communication, as well as the average duration of signal fading for two user movement scenarios.

2 System Model

We consider a network model of M UAVs deployed to serve N users. Let the service area be a square with the side $S_A = 2R$ (see Fig. 1).

All UAVs move inside the simulation area, and at each moment, the nearest UAV is selected for user service. The distance from the user to the nearest UAV is indicated by ρ. The user is intercepted by another UAV when it becomes the closest to the user. All other UAVs are considered as interference sources for the users.

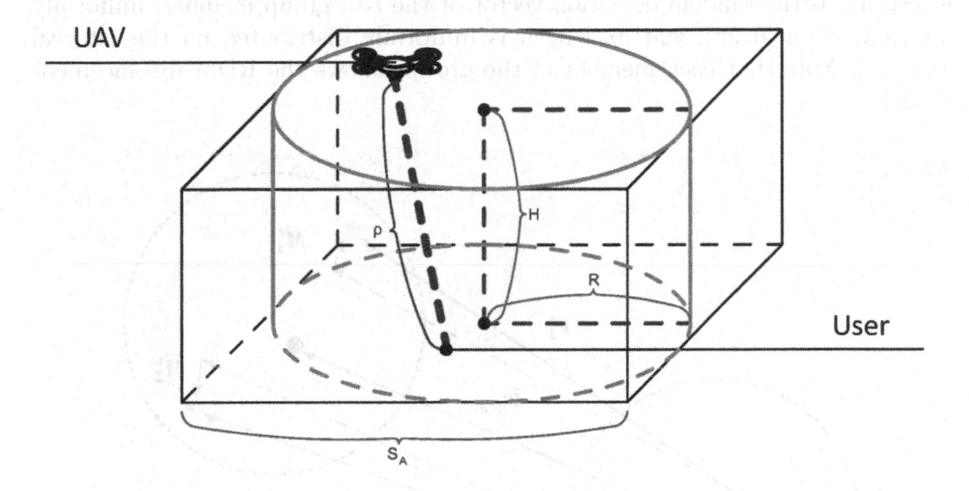

Fig. 1. System model

2.1 Users Mobility

In this paper, as a model of user movement, we consider two models – Random Direction Mobility (RDM) [18] and Reference Point Group Mobility (RPGM) [4]. Also, these two models were considered in [25].

Random Direction Mobility. In the random direction model model, at each moment of time, the i-th user ($i = 1, \ldots, N$) selects a random direction uniformly distributed over $(0, 2\pi)$, and moves in this direction at a constant speed v_i over a period of time that has an exponential distribution with the parameter $1/E(\tau_i)$, where $E(\tau_i)$ is the average movement time. Note that the speed and direction of the movement of one user are independent of the speed and direction of other users.

Reference Point Group Mobility. According to the reference point group mobility model, users are divided into groups, each group has its own leader. The leader determines the direction of the movement of the remaining members of the group (see Fig. 2). In the RPGM model, the movement of the group leader at time t can be denoted by the vector $\overrightarrow{V}^t_{group}$. The movement of i-th group member deviates from $\overrightarrow{V}^t_{group}$ on $\overrightarrow{M}^t_i$, and its movement vector can be calculated as:

$$\overrightarrow{V}^t_i = \overrightarrow{V}^t_{group} + \overrightarrow{M}^t_i,$$

where $\overrightarrow{M}^t_i$ is the random deviation vector of the i-th group member, uniformly distributed on $[0, 2\pi]$, and its length is uniformly distributed on the interval $[0, r_{max}]$. Note that each member of the group follows the RDM displacement model.

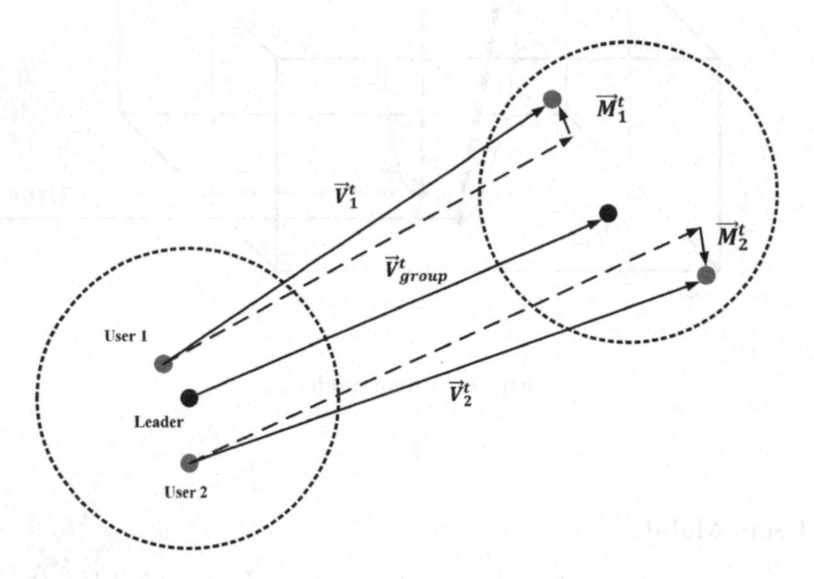

Fig. 2. RPGM trajectory (repeated from [25])

2.2 UAVs Mobility

In this paper, we use two models for UAVs movement: the random patrol model and the ring patrol model. These patrol models, as the most accessible for analysis and deployment, were chosen to assess the quality of user service when UAVs are not aware of their behavior.

Random Patrol Model. The random patrol model involves alternating the vertical and horizontal movement of the i-th ($i = 1, \ldots, M$) UAV, changing it's height $h_i(t)$ and the horizontal position $z_i(t)$ in accordance with the Random Waypoint Mobility (RWP) model (Fig. 3). Initially, the UAV is launched at a random height H_1 uniformly distributed in the interval $[H_{min}; H_{max}]$. Then the UAV selects a random point at a height of H_2, also uniformly distributed in the interval $[H_{min}; H_{max}]$, and moves toward it at a speed v (the same for all UAVs) for several time intervals, without changing direction, until it reaches its destination. Upon reaching a point at a height of H_2, the UAV remains at this point for the delay time T_s, uniformly distributed in the interval $[\tau_{min}; \tau_{max}]$, where τ_{min} is the minimum, and τ_{max} is the maximum delay time. During this time, the UAV moves in the horizontal plane, choosing a point uniformly distributed on the segment $[z_i(t) - R, z_i(t) + R]$ and moves toward it with a speed v.

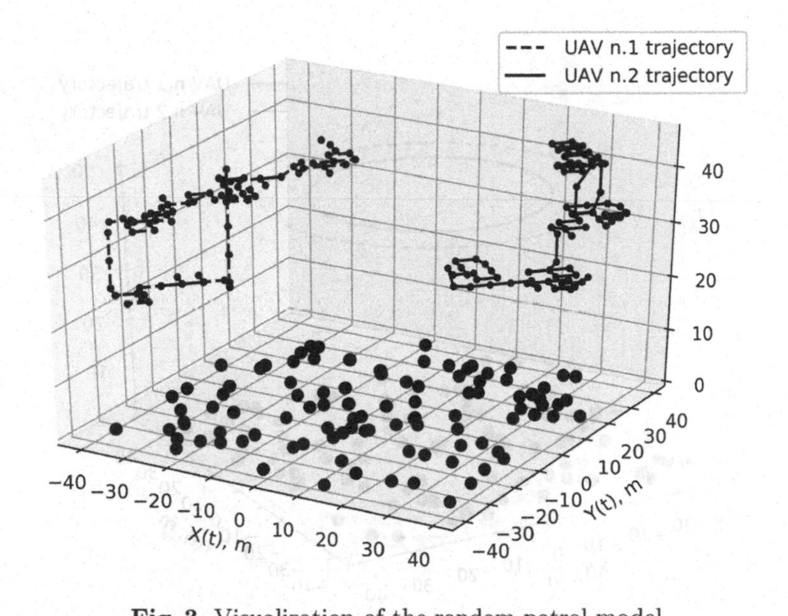

Fig. 3. Visualization of the random patrol model

Note that a change in the horizontal position of the UAV at the current height may occur several times during the delay, and the UAV may not have time to reach a point. In this case, UAV movement is interrupted and it remains in the current horizontal position. After the time T_s, the UAV selects the vertical point again and flies to it, further performing the same iterations.

Ring Patrol Model. This model is based on a deterministic ring trajectory (Fig. 4). All UAVs fly at the same height H_D. To achieve uniform coverage of the patrol area with a network of M UAVs in the case of a ring path, the following conditions must be met:

– The i-th UAV moves in a circle with a radius:

$$R_i = \sqrt{\frac{i}{M+1}}, \; i = \overline{1, M};$$

– The i-th UAV makes a full turn in τ seconds with a constant speed:

$$v_i = 2\pi\frac{R_i}{\tau}, \; i = \overline{1, M};$$

– All UAVs fly at the same angular velocities, so the angle between adjacent UAVs is $\frac{2\pi}{M}$ and is kept at any time.

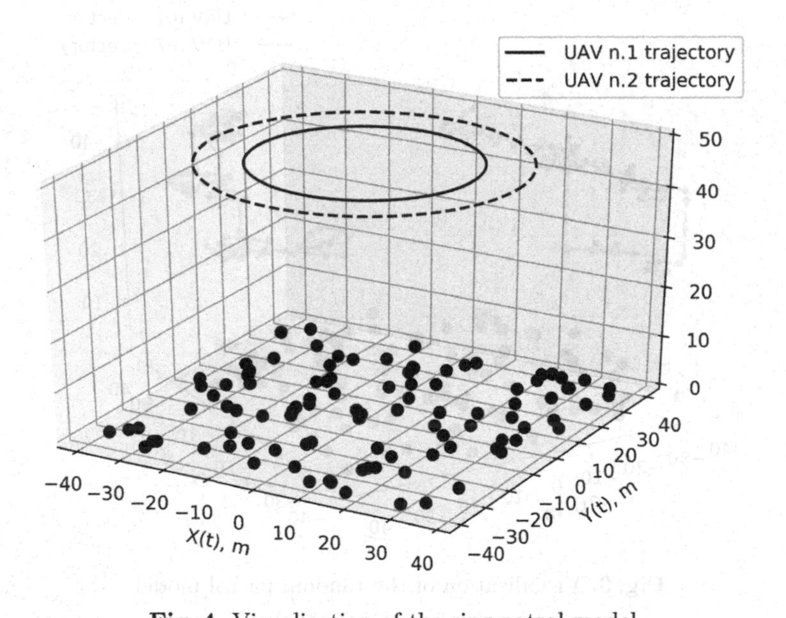

Fig. 4. Visualization of the ring patrol model

However, in the random patrol model, all UAVs have the same speed v. Therefore, in order to get results that are fair for comparison, we will calculate the speed of each UAV in its orbit for the ring patrol model based on the average speed v_{av}, which should be equal to the speed of each UAV in the random spatial patrol model ($v_{av} = v$):

$$\tau_{av} = \frac{2\pi \sum_{i=1}^{M} R_i}{v_{av} M},$$

$$v_i = 2\pi\frac{R_i}{\tau_{av}}, \quad i = \overline{1, M}.$$

3 Metrics of Interest

The paper considers the UAVs patrolling models, the effectiveness of which is estimated from the point of the user coverage probability. In addition, an indicator of the average duration of signal fading is estimated.

3.1 Coverage Probability

To assess the applicability of the UAVs mobility model in the proposed scenario, the coverage probability indicator is estimated, which is defined as the proportion of users, for whom the signal-to-noise ratio (SNR) at the receiver exceeds a predetermined threshold (γ) necessary for the communication:

$$P_c = \frac{1}{N}\sum_{j=1}^{N} 1\{SNR_j \geq \gamma\}.$$

where SNR_j is the SNR for the j-th user from the nearest UAV, and it is calculated as:

$$SNR_j = 10 \cdot \log_{10} \frac{P_{rx_j}}{N_0},$$

where P_{rx_j} is the power of the received signal [W] at the receiving antenna (device of the j-th user); N_0 is the noise power [W]. The power of the signal received by the user device is modeled using the Friis transfer formula:

$$P_{rx_j} = P_{tx} + G_{tx} + G_{rx} + 20 \cdot \log_{10}\left(\frac{c}{4\pi r_j f_c}\right).$$

Here, P_{tx} denotes the transmit power (antenna power) [dBm], G_{tx} and G_{rx} are the transmit and receive antenna gains [dBi], c is the speed of light [m/s], f_c is the frequency [GHz], r_j is the distance between user j and the nearest UAV [m]. Noise power $N_0 = -174 + 10 \cdot \log_{10}(B)$ [dBm], where B is the bandwidth [Hz], -174 is the noise power [dBm], emitted for 1 Hz. The conversion from dBm to W is performed as

$$W = \frac{10^{\frac{dBm}{10}}}{1000}.$$

If the value of SNR_j exceeds a predetermined threshold value ($\gamma = 20$ dBm) [1], then the connection is available and the user is considered covered.

3.2 Average Fade Duration

Consider the stochastic process $S_j(t)$, which characterizes the quality of the connection for the j-th user at time t, so that the value of the $S_j(t)$ will coincide with the SNR at the receiver of the j-th user. Let the threshold γ exist for the process under consideration.

The random variables (RVs) τ_{ij}^{-} – the duration of the i-th signal absence period and τ_{ij}^{+} – the duration of the i-th communication period between the devices are shown in Fig. 5, together with the moments t_{ij}^- and t_{ij}^+ as the intersections of the threshold value of the ratio $\gamma = SNR^*$ by the stochastic process $S_j(t)$.

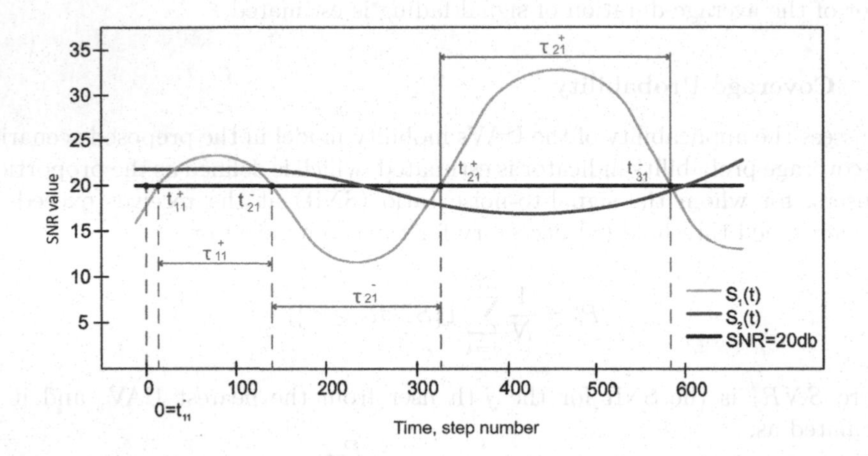

Fig. 5. An example of two trajectories of a stochastic process $S_j(t)$

Our study interest is the estimation of the means of the RV τ_{ij}^-. Note that such an indicator of service quality as the average fade duration (AFD), studied in [7,9,20], corresponds to the average of RV τ_{ij}^-.

Let $N_j(T)$ coincides with the number of periods of lack of communication between the interacting receiver and transmitter during the time T, i.e. periods during which the level of the signal received by the user was below enough to provide a service with the required quality. For example, for the green curve in Fig. 5 $N_1(600) = 3$.

Here $\tau_{ij}^-(T)$ is a random value of i-th lack period duration of communication for the j- th user on the modeling interval with length T, $i = 1, ..., N_j(T)$. Then, for $N_j(T) > 0$ average signal fading duration $\overline{\tau}_j$ for the j-th user during the simulation time T can be found by the formula:

$$\overline{\tau}_j = \frac{1}{N_j(T)} \sum_{i=1}^{N_j(T)} \tau_{ij}^-(T).$$

To calculate the average AFD $\overline{\tau}$ for N users, we use the formula:

$$\overline{\tau} = \frac{1}{N} \sum_{j=1}^{N} \overline{\tau}_j.$$

4 Numerical Results

For numerical analysis, the simulator was developed, consisting of two models of user movement and two models of UAV movement. As the initial data for modeling, we used the values given in the Table 1, which were chosen as the most consistent with the real scenarios of UAV user coverage [12].

Table 1. Simulation parameters

Symbol	Value	Description
P_{tx}	24 dBm	Transmitting power
G_{tx}	3 dBm	Transmitting antenna gain
G_{rx}	3 dBm	Receiving antenna gain
N_0	27.434 dBm	Noise power
B	0.56 GHz	Bandwidth
γ	20 dB	SNR threshold
M	3 or 5	Number of UAVs
N	100	Number of users
S_A	100*100 m^2	Area of interest
H_D	20 m	UAV altitude for the ring patrol model
$[H_{min}, H_{max}]$	[15; 20] m	UAV altitude Range for the random patrol model
v_i	1.4 m/s	Users speed
v_{av}	5 m/s	UAVs average speed

In Figs. 6 and 7, the curves indicated by RING correspond to the ring patrol model, and the RPM corresponds to the random patrol model. In addition, the data in Figures are averaged over the number of runs of the simulation model. They display graphs of the coverage probability and the average duration of signal fading for different models of user movement and UAVs; different number of user groups and UAVs.

In the following figures, on the plots of coverage probability, we can observe that the curve corresponding to the random patrol model has the most stable character. Due to the features of this model, UAVs can "hang" over a group of users or a large number of them for a long time, and therefore the smallest deviation from the coverage probability averaged over launches is ensured. In addition, this result is facilitated by the fact that in the RPM model, UAVs can change their height, and with UAVs height is decreasing, the radius of users coverage is increasing. At the same time, the coverage probability indicator for the ring patrol model is spasmodic in nature with the largest discrepancy interval compared to the random patrolling model. Since the model involves constant movement, at certain points in time, UAVs can be located above the smallest congestion of users.

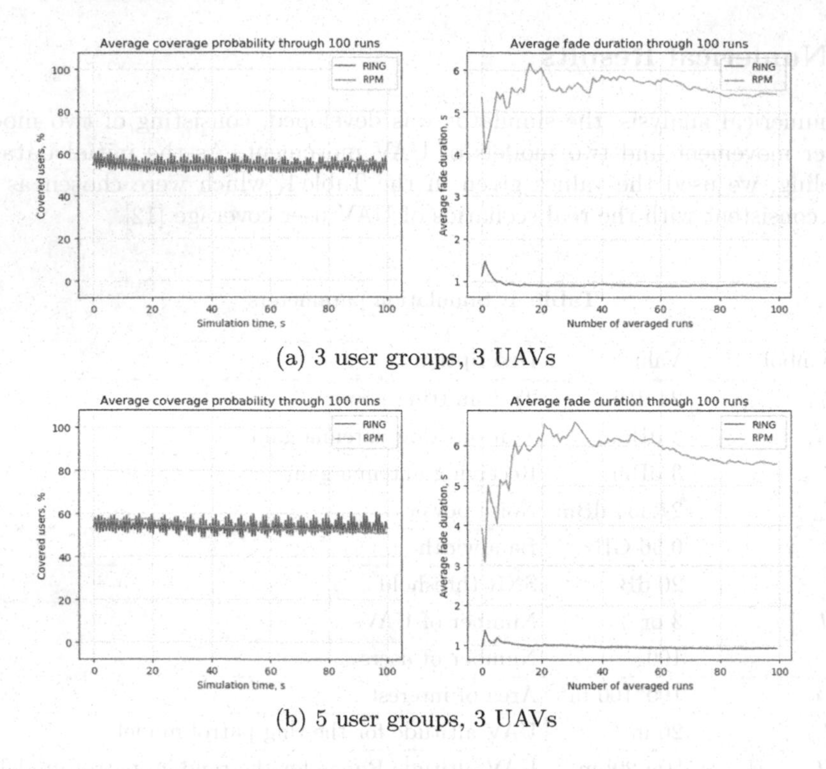

(a) 3 user groups, 3 UAVs

(b) 5 user groups, 3 UAVs

Fig. 6. User mobility model: Reference Point Group Mobility

However, for the random patrol model, the AFD behaves worse than for the ring patrolling model. As mentioned above, UAVs can be located over one cluster of users or a group of users for a long time, while not serving another cluster or group of users for a longer time.

After analyzing the results, we can conclude that both models show better performance in the case of users moving according to random direction mobility both in terms of coverage probability and in terms of average fade duration. This conclusion gives the right to believe that in a scenario where there are several user groups that are following their group leaders, other UAV moving models should be used to ensure high-quality communications with low latencies, which is the main requirement for network parameters in the new 5G paradigm. Models of random and ring patrolling are more suitable when applied to serve a large crowd of people concentrated in one area.

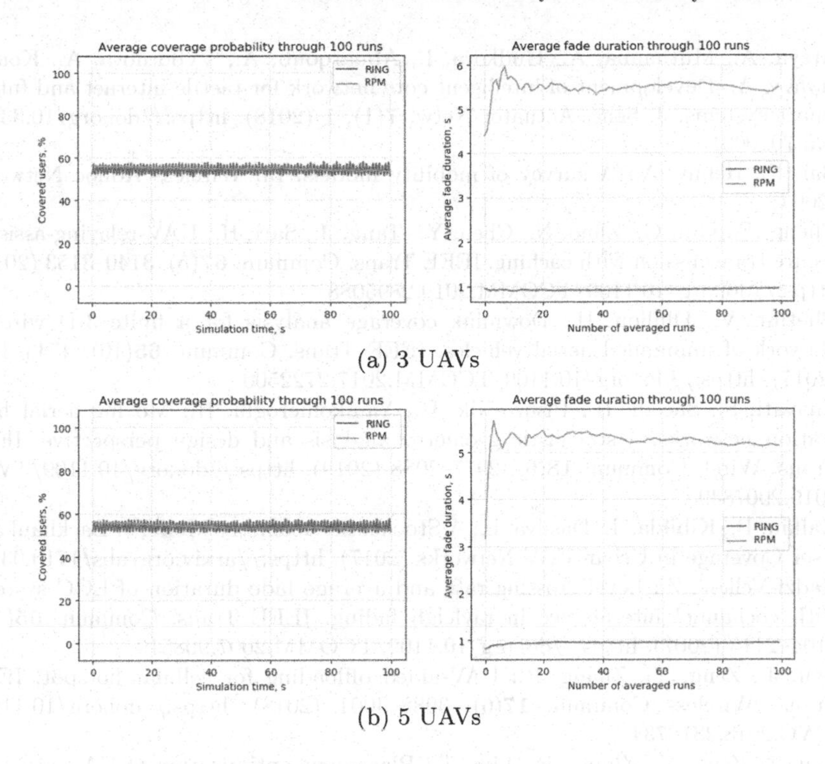

(a) 3 UAVs

(b) 5 UAVs

Fig. 7. User mobility model: Random Direction Mobility

5 Conclusion

The results obtained in the article showed that none of the models is universal from the point of view of the coverage probability. The advantage of the ring patrol model is the uniform coverage of the region at a low average fade duration. Also, such a model gives good results for non-feedback scenarios, when the drones have no information about the movement of users. In the latter case, the random patrol model yields results that are superior to the ring patrol. It is planned to continue the investigation of the random patrol model for use in the 3GPP 5G Integrated Access and Backhaul (IAB).

References

1. How to: Define Minimum SNR Values for Signal Coverage. http://www.wireless-nets.com/resources/tutorials/define_SNR_values.html
2. Aslan, S., Demirci, S.: Solving UAV localization problem with artificial bee colony (ABC) algorithm. In: 2019 4th International Conference on Computer Science and Engineering (UBMK), pp. 735–738 (2019). https://doi.org/10.1109/UBMK.2019.8907034

3. Ateya, A., Muthanna, A., Gudkova, I., Abuarqoub, A., Vybornova, A., Koucheryavy, A.: Development of intelligent core network for tactile internet and future smart systems. J. Sens. Actuator Netw. **7**(1), 1 (2018). https://doi.org/10.3390/jsan7010001
4. Bai, F., Helmy, A.: A survey of mobility models. In: Wireless Adhoc Networks (2004)
5. Cheng, F., Gui, G., Zhao, N., Chen, Y., Tang, J., Sari, H.: UAV-relaying-assisted secure transmission with caching. IEEE Trans. Commun. **67**(5), 3140–3153 (2019). https://doi.org/10.1109/TCOMM.2019.2895088
6. Chetlur, V., Dhillon, H.: Downlink coverage analysis for a finite 3-D wireless network of unmanned aerial vehicles. IEEE Trans. Commun. **65**(10), 4543–4558 (2017). https://doi.org/10.1109/TCOMM.2017.2722500
7. Enayati, S., Saeedi, H., Pishro-Nik, H., Yanikomeroglu, H.: Moving aerial base station networks: a stochastic geometry analysis and design perspective. IEEE Trans. Wirel. Commun. **18**(6), 2977–2988 (2019). https://doi.org/10.1109/TWC.2019.2907849
8. Galkin, B., Kibiłda, J., Dasilva, L.: A Stochastic Geometry Model of Backhaul and User Coverage in Urban UAV Networks (2017). https://arxiv.org/abs/1710.03701
9. Hadzi-Velkov, Z.: Level crossing rate and average fade duration of EGC systems with cochannel interference in rayleigh fading. IEEE Trans. Commun. **55**(11), 2104–2113 (2007). https://doi.org/10.1109/TCOMM.2007.908519
10. Lyu, J., Zeng, Y., Zhang, R.: UAV-aided offloading for cellular hotspot. IEEE Trans. Wireless Commun. **17**(6), 3988–4001 (2018). https://doi.org/10.1109/TWC.2018.2818734
11. Lyu, J., Zeng, Y., Zhang, R., Lim, T.: Placement optimization of UAV-mounted mobile base stations. IEEE Commun. Lett. **21**(3), 604–607 (2017). https://doi.org/10.1109/LCOMM.2016.2633248
12. Mezzavilla, M., et al.: End-to-end simulation of 5G mm wave networks. IEEE Commun. Surv. Tutor. **20**(3), 2237–2263 (2018). https://doi.org/10.1109/COMST.2018.2828880
13. Mozaffari, M., Saad, W., Bennis, M., Debbah, M.: Drone small cells in the clouds: design, deployment and performance analysis. In: 2015 IEEE Global Communications Conference (GLOBECOM), pp. 1–6 (2015). https://doi.org/10.1109/GLOCOM.2015.7417609
14. Mozaffari, M., Saad, W., Bennis, M., Debbah, M.: Unmanned aerial vehicle with underlaid device-to-device communications: performance and tradeoffs. IEEE Trans. Wireless Commun. **15**(6), 3949–3963 (2016). https://doi.org/10.1109/TWC.2016.2531652
15. Mozaffari, M., Saad, W., Bennis, M., Debbah, M.: Mobile unmanned aerial vehicles (UAVs) for energy-efficient internet of things communications. IEEE Trans. Wireless Commun. **16**(11), 7574–7589 (2017). https://doi.org/10.1109/TWC.2017.2751045
16. Mozaffari, M., Saad, W., Bennis, M., Debbah, M.: Communications and control for wireless drone-based antenna array. IEEE Trans. Commun. **67**(1), 820–834 (2019). https://doi.org/10.1109/TCOMM.2018.2871453
17. Mozaffari, M., Kasgari, A.T.Z., Saad, W., Bennis, M., Debbah, M.: Beyond 5G With UAVs: foundations of a 3D wireless cellular network. IEEE Trans. Wireless Commun. **18**(1), 357–372 (2019). https://doi.org/10.1109/TWC.2018.2879940

18. Nain, P., Towsley, D., Liu, B., Liu, Z.: Properties of random direction models. In: Proceedings IEEE 24th Annual Joint Conference of the IEEE Computer and Communications Societies, vol. 3, pp. 1897–1907 (2005). https://doi.org/10.1109/INFCOM.2005.1498468
19. Ostrikova, D., Beschastnyi, V., Gudkova, I., Zeifman, A.: Optimal multicast subgrouping in mobility-aware 5g systems: challenges, modeling, and opportunities. In: 12th International Workshop on Applied Problems in Theory of Probabilities and Mathematical Statistics (Summer Session) in the Framework of the Conference on Information and Telecommunication Technologies and Mathematical Modeling of High-Tech Systems, Lisbon, Portugal, pp. 13–22 (2018)
20. Shankar, P.M.: Fading and Shadowing in Wireless Systems. Springer, Heidelberg (2014). https://doi.org/10.1007/978-1-4614-0367-8
21. Sharma, P., Kim, D.: UAV-enabled downlink wireless system with non-orthogonal multiple access. In: 2017 IEEE Globecom Workshops (GC Wkshps), pp. 1–6 (2017). https://doi.org/10.1109/GLOCOMW.2017.8269066
22. Sharma, P., Kim, D.: Random 3D mobile UAV networks: mobility modeling and coverage probability. IEEE Trans. Wireless Commun. **18**(5), 2527–2538 (2019). https://doi.org/10.1109/TWC.2019.2904564
23. Sopin, E., Daraseliya, A., Correia, L.: Performance analysis of the offloading scheme in a fog computing system. In: 2018 10th International Congress on Ultra Modern Telecommunications and Control Systems and Workshops (ICUMT), pp. 1–5 (2018). https://doi.org/10.1109/ICUMT.2018.8631245
24. Sun, J., Masouros, C.: Deployment strategies of multiple aerial BSs for user coverage and power efficiency maximization. IEEE Trans. Commun. **67**(4), 2981–2994 (2019). https://doi.org/10.1109/TCOMM.2018.2889460
25. Tafintsev, N., et al.: Improved network coverage with adaptive navigation of mmwave-based drone-cells. In: 2018 IEEE Globecom Workshops (GC Wkshps), pp. 1–7 (2018). https://doi.org/10.1109/GLOCOMW.2018.8644097
26. Zeng, Y., Zhang, R.: Energy-efficient UAV communication with trajectory optimization. IEEE Trans. Wirel. Commun. **16**(6), 3747–3760 (2017). https://doi.org/10.1109/TWC.2017.2688328
27. Zeng, Y., Zhang, R., Lim, T.: Throughput maximization for UAV-enabled mobile relaying systems. IEEE Trans. Commun. **64**(12), 4983–4996 (2016). https://doi.org/10.1109/TCOMM.2016.2611512
28. Zhang, S., Zhang, H., Di, B., Song, L.: Cellular UAV-to-X communications: design and optimization for multi-UAV networks. IEEE Trans. Wireless Commun. **18**(2), 1346–1359 (2019). https://doi.org/10.1109/TWC.2019.2892131

Structures and Deployments of a Flying Network Using Tethered Multicopters for Emergencies

Truong Duy Dinh[1], Vladimir Vishnevsky[2], Andrey Larionov[2], Anastasia Vybornova[1], and Ruslan Kirichek[1,2](✉)

[1] The Bonch-Bruevich Saint-Petersburg State University of Telecommunications, St.Petersburg, Russia
din.cz@spb.ru, a.vybornova@gmail.com, kirichek@sut.ru

[2] V. A. Trapeznikov Institute of Control Sciences of RAS, Moscow, Russia
vishn@inbox.ru, larioandr@gmail.com

Abstract. In recent years, the interest of tethered UAVs high-altitude platforms has been widely constantly increasing in many fields. The long-time operating possibility is one of the main advantages of tethered unmanned high-altitude platforms compared to autonomous UAVs. In the paper, a flying network for emergencies using tethered multicopters is proposed. The combination of tethered unmanned high-altitude platforms and groups of UAVs in flying network for emergencies is expected to enhance the effectiveness of search and rescue operation in the wilderness as well as after natural disasters.

Keywords: UAV · Flying network · Tethered multicopters · Search and rescue

1 Introduction

Over the past decade, the emergence of new technologies as well as the development of science and technology has greatly assisted search and rescue operation in emergencies. The dissemination of UAVs for civilian purposes has turned them into a useful search and rescue (SAR) tool in different situations, such as for emergency prevention, monitoring emergencies, searching for missing people after natural disasters, or urgently delivering the necessary cargo to places where it is needed in an emergency. In addition, UAVs are used for environmental purposes, such as to protect beaches, study the melting of polar ice, monitor forests, monitor the coast and water areas, determine the effects of various pollutants, etc. [1,2].

The reported study was funded by RFBR, project number 20-37-70059.

V. M. Vishnevskiy et al. (Eds.): DCCN 2020, LNCS 12563, pp. 28–38, 2020.
https://doi.org/10.1007/978-3-030-66471-8_3

Multifunctional complexes with UAVs in control and communication systems are utilized for relaying signals or in studies of the pattern of radio signals transmission, and for inspection of cell towers [8]. In some cases, UAVs can work as "network nodes" to connect the network to the Internet (Internet of Drones - IoD) [9]. Moreover, in order to expand the working area, cellular networks can also be employed as an additional communication channel to UAVs, along with conventional P2P (point to point) networks, for example, in automated air traffic control systems. In many scientific articles [3–7], using groups of UAVs had been proven to be much more effective than using only one UAV. However, the main disadvantage of UAVs is a limited time of operation due to the small battery resource of UAVs equipped with electric motors or the fuel reserve for internal combustion engines. In order to solve this problem, tethered UAVs high-altitude platforms are consider. They can support long-term operation with power supply of engines and payload equipment is provided from the ground-based energy sources.

Due to above-mentioned features of UAV groups, in this study, the paper provides a Wi-Fi network based on groups of UAVs and tethered UAVs high-altitude platforms, called flying network for emergencies using tethered multicopters, that can help rescuers to communicate with victims or find their locations using Wi-Fi signals generated from their phone. In addition, the deployment of rescue operations in difficult or dangerous areas for rescuers is also addressed with the help of flying network.

2 Tethered UAVs High-Altitude Platform

At present, research centers in leading countries of the world are carrying out intensive scientific work on the design and implementation of tethered UAVs high-altitude platforms [10–17], given the wide spread of their practical application. The long-time operating possibility is one of the major advantages of tethered unmanned high-altitude platforms compared to autonomous UAVs. UAVs can be presented by two types: multicopter type and fixed wing type. Fixed wing type UAV has a high flight duration, maximum flight altitude, high speed, and high payload. On the other hand, multicopter type has the ability to stay stable in the air, as well as high maneuverability [21]. With these advantages, multicopter type is more suitable for tethered UAVs system due to its structural characteristics and missions (Fig. 1).

Tethered UAVs high-altitude platform consists of terrestrial and flying modules. The terrestrial module contains a ground control station for a high-altitude platform (TCS), a ground voltage converter, a winch of a tethered cable of a high-altitude platform and a mooring device (Fig. 2).

A tethered UAVs system consists of a multicopter, cable and flight platform. A new energy transfer technology will provide the multicopter with the ability of lifting to a height 300 m and with a payload of up to 50 kg, and a long working time (up to 24 h) which is limited only by the reliability characteristics of the multicopter. The cable, of either copper wires or optical fiber, ensures the transfer

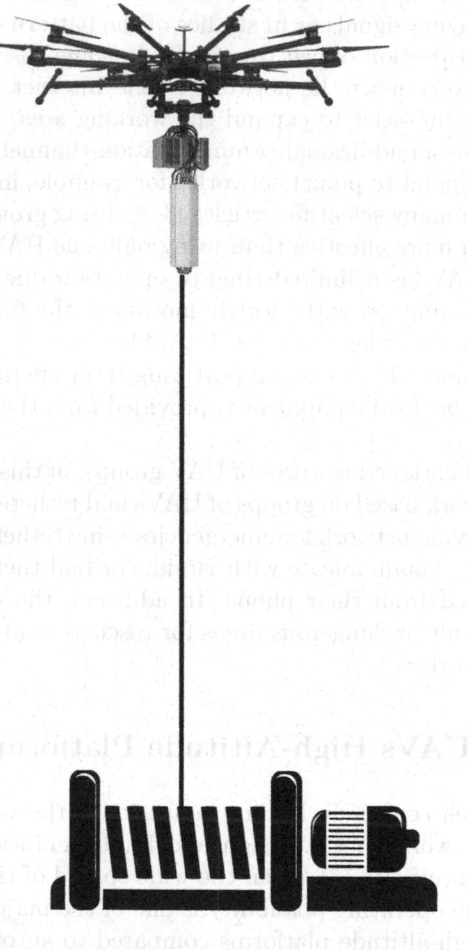

Fig. 1. Tethered UAV high-altitude platform application example

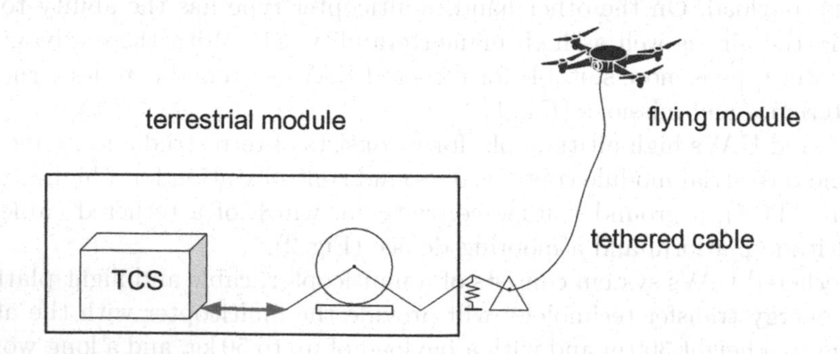

Fig. 2. Tethered UAV high-altitude platform components

of large amounts of information from board to ground and vice versa. Local navigation systems equipped in high-altitude platforms, provide high positioning accuracy and increase noise immunity compared to satellite navigation systems.

The architecture of the tethered high-altitude unmanned platform includes the following main components:

1. An unmanned multi-rotor aerial vehicle of large carrying capacity and a long operating time, designed to lift up to 300 m. and hold the telecommunications payload, video surveillance equipment, etc.
2. A high-power ground-to-board energy transmission system that provides power energy to the propulsion systems of the unmanned multi-rotor flight module and the payload equipment.
3. Control and stabilization system of the high-altitude platform, including a local navigation subsystem with ground-based radio beacons, providing increased positioning accuracy and noise immunity in the absence of signals from satellite navigation systems.
4. On-board payload equipment, including base station of the cellular network of the fourth generation (LTE); radar and radio relay equipment; equipment for video surveillance and environmental monitoring, etc.
5. Cable-rope on Kevlar base, including copper wires of small cross-section ($0.5\,\text{mm}^2$) for transmission of high-voltage (up to 2000 V), high-frequency (up to 200 kHz) signals and optical fiber for digital information transmission with a speed of up to 10 Gbit/s.
6. Ground control complex, which includes an AC voltage converter 380/2000 V, a system for diagnostics of the parameters of the high-altitude platform and an intelligent wrench with a microprocessor unit to control the cable-rope tension during lifting, descending and wind loads. In mobile configuration, the ground control center is located on a mobile platform with an electric generator installed on it, the output power of which is not less than 20 kW.

3 Flying Network for Emergency Using Tethered Multicopters

One of the important applications of UAVs in communication systems is the UAVs network or FANET (Flying Ad-Hoc Network). Nowadays, FANET is widely used in various fields: military, commercial, agricultural, etc. In particular, an application of FANET in the search and rescue operations was developed, named Flying network for emergencies (Fig. 3) [18].

Flying network for emergencies consists of two segments: a flying segment and a terrestrial segment. In the flying segment, UAVs are divided into groups, which are able to simultaneously communicate with each other and to the emergency services, victims or sensor nodes in the terrestrial segment without having any predefined and fixed infrastructure. In order to solve critical issues in FANET, such as communications and networking of the multiple UAVs, the modified of protocol IEEE 802.11p, called CMMpP, was presented in [19]. CMMpP - a

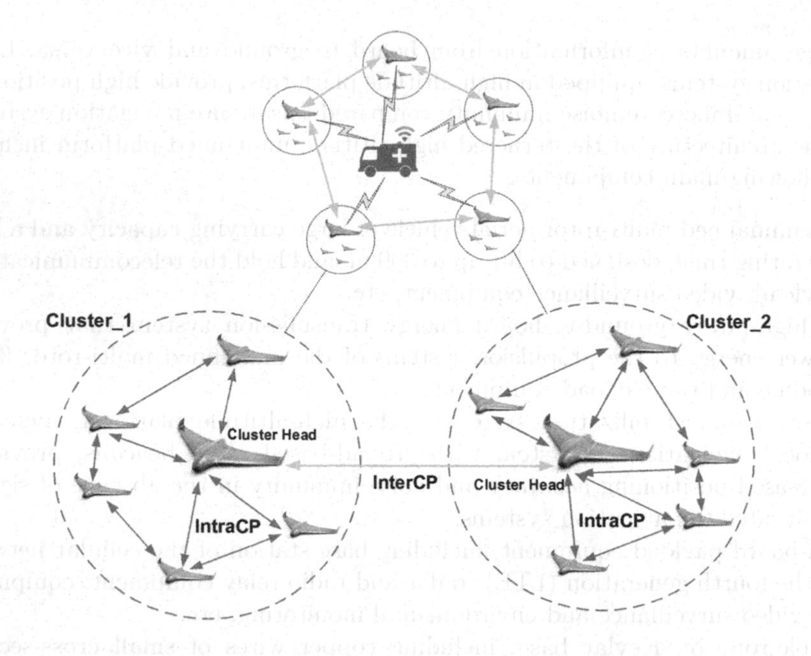

Fig. 3. An architecture of flying network for emergencies

new MAC protocol based on IEEE 802.11p and IEEE 1609.4 protocols to perform communications between UAVs as well as between groups of UAVs, which consists of cluster management protocol (CMP), intra-cluster communication protocol (IntraCP) and inter-cluster communication protocol (InterCP) (Fig. 4).

IEEE 802.11p was originally developed for VANET - Vehicular ad hoc networks [20]. It was adopted as the Medium Access Control (MAC) and Physical Layer (PHY) specifications for the lower-layer Dedicated Short-Range Communication standard (DSRC), which has advantageous characteristics, such as being able to operate in the frequency range of 5.9 GHz (less disturbed by other devices), wide coverage (up to 1000 m), fast transmission rate (up to 27 Mbps), self-organization and fast convergence.

Nevertheless, one of the weaknesses of UAVs in flying network for emergencies is the short working time. For multicopter type, the flight time is about 30–60 min, and for fixed wing type, it can reach 1–2 h. However, this is a relatively short time in the search and rescue missions, which often leads to inefficient operations or the need to replace UAVs many times. To increase the effectiveness of search and rescue operations, using tethered UAVs in flying network for emergencies is proposed, because of the following advantages:

- The effectiveness of the tethered UAVs system in various civilian areas, their mobility, compactness and cost-effectiveness compared to very expensive satellite systems;
- Super long working time, UAVs can operate up to 24 h powered by ground;

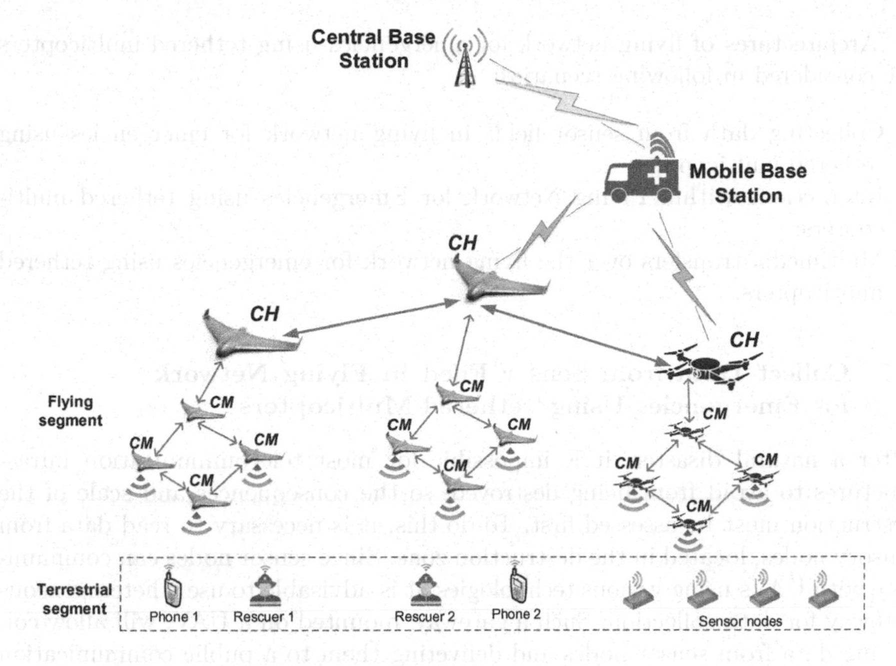

Fig. 4. Interaction of UAV groups

- Possibility of lifting the platform to a height of up 500 m with a payload of up to 50 kg;
- Ultra-wide bandwidth for data transmission through optical fiber inside the cable;
- The ability to shoot high-definition video and images acquired by the camera mounted on the UAV and then they are sent back to the ground through an optical fiber inside the cable;
- The system either can be freestanding or mounted on the rear of the vehicle. It is suitable for various industrial applications, such as television broadcasting, alarm relays, video surveillance, etc. When the system is installed on a car, the attached UAVs themselves can follow the car within a speed 25 km per hour;
- Local navigation system based on tethered UAVs system provides high positioning accuracy and increased noise immunity compared to satellite navigation systems;
- A relatively short time of deployment of tethered UAVs system, approximately no more than 10 min;
- Tethered UAVs system provides the possibility of its operation at temperatures from −50 to +50 degrees Celsius, and the UAV itself can perform a flight with wind up to 15 meters per second;
- The ability to expand the operating range of the tethered UAVs system by using a chain of UAVs tethered one to the other. The first UAV in the chain is tethered to a ground station, while the last one serves as end effector.

Architectures of flying network for emergencies using tethered multicopters are considered in following scenarios:

- Collecting data from sensor fields in flying network for emergencies using tethered multicopters;
- Interactions within Flying Network for Emergencies using tethered multicopters;
- Multimedia transfers over the flying network for emergencies using tethered multicopters.

3.1 Collect Data from Sensor Filed in Flying Network for Emergencies Using Tethered Multicopters

After a natural disaster, it is impossible for most telecommunication infrastructures to avoid from being destroyed, so the consequences and scale of the destruction must be assessed first. To do this, it is necessary to read data from sensory nodes, located in the destruction zone. Since sensor nodes can communicate with UAVs using various technologies, it is advisable to use a heterogeneous gateway for data collection. Such a gateway, mounted on a UAV, will allow collecting data from sensor nodes and delivering them to a public communication network [23,24].

In a SAR operation, mobile base stations will deploy groups of UAVs, including tethered UAVs, to areas around MBS to gather information. All UAVs are equipped with a heterogeneous gateway, which is a network device or a relay system designed to ensure the interaction of two information networks that have different characteristics, using different sets of protocols and supporting different transmission technologies [22]. The gateway can support technologies, such as ZigBee, 6LoWPAN, LoRa, Bluetooth, NB-IoT, and act as a connecting link between sensor nodes and mobile base station. With tethered UAVs, more data can be collected and the coverage is also extended. An architecture of collecting data from sensor fields in flying network for emergencies using tethered multicopters is shown in Fig. 5.

3.2 Interaction of Flying Network for Emergencies Using Tethered Multicopters

In the flying network for emergencies, communication among UAVs in a group and among groups of UAVs is of paramount importance. Technology IEEE 802.11p with the modified protocol CMMpp in [19] was developed to solve this issue. Moreover, tethered UAVs can be used in this network, which is presented in Fig. 6. Tethered UAVs become super cluster nodes, which can receive information from cluster nodes of the groups or can replace cluster nodes when all of UAVs in the group can not be the cluster head. With the advantage of having a very long flight time, tethered UAVs will keep the network stable and reliable during the mission.

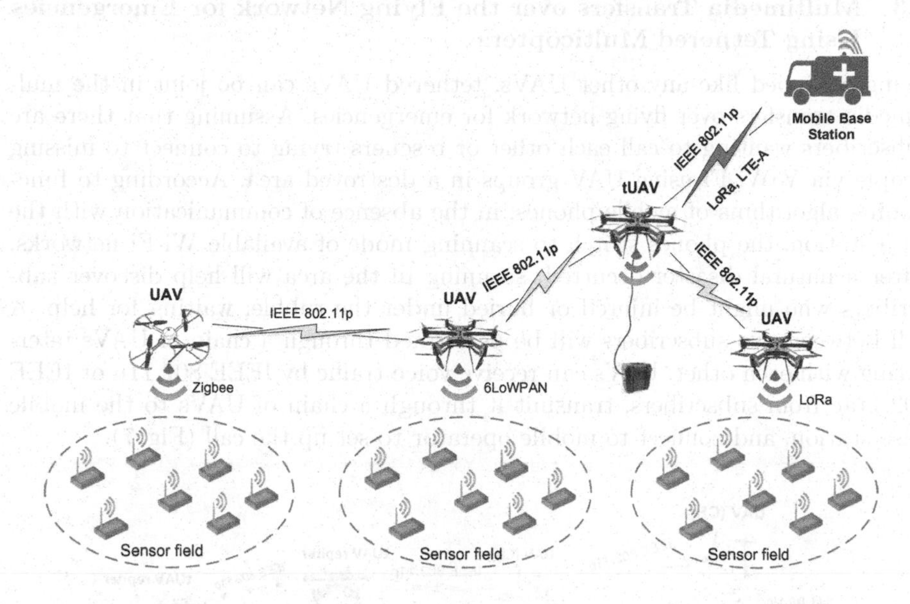

Fig. 5. Collect data from sensor filed in flying network for emergencies using tethered multicopters

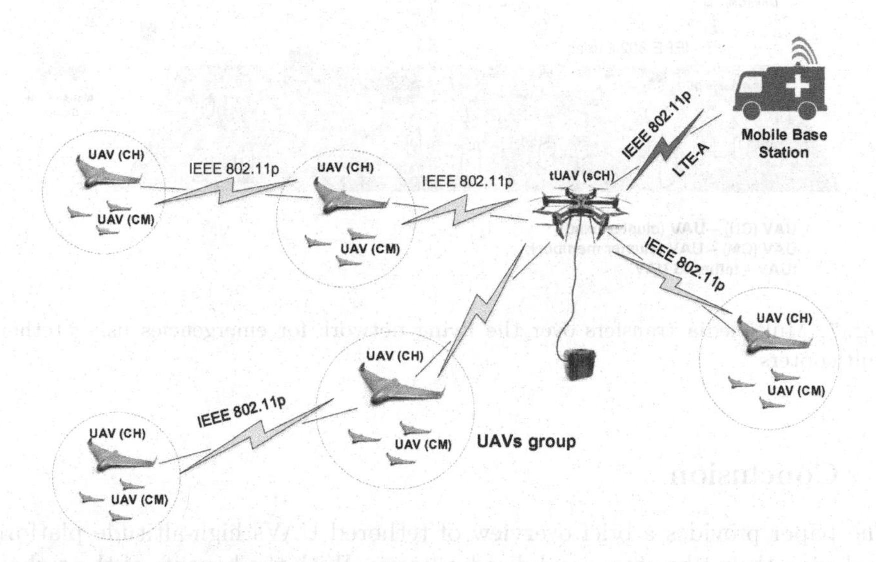

Fig. 6. Interaction of Flying Network for Emergencies using tethered multicopters

3.3 Multimedia Transfers over the Flying Network for Emergencies Using Tethered Multicopters

Being equipped like any other UAVs, tethered UAVs can be joint in the multimedia transfers over flying network for emergencies. Assuming that there are subscribers wanting to call each other or rescuers trying to connect to missing people via VoWi-Fi using UAV groups in a destroyed area. According to functioning algorithms of mobile phones, in the absence of communication with the base station, the phones switch to scanning mode of available Wi-Fi networks. After a natural disaster occurred, scanning in the area will help discover subscribers who might be injured or buried under the rubble, waiting for help. A call between two subscribers will be performed through a chain of UAVs interacting with each other. UAVs can receive voice traffic by IEEE 802.11n or IEEE 802.11ac from subscribers, transmit it through a chain of UAVs to the mobile base station, and connect to mobile operator to set up the call (Fig. 7).

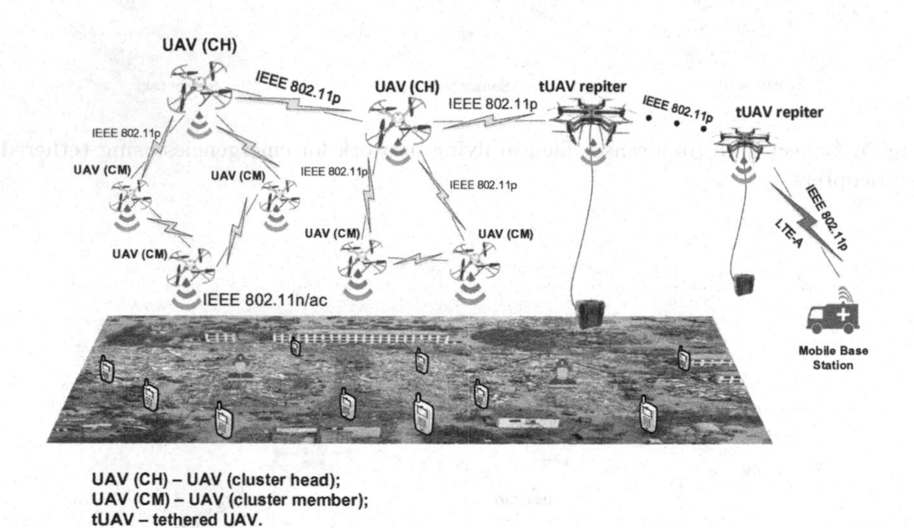

Fig. 7. Multimedia transfers over the flying network for emergencies using tethered multicopters

4 Conclusion

The paper provides a brief overview of tethered UAVs high-altitude platforms analyzing their advantages and disadvantages. With the benefits of the tethered UAVs high-altitude platform, flying network using tethered multicopters was proposed for emergency situations. Different architectures of this network were presented in order to enhance the effectiveness of search and rescue operations. In the future work, the research will prioritize in conducting a series of experiments and simulations to evaluate the performance of the proposed architectures.

Acknowledgment. The reported study was funded by RFBR, project number 20-37-70059.

References

1. Scherer, J., et al.: An autonomous multi-UAV system for search and rescue. In: Proceedings of the First Workshop on Micro Aerial Vehicle Networks, Systems, and Applications for Civilian Use, pp. 33–38 (2018)
2. Cubber, G.D., et al.: Introduction to the use of robotic tools for search and rescue, pp. 1–17 (2017)
3. Dinh, T.D., et al.: Unmanned aerial system-assisted wilderness search and rescue mission. Int. J. Distrib. Sensor Netwo. **15**(6), 1–15 (2019)
4. Kirichek, R., Paramonov, A., Koucheryavy, A.: Swarm of public unmanned aerial vehicles as a queuing network. In: Vishnevsky, V., Kozyrev, D. (eds.) DCCN 2015. CCIS, vol. 601, pp. 111–120. Springer, Cham (2016). https://doi.org/10.1007/978-3-319-30843-2_12
5. Koucheryavy, A., Vladyko, A., Kirichek, R.: State of the art and research challenges for public flying ubiquitous sensor networks. In: Balandin, S., Andreev, S., Koucheryavy, Y. (eds.) ruSMART 2015. LNCS, vol. 9247, pp. 299–308. Springer, Cham (2015). https://doi.org/10.1007/978-3-319-23126-6_27
6. Ding, X.C., Rahmani, A.R., Egerstedt, M.: Multi-UAV convoy protection: an optimal approach to path planning and coordination. IEEE Trans. Rob. **26**(2), 256–268 (2010)
7. Amelin, K., Amelina, N., Granichin, O., Granichina, O., Andrievsky, B.: Randomized algorithm for UAVs group flight optimization. IFAC Proc. Vol. **46**(11), 205–208 (2013)
8. Shakhatreh, H., et al.: Unmanned aerial vehicles (UAVs): a survey on civil applications and key research challenges. IEEE Access **7**, 48572–48634 (2019)
9. Gharibi, M., Boutaba, R., Waslander, S.L.: Internet of drones. IEEE Access **4**, 1148–1162 (2016)
10. Fagiano, L.: Systems of tethered multicopters: modeling and control design. IFAC-PapersOnLine **50**(1), 4610–4615 (2017)
11. Al-Radaideh, A., Sun, L.: Self-localization of a tethered quadcopter using inertial sensors in a GPS-denied environment. In 2017 International Conference on Unmanned Aircraft Systems (ICUAS), pp. 271–277. IEEE (2017)
12. Vishnevsky, V., Meshcheryakov, R.: Experience of developing a multifunctional tethered high-altitude unmanned platform of long-term operation. In: Ronzhin, A., Rigoll, G., Meshcheryakov, R. (eds.) ICR 2019. LNCS (LNAI), vol. 11659, pp. 236–244. Springer, Cham (2019). https://doi.org/10.1007/978-3-030-26118-4_23
13. Kozyrev, D.V., Phuong, N.D., Houankpo, H.G.K., Sokolov, A.: Reliability evaluation of a hexacopter-based flight module of a tethered unmanned high-altitude platform. In: Vishnevskiy, V.M., Samouylov, K.E., Kozyrev, D.V. (eds.) DCCN 2019. CCIS, vol. 1141, pp. 646–656. Springer, Cham (2019). https://doi.org/10.1007/978-3-030-36625-4_52
14. Perelomov, V.N., Myrova, L.O., Aminev, D.A., Kozyrev, D.V.: Efficiency enhancement of tethered high altitude communication platforms based on their hardware-software unification. In: Vishnevskiy, V.M., Kozyrev, D.V. (eds.) DCCN 2018. CCIS, vol. 919, pp. 184–200. Springer, Cham (2018). https://doi.org/10.1007/978-3-319-99447-5_16

15. Vishnevsky, V., Tereschenko, B., Tumchenok, D., Shirvanyan, A.: Optimal method for uplink transfer of power and the design of high-voltage cable for tethered high-altitude unmanned telecommunication platforms. In: Vishnevskiy, V.M., Samouylov, K.E., Kozyrev, D.V. (eds.) DCCN 2017. CCIS, vol. 700, pp. 240–247. Springer, Cham (2017). https://doi.org/10.1007/978-3-319-66836-9_20
16. Vishnevsky, V.M., Tereschenko, B.N.: Russian Federation Patent - 2572822 "Method of remote power for objects by wire". The patent is registered in the state register of inventions of the Russian Federation. Accessed 16 Dec 2015
17. Ferreira de Castro, D., Santos, J.S., Batista, M., Antônio dos Santos, D., Góes, L.: Modeling and control of tethered unmanned multicopters in hovering flight. In: AIAA Modeling and Simulation Technologies Conference, pp. 2333–2342 (2015)
18. Dinh, T.D., Pham, V.D., Kirichek, R., Koucheryavy, A.: Flying network for emergencies. In: Vishnevskiy, V.M., Kozyrev, D.V. (eds.) DCCN 2018. CCIS, vol. 919, pp. 58–70. Springer, Cham (2018). https://doi.org/10.1007/978-3-319-99447-5_6
19. Dinh, T.D., Le, D.T., Tran, T.T.T., Kirichek, R.: Flying ad-hoc network for emergency based on IEEE 802.11p multichannel MAC protocol. In: Vishnevskiy, V.M., Samouylov, K.E., Kozyrev, D.V. (eds.) DCCN 2019. LNCS, vol. 11965, pp. 479–494. Springer, Cham (2019). https://doi.org/10.1007/978-3-030-36614-8_37
20. IEEE Standards Association. 802.11 p-2010-IEEE standard for information technology-local and metropolitan area networks-specific requirements-part 11: Wireless lan medium access control (mac) and physical layer (phy) specifications amendment 6: Wireless access in vehicular environments (2010). http://standards.ieee.org/findstds/standard/802.11p-2010.html
21. Paredes, J. A., Saito, C., Abarca, M., Cuellar, F.: Study of effects of high-altitude environments on multicopter and fixed-wing UAVs' energy consumption and flight time. In: 2017 13th IEEE Conference on Automation Science and Engineering (CASE), pp. 1645–1650. IEEE (2017)
22. Kirichek, R., Kulik, V.: Long-range data transmission on flying ubiquitous sensor networks (FUSN) by using LPWAN protocols. In: Vishnevskiy, V.M., Samouylov, K.E., Kozyrev, D.V. (eds.) DCCN 2016. CCIS, vol. 678, pp. 442–453. Springer, Cham (2016). https://doi.org/10.1007/978-3-319-51917-3_39
23. Goudarzi, S., Kama, N., Anisi, M.H., Zeadally, S., Mumtaz, S.: Data collection using unmanned aerial vehicles for internet of things platforms. Comput. Electric. Eng. **75**, 1–15 (2019)
24. Kirichek, R., Vladyko, A., Paramonov, A., Koucheryavy, A.: Software-defined architecture for flying ubiquitous sensor networking. In: 2017 19th International Conference on Advanced Communication Technology (ICACT), pp. 158–162. IEEE (2017)

Multipath Redundant Network Protocol Without Delivery Guarantee

Ilya Noskov[1](✉) and Vladimir Bogatyrev[1,2]

[1] ITMO University, Saint Petersburg, Russia
noskovii@mail.ru

[2] Saint-Petersburg State University of Aerospace Instrumentation, Saint Petersburg, Russia
vladimir.bogatyrev@gmail.com

Abstract. Timely and faultless packets delivery problem in real-time systems is described in the paper. New protocol based on multipath redundant transmissions with high probability of faultless and timely delivery is presented. This approach is based on UDP protocol and using redundant transmissions via multipath reserve channels between a client and a server. The multiplicative criteria based on faultless and timely packets delivery probability and average delivery time reserve relatively delivery time restriction defined in the real-time computer system is used for efficiency evaluation for time sensitive systems in this paper. Faultless and timely delivery packet probability and faultless probability were used as a criterion for developed protocol efficiency analyzing in systems without strong delivery packet time limitations or packet delivery time insensitive systems. Effective using areas of developed protocol are described in the paper. Developed protocol provides faultless and timely delivery packets improvements for different type of systems (real-time and time insensitive). The efficiency of redundant multipath transmissions is analyzed and researched using obtained results from experiments with simulation models developed in OMNeT++ environment. This paper can be useful for network engineers who develop new transport or application layer protocols to provide reliable network transmissions in computer networks. This paper can be considered as a theoretical base for developing new multipath redundant protocol implementation and using it in real computer networks.

Keywords: Multipath transmissions · Network protocol · Faultless and timely delivery probability · Redundancy coefficient · Real-time and high availability systems · Critical to delays packets · OMNeT++ · UDP

1 Introduction

Nowadays there are many research works and papers [1–8] described modern computer networks problems. Authors provide new ways and solutions that help

V. M. Vishnevskiy et al. (Eds.): DCCN 2020, LNCS 12563, pp. 39–51, 2020.
https://doi.org/10.1007/978-3-030-66471-8_4

to improve quality of network interconnection between nodes. Fundamental problems of designing and developing information and communication systems are represented in [9–13]. Security issues of computer networks are considered in [14,15]. At the present time computer modelling is widely used in many areas of engineering and researching because this approach helps us to save resources for building real expensive systems. Authors provide different models for network traffic analyzing and acceleration of simulation modelling in [16,17]. Fault tolerance of cluster configurations and exchanging problems between systems are described in [18,19]. But these approaches don't provide real network protocol solutions for solving fault tolerance problems in cluster systems which are built with using computer network infrastructure. In papers [20–22] authors provide researching connected with redundant multipath transmissions, but they didn't consider influence of real network protocols in developed models and didn't provide suggestions for improving or developing new reliable multipath network protocols. Real-time communication system reliability is associated not only with supporting the availability, fault tolerance and reliability of the system structure, but also with timely delivery of critical to delays packets in real-time computer networks which provide computer communications in the client-server architecture [23]. Network engineers change physical network topology (add new links between routers/switches, add new network equipment etc.) or/and use different network protocols (FHRP, MPTCP, SCTP etc.) to achieve better reliability in computer networks. Approach with using new protocols is more suitable because it does not require new equipment and provides only software devices upgrading in order to achieve better network characteristics and user experience. It is important from economy point of view because we don't need to buy new equipment.

In the paper [24] authors provide survey of recent transport layer protocols. They describe new congestion control algorithms used in transport layer for better network performance, provide information about new transport layer protocols which control of packets delivery and use connection establishment between nodes before sending data. These protocols are developed based on TCP protocol architecture. Also they mentioned about multipath modification of TCP - MPTCP. But there are no papers about multipath redundant transmissions based on UDP protocol without delivery guarantee. But these protocols are widely used in real-time computer systems and delay sensitive applications.

UDP is transport layer network protocol which does not use handshaking for connection establishment and does not ensure of packets delivery. Time sensitive applications often use UDP for real-time traffic sending scenario because new packets have bigger priority and loss out of date packets and receiving new packets is more preferable than waiting retransmissions for lost packets. In this cases packets are becoming outdated very quickly, and retransmissions of lost packets (like in TCP protocol) is not suitable for such systems. That is why it is important to research and improve UDP transport protocol for time sensitive systems and applications.

Developing and using new transport layer protocols needs to change kernel source code of operating system on communication nodes to provide opportunity to using new protocol. In most cases we cannot upgrade kernel without shutdown, also some operating systems are proprietary and we have no access to source code for its modification. That is why developing new reliable protocol over existing transport protocols is more suitable and scalable solution. New modification will be able to easy integrated to different systems and we shouldn't change kernel source code.

The main aim of this work is the developing simulation model of new multipath redundant transmissions protocol prototype based on UDP in the OMNeT++ environment and find out effective using areas of this protocol.

2 Developing Model of UDP Protocol Multipath Extension

Simulation environment OMNeT++ is modern specialized tool for simulating and researching computer network models. This environment contains a large library of real network protocols and equipment models [25]. There are many models developed in this environment [26–29]. It shows that it is flexible tool with different layers network protocols implementations.

UDP is used as a transport protocol for our prototype. The OMNeT++ environment has implementations of various types of generator and sink applications which use UDP transport protocol. There are UDPBasicApp class and UDPSink class implementations in OMNeT++ environment. These classes provide functionality of client and server UDP-based applications. But these applications don't support redundant multipath transmissions. New classes of generator and sink applications have been developed in OMNeT++ environment using C++ programming language in order to provide UDP-based protocol with a redundant transmissions opportunity.

New UDP application classes extend of base classes and allow to specify several addresses in configuration file to provide redundant transmission via sending copy of datagrams to these addresses. Port number and packets length

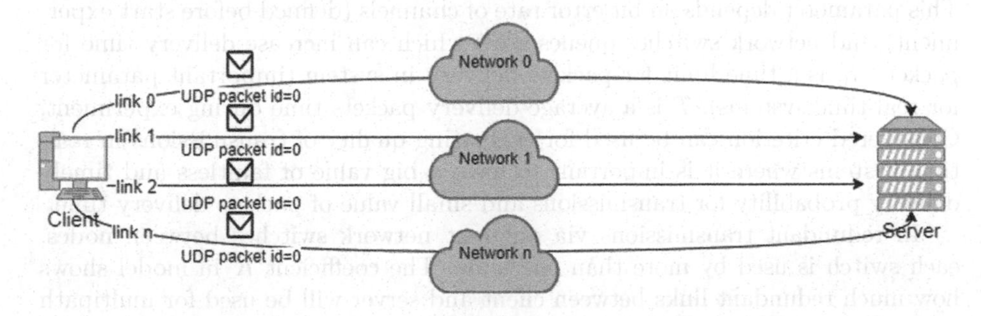

Fig. 1. Multipath transmissions using UDP-based protocol.

are set in configuration file. In Fig. 1 you can see example of the traffic flow session between client and server using developed protocol based on UDP multipath transmissions.

User should set n interfaces which will be used for multipath data transmissions in new protocol. After that UDP sockets are created for n selected interfaces on the server and the client sides. UDP-based protocol is using all created sockets in one session for providing multipath data transmissions of user data flow. The protocol is creating copy of each datagram and sending copies via all sockets for current session. UDP datagrams are being transmitted from the client to the server via different physical channels simultaneously. New protocol header has ID to identify duplicates on the receiver side. The server side application recognizes copies in received data and drop extra packets. Fastest arrived datagram will be saved and transferred to application for further processing and providing data to user.

3 Developing Computer Network Model Using Modified UDP Protocol

The simulation model of system with server, five network switches and five clients was developed. This model is based on a EtherSwitch model which represents model of network switch and a StandardHost model which simulates client and server behaviour. Figure 2 shows this model in the OMNeT++ environment.

In this model network switches with connected to them clients and server present different network segments which can be used for redundant transmissions.

The criterion M has been used for efficiency evaluation of redundant transmissions.

$$M = P(t_0 - T) \tag{1}$$

This (1) represents the multiplicative criteria based on faultless and timely packets delivery probability and average delivery time reserve relatively delivery time restriction. P is a probability of timely (packet should be received before t_0 comes) and faultless packets delivery. Value of P is calculated after experiment. This parameter depends on bit error rate of channels (defined before start experiment) and network switches queues sizes which can increase delivery time for packets. t_0 is a time limit for packets delivery in system (important parameter for real-time systems). T is a average delivery packets time during experiment. Considered criterion can be used for evaluating quality of transmissions in real-time systems where it is important to have a big value of faultless and timely delivery probability for transmissions and small value of packets delivery time.

In redundant transmissions via different network switches between nodes, each switch is used by more than one node. The coefficient K in model shows how much redundant links between client and server will be used for multipath transmissions. We carried out simulation experiments with different values of the redundancy coefficient K.

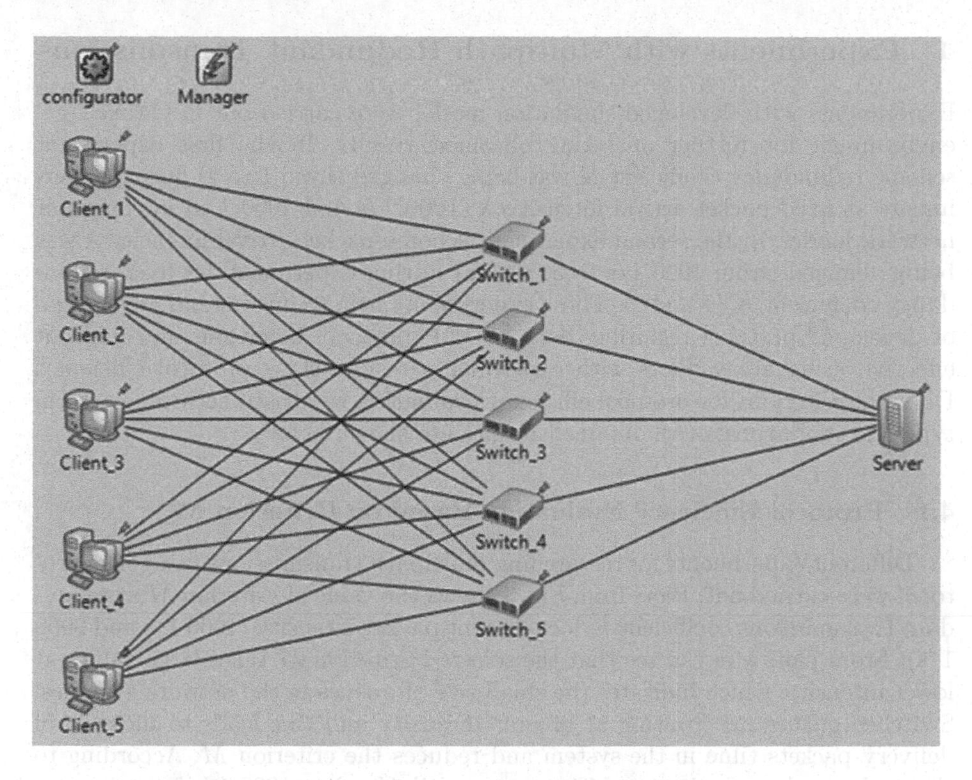

Fig. 2. Redundant computer network model in OMNeT++ environment.

These parameters have been used in simulation experiments with redundant transmissions: $L = 10$ Mbit/s - communication channels throughput. $B = 0.0001$ - bit error probability for channel, $\lambda = 1000$ 1/s - packet arrival rate, packets length for this simulation process is 100 B. Network delay time for 100 B packet for two links (client-switch and switch-server) has been calculated. This value includes data sending time according to throughput value and propagation delay in channel. Processing packet time on nodes (client, switch and server) was added to the total time. Obtained value was multiplied by two times and considered as a delivery time limit $t_0 = 0.0004$ s for our real-time system. This value can be different for other real-time systems and depends on limitations for delivery packet time.

All described parameters are set in OMNeT++ environment in $*.ini$ (general simulation experiment parameters), $*.ned$ (contains network model topology description) and $*.xml$ (equipment parameters) configuration files.

4 Experiments with Multipath Redundant Transmissions

Experiments with developed simulation model were carried out in OMNeT++ environment for further analyzing obtained results. In the first experiment scheme redundancy coefficient K was being changed (from 1 to 4) during experiments at fixed packet arrival intensity λ (1000 1/s and 2000 1/s) in computer network model. In the second experiment scheme packet arrival intensity λ was being changed (from 1000 1/s to 5000 1/s) during experiments at fixed redundancy coefficient K (1 and 2). These experiments help to understand advantages of developed protocol regarding default UDP protocol behaviour and find out effective using areas. Plots with experiments results show protocol efficiency. Different criterions for protocol efficiency evaluation were used according system type and its purposes. All obtained results are shown below.

4.1 Protocol Efficiency Evaluation Based on Criterion M

Different experiments for researching multipath transmissions using new protocol were carried out. Plots from Fig. 3 shows the value of criterion M at redundant transmissions coefficient K for different packet intensity (1000 1/s and 2000 1/s). From plots you can see that the selected criterion M takes large values at lower intensity, which indicates the small size of queues in the network switches. Switches queues are growing at greater intensity and this leads to increase of delivery packets time in the system and reduces the criterion M. According to these plots value of criterion M increases until $K=3$ at 1000 1/s intensity and

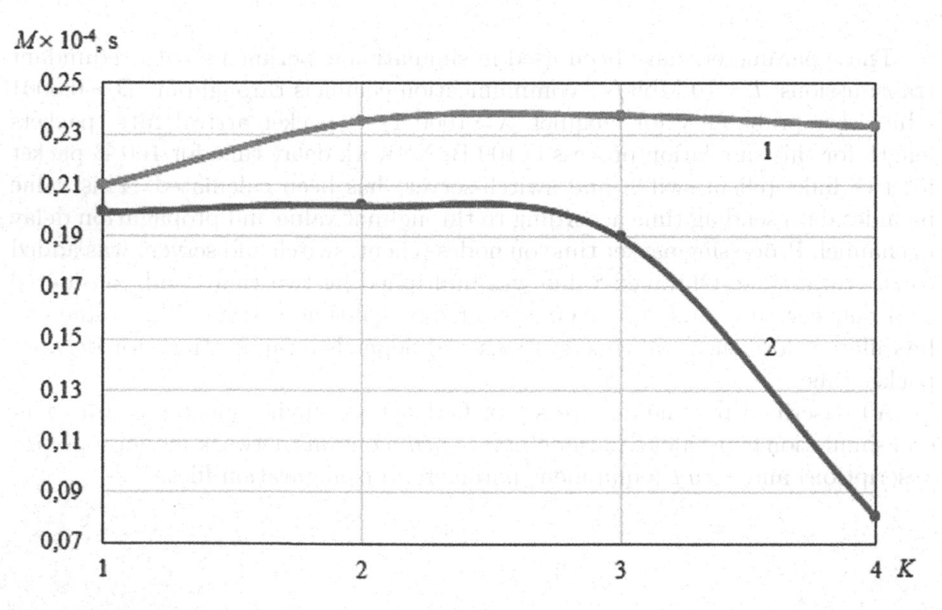

Fig. 3. The dependence of the efficiency criterion M on the redundancy coefficient K: at an intensity of 1000 1/s (curve 1); at an intensity of 2000 1/s (curve 2).

increases until $K = 2$ at 2000 1/s intensity. It says that we increase of faultless and timely delivery probability in developed real-time system simulation model using new protocol. In this network configuration we can conclude that packets transmissions with lower intensity is more efficient on all values of redundancy coefficient K. In the curve 2 you can see sharp decline at $K = 4$. It allows to make conclusion that switch buffers are overflowed and packets were being dropped.

Increasing the number of redundant transmissions helps us to reduce packets lost probability. But the end to end packets delay in the system increases because queues in network switches are growing. Delays influence on probability of timely delivery packets and reduce value of criterion M. Efficiency criterion represents multiply probability of timely and faultless packet delivery and average delivery time reserve relatively delivery time restriction. This approach helps us to consider packet delivery time as a part of our criterion. It is needed for real-time systems where delivery time is very important characteristic of interaction because in such systems information is being out of date very fast. For systems where delivery time is not critical we can consider probability of faultless packet delivery and probability of faultless and timely delivery packets as a main efficiency criterion. In this case we don't consider delivery time reserve relatively delivery time restriction. It can be useful for system without strong packets delivery time limitations.

Figure 4 shows the criterion M at different packets arrival intensity for different values of redundant transmissions coefficient K (1 and 2). As you can see from plots increasing of redundancy coefficient is not effective for all area.

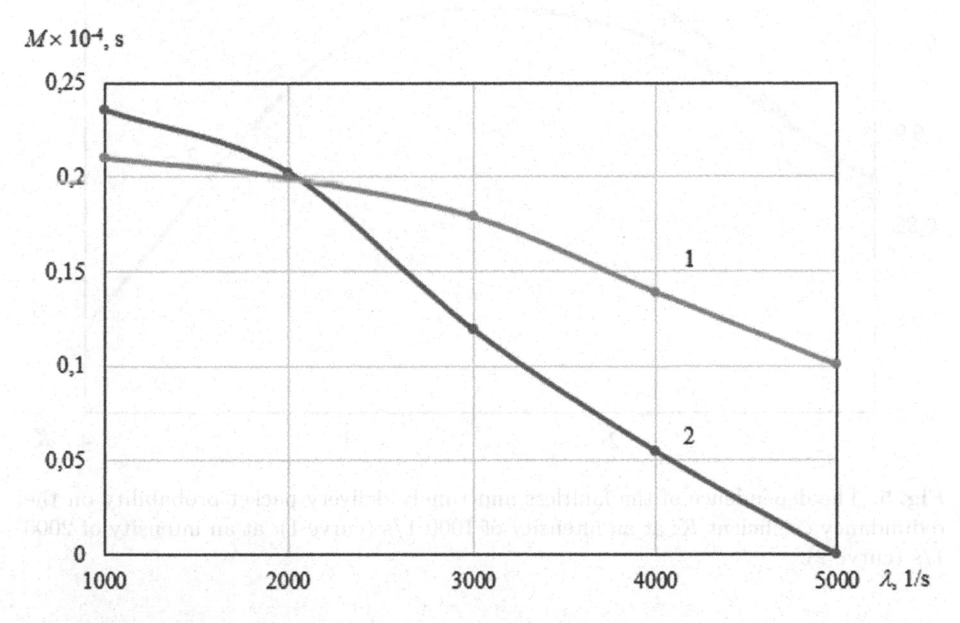

Fig. 4. The dependence of the efficiency criterion M on the intensity of packet arrival: with the redundancy coefficient $K = 1$ (curve 1); with the redundancy coefficient $K = 2$ (curve 2).

There is area in which transmissions with a high redundancy coefficient is more efficient on the same network configuration. After overcoming the intensity threshold of 2000 1/s, the redundant model is becoming less efficient than model without reservation. Increasing traffic in the network in redundant model leads to the queues growing in network switches. That is why delivery time of packets is growing. This case is unacceptable for real-time systems. But the probability of delivery packets increases using redundant transmissions. It is important for delivery time insensitive systems.

4.2 Protocol Efficiency Evaluation Based on Faultless and Timely Delivery Probability as a Efficiency Criterion

Faultless and timely delivery probability as a efficiency criterion can be used for protocol evaluation in systems without strong time limitations to packet delivery. For delivery time insensitive systems faultless and timely delivery probability is more important criteria than M. At the Fig. 5 you can see the dependence of the faultless and timely delivery packet probability on the redundancy coefficient K at an packet arrival intensity $\lambda = 1000$ 1/s and 2000 1/s. In these cases faultless and timely delivery packet probability increases for all K value excluding $K = 4$.

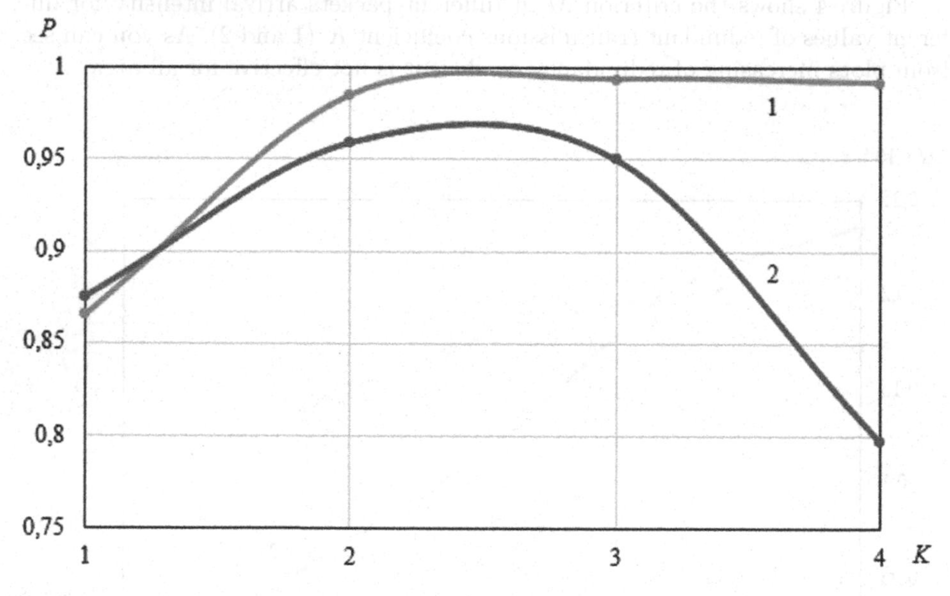

Fig. 5. The dependence of the faultless and timely delivery packet probability on the redundancy coefficient K: at an intensity of 1000 1/s (curve 1); at an intensity of 2000 1/s (curve 2).

Figure 6 shows faultless and timely delivery packet probability at packets arrival intensity for different value of redundancy coefficient K (1 and 2).

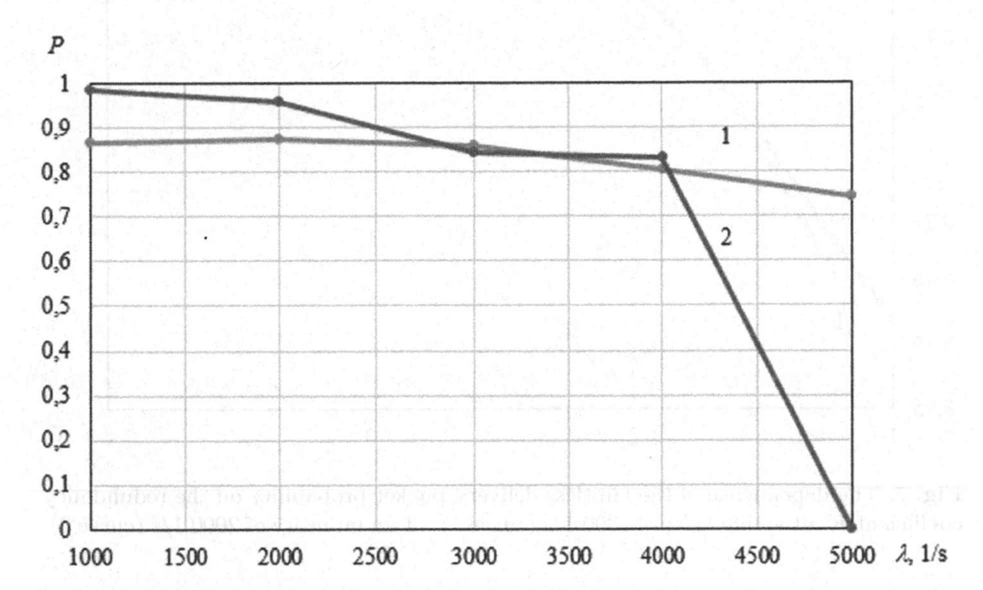

Fig. 6. The dependence of the faultless and timely delivery packet probability on the intensity of packet arrival: with the redundancy coefficient $K = 1$ (curve 1); with the redundancy coefficient $K = 2$ (curve 2).

From plots above we can see that redundant transmissions help to achieve bigger packet delivery probability in most cases until $\lambda = 4000$ (after that packets are dropped because switches queues are overflowed). It helps us to transmit data with more faultless and timely delivery probability in computer networks without strong delivery time limitations using new developed protocol. This approach is useful in data centers which backup big amount of user data or servers where faultless data transmissions more important criteria than data delivery time.

4.3 Protocol Efficiency Evaluation Based on Faultless Delivery Probability as a Efficiency Criterion

For systems without packet delivery time limitations we can use faultless delivery probability as a efficiency criterion for such systems. Figure 7 shows that developed protocol allows to increase faultless delivery probability in the all area.

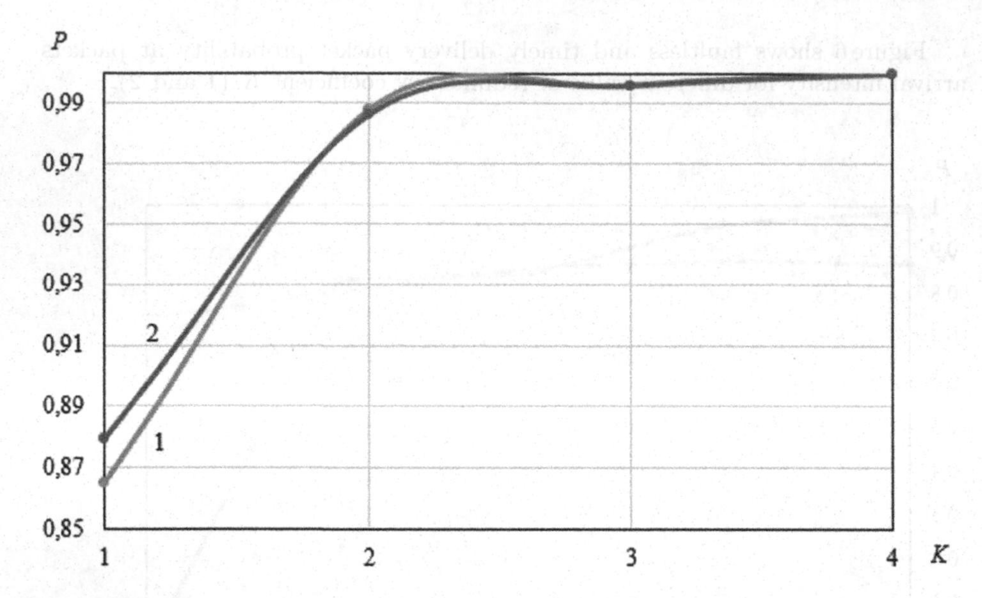

Fig. 7. The dependence of the faultless delivery packet probability on the redundancy coefficient K: at an intensity of 1000 1/s (curve 1); at an intensity of 2000 1/s (curve 2).

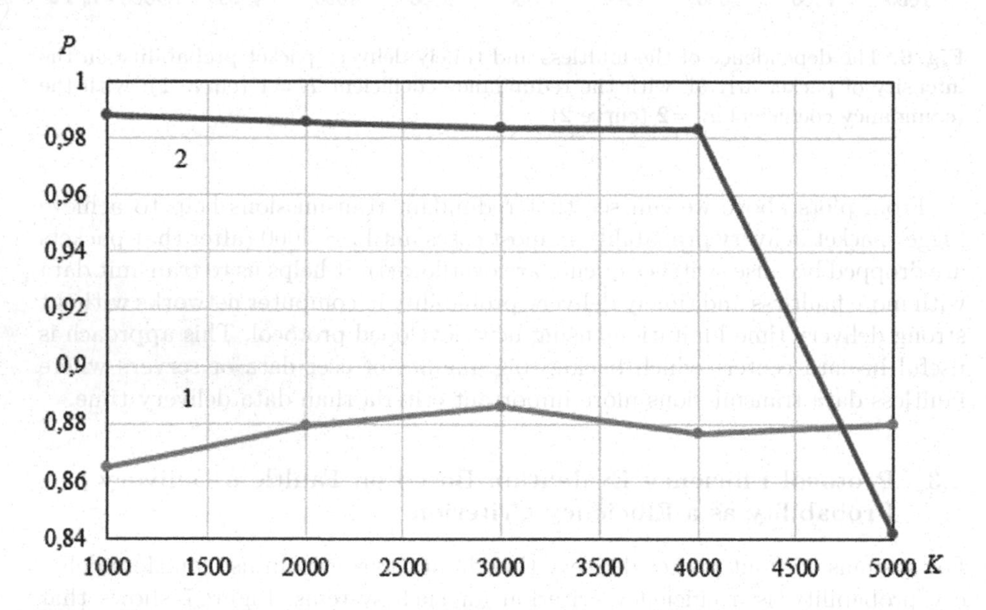

Fig. 8. The dependence of the faultless delivery packet probability on the intensity of packet arrival: with the redundancy coefficient $K = 1$ (curve 1); with the redundancy coefficient $K = 2$ (curve 2).

Figure 8 shows advantages of new protocol in wide range of packet arrival intensity. But if λ more than 4000 1/s we can see decrease of faultless probability because queues in network equipment are full. That is why we can conclude that developed protocol helps us to increase faultless delivery packet probability and may be used as a main protocol for time insensitive systems and applications.

5 Conclusions

The simulation model of computer network with the possibility of increasing the redundancy of packets transmissions has been developed in the OMNeT++ environment. Experiments to assess the effectiveness of packets transmissions with different intensity and redundancy coefficient was carried out. Areas with effective using of redundant transmissions in computer networks have been described for systems with packet delivery time restrictions and delivery time insensitive systems. Developed model allows to transmit packets via several physical channels and provides redundant data transfer. Modifications of UDP protocol on the layer above are proposed. The presented results can be used in the design of high-reliable real-time computer network systems based on UDP protocol with strong restrictions to packet delivery time. This research can be used as a theoretical and practice base to develop new transport or application layer protocols based on UDP and provides more reliable and timely transmissions of important data in real-time computer systems.

Found out areas of effective using of developed protocol prototype based on UDP multipath transmissions. Redundancy coefficient is should be chosen according buffers sizes of network switches and their throughput. For delay insensitive applications we can use multipath redundant transmissions more widely (using faultless and timely delivery probability as a criterion instead of M).

We are planning to carry out more experiments in different scenarios and compare developed protocol performance with other protocols. After that proof of concept will be developed to demonstrate protocol work in real computer networks.

References

1. Birman, K.P., Joseph, T., Raeuchle, T., El Abbadi, A.: Implementing fault-tolerant distributed objects. IEEE Trans. Softw. Eng. **11**(6), 502–508 (1985)
2. Coulouris, G., Dollimore, J., Kindberg, T., Blair, G.: Distributed Systems: Concepts and Design, 5th edn. Addison-Wesley, Boston (2011)
3. Defago, X., Schiper, A., Sergent, N.: Semi-passive replication. In: Proceedings of the 17th IEEE Symposium on Reliable Distributed Systems (SRDS), West Lafayette, IN, USA, pp. 43–50 (1998)
4. Gunnar, A., Johansson, M.: Robust load balancing under traffic uncertainty-tractable models and efficient algorithms. Telecommun. Syst. **48**(1–2), 93–107 (2011)
5. Kim, Y., Righter, R., Wolff, R.: Job replication on multiserver systems. Adv. Appl. Prob. **41**, 546–575 (2009)
6. Kurose, J.F., Ross, J.F.: Computer Networking: A Top-Down Approach, 6th edn. Pearson, Boston (2013)

7. Malichenko, D.: Optimization of network overhead for transport layer coding. In: 9th Conference of Open Innovations Community FRUCT, pp. 92–95 (2011)
8. Lee, M.H., Dudin, A.N., Klimenok, V.I.: The SM/M/N queueing system with broadcasting service. Math. Probl. Eng., Article ID 98171 (2006)
9. Sorin, D.: Fault Tolerant Computer Architecture. Morgan Claypool, San Rafael (2009)
10. Kopetz, H.: Real-Time Systems: Design Principles for Distributed Embedded Applications. Springer, Heidelberg (2011). https://doi.org/10.1007/978-1-4419-8237-7
11. Zakoldaev, D.A., Korobeynikov, A.G., Shukalov, A.V., Zharinov, I.O.: Workstations industry 4.0 for instrument manufacturing. In: IOP Conference Series: Materials Science and Engineering, vol. 665, p. 012015. IOP Publishing (2019). https://doi.org/10.1088/1757-899X/665/1/012015
12. Zakoldaev, D.A., Korobeynikov, A.G., Shukalov, A.V., Zharinov, I.O.: Cyber and physical systems technology classification for production activity of the Industry 4.0 smart factory. In: IOP Conference Series: Materials Science and Engineering, vol. 582, no. 1, p. 012007 (2019)
13. Astakhova, T.N., Verzun, N.A., Kasatkin, V.V., Kolbanev, M.O., Shamin, A.A.: Sensor network connectivity models. Informatsionno-upravliaiushchie sistemy (5), 38–50 (2019). https://doi.org/10.31799/1684-8853-2019-5-38-50
14. Vishnevskii, V.M.: Teoreticheskie osnovy proektirovaniya (Theoretical Foundations of Design). Tekhnosfera, Moscow (2003)
15. Aliev, T.: The synthesis of service discipline in systems with limits. In: Vishnevsky, V., Kozyrev, D. (eds.) DCCN 2015. CCIS, vol. 601, pp. 151–156. Springer, Cham (2016). https://doi.org/10.1007/978-3-319-30843-2_16
16. Poymanova, E.D., Tatarnikova, T.M.: Models and methods for studying network traffic. In: 2018 Wave Electronics and its Application in Information and Telecommunication Systems (WECONF), pp. 1–5 (2018). https://doi.org/10.1109/WECONF.2018.8604470
17. Kutuzov, O., Tatarnikova, T.: On the acceleration of simulation modeling. In: Proceedings of 2019 22nd International Conference on Soft Computing and Measurements, SCM 2019, pp. 45–47 (2019). https://doi.org/10.1109/SCM.2019.8903785
18. Bogatyrev, V.A.: Fault tolerance of clusters configurations with direct connection of storage devices. Autom. Control Comput. Sci. **45**(6), 330–337 (2011)
19. Bogatyrev, V.A.: Exchange of duplicated computing complexes in fault-tolerant systems. Autom. Control Comput. Sci. **45**(5), 268–276 (2011)
20. Bogatyrev, V.A., Parshutina, S.A.: Redundant distribution of requests through the network by transferring them over multiple paths. In: Vishnevsky, V., Kozyrev, D. (eds.) DCCN 2015. CCIS, vol. 601, pp. 199–207. Springer, Cham (2016). https://doi.org/10.1007/978-3-319-30843-2_21
21. Bogatyrev, V.A., Parshutina, S.A.: Efficiency of redundant multipath transmission of requests through the network to destination servers. In: Vishnevskiy, V.M., Samouylov, K.E., Kozyrev, D.V. (eds.) DCCN 2016. CCIS, vol. 678, pp. 290–301. Springer, Cham (2016). https://doi.org/10.1007/978-3-319-51917-3_26
22. Bogatyrev, A.V., Bogatyrev, V.A., Bogatyrev, S.V.: Transmission, multipath redundant, with packet segmentation. In: 2019 Wave Electronics and its Application in Information and Telecommunication Systems (WECONF), Saint-Petersburg, Russia, pp. 1–4 (2019). https://doi.org/10.1109/WECONF.2019.8840643

23. Noskov, I.I., Bogatyrev, V.A.: Interaction model of computer nodes based on transfer reservation at multipath routing. In: 2019 Wave Electronics and its Application in Information and Telecommunication Systems (WECONF), p. 8840607 (2019). https://doi.org/10.1109/WECONF.2019.8840607. https://ieeexplore.ieee.org/document/8840607
24. Polese, M., Chiariotti, F., Bonetto, E., Rigotto, F., Zanella, A., Zorzi, M.: A survey on recent advances in transport layer protocols. CoRR, abs/1810.03884 (2018)
25. Varga, A., Hornig, R.: An overview of the OMNeT++ simulation environment. In: Simulation Tools and Techniques for Communications, Networks and Systems Workshops, Simutools 2008 (2008)
26. Vesely, V., Rek, V., Rysavy, O.: Enhanced interior gateway routing protocol with IPv4 and IPv6 support for OMNeT++. In: Advances in Intelligent Systems and Computing, vol. 2015, no. 1, pp. 65–85 (2016). ISSN 2194-5357
27. Vesely, V., Sveda, M.: L2 Protocols in OMNeT++. IP Networking 1 - Theory and Practice, pp. 37–40. Zilina University Publisher, Zilina (2012)
28. Vesely, V., Rysavy, O., Sveda, M.: Protocol independent multicast in OMNeT++. In: The International Academy, pp. 132–137. Research and Industry Association (2014)
29. Noskov, I.I., Bogatyrev, V.A.: Simulating of fault-tolerant gateway based on VRRP protocol in OMNeT++ environment. In: CEUR Workshop Proceedings - 2019, vol. 2522 (2019). https://www.scopus.com/record/display.uri?eid=2-s2.0-85077504578&origin=inward&txGid=702b35ab02dafa1f36315620c44296b4. ISSN 16130073

Modelling Multi-connectivity in 5G NR Systems with Mixed Unicast and Multicast Traffic

Vitalii Beschastnyi[1(✉)], Daria Ostrikova[1], Sergey Melnikov[1], and Yuliya Gaidamaka[1,2]

[1] Peoples' Friendship University of Russia (RUDN University), 6 Miklukho-Maklaya St, Moscow 117198, Russian Federation
{beschastnyy-va,ostrikova-dyu}@rudn.ru, melnikov@linfotech.ru

[2] Federal Research Center "Computer Science and Control" of the Russian Academy of Sciences (FRC CSC RAS), 44-2 Vavilov St, Moscow 119333, Russian Federation
gaydamaka-yuv@rudn.ru

Abstract. 3GPP New Radio (NR) radio access technology operating in millimeter wave (mmWave) frequency band is considered as key enabler for Fifth-generation (5G) mobile system. Despite the enormous available bandwidth potential, mmWave signal transmissions suffer from fundamental technical challenges like severe path loss, sensitivity to blockage, directivity, and narrow beamwidth, due to its short wavelengths. To address the problem of quality degradation due to the line-of-sight (LoS) blockage by various objects in the channel, 3GPP is currently working on multi-connectivity (MC) mechanisms that allow a user to remain connected to several mmWave access points simultaneously as well as switch between them in case its active connection drops. In this paper, exploiting the methods of stochastic geometry and queuing theory we propose a model of 5G NR base station (BS) serving a mixture of unicast and multicast traffic. MC techniques is proposed to be used for cell-edge users. The proposed model is validated against computer simulations in terms of session drop probabilities and system resource utilization metrics. Our findings are illustrated with a numerical example.

Keywords: 5G NR · mmWave · Multi-connectivity · simulation

1 Introduction

Every year, user needs and the number of device connections are increasing. According to the Cisco statistics [1], mobile video traffic has been believed to

The publication has been prepared with the support of the "RUDN University Program 5–100" (V.A. Beschastnyi, original draft preparation; D.Yu. Ostrikova, visualization and validation). The reported study was funded by RFBR, projects Nos. 18-07-00156 (Yu.V. Gaidamaka, conceptualization and methodology) and 18-07-00576 (Yu.V. Gaidamaka, supervision and project administration).

V. M. Vishnevskiy et al. (Eds.): DCCN 2020, LNCS 12563, pp. 52–63, 2020.
https://doi.org/10.1007/978-3-030-66471-8_5

increase by 50 percent annually. Due to the growing popularity of multifunctional multimedia applications and services, wireless mobile networks must constantly evolve, offering higher data rates, as well as reducing data transfer delays and increasing energy efficiency, which leads to an increase in the quality of service for end users. 3GPP New Radio (NR) radio access technology standardized by 3rd Generation Partnership Project (3GPP) [2] is expected to play a key role in 5G systems. NR systems promise multi-gigabit rates together with reduced latency at the air interface compared to 4G Networks. Despite the enormous available bandwidth potential, mmWave signal transmissions suffer from fundamental technical challenges like severe path loss, sensitivity to blockage, directivity, and narrow beamwidth, due to its short wavelengths. To effectively support system design and deployment, accurate channel modeling comprising several 5G technologies and scenarios is essential [3,4]. Though mmWave frequency band offers huge amount of available resources, the lack of PTM capabilities may imply a future limitation of 5G networks, leading to inefficient service provisioning and utilization of the network and spectrum resources [5–7].

In indoor scenarios 5G NR systems mostly suffer from obstacles such as humans and vehicles, which generally tend to be mobile and are often termed "blockers" [8]. For the sake of block error probability mitigation at the air interface a user equipment (UE) experiencing such type of blockage may lower its modulation and coding scheme (MCS) that depends on propagation environment and the distance to NR base station (BS), or enter outage conditions in case further lowering of MCS is impossible [9]. Multi-connectivity operation recently proposed by 3GPP is aimed at solving the problem of outage conditions by maintaining several simultaneously active links for adjacent NR BSs so that the connection is redirected between them in case of blockage [10,11]. However, when lowering MCS, additional physical resources are utilized to support the required rate at the air interface. Once the required amount of resource in not available, a session should either reduce its rate, or it should be dropped [3].

In [12] authors introduce a joint scheduling framework based on dynamic MC to satisfy the different requirements of enhanced mobile broadband (eMBB) and ultra-reliability low-latency communication (URLLC) traffics in 5G. The framework aims to optimize multiple traffics in independent link dynamically. BSs will form MC clusters, which vary with network characteristics to accomplish high rate. Maximum capacity of eMBB traffic is achieved while the latency of the uRLLC traffics is guaranteed. The authors of [13] characterize the outage probability and spectral efficiency associated with different degrees of MC in a typical 5G urban scenario, where the line-of-sight propagation path can be blocked by buildings as well as humans. These results demonstrate that the degrees of MC of up to 4 offer higher relative gains.

In this paper, we consider a 5G NR BS deployment serving mixture of unicast and multicast sessions. MC techniques is proposed to be used for cell-edge users. First, by means of stochastic geometry we derive the amount of required resource in terms of average values that will further simplify our computations. Then, we present our simulation approach and finally validate our model against computer simulations. Our findings showed the advantages of multicast sessions over unicast ones in terms of reliability and resource consumption.

The remainder is organized as follows. First, our system model of NR BS operation under unicast and multicast traffic load introduced in Section 2. The simulation approach is offered in Sect. 3. Numerical results illustrating the resource occupation effect of multicast sessions are presented in Section 4. Conclusions are drawn in the last section.

2 System Model

This section provides an overview of the considered scenario with 5G NR Base Station that serves unicast (point-to-point, PTP) and multicast (point-to-multipoint, PTM) sessions[14]. As we assume only rather small mobile obstacles in our model, the BS has circularly-shaped coverage area of radius d^E_{LoS}. We assume random distribution of users that follows Poisson Point Process (PPP) with density parameter ρ. The coverage area radius can be evaluated using the mmWave propagation model mentioned below together with MCSs what gives us the SNR threshold at which the session can be no longer served at current BS. The mapping between CQIs and spectral efficiency for 3GPP NR systems is given in Table 1 where LoS ad n nLoS ranges are estimated for the default system paramenters presented in Sect. 4. This approach follows [3], where authors compute mean demand for the model without multiconnectivity feature.

Table 1. CQI, MCS, and distance mapping.

CQI, i	MCS	$E(i)$	SNR, [dB]	LoS range, [m]	nLoS range [m]
1	QPSK, 78/1024	0,15237	-9,478	758–1033	146–199
2	QPSK, 120/1024	0,2344	-6,658	572–758	111–146
3	QPSK, 193/1024	0,377	-4,098	445–572	86–111
4	QPSK, 308/1024	0,6016	-1,798	350–445	68–86
5	QPSK, 449/1024	0,877	0,399	280–350	54–68
6	QPSK, 602/1024	1,1758	2,424	223–280	43–54
7	16QAM, 378/1024	1,4766	4,489	182–223	35–43
8	16QAM, 490/1024	1,9141	6,367	145–182	28–35
9	16QAM, 616/1024	2,4063	8,456	119–145	23–28
10	16QAM, 466/1024	2,7305	10,266	96–119	18–23
11	16QAM, 567/1024	3,3223	12,218	78–96	15–18
12	16QAM, 666/1024	3,9023	14,122	64–8	12–15
13	16QAM, 772/1024	4,5234	15,849	52–64	10–12
14	16QAM, 873/1024	5,1152	17,786	42–64	8–10
15	16QAM, 948/1024	5,5547	19,809	0–42	0–8

When establishing a user session, BS allocates the amount of physical resource that is generally a random variable and depends on propagation model and the distance from UE to the serving BS. According to [15,16], the mmWave linear path loss L in dBs for UEs in LoS and nLoS conditions is given by:

$$L(x) = \begin{cases} 32.4 + 21\log(x) + 20\log f_c, \text{non-blocked}, \\ 47.4 + 21\log(x) + 20\log f_c, \text{blocked}, \end{cases}$$

where f_c is operational frequency measured in GHz, and x is the distance between BS and UE. These expressions allow us deriving maximum distances d^E_{nLoS} and d^E_{LoS} at which a UE is able to start a session in blocked and non-blocked conditions correspondingly. This can be done by taking the lowest SNR defined by MCS as the value of L threshold [9,15].

In our model the coverage area is divided into two sectors: "inner sector (zone)" with radius $R_I = d^E_{nLoS}$; and the "outer sector (zone)" with width $R_O = R_C - R_I$, where R_O is restricted by the radius of the coverage area R_C. This division allows us for improving resource demand estimation as demands of sessions from inner zone that are always served at the same BS considerably different from sessions from outer zone that are subject of MC mechanism.

Then, to define the resource demands $b^{(n)LoS}_{I/O}$ that depend on both UE location (inner or outer zone) and LoS conditions we use approximations of spectral efficiency for each of the four propagation scenarios. The approximations are performed as the following: first, using the conventional MCS we define the longest possible distances x_i for each given i-CQI in LoS and nLoS conditions (1); then we calculate the coverage area for the CQIs as $\Omega_i = \pi(x_i^2 - x_{i-1}^2)$ for i-CQI, $x_0 = 0$; finally, we derive the average resource demand for each of the scenarios (2).

$$x_i = \begin{cases} \left(\dfrac{P_A G_A G_U}{S_i N_0 W\left(10^{2\log_{10} f_c + 3.24}\right)}\right)^{\frac{1}{\gamma}}, \text{non-blocked}, \\ \left(\dfrac{P_A G_A G_U}{S_i N_0 W\left(10^{2\log_{10} f_c + 4.74}\right)}\right)^{\frac{1}{\gamma}}, \text{blocked}, \end{cases} \tag{1}$$

where P_A is the NR BS transmit power, S_i is the worst possible SNR given by MCS for i-CQI, G_A and G_U are the antenna array gains at the NR BS and the UE ends, respectively, N_0 is the power spectral density of noise, W is the given bandwidth, γ – path loss exponent.

$$b^{(n)LoS} = \begin{cases} \left\lceil v \cdot \left(s_A \cdot \sum\limits_{i:0<x_i\leq R_I} \dfrac{\Omega_i^{(n)LoS} \cdot E_i}{R_I} \right)^{-1} \right\rceil, \text{inner zone}, \\ \left\lceil v \cdot \left(s_A \cdot \sum\limits_{i:R_I<x_i\leq R_C} \dfrac{\Omega_i^{(n)LoS} \cdot E_i}{R_C} \right)^{-1} \right\rceil, \text{outer zone}, \end{cases} \tag{2}$$

where s_A is the physical resource block (PRB) measured in MHz, E_i is the spectral efficiency corresponding to i-CQI, v is the required service data rate.

Mobile obstacles may appear on LoS towards the BS and thus reduce the SNR of the established connection. In this paper we consider the typical type of blockers for indoor scenarios which is humans with their blockage radius r_B approximating the width of human body. We model blockers' movement with two exponentially distributed random variables with parameters θ_{LoS} and θ_{nLoS} that correspond to the time intervals when a blocker is passing the LoS and periods between two consecutive blockage events [3]. It should be noted that the θ_{LoS} and θ_{nLoS} intensities generally depend on density of users and their behaviour, i.e. their velocity and mobility model. It means that in case of dense networks, system performance may be severely compromised not only by the number of active devices, but also the increased LoS blocking probability.

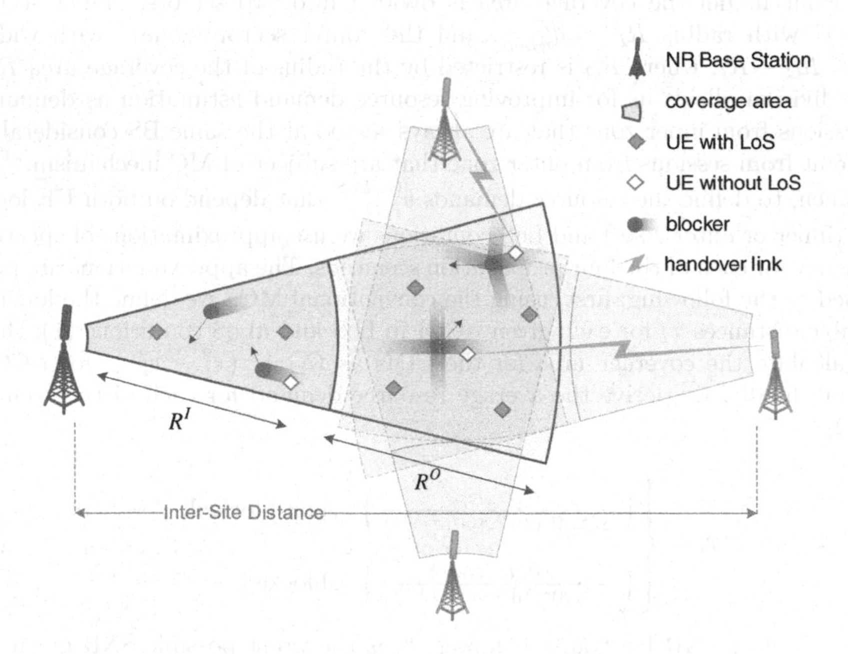

Fig. 1. Communication scenario.

Figure 1 illustrates the communication scenario considered in our paper. We observe the resource allocation process at a single NR BS. We suppose there are no interfering beams that could worsen channel quality. Each blocker has its own radio shadow area with length $d_B = \frac{d(h_B - h_U)}{(h_A - h_U)} + r_B$ [3]. Inside inner zone UE has only one active link with the target NR BS, resources are allocated even if UE in the radio shadow area. As the MCS does not support connection in nLoS conditions whenever UE at the distance $d > R^I$, multiple links are maintained towards the target and nearby BSs; once it gets into the shadow, user session is handed over to the BS with the greatest SNR at the moment.

As we assume that NR BS maintains constant bitrate for an established session, in our model session is only dropped when its new demand is more than remaining resource. Session requires additional resource in the following cases: (i) new session is initiated; (ii) LoS towards UE is blocked in inner zone; (iii) ongoing session is handed over back to original NR BS.

We also assume that there is no prioritization among traffic types. Whenever a session is attempting for establishment, NR BS allocates unoccupied resource without any reservation mechanisms [17].

3 Simulation Approach

In this section we present the system-level tool that we used to validate the above-mentioned model. As we have limited system capacity in terms of resource, this guarantees accessibility of steady-state conditions [18] that are detected using exponentially-weighted moving average technique with smoothing parameter set to 0.05.

The simulation tool utilizes an event-driven engine that allows for processing predefined set of events [19,20], such as session establishment and aborting, blockages, handover, resource management, etc. This approach provides both much flexibility and powerful capabilities for statistic data collection. High-level algorithm of the simulation is described as UML activity diagram presented in Fig. 2.

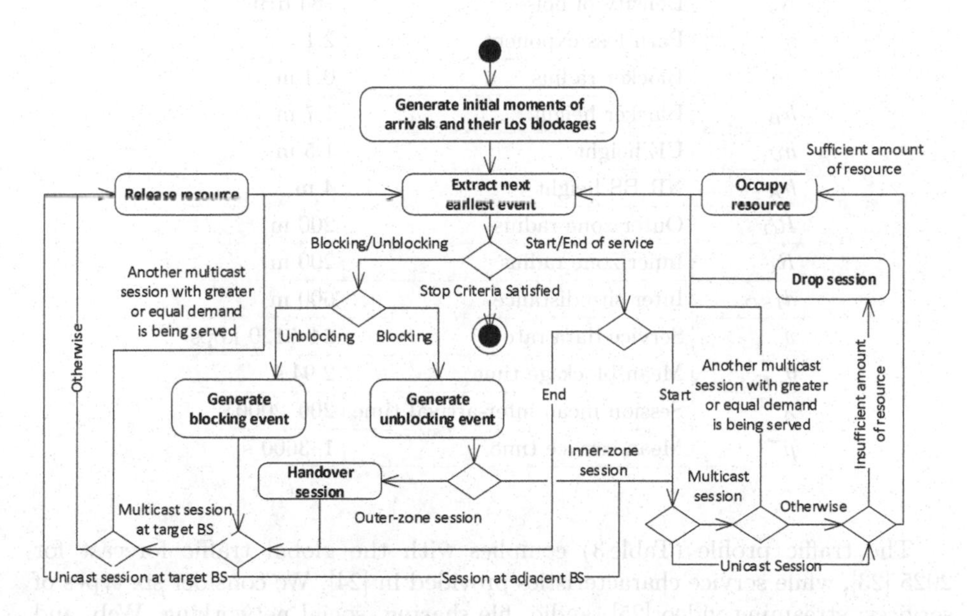

Fig. 2. Simulation algorithm.

Data is collected during the steady-state period only providing better accuracy compared to analytical results. In what follows, we demonstrate only point estimates of the metrics of interest.

4 Numerical Assessment

In this paper, we analyze NR BS resource allocation process needed for establishing and maintaining unicast and multicast [21] sessions composing mixed traffic load. Our metrics of interest are unicast and multicast session drop probabilities and mean NR BS resource utilization which are the key indicators of system reliability. The input system parameters [22] are given in Table 2.

Table 2. Input data

Notation	Description	Values
P_A	Transmit power	0.2 W
G_A	BS antenna array gain	2.58 dBi
G_U	UE antenna array gain	2.58 dBi
f_c	Operational frequency	28 GHz
W	Bandwidth	1 GHz
s_A	Service unit	1.44 MHz
N_0	Density of noise	–84 dBi
γ	Path loss exponent	2.1
r_B	Blocker radius	0.4 m
h_B	Blocker height	1.7 m
h_U	UE height	1.5 m
h_A	NR BS height	4 m
R_O	Outer zone radius	200 m
R_I	Inner zone radius	200 m
d_{ISD}	Inter site distance	600 m
v	Service data rate	64–5120 kbps
θ_{nLoS}^{-1}	Mean blockage time	2.94 s
λ^{-1}	Session mean inter-arrival time	200–2000 s
μ^{-1}	Mean service time	1–3600 s

The traffic profile (Table 3) complies with the global traffic forecast for 2025 [23], while service characteristics provided in [24]. We consider six types of services: streaming video [25], audio, file sharing, social networking, Web, and Machine-to-Machine [26] traffic. Data is transmitted by unicast sessions for all the services, but video can be also streamed via multicast sessions to reduce the amount of utilized resource.

Table 3. Service characteristics.

Service	Data rate, [kbps]	Duration, [s]	Share, [%]
Video	5120	3600	76
Audio	64	90	1
File sharing	1024	2	3
Social networking	384	8	8
Web browsing	500	3	2
M2M	200	1	10

In Fig. 3 and Fig. 4 we present session drop probabilities for different density values. One may observe extremely rapid degradation of unicast video service compared to other services. This can be explained by the huge resource demand of video traffic coupled with long-lasting service delivery time. It is much easier for services with smaller demands to fit into the remainder of the resources in case of high traffic load at NR BS.

Figure 4 shows that in case of handover sessions the gap between video and other types of services in terms of session continuity becomes even greater.

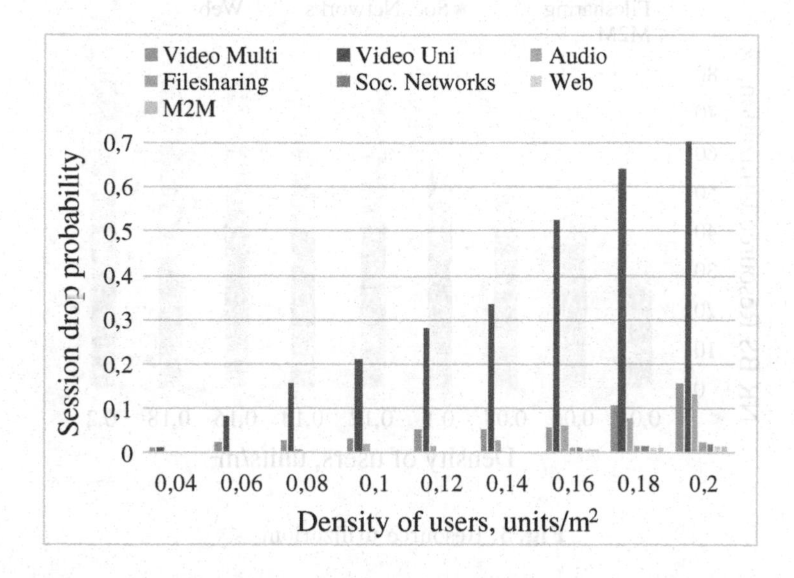

Fig. 3. Session drop probabilities.

In Fig. 5 we demonstrate resource distribution among the above-mentioned services. Along with the network densification the share of resource occupied by video traffic declines, ceding it to the more light-weight and short-time services.

We also present numerical results for session drop probabilities (Fig. 6) as function of inter-site distance for different ratio between unicast and multicast

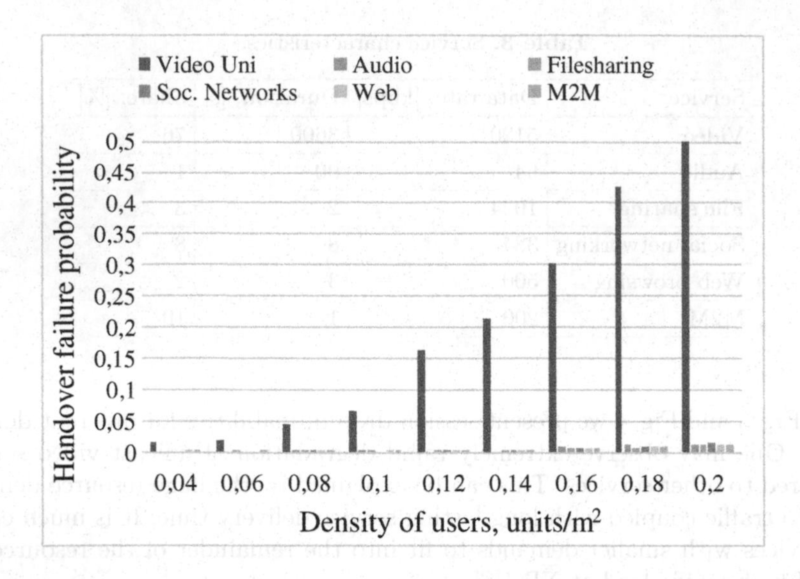

Fig. 4. Handover failure probabilities.

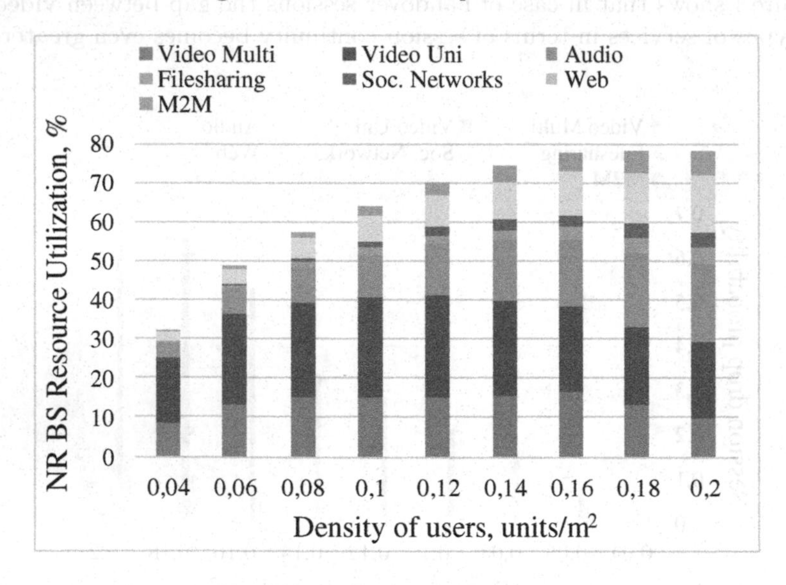

Fig. 5. Resource utilization.

traffic shares. As may be observed, the trend of unicast numerical superiority over multicast in drop probabilities is mainly kept within all the variety of traffic shares. However, when the number of multicast sessions becomes insufficient to continuously occupy allocated amount of resource, the multicast session drop probabilities start to outnumber unicast ones.

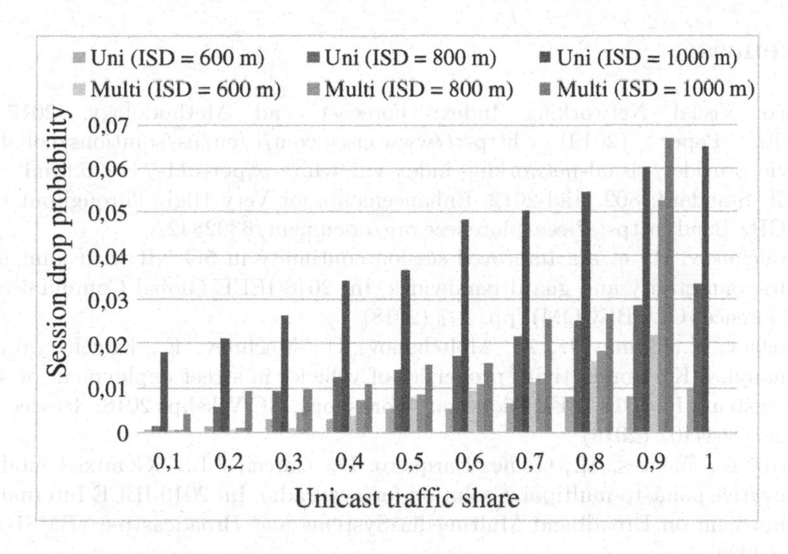

Fig. 6. Session drop probabilities as function of ISD.

By setting a threshold at some maximum allowed value of drop probability, it is possible to estimate the longest possible distance between adjacent NR BSs. Such approach allows for optimization of NR BS deployment minimizing the number of BSs to provide acceptable coverage within a given area. This may significantly reduce capital and operations expenses of network operator, but in this case one should give thorough consideration to the network scalability.

Our findings show that multicast sessions offer more reliability compared to unicast communications providing four-times less drop rates while occupying less amount of resource. This can be explained by the concept of resource occupation while organizing a multicast session. Once amount of resource is allocated to a multicast session, all the successive multicast sessions providing the access to the same content will keep on capturing this resource until the last session comes to its end.

5 Conclusion

In this paper, we considered the radio frequency resource allocation by a 5G NR BS with multi-connectivity mechanism that allows for reducing density of BS deployment. By means of the developed simulation tool we provided the numerical analysis of 5G NR BS deployment serving mixture of unicast and multicast sessions. Our findings revealed the advantages of multicast sessions over unicast ones in terms of reliability and resource consumption.

Our future work is to study fair radio resources distribution policies between unicast and multicast multimedia services in next-generation mobile networks that allow to protect disadvantaged types of services to fulfil QoE requirements.

References

1. Cisco Visual Networking Index: Forecast and Methodology, 2017–2022 White Paper (2019). https://www.cisco.com/c/en/us/solutions/collateral/serviceprovider/visual-networking-index-vni/white-paper-c11-741490.html
2. IEEE Standard: 802.11ad-2012: Enhancements for Very High Throughput in the 60 GHz Band. https://ieeexplore.ieee.org/document/6392842/
3. Kovalchukov, R., et al.: Improved session continuity in 5G NR with joint use of multi-connectivity and guard bandwidth. In: 2018 IEEE Global Communications Conference (GLOBECOM), pp. 1–7 (2018)
4. Begishev, V., Samuylov, A., Moltchanov, D., Machnev, E., Koucheryavy, Y., Samouylov, K.: Connectivity properties of vehicles in street deployment of 3GPP NR systems. In: 2018 IEEE Globecom Workshops, GC Wkshps 2018 - Proceedings, art.no. 8644102 (2019)
5. Garro, E., Fuentes, M., Gomez-Barquero, D., Carcel, J.L.: 5G mixed mode: an innovative point-to-multipoint solution for new radio. In: 2019 IEEE International Symposium on Broadband Multimedia Systems and Broadcasting (BMSB), pp. 1–5 (2020)
6. Wang, J., Hongbo, X., Zhu, B., Fan, L., Zhou, A.: Hybrid beamforming design for mmWave joint unicast and multicast transmission. IEEE Commun. Lett. **22**(10), 2012–2015 (2018)
7. Säily, M., Barjau, C., Navrátil, D., Prasad, A., Gómez-Barquero, D., Tesema, F.B.: 5G radio access networks: enabling efficient point-to-multipoint transmissions. IEEE Vehicular Technol. Mag. **14**(4), 29–37 (2019)
8. Samuylov, A.: Characterizing resource allocation trade-offs in 5G NR serving multicast and unicast traffic. IEEE Trans. Wirel. Commun. **19**(5), 3421–3434 (2020)
9. 3GPP, "NR: Physical channels and modulation (Release 15)", 3GPP TR 38.211 (2017)
10. 3GPP, "NR: Multi-connectivity; Overall description (Release 15)", 3GPP TS 37.34 V15.2.0 (2018)
11. Petrov, V., et al.: Dynamic multi-connectivity performance in ultra-dense urban mmWave deployments. IEEE J. Selected Areas Commun. **35**(9), 2038–2055 (2017)
12. Zhang, K., Zhang, J., Tao, X.: Dynamic multiconnectivity based joint scheduling of eMBB and uRLLC in 5G networks. IEEE Syst. J., 1–11 (2020)
13. Gapeyenko, M., Petrov, V., Riza, M., Andreev, S., Himayat, N., Koucheryavy, Y.: On the degree of multi-connectivity in 5G millimeter-wave cellular urban deployments. IEEE Trans. Vehicular Technol. **6**(2), 1973–1978 (2018)
14. Feng, W., Li, Y., Niu, Y., Su, L., Jin, D.: Multicast spatial reuse scheduling over millimeter-wave networks. In: Wireless Communications and Mobile Computing Conference (IWCMC), pp. 317–322. IEEE (2017)
15. Polese, M., Zorzi, M.: Impact of channel models on the end-to-end performance of Mmwave cellular networks. In: IEEE 19th International Workshop on Signal Processing Advances in Wireless Communications (SPAWC), pp. 1–5 (2018)
16. Biason, A., Zorzi, M.: Multicast via Point to multipoint transmissions in directional 5G mmWave communications. IEEE Commun. Mag. **57**(2), 88–94 (2019)
17. Moltchanov, D., et al.: Improving session continuity with bandwidth reservation in mmwave communications. IEEE Wirel. Commun. Lett. **8**(1), 05–108 (2019)
18. Naumov, V., Samouylov, K.: Analysis of multi-resource loss system with state-dependent arrival and service rates. Prob. Eng. Inf. Sci. **31**(4), 413–419 (2017)

19. Mezzavilla, M., et al.: End-to-end simulation of 5G mmWave networks. IEEE Comm. Surv. Tutor. **20**(3), 2237–2263 (2018)
20. Samouylov, K., Naumov, V., Sopin, E., Gudkova, I., Shorgin, S.: Sojourn time analysis for processor sharing loss system with unreliable server. In: Wittevrongel, S., Phung-Duc, T. (eds.) ASMTA 2016. LNCS, vol. 9845, pp. 284–297. Springer, Cham (2016). https://doi.org/10.1007/978-3-319-43904-4_20
21. Vikhrova, O., Pizzi, S., Sinitsyn, I., Molinaro, A., Iera, A., Samuylov, K., Araniti, G.: An analytic approach for resource allocation of IoT multicast traffic. In: Proceedings of the International Symposium on Mobile Ad Hoc Networking and Computing (MobiHoc), pp. 25–30 (2019)
22. Samuylov, A., Moltchanov, D., Krupko, A., Kovalchukov, R., Moskaleva, F., Gaidamaka, Yu.: Performance analysis of mixture of unicast and multicast sessions in 5G NR systems. In: 10th International Congress on Ultra Modern Telecommunications and Control Systems and Workshops (ICUMT - 2018), Russia, Moscow (2018)
23. Ericsson Mobility Report (2020). https://www.ericsson.com/49da93/assets/local/mobility-report/documents/2020/june2020-ericsson-mobility-report.pdf
24. Khatibi, S., Correia, L.M.: Modelling virtual radio resource management in full heterogeneous networks. EURASIP J. Wirel. Commun. Netw. **2017**(1), 1–17 (2017). https://doi.org/10.1186/s13638-017-0858-7
25. Borodakiy, V.Y., Samouylov, K.E., Gudkova, I.A., Markova, E.V.: Analyzing mean bit rate of multicast video conference in LTE network with adaptive radio admission control scheme. J. Math. Sci. **218**(3), 257–268 (2016)
26. Gudkova, I., et al.: Analyzing impacts of coexistence between M2M and H2H communication on 3GPP LTE system. In: Mellouk, A., Fowler, S., Hoceini, S., Daachi, B. (eds.) WWIC 2014. LNCS, vol. 8458, pp. 162–174. Springer, Cham (2014). https://doi.org/10.1007/978-3-319-13174-0_13

IoT Traffic Prediction with Neural Networks Learning Based on SDN Infrastructure

Artem Volkov[2], Ali R. Abdellah[1,2], Ammar Muthanna[2,3(✉)], Maria Makolkina[2,3], Alexander Paramonov[2,3], and Andrey Koucheryavy[2]

[1] Electronics and Communications Engineering, Electrical Engineering Department, Al-Azhar University, Qena 83513, Egypt
alirefaee@azhar.edu.eg

[2] The Bonch-Bruevich Saint-Petersburg State University of Telecommunications, Pr. Bolshevikov, 22, St.Petersburg 193232, Russia
artemanv.work@gmail.com, ammarexpress@gmail.com, akouch@mail.ru

[3] Peoples' Friendship University of Russia (RUDN University), 6 Miklukho-Maklaya Street, Moscow 117198, Russia

Abstract. In recent time, there are more achievements in technologies for 5G/MT-2020 networks the international research area have. Any way, one of the main important task in this area is the IoT traffic recognition and prediction. Currently, researchers face a new challenge to their experience, talents and desire to reach new heights in information and communication technologies. The new challenge is IMT-2030 networking technologies and services. In this case, the question with effective IoT traffic prediction methods is still relevant during transition to the next IMT-2030 network and services. IoT it is the ubiquitous conception, on which the new IMT-2030 services also based. For example, Tactile Internet, part of the solutions in digital avatars, and others. There more, these algorithms have to be more efficient and fastly in work with huge data capacity, which characterized the different services of Internet of Things. Recently, there are Machine Learning and Big Data algorithms are took this place of new algorithms for efficient and complex algorithms. In this paper, we implement IoT traffic prediction approaches using single step ahead and multi-step ahead prediction with NARX neural network. As a data we used the metadata of flows which were received through the northern interface. The prediction accuracy has been evaluated using three neural network traing algorithms: Traincgf, Traincgp, Trainlm, with MSE as performance function in term of using mean absolute percent of error (MAPE) as prediction accuracy measure IoT.

Keywords: Internet of Things · 5G/IMT-2020 · Neural networks · Traffic prediction

V. M. Vishnevskiy et al. (Eds.): DCCN 2020, LNCS 12563, pp. 64–76, 2020.
https://doi.org/10.1007/978-3-030-66471-8_6

1 Introduction

The concept of fifth generation communication networks includes a whole range of other concepts and technologies, and not only describes the principles of organizing a mobile access network [1]. Key changes are being made to both the network and computing infrastructure and the service level. To systematize and structure technological solutions in IMT-2020 communication networks, international standardizing organizations are developing a stack of recommendations and standards. These recommendations and standards regulate the structure, requirements and protocols for the interaction of systems in IMT-2020 communication networks. A number of organizations work on this task, such as ITU (International Telecommunication Union), IEEE (The Institute of Electrical and Electronic Engineers), ISO/MEC, etc. However, the general development trends of technologies within the framework of 5G/IMT-2020 communication networks can be determined by the following characteristics:

* Trends that concern users and applications. Mobile devices play a diverse, continuously changing role in the life of a modern person. Future IMT-systems should support new usage scenarios, including applications requiring high-speed data transfer, connecting a large number of devices, as well as ultra-low latency and high reliability.
* network topology changes very fast so fragmentation regularly happens.
* User-oriented communication systems with ultra-low latency and high reliability. Modern users are used to receiving services very quickly, one might say instantly. Accordingly, current and future systems should instantly establish the necessary connections and provide an instant response from the applications used in the same way, instant feedback should be a key factor in the successful development of cloud services.
* Support for high user density environments. It is necessary that modern and future communication networks can provide all the necessary services with a high density of device connection without loss of quality of the services provided.
* Providing high quality while high mobility. Modern and future communication networks in the process of providing services should ensure that the level of quality of service remains unchanged in conditions of high user mobility (for example, a user travels on a high-speed train at a speed of about 300–400 km/h).
* IoT applications and private IoT concepts. The concept of the Internet of Things played a key role in the breakthrough in the development of a new generation of communication networks. The essence of the concept is both simple and at the same time complicated in systemic issues, both from the technology side and from the side of ensuring the safety of people, citizens, enterprises, states. The concept of the Internet of Things is not new, and at the moment, as part of its development, it has stood out and isolated by separate concepts for its implementation. Each of the isolated IoT concepts carries its own achievements in the development of society, business, science. For example, the concept of "Smart City" can be distinguished as a separate

concept, which has degenerated from the main IoT concept, and carries its own requirements for implementation, for example, to such a parameter as security and the like [2].

However, as already noted in the abstract, IMT-2020 communication networks in the course of their development brought new ideas and new goals in technological solutions at all levels, formed requirements that at the time of 2020 could not be implemented for a number of reasons. For new technologies and services, it is necessary to carry out additional large-scale research, thus, a vision of new services and the capabilities of the next generation communication networks - IMT-2030. For example, the concept of Tactile Internet, which was originally supposed to be implemented as part of the 5G/IMT-2020 communication networks, is shifting in its entry into the market, as a result of research and development it turned out that a necessary requirement is a delay in this type of service, which is 1 ms. This indicator is at least not achievable in the framework of modern approaches to building networks and organizing data processing. So for other services based on the Internet of Things, new approaches to the implementation of the infrastructure component of networks and cloud computing have become necessary [3].

As a solution to the problem, the International Telecommunication Union proposed the use of the following technologies for building communication networks: SDN/NFV, which will allow to achieve new opportunities in the effectiveness of traffic management and switching at different levels of communication networks. For the implementation of applications demanding on time characteristics, such solutions as MEC (Multi access Edge Computing) began to appear and one of the latter - the concept of fog computing. Thus, at the time of 2020, 5G/IMT-2020 communication networks brought a new technological breakthrough both in infrastructure and in service, while setting and creating new, complex tasks, with new requirements, the implementation of which requires unification in terms of solutions, as well as transforming approaches to development and implementation.

However, given the smooth transition in the development of network generations, it is worth noting that since most of the new services are based on the concept of the Internet of Things, and still require the fulfillment of the set requirements, as well as the development of new approaches to their implementation, the issue of identifying the types of the Internet traffic of Things also its forecasting remains very relevant now and in the near future. In this case, taking into account the general vector of development of networks and services, new approaches and methods should be consistent with those technological solutions of communication networks and services that brought 5G/IMT-2020 communication networks. Thus, the software-defined Networking & Network Function Virtualization is taken as a network basis. It should also be borne in mind that the task of identifying and predicting the traffic of various types of applications of the Internet of Things should be solved by fulfilling the following requirements: high speed of computing, self-learning system and the ability to adapt to the geographical specifics of the communication network and applications, as well as work with traffic should occur without direct intervention into the structure and essence of traffic in order to provide high-quality services for IoT services, as well as maintaining user privacy.

Under the above requirements, algorithms that belong to the class of Artificial Intelligence are suitable, namely, Machine Learning algorithms. Today, artificial intelligence (AI) technologies are positioned as the main factors stimulating the economic growth of states in modern conditions. Specialists have high hopes for the introduction of Artificial Intelligence technologies in the communication network, since with the advent of new services, new architectures, the efficiency of decision-making is much greater than what a person can provide. In order to make a decision, taking into account a set of optimality criteria, in modern communication networks it is necessary to process a large amount of data, as well as take into account the specifics of decisions and implementation features, taking into account the geographical location of the network and so on. Thus, the research and development of new methods in monitoring and managing communication networks based on Artificial Intelligence technologies is a necessary solution. At the same time, SDN/NFV communication network building technologies, due to their technical capabilities, provide the necessary level of abstraction from network devices and the possibility of easy and programmer-independent device programmability. In these solutions, the principles of openness of platform solutions, openness in protocol interactions of the inherent flexible scalability were originally laid down [4,5].

Mobile and wireless communication is a driving force in strengthening the economical, technological, financial, social, health, education and research environments. In order to support the growth rate and to match with ever-changing requirements, industry and academia always strive to devise new mechanisms and techniques for mobile and wireless communication to ease human life, e.g.., evolution of Internet of Things, 5G, Vehicular Networks, Tactile network etc. Recently, due to the advancement in the computational power and development in the deep learning algorithm [5–8], AI are showing promising results in various applications viz. Security, Networks, Information Retrieval etc. and there is no doubt that Artificial Intelligence (AI) reduces the human intervention and provides reliable results, irrespective of the fields [9–12]. Henceforth, the development in the AI can also provide new ideas to the 5G networks for efficient traffic control and management. In the field of artificial intelligence in communication networks, the greatest attention is currently paid to the use of machine learning methods Machine Learning (ML) [8,18,19]. This is reflected in the activities of standardizing organizations [5]. There are several reasons why machine learning algorithms are in demand in solving problems for communication networks of the 5G and future generations:

- ⋆ Trends that concern users and applications. Mobile devices play a diverse, continuously changing role in the life of a modern person. Future IMT-systems should support new usage scenarios, including applications requiring high-speed data transfer, connecting a large number of devices, as well as ultra-low latency and high reliability.

- ⋆ network topology changes very fast so fragmentation regularly happens.
- ⋆ Machine learning allows network and communication systems to meet the individual needs of users.
- ⋆ Machine learning allows you to gain new knowledge from large databases.
- ⋆ Machine learning can replace a person in solving monotonous problems.
- ⋆ Machine learning allows to implement systems that are difficult and expensive to create manually because of the need to have special skills or specific knowledge.
- ⋆ Machine learning allows you to get adequate forecasts for dynamic traffic changes across various applications, which is especially important in heterogeneous networks.

Deep Learning (DL) is currently presented as a breakthrough technology in machine learning [11]. Deep learning allows you to train a model to predict the outcome of a set of input data. At the same time, multilayer neural networks are used, which are constantly updated with new volumes of data.

This article will discuss two architectural approaches in the development of artificial Neural Networks for predicting the traffic of the Internet of Things, based on data on flows (metadata) obtained through the management level of a software-configured network.

2 Related Works

Several researchers have focused on the study traffic modelling, analysis, and prediction, using machine learning and in the field of IoT technology. Our focus in this paper is on the IoT traffic prediction using single step and multistep prediction with Time Series NARX recurrent Neural Networks based SDN infrastructure, and, hence, in the rest of this section, we review the notable works relevant to our focus area in this paper.

Ali R. Abdellah et al. [8] introduced an approach for IoT traffic time series prediction using a multistep ahead prediction with Time Series NARX recurrent Neural Networks, he evaluated a prediction accuracy has been evaluated using the performance functions MSE, SSE, and MAE, besides, another measure of prediction accuracy the mean absolute percentage of error (MAPE).

Filip Pilka et al. [13] presented three different approaches to multi-step ahead prediction using neural networks for video traffic prediction and described basic principles of two types of neural networks, he used for all the three approaches: the multilayer feedforward artificial neural network and the Nonlinear AutoRegressive model with eXogeneous inputs neural network (NARX). Then he briefly described the structure of the video trace files.

Shiva et al. [14] introduced different approaches for multi-step ahead traffic prediction based on modified ANFIS for future traffic prediction he presented comparitive between three different approaches for multi-step ahead prediction they applied them for data gathered fro tehran highway by modifying the structure od adaptive neuro-fuzzy inference system (ANFIS).

Mohamed et al. [15] introduced efficient prediction of network traffic for real-time applications. They explored a number of predictors and searched for a predictor which has high accuracy and low computation complexity and power consumption.

Artem Volkov et al. [16] investigated SDN load prediction algorithm based on artificial intelligence they proposed a novel approach for SDN load prediction based on artificial intelligence algorithms and totally monitoring of OpenFlow channels activities.

3 Problem Statement

This article was tasked with the study of two architectural approaches in Artificial Neural Networks to predict the Internet of Things traffic. As a IoT traffic prediction approaches were used single step ahead and multi-step ahead prediction with NARX neural network.

4 The System Structure

4.1 Architecture of Infrastructure and Structure of Data

At the time of the research, in the scientific world you can find many works that are aimed at detecting the types of traffic, developing forecasting models, [12] both traffic and the load of telecommunication systems [16,17]. As previously mentioned in this article, at present, the communication network, from the point of view of generations, is in transition, some of the technologies of 5G/IMT-2020 communication networks are being introduced, however, in the process of developing this generation of networks, new requirements and new ones have been developed visions in services, for the implementation of which, a certain amount of time will be required. Thus, the concept of IMT-2030 communication networks appeared, which in turn includes a number of technologies and solutions, the development and implementation of which were initially considered within the framework of the IMT-2020 generation (for example, Tactile Internet), as well as new types of services, for example such as, a servant of virtual telepresence, etc. At the same time, one of the fundamental concepts is the concept of the Internet of Things, on the basis of which private concepts were also proposed (for example, Smart City, Smart Home, and so on). However, the issue of developing methods and algorithms for the effective identification and forecasting of traffic flows of various Internet services of Things remains very relevant.

Thus, as previously mentioned, this article proposes the use of Machine Learning algorithms with Deep Learning to generate forecasts of the activity of the Internet of Things traffic. These algorithms are a certain kind of software modules that are integrated into a common analytical system according to deep classification and predictive analytics, which in turn is implemented at the service level of SDN/NFV networks. One of the requirements for this system was to propose a solution that would not in any way introduce delays in the Internet of Things and others, as well as be portable and scalable.

Thus, in order to fulfill these requirements, a hypothesis was expressed about the possibility of analyzing data streams, with the aim of classifying and predicting them, based on the metadata stream, which can be obtained through interaction with the controller of a software-configured network, through the northern interface. In turn, the controller receives all monitoring data about all the controlled flows through the southern network interface, thanks to the OpenFlow protocol.

This hypothesis was already tested in our previous work [12] on the development of a method for classifying the streams of the Internet of Things based on the approach on the analysis of stream metadata. Based on this approach and the existing level of development of the analytical system, as already mentioned above, the task was set to study two architectural approaches in Artificial Neural Networks to perform the difficult task of predicting flows. For this, in order to study and compare methods among themselves, we used real metadata of the Internet of Things, which were aggregated using the software of the previous work [3,12], as a tool for developing and comparing two solutions, we used special software for modeling of complex processes - Matlab.

The architecture of the laboratory bench, on the basis of which data was collected on the flows of the Internet of Things, is shown in Fig. 1. Also in Fig. 1, the corresponding location of the analytical service is reflected, in which the corresponding algorithm for predicting the traffic of the Internet of Things will be implemented.

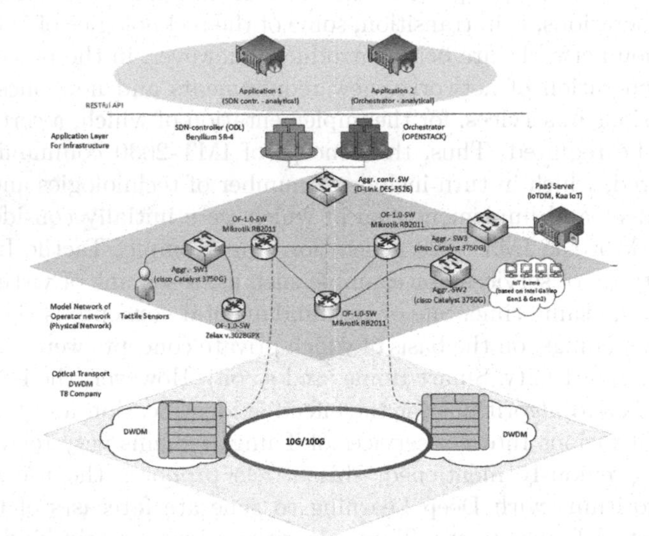

Fig. 1. Stand architecture

In [12], a method was proposed for compiling a meta-model of flows based on two global parts of the SDN stream table, namely: Match Field and Actions. In order to ensure portability and flexible scaling of the system, regardless of the data transmission medium, it is necessary that the load forecasting algorithms be implemented in the form of high-level service analytic modules, as shown in Fig. 1. As a result, the analytical system will work with all devices and streams in the form of work with "digital/software objects" that have a certain set of parameters and functions for actions on these parameters. This parameter will be processed by Nero networks. This method and the possibility of an abstraction level allows us to implement an analytical system that will work with metadata flow on-the-fly, without direct intervention. As a result, this approach will allow not to introduce additional delays in the traffic and change its profile.

To test the previously put forward hypothesis, we used the infrastructure of the Laboratory for Programmable Networks (SDNLab) of the Department of Communication and Data Networks of St. Petersburg State Telecommunications University named after prof. M.A. Bonch-Bruevich. Based on this infrastructure, the corresponding training and test data sets were formed, which were then used in the developed software in the Matlab simulation environment. A mathematical method for processing primary data from SDN switches received through the northern interface of a software-configured network controller was given in [3]. Thus, the final structure of these flows for research is presented in the form of the following matrix (1) (Fig. 2):

$$
\begin{array}{cccc}
 & \textbf{(T\_Delta)} & \textbf{(Byte\_Count)} & \textbf{(Packet\_Count)} \\
 & T_{delta} & B_{delta-12} & P_{delta-13} \\
DataSet_{ML} = & T_{delta} & B_{delta-22} & P_{delta-23} \\
 & \ldots & \ldots & \ldots \\
 & T_{delta} & B_{delta-N2} & P_{delta-N3}
\end{array}
$$

Fig. 2. The final structure

Where T_{delta} - represents the value of the data acquisition period, in this work - 1 sec. And the parameters B_{delta} and P_{delta} indicate the number of bits transmitted in a set time interval and, accordingly, packets, based on the counters of the studied flows. Further, as already mentioned above, the obtained data on the flows of the Internet of Things were used in the considered models of neural networks. It is worth considering that for each type of traffic, in order to form training samples of Neural Networks, the corresponding data sets (matrix 1) were formed. Further, during the operation of this system online and during testing, unallocated data of the studied flows were used.

4.2 Artificial Neural Networks with Multi-step and Single Step Prediction

Recently, Artificial Neural networks with multi-step and single step prediction Recently, Artificial Neural Network technologies are used in various directions, however, all tasks can be classified into several classes, for example, objects classification, data classification, forecasting and so on. The use of these algorithms is determined by a number of their advantages, for example, such as high recognition rates of the studied objects, prediction of complex systems within a limited time, and complexity, from the point of view of systematic objects.

A data stream from the generated $DataSet_{ML}$ is fed to the input layer of neurons (the so-called placeholders). Placeholders are connected to the first neural network layer by a fully connected architecture. In this work, we implement IoT traffic prediction approaches using a single step ahead and multi-step ahead prediction with NARX neural network. The prediction accuracy has been evaluated using three neural network training algorithms Traincgf, Traincgp, Trainlm, with MSE as performance function in terms of using mean absolute percent of error (MAPE) as prediction accuracy measure. The Input and output time series rely upon predicting the following value of one time-series is given another time-series. The previous values of both series (for best accuracy), or only one of the series (for a simpler system) may be used to predict the target series.

5 Simulation Results

In this work, we implement IoT traffic prediction approaches using single step ahead and multi-step ahead prediction with NARX neural network. The prediction accuracy has been evaluated using three neural network training algorithms Traincgf, Traincgp, Trainlm, with MSE as performance function in term of using mean absolute percent of error (MAPE) as prediction accuracy measure.

The Input and output time series rely upon predicting the following value of one time-series is given another time-series. The previous values of both series (for best accuracy), or only one of the series (for a simpler system) may be used to predict the target series. The dataset can be used to show how a neural network can be learned to make predictions. The Data (matrix 1) after the collection and preprocessing of the primary SDN's data set that was randomly divided into 75%, 15% and 15% for training, validation, and testing, respectively. The implementation of the recurrent neural network to predict the performance accuracy of IoT traffic. Table 1 shows the prediction accuracy for the IoT traffic in cases of using the above-mentioned training algorithms with the measure of performance accuracy MAPE.

Table 1. The measure of prediction accuracy for the predicted model validation using RMSE and MAPE.

Performance function	Training algorithm	Single step prediction	Multistep prediction
MSE	Trainlm	1.4	3.7
	Traincgf	2.5	7.3
	Traincgp	1.24	1.66

Table 1 displays the prediction accuracy of IoT traffic in case of the number of hidden units in LSTM layer are 500, 100,50, respectively in order to estimate the error of prediction we use the RMSE and another measure for performance accuracy MAPE.

From the tabulated results, the model predicted in the case of number hidden units 500 has the best prediction accuracy with RMSE value equal 0.93432 and MAPE equal 2.4% in comparison to its peers. The maximum average prediction accuracy improvement in the case of 500 hidden units is 14.35%. Also, the model predicted in the case of 100 hidden units dropped to has a prediction accuracy with RMSE value 2.4163 and the MAPE dropped to 9.32% and the maximum average of prediction accuracy in this case is 7.43%.

On the other hand, the model predicted in the cases of a number of hidden units 50 has the lowest prediction accuracy with RMSE value 6.5104 and MAPE dropped to 16.75% comparison of them.

Figure 3, Fig. 4 and Fig. 5 shows the prediction accuracy for the IoT throughput using RMSE.

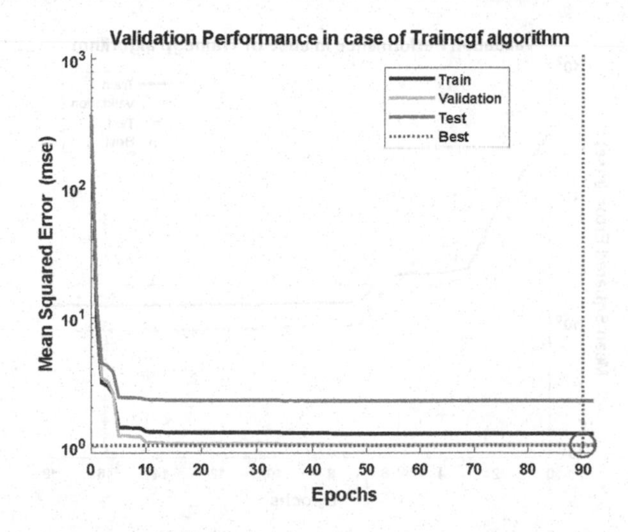

Fig. 3. The best performance validation in case Traingf algorithm

Figure 3 display the performance validation of network in case of using Traincgf algorithm as shown in figure the validation performance start increases at epoch 1 and decrease until 10 epochs then become constant until the 90 epoch which the best validation performance.

Figure 4 display the performance validation of network in case of using Trainlm algorithm as shown in figure the validation performance start increases at epoch 1 and decrease until 5 epoch then become constant until the 150 epoch which the best validation performance.

Figure 5 display the performance validation of network in case of using Traincgp algorithm as shown in figure the validation performance start increases

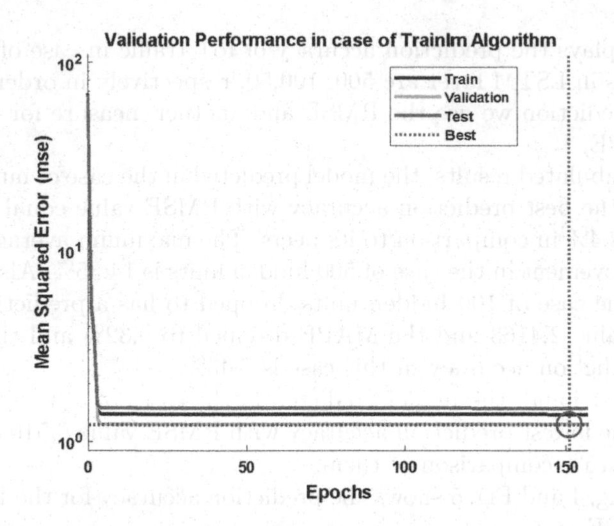

Fig. 4. The best performance validation in case Trainlm algorithm

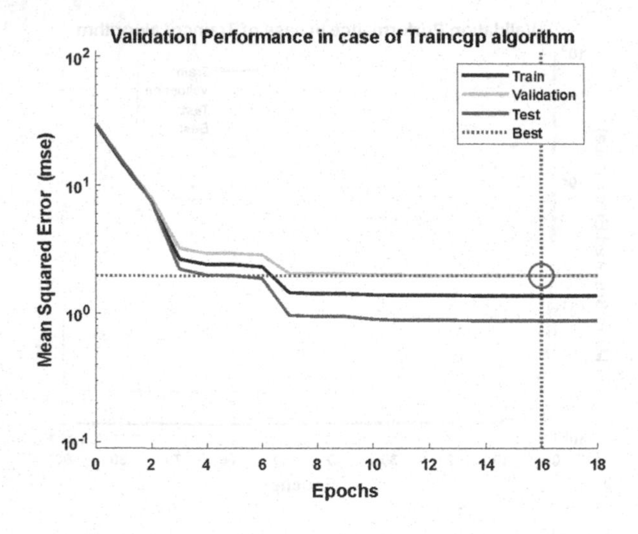

Fig. 5. The best performance validation in case Traincgp algorithm

at epoch 1 and decrease until 7 epochs then become constant until the 16 epoch which the best validation performance.

6 Conclusion

This paper was proposed a single step ahead and multi-step prediction method for IoT traffic based on NARX neural network. Based on SDN model network at SDN Laboratory we provide experiment for different neural network training algorithms (Traincgf, Traincgp, Trainlm, and MSE). Our theoretical and experimental results show the prediction accuracy for the IoT traffic in cases of using all these training algorithms with the measure of performance accuracy mean absolute percent of error MAPE.

Acknowledgments. The publication has been prepared with the support of the "RUDN University Program 5-100".

References

1. Muthanna, A., et al.: Analytical evaluation of D2D connectivity potential in 5G wireless systems. In: Galinina, O., Balandin, S., Koucheryavy, Y. (eds.) NEW2AN/ruSMART -2016. LNCS, vol. 9870, pp. 395–403. Springer, Cham (2016). https://doi.org/10.1007/978-3-319-46301-8_33
2. Hammoudeh, M., Newman, R., Dennett, C., Mount, S., Aldabbas, O.: Map as a service: a framework for visualising and maximising information return from multi-modal wireless sensor networks. Sensors **15**(9), 22970–23003 (2015)
3. Volkov, A., Khakimov, A., Muthanna, A., Kirichek, R., Vladyko, A., Koucheryavy, A.: Interaction of the IoT traffic generated by a smart city segment with SDN core network. In: Koucheryavy, Y., Mamatas, L., Matta, I., Ometov, A., Papadimitriou, P. (eds.) WWIC 2017. LNCS, vol. 10372, pp. 115–126. Springer, Cham (2017). https://doi.org/10.1007/978-3-319-61382-6_10
4. Mohammed, R., Mohammed, S.A., Shirmohammadi, S.: Machine learning and deep learning based traffic classification and prediction in software defined networking. In: IEEE International Symposium on Measurements & Networking (M&N), Catania, Italy, vol. 2019, pp. 1–6 (2019). https://doi.org/10.1109/IWMN.2019.8805044
5. Jiang, C., Zhang, H., Ren, Y., Han, Z., Chen, K., Hanzo, L.: Machine learning paradigms for next-generation wireless networks. IEEE Wirel. Commun. **24**(2), 98–105 (2017)
6. Carlin, A., Hammoudeh, M., Aldabbas, O.: Intrusion detection and countermeasure of virtual cloud systems-state of the art and current challenges. Int. J. Adv. Comput. Sci. Appl. **6**(6), 1–15 (2015)
7. Le, L.V., Lin, B.S., Do, S.: Applying big data, machine learning, and SDN/NFV for 5G early-stage traffic classification and network QoS control. Trans. Netw. Commun. **6**(2), 36 (2018)
8. Abdellah, A.R., Mahmood, O.A.K., Paramonov, A., Koucheryavy, A.: IoT traffic prediction using multi-step ahead prediction with neural network. In: IEEE 11th International Congress on Ultra-Modern Telecommunications and Control Systems and Workshops (ICUMT) (2019)
9. Abdellah, A.R., Muthanna, A., Koucheryavy, A.: Robust estimation of VANET performance-based robust neural networks learning. In: Galinina, O., Andreev, S., Balandin, S., Koucheryavy, Y. (eds.) NEW2AN/ruSMART -2019. LNCS, vol. 11660, pp. 402–414. Springer, Cham (2019). https://doi.org/10.1007/978-3-030-30859-9_34
10. Abdellah, A.R., Muthanna, A., Koucheryavy, A.: Energy estimation for VANET performance based robust neural networks learning. In: Vishnevskiy, V.M., Samouylov, K.E., Kozyrev, D.V. (eds.) DCCN 2019. CCIS, vol. 1141, pp. 127–138. Springer, Cham (2019). https://doi.org/10.1007/978-3-030-36625-4_11

11. Kato, N., Fadlullah, Z.M., Mao, B., Tang, F., Akashi, O., Inoue, T., Mizutani, K.: The deep learning vision for heterogeneous network traffic control: proposal, challenges, and future perspective. IEEE Wirel. Commun. **24**(3), 146–153 (2017)
12. Artem, V., Ateya, A.A., Muthanna, A., Koucheryavy, A.: Novel AI-based scheme for traffic detection and recognition in 5G based networks. In: Galinina, O., Andreev, S., Balandin, S., Koucheryavy, Y. (eds.) NEW2AN/ruSMART -2019. LNCS, vol. 11660, pp. 243–255. Springer, Cham (2019). https://doi.org/10.1007/978-3-030-30859-9_21
13. Pilka, F., Oravec, M.: Multi-step ahead prediction using neural networks. In: IEEE Proceedings ELMAR-2011, Zadar, pp. 269–272 (2011)
14. Rahimipour, S., Agha-Mohaqeq, M., Hashemi, S.M.T.: Different approaches for multi-step ahead traffic prediction based on modified ANFIS. In: Proceedings of the Third International Conference on Contemporary Issues in Computer and Information Sciences (CICIS 2012), pp. 156–161 (2012)
15. Iqbal, M.F., Zahid, M., Habib, D., John, L.K.: Efficient prediction of network traffic for real-time applications. J. Comput. Netw. Commun. **2019**, 1–11 (2019). https://doi.org/10.1155/2019/4067135
16. Volkov, A., Proshutinskiy, K., Adam, A.B.M., Ateya, A.A., Muthanna, A., Koucheryavy, A.: SDN load prediction algorithm based on artificial intelligence. In: Vishnevskiy, V.M., Samouylov, K.E., Kozyrev, D.V. (eds.) DCCN 2019. CCIS, vol. 1141, pp. 27–40. Springer, Cham (2019). https://doi.org/10.1007/978-3-030-36625-4_3
17. Muthanna, A., Volkov, A., Khakimov, A., Muhizi, S., Kirichek, R., Koucheryavy, A.: Framework of QoS management for time constraint services with requested network parameters based on SDN/NFV infrastructure. In: International Congress on Ultra Modern Telecommunications and Control Systems and Workshops, vol. 2018. IEEE Computer Society, November 2019. https://doi.org/10.1109/ICUMT.2018.8631274
18. Ellah, A.R.A., Essai, M.H., Yahya, A.: Robust Backpropagation Learning Algorithm Study for Feed Forward Neural Networks. Thesis, Al-Azhar University, Faculty of Engineering (2016)
19. Vikhrova, O., Pizzi, S., Iera, A., Molinaro, A., Samuylov, K., Araniti, G.: Performance analysis of paging strategies and data delivery approaches for supporting group-oriented IoT traffic in 5G networks. In: IEEE International Symposium on Broadband Multimedia Systems and Broadcasting, BMSB-2019, vol. 2019, art. no. 8971897 (2019)

Transmission Latency Analysis in Cloud-RAN

Eduard Sopin[1,2(✉)], Anatoly Botvinko[1], Alexandra Darmolad[1], Dinara Bixalina[1], and Anastasia Daraseliya[1]

[1] Peoples' Friendship University of Russia (RUDN University), 6 Miklukho-Maklaya St, Moscow 117198, Russian Federation
{sopin-es,bottvinko-ab,1032162870,1032163087,daraseliya-av}@rudn.ru

[2] Institute of Informatics Problems, FRC CSC RAS, Vavilova 44-2, Moscow 119333, Russian Federation

Abstract. Cloud-based Radio Access Network (C-RAN) is a centralized cloud computing architecture for radio access networks (RANs) that provides large-scale deployment, joint support for radio technologies, and real-time virtualization capabilities. By moving signal processing functions to a data center, C-RAN significantly reduces power consumption and deployment cost. The architecture of the cloud radio access network consists of three main components: a pool of base-band units (BBU pool), remote radio heads (RRHs), and a transport network. In C-RAN, base stations are replaced by remote radio heads: data blocks are digitized, transmitted through the fiber-optical infrastructure, and remotely processed in BBU pool. One of the main issues is to control the round-trip delay between the remote radio heads and the BBU pool. In the paper, we describe a C-RAN in terms of queuing network and accurately evaluate all delay components. Besides, we analyze the required computational resources of the BBU pool required to satisfy the strict round-trip delay budget in C-RAN.

Keywords: Cloud-RAN · Queuing network · Round-trip delay · Cloud computing

1 Introduction

Cloud-based Radio Access Network (C-RAN) is a new architectural concept for mobile communication networks designed to support high data rates with lower costs and are expected to provide low latency, high flexibility, and low power consumption to meet 5G requirements. In the traditional RAN architecture, the baseband processing and radio functions are located inside the base station (BS), while in the C-RAN, the functionality of the base station is separated

The publication has been prepared with the support of the "RUDN University Program 5-100" (recipient E. Sopin). The reported study was funded by RFBR, project number 20-07-01052 (recipient A. Daraseliya).

V. M. Vishnevskiy et al. (Eds.): DCCN 2020, LNCS 12563, pp. 77–86, 2020.
https://doi.org/10.1007/978-3-030-66471-8_7

from the cellular node and distributed between the Remote Radio Head (RRH) and the BaseBand Unit pool (BBU pool), which located far from each other [10]. Processing functions in the main frequency band are virtualized by means of Network Function Virtualization (NFV) and moved to the BBU pool in the central cloud through a Software Defined Network [9,13,14]. RRH is located in the BS and contains low power antennas and performs all the radio frequency functions necessary to emit a signal in a cell. They perform amplification, analog-to-digital conversion of radio signals, and send the digitized radio signals to the central BBU pool, where the received signals are processed, and cloud resources are dynamically allocated on demand [12].

Flexible distribution of computing resources across all RRHs and centralized processing of radio signals in the BBU pool improves the statistical multiplexing coefficient and simplifies the maintenance of cellular networks. Besides, the usage of a centralized cloud for BBU functions allows flatening of the daily and weekly peaks in traffic demand. For example, in the middle of the day, the load on the BSs located in the city center is increasing. But in the evening peole are coming home, so the load shifts to the uptown. Since the BBU pool serves the traffic from all BSs in the city, dynamic resource allocation avoids congestion. In addition, the C-RAN architecture allows deploying a large number of RRH's at low interference using coordinated multi-point (CoMP) techniques such as coordinated transmission and reception, which reduces the number of base stations required, resulting in reduced operational and capital expenditures.

However, the described architecture may suffer from increased delays due to signal processing in the BBU pool. In practical deployments, the delay incurred by transmitting the signal and processing it, limits the maximum area that a BBU can serve [12]. In the literature, the maximum distance between RRH and the BBU pool is considered between 20 and 40 km [10]. In [15], authors propose a hybrid cloud-fog RAN architecture to deal with the increased transmission latencies. In [7], machine learning techniques are proposed to predict the required resources in the BBU pool. However, despite a large amount of technical works in the field, there are few of them that accurately analyze the total round-trip delay incurred in the C-RAN deployments. In this paper, we carefully indicate all delay components of the round trip delay, formalize the process in terms of queuing theory and provide formulas for the mean response time and amount of computational resources required to satisfy the delay budget. Unlike the work [11], we do not assume fixed queuing delays in the BBU pool, but use queuing theory methods to deduce more accurate formulas. Besides, we consider more realistic scenario, in which RRHs are not connected directly to the BBU pool, but there is a router between them, which gathers code blocks from several RRHs and sends the aggregated traffic flow to the BBU pool.

The rest of the paper is organized as follows. The system model with the scenario description is presented in Sect. 2. Then, we deduce formulas for evaluation of all delay components in Sect. 3. Section 4 is devoted to the case study with the detailed discussion. Section 5 concludes the paper.

2 System Model

Consider N_c RRH clusters are served by the BBU pool located in the central cloud. Each cluster consist of N_r RRHs, and each RRH produce a flow of code blocks with rate λ_r.

The components of the round-trip delay are shown in the Fig. 1.

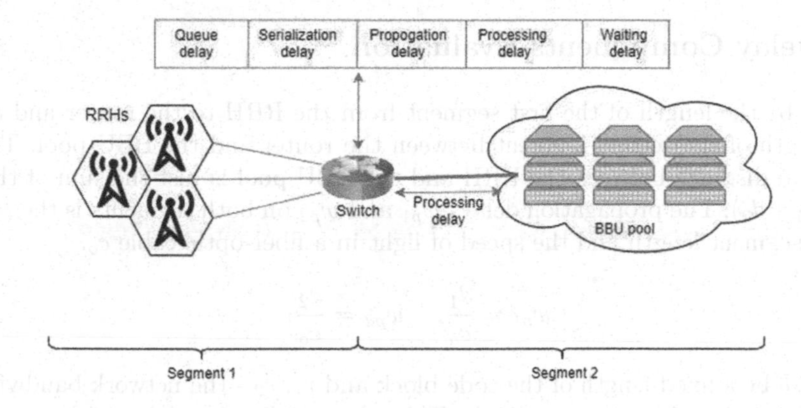

Fig. 1. Round-trip delay components in a C-RAN

User Equipment (UE), sends its signal to the RRH, where it is digitized, grouped into code blocks, and then transmitted for further processing to the BBU pool. Code blocks pass through two network segments between the RRH and the BBU pool. Delays may occur in each network segment, which include a propagation delay w_p and a serialization delay w_t for sending a code block over the network. The transmission delay w_t at each intermediate node between the BBU pool and RRH can be expressed as the ratio of the number of sent bits to the bandwidth of the communication line. A number of articles consider the ideal case where each RRH is connected by a separate fiber-optic channel directly to the BBU pool, but this is not a realistic scenario. We assume the presence of a router that combines several RRH signal streams, and then sends it in one channel. Hence, there is additional queuing delay w_q on this router.

Also, processing delay of w_s occurs in the BBU pool. The largest component of processing delay is due to forward error correction [3]. Forward Error Correction (FEC) is a signal coding/decoding technique with the ability to detect errors and correct information by the forward method. Thus, the receiving equipment can detect and correct errors that occur in the transmission channel. FEC dramatically reduces the bit error rate (BER), which allows increasing of the signal transmission distance without regeneration. Besides, when there is a shortage of resources in the BBU pool, the code blocks are stored in a buffer, so there is additional waiting delay w_b occurs. After that, the BBU sends the response back to the UE. Consequently, the round-trip delay includes twice all the described delays except waiting and processing latencies in the BBU pool. The total delay ξ_T can be expressed by formula (1), which is the sum of all the delay components:

$$\xi_T = 2\left(w_{p_1} + w_{t_1} + w_q + w_{p_2} + w_{t_2}\right) + w_s + w_b. \quad (1)$$

The total delay ξ_T must meet the strict latency requirements. In the case of LTE, the delay budget is about 3 ms [1].

3 Delay Components Evaluation

Let d_1 be the length of the first segment from the RRH to the router and d_2 - the length of the second segment between the router and the BBU pool. Then the total distance between the RRH and the BBU pool is just the sum of them ($d = d_1 + d_2$). The propagation delays w_{p1} and w_{p2} on both segments is the ratio of the segment length and the speed of light in a fiber-optic cable c_o.

$$w_{p1} = \frac{d_1}{c_o}, \quad w_{p2} = \frac{d_2}{c_o}. \quad (2)$$

Let b be a fixed length of the code block and r_1, r_2 - the network bandwidth in segments 1 and 2 correspondingly. Then the serialization delay is the ratio of the code block length and the bandwidth.

$$w_{t1} = \frac{b}{r_1}, \quad w_{t2} = \frac{b}{r_2}. \quad (3)$$

For the queuing delay at the intermediate router, we employ the queuing theory methods. Since the arrival of code blocks cannot be approximated by Poisson process, and the serving time at the router is also nonexponential, but close to deterministic, we use the $G/G/1$-type queuing system. Analytical review in work [2] showed that one of the most successful approximations for calculating the average waiting time Mw_q in this type of queuing system is the following formula:

$$Mw_q = \frac{\rho_1 \frac{b}{r_2}(v_a^2 + v_b^2)}{2(1-\rho_1)} f(\rho_1, v_a, v_b). \quad (4)$$

Here ρ_1 is the offered load and may be evaluated as the product of the total arrival intensity from all N_r RRHs in a cluster and the average service time, which makes $\rho_1 = \frac{N_r \lambda_r b}{r_2} < 1$. Besides, v_a is the variation coefficient of the interarrival times and v_b is the variation coefficient of the service times. Finally, $f(\rho_1, v_a, v_b)$ is the correction function, which depends on the value of the offered load ρ_1 and variation coefficients v_a and v_b:

$$f(\rho_1, v_a, v_b) = \begin{cases} e^{-\frac{2(1-\rho_1)}{3\rho_1}\frac{(1-v_a^2)^2}{v_a^2+v_b^2}}, & v_a < 1; \\ e^{-(1-\rho_1)\frac{v_a^2-1}{v_a^2+4v_b^2}}, & v_a \geq 1. \end{cases} \quad (5)$$

For the evaluation of the delays in the BBU pool, we also use the queuing theory methods. First, we deduce formula for the average processing delay w_s. The processing delay in the BBU pool is the time to process the radio signal, including demodulation and coding. The decoding calculation has its own performance directly related to the number of cycles performed by the FEC algorithm, and the average processing delay Mw_s can be expressed as

$$Mw_s = \frac{kbF}{pO} + J. \tag{6}$$

Formula (6) can be obtained using the following considerations. The BBU pool executes k cycles of the FEC algorithm for each code block. Parameter b, as before, is the length of the code block in bits, which may vary, depending on the technology in use, the coding rate, and the puncturing rate adjustment algorithm [5]. Each bit of the code block is usually processed by decoder with complexity F, expressed in bitwise operations. The processor clock speed allocated for BBU is denoted as p (in Hz) and O is the processor efficiency in operations per clock cycle, which is defined by the number of processor cores. In addition, we denote J - the time required to process other wireless functions [3].

To calculate the waiting time w_b in the BBU pool, we model it in terms of $G/G/m$ queuing system. Since the amount of computational resources in the BBU pool are optimized to serve the offered load, we use the heavy traffic asymptotic approximation for the $G/G/m$ queuing system [4]. Let m be the number of BBUs in the BBU pool, the average service time is given by formula (6). Besides, let σ_T be the variance of the interarrival times and σ_X is the variance of the service times. Then, the average waiting time Mw_b in the BBU pool can be expressed by the following formula:

$$Mw_b = \frac{\lambda_r N_r N_c(\sigma_T^2 + \frac{\sigma_X^2}{m})}{2(1-\rho_2)} \tag{7}$$

Here $\lambda_r N_r N_c$ is the total arrival intensite of code blocks from all RRHs from all clusters and ρ_2 is the load at the BBU pool:

$$\rho_2 = \frac{\lambda_r N_r N_c M w_s}{m}. \tag{8}$$

Combining all the delay components discussed above, the total average delay $M\xi_T$ between the BBU pool and the RRH can therefore be expressed as:

$$M\xi_T = 2\left(\frac{d_1}{c_0} + \frac{b}{r_1} + \frac{\rho_1 \frac{b}{r_2}(v_a^2 + v_b^2)}{2(1-\rho_1)} f(\rho_1, v_a, v_b). + \frac{d_2}{c_0} + \frac{b}{r_2}\right) \tag{9}$$

$$+ \frac{kbF}{pO} + J + \frac{\lambda_r N_r N_c(\sigma_T^2 + \frac{\sigma_X^2}{m})}{2(1-\rho_2)}.$$

Table 1. Input parameters

N_r	50
N_c	40
b	6144 bits
c_0	$2 \cdot 10^8$ m/s
$r_1 = r_2$	10 Gbit/s
k	7 cycles
F	200 operations per bit
p	3,47 GHz
O	2 operations per cycle
J	0,3 ms
v_a	0,3
v_b	0,4

4 Numerical Analysis

In this section we present the results of the numerical analysis. We consider the scenario from [3], where the BBU pool must decode the code block from RRH and by setting the average round-trip delay $M\xi_T$ less than or equal to the delay budget, we find the maximum distance $d = d_1 + d_2$ between the BBU pool and RRH. Take $d_1 = \frac{1}{3}d$ and $d_2 = \frac{2}{3}d$. The length of the code block b is set to 6144 bits as the maximum code block size in LTE [6]. The variation coefficients of of interarrival and service times are assumed equal on both intermediate router and the BBU pool. Their variances, which are used in formula (7), are deduced from variation coefficients and the averages by the well-known formula. The complexity of the decoder F is 200 operations per bit [8]. We consider dual core processor, so the processor efficiency $O = 2$ operations per cycle. The delay budget is assumed 3 ms [1]. The input parameters for the evaluation are summarized in Table 1.

First of all, we analyze the dependence of the round trip delay in a C-RAN on the computational resources of the BBU pool.

Figures 2 and 3 show the calculated average round-trip delay with total distance $d = 20$ km in cases $\lambda_r = 3 \cdot 10^4$ and $\lambda_r = 2.5 \cdot 10^4$ code blocks per second respectively, as a function of number of BBUs m in the BBU pool. One can note that the dependence is nearly linear. It may be conluded from the figures, that for the case of $\lambda_r = 3 \cdot 10^4$, the minimum required number of BBUs to satisfy the delay budget is approximately 101.5 thousands, while in case $\lambda = 2.5 \cdot 10^4$ it is 84.6 thousands.

Next, we analyze the effect of the code block arrival intensity on the average round-trip delay. Here we assume that there are $m = 10^5$ BBUs in the BBU pool and range the arrival intensity λ_r from $2.5 \cdot 10^4$ to $3 \cdot 10^4$ code blocks per second. Note that $\lambda_r = 2.5 \cdot 10^4$ corresponds to data flow through a RRH with

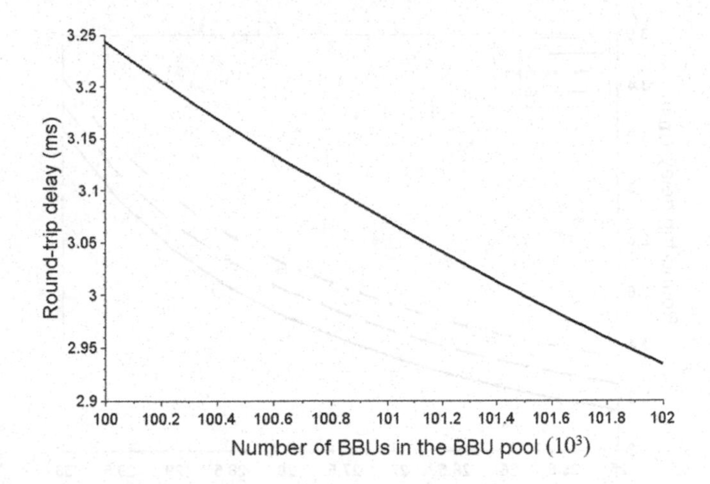

Fig. 2. The round-trip delay as a function of the number of BBUs in the BBU pool, $\lambda_r = 3 \cdot 10^4$

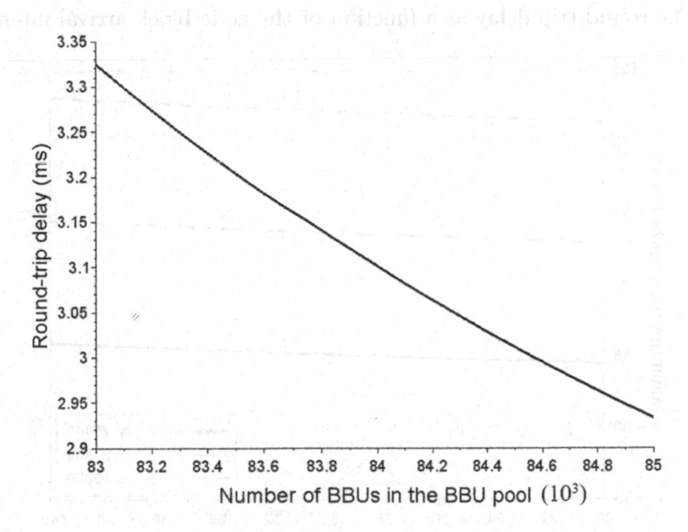

Fig. 3. The round-trip delay as a function of the number of BBUs in the BBU pool, $\lambda_r = 2.5 \cdot 10^4$

approximate bitrate 154 Mbps, and $\lambda_r = 3 \cdot 10^4$ - to the bitrate 190 Mbps. The results are depicted on Fig. 4 for various distances $d = 20; 30; 40$. As it is seen from the figure, the increase of the distance leads to a small increase of average round-trip delay. However, even small increase may be critical if there are not enough computational resources in the BBU pool.

Finally, we characterize the relation between the distance between RRHs and the BBU pool on the one hand, and the minimum required number of BBUs to satisfy the strict delay budget on the other hand. The Fig. 5 depicts the numerical

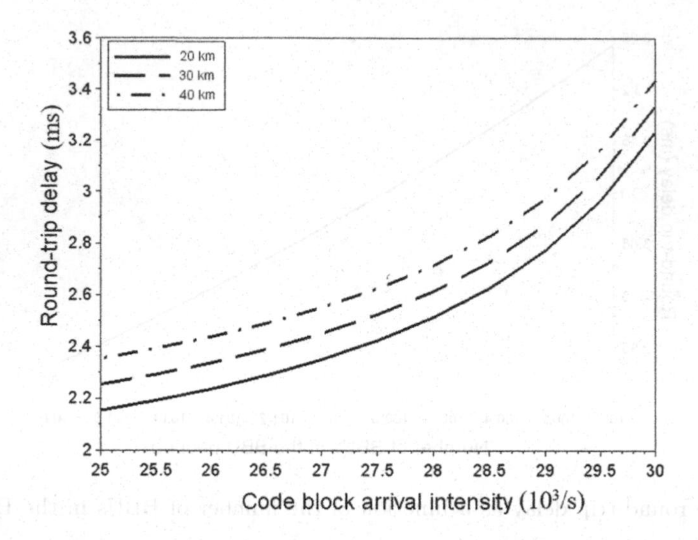

Fig. 4. The round-trip delay as a function of the code block arrival intensity λ_r

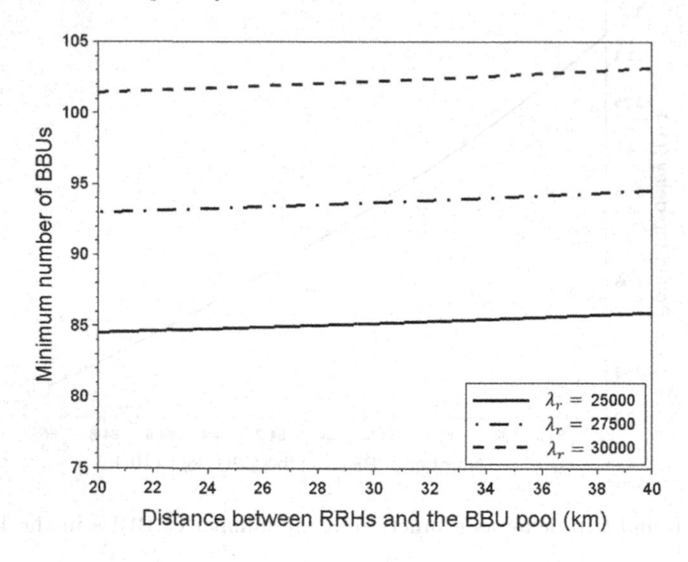

Fig. 5. Minimum number of BBUs required as a function of the distance d between RRH and the BBU pool

results for three values of the arrival intensity λ_r, namely 25000, 27500 and 30000 code blocks per second.

As one may conclude from Fig. 5, the distance between RRHs and the BBU pool may be even longer than 40 km, but it makes the system less tolerate to the queuing delays at the BBU pool. As a result, if a network operator plans its C-RAN with larger coverage area, then it should provide the BBU pool enough computational resources.

5 Conclusion

In the paper, we developed a mathematical model of the Cloud RAN to analyze the incurred round-trip delay. We carefully indicated components of the round-trip delay that reveal the interdependencies between computational resources of the BBU pool, distance between the RRHs and the BBU pool and the round-trip delay. To quantify the average delays we used well-known methods of queuing theory. The developed model was used to evaluate the required computational resources in the BBU pool to satisfy the strict delay budget in C-RAN. The analysis showed that the coverage radius of the BBU pool may be bigger than 40 km, but it leads to increased capital expenditures for providing the BBU pool additional computational resources.

References

1. Alyafawi, I., Schiller, E., Braun, T., Dimitrova, D., Gomes, A., Nikaein, N.: Critical issues of centralized and cloudified LTE-FDD radio access networks. In: 2015 IEEE International Conference on Communications (ICC), pp. 5523–5528. IEEE (2015)
2. Basharin, G., Bocharov, P., Kogan, Y.: Analiz ocheredey v vychislitelnyh setyah. Nauka, Teoriya i metody rascheta (1989)
3. Bhaumik, S., et al.: CloudIQ: a framework for processing base stations in a data center. In: Proceedings of the 18th Annual International Conference on Mobile Computing and Networking, pp. 125–136 (2012)
4. Bose, S.K.: Simulation techniques for queues and queueing networks. In: An Introduction to Queueing Systems, pp. 257–281. Springer, Boston (2002). https://doi.org/10.1007/978-1-4615-0001-8_7
5. Brejza, M.F., Li, L., Maunder, R.G., Al-Hashimi, B.M., Berrou, C., Hanzo, L.: 20 years of turbo coding and energy-aware design guidelines for energy-constrained wireless applications. IEEE Commun. Surv. Tutor. **18**(1), 8–28 (2015)
6. 3rd Generation Partnership Project (3GPP): LTE; evolved universal terrestrial radio access (E-UTRA); multiplexing and channel coding, (3GPP TS 36.2. 212 version 12.6. 0 release 12)", Technical Specification (2015)
7. Guerra-Gomez, R., Boque, S.R., García-Lozano, M., Bonafe, J.O.: Machine-learning based traffic forecasting for resource management in C-RAN. In: 2020 European Conference on Networks and Communications (EuCNC), pp. 200–204 (2020)
8. Holma, H., Toskala, A.: LTE for UMTS: OFDMA and SC-FDMA Based Radio Access. Wiley, Hoboken (2009)
9. Khakimov, A., Ateya, A.A., Muthanna, A., Gudkova, I., Markova, E., Koucheryavy, A.: IoT-fog based system structure with SDN enabled. In: ICFNDS 2018 (2018)
10. Kuilin, C., Ran, D.: C-RAN the road towards green ran. China Mobile Research Institute, White Paper (2011)
11. Marotta, M.A., Ahmadi, H., Rochol, J., DaSilva, L., Both, C.B.: Characterizing the relation between processing power and distance between BBU and RRH in a cloud RAN. EURO-COST, IRACON, TD(19)11040 (2019)
12. Musumeci, F., Bellanzon, C., Carapellese, N., Tornatore, M., Pattavina, A., Gosselin, S.: Optimal BBU placement for 5G C-RAN deployment over WDM aggregation networks. J. Lightwave Technol. **34**(8), 1963–1970 (2015)

13. Muthanna, A., et al.: SDN multi-controller networks with load balanced. In: ICFNDS 2018 (2018)
14. Muthanna, A., et al.: Secure and reliable IoT networks using fog computing with software-defined networking and blockchain. J. Sensor Actuator Netw. **8**(1), 15 (2019)
15. Tinini, R.I., Batista, D.M., Figueiredo, G.B., Tornatore, M., Mukherjee, B.: Low-latency and energy-efficient BBU placement and VPON formation in virtualized cloud-fog RAN. IEEE/OSA J. Opt. Commun. Networking **11**(4), B37–B48 (2019)

Resource Queuing System with Preemptive Priority for Performance Analysis of 5G NR Systems

Eduard Sopin[1](✉), Vyacheslav Begishev[1], Dmitri Moltchanov[1,2], and Andrey Samuylov[1,2]

[1] Peoples' Friendship University of Russia (RUDN University), 6 Miklukho-Maklaya Street, Moscow 117198, Russian Federation
{sopin-es,begishev-vo,molchanov-da,samuylov-ak}@rudn.ru

[2] Tampere University, 33720 Tampere, Finland
{dmitri.moltchanov,andrey.samuylov}@tuni.fi

Abstract. One of the ways to enable smooth coexistence of ultra reliable low latency communication (URRLC) and enhances mobile broadband (eMBB) services at the air interface of perspective 5G New Radio (NR) technology is to utilize preemptive priority service. In this paper, we provide approximate analysis of the queuing system with random resource requirements, two types of customers and preemptive priority service procedure. The distinctive feature of the systems – the random resource requirements – allows to capture the essentials of 5G NR radio interface but inherently increases the complexity of analysis. We present the main performance metrics of interest including session drop probability and system resource utilization as well as assess their accuracy by comparing with computer simulations. The developed model is not inherently limited to URLLC and eMBB coexistence and can be utilized in performance evaluation of 5G NR systems with priority-based service discipline at the air interface, e.g., in context of network slicing. Among other conclusions we explicitly show that both session drop and interruption probabilities of low priority traffic heavily depend not only on the intensity of high priority traffic but on stochastic characteristics of the resource request distribution.

Keywords: Resource queuing system · Preemptive priority · Blocking probability · Interruption probability · URLLC traffic · Network slicing

1 Introduction

Fifth generation (5G) cellular systems promises to bring not only extreme bandwidth at the air interface but to introduce advanced functionalities related to

The publication has been prepared with the support of the "RUDN University Program 5-100" (recipient V. Begishev). The reported study was funded by RFBR, project numbers 18-07-00576 (recipients D. Moltchanov and A. Samuylov) and 19-07-00933 (recipient E. Sopin).

V. M. Vishnevskiy et al. (Eds.): DCCN 2020, LNCS 12563, pp. 87–99, 2020.
https://doi.org/10.1007/978-3-030-66471-8_8

quality of service differentiation to mobile virtual mobile network operators (MVNO, [5,8]) as well as traffic coexistence at the air interface [2,6]. As the standardization of 5G New Radio (NR) interface is close to its completion the research community continue to explore the way to support these advanced functionalities.

In recent years, queuing systems with random resource requirements, where customers require not only a server but also a random volume of resources, have drawn significant attention for their ability to capture specifics of session serving process in prospective cellular systems including 5G NR technology [3,11]. These systems are often used on conjunction with stochastic geometry models to capture the stochastic properties induced by user locations with respect to base stations (BS) as well as traffic service dynamics at BS. Recently, these models have been utilized to investigate various session continuity mechanisms in 5G NR systems including multiconnectivity [17,18], resource reservation [3] as well as their joint functionality [9]. However, despite many research activities in the field, resource queuing systems with priorities have not been addressed yet.

The model presented in this paper has a large scope of applications in 5G NR systems. Of particular interest is provisioning of ultra reliable low latency service (URLLC) at New Radio (NR) base stations in industrial applications in context of network slicing [6,16]. Recall that URLLC service requires extremely small delays and loss guarantees at the air interface. To ensure it when mixed with conventional enhanced mobile broadband (eMBB, [4,12]) service at a single NR BS, several approaches ranging from the use of intentional overlapping by using non-orthogonal multiple access (NOMA, [8]) to explicit static bandwidth reservations have been proposed in the past. In this context, explicit prioritization may provide an alternative approach to maintain extreme service characteristics of URLLC traffic.

Another application area of the queuing system with priorities is network slicing functionality of 5G systems [6]. The concept of network slicing has been originally introduced in 3GPP in Release 14 [1] in context of vehicular-to-everything (V2X) communications. Later, in 3GPP Releases 15 and 16 it has been extended to include MVNOs and bundling of services with similar QoS requirements. The use-cases are flexible and include service provisioning for third parties such as MVNOs, content providers, etc. In the context of 5G system, a network slice instance is a unification of virtual resources in end-to-end fashion wherein a set of virtual network functions are instantiated and connected via a virtual network. One of the ways to support network slices with different service requirements in 5G NR systems is to use priority-based service discipline.

The specified model allows to account for random resource requirements at the air interface of both eMBB and URLLC service induced by random locations of user equipment in the coverage area of NR BS [3,10]. Preemptive priority discipline simultaneously accounts for efficient use of resources at the NR BS and ensures that URLLC traffic receives absolute priority over conventional eMBB traffic reaching the prescribed loss guarantees. Supplementing the model with a certain deployment of NR BSs and UEs in the considered area one may characterize the required density of NR BSs needed to maintain the prescribed performance provided to both URLLC and eMBB traffic types.

The rest of the paper is organized as follows. The resource queuing system model is introduced and analyzed in Sect. 2. Applications and numerical results are discussed in Sect. 3. Conclusions are drawn in the last section.

2 Model Description and Analysis

In this section we first introduced the queuing model by specifying their parameters and priority mechanism. Then, we proceed analyzing the system and proposing an approximate approach to the system analysis. Metrics of interest are finally derived.

2.1 Description of the System

We consider a multiserver queuing system with N servers and resource volume R. In practical context N represents the number of simultaneously supported sessions and can be infinite while R characterizes the amount of resources measures in, e.g., resource blocks (RB) or Hz. Two types of customers are served in the queuing system: first type are the preemptive priority customers and the second type are non-priority customers. Customers arrive according to Poisson process with intensities λ_1 and λ_2 correspondingly. These flows may represent, e.g. URLLC and eMBB traffic flows or slices with different service requirements. An arriving customer of type l requires discrete random volume of resources according to probability distribution $\{p_{l,r}\}$, $l = 1, 2$, $r = 1, 2, ...$, where $\sum\limits_{r=1}^{\infty} p_{l,r} = 1$. The randomness of resource request distribution allows to capture stochastic nature of user locations in the service area of a BS. The service times are exponentially distributed with intensities μ_1 and μ_2, correspondingly.

Assume that there are n_1 customers of the first type totally occupying r_1 resources and n_2 customers of the second type occupying r_2 resources. Upon arrival of a second type customer that requires j resources, if there is free server ($n_1 + n_2 < N$) and total volume of unoccupied resources is greater than the required volume of the customer ($R - r_1 - r_2 \geq j$), then the customer is accepted and the required resource volume is allocated to the customer. If arriving customer is the priority customer, then it is still accepted if $n_1 < N$ and $R - r_1 \geq j$, but the service of one or more customers of second type is interrupted. In this case, the system chooses the most "heavy" customer (the one that occupies the biggest part of resources) and terminates it. If there are still not enough resources, then the termination procedure continues until the resources can be allocated to the arriving priority customer $R - r_1 - r_2^* \geq j$.

Since the evaluation of the number of customers to be terminated upon arrival of a priority customer is the most complex part of the analysis, further we assume the following assumption. If termination of the two most "heavy" customers is not enough for the priority customer, then all non-priority customers are terminated. The applicability of the assumption is verified in Sect. 3.

The behavior of preemptive priority customers is equivalent to the behavior of these customers in the same queuing system without non-priority customers. Thus, we focus on the metrics of interest associated with non-priority customers.

2.2 System Analysis

To decrease the complexity of the stochastic process that describes the behavior of the system, we employ the technique originally proposed in [15]. According to it, instead of keeping track of resources allocated to all the customers, we follow only a total amount of occupied resources for each type of customers. Then, the behavior of the system can be described by a simplified process, $X(t) = (\xi_1(t), \gamma_1(t), \xi_2(t), \gamma_2(t))$, where $\xi_l(t)$ is the number of l-type customers at time t and $\gamma_l(t)$ is total resource volume occupied by l-type customers.

Let $q(n_1, r_1, n_2, r_2)$ be the stationary distribution of $X(t)$, $Q(n,r)$ and $P(n,r)$ – marginal stationary distributions of first and second type customers, respectively. For the resource queuing systems with priorities, it is impossible to obtain analytical expressions of the stationary distribution in the product form [13,14], that is why we concentrate on evaluation of the marginal distributions $Q(n,r)$ and $P(n,r)$. Since preemptive priority customers are not affected by non-priority customers, then, according to [15] we have

$$Q(n,r) = Q(0,0)\frac{\rho_1^n}{n!}p_{1,r}^{(n)}, \tag{1}$$

where $\rho_l = \frac{\lambda_l}{\mu_l}$, $p_{1,r}^{(n)}$ is the probability that n first type customers totally occupy r resources and $Q(0,0)$ is calculated using the normalizing condition. Particularly, the normalizing constant and all characteristics may be evaluated using functions $G(n,r)$, which are introduced in [19] together with efficient recurrent algorithm for their calculation.

$$G(n,r) = \sum_{k=0}^{n}\sum_{j=0}^{r}\frac{\rho^k}{k!}p_{1,j}^k. \tag{2}$$

Note that probabilities $p_{1,r}^{(n)}$ are evaluated from distribution $\{p_{1,r}\}$ by n-fold convolution.

To derive equations for approximation of stationary probabilities $P(n,r)$, we introduce additional notation. Let $\pi(k,j)$ be the probability that arriving second type customers cannot be accepted to the system conditional to k second type customers are in the system with j resources occupied, i.e.,

$$\pi(k,j) = 1 - \sum_{i=1}^{R-j} p_{2,i}\frac{G(N-k-1, R-j-i)}{G(N-k, R-j)}. \tag{3}$$

The fraction inside the sum represent conditional probability that there are no more than $N-k-1$ priority customers with no more than $R-j-i$ resource occupied under condition that there is no more than $N-k$ of them occupying no more than $R-j$ resources.

Further, let $\gamma_i(k,j)$ be the probability that one nonpriority customer occupies i resources under condition that k of them occupy j resources, $1 \le i \le j-k+1, 1 \le k \le j$,

$$\gamma_i(k,j) = \frac{p_{2,i} \cdot p_{2,j-i}^{(k-1)}}{p_{2,j}^{(k)}}. \quad (4)$$

Employing the order statistics approach, we derive the probability distribution of the most "heavy" customers. Let $\gamma_{max,i}(k,j)$ be the probability that the most "heavy" customer occupy i resources under condition that there are totally k nonpriority customers occupying j resources:

$$\gamma_{max,i}(k,j) = (\Gamma_{k,j}(i+1))^k - (\Gamma_{k,j}(i))^k, 1 \le i \le j-k+1, \quad (5)$$

where $\Gamma_{k,j}(x)$ is the cumulative distribution function (CDF), which corresponds with distribution $\{\gamma_i(k,j)\}_{i>0}$.

One of the most challenging parts is to analyze how many nonpriority customers should be interrupted upon arrival of a priority customer. So, we introduce $\beta_i(k,j)$ - the probability that upon arrival of a priority customer, exactly i resources should be released by second type customers under condition that there are k of them in the system occupying totally j resources.

$$\beta_i(k,j) = \sum_{r=0}^{R-j} \frac{G(N-k,r) - G(N-k,r-1)}{G(N-k,R-j)} p_{1,R+i-r-j}, \quad (6)$$
$$1 \le k \le N-1, k \le j \le R, 1 \le i \le j,$$

$$\beta_0(k,j) = \sum_{r=0}^{R-j-1} \frac{G(N-k,r) - G(N-k,r-1)}{G(N-k,R-j)} \sum_{s=1}^{R-r-j} p_{1,s}, \quad (7)$$
$$1 \le k \le N-1, 1 \le k \le R-1.$$

These probabilities will be used to analyze the number of interrupted nonpriority customers. Further we assume that upon arrival of a priority customer, there are four possibilities, namely i) no interruption, ii) interruption of one nonpriority customer, iii) interruption of two customers and iv) interruption of all customers. Since in the considered system, the most heavy customers are interrupted in the first place, the proposed assumption is expected to give good approximation. So, let $\delta_{s,i}(k,j)$, $s=1,2$ be the probability that exactly s customers are interrupted, which occupied totally i resources under condition that there were k customers with j resources occupied. These probabilities are evaluated as follows:

$$\delta_{1,i}(k,j) = \gamma_{max,i}(k,j) \sum_{l=1}^{i} \beta_l(k,j), 1 \le i \le j-k+1, 1 \le k \le N-1, \quad (8)$$

$$\delta_{2,i}(k,j) = \sum_{m=1}^{i-1} \gamma_{max,m}(k,j) \cdot \gamma_{max,i-m}(k-1,j-m) \sum_{l=m+1}^{i} \beta_l(k,j), \quad (9)$$
$$2 \le i \le j-k+2,$$

$$\delta_{all}(k,j) = 1 - \delta_0(k,j) - \sum_{i=1}^{j-k+1} \delta_{1,i}(k,j) - \sum_{i=2}^{j-k+2} \delta_{2,i}(k,j), \quad (10)$$

where $\delta_0(k,j)$ is the probability that no interruption occurs upon arrival of a first type customer:

$$\delta_0(k,j) = G(N-k, R-j)^{-1} \sum_{r=1}^{R-j-N+k+1} p_{1,r} \cdot G(N-k-1, R-j-r) \quad (11)$$

Finally, let $\alpha_{s,i}(j)$, $s = 1,2$, be the probability that exactly s non-priority customers occupying i resources are interrupted upon arrival of a priority customer under condition that there are N customers of second type with j resources occupied, i.e.,

$$\alpha_{1,i}(j) = \gamma_{max,i}(N,j) \sum_{m=1}^{R-j+i} p_{1,m}, 1 \le i \le j-N+1 \quad (12)$$

$$\alpha_{2,i}(j) = \sum_{l=1}^{i-1} \gamma_{max,l}(N,j) \cdot \gamma_{max,i-l}(N-1,j-l) \sum_{m=R-j+l+1}^{R-j+i} p_{1,m}, \quad (13)$$
$$2 \le i \le j-N+2$$

$$\alpha_{all}(j) = 1 - \sum_{i=1}^{j-N+1} \alpha_{1,i}(j) - \sum_{i=2}^{j-N+2} \alpha_{2,i}(j), 1 \le j \le R. \quad (14)$$

Utilizing the introduced probabilities, we can derive the system of equilibrium equations as follows:

$$\lambda_2(1-\pi(0,0))P(0,0) = \mu_2 \sum_{j=1}^{R} P(1,j) + \lambda_1 \sum_{k=1}^{N-1} \sum_{j=k}^{R} P(k,j)\delta_{all}(k,j)$$
$$+ \lambda_1 \sum_{j=N}^{R} P(N,j)\alpha_{all}(j) + \lambda_1 \sum_{j=1}^{R} P_1(j)\delta_j + \sum_{j=2}^{R} P_2(j)\delta_j \quad (15)$$

$$P(k,j)\left[\lambda_2(1-\pi(k,j))+k\mu_2+\lambda_1(1-\pi_{b1})(1-\delta_0(k,j))\right]$$
$$=(k+1)\mu_2\sum_{i=1}^{R-j}P(k+1,j+i)\cdot\gamma_i(k+1,j+i)$$
$$+\lambda_2\sum_{i=1}^{j-k+1}p_{2,i}P(k-1,j-i)\cdot\frac{G(N-k,R-j)}{G(N-k+1,R-j+i)}$$
$$+\lambda_1\sum_{i=1}^{\min(j+i-k,R-j)}P(k+1,j+i)\delta_{1,i}(k+1,j+i) \quad (16)$$
$$+\lambda_1\sum_{i=2}^{\min(R-j,j+i-k)}P(k+2,j+i)\delta_{2,i}(k+2,j+i),$$
$$1\leq k\leq N-1, k\leq j\leq R,$$

$$P(N,j)\left[N\mu_2+\lambda_1\right]=\lambda_2\sum_{i=1}^{j-N+1}p_{2,i}P(N-1,j-i)\cdot\frac{G(0,0)}{G(1,R-j+i)}. \quad (17)$$

By solving the system (15)–(17) we obtain the marginal stationary probabilities $P(n,r)$.

2.3 Performance Metrics of Interest

Now, we proceed with performance metrics of interest. First, we analyze the blocking probabilities upon arrival of both types of customers, namely π_{b1} and π_{b2}. For priority customers, the blocking probability is obtained in [19], while π_{b2} is deduced using similar logic

$$\pi_{b1}=1-G(N,R)^{-1}\sum_{r=1}^{R}p_{1,r}G(N-1,R-r); \quad (18)$$

$$\pi_{b2}=\sum_{j=0}^{R}P(N,j)+\sum_{k=0}^{N-1}\sum_{j=0}^{R}P(k,j)\pi(k,j). \quad (19)$$

Finally, we proceed with the probability π_i that a non-priority customer is interrupted. The intensity of customer interruption is obtained as $\lambda_2(1-\pi_{b,2})-\bar{N}_2\mu_2$ from the equality of arrival intensity and intensity of leaving the system. Here $\bar{N}_2$ is the average number of 2-type customers in the system. Then, the interruption probability is given by the ratio of interruption and arrival intensities, that is,

$$\pi_i=1-\frac{\bar{N}_2\mu}{\lambda_2(1-\pi_{b,2})}. \quad (20)$$

3 Applications and Numerical Results

In this section, we first review the applications scope of developed model. Then, we proceed providing numerical example characterizing the performance of low priority arrival flow in the considered system.

3.1 Applications Scope

The developed model can be utilized as a service model NR BSs with multiple arrival flows having different priorities in composite performance evaluation frameworks, e.g., [3,20]. These frameworks are logically divided into two parts: (i) a queuing part and (ii) queuing parametrization. The former captures the features of radio resource allocation at NR BSs. At the input, it accepts the CDF of the amount of requested resources, $F_R(x)$, and the temporal intensity of the UE stage changes, α, and delivers the metrics of interest depending on these intermediate metrics as well as parameters of additional quality of service provisioning mechanisms.

The distinctive feature of the developed queuing model is that it simultaneously captures the randomness of resource request distribution and priority service. The former captures the effect of randomness on user locations in BS service area. Thus, it is applicable to 5G NR systems, where one has to prioritized arrivals traffic flows, e.g., systems supporting two or more types of traffic with different priorities or systems supporting network slicing capabilities at the air interface.

To utilize the proposed service model in realistic environment one needs to provide the following metrics as a function of NR BS parameters, propagation characteristics and environment: (i) coverage of NR BS, (ii) mean resource request of sessions (or slices, depending on the application). The former can be determined using the procedure offered in [3,20] while the latter can be found as discussed in [10]. Note that we also do not explicitly specify the operational frequency of NR technology. In case of millimeter wave band one also needs to account for blockage phenomena using the models proposed in [7].

3.2 Accuracy Assessment

We start assessing the accuracy of the developed approximation. To this aim, we have developed a single-purpose simulation tool that models the considered resources queuing system with two types of customers and preemptive priority service discipline. Here, we assume that resource requirements of both priority and non-priority customers have geometric distribution with parameter 0.5, binomial distribution with parameter 0.25 and Poisson distribution with parameter 1. Further, we assume that $N = 60$, $R = 100$, the arrival intensity of priority customers is $\lambda_1 = 35$, and service intensities are $\mu_1 = \mu_2 = 1$. The arrival intensity of non-priority customers λ_2 varies from 10 to 20.

Figure 1 shows the comparison results. One can note that the blocking probability of the priority customers shows almost perfect match with the simulations,

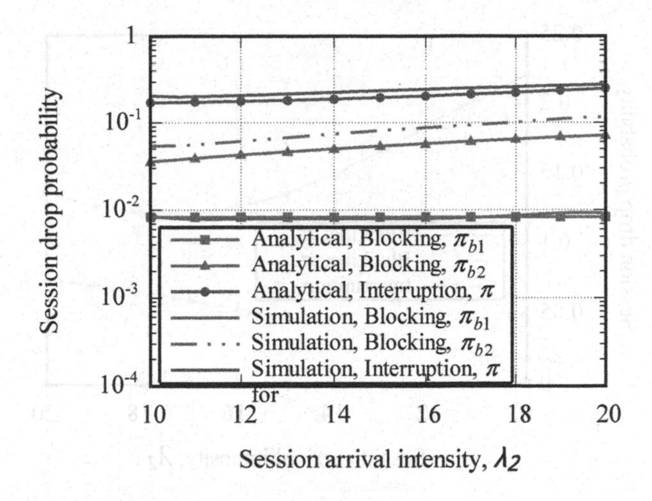

Fig. 1. Comparison of analytical and simulation results

while other two probabilistic metrics have bigger relative error. Notably, the relative error of all probabilities decreases with the increase of the non-priority customers intensity.

3.3 Performance Evaluation

We now proceed analyzing the response of the performance metrics of interest to the system parameters. Recall, that in the considered system we are primarily interested in service metrics related to the low priority traffic as high priority traffic receives exclusive access to the system resources.

First, we complement the dependence illustrated in Fig. 1 with Fig. 2 showing the considered performance metrics as a function of higher priority arrivals, λ_1. As one may observe, logical the increase in both, λ_1 and λ_2, results in worse performance of lower priority traffic in terms of session drop (blocking) probability upon arrival. However, the magnitude of the associated increase is different. On the other hand, the behavior of session interruption probability is drastically different. Particularly, when λ_2 increases the interruption probability always increases as well. However, the response to λ_1 is different – there is a bending point when the session interruption probability of low priority traffic starts to increase. This is explained by the fact that under these conditions more low priority sessions are blocked upon arrivals increasing the amount of resources left for those low priority sessions that are currently in the system.

The developed models accepts the CDF of resource requirements as a part of the model specifications. Depending on the application rate requirements as well as the service area of BSs, CDF may substantially differ. We thus investigate the effect of the structure of CDF of resource requirements in Fig. 3 and Fig. 4 by showing the metrics of interest as a function of geometric, Poisson and binomial distributions with the same mean value. As one may observe, Binomial

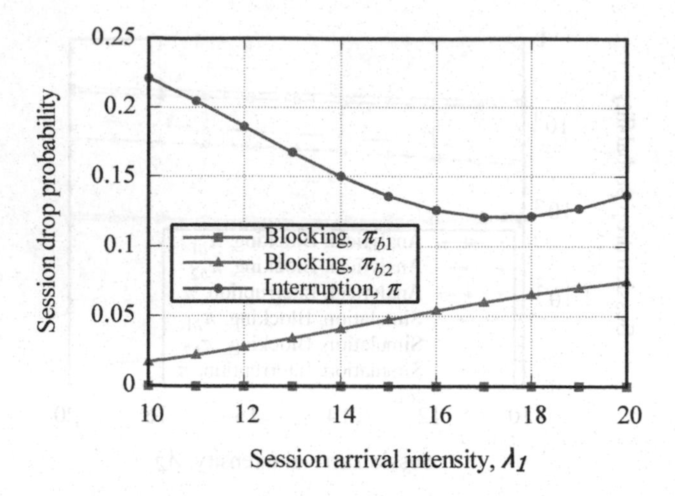

Fig. 2. Drop as a function of λ_1 for binomial, geometric and Poisson distribution

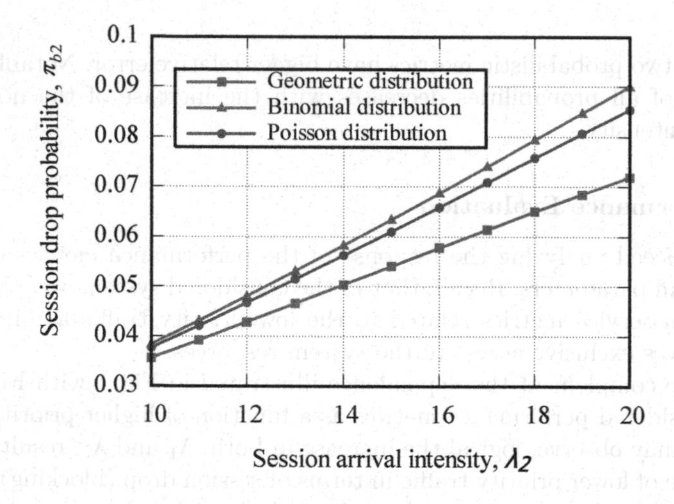

Fig. 3. π_{b2} as a function of λ_2 for binomial, geometric and Poisson distribution

and Poisson distributions lead to rather similar results with 3–5% difference in absolute values of considered metrics of interest. One the other hand, Geometric distribution results in drastically different performance measures. Observe that the variance of the considered Binomial, Poisson and Geometric distributions are 0.75, 1 and 2, respectively. This implies that not only mean value of the resource request distribution but higher order moments affect key performance indicators in 5G NR systems.

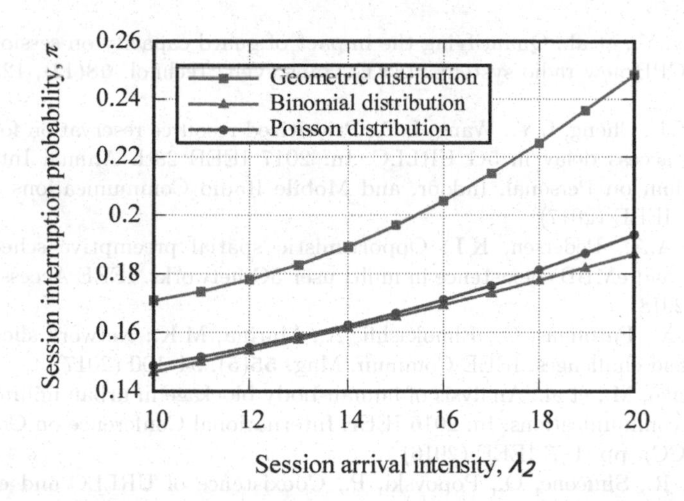

Fig. 4. π as a function of λ_2 for binomial, geometric and Poisson distribution

4 Conclusion

Motivated by coexistence of high-priority URLLC and low priority eMBB traffic at 5G NR air interface, in the paper, we have analyzed the limited resources queuing system with two types of customers and preemptive priority service discipline. We have developed an approach for approximate analysis performance measures for both type of customers.

The application scope of the proposed system is not limited by URLLC and eMBB coexistence at the air interface. Basically, the model can be applied to 5G NR systems, where prioritization is required at the air interface, e.g., network slicing. The developed model can be utilized in composite performance evaluation frameworks, where the service process at BSs is separated from the model parameterization.

The reported results of direct comparison with computer simulations indicate that the approximation accuracy lies within 5–10%. Furthermore, although the complexity of the proposed approach is lower compared to the direct solution of equilibrium equations, it is still high requiring significant computational efforts. Thus, the focus of our future work is to improve the approximation accuracy and decrease the computational requirements of the procedure.

References

1. 3GPP: Service requirements for V2X services (Release 14). 3GPP TS 22.185 V14.3.0, March 2017
2. Afolabi, I., Taleb, T., Samdanis, K., Ksentini, A., Flinck, H.: Network slicing and softwarization: a survey on principles, enabling technologies, and solutions. IEEE Commun. Surv. Tutor. **20**(3), 2429–2453 (2018)

3. Begishev, V., et al.: Quantifying the impact of guard capacity on session continuity in 3GPP new radio systems. IEEE Trans. Veh. Technol. **68**(12), 12345–12359 (2019)
4. Chen, Y.J., Cheng, L.Y., Wang, L.C.: Prioritized resource reservation for reducing random access delay in 5G URLLC. In: 2017 IEEE 28th Annual International Symposium on Personal, Indoor, and Mobile Radio Communications (PIMRC), pp. 1–5. IEEE (2017)
5. Esswie, A.A., Pedersen, K.I.: Opportunistic spatial preemptive scheduling for URLLC and eMBB coexistence in multi-user 5G networks. IEEE Access **6**, 38451–38463 (2018)
6. Foukas, X., Patounas, G., Elmokashfi, A., Marina, M.K.: Network slicing in 5G: survey and challenges. IEEE Commun. Mag. **55**(5), 94–100 (2017)
7. Gapeyenko, M., et al.: Analysis of human-body blockage in urban millimeter-wave cellular communications. In: 2016 IEEE International Conference on Communications (ICC), pp. 1–7. IEEE (2016)
8. Kassab, R., Simeone, O., Popovski, P.: Coexistence of URLLC and eMBB services in the C-RAN uplink: an information-theoretic study. In: 2018 IEEE Global Communications Conference (GLOBECOM), pp. 1–6. IEEE (2018)
9. Kovalchukov, R., et al.: Improved session continuity in 5G NR with joint use of multi-connectivity and guard bandwidth. In: 2018 IEEE Global Communications Conference (GLOBECOM), pp. 1–7. IEEE (2018)
10. Kovalchukov, R., Moltchanov, D., Gaidamaka, Y., Bobrikova, E.: An accurate approximation of resource request distributions in millimeter wave 3GPP new radio systems. In: Galinina, O., Andreev, S., Balandin, S., Koucheryavy, Y. (eds.) NEW2AN/ruSMART 2019. LNCS, vol. 11660, pp. 572–585. Springer, Cham (2019). https://doi.org/10.1007/978-3-030-30859-9_50
11. Lu, X., et al.: Integrated use of licensed-and unlicensed-band mmWave radio technology in 5G and beyond. IEEE Access **7**, 24376–24391 (2019)
12. Makeeva, E., Polyakov, N., Kharin, P., Gudkova, I.: Probability model for performance analysis of joint URLLC and eMBB transmission in 5G networks. In: Galinina, O., Andreev, S., Balandin, S., Koucheryavy, Y. (eds.) NEW2AN/ruSMART -2019. LNCS, vol. 11660, pp. 635–648. Springer, Cham (2019). https://doi.org/10.1007/978-3-030-30859-9_55
13. Naumov, V.A., Samuilov, K.E.: Analysis of networks of the resource queuing systems. Autom. Remote Control **79**(5), 822–829 (2018). https://doi.org/10.1134/S0005117918050041
14. Naumov, V.A., Samouylov, K.E.: Conditions for the product form of the stationary probability distribution of Markovian resource loss systems. Tomsk State Univ. J. Control Comput. Sci. **46**, 64–72 (2019)
15. Naumov, V.A., Samuilov, K.E., Samuilov, A.K.: On the total amount of resources occupied by serviced customers. Autom. Remote Control **77**(8), 1419–1427 (2016). https://doi.org/10.1134/S0005117916080087
16. Orsino, A., et al.: Caching-aided collaborative D2D operation for predictive data dissemination in industrial IoT. IEEE Wirel. Commun. **25**(3), 50–57 (2018)
17. Petrov, V., et al.: Achieving end-to-end reliability of mission-critical traffic in softwarized 5G networks. IEEE J. Sel. Areas Commun. **36**(3), 485–501 (2018)
18. Petrov, V., et al.: Dynamic multi-connectivity performance in ultra-dense urban mmWave deployments. IEEE J. Sel. Areas Commun. **35**(9), 2038–2055 (2017)

19. Samouylov, K., Sopin, E., Vikhrova, O., Shorgin, S.: Convolution algorithm for normalization constant evaluation in queuing system with random requirements. In: AIP Conference Proceedings, vol. 1863, p. 090004. AIP Publishing LLC (2017)
20. Samuylov, A., et al.: Characterizing resource allocation trade-offs in 5G NR serving multicast and unicast traffic. IEEE Trans. Wireless Commun. **19**(5), 3421–3434 (2020)

Redundant Servicing of a Flow of Heterogeneous Requests Critical to the Total Waiting Time During the Multi-path Passage of a Sequence of Info-Communication Nodes

V. A. Bogatyrev[1,2,3](✉), A. V. Bogatyrev[1], and S. V. Bogatyrev[1]

[1] NEO Saint Petersburg Competence Center, Saint Petersburg, Russia
{anatoly,stanislav}@nspcc.ru

[2] ITMO University, Saint Petersburg, Russia

[3] Saint Petersburg State University of Aerospace Instrumentation, Saint Petersburg, Russia
vladimir.bogatyrev@gmail.com
https://nspcc.ru/en

Abstract. The possibilities are investigated and analytical models of redundant multiway servicing of a heterogeneous request flow with their replication rate depending on the maximum permissible waiting time for replicas accumulated in the queues of nodes that make up the path for real-time information and communication systems are proposed. Two options are considered for redundant servicing of a heterogeneous flow during the sequential passage of copies of requests through parallel-connected nodes grouped in groups. For the first option, when generating a request, a certain number of copies are created, for each of which a path is predefined as a sequence of nodes of different groups involved in servicing this copy. For the second option, the paths are formed dynamically at each stage, and a copy of the request, executed first at some stage of the sequential passage of groups of redundant nodes, is transferred for redundant service to the next group of nodes. At various stages of service, the redundancy ratio can vary.

Keywords: Redundant service · Latency · Heterogeneous flow · Distribution of requests copies · Real-time

1 Introduction

Increased requirements for reliability [1,2], continuity, and timeliness of data processing and transmission processes are imposed on real-time systems, including cyber-physical systems [3–5], in which communication is often based on wireless technologies [6,7].

For info-communication, including cluster, systems operating in real-time, especially as part of a cyber-physical system, operability is determined not only

V. M. Vishnevskiy et al. (Eds.): DCCN 2020, LNCS 12563, pp. 100–112, 2020.
https://doi.org/10.1007/978-3-030-66471-8_9

by the reliability of the system structure but also by the functional reliability characterized by the probability of timely error-free fulfillment of requests that are critical to delays [8–10].

In the well-known works [11–15] related to ensuring the reliability of distributed computer systems including cluster ones [16–18], questions of assessing the reliability of real-time cluster systems are not addressed, for which strict requirements are imposed on service request delays and on the continuity of the computing process, including when the recovery time after failures of reserved resources can exceed the maximum allowable time of interruption of the computing process [8,19].

To reduce the average network transmission delays, transport coding allows [20,21], in which message fragments are transmitted along different routes, and in case of loss or error of frame transmissions, the entire message can be restored without retransmissions.

Multi-path routing allows for increased network availability and reduced reconfiguration time. With multi-path transmission, the main and several backup routes (paths) are formed, and in case of failure of the nodes making up the path, a switch to a backup, the pre-registered path is performed, which allows to speed up reconfiguration, and in some cases to ensure that the permissible delays in real-time systems are not exceeded. The efficiency of multi-path transmissions is achieved with prioritization of traffic and load balancing, including network reconfiguration after failures [22–24].

Reliability and timeliness of query execution in a redundant information and communication system (server cluster or switching nodes) can be improved as a result of redundant servicing of copies of requests with the issuance of one completed copy (for example, the first issued in time) [12–14].

Application of the concept of redundant maintenance to systems with multi-path routing can increase not only the structural reliability and availability of the network but also (as shown in [9,10,25]) the probability of timely service of requests that are critical to latency, which is important for real-time systems, including cyber-physical systems [26].

The direction of redundant multipath service with query replication [8–10], researched in this article, is the development of the concepts of multipath routing [22–24], multicast transmissions [7], broadcast service [27,28] and dynamic distribution of requests [29,30]. The solutions proposed in the article are focused on the possible application in the concept of Ultrareliable and Low-Latency Wireless Communication [31–34] that is currently being intensively developed.

The redundant multi-way service of a request by a sequence of groups of redundant nodes (clusters) is considered to be successful in real-time if at least one copy of it is executed correctly without exceeding the maximum permissible total waiting time in queues of the nodes sequentially executing it. The effectiveness of redundant maintenance for multi-level cluster systems (sequentially connected groups of redundant nodes, multicluster) is estimated by the probability of the timeliness of multi-stage servicing of at least one copy of the request [8]. Multistage maintenance involves passing a request through all the nodes that make up the active path. A path is a set of nodes whose operability ensures the fulfillment of a functional task of sequentially performing a request service. The path is active if the nodes included in it are involved in servicing the request or its copies. Potentially possible paths that are not used in the execution of the request are reserved.

For requests serviced in real-time, it is important not only to maintain the operability of the nodes that make up the path, but also the timeliness of the service and requests passing through it. The task is complicated by the heterogeneity of the flow when different requests may have different criticality to service time. The aim of the work is to investigate the possibility of increasing the functional reliability of a distributed system while increasing the probability of timely execution of time-critical requests of a heterogeneous stream as a result of their replication and redundant servicing by a sequence of nodes included in the path, taking into account the accumulation of latency on all nodes of the path.

2 Options for the Formation of the Path for Phased Redundant Service of Requests in a Multi-level Cluster

As the object of research consider m levels computer cluster [10] comprising at i-th level of n_i parallel-connected servers [8], each of them can be represented as single-channel queuing systems of the $M/M/1$ type with infinite queues [35]. With redundant maintenance at each level, multiple copies of the request are executed.

The following options for organizing redundant services in a multi-level cluster are researched:

Option S_1: k copies of request (replicas) are created in the node of the request source, and for each copy, the service path (route) is specified with the servers that execute the request (copy of the request) at each stage (level) [10]. For requests of different criticality to the total waiting time in the nodes making up the path, their number (paths' redundancy ratio) can be set various.

Option S_2: When a request is generated by a source, it's k_1 copies are created, distributed for nodes in k_1 first-level servers. When one of the copies is serviced at the first level, k_2 copies of requests are created, which are transferred to services in selected k_2 second-level nodes, and so on, until the request is serviced all m levels of the cluster [10]. The redundancy ratio of requests is set depending on their criticality to the total waiting time in the nodes that make up the path and can be different for different levels.

For option S_1: k paths of request execution are generated from the source, the change of which, as well as the multiplicity of backup copies, does not occur during sequential servicing of the request. The number of copies of requests (redundancy ratio) at different levels is the same. For a non-uniform flow of requests, the redundancy ratio may vary depending on the criticality of the requests to the total waiting time in path. If the number of redundant service paths generated when a request arrives in the system is greater than the multiplicity of structural reservation of nodes at any level, then several copies of the request are sequentially serviced.

For option S_2: servicing paths are dynamically formed during the transfer of a request between cluster levels. The multiplicity of backup copies at different levels of the cluster may vary [12,13] and for a heterogeneous flow, if depends on the criticality of the requests to the allowable waiting time.

In this paper, we set the task of constructing a new model that allows, for a heterogeneous request flow, of different criticality, to wait delays in the nodes making up the path, to take into account the requirements for not exceeding the maximum allowable accumulated waiting time for sequential redundant request servicing by nodes included in the path.

3 Redundant Service for a Homogeneous Request Flow in a Group of Parallel Connected Nodes

Consider a single-level system of n redundant nodes connected in parallel to a cluster with an inhomogeneous flow of requests with allocation of z types of requests by the allowable waiting time in queues t_1, t_2, ..., t_z. The fractions of flows of heterogeneous requests are equal to g_1, g_2, ..., g_z respectively, and their intensities λg_1, λg_2, ..., λg_z, and

$$\sum_{i=1}^{z} g_i = 1.$$

The probability that the waiting time in the queue of an non redundant request of the i-th flow is less than the time t_i with unattended servicing and assuming that the average execution time of requests of all types v is calculated as

$$P_i = 1 - \frac{\Lambda_0 v}{n} e^{(\frac{\Lambda_0}{n} - \frac{1}{v})t},$$

where

$$\Lambda_0 = \sum_{i=1}^{z} g_i \Lambda.$$

The probability that the waiting time in the queue of requests of all types is less than the maximum permissible time for each of them is defined as:

$$P_c = \prod_{i=1}^{z} P_i.$$

Consider the case of redundant servicing of requests of z types, with a reservation ratio of k_1, k_2, ..., k_z. The probability of timely servicing of at least one of the copies of the request k_i of the i-th flow is calculated as [12,13]

$$P_i = 1 - [\frac{\Lambda_0 v}{n} e^{(\frac{\Lambda_0}{n} - \frac{1}{v}) t_i}]^{ki},$$

where

$$\Lambda_0 = \sum_{i=1}^{z} k_i g_i \Lambda. \qquad (1)$$

It should be noted that the average residence time of a selected (specific) copy of a request in a certain node for unprioritized servicing of requests of all types is found as:

$$u = \frac{v}{(1 - \Lambda_0 v/n)}.$$

As an example, consider the case when there are two flows g with the probability of incoming requests of the first flow g and the second $1 - g$, respectively, when reserving the corresponding requests with redundancy, and we have the total intensity of the redundant flow:

$$\Lambda_0 = \Lambda(g k_1 + (1 - g) k_2).$$

The dependence of the probability of timely fulfillment of requests of the first and total flows on their intensity Λ (without redundancy) at different redundant ratios are shown in Fig. 1, in which curves 1–4 correspond to the probability of timely servicing of the first flow of requests with a redundancy $k1$ equals to 1–4 (in case of non-redundant service of requests of the second flow less critical to the waiting time). Curves 5–8 in Fig. 1 correspond to the probabilities of timely execution of requests of the first flow. The calculation was carried out with $n = 5$, $g = 0.5$ and the maximum allowable waiting time for the requests of the first and second flows are $t_1 = 2v$ and $t_2 = 5v$, v - the average time to complete the request is $v = 4 10^{-4}$ s (this query execution time takes place, for example, when transmitting a packet with a length of 4096 bits at a transmission speed of 10 Mbit / s). It can be seen from the presented graphs that with an increase in the redundancy ratio of requests of the first flow (more critical to the waiting time), especially with a low intensity of the flow of requests, it is advisable to increase the redundancy ratio of requests of the first flow to certain limits. However, an increase in the redundancy ratio of requests of the first flow leads to an increase in the total load of the system and to a decrease in the probability of timely service of requests of the second flow, which reduces the probability of timely service P_c of the total flow of requests.

The expediency and requests redundancy ratio depends not only on the intensity of the flow of requests, but also largely on the allowable waiting time for requests.

Figure 2 shows the dependence of the probability of timely execution of requests of the total flow on the intensity Λ of the input flow (prior to reservation requests) with different redundancy ratios and criticality to the allowable waiting time t_1 of requests of the first flow. Curves 1–4 represent cases when

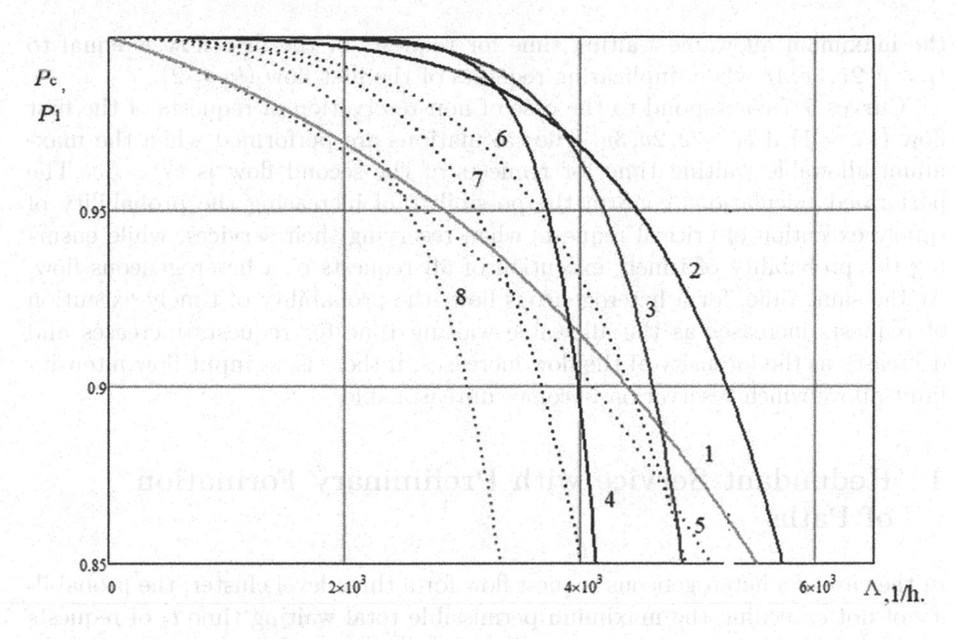

Fig. 1. Dependence of the probability of timely execution of requests of the first and total flow on the intensity of the total non redundant input flow Λ and the frequency of redundant requests.

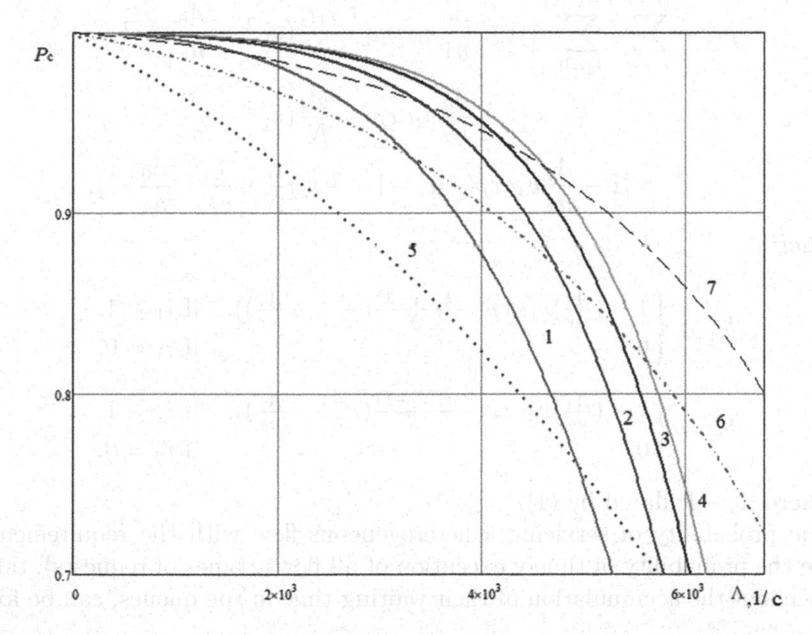

Fig. 2. Dependence of the probability of timely execution of requests of the total flow on the intensity of the input flow Λ at different criticality of the requests of the first flow to the allowable waiting time t_1 in case of duplication and not redundant for their maintenance.

the maximum allowable waiting time for requests of the first flow is equal to $t_1 = v, 2v, 3v, 4v$ when duplicating requests of the first flow ($k_1 = 2$).

Curves 5–7 correspond to the case of non-reservation of requests of the first flow ($k_1 = 1$) if $t_1 = v, 2v, 3v$. The calculations are performed when the maximum allowable waiting time for requests of the second flow is $t_2 = 5v$. The performed calculations confirm the possibility of increasing the probability of timely execution of critical requests when reserving their services, while ensuring the probability of timely execution of all requests of a heterogeneous flow. At the same time, for a heterogeneous flow, the probability of timely execution of requests increases as the allowable waiting time for requests increases and decreases as the intensity of the flow increases, if there is an input flow intensity limit above which reservation becomes unreasonable.

4 Redundant Service with Preliminary Formation of Paths

In the case of a heterogeneous request flow for a three-level cluster, the probability of not exceeding the maximum permissible total waiting time t_i of requests of the i-th type in queues of two levels t_0 for one (any) of the k_j assigned paths of the redundant execution of one copy of the request is calculated as:

$$\begin{aligned} p_{1i} = \sum_{i_1=0}^{N-1} \sum_{i_2=0}^{N-1-i_1} \{[1 - \frac{\Lambda_0}{n1} v_1 exp(-\frac{i_1 t_i}{N}(v_1^{-1} - \frac{\Lambda_0}{n_1}) - b_{i_1}] \\ \times [1 - \frac{\Lambda_0}{n_2} v_2 exp(-\frac{i_2 t_i}{N}(v_2^{-1} - \frac{\Lambda_0}{n_2}) - b_{i_2}] \\ \times [1 - \frac{\Lambda_0}{n_3} v_3 exp(-(t_i - (i_1 + i_2)\frac{t_i}{N})(\frac{1}{v_3} - \frac{\Lambda_0}{n_3}))]\}, \end{aligned} \quad (2)$$

where

$$b_{i_1} = \begin{cases} 1 - (\frac{\Lambda_0}{n_1}) v_1 exp(-\frac{t_i(i_1-1)}{N}(v_1^{-1} - \frac{\Lambda_0}{n_1})), & \text{if } i_1 \geq 1, \\ 0, & \text{if } i_1 = 0. \end{cases}$$

$$b_{i_2} = \begin{cases} 1 - (\frac{\Lambda_0}{n_2}) v_2 exp(-\frac{t_i(i_2-1)}{N}(v_2^{-1} - \frac{\Lambda_0}{n_2})), & \text{if } i_2 \geq 1, \\ 0, & \text{if } i_2 = 0. \end{cases}$$

where Λ_0 calculated by (1).

The probability of servicing a heterogeneous flow with the requirement to ensure the probability of timely execution of all flows (types of requests), taking into account the accumulation of their waiting time in the queues, can be found as:

$$P_1 = \prod_{i=1}^{z} [1 - (1 - p_{1i})^{k_i}],$$

5 Redundant Service of Heterogeneous Flow with the Formation of Copies of Requests

The probability of a request flow of different criticality for service delays being evaluated will be evaluated. z types of requests according to the criticality of the total waiting time at nodes sequentially receiving a service request will be allocated, while for the i-th type of requests, the maximum allowable accumulated waiting time by all nodes sequentially serving a request is set to t_i.

For redundant service according to option S_2, the formation of a given number of copies of nodes transmitted to the next level is carried out by a copy of the request executed first in time.

As a result, the intensity of requests arriving at the j-th level, taking into account the multiplicity of reservation of k_{ji} requests of the i-th type will be equal to

$$\Lambda_j = \sum_{i=1}^{z} g_i k_{ji} \Lambda.$$

For a three-level cluster, the probability of timely execution of at least one copy of the i-th type request, taking into account the total delay at all levels, is calculated as:

$$\begin{aligned} P_{2i} = \sum_{i_1=0}^{N-1}\sum_{i_2=0}^{N-1-i_1} \{ & [1 - \{(\Lambda_1/n_1)v_1 exp(-i_1\frac{t_i}{N}(v_1^{-1} - \frac{\Lambda_1}{n_1}))\}^{k_{1i}} - b_{i_1}] \\ & \times [1 - \{(\Lambda_2/n_2)v_2 exp(-i_2\frac{t_i}{N}(v_2^{-1} - \frac{\Lambda_2}{n_2}))\}^{k_{2i}} - b_{i_2}] \\ & \times [1 - [(\Lambda_3/n_3)v_3 exp(-(t_0 - (i_1 + i_2)\frac{t_i}{N})(\frac{1}{v_3} - \frac{\Lambda_3 k_3}{n_3})]^{k_3}]\}, \end{aligned}$$

where

$$b_{i_1} = \begin{cases} 1 - \{(\Lambda_1/n_1)v_1 exp(-\frac{t_i}{N}(i_1 - 1)(v_1^{-1} - (\Lambda_1/n_1)))\}^{k_{1i}}, & \text{if } i_1 \geq 1, \\ 0, & \text{if } i_1 = 0. \end{cases}$$

$$b_{i_2} = \begin{cases} 1 - \{(\Lambda_2/n_2)v_2 exp(-\frac{t_i}{N}(i_2 - 1)(v_2^{-1} - \frac{\Lambda_2}{n_2}))\}^{k_{2i}}, & \text{if } i_2 \geq 1, \\ 0, & \text{if } i_2 = 0. \end{cases}$$

For option S_2 of redundant service, the efficiency of servicing an inhomogeneous flow with the requirement to ensure the probability of timely execution of all types of requests taking into account the accumulation of their waiting time in queues is defined as

$$P_2 = \prod_{i=1}^{z} P_{2i}.$$

For option S_2 of the redundant service, the total average residence time of the request of the i-th flow (type) at all m levels of the system calculated as

$$T_i = \sum_{j=1}^{m} \{ \int_0^{\infty} [\frac{\Lambda_j v_j}{n_j} e^{(\frac{\Lambda_0}{n} - \frac{1}{v})t}]^{k_{ji}} dt \}.$$

For a heterogeneous flow, the efficiency criterion can be taken as the mathematical expectation of the total delays in the request queues of all z types

$$T = \sum_{i=1}^{z} g_i \sum_{j=1}^{m} \{ \int_0^{\infty} [\frac{\Lambda_j v_j}{n_j} e^{(\frac{\Lambda_0}{n} - \frac{1}{v})t}]^{k_{ji}} dt \}.$$

The implementation of multi-path redundant service with the formation of copies of requests at each stage requires the interaction of group nodes, which can be quite simply organized in a server cluster, but the implementation of such interaction in a group of communication nodes is difficult, since it requires interaction through a network, which can significantly slow down the maintenance process.

6 Comparison of Reserved Service Options

Consider a two-level cluster. In the calculations, we assume that, the number of reserved servers at each level is the same and equal to $n = 8$ pcs, the average query execution time by the servers of the first and second level is $v_1 = v_2 = 0.4$ s, and the total allowable wait time is $t_0 = 0.2$ s.

In Fig.3 shows the dependencies of the probabilities of timely service in a two-level cluster on the intensity of the input request flow Λ. In Fig. 3, curve 1 corresponds to non-redundant service $k = 1$, curves 2–3 correspond to service option S_1 with a request redundancy ratio of $k = 2, 3$, and curves 4, 5 correspond to option S_2 with a redundancy ratio of $k = 2, 3$.

In Fig. 4 shows the dependencies of the probability of timely service of requests on their redundancy ratio (number of generated copies) k. In Fig. 4, curves 1–3 represent service option S_1, and curves 4–6 option S_2 for request flow intensities corresponding to $\Lambda = 3.3; 3; 3.5$ 1/s.

These graphs allow to conclude that there is an optimal multiplicity of redundant service, and the smaller the system load and the permissible total waiting time, the greater the redundancy ratio at which the maximum probability of multistage service timeliness is achieved.

Calculations show that the S_2 redundant service option improves the probability of timely service requests, however, its implementation is more complicated and requires additional research on the organization of interaction between nodes included in the redundant group (cluster).

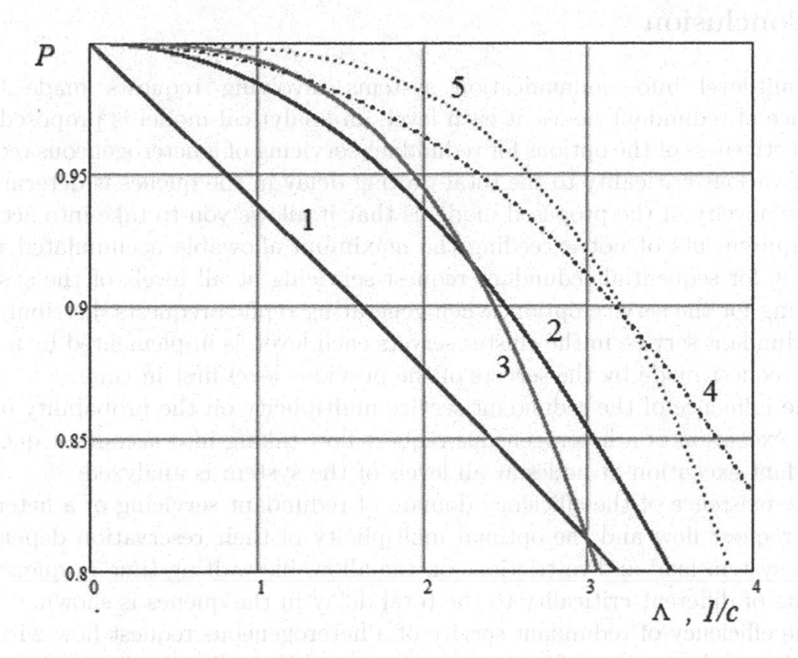

Fig. 3. Dependencies of the probabilities of timely service in a two-level cluster on the intensity of the input request flow Λ.

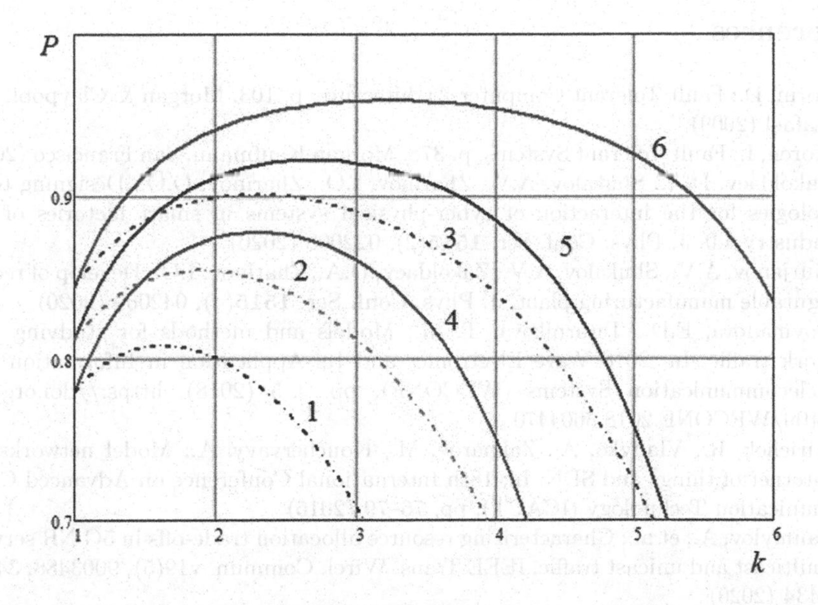

Fig. 4. Dependence of the probability of timely service requests on the frequency of their reservation.

7 Conclusion

For multilevel info-communication systems involving requests made by a sequence of redundant nodes at each level, an analytical model is proposed and the effectiveness of the options for redundant servicing of a heterogeneous request flow of various criticality to the total waiting delay in the queues is determined.

The novelty of the proposed model is that it allows you to take into account the requirements of not exceeding the maximum allowable accumulated waiting time for sequential redundant request servicing at all levels of the system. Including for the service option, when generating replica requests distributed to the redundant service in the cluster servers each level, is implemented by a copy of the request made by the servers of the previous level first in time.

The influence of the redundant service multiplicity on the probability of the timely execution of a heterogeneous request flow taking into account sequential redundant execution at nodes at all levels of the system is analyzed.

The existence of the efficiency domain of redundant servicing of a heterogeneous request flow and the optimal multiplicity of their reservation depending on the system load and restrictions on the allowable waiting time in queues for requests of different criticality to the total delay in the queues is shown.

The efficiency of redundant service of a heterogeneous request flow with the formation of the number of copies of requests and their distribution in the queue of servers at each system level, taking into account the criticality to the delay of requests of different flows, is shown.

References

1. Sorin, D.: Fault Tolerant Computer Architecture, p. 103. Morgan & Claypool, San Rafael (2009)
2. Koren, I.: Fault Tolerant Systems, p. 378. Morgan Kaufmann, San Francisco (2009)
3. Zakoldaev, D.A., Shukalov, A.V., Zharinov, I.O., Zharinov, O.O.: Designing technologies for the interaction of cyber-physical systems in smart factories of the industry 4.0. J. Phys. Conf. Ser. **1515**(2), 022008 (2020)
4. Gurjanov, A.V., Shukalov, A.V., Zakoldaev, D.A., Zharinov, I.O.: Develop of reconfigurable manufacturing plant. J. Phys. Conf. Ser. **1515**(4), 042060 (2020)
5. Poymanova, E.D., Tatarnikova, T. M.: Models and methods for studying network traffic. In: 2018 Wave Electronics and Its Application in Information and Telecommunication Systems (WECONF), pp. 1–5 (2018). https://doi.org/10.1109/WECONF.2018.8604470
6. Kirichek, R., Vladyko, A., Zakharov, M., Koucheryavy, A.: Model networks for internet of things and SDN. In: 18th International Conference on Advanced Communication Technology (ICACT), pp. 76–79 (2016)
7. Samuylov, A., et al.: Characterizing resource allocation trade-offs in 5G NR serving multicast and unicast traffic. IEEE Trans. Wirel. Commun. v19(5), 9003488, 3421–3434 (2020)
8. Bogatyrev, V.A., Bogatyrev, A.V.: Functional reliability of a real-time redundant computational process in cluster architecture systems. Autom. Control Comput. Sci. **49**(1), 46–56 (2015). https://doi.org/10.3103/S0146411615010022

9. Bogatyrev, A.V., Bogatyrev, V.A., Bogatyrev, S.V., Transmission, multipath redundant, with packet segmentation. In: 2019 Wave Electronics and its Application in Information and Telecommunication Systems (WECONF), Saint-Petersburg, Russia, pp. 1–4 (2019). https://doi.org/10.1109/WECONF.2019.8840643
10. Bogatyrev, V.A., Bogatyrev, A.V., Bogatyrev, S.V.:Redundant, maintenance of a non-uniform query stream by a sequence of nodes that are grouped together in groups. In: 2020 Wave Electronics and its Application in Information and Telecommunication Systems, WECONF 2020 9131463 (2020)
11. Jin, H., Li, D., Wu, S., Shi, X., Pan, X.: 2009 Live virtual machine migration with adaptive memory compression. In: Proceedings of the IEEE International Conference on Cluster Computing, CLUSTER 2009, New Orleans, USA (2009). Art. 5289170. https://doi.org/10.1109/CLUSTR.2009.5289170
12. Sahni, S., Varma, V.: A hybrid approach to live migration of virtual machines. In: 2012 Proceedings of the IEEE International Conference on Cloud Computing for Emerging Markets, CCEM 2012, Bengaluru, India, pp. 12–16 (2012). https://doi.org/10.1109/CCEM.2012.6354587
13. Machida, F., Kawato, M., Maeno, Y.: Redundant virtual machine placement for fault-tolerant consolidated server clusters. In: 2010 IEEE Network Operations and Management Symposium - NOMS 2010, Osaka, pp. 32–39 (2010). https://doi.org/10.1109/NOMS.2010.5488431
14. Kim, S., Choi, Y.R.: Constraint-aware VM placement in heterogeneouscomputing clusters. Clust. Comput. **23**, 71–85 (2020). SI
15. Yang, C., Liu, J., Hsu, C., et al.: On improvement of cloud virtual machine availability with virtualization fault tolerance mechanism. J. Supercomput. **69**, 1103–1122 (2014). https://doi.org/10.1007/s11227-013-1045-1
16. Jo, C., Cho, Y., Egger, B.: A machine learning approach to live migration modeling. In: Proceedings of the 2017 Symposium on Cloud Computing, SoCC, vol. 17, pp. 351–364 (2017)
17. Keller, G., Lutyya, H.: Dynamic management of applications with constraints in virtualized data centres. In: Proceedings of IFIP/IEEE International Symposium on Integrated Network Management (IM) (2015)
18. Wang, Y.B., et al.: Markov process-based availability analysis of rendering cluster systems. In: Advanced Materials Research, vol. 225–226, pp. 1024–1027. Trans Tech Publications Ltd., April 2011. https://doi.org/10.4028/www.scientific.net/amr.225-226.1024
19. Bogatyrev, V.A., Bogatyrev, S.V., Derkach, A.N.: Timeliness of the Reserved Maintenance by Duplicated Computers of Heterogeneous Delay-Critical Stream, vol. 2522, pp. 26–36. CEUR Workshop Proceedings (2019)
20. Kabatiansky, G., Krouk, E., Semenov, S.: Error Correcting Coding and Security for Data Networks. Analysis of the Superchannel Concrete EPT, p. 288. Wiley, New York (2005)
21. Krouk, E., Semenov, S.: Application of coding at the network transport level to decrease the message delay. In: Proceedings of 3rd International Symposium on Communication Systems Networks and Digital Signal Processing, Staffordshire University, UK, pp. 109–112 (2002)
22. Merindol, P.: Improving load balancing with multipath routing. In: Merindol, P., Pansiot, J., Cateloin, S. (eds.) Proceedings of the 17th International Conference on Computer Communications and Networks, ICCCN 2008, pp. 54–61. IEEE (2008)

23. Chanak, P., Samanta, T.: Indrajit Banerjee Fault-tolerant multipath routing scheme for energy efficient wireless sensor networks. Int. J. Wirel. Mob. Netw. (IJWMN) **5**(2), 33–45 (2013)
24. Rajeev, V., Muthukrishnan, C.R.: Reliable backup routing in fault tolerant real-time networks. In: Proceedings Ninth IEEE International Conference on Networks, ICON (2001)
25. Bogatyrev, V.A., Bogatyrev, S.V., Bogatyrev, A.V.: Timely Redundant Service of Requests by a Sequence of Cluster, vol. 2590, pp. 1–12. CEUR Workshop Proceedings (2020)
26. Zakoldaev, D.A., Korobeynikov, A.G., Shukalov, A.V., Zharinov I.O.: Cyber and physical systems technology classification for production activity of the industry 4.0 smart factory. In: IOP Conference Series: Materials Science and Engineering, vol. 582, No. 1, pp. 012007 (2019)
27. Lee, M.H., Dudin, A.N., Klimenok, V.I.: The SM/V/N queueing system with broadcasting service. Math. Probl. Eng. **2006**, 18 (2006). V. Article ID 98171
28. Dudin, A.N., Sun, B.: A multiserver MAP/PH/N system with controlled broadcasting by unreliable servers. Autom. Control Comput. Sci. **5**, 32–44 (2009)
29. Bogatyrev, V.A.: Protocols for dynamic distribution of requests through a bus with variable logic ring for reception authority transfer. Autom. Control Comput, Sci. **33**(1), 57–63 (1999)
30. Bogatyrev, V.A., Bogatyrev, S.V., Golubev, I.Y.: Optimization and the process of task distribution between computer system clusters. Autom. Control Comput, Sci. **46**(3), 103–111 (2012). https://doi.org/10.3103/S0146411612030029
31. Siddiqi, M.A., Yu, H., Joung, J.: 5G ultra-reliable low-latency communication implementation challenges and operational issues with IoT devices. Electronics **8**, 981 (2019)
32. Ji, H., Park, S., Yeo, J., Kim, Y., Lee, J., Shim, B.: Ultra-reliable and low-latency communications in 5G downlink: physical layer aspectsUltra-reliable and low-latency communications in 5G downlink: physical layer aspects. IEEE Wirel. Commun. **25**, 124–130 (2018)
33. Sachs, J., Wikström, G., Dudda, T., Baldemair, R., Kittichokechai, K.: 5G radio network design for ultra-reliable low-latency communication. IEEE Netw. **32**, 24–31 (2018)
34. Bennis, M., Debbah, M., Poor, H.V.: Ultrareliable and low-latency wireless communication: tail. risk and scale. In: Proceedings of the IEEE, vol. 106, pp. 1834–1853 (2018)
35. Kleinrock, L.: Queueing Systems, Volume I – Theory, p. 417. Wiley, New York (1975)

Agriculture Management Based on LoRa Edge Computing System

Fatkhullokhodzha Sharofidinov[1], Mohammed Saleh Ali Muthanna[2,4], Van Dai Pham[1], Abdukodir Khakimov[3](✉), Ammar Muthanna[1], and Konstantin Samouylov[3]

[1] The Bonch-Bruevich Saint-Petersburg State University of Telecommunications, Prospekt Bolshevikov 22, St. Petersburg, Russia
fatkhullo0998@gmail.com, daipham93@gmail.com, ammarexpress@gmail.com

[2] Department of Automation and Control Processes, Saint-Petersburg Electrotechnical University "LETI", Saint-Petersburg, Russia
muthanna@mail.ru

[3] Peoples' Friendship University of Russia (RUDN University), 6 Miklukho-Maklaya Street, Moscow 117198, Russia
khakimov-aa@rudn.ru, ksam@sci.pfu.edu.ru

[4] School of Computer Science and Technology, Chongqing University of Posts and Telecommunications, Chongqing, China

Abstract. Internet of Things (IoT) technologies represent the future challenges of computing and communications. They can also be useful to improve traditional farming practices worldwide. Since the areas where agricultural land is located in remote places, there is a need for new technologies. These technologies must be suitable and reliable for communication over long distances and, at the same time, consume little energy. In particular, one of these relatively new technologies is the LoRa communication protocol, which uses long waves to work over long distances. This is extremely useful in agriculture, where the communicating areas are broad fields of crops and greenhouses. This study developed a greenhouse monitoring system based on LoRa technology, designed to work over long distances. The edge computing paradigms with a machine learning mechanism are proposed to analyze and control the state of the greenhouse, and in particular, to reduce the mount of data transmitted to the server.

Keywords: LoRa · Edge computing · Precision agriculture · Machine learning

1 Introduction

Agriculture has always been one of the most critical sectors in the life of every state. The foundation of human survival depends on agriculture in many

The publication has been prepared with the support of the "RUDN University Program 5-100" (recipients A. Khakimov). For the research, infrastructure of the 5G Lab RUDN (Russia) was used.

V. M. Vishnevskiy et al. (Eds.): DCCN 2020, LNCS 12563, pp. 113–125, 2020.
https://doi.org/10.1007/978-3-030-66471-8_10

respects, as the health of the population depends precisely on the quantity and quality of manufactured products for everyday consumption in the agricultural industry [1,2]. Today, there is a rather serious threat to the food security of the Earth's population due to the exponential growth of its population, urbanization, which suggests that the share of the population that is engaged in agriculture is significantly falling [3]. This problem is added to the process of global warming, which is overgrowing. This means that to prevent the global crisis associated with a lack of nutrition and the loss of a significant part of the crop, the introduction of modern technologies is necessary. The development of modern technologies can help us face these challenges. Agriculture has been free from the use of digital technologies [4,17]. Almost all jobs involved in people. All yield data were collected manually, and analytic was not relevant for agriculture since the amount of data on which conclusions and forecasts can be drawn was small. The work was planned based on weather conditions, data on soil quality, and environmental conditions, which were collected manually. Also, the standard processing schedule, which includes demoralization, fertilizer, continuous irrigation, led to cost overruns and undetected problems since local characteristic sand natural variability was not taken into account. With the growing interest and demand for intelligent agriculture, many attempts have been made to introduce Biotechnology on the farm [5]. IoT technology [6,7] allows farmers to control field plots using connected sensors via the Internet from anywhere in the world. Farmers and agronomists have already begun to use technology to improve their productivity. For example, sensors located in greenhouses allow you to receive detailed data in real time in the form of variables, such as the temperature of the soil and the environment, irrigation water and soil conductivity, the acidity level (PH) of the soil and irrigation water, the properties of irrigation water, data on nutrient composition of the soil, etc. These data can be transmitted and analyzed using communication technologies and artificial intelligence (AI) paradigms. Farmers will be able to remotely monitor crop sand equipment using their smartphones, as well as analyze some statistics. Therefore, our proposed management system based on modern communication technologies can be a solution for the introduction of smart (precision) agriculture.

The organization of the paper will be as follows: Sect. 2 presents the related work. Section 3 provides the precision agriculture system architecture based on edge computing. Section 4 proposed a monitoring system for precision agriculture. Section 5 experiment and results And the conclusion will be given in Sect. 6.

2 Related Works

Many works have been done regarding the use of IoT in agriculture. Authors in [8] analyzed the functionality of the most innovative and trending technologies in the agricultural sector, as well as the possibilities of the Internet of things in agriculture, this work also arguments are given, why the introduction of such technologies is not only useful but also cost effective. Since LoRa technology is one of the most attractive in terms of its application in agriculture, many studies have been cited in this area. For example, in [9,15], a study was carried out on the implementation of LoRa technology as a transmission protocol between nodes of a sensor network in a visual monitoring system for agricultural fields, the idea is to transfer only the changing parts of the image frame and transfer data using LoRa, when bandwidth is from 50 bits to 50 kbit/s. Authors in the [10] are developing the LoRaWAN gateway mobile device, which can be used to increase the productivity and accuracy of greenhouses. The presented development includes the use of the Halter LoRa mini gateway, managed by the Raspberry Pi model 3 B+. There are cases where machine learning has been used to optimize costs as well as the production process. For example, in [11] is a machine based, accurate, and intelligent irrigation system with LoRa P2P networks for automatically and seamlessly learning the irrigation experience from experienced farmers for organic greenhouse crops. The proposed system will first calculate the amount of water for each irrigation based on a trained irrigation model in combination with environmental data such as air temperature/humidity, moderate soil moisture, light intensity, etc., and then automatically water crops through long distance and lowpower wireless LoRa P2P network. Authors in the [12] present an analysis of the influence of the parameters of the variant physical layer on the performance of LoRa networks in a tree farm. The main goal of our study is to develop a system for monitoring and controlling the process of growing crops in year round greenhouses, where vegetables are grown for everyday consumption. Our system is based on the principles of data transmission using LoRa technology using edge computing paradigms to reduce the amount of traffic that will be further transmitted to the cloud. Cloud computing will be used to calculate, analyze, predict the state of culture and soil. Forecasting will be based on data received from soil moisture sensors in the greenhouses, room temperature and humidity sensors, crop status sensors, as well as a sensor that measures soil composition.

3 Precision Agriculture System Architecture Based on Edge Computing

This section provides details on the architecture system for four layers 1) sensor layer 2) edge layer 3) network/cloud layer and 4) application layer and results of the architecture system using edge computing are discussed in terms of delay time.

3.1 Layer-Based Architecture

The architecture of our system consists of 4 layers, as shown in Fig. 1. Sensor layer. This level of our system consists directly of sensors: temperature, soil, air humidity, CO2, and illumination. IoT devices are connected to the gateway using energy efficient wireless technology LoRa. At this level, data is being read from our greenhouse objects.

3.2 Edge Layer

This level solves the following tasks: data exchange with sensors, preliminary analysis of received data, diagnostics of sensor status, data compression. The edge calculation paradigm is used. The advantages of using edge computing for initial processing of data collected from sensors are in reducing the volume of data uploaded to cloud servers, and in rapid response in critical situations, which involves any action to resolve cases without waiting for a response from the server [14]. Also at this level is the averaging of data collected from sensors. Besides, the artificial intelligence paradigm is used at this level, which is to apply a machine learning mechanism based on data collected from sensors to predict the state of the greenhouse. This will significantly increase farm efficiency by helping to take action to improve the quality of the crop being grown.

3.3 Network/Cloud Layer

This level involves the transfer of data collected at the border through public communication networks to cloud servers. Cloud computing is used for in depth data analysis, as well as for their storage. At this level, all the advantages of cloud computing are used.

3.4 Application Layer

The collected data must be analyzed and converted into a readable form so that users can monitor and control the condition of the greenhouse in a way convenient for them. At this level, a policy of work is defined, and the process of control over the production of crops in the greenhouse takes place directly.

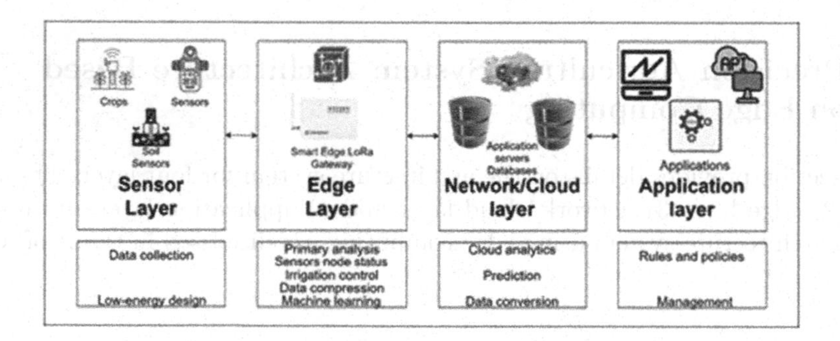

Fig. 1. Monitor and management system

Edge Computation and its Types

Today, edge computing determines the future of IoT. It ensures the stability of IoT devices and eliminates latency issues by providing data processing closer to the source.

IoT Edge Computing is a network of small data centers where missioncritical data is stored and processed locally. These data centers are connected in a grid, and they transmit the received data to a centralized data warehouse. This usually occurs within 10 square meters or less.

IoT Edge Computing is used to analyze and process data closer to the data source. The smart devices used in IoT Edge Computing are capable of processing critical data fragments and providing a fast realtime response. These devices prevent delays caused by sending data over the Internet to the cloud and delay response from the cloud.

These devices are designed to act as small data centers that provide nearzero latency. With this advanced capability, data processing is decentralized, and network traffic is significantly reduced. This data can then be collected by the cloud for further evaluation and processing.

Types of IoT Edge Computing
Generally, we can define three types of edge computations of IoT [13]:

- Local devices to accommodate a specific and welldefined target. They can be easily deployed and maintained.
- Local data centers to provide significant data processing and storage capabilities. They are usually preengineered and configured. They are assembled locally and provide good capital cost savings [16].
- Regional data centers, with a clear advantage in that they are closer to the data source. Although they have more processing and storage power than local data centers, they are expensive and require more maintenance. Such edge devices are designed either in prefabricated or custom versions.

The Benefits of Edge Computing
As IoT Edge Computing is implemented and widely distributed, a large number of industries are gaining potential benefits. In particular, edge computing offers seven major advantages in intelligent manufacturing:

1) Fast response time: Computing and storage power is local and distributed. Avoiding sending data back and forth in the cloud is key to reducing latency and getting more rapid responses. This helps prevent vital machine operations from being disrupted, or dangerous incidents occur
2) Sequential operations with sporadic communication: For many remote assets, controlling unpredictable areas of Internet connectivity, such as agricultural pumps, oil wells, windmills, or solar farms, can be challenging. The local storage and processing capabilities of edge devices ensure no data loss and prevent outages in the event of limited Internet connectivity.
3) Security: IoT Edge Computing has made data transfer between the cloud and devices redundant. We can filter sensitive data locally and move only the data model to the cloud.
4) Costeffective: A major practical issue with the introduction of the IoT was the cost of storage, processing power, and network bandwidth. IoT Edge

computing enables local data computing, allowing business organizations to distinguish between services that need to be performed locally and services that need to be sent to the cloud. This helps reduce the cost of developing a complete solution for the IoT.

5) Compatibility between modern and obsolete devices: the edge devices function as a link between current and obsolete machines. This allows outdated machines to interact with advanced tools for IoT solutions.
6) Increased fault tolerance: A decentralized peripheral computing architecture allows other network devices to become more resilient. This is a great advantage as a single server failure in the cloud can result in the defeat of thousands of IoT devices.
7) Reduced exposure: edge computing minimizes data transfer over the network. This, in turn, helps to reduce the impact of data during transit. In some cases, sensitive information such as personal identification information (PII) and the payment card industry (PCI) may not be sent at all. Such cases can help avoid some legal complications related to security and confidentiality. Also, data encryption and access control can make it highly secure from familiar threats.

3.5 Simulation Results

As a result of the simulation, it was found that the use of edge computing significantly reduces the load on the channels and can reduce delays in processing IoT traffic. Figure 3 shows a comparative analysis of the delay time in service when data passes through the system when the model is a traditional network without using edge computing and with a system using edge computing. As can

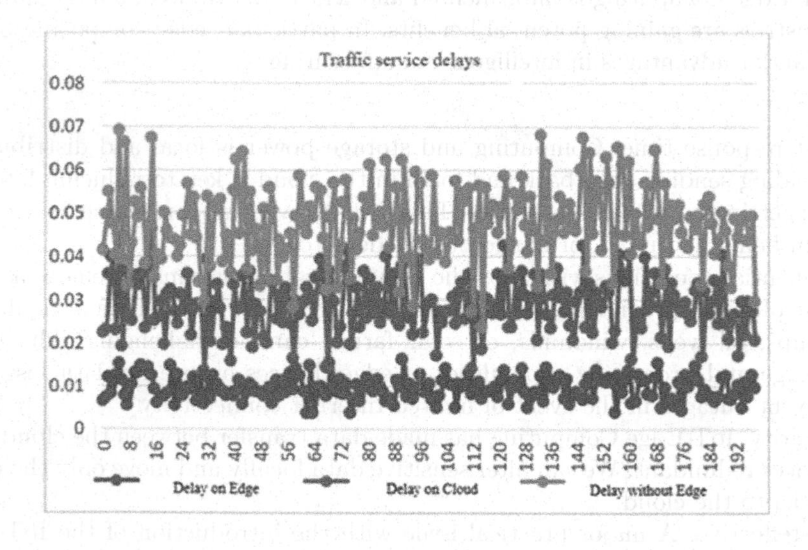

Fig. 2. Delay in service traffic

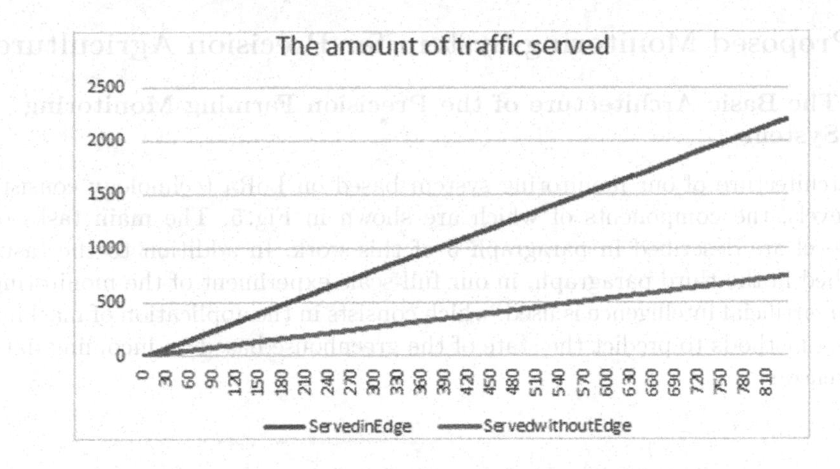

Fig. 3. The amount of traffic served

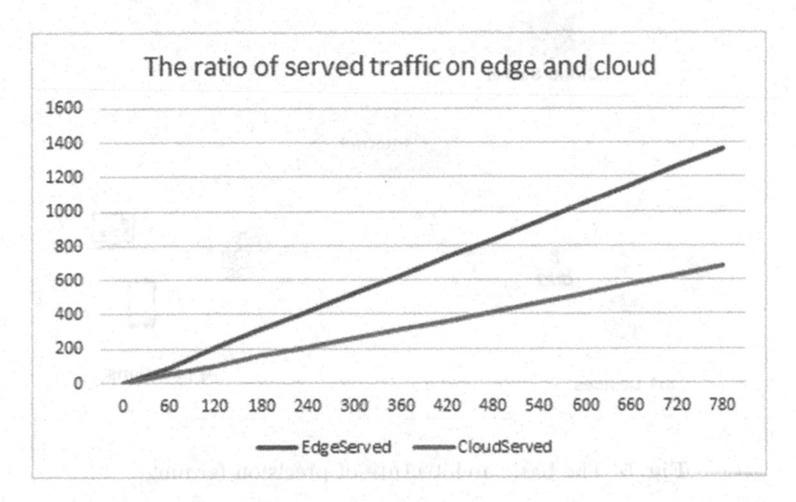

Fig. 4. Ratio of edge computing traffic service to the remote cloud

be seen from the Fig. 2, the delay in the system using edge computing is much lower compared to systems without the use of edge computing, when data is sent directly to the server without preliminary processing.

Without preprocessing, the amount of data uploaded to the server increases significantly, which increases the queue of data and service time. Since our precision farming system has very little traffic from sensors, most data can be maintained locally. This increases the amount of traffic being serviced and also reduces the uploading of data to the cloud server. As can be seen in Fig. 3 and Fig. 4 at Edge, up to half of the traffic can be serviced, and approximately three times more data can be serviced compared to systems where Edge is missing.

4 Proposed Monitoring System for Precision Agriculture

4.1 The Basic Architecture of the Precision Farming Monitoring System

The architecture of our monitoring system based on LoRa technology consists of 4 levels, the components of which are shown in Fig. 5. The main tasks of each level are described in paragraph 3 of this work. In addition to the tasks described in the third paragraph, in our full-scale experiment of the monitoring system, artificial intelligence is used, which consists in the application of machine learning methods to predict the state of the greenhouse based on incoming data from sensors.

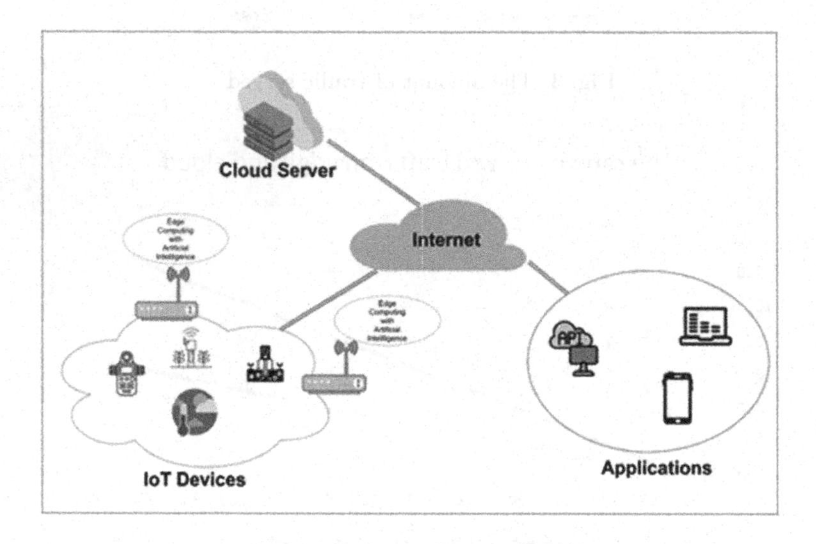

Fig. 5. The basic architecture of precision farming.

Using the Random Forest Algorithm to Classify a Greenhouse
Our prediction model based on machine learning is deployed at the edge level. This helps to identify actions to improve the plant's condition correctly. There are five types of sensors from which data is sent to a remote server. In our scheme, the data will be preprocessed at the gateway, which is the edge level. The Random Forest Classifier Model is used to classify the condition of a greenhouse. A random forest, as its name suggests, consists of a large number of individual decision trees that act as an ensemble. Each individual tree in a random forest gives a class forecast, and the class with the most votes becomes the forecast of our model. The fundamental concept underlying the random forest algorithm is simple, but strong is the wisdom of the crowd. Data science says that the reason why the random forest model works so well is that a large number of relatively uncorrected models (trees) acting as a committee will surpass any of the individual components of the models.

4.2 Sensor Node (Describe Sensor Node = MCU + Wireless Interface + Various Sensors)

Currently, there are many development boards to help create a prototype quickly. With our purpose, a development board equipped with LoRa interface is used to transmit sensor data and communicate with the gateway. As shown in Fig. 6, the development board TTGO LoRa32, which is based on modules such as ESP32 and LoRa SX1276, was used with some sensor modules to measure temperature, humidty, CO2 index, soil humidity, and light intensity.

Fig. 6. Sensor node

Moreover, the ESP32 module also supports the other communication interfaces such as WiFi and Bluetooth Low Energy [9,18]. Therefore, this development board can be used as a single-channel gateway for LoRa communication. A protocol is defined to transmit messages between the sensor nodes and the gateway. Before transmitting the sensor data, the node sends a request message to the gateway. After receiving the sensor data, the gateway also replies to an acknowledgment message to the sensor node. As shown in Fig. 7, the payload of a sensor message is specified with five bytes for measured data from sensors. The message type is defined in the header. By defining the message transmission protocol between the gateway and the sensor nodes, we can save the channel budget. In the payload, the data byte from 1 to 5 represent the measured data of temperature, humidity, CO2 index, soil humidity, and light intensity, respectively.

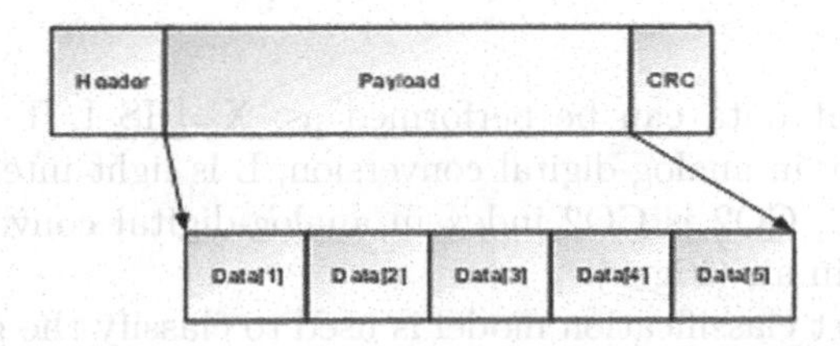

Fig. 7. Format of a sensor message

According to the identification of each device, the edge layer and cloud layer can access the device's data for further processing. In the next subsection, we will consider some techniques used to preprocess the sensor data and predict the status in the greenhouse.

4.3 Sensor Data Processing (Data Process on Edge, Prediction Method)

Describe methods of data processing on Edge. (Aggregation, averaging data, normalizing data) The approach of using the edge layer for preprocessing sensor data, we can save the bandwidth to the remote server, and reply action to actuator quickly without waiting server responses. However, some noises are added to measured data from sensor devices. Therefore, to increase accuracy sensor measurement, a simple filtering technique can be used at the edge layer. The averaging technique is easy to implement by adding several measures together, then dividing the total by several measurements. The last data coming from sensor devices are saved temporarily at the edge layer, with k measurements, the new data X_{inew} new are averaged: $X_{inew} = \frac{x_{i-k}+x_{i-k+1}+x_i}{k+1}$.

Moreover, the prediction model can be deployed at the edge layer to predict the condition in the greenhouses based on the received sensor data. It helps to identify actions correctly to improve the plant's state. There are five types of sensors sent to the remote server. In our scheme, the data will be preprocessed at the edge layer. Our prediction model is deployed to classify events into four classes: 1 – soil without water, 2 – environment correct, 3 – too much hot, 4 – very cold based on received data. The data frame for each device has a format in Table 1.

Table 1. caption

Soil Humidity (Analog)	Light (Analog)	Temperature (Celsius)	CO2 (Analog)	Air Humidity (%)	Class
550	280.83	21	154	86.7	4
773	695.83	22	150	87,6	1
568	764	34,4	169	61,1	3
130	0	22,1	34	72,9	2

Thus, the input data can be performed as: X=[HS L T CO2 HR] where HS is soil humidity in analog-digital conversion, L is light intensity in lux, T is temperature in °C, CO2 is CO2 index in analog-digital conversion and HR is humidity relative in air (Fig. 8).

A random forest classification model is used to classify the greenhouse state. A random forest, as its name implies, consists of a large number of individual decision trees that act as an ensemble. Each individual tree in a random forest gives a class forecast, and the class with the most votes becomes the forecast of our model. The fundamental concept underlying the random forest algorithm is the wisdom of the crowd. Data science says that the reason the random forest

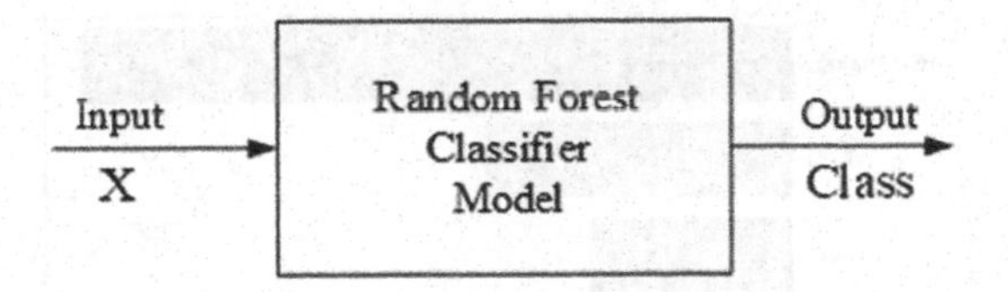

Fig. 8. SHEME

model works so well is this: a large number of relatively uncorrelated models (trees) acting as a committee will outperform any of the individual component models. The target output is class, For each event, the target output is followed as: $class_i = f(x_i) = f([HS_I\ L_I\ T_I\ CO2_I\ HR_I])$.

5 Experiment and Results

The learning model was evaluated with a different number of samples. By changing the number of samples used for training, we received the learning curve, as shown in Fig. 9. According to the graph, the training score increases when we have more samples to learn. With 300 samples, the model can predict the class of conditions in the greenhouse with high accuracy

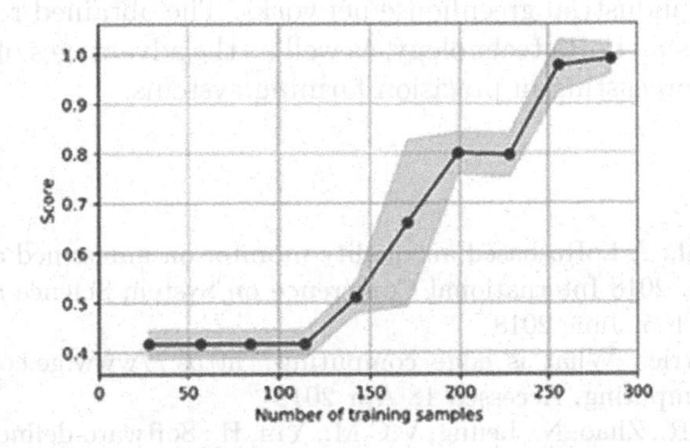

Fig. 9. Learning score based on the number of samples.

By monitoring the sensor data after a certain time, we can analyze the factor which is the most impact on plants. As shown in Fig. 10, the most important factor effected prediction result is soil humidity.

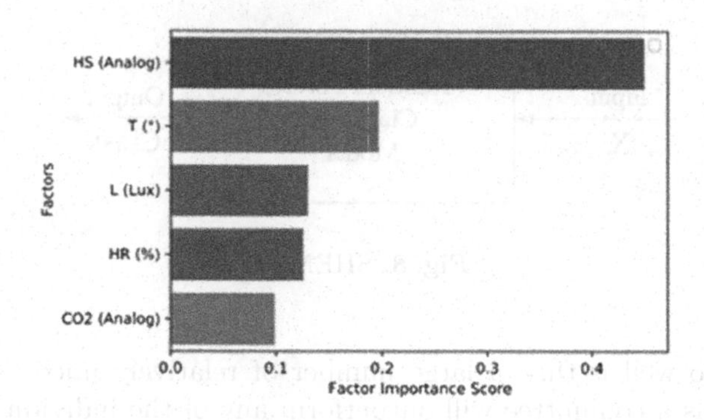

Fig. 10. Visualizing important factor

6 Conclusion

Using the paradigm of edge computing and artificial intelligence methods, we were able to create an effective model of the monitoring system for an agricultural facility. A monitoring system based on machine learning, which is taught at the edge level, helps farmers to monitor the condition of plants by predicting any case of deviation from normal factors, effectively irrigating plants, and optimally applying expensive chemicals. Our proposed monitoring system based on edge computing and machine learning can be used for deployment in both small home greenhouses and industrial greenhouse networks. The obtained results prove the effectiveness of using LoRa technology, as well as the advantages of using artificial intelligence for forecasting in precision farming systems.

References

1. Chen, L., et al.: A LoRa-based air quality monitor on unmanned aerial vehicle for smart city. In: 2018 International Conference on System Science and Engineering (ICSSE), pp. 1–5, June 2018
2. General Electric.: What is edge computing? https://www.ge.com/digital/blog/what-edge-computing. Accessed 18 Apr 2019
3. He, Y., Yu, F.R., Zhao, N., Leung, V.C.M., Yin, H.: Software-defined networks with mobile edge computing and caching for smart cities: a big data deep reinforcement learning approach. IEEE Commun. Mag. **55**(12), 31–37 (2017). https://doi.org/10.1109/MCOM.2017.1700246
4. Khakimov, A., Muthanna, A., Muthanna, M.S.A.: Study of fog computing structure. In: Proceedings of the 2018 IEEE Conference of Russian Young Researchers in Electrical and Electronic Engineering (EIConRus), Moscow, Russia, 29 January–1 February 2018, pp. 51–54 (2018)
5. Perera, C., Qin, Y., Estrella, J.C., Reiff-Marganiec, S., Vasilakos, A.V.: Fog computing for sustainable smart cities: a survey. ACM Comput. Surv. **50**(3) (2017). https://doi.org/10.1145/3057266
6. Taleb, T., Dutta, S., Ksentini, A., Iqbal, M., Flinck, H.: Mobile edge computing potential in making cities smarter. IEEE Commun. Mag. **55**(3), 38–43 (2017). https://doi.org/10.1109/MCOM.2017.1600249CM

7. Suresh, V.M., Sidhu, R., Karkare, P., Patil, A., Lei, Z., Basu, A.: Powering the IoT through embedded machine learning and LoRa. In: IEEE World Forum on Internet of Things, WF-IoT 2018 - Proceedings, vol. 2018, pp. 349–354. Institute of Electrical and Electronics Engineers Inc., January 2018. https://doi.org/10.1109/WF-IoT.2018.8355177
8. Carlin, A., Hammoudeh, M., Aldabbas, O.: Defence for distributed denial of service attacks in cloud computing. Procedia Comput. Sci. **73**, 490–497 (2015). https://doi.org/10.1109/NORCHIP.2014.7004716
9. Nguyen Gia, T., et al.: Energy efficient wearable sensor node for IoT-based fall detection systems. Microprocess. Microsyst. **56**, 34–46 (2018). https://doi.org/10.1016/j.micpro.2017.10.014
10. Rahmani, A.M., et al.: Exploiting smart e-Health gateways at the edge of healthcare internet-of-things: a fog computing approach. Future Gener. Comput. Syst. **78**, 641–658 (2018). https://doi.org/10.1016/j.future.2017.02.014
11. Muthanna, M.S.A., Wang, P., Wei, M., Ateya, A.A., Muthanna, A.: Toward an ultra-low latency and energy efficient LoRaWAN. In: Galinina, O., Andreev, S., Balandin, S., Koucheryavy, Y. (eds.) NEW2AN/ruSMART -2019. LNCS, vol. 11660, pp. 233–242. Springer, Cham (2019). https://doi.org/10.1007/978-3-030-30859-9_20
12. Dongare, A., et al.: Charm: exploiting geographical diversity through coherent combining in low-power wide-area networks. In: Proceedings of the International Conference on Information Processing in Sensor Networks, pp. 60–71 (2018)
13. Hammoudeh, M., Newman, R., Dennett, C., Mount, S., Aldabbas, O.: Map as a service: a framework for visualising and maximising information return from multi-modal wireless sensor networks. Sensors **15**(9), 22970–23003 (2015)
14. Carlin, A., Hammoudeh, M., Aldabbas, O.: Intrusion detection and countermeasure of virtual cloud systems-state of the art and current challenges. Int. J. Adv. Comput. Sci. Appl. **6**(6), 1–15 (2015)
15. Gia, T.N., Thanigaivelan, N.K., Rahmani, A.M., Westerlund, T., Liljeberg, P., Tenhunen, H.: Customizing 6LoWPAN networks towards internet-of-things based ubiquitous healthcare systems. In: NORCHIP 2014–32nd NORCHIP Conference: The Nordic Microelectronics Event. Institute of Electrical and Electronics Engineers Inc. (2015). https://doi.org/10.1109/NORCHIP.2014.7004716
16. Daraseliya, A.V., Sopin, E.S., Samuylov, A.K., Shorgin, S.Y.: Comparative analysis of the mechanisms for energy efficiency improving in cloud computing systems. In: Galinina, O., Andreev, S., Balandin, S., Koucheryavy, Y. (eds.) NEW2AN/ruSMART -2018. LNCS, vol. 11118, pp. 268–276. Springer, Cham (2018). https://doi.org/10.1007/978-3-030-01168-0_25
17. Sopin, E.S., Daraseliya, A.V., Correia, L.M.: Performance analysis of the offloading scheme in a fog computing system. In: International Congress on Ultra Modern Telecommunications and Control Systems and Workshops, vol. 2018, art. no. 8631245 (2019)
18. Daraseliya, A.V., Sopin, E.S., Rykov, V.V.: On optimization of energy consumption in cloud computing system. In: CEUR Workshop Proceedings, vol. 2332, pp. 23–31 (2018)

Dynamic Algorithm for Building Future Networks Based on Intelligent Core Network

Abdukodir Khakimov[1(✉)], Ammar Muthanna[2], Ibrahim A. Elgendy[3], and Konstantin Samouylov[1]

[1] Peoples' Friendship University of Russia (RUDN University), Miklukho-Maklaya Street, 6, Moscow 117198, Russia
khakimov-aa@rudn.ru, ksam@sci.pfu.edu.ru

[2] The Bonch-Bruevich Saint-Petersburg State University of Telecommunications, Pr. Bolshevikov, 22, St. Petersburg 193232, Russia
ammarexpress@gmail.com

[3] School of Computer Science and Technology, Harbin Institute of Technology, Harbin, China
ibrahim.elgendy@hit.edu.cn

Abstract. 6G/IMT-2030 is designed to provide users with innovative speeds of terabit per second, which are proposed to be achieved using a number of advanced technologies, such as Mobile Edge Computing (MEC), Internet of Things (IoT), millimeter wave (mmWave), new radio and software defined networking. It is necessary to solve several important aspects in order to satisfy Quality of Service (QoS), first of all, to ensure network coverage density even in sparsely populated areas. In this paper we proposed software defined network based mobile edge computing dynamic algorithm for improving network performance. In addition, this algorithm can help the service provided to adapt with a required load on the radio links. Furthermore, local content caching and Local Internet Breakout (LIB) can be utilized to reduce the transport network requirements. Finally, the proposed algorithm is analyzed using some use cases and we developed testbed to emulate operator infrastructure.

Keywords: Dynamic algorithm · Edge computing · 5G · Core network

1 Introduction

Mobile and radio communications will continue to play an important role in strengthening economic, technological, financial, social growth, healthcare, education, and research. To maintain growth and to meet ever-changing requirements, industry and research industries are always striving to develop and

The publication has been prepared with the support of the "RUDN University Program 5-100" (recipient K. Samouylov). For the research, infrastructure of the 5G Lab RUDN (Russia) was used.

V. M. Vishnevskiy et al. (Eds.): DCCN 2020, LNCS 12563, pp. 126–136, 2020.
https://doi.org/10.1007/978-3-030-66471-8_11

implement new mechanisms and methods in mobile and radio communications, to facilitate people's lives, for example, the development of the Internet of Things [1], 5G network [2,3], decentralized transport networks [4], tactile internet [5], etc..

Currently, the main direction in the communication networks and systems development is the creation of 5G communication networks. Initially, the 5G communication network creation was based on the Internet of Things concept, for the implementation of which required the creation of superdense communication networks with the number of devices around 1 million per 1 Km^2 [6]. This non-trivial task has recently been made even more complicated by the advent of the Tactile Internet concept, which required transmitting tactile sensations to limit the round-trip latency over the network to 1ms. This leads to a review of the fundamentals of building communication networks and due to fundamental restrictions on the speed of light transmission to the decentralization of the network and its resources. These requirements are not only for the transmission of tactile sensations but also for unmanned vehicles with network support, augmented reality applications, etc..

Therefore, as part of the implementation of the 5G communication network, the concept of ultra-low latency communication networks appears, which based on available research in the field of communication networks in 2030, will determine the basis for building communication networks in the long term. In the 2030 networks, and, in our opinion, much earlier, human-machine interactions H2M and, in particular, human-robot avatars will play an exceptional role. The requirements for communication networks will be largely determined by the implementation of Industry 4.0, which already today takes on a very definite form as part of the work on the industrial Internet of Things. Here, the interaction of the network and production requires a new look at the problems of compatibility of technical means, services, classes, and parameters of quality of service and quality of perception [7].

To achieve very high availability and reliability of the system, as well as the ultra-low latency required for a tactile Internet system, system components create significant limitations in terms of context, content and mobility [8,9]. Thus, there must be installed the appropriate design for test compound required between the primary and secondary domain with the required specifications. The main problem facing checking the network domain is the 1ms end-to-end latency. All the proposed work for the network area proves that it must use new technologies such as software-defined networks (SDN), network function virtualization (NFV), mobile edge computing (MEC) and software-defined radio (SDR). 5G communication system will facilitate the implementation of the network domain for the Tactile Internet system, as it will determine the basic structure. The main problem is adapting the 5G structure to achieve a 1ms end-to-end delay. A Tactile Internet system can usually be considered as a three-layer system based on a 5G system structure proposed by NGMN [14]. This paper discusses the benefits of edge computing in mobile networks and there use cases that can be deployed in the near future. This helps consider architectures and deployment options for MEC and evaluates the long-term role of MEC in evolving 4G and 5G networks, In this work, we propose a dynamic algorithm to ensure network performance.

After studied other works on this topic [10–12], we were able to build our stand of the 5G network:

- EPC virtualization using network function virtualization (NFV) technology (vEPC),
- Separation of control planes and data in vEPC using SDN/NFV technology

1.1 NFV-based EPC Architecture

As part of NFV-based EPC concept, EPC nodes are transferred from dedicated hardware platforms to virtual machines (VMs) or containers and implemented as virtual network functions (VNFs) operating as software in a cloud environment (for example, OpenStack). However, there are no functional changes: the interfaces and protocols (for example, GTP) used for communication between nodes are still standardized by 3GPP. Network functions are created and managed by the cloud controller or with some of the latest MANO tools such as OpenStack Tacker, OpenBaton or OpenMANO. Resources for VNFs are provided by NFV infrastructure (NFVI).

NFV provides benefits such as cost reduction and flexibility. This task of redesigning the EPC architecture seems to be the most feasible in life, because it does not require any major changes in the current deployment of EPC. Each node can be virtualized using multiple virtual machines, which makes it possible to deploy multiple NFV-based EPCs (vEPCs) simultaneously. Support for generic interfaces and protocols allows mobile operators to employ interaction of vEPCs with existing EPCs.

However, this approach has a number of limitations. Keeping all VNFs within 3GPP standards creates problems when adding new VNFs, since the latter must be configured and created in a consistent manner. In addition, the scaling and initialization process are not efficient due to the close relations between the control and data planes in the P-GW and S-GW due to the fact that the planes have different requirements for processing resources, as was mentioned earlier. In addition, the transfer of information from EPC nodes to VNFs can result in loss of information during scaling procedures, such as removing VNFs from the system. This rises problems of reliability and fault tolerance. With this approach, EPC gateways (SGW, PGW) are separated into the control plane - SGW-C and PGW-C and the data plane - SGW-U and PGW-U. Then, using the NFV technology, the control plane functions (SGW-C, PGW-C) become VNFs. In this case, the data plane functions (SGW-U, PGW-U) remain either on dedicated hardware platforms or also become virtual.

After separation of the data and control planes, SGW and PGW can be combined: SGW-C and PGW-C - into a single control object called GW-C; SGW-U and PGW-U - into a single data plane object called GW-U. In all cases, an SDN controller, that can also be virtualized, is used to communicate between the planes. Its functions include interpreting signal messages received from the control plane, and setting up forwarding rules in the data plane through an open API. The data plane (SGW-U, PGW-U) can be implemented as OpenFlow switches capable of encapsulating and decapsulating GTP.

Compared to the first method, this approach not only has advantages such as flexibility and backward compatibility, but also overcomes the shortcomings of the vEPC architecture with the implementation of SDN. The SDN separates control and data planes, which can now be scaled independently in a cost-effective way. The separation of planes also makes it possible to move them across the network, e.g., closer to the user to reduces end-to-end delay, and also contributes to the development of mobile edge computing. In addition, SDN provides better UE mobility management through flexible distribution of traffic flows in the infrastructure.

1.2 SDN/NFV-based EPC Architecture

With this approach, EPC gateways (SGW, PGW) are separated into the control plane - SGW-C and PGW-C and the data plane - SGW-U and PGW-U. Then, using the NFV technology, the control plane functions (SGW-C, PGW-C) become VNFs. In this case, the data plane functions (SGW-U, PGW-U) remain either on dedicated hardware platforms or also become virtual.

After separation of the data and control planes, SGW and PGW can be combined: SGW-C and PGW-C - into a single control object called GW-C; SGW-U and PGW-U - into a single data plane object called GW-U. In all cases, an SDN controller, that can also be virtualized, is used to communicate between the planes. Its functions include interpreting signal messages received from the control plane, and setting up forwarding rules in the data plane through an open API. The data plane (SGW-U, PGW-U) can be implemented as OpenFlow switches capable of encapsulating and decapsulating GTP.

Compared to the first method, this approach not only has advantages such as flexibility and backward compatibility, but also overcomes the shortcomings of the vEPC architecture with the implementation of SDN. The SDN separates control and data planes, which can now be scaled independently in a cost-effective way. The separation of planes also makes it possible to move them across the network, e.g., closer to the user to reduces end-to-end delay, and also contributes to the development of mobile edge computing. In addition, SDN provides better UE mobility management through flexible distribution of traffic flows in the infrastructure.

2 Related Works

The new generation of communication technologies 5G offers many new applications and services in terms of people and devices (things). Consequently, the growing demand for bandwidth in 5G networks can lead to a shortage of wireless spectrum, despite its under-utilization in urban areas. The smart city paradigm implies the widespread use of inter-machine interaction devices, which requires the development of a more flexible spectrum management system. The advent of 5G networks leads to new issues, in particular, to the expansion of spectrum capabilities. It was estimated that the spectrum limit will be reached by 2025 [13–15]. Therefore, new strategies for the efficient use of the wireless spectrum [16] are required.

A brief overview of the basic concepts of machine learning and their involvement in the most popular 5G applications, including cognitive radio systems, Massive MIMO systems, femto/small cells, heterogeneous networks, intelligent networks, energy storage, device-to-device communications, were covered in [17,18]. This work determines the relevance of this issue, the problems of the formulation and methodology of effective machine learning in the context of developing networks in order to touch hitherto unattainable resources and ways of applications.

In [19] authors offer relevant input and output characteristics of heterogeneous network traffic and systems of controlled deep neural networks. In this work, the authors describe how the proposed system works and how it differs from traditional neural networks. In addition, preliminary results demonstrate the promising performance of the proposed deep machine learning system compared to the Open Shortest Path First reference strategy in terms of signal transmission, throughput and delay.

A detailed literature review was performed in [20], including a list of literature on large data sets, machine learning methods, SDN and NFV technologies and methods for their development using 5G systems. The work mainly considers the deployment of these technologies in the areas of organizing and managing data traffic.

In [21], the authors proposed a practical platform and traffic prediction processes based on large data arrays, machine learning, and network Key Performance Indicators (KPIs), which are flexible enough to provide accurate prediction of data flow parameters of various types of cells (GSM, 3G, 4G) for long-term and short-term prediction. The performance of the proposed model was evaluated, applicable to a real set of data taken from KPI - more than 6,000 hundred real networks during 2016 and 2017. While the authors in [22] proposed the Internet of Vehicle (IoV) system, where city maps are segmented into a small number of unique maps. They apply the Ant Colony algorithm to each map to find the best route. In addition, a function for calculating the intensity of data streams for the model of intense data streams, based on the principles of fuzzy logic, is proposed. The proposed IoV-based route selection method is compared with existing shortest-path selection algorithms such as Dijikstra algorithm, Kraskel algorithm and Prima algorithm. The experimental results showed a good performance of the proposed method of route selection based on IoV.

3 Proposed Algorithm

The principle of operation in our proposed algorithm is to move and arrange on the network in such a way as to reduce the load on one node dynamically. That is mean, migrate only when the load on the central node increases. The diagram in Fig. 2 shows the step-by-step EPC migration process. The controller initiates the migration thereby notifying the application itself (epc). The essence of the algorithm is to first clone the application and only then transfer sessions, this way we minimize application downtime.

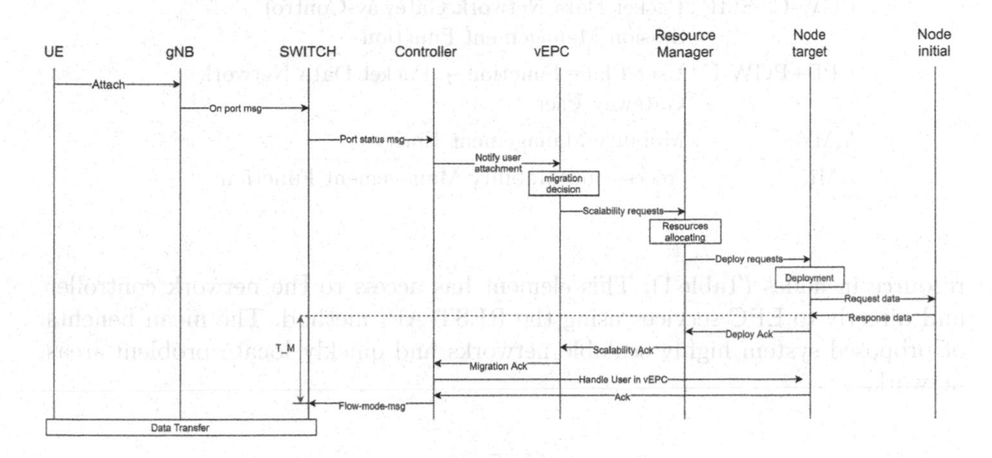

Fig. 1. Diagram of proposed algorithm

4 Testbed Based on 5G Model Network

To implement the proposed methodology for the dynamic distribution of services, we have assembled a 5G NSA testbed shown in Fig. 2 based on the 5G laboratory at RUDN University. Each network element was packed in a separate container with its dependencies, and launched in one virtual plane. As RAN we implanted working on FPGA. With the following features:

Frequency range: 10 MHz to 6 GHz.
Instant bandwidth: up to 160 MHz;
Power: 50–100 mW (17–20 dBm)
Noise range: 0–31.5 dBm
DAC resolution: 16 bits;
Number of channels: 2

As an FPGA controller, OAI (OpenAirInterface) is used.

Our goal is to distribute the elements of the NSA network to reduce resource costs while maintaining the required quality of QoS. To do this, we offer a mechanism for the seamless transfer of the service to other nodes. The sequence diagram is presented in Fig. 1. It describes how the service is transferred using the SDN with the least loss with active users. For this procedure, an additional element is introduced that will be responsible for allocating and deploying the

Table 1. Network elements

HSS+UDM	Home Subscriber Serve + Unified Data Management
PCF+PCRF	Policy Control Function + Policy and Charging Rules Function
PGW-C+SMF	Packet Data Network Gateway-Control + Session Management Function
UPF+PGW-U	User Plane Function + Packet Data Network Gateway-User
MME	Mobility Management Entity
AMF	Access and Mobility Management Function

resource in nodes (Table 1). This element has access to the network controller and directly to EPC services using the REST API method. The mean benefits of proposed system highly scalable networks and quickly locate problem areas network.

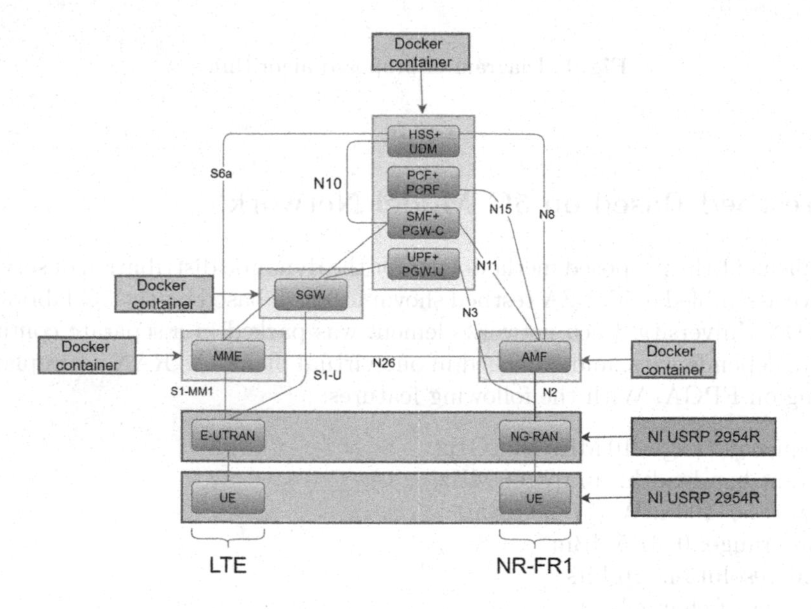

Fig. 2. Testbed based on model network

In the experiment, we used a different number of active user sessions to study the possible effects on the migration process. Since active sessions have a direct connection to HSS + UDM. It was these elements that we use to the migration process in order to get the worst case scenario.

5 Results

As mentioned earlier, for the migration process, we subjected the EPC as part of HSS, UDM, PCF, PCRF, PGW-C, SMF, UPF, PGW-U in one container. And we watched the duration of the migration depending on the sessions and packet loss. The migration duration to consider the time T_M shown in Fig. 2.

Further, the following results were obtained. Figure 3 shows that with 640 active sessions, migration occurs in 4 min. And with small amounts of a session in less than one minute. And it is clear that then there is a sharp increase in migration time. This is due to the fact that the system for preserving the network is trying to transfer sessions block by block because of this, such indicators of the migration duration.

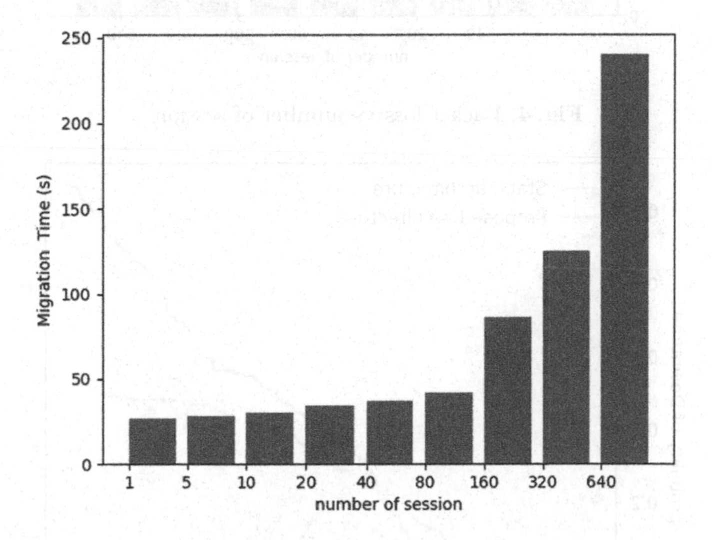

Fig. 3. Migration time vs number of session

The Fig. 4 shows packet loss during migration with the same number of sessions. As a result, we get that with more user migration, the migration time increases but the number of patch packets per user decreases.

The next stage of the experiment was to study the probability of network failure with large amounts of session. To do this, we reproduced a large traffic generator up to 14,000 subscribers with a simultaneous connection to the EPC under standard conditions. Since the testbed is in a laboratory, so these values may differ from the actual network performance. Our task was to study the relative comparison of our proposed distributed network from traditional centralized ones. As a result, when we distributed services on edge computing, we get a more reliable network. Where at 14,000 subscribers the probability of service failure is reduced by ~0.25, Fig. 5.

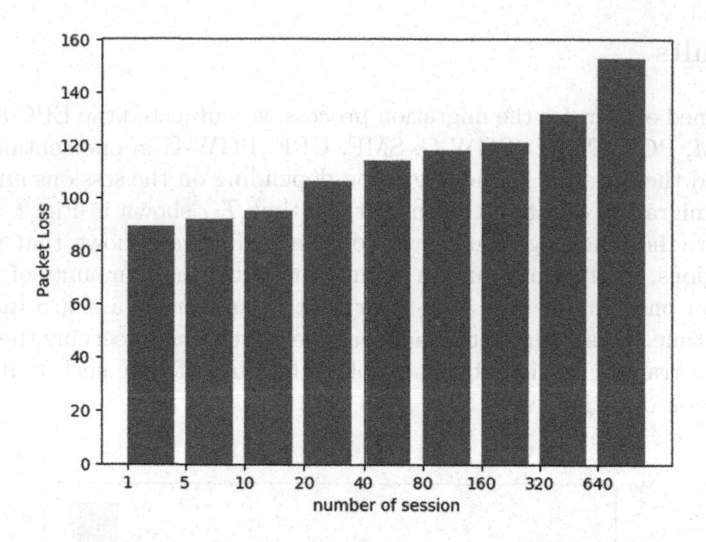

Fig. 4. Packet loss vs number of session

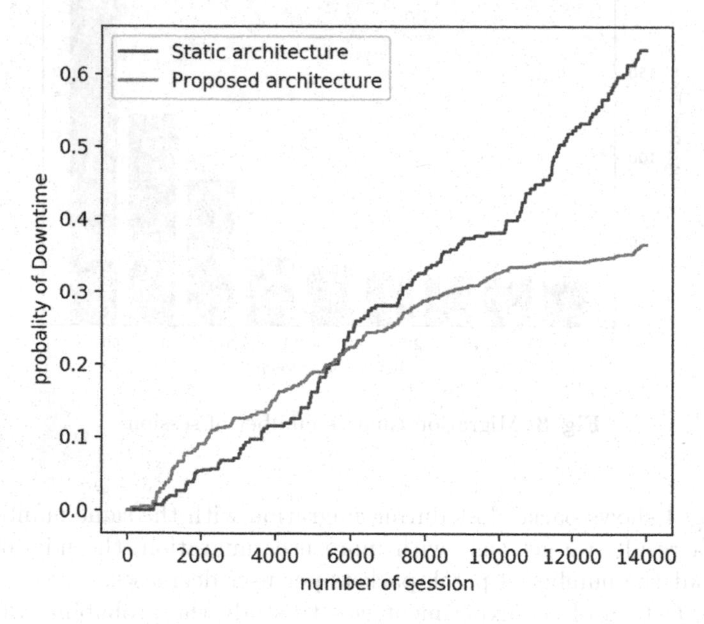

Fig. 5. The probability of downtime for static and proposed system structure

6 Conclusion

In 2020, we can already talk about the gradual introduction of new network technologies and computing infrastructure, as well as new services. At the same time, the 5G concept is replaced by the concept of the 6G network. In this paper, we proposed an algorithm for dynamically network rebuilding by distributing services (in our case, epc). We had modeled a real mobile network NSA 5G, based on 5G laboratory at RUDN University. This paper reproduced the distribution of HSS + UDM in MEC nodes. Finally, we received that the overall probability of service failure decreased by $\sim$0.3 at 14000 sessions.

In ongoing and future work, we plan to integrate the genetic algorithm to determine the optimal resolution of services which will reduce packet loss during migration. In addition, we also plan to do experiments not only with EPC but also with other network elements even outside of operator services.

References

1. Muthanna, A., et al.: Secure and reliable IoT networks using fog computing with software-defined networking and blockchain. J. Sens. Actuator Netw. **8**(1), 15 (2019)
2. Ateya, A.A., Muthanna, A., Vybornova, A., Darya, P., Koucheryavy, A.: Energy - aware offloading algorithm for multi-level cloud based 5G system. In: Galinina, O., Andreev, S., Balandin, S., Koucheryavy, Y. (eds.) NEW2AN/ruSMART -2018. LNCS, vol. 11118, pp. 355–370. Springer, Cham (2018). https://doi.org/10.1007/978-3-030-01168-0_33
3. Khayyat, M., Alshahrani, A., Alharbi, S., Elgendy, I., Paramonov, A., Koucheryavy, A.: Multilevel service-provisioning-based autonomous vehicle applications. Sustainability **12**(6), 2497 (2020)
4. Jaiswal, R.K., Jaidhar, C.: Location prediction algorithm for a nonlinear vehicular movement in VANET using extended Kalman filter. Wireless Netw. **23**(7), 2021–2036 (2017). https://doi.org/10.1007/s11276-016-1265-4
5. Ateya, A.A., et al.: Model mediation to overcome light limitations-toward a secure tactile internet system. J. Sens. Actuator Netw. **8**(1), 6 (2019)
6. Chang, K.-C., Chu, K.-C., Wang, H.-C., Lin, Y.-C., Pan, J.-S.: Energy saving technology of 5g base station based on internet of things collaborative control. IEEE Access **8**, 32935–32946 (2020)
7. Nørgaard, B., Guerra, A.: Engineering 2030: conceptualization of industry 4.0 and its implications for engineering education. In: 7th International Research Symposium on PBL, p. 34 (2018)
8. Aijaz, A., Simsek, M., Dohler, M., Fettweis, G.: Shaping 5G for the tactile internet. In: Xiang, W., Zheng, K., Shen, X.S. (eds.) 5G Mobile Communications, pp. 677–691. Springer, Cham (2017). https://doi.org/10.1007/978-3-319-34208-5_25
9. Daraseliya, A.V., Sopin, E.S., Samuylov, A.K., Shorgin, S.Y.: Comparative analysis of the mechanisms for energy efficiency improving in cloud computing systems. In: Galinina, O., Andreev, S., Balandin, S., Koucheryavy, Y. (eds.) NEW2AN/ruSMART -2018. LNCS, vol. 11118, pp. 268–276. Springer, Cham (2018). https://doi.org/10.1007/978-3-030-01168-0_25
10. Parvez, I., Rahmati, A., Guvenc, I., Sarwat, A.I., Dai, H.: A survey on low latency towards 5G: RAN, core network and caching solutions. IEEE Commun. Surv. Tutor. **20**(4), 3098–3130 (2018)
11. Alvarez, F., et al.: An edge-to-cloud virtualized multimedia service platform for 5G networks. IEEE Trans. Broadcast. **65**(2), 369–380 (2019)
12. Carlin, A., Hammoudeh, M., Aldabbas, O.: Defence for distributed denial of service attacks in cloud computing. Procedia Comput. Sci. **73** (2015). https://doi.org/10.1016/j.procs.2015.12.037

13. Ericsson mobility report: on the pulse of the networked society. http://www.abc.es/gestordocumental/uploads/internacional/EMR-June-2016-D5201.pdf
14. Cisco, C.V.N.I.: Global mobile data traffic forecast update, 2016–2021, white paper, pp. 0018–9545 (2017)
15. Sopin, E.S., Daraseliya, A.V., Correia, L.M.: Performance analysis of the offloading scheme in a fog computing system, pp. 1–5 (2018)
16. Palola, M., et al.: Live field trial of licensed shared access (LSA) concept using LTE network in 2.3 GHz band. In: 2014 IEEE International Symposium on Dynamic Spectrum Access Networks (DYSPAN), pp. 38–47. IEEE (2014)
17. Jiang, C., Zhang, H., Ren, Y., Han, Z., Chen, K.-C., Hanzo, L.: Machine learning paradigms for next-generation wireless networks. IEEE Wirel. Commun. **24**(2), 98–105 (2016)
18. Daraseliya, A., Sopin, E., Rykov, V.: On optimization of energy consumption in cloud computing system, October 2018
19. Kato, N., et al.: The deep learning vision for heterogeneous network traffic control: proposal, challenges, and future perspective. IEEE Wirel. Commun. **24**(3), 146–153 (2016)
20. Le, L.-V., Lin, B.-S., Do, S.: Applying big data, machine learning, and SDN/NFV for 5G early-stage traffic classification and network QoS control. Trans. Netw. Commun. **6**(2), 36 (2018)
21. Le, L.-V., Sinh, D., Tung, L.-P., Lin, B.-S.P.: A practical model for traffic forecasting based on big data, machine-learning, and network KPIs. In: 2018 15th IEEE Annual Consumer Communications & Networking Conference (CCNC), pp. 1–4. IEEE (2018)
22. Kumar, P.M., Manogaran, G., Sundarasekar, R., Chilamkurti, N., Varatharajan, R., et al.: Ant colony optimization algorithm with internet of vehicles for intelligent traffic control system. Comput. Netw. **144**, 154–162 (2018)

Methods and Models for Using Heterogeneous Gateways in the Mesh LPWANs

Viacheslav Kulik[1(✉)], Van Dai Pham[1], and Ruslan Kirichek[1,2]

[1] Bonch-Bruevich Saint-Petersburg State University of Telecommunications, 193232 Saint-Petersburg, Russia
vslav.kulik@gmail.com, fam.vd@spbgut.ru, kirichek@sut.ru
[2] V.A. Trapeznikov Institute of Control Sciences of Russian Academy of Sciences, 117997 Moscow, Russia

Abstract. Nowadays, a group of low-power wide-area networks (LPWANs) is one of the options providing the communication infrastructure for the Internet of Things applications. Most of LPWANs use a star topology model between end-nodes and a base station or a gateway. However, the mesh topology has shown the mobility and the dynamic in the network deployment. This article considers a question for constructing mesh LPWAN networks. In this article, the authors investigated existing standards of the developing gateways in the LPWAN networks, creating a structure of the mesh LPWAN with a conversion structure – heterogeneous gateway. We used this structure to create a simulation model of this network. The simulation model was mapped with a model based on the more traditional star topology with a simple gateway and an edge server. The results of the simulation can be used to design new mesh LPWAN networks.

Keywords: Heterogeneous gateways · Internet of Things · Mesh networks · Time analysis · Simulation · LPWAN

1 Introduction

Modern society is undergoing rapid changes due to the occurrence and implementation of modern technologies in everyday life. These technologies have both positive and negative impacts on urban infrastructure and society. It is proposed to use special tools to monitor and manage urban infrastructure based on the Internet of things (IoT) technologies to minimize negative aspects. Together with artificial intelligence, unmanned car/air transport management, environmental control, and emergency response systems are part of the concept of urban space organization – smart city (SC) [1,2,15]. In the process of developing SC, it is expected to connect things to the global network and also collect the information augmented to citizens. Wireless sensor and actuator networks consist of many sensors for data collection and actuators to control remote devices. There

V. M. Vishnevskiy et al. (Eds.): DCCN 2020, LNCS 12563, pp. 137–148, 2020.
https://doi.org/10.1007/978-3-030-66471-8_12

is a requirement of providing communication infrastructure in the smart city to organize these such networks.

When smart city technologies were implemented in the urban infrastructure, many problems related to ensuring connectivity of the system's endpoints arise. It is expected that many things can connect to the global network Internet and communicate with each other. One of the main problems is providing a high data transmission distance in urban conditions. Modern mobile networks are not well suited for transmitting small amounts of data from low-performance computing devices with sensors/actuators connected to them due to the high level of cellular modules' energy consumption and a large amount of service traffic transmitted over these networks. To connect this type of device to the SC network infrastructure – low-power wide-area networks (LPWAN) are used [11,14]. The LPWAN networks provide the ability of long-range communication and require a low power consumption. The range of data transmission in LPWAN networks varies from 1 to 10 km in urban conditions, depending on the transceiver's power and the type of power supply of the end-node (EN). According to the LPWAN switch characteristics, up to 50 thousand nodes can be working in each subnet, simultaneously transmitting data to the switch at speed from 0.3 to 30 Kbit/s [4]. These technical characteristics allow us to implement a network with a high density of EN placement.

In most cases, LPWAN systems are based on the star network topology. However, when it is not possible to provide a guaranteed high-quality connection EN with the network switch or the ENs are located outside of the gateway's coverage, a mesh or multi-hop communication is proposed to ensure the connection between ENs and the gateway [8,13]. This topology allows organizing a network connection between the EN and the switch and between the EN itself. This solution theoretically allows increasing the number of devices in the network per hub and increasing data transmission reliability in the network with multiple routes for data delivery to the destination. Thus, the network can be expanded without the requirement of adding new gateways. Since the long-range communication in the LPWAN networks, one gateway might serve and cover many devices.

In most cases, ENs transmit data to devices that are located outside the local LPWAN network. Devices called LPWAN gateways are used to solve these problems. These devices are necessary for ensuring interaction between EN and devices located in the external network. In most existing systems, the gateway and switch functions are combined into a single device, and the incoming data are processed on remote servers. This network structure is acceptable if the system does not have strict system response time requirements, which are achieved by reducing delivery and data processing time. This problem is most often solved using two approaches: organizing a local edge or fog server in a local network [3], or by extending the functionality of the LPWAN network gateway. A gateway with extended functionality is called a heterogeneous gateway (HG) [9,10]. This gateway allows us to dynamically add new software (SW) for data processing, depending on the requirements. Currently, the application of HG within LPWAN networks is a new unconventional task. In this paper, the authors study models and methods for ensuring interaction between LPWAN networks and other communication networks. Based on the research, a simulation model of the LPWAN mesh network with heterogeneous gateways is proposed and compared with a network with a more traditional star topology and without using HG.

2 Structure of the LPWAN Networks

Typically, the LPWAN networks consist of the following elements [11]:

- end-node – the low-performance device that is used to interact with sensors and/or actuators connected to them;
- switch – the device that acts as routers in the LPWAN networks;
- gateway – the device that receives and extracts useful data from packets of the LPWAN network format and then encapsulates them and sends it to the target network (usually the IP).

This network supports a star topology in its classic form, where multiple nodes connect to a single Central device that serves as a gateway and a switch. A group of such devices can be combined using external network technologies, such as IP, into a single network that allows them to interact with each other.

However, there are already existing such solutions that support the mesh network topology. In these networks, users can interact with the switch/gateway and each other. The function of dynamic mesh routing is experimental in WAN networks due to the high level of requirements for the service traffic bandwidth of the communication channel. LPWAN systems with pre-configured routing tables for each EN are a more common solution. Such tables can be generated manually by developers or network administrators, or by a single switch in the local network before the system running.

LPWAN systems are more often part of more complex network infrastructure, such as smart city systems [5]. Within these systems, LPWAN devices interact with remote and local SC platforms, computer appliance (CA) for receiving, storing, and analyzing data from LPWAN systems. Based on the data analysis results, the CA either independently decides on the SC system's further functioning or transmit information to the SC system operator if the situation is sufficiently critical. Special devices are used in the operator's network (ON) to transmit data through the operator's network, which provides the SC systems' infrastructure. The structure of the LPWAN network functioning within the SC system is shown in Fig. 1.

LPWAN structure can be simplified and reduced the cost of implementing many devices by using special devices that combine the functions of the switch, gateway, and SC edge platform that called the heterogeneous gateways defined in ITU-T recommendations Q. 4060 "The structure of the testing of heterogeneous Internet of Things gateways in a laboratory environment" [6] and Q. 3055 "Signaling protocol for heterogeneous Internet of Things gateways" [7]. The HG can be used in LPWAN networks only if there is a network interface that supports network data transmission technologies and implements all the functions

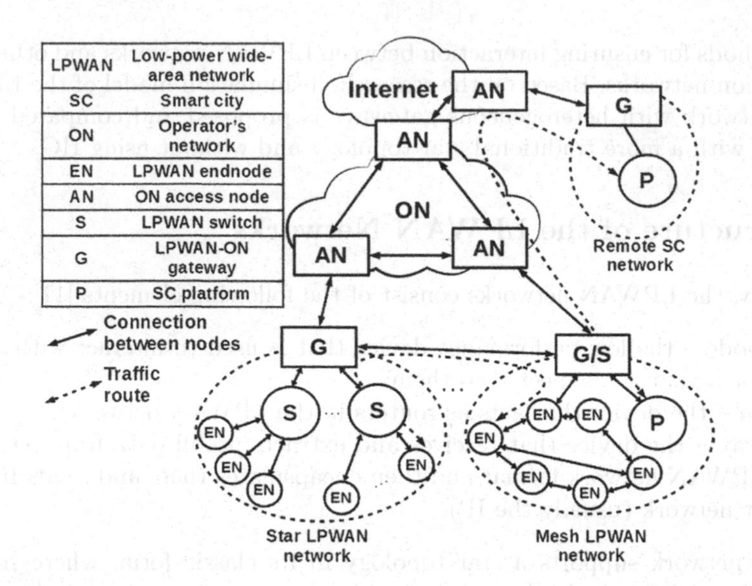

Fig. 1. The structure of the work LPWAN network within the SC system

typical for the hub, gateway, and edge platform. All the described features could be implemented using special software that operates on the operating system's service and application level. In Fig. 2 shows the structure of the LPWAN network, which includes a heterogeneous gateway for routing, edge processing, and packets transmission from LPWAN devices to the target network.

The HG can be used in the LPWAN networks only if there is a network interface that supports network information exchange technology and implements all the functions typical for the switch, gateway, and edge platform. All the described functions can be implemented using a semantic gateway, which works in the user's virtual workspace. Figure 3a shows the structure of the LPWAN network, which includes a heterogeneous gateway for routing, edge processing, and further sending network packets from LPWAN devices to the target network, and Fig. 3b shows the LPWAN network with the star topology and without using the HG.

As can be seen in 4, the difference in two structures is the ES and S/G are combined into the HG. The advantages and disadvantages of the proposed LPWAN network topology using heterogeneous gateways should be determined. In the next section, we consider analytical and simulation models to compare the operation of this network with the operation of a traditional LPWAN network with a star topology.

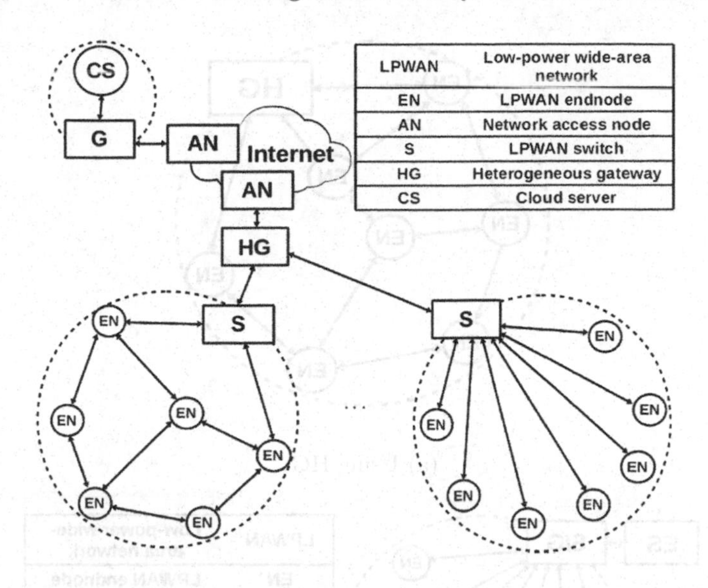

Fig. 2. The example of the LPWAN network structure using HG

3 Analytical and Simulation Models

Based on the proposed structure, it is possible to describe the operation of a simulation model that can be used to study the properties of the proposed network according to various parameters (for example, message service time, the amount of transmitted useful data per unit of time, the power consumption of devices in the network, etc.) and the model to which the model will be compared, including the heterogeneous gateway. Figure 4a represents the multi-hop transmission between end-nodes to a heterogeneous gateway. Each node is considered as a queueing system. In this case, the HG is used to serve the mesh/multi-hop networks. Besides, Fig. 4b shows a communication model from the end-nodes via a switch/gateway to an edge server. This model represents a direct connection between the end-nodes and the switch/gateway in the star topology model. In this traditional structure, an additional sever is used to process data at the edge.

It is necessary to take into account the features of LPWAN devices functioning to develop this model. The model network was developed, consisting of the LoRaWAN - CubeCell Dev-Board end node (Fig. 5a) and the Dragino LG02 gateway (Fig. 5b) for this purpose. These devices were used to measure the intensity of message receipt without using software delay for the EN and the intensity of message service at the gateway. The end-nodes continuously send a data packet with a size of 30 bytes. The following parameters are configured for LoRa transceivers:

- Center frequency = 868 Mhz.
- Transmission power = 14 dBm.
- Bandwidth = 125 kHz.

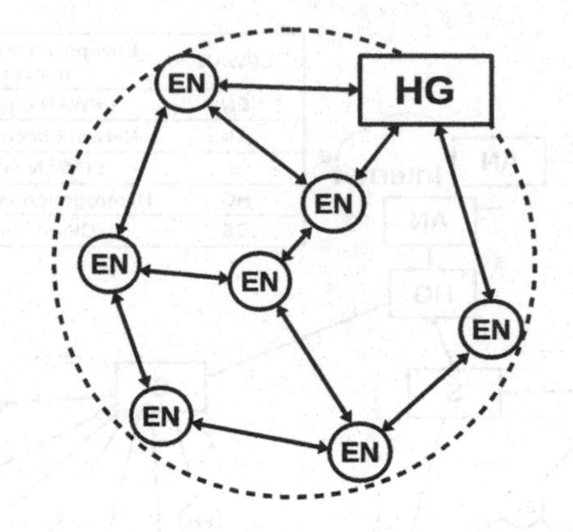

(a) Using HG

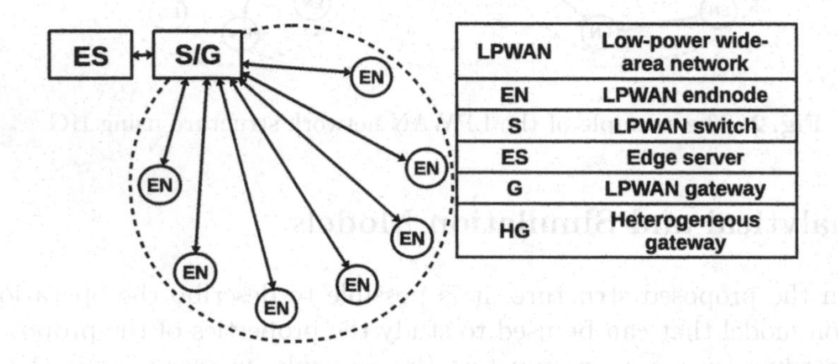

(b) Without using HG

Fig. 3. LPWAN structure

- Spreading factor = 10.
- Coding rate = 4/5.

As a result of the measurements, we have received analytical models describing the intensity of receiving and servicing messages for LoRaWAN devices based on the method of least squares, the method of generalized reduced gradients, and the Kolmogorov-Smirnov test. Figure 6 shows the ratio of the obtained analytical models to the original empirical distributions.

Thus, the intensity of the message service for the EN (2) and the server (4), and delay rates for the network interface (1) and for communication channel (3), can be described using the following probability density expressions:

$$f_a(x) = p_1\lambda_1 e^{\lambda_1 x} + p_2\frac{\lambda_2^{a_2}}{\Gamma(a_2)}xe^{-\lambda_2 x} + p_3\frac{\lambda_3^{a_3}}{\Gamma(a_3)}xe^{-\lambda_3 x} \tag{1}$$

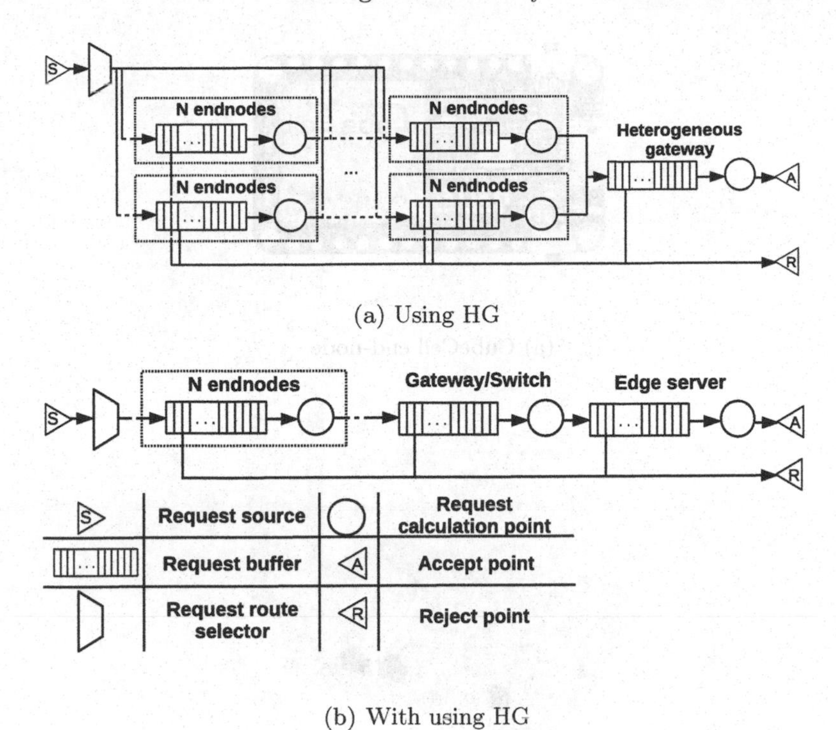

(a) Using HG

(b) With using HG

Fig. 4. Structure of the simulation model

$$f_c(x) = G_c(x, p_1, a_1, \lambda_1, c_1) + G_c(x, p_2, a_2, \lambda_2, c_2) + ... + G_c(x, p_{10}, a_{10}, \lambda_{10}, c_{10}) \quad (2)$$

$$f_{cd}(x) = p\frac{\lambda}{\Gamma(a)}xe^{-\lambda x} \quad (3)$$

$$f_s(x) = p\lambda e^{\lambda x} \quad (4)$$

$$G_c(x, p, a, \lambda, c) = p\frac{\lambda^a}{\Gamma(a)}(x - c)e^{-\lambda(x-c)} \quad (5)$$

Where:

- $f_a(x)$ – probability density of message arrival intensity,
- $f_c(x)$ – message service intensity,
- $G_c(x, p, a, \lambda, c)$ – bias Gamma probability density distribution,
- p_i – probability of falling into a given distribution,
- a_i – scale coefficient,
- λ_i – form coefficient,
- c_i – displacement coefficient.

(a) CubeCell end-node

(b) Dragion gateway

Fig. 5. Used LPWAN hardware

These analytical models, together with the probabilistic distributions obtained during the experiment for the time intervals between message arriving, the latency indicators of the LoRaWAN network interface, the service time of messages on the EN, the latency on the communication channel, and the service time of messages on the server were used to model a network of 1000 LoRaWAN end-nodes, according to Fig. 3a and 4a for the network using HG and Fig. 3b and 4b for the network without HG.

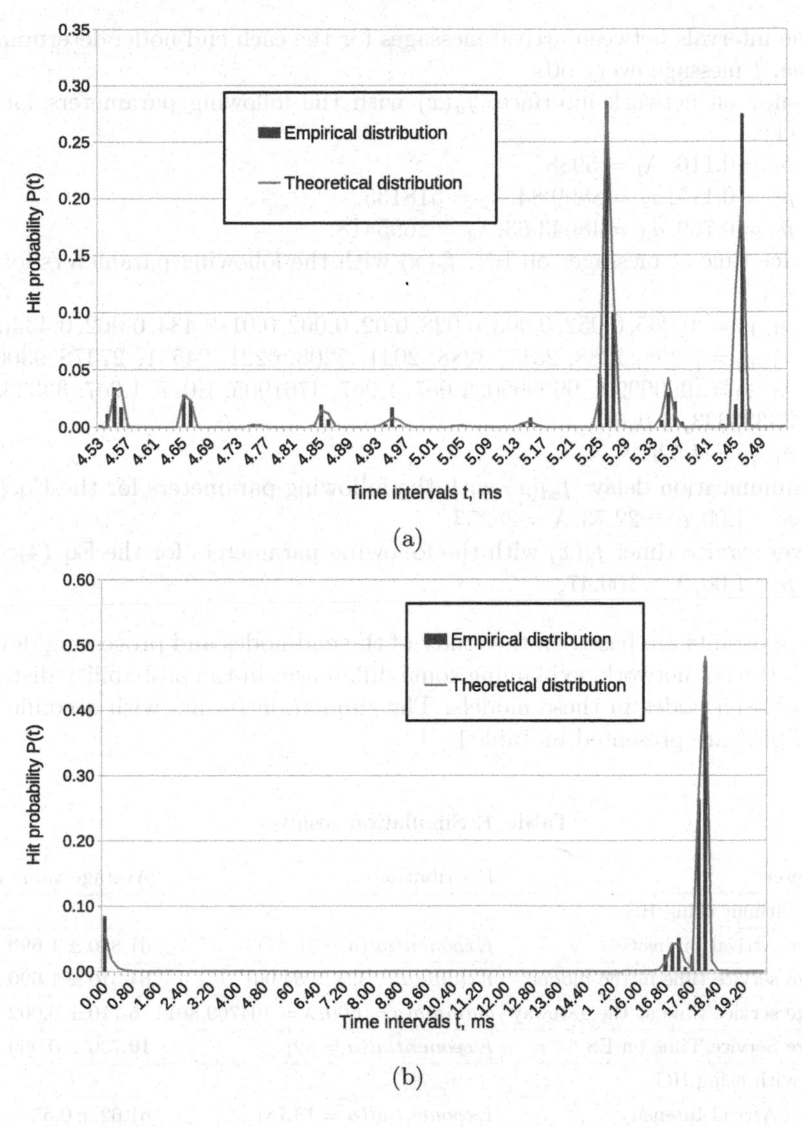

Fig. 6. The ratio of the empirical distribution to the theoretical distribution for: a) the network interface latency; b) the message service intensity

4 Simulation Results

The obtained experimental network's probability distributions allow us to simulate the work of the previously described networks (Fig. 3a and Fig. 3b). The probability distributions used in the simulation using the Python library Ciw [12] are presented below:

1. Time intervals between arrival messages for the each end node: deterministic value, 1 message every 60 s.
2. Latency on network interface: $f_a(x)$ with the following parameters for the Eq. (1):
   - $p_1 = 0.116$, $\lambda_1 = 5988$,
   - $p_2 = 0.115, a_2 = 8899.84, \lambda_2 = 518135$,
   - $p_3 = 0.769, a_3 = 48643.63, \lambda_3 = 2695418$.
3. Service time of messages on EN: $f_c(x)$ with the following parameters for the Eq. (2):
   - $p_{1..10} = [0.065, 0.052, 0.003, 0.028, 0.02, 0.002, 0.01, 0.434, 0.062, 0.434]$,
   - $a_{1..10} = [\ 298,\ 1288,\ 2317,\ 3288,\ 2041,\ 5208,\ 6221,\ 24551,\ 27478, 93005]$,
   - $\lambda_{1..10} =$ [9900990, 9900990, 1.0e7, 1.0e7, 4761905, 1.0e7, 1.0e7, 33333333, 33333333, 1.0e8],
   - $c_{1..10} = 4.52e^{-3}$.
4. Communication delay: $f_{cd}(x)$ with the following parameters for the Eq. (3):
   - $p = 1.00, a = 22.73, \lambda = 25253$.
5. Server service time: $f_s(x)$ with the following parameters for the Eq. (4):
   - $p = 1.00, \lambda = 100.47$.

These results are based on the study of the end-nodes and processing devices for each type of network, explaining some differences in the probability distributions between nodes in these models. The simulation results with a confidence level of 95% are presented in Table 1.

Table 1. Simulation results

Parameters	Distribution	Average value (ms)
Model without using HG		
Message Arrival Intensity	$Exponential(a = 16.57)$	61.850 ± 1.690
Message service time in the model	$Erlang(a = 3, \lambda = 44.06)$	84.310 ± 1.690
Message service time at the gateway	$Gamma(a = 600, \lambda = 104769.80)$	5.710 ± 0.002
Message Service Time on ES	$Exponential(a = 82)$	10.730 ± 0.060
Model with using HG		
Message Arrival Intensity	$Exponential(a = 13.78)$	61.62 ± 0.55
Message service time in the model	$Erlang(a = 7, \lambda = 265.86)$	27.40 ± 0.07
Message service time at the HG	$Erlang(a = 4, \lambda = 307.61)$	15.52 ± 0.06

According to the simulation results, the service time was reduced in the model using HG. Moreover, the service time decreased and smaller than the sum of the service time at the gateway and server. Thus, for a more complex assessment of the need to use a heterogeneous gateway, it is necessary to conduct experiments, both based on the server's energy consumption, the gateway, and HG, and based on a real model network. Also, a study of emergency cases in which each end node sends messages should be much more intensive than one message per second did not conduct in this paper. The authors assume that in this case, the network topology with a heterogeneous gateway will not have significant advantages, since both the traditional model (S/G) and the proposed model (HG) will have the narrowest and most critical place is the node providing access to the external network. However, this issue may have unexpected results at the radio communication channel level and requires further research.

5 Conclusion

This paper has analyzed the model and methods for ensuring LPWANs and other communication networks' interaction. Based on this analysis, we have proposed the structure and simulation model of the mesh LPWAN network using heterogeneous gateways. With the presented network structure, we performed the simulation series. The mesh network results using HG were compared with the star topology network and without using HG.

Looking into further study, we intend to develop a model mesh network based on the developed structure, including HG and dozens of ENs. This model's functional results will be compared with the network's properties with the star topology and without using HG. It is also proposed to study the energy consumption of LPWAN network elements: gateway, edge server, EN, and HG.

Acknowledgment. The publication has been prepared with the support of the grant from the President of the Russian Federation for state support of leading scientific schools of the Russian Federation according to the research project SS-2604.2020.9.

References

1. Abu-Matar, M., Mizouni, R.: Variability modeling for smart city reference architectures. In: 2018 IEEE International Smart Cities Conference (ISC2). IEEE, September 2018. https://doi.org/10.1109/isc2.2018.8656967
2. Alcatel-Lucent: Smart city network architecture guide p. 34 (2019). https://www.al-enterprise.com/-/media/assets/internet/documents/smart-city-network-architecture-guide-en.pdf
3. Alshahrani, A., Elgendy, I.A., Muthanna, A., Alghamdi, A.M., Alshamrani, A.: Efficient multi-player computation offloading for VR edge-cloud computing systems. Appl. Sci. **10**(16), 5515 (2020). https://doi.org/10.3390/app10165515
4. Corporation, S.: Sx1276/77/78/79 lora modules datasheet p. 132. https://www.semtech.com/products/wireless-rf/lora-transceivers/sx1276
5. Ferreira, C.M.S., Oliveira, R.A.R., Silva, J.S.: Low-energy smart cities network with LoRa and bluetooth. In: 2019 7th IEEE International Conference on Mobile Cloud Computing, Services, and Engineering (MobileCloud). IEEE, April 2019. https://doi.org/10.1109/mobilecloud.2019.00011
6. ITU-T: Q.4060 the structure of the testing of heterogeneous internet of things gateways in a laboratory environment (2018). https://www.itu.int/rec/T-REC-Q.4060-201810-I
7. ITU-T: Q.3055 signalling protocol for heterogeneous internet of things gateways p. 29 (2019). https://www.itu.int/rec/T-REC-Q.3055-201912-I

8. Kirichek, R., Vishnevsky, V., Pham, V.D., Koucheryavy, A.: Analytic model of a mesh topology based on LoRa technology. In: 2020 22nd International Conference on Advanced Communication Technology (ICACT). IEEE, February 2020. https://doi.org/10.23919/icact48636.2020.9061519
9. Lamberti, R., et al.: An open multimodal mobility platform based on distributed ledger technology. In: Galinina, O., Andreev, S., Balandin, S., Koucheryavy, Y. (eds.) NEW2AN/ruSMART -2019. LNCS, vol. 11660, pp. 41–52. Springer, Cham (2019). https://doi.org/10.1007/978-3-030-30859-9_4
10. Kulik, V., Kirichek, R.: The heterogeneous gateways in the industrial internet of things. In: 2018 10th International Congress on Ultra Modern Telecommunications and Control Systems and Workshops (ICUMT). IEEE, Novemeber 2018. https://doi.org/10.1109/icumt.2018.8631232
11. LoRa-Alliance: Lorawan specification v. 1.0.2 p. 70. https://lora-alliance.org/sites/default/files/2018-05/lorawan1_0_2-20161012_1398_1.pdf
12. Palmer, G.I., Knight, V.A., Harper, P.R., Hawa, A.L.: Ciw: an open-source discrete event simulation library. J. Simul. **13**(1), 68–82 (2018). https://doi.org/10.1080/17477778.2018.1473909
13. Pham, V.D., Dinh, T.D., Kirichek, R.: Method for organizing mesh topology based on LoRa technology. In: 2018 10th International Congress on Ultra Modern Telecommunications and Control Systems and Workshops (ICUMT). IEEE, November 2018. https://doi.org/10.1109/icumt.2018.8631270
14. Sanchez-Iborra, R., Cano, M.D.: State of the art in LP-WAN solutions for industrial IoT services. Sensors **16**(5), 708 (2016). https://doi.org/10.3390/s16050708
15. Tolcha, Y.K., et al.: Oliot-OpenCity: open standard interoperable smart city platform. In: 2018 IEEE International Smart Cities Conference (ISC2). IEEE, Septemebr 2018. https://doi.org/10.1109/isc2.2018.8656763

Research on Using the AODV Protocol for a LoRa Mesh Network

Van Dai Pham[1], Duc Tran Le[2](✉), Ruslan Kirichek[1,3], and Alexander Shestakov[1]

[1] Bonch-Bruevich Saint-Petersburg State University of Telecommunications, Saint Petersburg 193232, Russia
fam.vd@spbgut.ru, kirichek@sut.ru, alexandr.shestakov01@yandex.ru

[2] The University of Danang - University of Science and Technology, Danang, Vietnam
letranduc@dut.udn.vn

[3] V.A. Trapeznikov Institute of Control Sciences of Russian Academy of Sciences, Moscow 117997, Russia

Abstract. In this paper, we consider the LoRa technology to expand sensor network coverage in smart sustainable cities. A model of a LoRa mesh network is proposed using the AODV protocol in packet routing. With a simulation model developed based on OMNET++, a series of computer experiments was carried out with changing various parameters. In the experiments results, the end-to-end delay and packet loss ratio were analyzed in the dependence on the number of nodes and packet size in the network. The simulation results show that the latency is relatively high in the LoRa mesh network, but it might be accepted for some applications.

Keywords: IoT · LoRa · Mesh network · AODV · Delay · Packet loss

1 Introduction

Over the past decade, the Internet of Things (IoT) has received significant attention in the scientific and industrial fields. People can control, monitor, and do a lot more from a remote distance. It is done by connecting various objects reducing physical distance. The IoT movement creates the need for new wireless technologies, capable of supporting the large numbers of devices in the IoT space. These systems require a technology that consumes less power and also covers long distances. However, many technologies such as ZigBee, WiFi, Bluetooth popularly used at present consumes high power and is not suitable for battery-operated systems. To fulfill the communication requirements of IoT, we need new technology. Low-Power Wide Area Network (LPWAN) offers radio coverage over a large area by way of base stations and adapting transmission rates, transmission power, modulation, duty cycles, where end-devices incur a very low energy consumption due to their being connected.

V. M. Vishnevskiy et al. (Eds.): DCCN 2020, LNCS 12563, pp. 149–160, 2020.
https://doi.org/10.1007/978-3-030-66471-8_13

According to the ITU-T Y.4903/L.1603 recommendation for the development of Smart Sustainable Cities, it is necessary to provide full Internet access covered in the city. One of the well-known technologies included in the LPWAN network group is LoRa (Long-Range) technology developed by Semtech. The long-range and low-power nature of LoRa makes it a promising candidate for smart sensing technology in civil infrastructures (such as health monitoring, smart metering, environment monitoring, etc.), as well as in industrial applications.

Many technologies are proposed to use in IoT applications. Every technology has its features, advantages, and disadvantages. However, no single technology can serve all IoT applications because different applications will have different requirements. Based on the requirement, we can only choose a technology that is best suited for the specific application from the existing technologies.

WiFi is widely used in many IoT applications, the most common being the link from the gateway to the router that connects to the Internet. However, it is also used for primary wireless links requiring high speed and medium distance.

The WiFi is usually confined within a small area such as a building, home office. It can span over a limited range that is 300 m in radius. However, Lora technology provides long-range communication. A single LoRa gateway can span an area of 100 km^2. In addition, the LoRa is based on the chirp spread spectrum (CSS), which is highly resistant to multipath and fading.

ZigBee is an IEEE 802.15.4-based specification for a suite of high-level communication protocols used to create personal area networks. It consists of small, low-power digital radios. These are best suited for home automation, medical device data collection, small scale projects that need to transfer data over small distances. Its range is 10–100 m. Zigbee also consumes power more than LoRa and is not suitable for outdoor applications with low power requirements. It also does not have the long-range capability of LoRa technology.

Bluetooth is a wireless transmission technology that theoretically allows short-distance connections between devices up to 100 m away but is only about 10 m in actual use. It is used mainly for speakers, health monitors, and other short-range applications. That is the reason why in the context of long-range transmission, LoRa proves superior to Bluetooth.

In summary, LoRa technology will revolutionize IoT by enabling data communication over a long-range distance while using very little power. LoRa fills the technology gap of Cellular and WiFi/BLE based networks that require either high bandwidth or high power or have a limited range or inability to penetrate deep indoor environments. In effect, LoRa Technology is flexible for rural or indoor use cases in smart cities, smart homes and buildings, smart agriculture, smart metering, and smart supply chain and logistics.

2 LoRa Overview

LoRa commonly refers to two distinct layers: LoRa physical layer, which is based on the Chirp Spread Spectrum (CSS) [2] radio modulation technique, provides long-range, low-power, and low-throughput communications. LoRaWAN (Long

Range Wide Area Network), a MAC layer protocol, provides a medium access control mechanism for many end-devices to communicate with a gateway using the LoRa modulation. The LoRa physical layer is based on the Chirp Spread Spectrum (CSS) with integrated Forward Error Correction (FEC) that provides low sensitivity for long communication ranges. CSS has many advantages:

- Great link budget, low-power transmission.
- Resistant to multipath and other interference.
- Orthogonality of spreading factors.
- Simplified electronic for receiver synchronization.
- Robust against Doppler shift (mobile application).

With this design, devices using different data rates do not interfere with each other, and communication between devices and gateways can be operated over multiple frequency channels, increasing the network capacity.

LoRa networks operate in unlicensed ISM frequency bands, which for Europe is the frequency band 867–869 MHz with a central frequency of 868 MHz. For this band, the LoRa specifications define 10 channels of 125/250 kHz uplink and 125 kHz downlink. On the other hand, for North America, the frequency band is 902–928 MHz with a central frequency of 915 MHz, and it defines 64 channels of 125/500 kHz uplink and 500 kHz downlink.

Besides the frequency range specifying possibility, several parameters are available for the customization of the LoRa technology [17]:

- Bandwidth: The Bandwidth (BW) is the distance between the lowest and highest frequency in each chirp. A higher BW will increase the data rate and decrease the transmission time on-air for a packet. It will also decrease the decoding sensitivity since the radio signal is exposed more to noise. Using a low BW for the same size packet means longer transmission time and a higher risk that the receiver will fall out of synchronization due to imperfect receiver clock drift.
- Spreading Factor (SF): The spreading Factor (SF) is the ratio between the chip rate and the underlying symbol rate. If we increase the SF (i.e. more bits per symbol), the Signal to Noise Ratio (SNR) will be increased, and consequently, it will increase the range and the sensitivity, but also the time on-air to send a symbol.
- Coding Rate (CR) LoRa uses Forward Error Correction (FEC) for the payload. The level of FEC is set by the Coding Rate (CR) parameter. CR increases robustness against interference but increases the time on-air when more redundant bits are used for corrections.

By varying these parameters, it is possible to determine the optimal mode of the operation for different data transmission conditions between the transmitter and the receiver. In LoRa, the chirp rate depends only on the bandwidth: the chirp rate is equal to the bandwidth (one chirp per second per Hertz of bandwidth). The symbol rate and the bit rate at a given spreading factor are proportional to the frequency bandwidth, so a doubling of the bandwidth will effectively double the transmission rate. The duration of a symbol (Ts) to the bandwidth and the spreading factor is shown in formula (1).

$$T_S = \frac{2^{SF}}{BW} \tag{1}$$

Moreover, LoRa includes a forward error correction code. The code rate (CR) equals $\frac{4}{4+n}$, with $n \in \{1, 2, 3, 4\}$.

In the formula (2), the relationship between parameters BW, SF, and CR is determined:

$$DR = SF \times \frac{BW}{2^{SF}} \times CR \tag{2}$$

These parameters also influence decoder sensitivity. Therefore, increasing BW will decrease receiver sensitivity, and increasing the spreading factor will raise the receiver sensitivity.

3 LoRa Mesh Networks

In this work, we consider the use of LoRa to expand the coverage of the sensor network. The proposed network model is a mesh network. Why the topology is mesh?

In the IoT networks, we have two popular types of topology: star and mesh.

Star topology is used in the IoT networks. The advantage of star topology is that all the network complexity is driven to a central node, so all the other nodes only need to communicate in their time or frequency slot. How they communicate depends on whether wireless multiplexing is done through frequency-division multiple access, time-division multiple access, or code-division multiple access. The primary disadvantage of star topology is that the radio link between the gateway and the end node or terminal can be very long, which means the further a node is away from the gateway, the more energy it has to expend relaying a message. Also, LoRa devices may still be unable to communicate wirelessly with a nearby gateway due to physical obstacles that can attenuate wireless signal strength and result in data losses and communication errors [9].

That is the reason why most of the existing technologies, nowadays, are based on mesh networks [5]. Mesh networking is a solution for increasing the communication range and packet delivery ratio without installing additional LoRa gateways. In the mesh network, the infrastructure node can relay a packet and cooperate with other nodes to efficiently route a packet to the gateways. Each node acts as a repeater. Mesh networks dynamically connect end-nodes and self-configure the routing paths [1]. The primary advantage of mesh topology is that it has low transmitted power and shorter links, which allows for a fairly long battery life and enables us to move much data around the network. Moreover, routing and configuration functionalities are dedicated to wireless mesh networks, and the load on mesh clients is decreased considerably due to this dedication. This makes LoRaWAN mesh one of the best ways to collect data from many remote sensors at once.

4 Routing in LoRa Mesh Networks

Designing and developing routing protocols in wireless mesh networks is a challenging issue that should cover multiple performance metrics such as minimum hop count, preventing disruption of the service methods according to robustness concepts. Using the mesh infrastructure to perform routing processes as efficiently as possible, and increasing the scalability of routing protocols to install or maintain routing paths in a mesh network with large capacity [6]. To implement a LoRa mesh network model with the LoRa gateways, we need to choose the appropriate routing protocol.

Typically, ad-hoc routing protocols are divided into three categories based on the network topology information used for route discovery: proactive, reactive, or hybrid.

- Proactive Routing Protocols: This kind of protocol periodically exchanges the topology information between all the network nodes. Therefore, proactive routing protocol has no route discovery since the destination route is saved and maintained within a table. The tables usually must be updated. These protocols are used where the route requirements are frequent. However, the drawback of this protocol is that it gives low idleness to constant application [12]. DSDV (Destination Sequenced Distance Vector), OLSR (Optimized Link State Routing) are examples.
- Reactive Routing Protocols: These routing protocols choose routes to other nodes only when they are needed. A route discovery process is launched when a node wants to communicate with another station for which it does not possess any route table access. AODV (Ad-hoc On-Demand Distance Vector routing protocol), DSR (Dynamic Source Routing) are examples.
- Hybrid Routing Protocols: Hybrid Routing joining nearby proactive routing protocols and global reactive routing protocols to reduce routing overhead and delay due to route disclosure process. The advantages of these protocols are higher efficiency and scalability. However, the disadvantage is high latency for locating new routes [4]. Examples of these protocols are ZRP (Zone Routing Protocol).

There are many protocols within each of these three categories, and a comprehensive review cannot be provided here. We will instead examine and evaluate a representative selection of protocols commonly used and referenced in the literature.

There are many kinds of research focused on the comparison of the protocols mentioned above. In [7,8], the authors found that AODV has shown better performance than other routing protocols (DSR, DSDV) with the increment of nodes number in all performance metrics (end-to-end delay, routing load, received packets, packet delivery ratio, dropped packets). In [19], Singh et al. showed that although DSR performs well in quick transmission, but it has high packet loss. Jogendra Kumar in [11] proved that AODV has the best execution as far as normal jitter and end-to-end delay in comparison with DYMO (Dynamic MANET On-Demand Routing Protocol), DSR, OSLR, ZRP.

In [14], Makodia et al. even pointed out that AODV is advised for secured communication. In addition, AODV also gives a good value of average throughput [21].

In addition to those evaluations, we can see that, due to its characteristics, OSLR is well-suited to large and dense networks with random and sporadic traffic. However, we need a few gateways to cover even a large area in the LoRa network. As such, the added overhead of choosing relays and updating topology information is unnecessary in our case [13].

Although DSDV has lower control overhead than OLSR [16], continuous updates are nonetheless unnecessary for networks with static nodes, as in a typical LoRa deployment scenario.

Since reactive protocols require sharing of topology information only when routes fail or a new route needs to be established, they allow for a reduced control overhead, and thus energy cost, in comparison to proactive protocols [15].

DSR protocol is designed for a network with potentially high mobility. Therefore it does not suit out case [13].

In summary, although the comparisons are made mainly in MANET and VANET networks, it also partly points out the advantages and disadvantages of the protocols. From those evaluations, perhaps, the suitable routing protocol for our LoRa mesh network is AODV. This is the reason why we choose to use the AODV protocol in the LoRa mesh network.

5 Network and Simulation Model

5.1 Network Model

Currently, mesh networks are popularly used in various IoT applications. In many cases, sensor networks are deployed far away from the access point to the Internet. In these cases, a model is proposed based on mesh networks in two segments. As shown in Fig. 1, sensor networks communicate with the gateways to connect to the server cloud. The gateways are able to communicate with each other in the mesh network while some gateways have access to the Internet. Moreover, using long-range communication as LoRa for the gateway network is the main idea to expand the coverage.

In particular, the mesh network can provide connectivity for IoT devices in smart sustainable cities applications. Using the LoRa technology as a communication method at the physical layer, devices may communicate with the others over a hundred meters. As shown in Fig. 2, a network of devices is deployed in an area of the city Saint-Petersburg. Each node is equipped with the LoRa interface configured with the same parameters. We can consider these devices used to collect data from the short-range communication sensors, then collected data are transmitted to the sink node connected to the external network to the remote server. The AODV protocol is proposed to establish routing paths to the sink node.

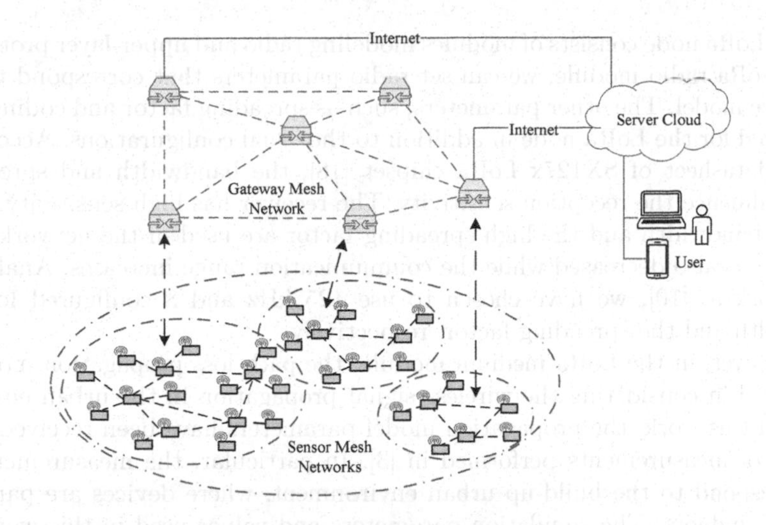

Fig. 1. Network model

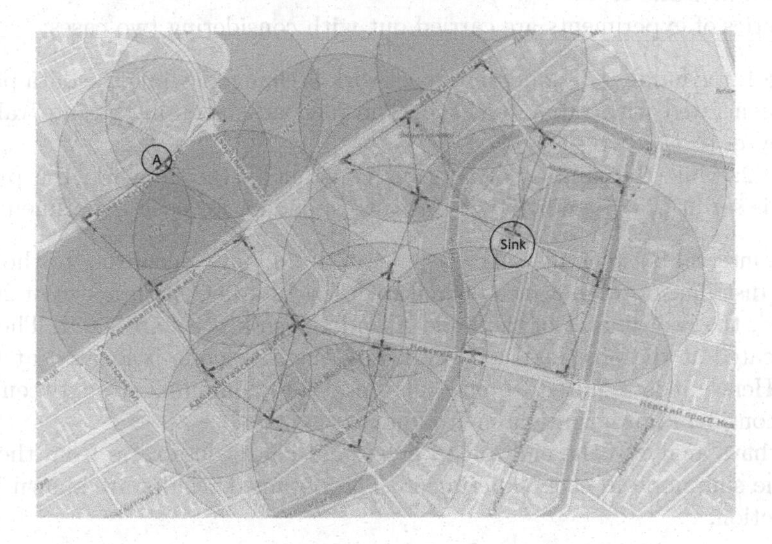

Fig. 2. Deployment area

5.2 Simulation Methodology and Parameter

The frameworks OMNET++ and inet are used to carry out the network simulation. They are known well to be used in numerous domains for simulating wired and wireless networks. Based on these frameworks and the work in [20], we have developed a module of the LoRa node. Since the OMNET ++ library and frameworks are designed based on modular and component-oriented principles, the LoRa node can be integrated with the build-in modules from the inet framework.

The LoRa node consists of modules modeling radio and upper-layer protocols. In the LoRa radio module, we can set radio parameters that correspond to our hardware model. The other parameters, such as spreading factor and coding rate configured for the LoRa node in addition to the usual configurations. According to the datasheet of SX127x LoRa chipset [18], the bandwidth and spreading factor influence the reception sensitivity. The receiver has high sensitivity when the low bandwidth and the high spreading factor are used in the network. The data rate is also decreased while the communication range increases. Analyzing from work in [10], we have chosen to use 125 kHz and 8 configured for the bandwidth and the spreading factor, respectively.

Moreover, in the LoRa medium module, the path loss propagation model is configured in considering the wireless signal propagation in the urban environment. In this work, the propagation model parameters have been received from a series of measurements performed in [3]. In particular, the measurements in [3] correspond to the build-up urban environment, where devices are partially deployed indoors. The simulation parameters and values used in this work are presented in Table 1.

A series of experiments are carried out with considering two cases:

- Case 1: the number of nodes in the network is changed, while the data packets are generated with random length. The payload size is in the interval from 20 bytes to 150 bytes.
- Case 2: while the number of nodes is constant in the network, the payload size is set in {20, 40, 60, 150} bytes corresponding for each experiment.

The interval between messages is generated randomly according to the exponential distribution with a mean equal to 120 s. In a field with a size of 2000 $\times$ 2000 m^2, the coordinates of the node A and the sink node are fixed. The node A is located at the position (400, 400), and the sink node is located at (1500, 1500). Hence, it is required to have relay nodes to communicate between them in the considered propagation environment.

We have analyzed the end-to-end delay and packet loss ratio from the node A to the sink node in each experiment. The obtained results are shown in the next section.

6 Result Analysis

6.1 Analysis by Varying the Number of Nodes

The experiments were performed with a changing number of nodes in the network. The coordinates of nodes were generated randomly according to the uniform distribution in the interval (0, 2000). Figure 3 shows a histogram of the delay required to deliver the data packet from the node A to the sink node. According to Fig. 3a, the delay does not change much when increasing the number of nodes in the network. Based on the probability density histogram, we can see that the delay varying in the interval to 2 s has a probability equal to

Table 1. Simulation parameters and values

Module	Parameter	Value
LoRa radio	Frequency, MHz	868
	Bandwidth (BW), kHz	125
	Spreading factor (SF)	8
	Coding rate (CR)	4/5
	Transmission power, dBm	20
	Antenna, dBi	5
LoRa medium	Log-distance path loss model	PL(d0) = 127.41 dB d0 = 40 m gamma = 2.08 sigma = 3.57
	Background noise, dBm	−96.616
Application	Payload length, byte	Case 1: randint(20, 150) Case 2: {20, 40, 60, 150}
	Send interval, s	Exponential(120s)
AODV	maxJitter, s	5
	helloInterval, s	60
	activeRouteTimeout, s	200
	rreqRetries	4
	timeoutBuffer, s	10
Others	Number of nodes	{15, 20, 25, 30}
	Field simulation size, m^2	2000 × 2000
	Simulation time, s	50000

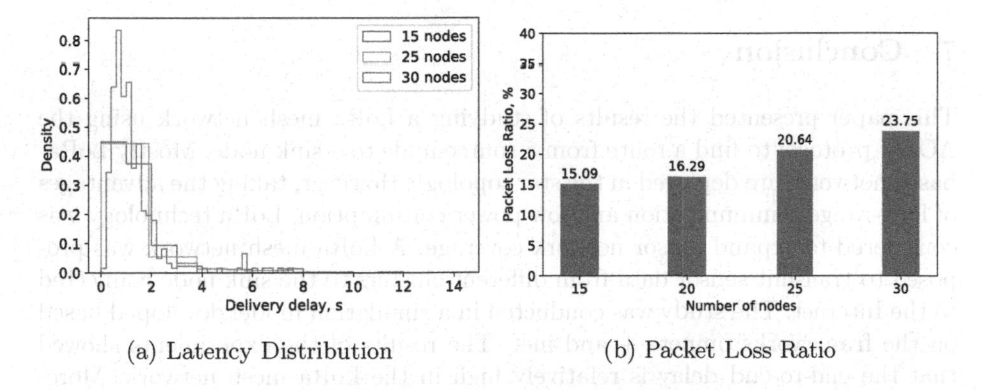

(a) Latency Distribution (b) Packet Loss Ratio

Fig. 3. Analysis by varying number of nodes

$2 \cdot 0.45 = 0.9$. However, the packet loss ratio increases when increasing the number of nodes in the network, as shown in Fig. 3b. Adding the number of nodes significantly affects the change of the packet loss ratio.

6.2 Analysis by Varying Payload Length

In the second case, with the fixed number of nodes in the network, a series of experiments was performed with changing payload length generated randomly in the first case. In this case, the delay distribution and packet loss ratio are presented in Fig. 4a and b, respectively. The delivery delay can reach up to several seconds via relay nodes to the destination. Besides, the packet loss ratio increases when increasing the payload length. However, the data packet size does not significantly affect the change of the packet loss ratio. Comparing the sending packets of size 20 and 150 bytes shows that the packet size does not greatly affect the probability of loss in this considered network.

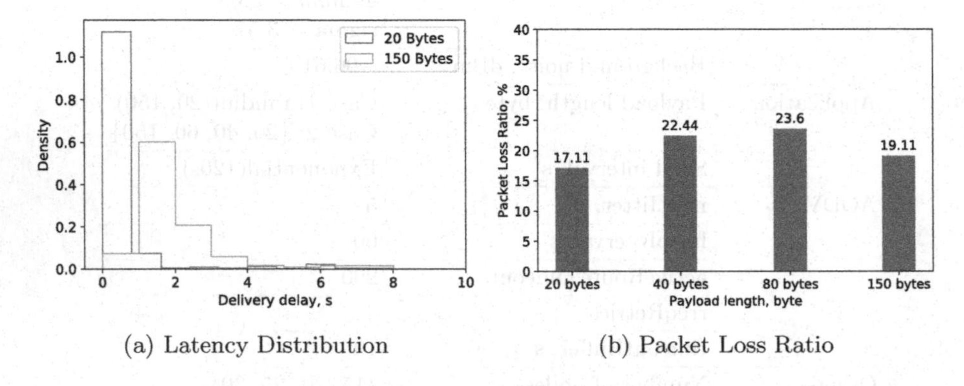

(a) Latency Distribution (b) Packet Loss Ratio

Fig. 4. Analysis by varying payload length

7 Conclusion

This paper presented the results of studying a LoRa mesh network using the AODV protocol to find a route from a source node to a sink node. Mostly LoRa-based networks are deployed in the star topology. However, taking the advantages of long-range communication and low power consumption, LoRa technology was considered to expand sensor network coverage. A LoRa mesh network was proposed to transmit sensor data from different clusters to the sink node connected to the Internet. The study was conducted in a simulation model developed based on the frameworks omnet++ and inet. The results of the experiments showed that the end-to-end delay is relatively high in the LoRa mesh network. Moreover, changing the number of nodes in the network and payload size affected the packet loss ratio.

Based on the study results, the AODV protocol might be used for the LoRa mesh network. However, the delay needs to be considered in such a network. Looking into the future, we intend to consider in developing the other protocol that has compatibility with both the LoRa mesh network and LoRaWAN.

Acknowledgment. The publication has been prepared with the support of the grant from the President of the Russian Federation for state support of leading scientific schools of the Russian Federation according to the research project SS-2604.2020.9.

References

1. Alotaibi, E., Mukherjee, B.: A survey on routing algorithms for wireless ad-hoc and mesh networks. Comput. Netw. **56**(2), 940–965 (2012)
2. Berni, A., Gregg, W.: On the utility of chirp modulation for digital signaling. IEEE Trans. Commun. **21**(6), 748–751 (1973)
3. Bor, M.C., Roedig, U., Voigt, T., Alonso, J.M.: Do LoRa low-power wide-area networks scale? In: Proceedings of the 19th ACM International Conference on Modeling, Analysis and Simulation of Wireless and Mobile Systems, pp. 59–67 (2016)
4. Chandra, A., Thakur, S.: Qualitative analysis of hybrid routing protocols against network layer attacks in MANET. Apoorva Chandra et al., Int. J. Comput. Sci. Mob. Comput. **4**(6), pp. 538–543 (2015)
5. Devalal, S., Karthikeyan, A.: LoRa technology-an overview. In: 2018 Second International Conference on Electronics, Communication and Aerospace Technology (ICECA), pp. 284–290. IEEE (2018)
6. Eslami, M., Karimi, O., Khodadadi, T.: A survey on wireless mesh networks: Architecture, specifications and challenges. In: 2014 IEEE 5th Control and System Graduate Research Colloquium, pp. 219–222. IEEE (2014)
7. Ferreiro-Lage, J.A., Gestoso, C.P., Rubiños, O., Agelet, F.A.: Analysis of unicast routing protocols for VANETs. In: 2009 Fifth International Conference on Networking and Services, pp. 518–521. IEEE (2009)
8. Haerri, J., Filali, F., Bonnet, C.: Performance comparison of AODV and OLSR in VANETs urban environments under realistic mobility patterns. In: Proceedings of the 5th IFIP mediterranean ad-hoc networking workshop, pp. 14–17 (2006)
9. Ke, K.H., Liang, Q.W., Zeng, G.J., Lin, J.H., Lee, H.C.: A LORA wireless mesh networking module for campus-scale monitoring: demo abstract. In: Proceedings of the 16th ACM/IEEE International Conference on Information Processing in Sensor Networks, pp. 259–260 (2017)
10. Kirichek, R., Vishnevsky, V., Pham, V.D., Koucheryavy, A.: Analytic model of a mesh topology based on LoRa technology. In: 2020 22nd International Conference on Advanced Communication Technology (ICACT), pp. 251–255. IEEE (2020)
11. Kumar, J.: Comparative performance analysis of AODV, DSR, DYMO, OLSR and ZRP routing protocols in manet using varying pause time. Int. J. Comput. Commun. Netw. (IJCCN) **3**(1), 43–51 (2012)
12. Kumar, R., Dave, M.: A comparative study of various routing protocols in VANET. arXiv preprint arXiv:1108.2094 (2011)
13. Lundell, D., Hedberg, A., Nyberg, C., Fitzgerald, E.: A routing protocol for LoRa mesh networks. In: 2018 IEEE 19th International Symposium on A World of Wireless, Mobile and Multimedia Networks (WoWMoM), pp. 14–19. IEEE (2018)
14. Makodia, B., Patel, T., Parmar, K., Hadia, S., Shah, A.: Implementing and analyzing routing protocols for self-organized vehicular adhoc network. In: 2013 Nirma University International Conference on Engineering (NUiCONE), pp. 1–6. IEEE (2013)
15. Pandey, K., Swaroop, A.: A comprehensive performance analysis of proactive, reactive and hybrid MANETs routing protocols. arXiv preprint arXiv:1112.5703 (2011)

16. Perkins, C.E., Bhagwat, P.: Highly dynamic destination-sequenced distance-vector routing (DSDV) for mobile computers. ACM SIGCOMM Comput. Commun. Rev. **24**(4), 234–244 (1994)
17. Pham, V.D., Dinh, T.D., Kirichek, R., et al.: Method for organizing mesh topology based on LoRa technology. In: 2018 10th International Congress on Ultra Modern Telecommunications and Control Systems and Workshops (ICUMT), pp. 1–6. IEEE (2018)
18. Semtech Corporation: Datasheet SX1276/77/78/79 LoRa Transciever, p. 132 (2019)
19. Singh, P.K., Lego, K., Tuithung, T.: Simulation based analysis of adhoc routing protocol in urban and highway scenario of VANET. Int. J. Comput. Appl. **12**(10), 42–49 (2011)
20. Slabicki, M., Premsankar, G., Di Francesco, M.: Adaptive configuration of LoRa networks for dense IoT deployments. In: NOMS 2018–2018 IEEE/IFIP Network Operations and Management Symposium, pp. 1–9. IEEE (2018)
21. Talooki, V.N., Ziarati, K.: Performance comparison of routing protocols for mobile ad hoc networks. In: 2006 Asia-Pacific Conference on Communications, pp. 1–5. IEEE (2006)

On the Algebraic Theory of Loop Free Routing

Hussein Khayou(✉), Margarita A. Rudenkova(✉), and Leonid I. Abrosimov(✉)

National Research University "Moscow Power Engineering Institute", Krasnokazarmennaya 14, 111250 Moscow, Russia
hussein.khayou@gmail.com, {RudenkovaMA,AbrosimovLI}@mpei.ru

Abstract. Validation models not only provide a better understanding of the system, but can also help in improving the reliability and robustness of the design. EIGRP metric can be modeled algebraically using bi-semigroups and semirings in case the cost function is homomorphic [1]. EIGRP uses DUAL algorithm which is the basis for loop-free distance vector routing with non lexical metric. DUAL was validated using the classical shortest path problem [13,20]. However, it was shown that DUAL does not perform as expected in the absence of monotonicity [1,23]. This article approaches loop free routing from an algebraic perspective. Conditions for loop free routing and the relations between them were presented algebraically and proved correct. Then, we investigate loop free routing in the presence and the absence of monotonicity. Our goal is to provide theory for loop free routing with an arbitrary metric.

Keywords: Semirings · Routing algebra · Loop free routing · Bellman Ford · Diffusing computation

1 Introduction

A routing loop is a common problem in computer networks. This happens when the computed path towards a particular destination contains a loop due to inaccurate routing tables, thus packets destined to this destination will loop endlessly unless they are eventually dropped. This is especially true in early distance vector protocols such as routing information protocol (RIP) which is prone to counting to infinity. In link state protocols such as open shortest first (OSPF) and intermediate system to intermediate system (IS-IS), routing loops can still occur [20], however, they are short lived, as they disappear as soon as the information about the new topology is flooded across the network, and all the routers synchronize their link state databases.

New distance vector protocols such as border gateway protocol (BGP) and enhanced interior gateway routing protocol (EIGRP) are designed to prevent loops before they occur. EIGRP protocol is a Cisco proprietary protocol based on the interior gateway protocol (IGRP). EIGRP was converted to an open

V. M. Vishnevskiy et al. (Eds.): DCCN 2020, LNCS 12563, pp. 161–175, 2020.
https://doi.org/10.1007/978-3-030-66471-8_14

standard in 2013 and was published in RFC 7868 in 2016 [22]. EIGRP employs diffusing update algorithm (DUAL). The convergence time with DUAL rivals that of any other existing routing protocol [3]. EIGRP employs SNC (Source Node Conditions), which is one of the sufficient conditions for loop freedom. This condition is met when the neighbor's advertised distance for a particular destination is strictly less than the feasible distance for that destination [13,20]. Other sufficient conditions includes DIC (Distance Increase Condition), and CSC (Current Successor Condition).

DUAL is proved to be loop free at every instance of time, and to converge in a finite time after the occurrence of link-cost changes [20]. However, EIGRP uses the same composite metric in IGRP which utilizes the available bandwidth, delay, load utilization, and link reliability for metric calculations. Composite metric is modeled in [1] using an algebraic construct called the functional product, where the authors showed that EIGRP's metric is non monotonic resulting that EIGRP solves a local optimal solution as opposed to the global optimal solution, which is solved by the classical shortest path problem. To the best of our knowledge, no formal proof of correctness of DUAL with a non monotonic metric has been given yet.

In this paper we investigate the concept of loop free routing with a generic metric using the matrix model with semirings. We provide an algebraic representation for the sufficient conditions of loop free routing in the semiring model. We also explore the relation between them and provide an algebraic proof for their correctness. We also introduce the concept of monotone routing. It was seen that if the routing was decreasing (or increasing) in one iteration, it will continue doing so in the next iterations as long as the topology is fixed. The generality of the model helps in showing some theory and guidelines on the design of new loop free routing protocols in the presence and absence of monotonicity.

This paper is organized as follows: next section surveys the literature in this field, an overview of the semiring model is presented in Sect. 3, sufficient conditions for loop free routing are presented in Sect. 4, monotone routing is introduced in Sect. 5, and loop free routing in a non-monotone algebra is discussed in Sect. 6, and Sect. 7 concludes the paper.

2 Related Works

Recently, there have been some efforts to apply formal methods specifically algebraic specifications to existing routing protocols. In [5] and [2], Sobrinho developed an algebraic framework for investigating the convergence properties of distance vector and path vector protocols. It is shown that "monotonicity" (a property related to the inflationary property presented in Sect. 3) implies protocol convergence in every network but not necessarily to a "global optimal" (the notion of optimality is defined in Sect. 3 also). However, "isotonicity" (which is a property related to distributivity, we refer to it as monotonicity in this article) assures convergence onto global optimal paths when the protocol converges. For link state protocols Sobrinho presented a more specific less general algebraic frame work [4]. It was seen that local optimality rather than global optimality is more appropriate for modeling inter-domain routing protocol such as BGP [2,5–7].

Griffin and Sobrinho proposed metarouting as a means of defining routing protocols in a high-level and declarative manner [8]. Metarouting was based on Sobrinho's algebra. Sobrinho's model uses a partial order relation to construct the algebra. There is another approach to model path problems using algebraic structures called *semirings* ([9], and [10] contains modern surveys of this area). The two models can be related to each other [11]. Routing operations in the semiring model become matrix operations.

Diffusing Computation concept was first proposed by Dijkastra and Scholten [12]. After that, most of the work on loop free routing has been done by Garcia who presented DUAL [13,20] and proved its loop freedom. DUAL was adopted later in EIGRP [3]. DIC was discussed in the literature prior to the work of Jaffe and Moss, however, they were first to prove that DIC is sufficient for loop freedom [14]. CSC and SNC were proposed and proved correct by Garcia [13,20]. Gouda and Schneider [23] have provided a graph theoretical approach to show that IGRP and EIGRP protocol do not behave as expected because of the composite metric which is not nonmonotonic in general. Algebraic theory of routing was also used in more recent studies in [1,15,17,24]. In [25] the author has developed algorithms to solve for optimal solutions of the shortest path problem for IGRP's like metrics.

3 Semirings and Graphs

We briefly describes in this section how semirings can be related to the shortest path problem. More details can be found in [21]. Semirings are structures of the form $(S, \oplus, \otimes, \overline{0}, \overline{1})$, where S is a non-empty and non-trivial set and the axioms in Table 1 hold. Semirings differ from rings in that the additive operation do not need to admit inverses. That is, $(S, \oplus)$ is only required to be a monoid, not a group [9]. This allows us to define a non-trivial "natural order" on S:

$$a \leq_{\oplus} b \equiv a = a \oplus b$$

We can also define the strict version of this order

$$a <_{\oplus} b \equiv a = a \oplus b \neq b$$

We need $\oplus$ to be idempotent ($\forall a \in S : a \oplus a = a$) so the relation $\leq_{\oplus}$ becomes reflexive. In this case $\leq_{\oplus}$ becomes a partial order. If $\oplus$ is also selective ($\forall a, b \in S : a \oplus b \in \{a, b\}$) then $\leq_{\oplus}$ becomes total order. In addition, we define a "canonical order" on the semi-group $(S, \oplus)$

$$a \trianglelefteq_{\oplus} b \equiv \exists c \in S | a = b \oplus c$$

If $\oplus$ is commutative and idempotent then

$$\forall a, b \in S : a \trianglelefteq_{\oplus} b \Leftrightarrow a \leq_{\oplus} b$$

We define monotonicity from the left and the right respectively

$$a \leq_{\oplus} b \Rightarrow c \otimes a \leq_{\oplus} c \otimes b \tag{3.1}$$

$$a \leq_{\oplus} b \Rightarrow a \otimes c \leq_{\oplus} b \otimes c \tag{3.2}$$

Table 1. Axioms for semirings

Axiom	Explanation
$\oplus$ Associative	$a \oplus (b \oplus c) = (a \oplus b) \oplus c$
$\oplus$ Commutative	$a \oplus b = b \oplus a$
$\oplus$ Identity	$a \oplus \overline{0} = \overline{0} \oplus a = a$
$\otimes$ Associative	$a \otimes (b \otimes c) = (a \otimes b) \otimes c$
$\otimes$ Identity	$a \otimes \overline{1} = \overline{1} \otimes a = a$
$\otimes$ Annihilator	$a \otimes \overline{0} = \overline{0} \otimes a = \overline{0}$
Left distributivity	$a \otimes (b \oplus c) = (a \otimes b) \oplus (a \otimes c)$
Right distributivity	$(a \oplus b) \otimes c = (a \otimes c) \oplus (b \otimes c)$

Monotonicity holds in semirings due to distributivity. In addition, we define the left and right strict inflationary property respectively

$$\forall a, b \in S, a \neq \overline{0}, b \neq \overline{1} : a <_{\oplus} b \otimes a \tag{3.3}$$

$$\forall a, b \in S, a \neq \overline{0}, b \neq \overline{1} : a <_{\oplus} a \otimes b \tag{3.4}$$

The semiring $(S, \oplus, \otimes, \overline{0}, \overline{1})$ can be used to define a semiring $(\mathbb{M}_n(S), \oplus, \otimes, J, I)$ of $n \times n$ matrices. The J and I matrices are:

$$J(i,j) = \overline{0} \tag{3.5}$$

$$I(i,j) = \begin{cases} \overline{1}, & \text{if } i = j; \\ \overline{0}, & \text{otherwise.} \end{cases} \tag{3.6}$$

The classical shortest path problem can be modeled with semirings [16]. Given a semiring $(S, \oplus, \otimes, \overline{0}, \overline{1})$ and a graph $G = (V, E)$, a *weight function* is a mapping $w : E \rightarrow S - \{\overline{0}\}$. The graph can be presented using what we call *weighted adjacency matrix*.

$$A(i,j) = \begin{cases} w(e), & \text{if } e = (i,j) \in E; \\ \overline{0}, & \text{otherwise.} \end{cases} \tag{3.7}$$

A path $p = v_1, v_2, \cdots, v_{k+1}$ of length k is a sequence of nodes such that $(v_m, v_{m+1}) \in E$ for each $m, 1 \leq m \leq k$. The weight of the path p is given by

$$w(p) = w(v_1, v_2) \otimes w(v_2, v_3) \otimes \cdots \otimes w(v_k, v_{k+1}) \tag{3.8}$$

Null paths are denoted by ϵ and are given the weight $\overline{1}$. Non-existing paths are given the weight $\overline{0}$. The power of a matrix A is defined inductively:

$$A^0 = I \tag{3.9}$$

$$A^{k+1} = A \otimes A^k \tag{3.10}$$

$$A^{(0)} = I \tag{3.11}$$

$$A^{(k+1)} = A^{k+1} \oplus A^{(k)} \tag{3.12}$$

Let $P^k(i,j)$ be the set of paths from node i to j which has exactly k arcs. $P^{(k)}(i,j)$ is the set of paths from node i to j with k arcs at most. $P(i,j)$ is the set of all possible paths from node i to j. We say that p is a simple path if it does not have loops. We denote by $SP(i,j)$ the set of all possible simple paths from node i to j. Additionally, $SP^{(k)}(i,j)$ is the set of all simple paths from i to j with at most k arcs of length.

Theorem 1.

$$A^k(i,j) = \bigoplus_{p \in P^k(i,j)} w(p) \tag{3.13}$$

$$A^{(k)} = \bigoplus_{p \in P^{(k)}(i,j)} w(p) \tag{3.14}$$

If there exists a $q \geq 0$ such that $A^{(q+1)} = A^{(q)}$ then $\forall k \geq q : A^{(k)} = A^{(q)}$. We say then that A is q-stable. We say also that $A^{(k)}$ **converges** to $A^* = A^{(q)}$, and we call A^* the **closure** matrix of A.

$$A^* = \bigoplus_{k \geq 0} A^{(k)} = \bigoplus_{p \in P(i,j)} w(p) \tag{3.15}$$

We interpret (3.15) as: A^* is the solution to the **global optimal** path problem. It remains to see when this solution exists. Theorem 2 gives that the inflationary property is sufficient for global optimality in semirings, and the solution is reached with at most $n-1$ iterations. Theorem 3 states that the solution will be the same when constructed using only simple paths. Since only single paths are used then we will reach the solution after $d \leq n-1$ iterations, where d is the diameter of the graph (the number of arcs in the longest path in the graph).

Theorem 2. *Let* $(S, \oplus, \otimes, \overline{0}, \overline{1})$ *be an idempotent semiring, and let* $\overline{1}$ *be an annihilator for* $\oplus$*. Then, all weighted adjacency matrices in* $(\mathbb{M}_n(S), \oplus, \otimes, J, I)$ *are* $(n-1)$*-stable [21].*

Proof. Let $A \in \mathbb{M}_n(S)$ where A is the weighted adjacency matrix of graph $G = (V, E)$. It is sufficient to show that $A^{(n)} = A^{(n-1)}$.

Let $(a,b) \in S^2$ then

$$a \oplus (a \otimes b) = (a \otimes \overline{1}) \oplus (a \otimes b) = a \otimes (\overline{1} \oplus b) = a \otimes \overline{1} = a$$

This means that $\forall(a,b) \in S^2 : a \leq_{\oplus} a \otimes b$, which is the left inflationary property. Similarly, $\forall(a,b) \in S^2 : a \leq_{\oplus} b \otimes a$—right inflationary property holds. For the inflationary property to hold in the classical shortest path problem (semiring $(\mathbb{R} \cup \{\infty\}, \min, +, \infty, 0)$), we should restrict the set to $\mathbb{R}^+$, so we use only positive number for path weights.

Let $p \in P^n(i,j), i \neq j$. Since all the nodes in the graph which are different from i and j are $n-2$. Then any simple path will have at most $n-1$ arc. This means that p is not a simple path. So it contains at least one loop. Let the loop be at node k, $p_{i,k}$ is the sub-path from i to k, $p_{k,k}$ the loop at node k, $p_{k,j}$ the sub-path from k to j, and finally p' be the path after removing the loop, which can be obtained from joining $p_{i,k}$ and $p_{k,j}$. We have

$$
\begin{aligned}
w(p) &= w(p_{i,k}) \otimes w(p_{k,k}) \otimes w(p_{k,j}) \\
w(p') &= w(p_{i,k}) \otimes w(p_{k,j}) \\
w(p) \oplus w(p') &= (w(p_{i,k}) \otimes w(p_{k,k}) \otimes w(p_{k,j})) \oplus (w(p_{i,k}) \otimes w(p_{k,j})) \\
&= w(p_{i,k}) \otimes ((w(p_{k,k}) \otimes w(p_{k,j})) \oplus w(p_{k,j})) \\
&= w(p_{i,k}) \otimes w(p_{k,j}) \\
&= w(p')
\end{aligned}
$$

Thus, if $p \in P^n(i,j)$, there exists a path $p' \in P^{(n-1)}(i,j)$ so that $w(p) \oplus w(p') = w(p')$. By reordering the elements of the $\oplus$ sum—as $\oplus$ is commutative. We get

$$
\begin{aligned}
A^{(n)}(i,j) &= \bigoplus_{q \in P^{(n)}(i,j)} w(p) \\
&= \left(\bigoplus_{q \in P^{(n)}(i,j), q \neq p, q \neq p'} w(q) \right) (\oplus w(p) \oplus w(p')) \\
&= \left(\bigoplus_{q \in P^{(n)}(i,j), q \neq p, q \neq p'} w(q) \right) \oplus w(p') \\
&= \bigoplus_{q \in P^{(n)}(i,j), q \neq p} w(q)
\end{aligned}
$$

This means we can exclude p from the previous sum without affecting the result. Since we can do this for any $p \in P^n(i,j)$ then we get

$$
A^{(n)}(i,j) = \bigoplus_{q \in P^{(n)}(i,j), p \notin P^n(i,j)} w(p) \tag{3.16}
$$

But $P^{(n)}(i,j) - P^n(i,j) = P^{(n-1)}(i,j)$, then

$$
A^{(n)}(i,j) = \bigoplus_{q \in P^{(n-1)}(i,j)} w(p) = A^{(n-1)}(i,j)
$$

So $\mathbb{M}_n(S)$ is $(n-1)$-stable.

Theorem 3. *Let* $(S, \oplus, \otimes, \overline{0}, \overline{1})$ *be an idempotent semiring, and let* $\overline{1}$ *be an annihilator for* $\oplus$. *Then,*

$$A^{(k)}(i,j) = \bigoplus_{q \in SP^{(k)}(i,j)} w(q) \tag{3.17}$$

Proof. Let $p \in P^{(k)}(i,j) - SP^{(k)}(i,j)$. This means that p contains at least one loop. Therefore, we can find a path $p' \in P^{(k)}(i,j)$ same as we did in Theorem 2, which is formed by eliminating the loop from the path p, such that $w(p) \oplus w(p') = w(p')$.

$$A^{(k)}(i,j) = \bigoplus_{q \in P^{(k)}(i,j), q \neq p} w(q)$$

We can do the same for all paths in $P^{(k)}(i,j) - SP^{(k)}(i,j)$ because the set $P^{(k)} - SP(k)$ is finite. Then

$$A^{(k)}(i,j) = \bigoplus_{q \in SP^{(k)}(i,j)} w(q)$$

Bellman-Ford algorithm can be modeled using the following iteration [11,15, 17]

$$A^{\langle 0 \rangle} = I \tag{3.18}$$

$$A^{\langle k+1 \rangle} = A \otimes A^{\langle k \rangle} \oplus I \tag{3.19}$$

We call this iteration *distance vector iteration.* If A is the weighted adjacency matrix of graph $G = (V, E)$, then (3.19) can be written as

$$A^{\langle k+1 \rangle}(i,j) = I(i,j) \oplus \bigoplus_{q \in V} A(i,q) \otimes A^{\langle k \rangle}(q,j) \tag{3.20}$$

If $N(i) \subseteq V$ is the set of node i neighbors, then we can restrict the sum in (3.20) to the set $N(i)$ because $A(i,q) = \overline{0}$ when $q \notin N(i)$

$$A^{\langle k+1 \rangle}(i,j) = I(i,j) \oplus \bigoplus_{q \in N(i)} A(i,q) \otimes A^{\langle k \rangle}(q,j) \tag{3.21}$$

If we assume that the left strict inflationary property holds, and we do not allow links to have the weight $\overline{1}$, then we say the routing is **loop-free** for static topology i.e. only simple paths will be considered. This can be seen easily from Theorem 3. If $\otimes$ is cancellative (see (3.22) and (3.23)), and $\overline{1}$ is annihilator for $\oplus$ then the left and right strict inflationary properties hold.

$$a \neq \overline{0}, a \otimes b = a \otimes c \Rightarrow b = c \tag{3.22}$$

$$c \neq \overline{0}, a \otimes c = b \otimes c \Rightarrow a = b \tag{3.23}$$

We lose the loop free attribute in dynamic topology. If some of the links fail, then loops might occur causing the counting to convergence or counting to infinity problem. This can be explained easily algebraically [17]. Suppose that we start the distance vector iteration with an arbitrary matrix M rather than I. M represents the matrix of best path weights before the change in the topology, while A is the new weighted adjacency matrix after the change in the topology.

$$A_M^{\langle 0\rangle} = M \tag{3.24}$$

$$A_M^{\langle k+1\rangle} = A \otimes A_M^{\langle k\rangle} \oplus I, k \geq 1 \tag{3.25}$$

We find by induction that for $k \geq 1$

$$A_M^{\langle k\rangle} = A^k \otimes M \oplus A^{\langle k-1\rangle} \tag{3.26}$$

If A is $(n-1)$-stable and for $k \geq n$ we have

$$A_M^{\langle k\rangle} = A^k \otimes M \oplus A^* \tag{3.27}$$

While the term A^* may be reached soon, but it could be that the term $A^k \otimes M$ is preferred. So, the iteration will continue until $A^* \leq_{\oplus} A^k \otimes M$. This might not happen if there is no longer a path that connects the nodes i and j [17] – counting to infinity. Solutions to this problem includes limiting the number of hops to 15 like in the RIP protocol [18]. This hop limit make networks connected with more than 15 routers unreachable. In BGP the whole path is advertised and the algorithm is modified to consider only simple paths [19]. Another solution is by using the DUAL algorithm which is proved to be loop free [13,20].

4 Sufficient Conditions for Loop Freedom in Distance Vector Routing

We will assume in this section that $(S, \oplus, \otimes, \overline{0}, \overline{1})$ is a left strict inflationary, selective, and idempotent semiring. We call this a linear increasing semiring (LISR). In addition, $\overline{1}$ is not allowed as a link weight. We need selectivity to be able to define the successor node, which is the next hop node selected by a node i that corresponds to the best path toward a destination node j.

We will modify the model for distance vector routing in (3.18) and (3.19) to take account of topology changes. The change in the topology will be modeled as a change in the weighted adjacency matrix A. We will call the matrix computed by distance vector iteration "the routing matrix". We say that a node $q \neq i$ is a downstream node for i in routing toward j if the path selected by i passes through q. And we say that q is an upstream node for i if i is a downstream node for q. Equation (4.1) represents distance vector routing model in a dynamic topology. M_k represents the routing matrix (best paths weights) at stage k and A_{k+1} is the new weighted adjacency matrix that captures the new topology (it would be same as A_k if no topology changes occur at stage $k+1$ of routing)

$$M_{k+1} = A_{k+1} \otimes M_k \oplus I \tag{4.1}$$

The successor to i in routing toward j $(i \neq j)$ at stage $k+1$ is then a node s selected by i such that

$$M_{k+1}(i,j) = \bigoplus_{q \in V} A_{k+1}(i,q) \otimes M_k(q,j) = A_{k+1}(i,s) \otimes M_k(s,j) \qquad (4.2)$$

This is true because $\oplus$ is a selective operation. The importance of loop free routing is that it guarantees convergence in a dynamic topology as stated by Theorem 4.

Theorem 4. *If we arrive in routing to the matrix M. And the topology is settled on a weighted adjacency matrix A (no further change in the topology). If the routing is loop free afterwards (only simple paths are inspected), then the routing will converge to A^*.*

Proof. In (3.27), let $k \geq n$, then A^k will be equivalent to J as there is no simple path with a number of arcs greater than $n-1$ and the algorithm is assumed to consider only simple paths. Then, $A^{\langle k \rangle} = J \otimes M \oplus A^* = A^*$.

There are 3 sufficient conditions for loop free routing [20]. We express them algebraically in the following:

DIC *Distance Increase Condition* Node i is free to change its successor toward j to a node s that satisfies (4.2) if the distance is not increased i.e. $M_{k+1}(i,j) \leq_{\oplus} M_k(i,j)$. Otherwise node i must maintains its current successor.

CSC *Current Successor Condition* Node i is free to change its successor toward j to a node s that satisfies (4.2) if $M_k(s,j) \leq_{\oplus} M_{k-1}(s',j)$, where s' is the successor in the previous stage (step k). Otherwise node i must maintains its current successor.

SNC *Source Node Condition* Node i is free to change its successor toward j to a node s that satisfies (4.2) if $M_k(s,j) <_{\oplus} M_k(i,j)$. Otherwise node i must maintains its current successor.

In the rest of this section we will provide an algebraic proof for the correctness of these 3 conditions.

Theorem 5. *The SNC is a sufficient condition for loop free routing.*

Proof. In (4.1), suppose we arrive to a routing matrix M. Now, let i chooses a successor s_1 in the next step of routing toward j, so that the loop $(i, s_1, s_2, \cdots, s_k, i)$ will be formed. If all the nodes take into account the SNC when choosing their successors, then we have

$$\begin{aligned} M(s_1,j) &<_{\oplus} M(i,j) \\ M(s_2,j) &<_{\oplus} M(s_1,j) \\ &\vdots \\ M(s_k,j) &<_{\oplus} M(s_{k-1},j) \\ M(i,j) &<_{\oplus} M(s_k,j) \end{aligned}$$

We have

$$M(i,j) <_\oplus M(s_k,j) <_\oplus M(s_{k-1},j) <_\oplus \cdots <_\oplus M(s_2,j) <_\oplus M(s_1,j) <_\oplus M(i,j)$$

leading to the contradiction $M(i,j) <_\oplus M(i,j)$. Then no loop can be formed if all the nodes respect SNC conditions while choosing their successors.

Theorem 6. *DIC implies SNC.*

Proof. In (4.1), suppose we arrive to a routing matrix M and the weighted adjacency matrix is A. Suppose that node i chooses node s as a successor in routing toward j. Then, the new distance to j is $A(i,s) \otimes M(s,j)$, if node i takes into account the DIC when selecting its successor. Then we have $A(i,s) \otimes M(s,j) \leq_\oplus M(i,j)$. We have also $M(s,j) <_\oplus A(i,s) \otimes M(s,j)$ since $A(i,s) \neq \overline{1}$. Therefore, $M(s,j) <_\oplus M(i,j)$ and SNC holds.

Theorem 7. *CSC implies SNC.*

Proof. In (4.2), suppose that node i selects s as a successor in the $k+1$ step of routing toward j. Let s' be the successor in the k step. Now if CSC holds then $M_k(s,j) \leq_\oplus M_{k-1}(s',j)$. We have then $M_k(i,j) = A_k(i,s') \otimes M_{k-1}(s',j)$. Therefore, $M_{k-1}(s',j) <_\oplus M_k(i,j)$ and consequently $M_k(s,j) <_\oplus M_k(i,j)$.

Corollary 1. *DIC and CSC are sufficient conditions for loop free routing.*

5 Monotone Routing

Theorem 8 states that if the routing matrix increases (or decreases) in one distance vector iteration and the topology is fixed, then it will continue increasing (or decreasing) in the next iterations. Theorem 10 proves that the decreasing routing will eventually converge in a finite number of distance vector iterations.

Theorem 8. *Let $(S, \oplus, \otimes, \overline{0}, \overline{1})$ be an idempotent semiring. In (3.24) and (3.25)*

1. *If $A_M^{\langle 1 \rangle} \leq_\oplus M$ then $\forall k \geq 0 : A_M^{\langle k+1 \rangle} \leq_\oplus A_M^{\langle k \rangle}$*
2. *If $M \leq_\oplus A_M^{\langle 1 \rangle}$ then $\forall k \geq 0 : A_M^{\langle k \rangle} \leq_\oplus A_M^{\langle k+1 \rangle}$*

Proof. We will prove 1 by induction. 2 can be proven similarly. Let us assume the inequality in 1 holds for k then

$$A_M^{\langle k+1 \rangle} \oplus A_M^{\langle k \rangle} = A_M^{\langle k+1 \rangle}$$

We apply A on both sides. The distributivity of $\otimes$ on $\oplus$ implies

$$\left(A \otimes A_M^{\langle k+1 \rangle}\right) \oplus \left(A \otimes A_M^{\langle k \rangle}\right) = A \otimes A_M^{\langle k+1 \rangle}$$

Then we have

$$\left(\left(A \otimes A_M^{\langle k+1 \rangle}\right) \oplus I\right) \oplus \left(\left(A \otimes A_M^{\langle k \rangle}\right) \oplus I\right) = \left(A \otimes A_M^{\langle k+1 \rangle}\right) \oplus I$$

Because $I \oplus I = I$. Then

$$A_M^{\langle k+2\rangle} \oplus A_M^{\langle k+1\rangle} = A_M^{\langle k+2\rangle}$$

So the inequality holds for $k+1$ then it holds $\forall k \geq 0$.

In general, the routing might be decreasing for some destination nodes and increasing for some others, nevertheless, routing toward a node is independent from routing toward other nodes as stated by Observation 9.

Observation 9. *Routing toward a node j is independent from routing toward other nodes.*

This is true because the values of column $M_{k+1}(-,j)$ are constructed using only the values in column $M_k(-,j)$. We can, therefore, restrict our focus on one destination node j.

Theorem 10. *Let $(S, \oplus, \otimes, \overline{0}, \overline{1})$ be a left inflationary idempotent semiring. In (3.24) and (3.25) if $A_M^{\langle 1\rangle} \leq_\oplus M$ then the routing will converge to $A^* \otimes (A^n \otimes M \oplus I)$.*

Proof. From Theorem 2 we know that A will be $(n-1)$-stable. From (3.27) we have, for $k \geq 0$

$$A_M^{\langle k+n\rangle} = (A^{k+n} \otimes M) \oplus A^*$$

From Theorem 8 we have

$$A_M^{\langle k+n\rangle} \leq_\oplus A_M^{\langle k+n-1\rangle} \leq_\oplus \cdots \leq_\oplus A_M^{\langle 1+n\rangle} \leq_\oplus A_M^{\langle n\rangle}$$

Then

$$\begin{aligned}
A_M^{\langle k+n\rangle} &= A_M^{\langle n\rangle} \oplus A_M^{\langle 1+n\rangle} \oplus \cdots \oplus A_M^{\langle k+n\rangle} \\
&= ((A^n \otimes M) \oplus A^*) \oplus ((A^{n+1} \otimes M) \oplus A^*) \oplus \cdots \oplus ((A^{n+k} \otimes M) \oplus A^*) \\
&= (A^n \otimes M) \oplus (A^{n+1} \otimes M) \oplus \cdots \oplus (A^{n+k} \otimes M) \oplus A^* \\
&= ((I \oplus A \oplus A^2 \oplus \cdots \oplus A^k) \otimes (A^n \otimes M)) \oplus A^* \\
&= \left(A^{(k)} \otimes A^n \otimes M\right) \oplus A^*
\end{aligned}$$

Then for $k \geq n-1$ we have

$$\begin{aligned}
A_M^{\langle k+n\rangle} &= (A^* \otimes A^n \otimes M) \oplus A^* \\
&= A^* \otimes ((A^n \otimes M) \oplus I)
\end{aligned}$$

Corollary 2. *Let $(S, \oplus, \otimes, \overline{0}, \overline{1})$ be a left strict inflationary, selective, and idempotent semiring, and $\overline{1}$ is not allowed as a link weight. In (3.24) and (3.25) if $A_M^{\langle 1\rangle} \leq_\oplus M$ then the routing will converge to A^*.*

This is true because the routing will be then loop free because of DIC. This gives the idea that in order to make the routing loop free after a topology change we have to manipulate the routing matrix M in a way so that the resulting routing will be decreasing. Alternatively, in semirings, if A is stable then A^* is the "largest" solution (fixed point) of the equation $X = (A \otimes X) \oplus I$ [21]. This is because $X = (A^{k+1} \otimes X) \oplus A^{(k)}$ which can be proved inductively. Thus, after a topology change we need to increase the starting matrix M so that routing decrease to A^*. Noting that when the inflationary property holds in semirings, then only loops can cause an increase in routing.

6 Loop Free Routing in a Non-monotone Algebra

Monotonicity is not needed in the proof that SNC is sufficient condition for loop free routing. We can also prove the correctness of DIC and CSC without monotonicity, where we assume the strict inflationary property instead.

Decreasing routing is the foundation stone for loop free routing. However, if the underlying algebra is not monotone, then routing will not be necessarily decreasing even in a static topology. Let us consider the EIGRP metric as an example (see Eq. 6.1 [1,22]). We will use the default K values ($K_1 = K_3 = 1, K_2 = K_4 = K_5 = 0$). The metric then will take the form (bw, d). In the simple graph presented in Fig. 1, Let the distance from node k to j be $(2 \times 10^6, 1)$.

$$
\begin{aligned}
&f : \mathbb{N}^{\infty} \times \mathbb{N}^{\infty} \times \mathbb{N}[1, 255] \times \mathbb{N}[1, 255] \rightarrow \mathbb{R}_+^{\infty} \\
&f(bw, del, ld, rel) = 256 \times (K_1 \times \frac{10^7}{bw} + K_2 \times \frac{10^7}{bw \times (256 - ld)} + K_3 \times del) \\
&\qquad \times \begin{cases} 1, & \text{if } K_5 = 0 \\ \frac{K_5}{K_4 + rel}, & \text{otherwise} \end{cases}
\end{aligned} \tag{6.1}
$$

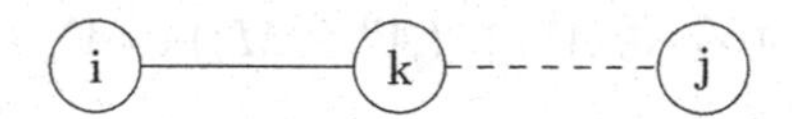

Fig. 1. Simple graph

After applying the cost function on this distance we get

$$f(2 \times 10^6, 1) = 256 \times (5 + 1) = 1530$$

Let us assume also that the cost of the link from i to k is $(2 \times 10^6, 1)$. The computed distance in node i to node j, in this case, is

$$(\min(2 \times 10^6, 2 \times 10^6), 1 + 1) = (2 \times 10^6, 2)$$

Applying the cost function on the above distance we get

$$f(2\times 10^6, 2) = 256 \times (5+2) = 1792$$

Now, let us assume that node k chooses another path towards j. Let the distance of the new path be $(5\times 10^6, 3)$. This distance is preferred to the previous distance of k, because

$$f(5\times 10^6, 3) = 256 \times (2+3) = 1280$$

The computed distance from i to j becomes

$$(\min(5\times 10^6, 2\times 10^6), 3+1) = (2\times 10^6, 4)$$

Applying the cost function on the above distance we get

$$f((2\times 10^6, 4)) = 256 \times (5+4) = 2304$$

Fig. 2. Simple graph 2

From the above discussion, the distance in node k decreased, however, the distance in i increased. It is clear, then, that the routing is not decreasing in this case even when the topology is static. In loop free routing, this will cause a diffusing computation in node i if we are using DIC as a feasibility condition. We can find similar scenarios for the other sufficient conditions for loop free routing. For example, if we use CSC, and there is a node u which uses node i as a successor in routing towards j (See Fig. 2). The distance increase in node i will cause a diffusing computation in u. If we use SNC, and the distance in i after the increase becomes "greater" than the distance in u, then there will be a diffusing computation in u. Note that if we are using the original distributed Bellman-Ford algorithm in the latter case, node i will choose node u as a successor causing a routing loop. This means that DBF algorithm is not necessarily loop free in a static topology if the underlying algebra is not monotone, which explains why we are not bounded by $n-1$ iteration to reach convergence in non-monotone algebra.

In DUAL the affected node after topology changes starts a diffusing computations. So, that all upstream nodes change their distance to a proper value. As a result, routing will decrease to convergence when no other topology change occurs, and no future diffusing computation will occur [20]. This is true if monotonicity holds, as Garcia used original shortest paths problem in his discussion. However, in non monotone algebra, like (EIGRP metric), there is no guarantee that no diffusing computation will occur after the original diffusing computation terminates. However, the new diffusing computation if it occurs, and assuming the new topology did not change, it will affect only a subset of nodes from the original upstream nodes that were affected in the first diffusing computation. This means that each subsequent diffusing computation will affect a smaller set of nodes. Then, after a while no diffusing computation will ever happen, and the protocol will reach an equilibrium point and converge. The only difference is that the protocol will converge to a local optimal solution rather than a global optimal one due to the loss of monotonicity. The optimality is in the sense that a node can not change its successor to a better path considering neighboring nodes' choices. Algebraically, the local optimal solution is a fixed point for the equation $L = (A \otimes L) \oplus I$.

7 Conclusion

We have used the matrix model with semirings to investigate loop free routing. We have shown that, when the strict inflationary property holds, Bellman-Ford will calculate loop-free routing paths in a static topology. However, in a dynamic topology, routing loops may occur.

Loop free routing conditions have been presented algebraically and proved to be correct. We have also shown the relations between these conditions. In Sect. 5 we have introduced the concept of monotone routing. We have shown that if the routing matrix decrease (or increase) in one distance vector iteration, it will continue decreasing (or increasing) in the next iterations as long as the topology is fixed. Thus, for loop free routing we have to manipulate the routing matrix in order to make the routing decreasing so that DIC holds.

Finally, we have discussed the effect of loss of monotonicity. We have shown that DBF will no longer be loop free even in a static topology, and we are no longer limited to $n - 1$ iterations to achieve optimality (in this case, local optimality). We have also demonstrated that diffusion computations can occur even in a static topology when using DUAL, however, the algorithm will converge to a local optimum.

References

1. Khayou, H., Sarakbi, B.: A validation model for non-lexical routing protocols. J. Netw. Comput. Appl. **98**, 58–64 (2017)
2. Sobrinho, J.L.: An algebraic theory of dynamic network routing. IEEE/ACM Trans. Netw. **13**(5), 1160–1173 (2005)
3. Albrightson, R., Garcia-Luna-Aceves, J.J., Boyle, J.: EIGRP- a fast routing protocol based on distance vectors. Proc. Networld/Interop **94** (1994)
4. Sobrinho, J.L.: Algebra and algorithms for QoS path computation and hop-by-hop routing in the Internet. IEEE/ACM Trans. Netw. **10**(4), 541–550 (2002)
5. Sobrinho, J.L.: Network routing with path vector protocols: theory and applications. In: Proceedings of the 2003 Conference on Applications, Technologies, Architectures, and Protocols for Computer Communications (2003)
6. Gurney, A.J.T.: Construction and verification of routing algebras. Diss. University of Cambridge (2010)

7. Sobrinho, J.L., Griffin, T.G.: Routing in equilibrium. Math. Theory Netw. Syst. (2010)
8. Griffin, T.G., Sobrinho, J.L.: Metarouting. In: Proceedings of the 2005 Conference on Applications, Technologies, Architectures, and Protocols for Computer Communications (2005)
9. Gondran, M., Minoux, M.: Graphs, Dioids and Semirings: New Models and Algorithms, vol. 41. Springer, New York (2008). https://doi.org/10.1007/978-0-387-75450-5
10. Baras, J.S., Theodorakopoulos, G.: Path problems in networks. Synth. Lect. Commun. Netw. **3**(1), 1–77 (2010)
11. Griffin, T.G., Gurney, A.J.T.: Increasing bisemigroups and algebraic routing. In: Berghammer, R., Möller, B., Struth, G. (eds.) RelMiCS 2008. LNCS, vol. 4988, pp. 123–137. Springer, Heidelberg (2008). https://doi.org/10.1007/978-3-540-78913-0_11
12. Dijkstra, E.W., Scholten, C.S.: Termination detection for diffusing computations. Inf. Process. Lett. **11**(1), 1–4 (1980)
13. Garcia-Luna-Aceves, J.J.: A unified approach to loop-free routing using distance vectors or link states. ACM SIGCOMM Comput. Commun. Rev. **19**(4), 212–223 (1989)
14. Jaffe, J., Moss, F.: A responsive distributed routing algorithm for computer networks. IEEE Trans. Commun. **30**(7), 1758–1762 (1982)
15. Dynerowicz, S., Griffin, T.G.: On the forwarding paths produced by internet routing algorithms. In: 2013 21st IEEE International Conference on Network Protocols (ICNP). IEEE (2013)
16. Mohri, M.: Semiring frameworks and algorithms for shortest-distance problems. J. Autom. Lang. Comb. **7**(3), 321–350 (2002)
17. Alim, M.A., Griffin, T.G.: On the interaction of multiple routing algorithms. In: Proceedings of the Seventh COnference on Emerging Networking EXperiments and Technologies (2011)
18. Malkin, G.: RFC 2453: RIP Version 2 (1998)
19. Rekhter, Y., Li, T., Hares, S.: RFC 4271: A Border Gateway Protocol 4 (BGP-4) (2006)
20. Garcia-Lunes-Aceves, J.J.: Loop-free routing using diffusing computations. IEEE/ACM Trans. Netw. **1**(1), 130–141 (1993)
21. Griffin, T.: An Algebraic approach to internet routing. University of Cambridge. Lecture Notes (2010)
22. Savage, D., et al.: RFC 7868: Cisco's Enhanced Interior Gateway Routing Protocol (EIGRP) (2016)
23. Gouda, M.G., Schneider, M.: Maximizable routing metrics. IEEE/ACM Trans. Netw. **11**(4), 663–675 (2003). https://doi.org/10.1109/TNET.2003.815294
24. Sobrinho, J.L.: Correctness of routing vector protocols as a property of network cycles. IEEE/ACM Trans. Netw. **25**(1), 150–163 (2017). https://doi.org/10.1109/TNET.2016.2567600
25. Saad, M.: Non-isotonic routing metrics solvable to optimality via shortest path. J. Comput. Netw. **145**, 89–95 (2018)

Queueing System with Two Unreliable Servers and Backup Server as a Model of Hybrid Communication System

Valentina Klimenok[1], Alexander Dudin[1(✉)], and Vladimir Vishnevsky[2]

[1] Department of Applied Mathematics and Computer Science, Belarusian State University, 220030 Minsk, Belarus
{klimenok,dudin}@bsu.by

[2] Institute of Control Sciences of Russian Academy of Sciences and Closed Corporation "Information and Networking Technologies", Moscow, Russia
vishn@inbox.ru

Abstract. In this paper, we analyze a queueing system with two main unreliable servers and backup reliable server. The input flow is defined by the $BMAP$ (Batch Markovian Arrival Process). Heterogeneous breakdowns arrive to the main servers according to a $MMAP$ (Marked Markovian Arrival Process). Service times and repair times have PH (Phase type) distribution. The queueing system under consideration is an adequate model of operation of hybrid communication systems which combine the use of Free Space Optics and radio technologies. We derive a condition for the stable operation of the system, compute its stationary distribution and the key performance measures. Illustrative numerical examples give some insight into the behavior of the system.

Keywords: Unreliable queueing system · Heterogeneous servers · Backup server · Stationary distribution · Stationary performance measures

1 Introduction

The rapid and continuous increase in the number of users on the Internet, the increase in the volume and quality of information transmitted in broadband wireless networks requires a dramatic increase in the performance of multimedia communication channels. In this regard, in recent years, within the frame of the development of next generation networks, intensive research is being carried out to improve wireless performance. One of the directions for creating ultra-high-speed (up to 10 Gbit/s) and reliable wireless communications is the development of hybrid systems based on laser and radio technologies.

The FSO (Free Space Optics) technology has been widely used in recent times. This technology is based on the transmission of data by modulated radiation in the infrared (or visible) part of the spectrum through the atmosphere and their subsequent detection by an optical photo-detector device. The main advantages of atmospheric optical communication lines are as follows.

V. M. Vishnevskiy et al. (Eds.): DCCN 2020, LNCS 12563, pp. 176–195, 2020.
https://doi.org/10.1007/978-3-030-66471-8_15

- High bandwidth and quality of digital communication. Modern FSO-solutions can provide a transmission speed of digital flows up to 10 Gbit/s with a bit error rate of 10^{-12}, which is currently impossible to achieve with any other wireless technologies;
- High security of the channel from unauthorized access and stealth. No wireless transmission technology can offer such confidentiality of communication as laser. Absence of pronounced external signs (basically, this electromagnetic radiation) allows to hide not only the transmitted information, but also the very fact of information exchange. Therefore, laser systems are often used for a variety of applications where high confidentiality of data transmission is required, including financial, medical and military organizations;
- High level of noise immunity and noninterference. FSO-equipment is immune to radio interference and does not create an interference itself;
- Speed and easiness of deployment of the FSO network.

Along with these advantages of wireless optical systems, their main disadvantages are also known: the dependence of the accessibility of the communication channel on weather conditions and the need to provide direct visibility between the optical transmitter and the receiver. Unfavorable weather conditions, such as snow, fog, can significantly reduce the effective range of operation of laser atmospheric communication lines. Thus, the attenuation of a signal in an optical channel in a strong fog can reach a critical value of 50–100 dB/km. Therefore, in order to achieve operator reliability values of the FSO communication channel, it is necessary to resort to the use of hybrid solutions.

Hybrid radio-optical equipment is based on the use of redundant radio channels (centimeter and/or millimeter range of radio waves) together with an optical channel. Note that the operation of the radio channel of the centimeter range of radio waves is practically independent of the weather. The performance of a millimeter-wave wireless channel is not affected by fog. At the same time, the signal/noise ratio, which determines the quality of channel's operation, is greatly reduced with heavy rain. This complementary behavior of optical and broadband radio channels has made it possible to put forward the concept of hybrid carrier-class systems that function reliably in all weather conditions.

Due to the high need for high-speed and reliable communication channels, the following architectures of hybrid systems are currently being used to solve the "last mile" problem (see [1–6]): a high-speed laser channel is reserved by a broadband radio channel operating under the IEEE 802.11n protocol in the centimeter band of radio waves ("cold" or "hot" reserve); The FSO channel is reserved by the radio channel of the millimeter-wave E-band of radio waves (71–76 GHz, 81–86 GHz); The FSO channel and the radio channel of the millimeter band operate in parallel and are reserved by the channel IEEE 802.11n, which is in the cold reserve. In [7] the authors consider a hybrid communication system consisting of FSO links supported by terrestrial optical fiber connections.

Practical needs stimulated theoretical studies on the performance evaluation and the selection of optimal modes for the operation of hybrid systems using queueing theory models with unreliable service channels. Initially, these papers, see, e.g. [8–11] used simplified assumptions about the Poisson character of the input flow, flow of breakdowns, the exponential distribution of packet service time and repair time of communication channels.

Papers from [8] are mainly focused on the study of stationary reliability characteristics, methods and algorithms for optimal channel switching in hybrid communication systems by means of simulation. In the paper [10] the authors consider the hybrid communication system with so called hot redundancy where the FSO channel and backup IEEE 802.11n channel transmit data in parallel. In the paper [9], the hybrid communication system with cold redundancy is investigated. Here the radio-wave link is assumed to be absolutely reliable and backs up the FSO channel in cases when the latter interrupts its functioning because of the unfavorable weather conditions. In the paper a statistical analysis of meteorological data for duration of the periods of favorable and unfavorable weather conditions is also carried out. The paper [11] deals with hybrid communication system consisting of the FSO channel and millimeter-wave radio channel. It is assumed that periods of favorable weather conditions for both channels alternate with periods of unfavorable weather conditions for one of the channels. To model this system, the authors consider two-channel queueing system with unreliable heterogeneous servers which fail alternately.

In further works [12–14], more complicated models of unreliable single-server queues are considered. They generalize models of [9–11] to the case of much more adequate processes describing the operation of corresponding hybrid communication systems. The input flow and the flow of breakdowns are described by Markovian Arrival Processes ($BMAP$ and MAP), see [15], and packet transmission time via communication channels and repair time are assumed to have Phase type (PH)-distributions, see [16]. Although these assumptions complicate the study of models that adequately describe the operation of hybrid systems, but they allow to take into account the non-stationary, correlated nature of information flows in modern and future 5G networks.

Almost in all previous papers, the subject for study were hybrid communication systems consisting of the main FSO channel and backup low speed radio channel. Such systems were modeled by single-server queues with a backup server. One way to increase the reliability and speed of information transmission is to create hybrid system consisting of two main unreliable but high-speed channels (FSO channel and a radio channel of millimeter-wave) which are reserved by reliable but low-speed radio channel IEEE 802.11n which is in the cold reserve. The unreliability of the main channels is due to the lack of favorable weather conditions: the FSO channel can not transmit information in poor visibility conditions and the millimeter-wave channel can not transmit information when precipitation occurs. Such a hybrid system can be modeled by unreliable two-server queueing system with a backup server. In the present paper, we consider such a queueing system. We assume that customers arrive into the system in batch correlated flow $BMAP$, heterogeneous breakdowns arrive to the main servers in $MMAP$ (Marked Markovian Arrival Process, see [17]), service and repair times have PH distribution. We investigate the operation of the system in steady state, derive a stability condition, compute the stationary distribution and performance measures of the system. We also present illustrative numerical examples which give some insight into the behavior of the system.

2 Mathematical Model

We consider a queueing system with infinite waiting room and two unreliable heterogeneous servers, which model FSO and millimeter-wave channels, and backup reliable server which models radio-wave IEEE802 channel. In the following, FSO channel will be called as server 1, millimeter-wave channel as server 2 and radio-wave IEEE802 channel as server 3. Customers arrive into the system in accordance with a $BMAP$. The $BMAP$ is a very general arrival process which is able to capture correlation and burstiness that are commonly seen in the traffic of modern communication networks. The $BMAP$ is defined by the underlying process ν_t, $t \geq 0$, which is an irreducible continuous-time Markov chain with finite state space $\{0, \dots, W\}$, and the matrix generating function $D(z) = \sum\limits_{k=0}^{\infty} D_k z^k$, $|z| \leq 1$. The batches of customers enter the system only at the epochs of the chain ν_t, $t \geq 0$, transitions. The $(W+1) \times (W+1)$ matrices D_k, $k \geq 1$, and non-diagonal entries of the matrix D_0 define the rates of the process ν_t, $t \geq 0$, transitions which are accompanied by generating the k-size batch of customers, $k \geq 0$. The intensity (fundamental rate) of the $BMAP$ is defined as $\lambda = \boldsymbol{\theta} D'(1)\mathbf{e}$ where the vector $\boldsymbol{\theta}$ is the unique solution of the system $\boldsymbol{\theta} D(1) = \mathbf{0}$, $\boldsymbol{\theta}\mathbf{e} = 1$. Hereinafter $\mathbf{e}$ is a column vector of 1's and $\mathbf{0}$ is a row vector of 0's. For more information about the $BMAP$ see, e.g. [15].

If the arriving customer meets both servers 1 and 2 idle, it starts service at server 1. If the arriving customer meets one of the servers 1 or 2 idle, it starts service at the idle server. If both servers are busy at the customers arrival moment, the customer moves to the buffer.

The service time of a customer by the jth server, $j = 1, 2, 3$, has PH type distribution with irreducible representation $(\boldsymbol{\beta}_j, S_j)$. Such a service time can be interpreted as the time until the continuous-time Markov chain $m_t^{(j)}, t \geq 0$, reaches the single absorbing state $M_j + 1$ if it has the state space $\{1, .., M_j, M_j + 1\}$, initial state is selected according to the vector $\boldsymbol{\beta}_j$, transitions within non-absorbing states are governed by the sub-generator S and the rates of transitions into the absorbing state are given by the vector $\mathbf{S}_0^{(j)} = -S_j\mathbf{e}$. The mean service time is calculated as $b_1^{(j)} = \boldsymbol{\beta}_j(-S_j)^{-1}\mathbf{e}$ and the service rate is equal to $\mu_j = b_1^{(j)^{-1}}$. More detailed description of the PH type process can be found in [16].

Breakdowns of two types arrive to the servers 1, 2 according to a $MMAP$ which is defined by the underlying process η_t, $t \geq 0$, with state space $\{0, \dots, V\}$ and by the matrices H_0, H_1, H_2. The matrix H_0 defines the rates of the process η_t, $t \geq 0$, transitions which do not lead to generation of a breakdown. The matrix H_j defines the rates of the η_t, $t \geq 0$, transitions which are accompanied by generating a breakdown which is directed to the server j, for $j = 1, 2$. The rate of breakdowns directed to the jth server is calculated as $h_j = \boldsymbol{\vartheta} H_j \mathbf{e}$ where the vector $\boldsymbol{\vartheta}$ is the unique solution of the system $\boldsymbol{\theta}(H_0 + H_1 + H_2) = \mathbf{0}$, $\boldsymbol{\vartheta}\mathbf{e} = 1$.

When a breakdown breaks one of the main servers, the repair period at this server starts immediately and the other main server, if it is available, begins the service of the interrupted customer as a new. If the latter server is busy or under repair, the customer goes to the server 3 and starts its service as a new. However, if during the service time of the customer at the server 3 one of the main servers becomes fault-free, the customer restarts its service on this server. Service at the server 3 is terminated. This server can provide service only when both main servers are broken.

The repair period at the jth main server, $j = 1, 2$, has PH type distribution with an irreducible representation $(\boldsymbol{\tau}_j, T_j)$. The repair process at the jth server is directed by the Markov chain $r_t^{(j)}$, $t \geq 0$, with state space $\{1, \ldots, R_j, R_j + 1\}$ where $R_j + 1$ is an absorbing state. The rates of transitions into the absorbing state are given by the vector $\mathbf{T}_0^{(j)} = -T_j\mathbf{e}$. The repair rate is calculated as $\tau_j = (\boldsymbol{\beta}_j(-T_j)^{-1}\mathbf{e})^{-1}$. Breakdowns arriving to a server are ignored it the server is under repair at the moment of the breakdown arrival.

3 Process of the System States

Let, at the moment t,

i_t be the number of customers in the system, $i_t \geq 0$,

$n_t = 0$, if both main servers are fault-free (both ones are busy or idle); $n_t = 0_j$, if both main servers are fault-free, the jth server is busy and the other one is idle, $j = 1, 2$; $n_t = 1$, if the server 1 is under repair; $n_t = 2$, if the server 2 is under repair; $n_t = 3$, if both servers are under repair;

$m_t^{(j)}$ be the state of the directing process of the service at the jth busy server, $j = 1, 2, 3, m_t^{(j)} = \overline{1, M_j}$;

$r_t^{(j)}$ be the state of the directing process of the repair time at the jth server, $j = 1, 2, r_t^{(j)} = \overline{1, R_j}$;

ν_t and η_t be the states of the directing processes of the $BMAP$ and the $MMAP$, respectively, $\nu_t = \overline{0, W}$, $\eta_t = \overline{0, V}$.

The process of the system states is described by the regular irreducible continuous time Markov chain, $\xi_t, t \geq 0$, with state space

$$X = \{(0, n, \nu, \eta),\ i = 0, n = \overline{0, 3},\ \nu = \overline{0, W},\ \eta = \overline{0, V}\} \bigcup$$

$$\{(i, 0_j, \nu, \eta, m^{(j)}),\ i = 1, j = 1, 2, n = 0_j, \nu = \overline{0, W}, \eta = \overline{0, V},\ m^{(j)} = \overline{1, M_j}\} \bigcup$$

$$\{(i, 0, \nu, \eta, m^{(1)}, m^{(2)}),\ i > 1, \nu = \overline{0, W}, \eta = \overline{0, V}, m^{(k)} = \overline{1, M_k}, k = 1, 2\} \bigcup$$

$$\{(i, 1, \nu, \eta, m^{(2)}, r^{(1)}), i \geq 1, \nu = \overline{0, W}, \eta = \overline{0, V}, m^{(2)} = \overline{1, M_2}, r^{(1)} = \overline{1, R_1}\} \bigcup$$

$$\{(i, 2, \nu, \eta, m^{(1)}, r^{(2)}), i \geq 1, \nu = \overline{0, W}, \eta = \overline{0, V}, m^{(1)} = \overline{1, M_1}, r^{(1)} = \overline{1, R_2}\} \bigcup$$

$$\{(i,3,\nu,\eta,m^{(3)},r^{(1)},r^{(2)}),\ i>0, \nu=\overline{0,W}, \eta=\overline{0,V}, m^{(3)}=\overline{1,M_3},$$
$$r^{(j)}=\overline{1,R_j}, j=1,2\}.$$

Let $Q_{i,j}$, $i,j\geq 0$, be the matrices formed by the rates of the chain transitions from the states corresponding to the value i of the component i_t to the states corresponding to the value j of this component. The following statement is true.

Lemma 1. *Infinitesimal generator of the Markov chain ξ_t, $t\geq 0$, has the following block structure*

$$Q=\begin{pmatrix} Q_{0,0} & Q_{0,1} & Q_{0,2} & Q_{0,3} & Q_{0,4} & \cdots \\ Q_{1,0} & Q_{1,1} & Q_{1,2} & Q_{1,3} & Q_{1,4} & \cdots \\ O & Q_{2,1} & Q_1 & Q_2 & Q_3 & \cdots \\ O & O & Q_0 & Q_1 & Q_2 & \cdots \\ \vdots & \vdots & \vdots & \vdots & \vdots & \ddots \end{pmatrix},$$

where non-zero blocks are of the following form:

$$Q_{0,0}=\begin{pmatrix} D_0\oplus H_0 & I_{\bar W}\otimes H_1\otimes \boldsymbol{\tau}_1 & I_{\bar W}\otimes H_2\otimes\boldsymbol{\tau}_2 & O \\ I_a\otimes \boldsymbol{T}_0^{(1)} & D_0\oplus(H_0+H_1)\oplus T^{(1)} & O & I_{\bar W}\otimes H_2\otimes I_{R_1}\otimes\boldsymbol{\tau}_2 \\ I_a\otimes\boldsymbol{T}_0^{(2)} & O & D_0\oplus(H_0+H_2)\oplus T^{(2)} & I_{\bar W}\otimes H_1\otimes I_{R_2}\otimes\boldsymbol{\tau}_1 \\ O & I_{aR_1}\otimes\boldsymbol{T}_0^{(2)} & I_a\otimes\boldsymbol{T}_0^{(1)}\otimes I_{R_2} & D_0\oplus H\oplus T^{(1)}\oplus T^{(2)} \end{pmatrix},$$

$$Q_{0,1}=$$

$$\begin{pmatrix} D_1\otimes I_{\bar V}\otimes\boldsymbol{\beta}_1 & O_{a\times aM_2} & O & O & O \\ O & O & D_1\otimes I_{\bar V}\otimes\boldsymbol{\beta}_2\otimes I_{R_1} & O & O \\ O & O & O & D_1\otimes I_{\bar V}\otimes\boldsymbol{\beta}_1\otimes I_{R_2} & O \\ O & O & O & O & D_1\otimes I_{\bar V}\otimes\boldsymbol{\beta}_3\otimes I_{R_1R_2} \end{pmatrix},$$

$$Q_{0,k}=diag\{D_k\otimes I_{\bar V}\otimes\boldsymbol{\beta}_1\otimes\boldsymbol{\beta}_2,\ D_k\otimes I_{\bar V}\otimes\boldsymbol{\beta}_2\otimes I_{R_1},$$
$$D_k\otimes I_{\bar V}\otimes\boldsymbol{\beta}_1\otimes I_{R_2},\ D_k\otimes I_{\bar V}\otimes\boldsymbol{\beta}_3\otimes I_{R_1R_2}\},$$

$$Q_{1,0}=\begin{pmatrix} I_a\otimes\boldsymbol{S}_0^{(1)} & O & O & O \\ I_a\otimes\boldsymbol{S}_0^{(2)} & O & O & O \\ O & I_a\otimes\boldsymbol{S}_0^{(2)}\otimes I_{R_1} & O & O \\ O & O & I_a\otimes\boldsymbol{S}_0^{(1)}\otimes I_{R_2} & O \\ O & O & O & I_a\otimes\boldsymbol{S}_0^{(3)}\otimes I_{R_1}\otimes I_{R_2} \end{pmatrix},$$

$$Q_{1,1}=\left(Q_{1,1}^{(1)}\,Q_{1,1}^{(2)}\right),$$

$$Q_{1,1}^{(1)}=\begin{pmatrix} D_0\oplus H_0\oplus S_1 & O & I_{\bar W}\otimes H_1\otimes\mathbf{e}_{M_1}\otimes\boldsymbol{\beta}_2\otimes\boldsymbol{\tau}_1 \\ O & D_0\oplus H_0\oplus S_2 & I_{\bar W}\otimes H_1\otimes I_{M_2}\otimes\boldsymbol{\tau}_1 \\ O & I_a\otimes I_{M_2}\otimes\boldsymbol{T}_0^{(1)} & D_0\oplus(H_0+H_1)\oplus S_2\oplus T_1 \\ I_a\otimes I_{M_1}\otimes\boldsymbol{T}_0^{(2)} & O & O \\ O & O & I_a\otimes\boldsymbol{\beta}_2\otimes\mathbf{e}_{M_3}\otimes I_{R_1}\otimes\boldsymbol{T}_0^{(2)} \end{pmatrix},$$

$$Q_{1,1}^{(2)} = \begin{pmatrix} I_{\bar{W}} \otimes H_2 \otimes I_{M_1} \otimes \boldsymbol{\tau}_2 & O \\ I_{\bar{W}} \otimes H_2 \otimes \mathbf{e}_{M_2} \otimes \boldsymbol{\beta}_1 \otimes \boldsymbol{\tau}_2 & O \\ O & I_{\bar{W}} \otimes H_2 \otimes \mathbf{e}_{M_2} \otimes \boldsymbol{\beta}_3 \otimes I_{R_1} \otimes \boldsymbol{\tau}_2 \\ D_0 \oplus (H_0 + H_2) \oplus S_1 \oplus T_2 & I_{\bar{W}} \otimes H_1 \otimes \mathbf{e}_{M_1} \otimes \boldsymbol{\beta}_3 \otimes \boldsymbol{\tau}_1 \otimes I_{R_2} \\ I_a \otimes \boldsymbol{\beta}_1 \otimes \mathbf{e}_{M_3} \otimes \boldsymbol{T}_0^{(1)} \otimes I_{R_2} & D_0 \oplus H \oplus S_3 \oplus T_1 \oplus T_2 \end{pmatrix},$$

$$Q_{1,k} =$$

$$\begin{pmatrix} D_{k-1} \otimes I_{\bar{V}M_1} \otimes \boldsymbol{\beta}_2 & O & O & O \\ D_{k-1} \otimes I_{\bar{V}} \otimes \boldsymbol{\beta}_1 \otimes I_{M_2} & O & O & O \\ O & D_{k-1} \otimes I_{\bar{V}M_2R_1} & O & O \\ O & O & D_{k-1} \otimes I_{\bar{V}M_1R_2} & O \\ O & O & O & D_{k-1} \otimes I_{\bar{V}M_3R_1R_2} \end{pmatrix},$$

$$Q_{2,1} = \begin{pmatrix} I_a \otimes I_{M_1} \otimes \boldsymbol{S}_0^{(2)} & O \\ O & O \end{pmatrix} + \Big(O \mid diag\{I_a \otimes \boldsymbol{S}_0^{(1)} \otimes I_{M_2},$$

$$I_a \otimes \boldsymbol{S}_0^{(2)} \boldsymbol{\beta}_2 \otimes I_{R_1},\ I_a \otimes \boldsymbol{S}_0^{(1)} \boldsymbol{\beta}_1 \otimes I_{R_2},\ I_a \otimes \boldsymbol{S}_0^{(3)} \boldsymbol{\beta}_3 \otimes I_{R_1R_2}\}\Big),$$

$$Q_0 = diag\{I_a \otimes (\boldsymbol{S}_0^{(1)} \boldsymbol{\beta}_1 \oplus \boldsymbol{S}_0^{(2)} \boldsymbol{\beta}_2),\ I_a \otimes \boldsymbol{S}_0^{(2)} \boldsymbol{\beta}_2 \otimes I_{R_1},$$
$$I_a \otimes \boldsymbol{S}_0^{(1)} \boldsymbol{\beta}_1 \otimes I_{R_2},\ I_a \otimes \boldsymbol{S}_0^{(3)} \boldsymbol{\beta}_3 \otimes I_{R_1R_2}\},$$

$$Q_1 = \begin{pmatrix} Q_1^{(1,1)} & Q_1^{(1,2)} \\ Q_1^{(2,1)} & Q_1^{(2,2)} \end{pmatrix},$$

$$Q_1^{(1,1)} = \begin{pmatrix} D_0 \oplus H_0 \oplus S_1 \oplus S_2 & I_{\bar{W}} \otimes H_1 \otimes \mathbf{e}_{M_1} \otimes I_{M_2} \otimes \boldsymbol{\tau}_1 \\ I_a \otimes \boldsymbol{\beta}_1 \otimes I_{M_2} \otimes \boldsymbol{T}_0^{(1)} & D_0 \oplus (H_0 + H_1) \oplus S_2 \oplus T_1 \end{pmatrix},$$

$$Q_1^{(1,2)} = \begin{pmatrix} I_{\bar{W}} \otimes H_2 \otimes I_{M_1} \otimes \mathbf{e}_{M_2} \otimes \boldsymbol{\tau}_2 & O \\ O & I_{\bar{W}} \otimes H_2 \otimes \mathbf{e}_{M_2} \otimes \boldsymbol{\beta}_3 \otimes I_{R_1} \otimes \boldsymbol{\tau}_2 \end{pmatrix},$$

$$Q_1^{(2,1)} = \begin{pmatrix} I_a \otimes I_{M_1} \otimes \boldsymbol{\beta}_2 \otimes \boldsymbol{T}_0^{(2)} & O \\ O & I_a \otimes \boldsymbol{\beta}_2 \otimes \mathbf{e}_{M_3} \otimes I_{R_1} \otimes \boldsymbol{T}_0^{(2)} \end{pmatrix},$$

$$Q_1^{(2,2)} = \begin{pmatrix} D_0 \oplus (H_0 + H_2) \oplus S_1 \oplus T_2 & I_{\bar{W}} \otimes H_1 \otimes \mathbf{e}_{M_1} \otimes \boldsymbol{\beta}_3 \otimes I_{R_2} \otimes \boldsymbol{\tau}_1 \\ I_a \otimes \boldsymbol{\beta}_1 \otimes \mathbf{e}_{M_3} \otimes \boldsymbol{T}_0^{(1)} \otimes I_{R_2} & D_0 \oplus H \oplus S_3 \oplus T_1 \oplus T_2 \end{pmatrix},$$

$$Q_{k+1} = diag\{D_k \otimes I_{\bar{V}M_1M_2},\ D_k \otimes I_{\bar{V}M_2R_1},\ D_k \otimes I_{\bar{V}M_1R_2},\ D_k \otimes I_{\bar{V}M_3R_1R_2}\},\ k \geq 1,$$

where $H = H_0 + H_1$, $\otimes, \oplus$ *are the symbols of Kronecker's product and sum of matrices,* $diag\{\dots\}$ *denotes the block diagonal matrix with the diagonal blocks listed in the brackets,* $\bar{W} = W + 1$, $\bar{V} = V + 1$, $a = \bar{W}\bar{V}$, $\mathbf{e}_n$ *is a column vector of size* n, *consisting of 1's,* I (O) *is an identity (zero) matrix.*

Corollary 1. *The Markov chain* $\xi_t, t \geq 0$, *belongs to the class of continuous time quasi-Toeplitz Markov chains, see [18].*

Proof. The generator Q of the chain $\xi_t, t \geq 0$, has a block upper-Hessenberg structure and, starting from $i = 3$, the blocks $Q_{i,j}$ depend on i, j only through the difference $j-i$. Then, according the definition given in [18] the chain $\xi_t, t \geq 0$, is a quasi-Toeplitz Markov chain.

In what follows we need expressions for the matrix generating functions $Q^{(n)}(z) = \sum_{k=2}^{\infty} Q_{n,k}z^k$, $n = 0, 1$, and $Q(z) = \sum_{k=0}^{\infty} Q_k z^k$, $|z| \leq 1$. These expressions are given by the following

Corollary 2. *The matrix generation functions $Q(z)$, $Q^{(n)}(z)$, $n = 0, 1$, have the following form:*

$$Q^{(0)}(z) = diag\{D(z) - D_0 - D_1 z) \otimes I_{\bar{V}} \otimes \boldsymbol{\beta}_1 \otimes \boldsymbol{\beta}_2, (D(z) - D_0 - D_1 z) \otimes I_{\bar{V}} \otimes \boldsymbol{\beta}_2 \otimes I_{R_1},$$
$$(D(z) - D_0 - D_1 z) \otimes I_{\bar{V}} \otimes \boldsymbol{\beta}_1 \otimes I_{R_2}, (D(z) - D_0 - D_1 z) \otimes I_{\bar{V}} \otimes \boldsymbol{\beta}_3 \otimes I_{R_1 R_2}\}, \quad (1)$$

$$Q^{(1)}(z) = \quad (2)$$
$$z \begin{pmatrix} \bar{D}(z) \otimes I_{\bar{V}M_1} \otimes \boldsymbol{\beta}_2 & O & O & O \\ \bar{D}(z) \otimes I_{\bar{V}} \otimes \boldsymbol{\beta}_1 \otimes I_{M_2} & O & O & O \\ O & \bar{D}(z) \otimes I_{\bar{V}M_2R_1} & O & O \\ O & O & \bar{D}(z) \otimes I_{\bar{V}M_1R_2} & O \\ O & O & O & \bar{D}(z) \otimes I_{\bar{V}M_3R_1R_2} \end{pmatrix},$$

where $\bar{D}(z) = D(z) - D_0$,

$$Q(z) = Q_0 + \mathcal{Q}z + z diag\{D(z) \otimes I_{\bar{V}M_1M_2},$$
$$D(z) \otimes I_{\bar{V}M_2M_3R_1}, D(z) \otimes I_{\bar{V}M_1M_3R_2}, D(z) \otimes I_{\bar{V}M_3R_1R_2}\}, \quad (3)$$

where the matrix $\mathcal{Q}$ is of the form

$$\mathcal{Q} = \begin{pmatrix} \mathcal{Q}^{(1,1)} & \mathcal{Q}^{(1,2)} \\ \mathcal{Q}^{(2,1)} & \mathcal{Q}^{(2,2)} \end{pmatrix}, \quad (4)$$

$$\mathcal{Q}^{(1,1)} = \begin{pmatrix} I_{\bar{W}} \otimes H_0 \oplus S_1 \oplus S_2 & I_{\bar{W}} \otimes H_1 \otimes \mathbf{e}_{M_1} \otimes I_{M_2} \otimes \boldsymbol{\tau}_1 \\ I_{\bar{W}\bar{V}} \otimes \boldsymbol{\beta}_1 \otimes I_{M_2} \otimes \boldsymbol{T}_0^{(1)} & I_{\bar{W}} \otimes (H_0 + H_1) \oplus S_2 \oplus T_1 \end{pmatrix},$$

$$\mathcal{Q}^{(1,2)} = \begin{pmatrix} I_{\bar{W}} \otimes H_2 \otimes I_{M_1} \otimes \mathbf{e}_{M_2} \otimes \boldsymbol{\tau}_2 & O \\ O & I_{\bar{W}} \otimes H_2 \otimes \mathbf{e}_{M_2} \otimes \boldsymbol{\beta}_3 \otimes I_{R_1} \otimes \boldsymbol{\tau}_2 \end{pmatrix},$$

$$\mathcal{Q}^{(2,1)} = \begin{pmatrix} I_{\bar{W}} \otimes I_{\bar{V}M_1} \otimes \boldsymbol{\beta}_2 \otimes \boldsymbol{T}_0^{(2)} & O \\ O & I_{\bar{W}\bar{V}} \otimes \boldsymbol{\beta}_2 \otimes \mathbf{e}_{M_3} \otimes I_{R_1} \otimes \boldsymbol{T}_0^{(2)} \end{pmatrix},$$

$$\mathcal{Q}^{(2,2)} = \begin{pmatrix} I_{\bar{W}} \otimes (H_0 + H_2) \oplus S_1 \oplus T_2 & I_{\bar{W}} \otimes H_1 \otimes \mathbf{e}_{M_1} \otimes \boldsymbol{\beta}_3 \otimes I_{R_2} \otimes \boldsymbol{\tau}_1 \\ I_{\bar{W}\bar{V}} \otimes \boldsymbol{\beta}_1 \otimes \mathbf{e}_{M_3} \otimes \boldsymbol{T}_0^{(1)} \otimes I_{R_2} & I_{\bar{W}} \otimes H \oplus S_3 \oplus T_1 \oplus T_2 \end{pmatrix}.$$

4 Stationary Distribution

Let $\mathcal{Q}^-$ be a matrix obtained from the matrix $\mathcal{Q}$ given by (4) by formally deleting in its blocks the Kronecker cofactor $I_{\bar{W}}$ and Q_0^- be a matrix obtained from the matrix Q_0 by formally replacement in its blocks the cofactor I_a by the cofactor $I_{\bar{W}}$. Let also

$$\Psi = Q_0^- + \mathcal{Q}^-.$$

Theorem 1. *The necessary and sufficient condition for ergodicity of the Markov chain* ξ_t, $t \geq 0$, *is the fulfillment of the inequality*

$$\lambda < -\boldsymbol{\pi}_0(S_1 \oplus S_2)\mathbf{e} + \boldsymbol{\pi}_1\mathbf{S}_0^{(2)} + \boldsymbol{\pi}_2\mathbf{S}_0^{(1)} + \boldsymbol{\pi}_3\mathbf{S}_0^{(3)} \tag{5}$$

where $\boldsymbol{\pi}_0 = \mathbf{x}_0(\mathbf{e}_{V+1} \otimes I_{M_1M_2})$, $\boldsymbol{\pi}_1 = \mathbf{x}_1(\mathbf{e}_{V+1} \otimes I_{M_2} \otimes \mathbf{e}_{R_1})$, $\boldsymbol{\pi}_2 = \mathbf{x}_2(\mathbf{e}_{V+1} \otimes I_{M_1} \otimes \mathbf{e}_{R_2})$, $\boldsymbol{\pi}_3 = \mathbf{x}_3(\mathbf{e}_{V+1} \otimes I_{M_3} \otimes \mathbf{e}_{R_1R_2})$ *and the vectors* $\mathbf{x}_0, \mathbf{x}_1, \mathbf{x}_2, \mathbf{x}_3$ *are sub-vectors of the vector* $\mathbf{x} = (\mathbf{x}_1, \mathbf{x}_2, \mathbf{x}_3, \mathbf{x}_4)$, *which is the unique solution of the system of linear algebraic equations*

$$\mathbf{x}\Psi = 0, \ \mathbf{x}\mathbf{e} = 1. \tag{6}$$

Remark 1. Note that the ratio of the left part of inequality (5) and the right part of this inequality is the system load factor ρ, i.e.

$$\rho = \frac{\lambda}{-\boldsymbol{\pi}_0(S_1 \oplus S_2)\mathbf{e} + \boldsymbol{\pi}_1\mathbf{S}_0^{(2)} + \boldsymbol{\pi}_2\mathbf{S}_0^{(1)} + \boldsymbol{\pi}_3\mathbf{S}_0^{(3)}}.$$

Proof. It can be verified that the matrix $\sum_{k=0}^{\infty} Q_k$ is irreducible. Hence, from [18], the necessary and sufficient condition for ergodicity of the chain ξ_t is the fulfillment of the inequality

$$\mathbf{y}Q'(z)|_{z=1}\mathbf{e} < 0 \tag{7}$$

where the vector $\mathbf{y}$ is the unique solution of the system

$$\mathbf{y}Q(1) = \mathbf{y}, \quad \mathbf{y}\mathbf{e} = 1. \tag{8}$$

The theorem will be proven if we show that inequality (7) is equivalent to inequality (5). Let the vector $\mathbf{y}$ be of the form

$$\mathbf{y} = (\boldsymbol{\theta} \otimes \mathbf{x}_0, \boldsymbol{\theta} \otimes \mathbf{x}_1, \boldsymbol{\theta} \otimes \mathbf{x}_2, \boldsymbol{\theta} \otimes \mathbf{x}_3). \tag{9}$$

Substituting the vector $\mathbf{y}$ in the form (9) into (8) and using relation $\boldsymbol{\theta}D(1) = \mathbf{0}$, we reduce system (8) to the form (6).

Now we substitute into the inequality (7) the vector $\mathbf{y}$ in the form (9) and the expression for $Q'(1)$ calculated by formula (3). Taking into account that $\boldsymbol{\theta}D'(1)\mathbf{e} = \lambda$, we transform inequality (7) to the form

$$\lambda + \mathbf{x}\mathcal{Q}^-\mathbf{e} < 0. \tag{10}$$

Using the known relations $H\mathbf{e} = (H_0 + H_1 + H_2)\mathbf{e} = \mathbf{0}$, $S_n\mathbf{e} + \boldsymbol{S}_0^{(n)} = \mathbf{0}$, $T_n\mathbf{e} + \boldsymbol{T}_0^{(n)} = \mathbf{0}$, $n = 1, 2$, we reduce inequality (10) to the following form:

$$\lambda < \mathbf{x}_0(\mathbf{e}_{\bar{V}} \otimes I_{M_1M_2})(\mathbf{S}_0^{(1)} \oplus \mathbf{S}_0^{(2)})\mathbf{e} + \mathbf{x}_1(\mathbf{e}_{\bar{V}} \otimes I_{M_2} \otimes \mathbf{e}_{R_1})\mathbf{S}_0^{(2)}\mathbf{e}$$

$$+\mathbf{x}_2(\mathbf{e}_{\bar{V}} \otimes I_{M_1} \otimes \mathbf{e}_{R_2})\mathbf{S}_0^{(1)}\mathbf{e} + \mathbf{x}_3(\mathbf{e}_{\bar{V}} \otimes I_{M_3} \otimes \mathbf{e}_{R_1R_2})\mathbf{S}_0^{(3)}\mathbf{e}.$$

After using the notation introduced in the statement of the theorem, this inequality takes the form (5). □

Remark 2. In the physical interpretation of the inequality (5), we take into account that this inequality is derived under the system overload condition. Let us consider the physical meaning of the first term in the right-hand side of (5). The component $\boldsymbol{\pi}_0(m^{(1)}, m^{(2)})$ of the row vector $\boldsymbol{\pi}_0$ is the probability that servers 1 and 2 are fault-free and serve customers on the phases $m^{(1)}$ and $m^{(2)}$, respectively. The corresponding component of the column vector $(\mathbf{S}_0^{(1)} \oplus \mathbf{S}_0^{(2)})\mathbf{e}$ is the total service rate by servers 1 and 2 provided that the service on these servers is in the phases $m^{(1)}$ and $m^{(2)}$, respectively. Then the product $\boldsymbol{\pi}_0(\mathbf{S}_0^{(1)} \oplus \mathbf{S}_0^{(2)})\mathbf{e}$ represents the rate of the output flow in periods when customers are served by servers 1 and 2. The other summands of the sum on the right-hand side of inequality (5) are interpreted similarly: the second term is the rate of the output flow when customers are only served by server 2 (server 1 is under repair), the third term is the rate of the output flow when customers are only served by server 1 (server 2 is under repair), the fourth term is the rate of the output flow when customers are only served by server 3 (servers 1 and 2 are under repair). Then the right-hand side of inequality (5) expresses the total rate of the output flow under overload condition. Obviously, for the existence of a steady-state regime in the system, it is necessary and sufficient that the input rate λ be less than the rate of the output flow.

Corollary 3. *In the case of stationary Poisson flow of breakdowns and exponential distribution of service and repair times, ergodicity condition (5) is reduced to the following inequality:*

$$\lambda < \pi_0(\mu_1 + \mu_2) + \pi_1\mu_2 + \pi_2\mu_1 + \pi_3\mu_3, \tag{11}$$

where the vector $\boldsymbol{\pi} = (\pi_0, \pi_1, \pi_2, \pi_3)$ *is the unique solution to the system of linear algebraic equation*

$$\boldsymbol{\pi}\begin{pmatrix} -(h_1+h_2) & h_1 & h_2 & 0 \\ \tau_1 & -(h_2+\tau_1) & 0 & h_2 \\ \tau_2 & 0 & -(h_1+\tau_2) & h_1 \\ 0 & \tau_2 & \tau_1 & -(\tau_1+\tau_2) \end{pmatrix} = \mathbf{0}, \quad \boldsymbol{\pi}\mathbf{e} = 1.$$

In what follows we assume that the ergodicity condition given by Theorem 1 is satisfied, which ensures that there exist the stationary probabilities of the system states

$$p_0^{(n)}(\nu, \eta) = \lim_{t\to\infty} P\{i_t = 0, n_t = n, \nu_t = \nu, \eta_t = \eta\}, \; n = \overline{0,3}, \;\; \nu = \overline{0,W}, \eta = \overline{0,V};$$

$$p_1^{(0_n)}\{(\nu,\eta,m^{(k)})\} = \lim_{t\to\infty} P\{i_t = i, n_t = 0_n, \nu_t = \nu, \eta_t = \eta, m_t^{(n)} = m^{(n)}\},$$
$$n = 1,2,\ \nu = \overline{0,W},\ \eta = \overline{0,V},\ m^{(n)} = \overline{1,M^{(n)}};$$
$$p_i^{(0)}(\nu,\eta,m^{(1)},m^{(2)}) = \lim_{t\to\infty} P\{i_t = i, n_t = 0, \nu_t = \nu, \eta_t = \eta, m_t^{(1)} = m^{(1)},$$
$$m_t^{(2)} = m^{(2)}\}, i > 1,\ \nu = \overline{0,W},\ \eta = \overline{0,V},\ m^{(n)} = \overline{1,M^{(n)}}.$$
$$p_i^{(1)}(\nu,\eta,m^{(2)},r^{(1)}) = \lim_{t\to\infty} P\{i_t = i, n_t = 1,\ \nu_t = \nu, \eta_t = \eta,\ m_t^{(2)} = m^{(2)},$$
$$r_t^{(1)} = r^{(1)}\},\ i \ge 1, \nu = \overline{0,W}, \eta = \overline{0,V},\ m^{(2)} = \overline{1,M^{(2)}},\ r^{(1)} = \overline{1,R^{(1)}};$$
$$p_i^{(2)}(\nu,\eta,m^{(1)},r^{(2)}) = \lim_{t\to\infty} P\{i_t = i, n_t = 2,\ \nu_t = \nu, \eta_t = \eta,\ m_t^{(1)} = m^{(1)},$$
$$r_t^{(2)} = r^{(2)}\},\ i \ge 1,\ \nu = \overline{0,W},\ \eta = \overline{0,V},\ m^{(1)} = \overline{1,M^{(1)}},\ r^{(2)} = \overline{1,R^{(2)}};$$
$$p_i^{(3)}(\nu,\eta,m^{(3)},r^{(1)},r^{(2)}) = \lim_{t\to\infty} P\{i_t = i, n_t = 3,\ \nu_t = \nu, \eta_t = \eta,\ m_t^{(3)} = m^{(3)},$$
$$r_t^{(1)} = r^{(1)}, r_t^{(2)} = r^{(2)}\}, i \ge 1,\ \nu = \overline{0,W},\ \eta = \overline{0,V},\ m^{(3)} = \overline{1,M^{(3)}},$$
$$r^{(1)} = \overline{1,R^{(1)}}, r^{(2)} = \overline{1,R^{(2)}}.$$

Within each selected group, we order the probabilities in the lexicographic order of the components and form the vectors of these probabilities

$$\mathbf{p}_0^{(n)}, n = \overline{0,3};\ \mathbf{p}_1^{(0_1)}, \mathbf{p}_1^{(0_2)}, \mathbf{p}_i^{(n)}, n = \overline{0,3},\ i \ge 1.$$

Next, we form the vectors $\mathbf{p}_i$ of stationary probabilities corresponding to the values i of the denumerable component as follows:

$$\mathbf{p}_0 = (\mathbf{p}_0^{(0)}, \mathbf{p}_0^{(1)}, \mathbf{p}_0^{(2)}, \mathbf{p}_0^{(3)}),\ \mathbf{p}_1 = (\mathbf{p}_1^{(0_1)}, \mathbf{p}_1^{(0_2)}, \mathbf{p}_1^{(1)}, \mathbf{p}_1^{(2)}, \mathbf{p}_1^{(3)}),$$
$$\mathbf{p}_i = (\mathbf{p}_i^{(0)}, \mathbf{p}_i^{(1)}, \mathbf{p}_i^{(2)}, \mathbf{p}_i^{(3)}),\ i \ge 2.$$

These vectors are calculated using algorithm for calculating the stationary distribution of quasi-Toeplitz Markov chains, see [18].

5 Vector Generating Function of the Stationary Distribution. System Performance Characteristics

Calculating the stationary probability vectors $\mathbf{p}_i$, $i \ge 0$, we can also calculate various characteristics of the system performance. In the calculation process, the following result will be useful.

Lemma 2. *Vector generating function* $\boldsymbol{P}(z) = \sum_{i=0}^{\infty} \mathbf{p}_i z^i$, $|z| \le 1$, *satisfies the following equation:*

$$P(z)Q(z) = \mathcal{B}(z) \tag{12}$$

where

$$\mathcal{B}(z) = \mathbf{p}_0 Q(z) + z[\mathbf{p}_1 Q(z) - \mathbf{p}_0 Q^{(0)}(z) - \mathbf{p}_1 Q^{(1)}(z)] + z^2 \mathbf{p}_2 Q_0. \tag{13}$$

Formula (12) can be used to calculate the values of the function $\boldsymbol{P}(z)$ and its derivatives at the point $z = 1$ without calculating infinite sums. The obtained values allow to find the moments of the number of customers in the system and some other characteristics of the system. Note that it is not possible to calculate directly the value of $\boldsymbol{P}(z)$ and its derivatives at the point $z = 1$ from Eq. (12) since the matrix $Q(1)$ is singular. This difficulty can be overcome by using the recursion formulas given below in Corollary 4.

Let us denote $f^{(m)}(z)$ the mth derivative of the function $f(z)$, $m \geq 1$, and $f^{(0)}(z) = f(z)$.

Corollary 4. *The mth, $m \geq 0$, derivatives of the vector generating function $\boldsymbol{P}(z)$ at the point $z = 1$ are recursively calculated from the following system of linear algebraic equations:*

$$\begin{cases} \boldsymbol{P}^{(m)}(1)Q(1) = \mathcal{B}^{(m)}(1) - \sum\limits_{l=0}^{m-1} C_m^l \boldsymbol{P}^{(l)}(1)Q^{(m-l)}(1), \\ \boldsymbol{P}^{(m)}(1)Q'(1)\mathbf{e} = \frac{1}{m+1}[\mathcal{B}^{(m+1)}(1) - \sum\limits_{l=0}^{m-1} C_{m+1}^l \boldsymbol{P}^{(l)}(1)Q^{(m+1-l)}(1)]\mathbf{e}, \end{cases}$$

where the derivatives $\mathcal{B}^{(m)}(1)$ are calculated using formula (13) and expressions (1)–(3) for the vector generator functions $Q(z), Q^{(0)}(z), Q^{(1)}(z)$.

The proof of the corollary is parallel to the one outlined in [19] and is omitted here.

Having the stationary distribution $\boldsymbol{p}_i$, $i \geq 0$, been calculated we find a number of important stationary performance measures of the system and examine their behavior through the numerical experiments.

- Throughput of the system (the maximum rate of the flow that can be processed by the system)

$$\varrho = -\boldsymbol{\pi}_0(S_1 \oplus S_2)\mathbf{e} + \boldsymbol{\pi}_1\mathbf{S}_0^{(2)} + \boldsymbol{\pi}_2\mathbf{S}_0^{(1)} + \boldsymbol{\pi}_3\mathbf{S}_0^{(3)}.$$

- Mean number of customers in the system $L = \boldsymbol{P}'(1)\mathbf{e}$.
- Variance of the number of customers in the system $V = \boldsymbol{P}''(1)\mathbf{e} - L^2 + L$.
- Probability that i customers stay in the system $p_i = \mathbf{p}_i\mathbf{e}$.
- Probability $P_i^{(0)}$ that i customers stay in the system and both servers are fault- free

$$P_0^{(0)} = \mathbf{p}_0 \begin{pmatrix} \mathbf{e}_a \\ \mathbf{0}^T \end{pmatrix}, \; P_1^{(0)} = \mathbf{p}_1 \begin{pmatrix} \mathbf{e}_{a(M_1+M_2)} \\ \mathbf{0}^T \end{pmatrix}, \; P_i^{(0)} = \mathbf{p}_i \begin{pmatrix} \mathbf{e}_{aM_1M_2} \\ \mathbf{0}^T \end{pmatrix}, \; i \geq 2.$$

- Probability $P_i^{(1)} (P_i^{(2)})$ that i customers stay in the system and server 1 (server 2) is under repair

$$P_0^{(1)} = \mathbf{p}_0 \begin{pmatrix} \mathbf{0}_a^T \\ \mathbf{e}_{aR_1} \\ \mathbf{0}^T \end{pmatrix}, \quad P_0^{(2)} = \mathbf{p}_0 \begin{pmatrix} \mathbf{0}_{a(1+R_1)}^T \\ \mathbf{e}_{aR_2} \\ \mathbf{0}^T \end{pmatrix}$$

$$P_1^{(1)} = \mathbf{p}_1 \begin{pmatrix} \mathbf{0}^T_{a(M_1+M_2)} \\ \mathbf{e}_{aM_2R_1} \\ \mathbf{0}^T \end{pmatrix}, \quad P_0^{(2)} = \mathbf{p}_1 \begin{pmatrix} \mathbf{0}^T_{a(M_1+M_1+M_2R_1)} \\ \mathbf{e}_{aM_1R_2} \\ \mathbf{0}^T \end{pmatrix}.$$

$$P_i^{(1)} = \mathbf{p}_i \begin{pmatrix} \mathbf{0}^T_{aM_1M_2} \\ \mathbf{e}_{aM_2R_1} \\ \mathbf{0}^T \end{pmatrix}, \quad P_0^{(2)} = \mathbf{p}_0 \begin{pmatrix} \mathbf{0}^T_{aM_2(M_1+R_1)} \\ \mathbf{e}_{aM_1R_2} \\ \mathbf{0}^T \end{pmatrix}, i \geq 2.$$

- Probability $P_i^{(3)}$ that i customers stay in the system and both main servers are under repair

$$P_0^{(3)} = \mathbf{p}_0 \begin{pmatrix} \mathbf{0}^T \\ \mathbf{e}_{aR_1R_2} \end{pmatrix}, P_1^{(3)} = \mathbf{p}_1 \begin{pmatrix} \mathbf{0}^T \\ \mathbf{e}_{aM_3R_1R_2} \end{pmatrix}, P_i^{(3)} = \mathbf{p}_i \begin{pmatrix} \mathbf{0}^T \\ \mathbf{e}_{aM_3R_1R_2} \end{pmatrix}, i \geq 2.$$

- Probability that at an arbitrary time the servers are in the state n

$$P^{(n)} = \sum_{i=0}^{\infty} P_i^{(n)}, \ n = \overline{0,3}.$$

- Probability $P_{i,k}^{(0)}$ that an arriving batch of size k finds i customers in the system and both servers fault-free

$$P_{0,k}^{(0)} = \lambda^{-1}\mathbf{p}_0 \begin{pmatrix} I_{\bar{W}} \otimes \mathbf{e}_{\bar{V}} \\ O \end{pmatrix} D_k, \quad P_{1,k}^{(0)} = \lambda^{-1}\mathbf{p}_1 \begin{pmatrix} I_{\bar{W}} \otimes \mathbf{e}_{\bar{V}(M_1+M_2)} \\ O \end{pmatrix} D_k,$$

$$P_{i,k}^{(0)} = \lambda^{-1}\mathbf{p}_i \begin{pmatrix} I_{\bar{W}} \otimes \mathbf{e}_{\bar{V}M_1M_2} \\ O \end{pmatrix} D_k, \ i \geq 2.$$

- Probability $P_{i,k}^{(1)}(P_{i,k}^{(2)})$ that an arriving batch of size k finds i customers in the system and server 1 (server 2) under repair

$$P_{0,k}^{(1)} = \lambda^{-1}\mathbf{p}_0 \begin{pmatrix} O_{a\times\bar{W}} \\ I_{\bar{W}} \otimes \mathbf{e}_{\bar{V}R_1} \\ O \end{pmatrix} D_k, \quad P_{1,k}^{(1)} = \lambda^{-1}\mathbf{p}_1 \begin{pmatrix} O_{a(M_1+M_2)\times\bar{W}} \\ I_{\bar{W}} \otimes \mathbf{e}_{\bar{V}M_2R_1} \\ O \end{pmatrix} D_k,$$

$$P_{i,k}^{(1)} = \lambda^{-1}\mathbf{p}_i \begin{pmatrix} O_{aM_1M_2\times\bar{W}} \\ I_{\bar{W}} \otimes \mathbf{e}_{\bar{V}M_2R_1} \\ O \end{pmatrix} D_k, \ i \geq 2,$$

$$P_{0,k}^{(2)} = \lambda^{-1}\mathbf{p}_0 \begin{pmatrix} O_{a(1+R_1)\times\bar{W}} \\ I_{\bar{W}} \otimes \mathbf{e}_{\bar{V}R_2} \\ O \end{pmatrix} D_k, \quad P_{1,k}^{(2)} = \lambda^{-1}\mathbf{p}_1 \begin{pmatrix} O \\ I_{\bar{W}} \otimes \mathbf{e}_{\bar{V}M_1R_2} \\ O_{aM_3R_1R_2\times\bar{W}} \end{pmatrix} D_k,$$

$$P_{i,k}^{(2)} = \lambda^{-1}\mathbf{p}_0 \begin{pmatrix} O_{aM_2(M_1+R_1)\times\bar{W}} \\ I_{\bar{W}} \otimes \mathbf{e}_{\bar{V}M_1R_2} \\ O \end{pmatrix} D_k, \ i \geq 2.$$

- Probability $P_{i,k}^{(3)}$ that an arriving batch of size k finds i customers in the system and both main server under repair

$$P_{0,k}^{(3)} = \lambda^{-1}\mathbf{p}_0 \begin{pmatrix} O \\ I_{\bar{W}} \otimes \mathbf{e}_{\bar{V}R_1R_2} \end{pmatrix} D_k, \; P_{1,k}^{(3)} = \lambda^{-1}\mathbf{p}_1 \begin{pmatrix} O \\ I_{\bar{W}} \otimes \mathbf{e}_{\bar{V}M_3R_1R_2} \end{pmatrix} D_k,$$

$$P_{i,k}^{(3)} = \lambda^{-1}\mathbf{p}_i \begin{pmatrix} O \\ I_{\bar{W}} \otimes \mathbf{e}_{\bar{V}M_3R_1R_2} \end{pmatrix} D_k, \; i \geq 2.$$

6 Numerical Results

In this section, we present results of four numerical experiments. The purpose of the experiments is to study the behavior of the main performance characteristics of the system as functions of its parameters and illustrate the influence of the correlation in the input flow and variation in the repair process.

Experiment 1. In the experiment, we investigate the dependence of the mean number of customers in the system, L, on the input rate λ for different values of the breakdown rate h.

We suppose that the maximum size of batch in the $BMAP$ is 3. To specify the $BMAP$, we first define the matrices D_0 and D

$$D_0 = \begin{pmatrix} -1.349076 & 10^{-6} \\ 10^{-6} & -0.043891 \end{pmatrix}, \quad D = \begin{pmatrix} 1.340137 & 0.008939 \\ 0.0244854 & 0.0194046 \end{pmatrix}.$$

Now we express the matrices $D_k, k = \overline{1,3}$, in terms of the matrix D using the formula $D_k = Dq^{k-1}(1-q)/(1-q^3), k = \overline{1,3}$, where $q = 0.8$.

This $BMAP$ has the coefficient of correlation $c_{cor} = 0.407152$. The squared coefficient of variation is equal to $c_{var}^2 = 9.621426$.

$MMAP$ of breakdowns is defined by the matrices

$$H_0 = \begin{pmatrix} -8.110725 & 0 \\ 0 & -0.26325 \end{pmatrix},$$

$$H_1 = \frac{1}{3}\begin{pmatrix} 8.0568 & 0.053925 \\ 0.146625 & 0.116625 \end{pmatrix}, H_2 = \frac{2}{3}\begin{pmatrix} 8.0568 & 0.053925 \\ 0.146625 & 0.116625 \end{pmatrix}.$$

For this $MMAP$ $c_{cor} = 0.200504557$, $c_{var}^2 = 12.34004211$. We denote by h the total breakdown rate defined as $h = h_1 + h_2$ where, as defined above, h_1 and h_2 are the rates of breakdowns arriving at server 1 and server 2, respectively. It is evident from the form of the matrices H_1 and H_2 that $h_1 = \frac{1}{3}h$ and $h_2 = \frac{2}{3}h$.

PH service time distributions at server 1, server 2 and server 3 will be denoted as $PH_1^{(serv)}$, $PH_2^{(serv)}$, $PH_3^{(serv)}$, respectively. They are assumed to be Erlangian of order 2 with parameter 20, 15 and 4. These distributions are defined by the vectors $\boldsymbol{\beta}^{(1)} = \boldsymbol{\beta}^{(2)} = \boldsymbol{\beta}^{(3)} = (1, 0)$ and the matrices

$$S^{(1)} = \begin{pmatrix} -20 & 20 \\ 0 & -20 \end{pmatrix}, \quad S^{(2)} = \begin{pmatrix} -15 & 15 \\ 0 & -15 \end{pmatrix}, \quad S^{(3)} = \begin{pmatrix} -4 & 4 \\ 0 & -4 \end{pmatrix}.$$

PH repair time distributions at server 1 and server 2 coincide. They are assumed to be hyper-exponential of order 2 with the squared variation coefficient $c_{var}^2 = 25.07248$ and are defined by the following vector and matrix:

$$\boldsymbol{\tau}^{(1)} = \boldsymbol{\tau}^{(2)} = (0.05,\, 0.95),\;\; T^{(1)} = T^{(2)} = \begin{pmatrix} -0.003 & 0 \\ 0 & -0.245 \end{pmatrix}.$$

Figure 1 depicts the mean number L of customers in the system as a function of λ for different values of breakdown rate, $h = 0.0001, h = 0.001, h = 0.001$.

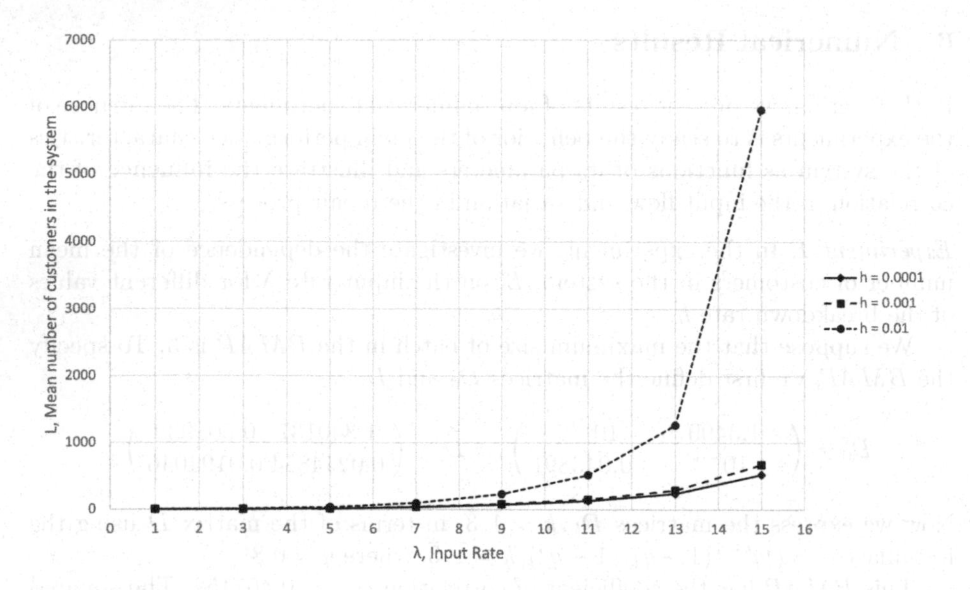

Fig. 1. Mean number L of customers in the system as a function of input rate λ for different values of breakdown rate: $h = 0.0001; h = 0.001; h = 0.001$.

It can be seen from Fig. 1 that the mean number of customers in the system, as expected, increases with increasing input rate λ and the rate of increase grows with increasing the breakdown rate h. You can also see that the mean queue length is rapidly increasing with the growth of load factor ρ.

Experiment 2. In the experiment, we investigate the dependence of the mean value L of the number of customers in the system on the input rate λ for different values of the coefficient of correlation in the $BMAP$. We consider three $BMAP$s having the same mean arrival rate but different coefficients of correlation. These $BMAP$s will be denoted as $BMAP_1, BMAP_2, BMAP_3$.

$BMAP_1$ is a stationary Poisson process with $D_0 = -\lambda$, $D_1 = \lambda$. For this process $c_{cor} = 0$, $c_{var} = 1$.

$BMAP_2$ is defined by the matrices D_0 and $D_k = Dq^{k-1}(1-q)/(1-q^3), k = \overline{1,3}$, where $q = 0.8$,

$$D_0 = \begin{pmatrix} -6.34080 & 10^{-6} \\ 10^{-6} & -0.13888 \end{pmatrix}, \quad D = \begin{pmatrix} 6.32140 \; 0.01939 \\ 0.10822 \; 0.03066 \end{pmatrix}$$

For this $BMAP$ $c_{var} = 3.5$, $c_{cor} = 0.1$.

$BMAP_3$ is the $BMAP$ defined in the Experiment 1. For this $BMAP$ $c_{var}^2 = 9.621425623$, $c_{cor} = 0.407152089$. We also fix the $MMAP$ of breakdowns and PH distributions of service and repair times the same as in Experiment 1.

Figures 2 depicts the mean number of customers in the system as a functions of λ for different $BMAP$s. We see that values of L depend, in some cases strongly, on the coefficient of correlation in the $BMAP$. Under the same value of input rate, λ, the mean number of customers in the system increases when the correlation increases. In addition, the difference in the values of L for different coefficients of correlation increases with increasing λ. In particular, we see that approximation of the $BMAP$ with coefficient correlation $c_{cor} = 0.4$ by the stationary Poisson process leads to a huge error in the calculation of L. Thus, under such an approximation, the mean estimates are too optimistic.

Experiment 3. In the experiment, we investigate the dependence of the number of customers in the system, L, on the breakdowns rate h for PH distributions of repair time with different coefficients of variation.

We take $BMAP$, $MMAP$ of breakdowns and PH distributions of service times from Experiment 1. We assume that PH distributions of repair times of server 1 and server 2 coincide and consider three different PH distributions (PH_1, PH_2, PH_3). The average repair time is equal to 20, however, the coefficients of variation of repair time are different.

PH_1 is the exponential distribution with the parameter 0.05 and coefficient of variation $c_{var} = 1$.

PH_2 is the hyper-exponential distribution of order 2. This distribution is defined by the following vector and matrix:

$$\boldsymbol{\tau} = (0.05, 0.95),\ T = \begin{pmatrix} -0.003 & 0 \\ 0 & -0.245 \end{pmatrix}.$$

In this case the repair time has coefficient of variation $c_{var} = 5$.

PH_3 is the hyper-exponential distribution of order 2. This distribution is defined by the following vector and matrix:

$$\boldsymbol{\tau} = (0.05, 0.95),\ T = \begin{pmatrix} -250000 & 0 \\ 0 & -0.05 \end{pmatrix}.$$

In this case the repair time has coefficient of variation $c_{var} = 9.9$.

Since in this experiment we choose the breakdown rates such that the following value differs from the previous one by an order of magnitude, it is reasonable to use a logarithmic scale (with a base of ten) on the X-axis. The resulting graph is shown in Fig. 3. It is seen from Fig. 3 that, under the same value of repair rate, h, the mean number of customers in the system L essentially varies for repair times with different coefficient of variation. In this example, the value of L increases when the variation increases and the difference in the value of L increases with increasing of h.

Experiment 4. In the experiment, we investigate the dependence of the throughput ϱ on the repair rate τ.

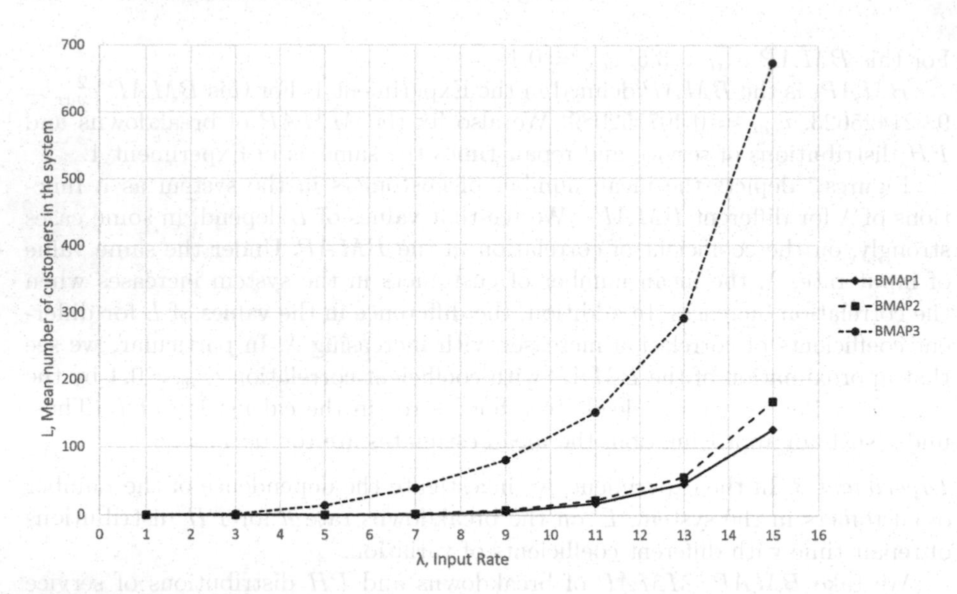

Fig. 2. Mean number of customers in the system as a function of input rate λ for $BMAP$s with different coefficients of correlation ($c_{cor} = 0; 0.1; 0.4$)

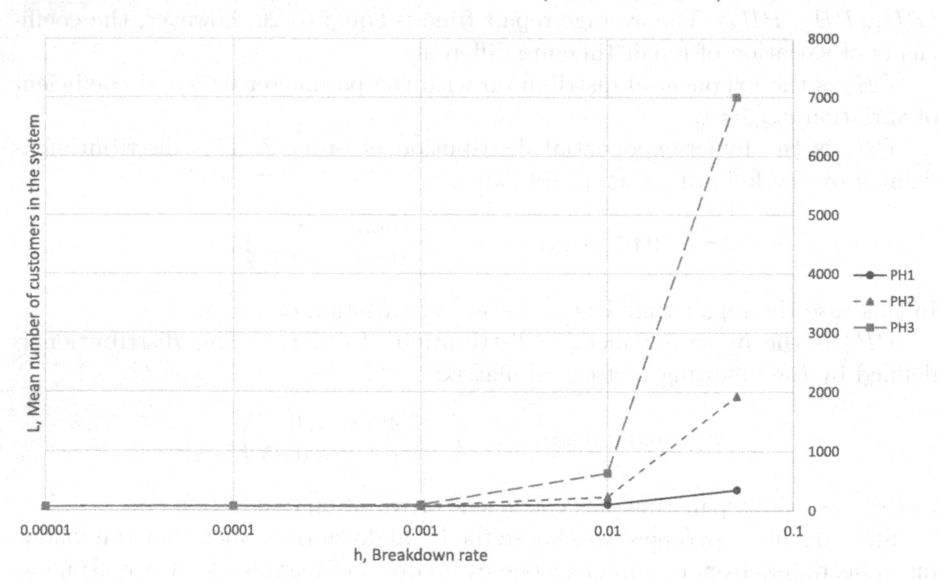

Fig. 3. Mean number of customers in the system as a function the breakdowns rate for repair times with different coefficients of variation

Here we use the same input data as in Experiment 1. Denote the identical repair rates of server 1 and server 2 as τ.

By analogy with the previous experiment, in this experiment we choose the breakdown rates such that the following value differs from the previous one by

Table 1. Values of throughput ϱ obtained in Experiment 4

$\tau \setminus h$	0.0001	0.0005	0.001	0.005	0.01	0.05	0.1	0.5
0.000001	2.627	2.177	2.094	2.020	2.010	2.002	2.001	2.000
0.00001	4.865	2.995	2.627	2.177	2.094	2.020	2.010	2.002
0.0001	11.95	6.49	4.865	2.994	2.627	2.176	2.094	2.020
0.001	16.68	14.06	11.95	6.494	4.864	2.994	2.626	2.176
0.01	17.42	17.08	16.68	14.06	11.94	6.489	4.858	2.986
0.1	17.49	17.46	17.42	17.08	16.67	14.05	11.92	6.439

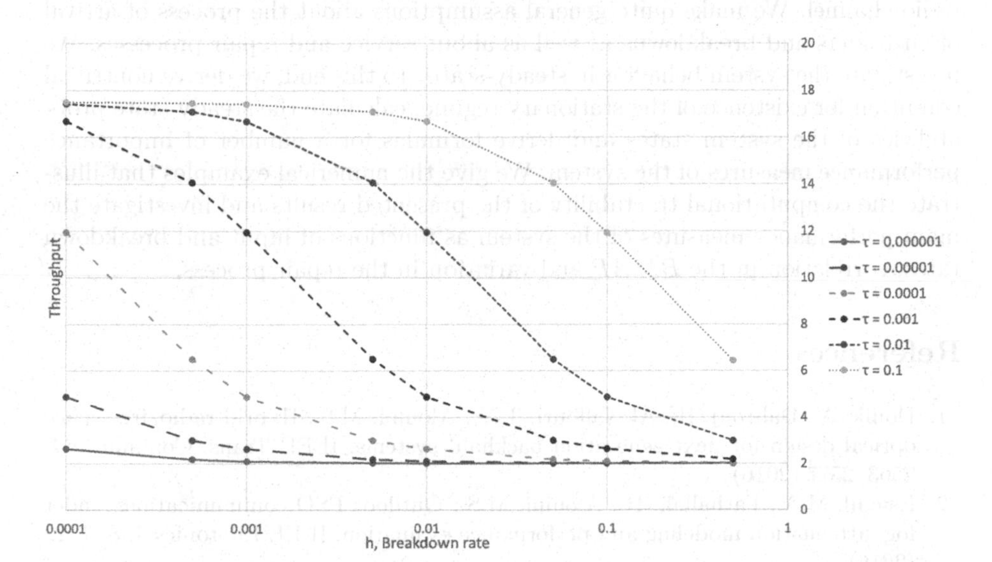

Fig. 4. Throughput of the system as a function of the breakdown rate h for different repair rates τ.

an order of magnitude. Thus, it is reasonable to use a logarithmic scale (with a base of ten) on the X-axis. The resulting graph is shown in Fig. 4. The data for the graph are also given in Table 1. As expectable, under the same value of repair rate τ, the throughput decreases when the breakdown rate increases. The behavior of the curves is more interesting. Let us consider the case when breakdowns rarely occur and repairs are fast. In this case we can assume that $\varrho \approx \mu_1 + \mu_2$, where μ_n is the service rate on server $n, n = 1, 2$. As it is seen from Table 1 this intuitive assumption is confirmed numerically: for $\tau = 0.1$ and $h = 0.0001$ $\varrho = 17.49$ while $\mu_1 + \mu_2 = 17.5$ in this experiment. It can also be seen from the Figure and the Table that, as expected, the throughput cannot exceed the horizontal asymptote $\varrho = \mu_1 + \mu_2$. Further, it can be noted that as repair rate is low and the breakdown rate increases, the curves tend to the horizontal asymptote $\varrho = \mu_3 = 2$, where μ_3 is the service rate on the backup server. Such a behavior coincides with the expected behavior of the system since, under the high breakdown rate, both main servers are almost always broken and only the backup reliable server serves customers.

7 Conclusion

In this paper, we have investigated the unreliable queueing system with backup server that can be used for modelling the hybrid communication system consisting of two main unreliable but high-speed channels (FSO channel and a radio channel of millimeter-wave) which are reserved by reliable but low-speed radio channel. We make quite general assumptions about the process of arrival of customers and breakdowns as well as about service and repair processes. We investigate the system behavior in steady-state. To this end, we derive nontrivial condition for existence of the stationary regime, calculate the steady-state probabilities of the system states and derive formulas for a number of importance performance measures of the system. We give the numerical examples that illustrate the computational tractability of the presented results and investigate the main performance measures of the system as functions of input and breakdown rates, correlation in the $BMAP$ and variation in the repair process.

References

1. Douik, A., Dahrouj, H., Al-Naffouri, T.Y., Alouini, M.S.: Hybrid radio/free-space optical design for next generation backhaul systems. IEEE Trans. Commun. **64**, 2563–2577 (2016)
2. Esmail, M.A., Fathallah, H., Alouini, M.S.: Outdoor FSO communications under fog: attenuation modeling and performance evaluation. IEEE Photonics J. **8**, 1–22 (2016)
3. Makki, B., Svensson, T., Eriksson, T., Alouini, M.S.: On the performance of RF-FSO links with and without hybrid ARQ. IEEE Trans. Wirel. Commun. **15**, 4928–4943 (2016)
4. Wu, Y., Yang, Q., Park, D., Kwak, K.S.: Dynamic link selection and power allocation with reliability guarantees for hybrid FSO/RF systems. IEEE Access **5**, 13654–13664 (2017)
5. Zhang, V, Chu, Y.J., Nguyen, T.: Coverage algorithms for WiFO: a hybrid FSO-WiFi femtocell communication system. In: Proceedings of the 26th International Conference on Computer Communication and Networks (ICCCN), pp. 1–6. IEEE (2017)
6. Zhou, K., Gong, C., Wu, N., Xu, Z.: Distributed channel allocation and rate control for hybrid FSO/RF vehicular ad hoc networks. J. Opt. Commun. Netw. **9**, 669–681 (2017)
7. Shehaj, M., Nace, D., Kalesnikau, I., Pioro, M.: Link dimensioning of hybrid FSO/fiber networks resilient to adverse weather conditions. Comput. Netw. **161**, 1–13 (2019)
8. Advanced Optical Wireless Communication Systems. Cambridge University Press (2012)

9. Sharov, S.Yu., Semenova, O.V.: Simulation model of wireless channel based on FSO and RF technologies. In: Proceedings of the 10th Distributed Computer and Communication Networks. Theory and Applications (DCCN-2010), pp. 368–374 (2010)
10. Vishnevsky, V.M., Semenova, O.V., Sharov, S.Y.: Modeling and analysis of a hybrid communication channel based on free-space optical and radio-frequency technologies. Autom. Remote Control **72**, 345–352 (2013)
11. Vishnevsky, V.M., Kozyrev, D.V., Semenova, O.V.: Redundant queuing system with unreliable servers. In: Proceedings of the 6th International Congress on Ultra Modern Telecommunications and Control Systems, pp. 383–386. IEEE Xplore Digital Library (2014)
12. Dudin, A., Klimenok, V., Vishnevsky, V.: Analysis of unreliable single server queuing system with hot back-up server. Commun. Comput. Inf. Sci. **499**, 149–161 (2015)
13. Klimenok, V.I.: Two-server queueing system with unreliable servers and Markovian arrival process. Commun. Comput. Inf. Sci. **800**, 42–55 (2017)
14. Vishnevsky, V.M., Klimenok, V.I.: Unreliable queueing system with cold redundancy. Commun. Comput. Inf. Sci. **522**, 336–347 (2015)
15. Lucantoni, D.M.: New results on the single server queue with a batch Markovian arrival process. Commun. Stat. Stoch. Model. **7**, 1–46 (1991)
16. Neuts, M.: Matrix-Geometric Solutions in Stochastic Models - An Algorithmic Approach. Johns Hopkins University Press, Baltimore (1981)
17. He, Q.M.: Queues with marked customers. Adv. Appl. Probab. **28**, 567–587 (1996)
18. Klimenok, V.I., Dudin, A.N.: Multi-dimensional asymptotically quasi-Toeplitz Markov chains and their application in queueing theory. Queueing Syst. **54**(245–259), 42–55 (2006)
19. Dudin, A., Klimenok, V., Lee, M.H.: Recursive formulas for the moments of queue length in the $BMAP/G/1$ queue. IEEE Commun. Lett. **13**, 351–353 (2009)

Flexible Random Early Detection Algorithm for Queue Management in Routers

Aminu Adamu[1(✉)], Vsevolod Shorgin[2], Sergey Melnikov[3], and Yuliya Gaidamaka[2,3]

[1] Umaru Musa Yar'adua University, Katsina, Nigeria
amadamum@gmail.com

[2] Federal Research Center "Computer Science and Control" of the Russian Academy of Sciences, Moscow, Russia
vshorgin@ipiran.ru, gaydamaka-yuv@rudn.ru

[3] Peoples' Friendship University of Russia (RUDN University), Moscow, Russia
melnikov@linfotech.ru

Abstract. With recent advancements in communication networks, congestion control remains a research focus. Active Queue Management (AQM) schemes are normally used to manage congestion in routers. Random Early Detection (RED) is the most popular AQM scheme. However, RED lacks self-adaptation mechanism and it is sensitive to parameter settings. Many enhancements of RED were proposed and are yet to provide stable performance under different traffic load situations. In this paper, AQM scheme called Flexible Random Early Detection (FXRED) is proposed. Unlike other RED's enhancements with static drop patterns, FXRED recognizes the state of the current network's traffic load and auto tune its drop pattern suitable to the observed load situation in order to maintain stable and better performance. Results of the experiments conducted have shown that regardless of traffic load's fluctuation, FXRED provides optimal performance and efficiently manages the queue.

Keywords: Network · AQM · RED · Flexible RED · Congestion control

1 Introduction

Congestion is one of the major phenomena that affect the quality of service (QoS) provided in networks, when congestion occurred, poor utilization, high delay, high packet loss rate and low throughput would be experienced [1]. Congestion remains a threat to performance in existing and especially future

The publication has been prepared with the support of the "RUDN University Program 5–100" (recipient Yu. Gaidamaka, conceptualization). The reported study was funded by RFBR, project numbers 18-07-00576 and 20-07-01064 (recipient V. Shorgin, validation and visualization).

V. M. Vishnevskiy et al. (Eds.): DCCN 2020, LNCS 12563, pp. 196–208, 2020.
https://doi.org/10.1007/978-3-030-66471-8_16

generation networks [2–6]. To ensure better network performance, congestion must be avoided. Active Queue Management (AQM) was introduced to control congestion and works by sending early congestion notification alert to end devices; subsequently end devices upon reception of such notification will learn to reduce their transmission rates before overflow occurs in the network routers. AQM also aimed at increasing throughput by reducing the number of packets dropped, managing queue lengths to absorb short-term congestion, providing a lower interactive delay and avoiding global synchronization by randomly dropping packets across all flows [7].

Random Early Detection (RED) algorithm proposed in [8] was one of the early generations of AQM schemes and it was recommended by Internet Engineering Task Force (IETF) to be used in Internet routers for congestion control [7], as such, some router vendors have implemented RED algorithm as a default congestion control algorithm in their product; Cisco implemented WRED [9]. RED algorithm controls congestion by actively dropping incoming packets probabilistically. Traditional RED algorithm is associated with four basic parameters, which are the queue's thresholds (min_{th} and max_{th}), average queue length (avg), maximum packet drop probability (max_p) and weighted parameter (w). The average queue length is computed using the Exponentially Weighted Moving Average (EWMA) method applied to the instantaneous queue length. If the computed avg is below the min_{th} threshold, no packet will be dropped and if avg is above max_{th} threshold, all the incoming packets will be dropped with probability 1, however, if avg is between thresholds min_{th} and max_{th}, then packet drop probability increases linearly to the maximum drop probability max_p [8].

Even though it was proven that RED is superior to conventional Tail drop algorithm [7], however, RED lacks self-adaptation mechanism and it is sensitive to parameter settings [10]. To address the shortcomings of RED, many enhanced versions of RED were proposed, such as Gentle RED [10], Dynamic RED [11], Double Slope RED [12], Nonlinear RED [13], Improved Nonlinear RED [14], Three Section RED [15], Adaptive Queue Management with Random Dropping [16], Change Trend Queue Management [17], etc. However, with RED or some of its enhanced versions, the effect of congestion management will be seriously affected once the traffic load changed [15–20].

In this paper, an AQM scheme termed as Flexible RED (FXRED) is proposed. Unlike RED and some of its enhanced variants, in addition to avg, FXRED considers the current traffic load as congestion indicator, i.e. FXRED recognizes the state of the current network's traffic load and adapts a drop pattern suitable to the observed load situation. FXRED increases its drop rate as load becomes high in order to avoid overflow and congestion, conversely, as load becomes low, FXRED decreases its drop rate in order to maximize link utilization and throughput. The scheme proposed in this paper could also be used to create queue-based service policies for network systems as in [21–23].

The rest of this paper is organized as follows. Section 2 presents the related works and the proposed Flexible RED algorithm is presented in Sect. 3. Results of the experiments conducted for the analysis of the proposed algorithm are presented in Sect. 4 and Sect. 5 concludes the paper.

2 Related Works

Traditionally, Active Queue Management (AQM) schemes are used to control congestion in routers. The most popular AQM is Random Early Detection (RED) algorithm proposed by S. Floyd and V. Jacobson in [8]. RED achieves the desired goals of AQM scheme by dividing the router's queue into three sections using threshold values min_{th} and max_{th} and performs its operations in two phases.

1st **Phase of RED operations:** computation of the average queue length *avg*.

Instead of using instantaneous queue length, RED continuously computes the average queue length (*avg*) using a simple exponentially weighted moving average.

2nd **Phase of RED operations:** decision on whether to accept or drop an incoming packet. In order to make a decision on whether to accept or drop an incoming packet, the computed *avg* in the first phase is compared with queue's threshold values, i.e. min_{th} and max_{th}. If $avg < min_{th}$, then all incoming packets will be accepted, if $min_{th} \leq avg < max_{th}$, then packets are dropped randomly across all the flows with probability which increases linearly as a function of *avg* to the maximum drop probability max_p, however, if $avg \geq max_{th}$, then all incoming packets will be dropped with probability 1.

Even though it was proven that RED is superior to conventional Tail drop algorithm, however, RED suffers from some shortcomings such as parameterization problem (i.e. setting RED parameters and their tuning in order to achieve better performance based on current load condition), low throughput, large delay/jitter, unfairness to connections, and induces instability in the network [15–20].

In order to improve the throughput of RED, Floyd in [10] proposed Gentle RED (GRED). In GRED, the region of the queue from min_{th} to max_{th} was further extended to $2max_{th}$. If the *avg* falls between min_{th} and max_{th}, packet dropping probability increases linearly from 0 to max_p (as done in traditional RED), further, if the *avg* falls within max_{th} and $2max_{th}$, then packet dropping probability increases linearly from max_p to 1, thereby making GRED more gentler than RED.

Zheng and Atiquzzaman in [12] proposed Double-Slope RED (DS-RED). In DS-RED, a mid-point (mid_{th}) was introduced between min_{th} and max_{th}, when *avg* is between min_{th} and mid_{th}, the drop probability is defined by a linear function, when *avg* is between mid_{th} and max_{th}, drop probability also increases linearly (with different slope). This is done in order to reduce the aggressiveness of RED and to improve link utilization. However, DS-RED's behavior is similar to that of Gentle RED [10].

The aggressiveness of RED and some of its enhanced versions made Zhou et al. to propose a Non-linear RED (NRED) in [13]. In NRED, the linear drop function of RED is replaced with a non-linear quadratic drop function. However, under high load conditions, NRED will cause overflow and forced packet drops.

Zhang et al. in [14] proposed MRED which is an improved nonlinear RED. MRED differs from GRED in the sense that if *avg* is between min_{th} and max_{th}, the drop probability is defined by a nonlinear function. However, if *avg* falls between max_{th} and $2max_{th}$, the drop probability increases linearly to the maximum of 1 as in GRED.

In [15], Feng et al. proposed Three Section RED (TRED). Unlike other versions of RED where a mid_{th} is introduce between min_{th} and max_{th}, in TRED, two mid-points $mid_{th}(1)$ and $mid_{th}(2)$ were introduced in order to capture three load scenarios, i.e. low, moderate and high. When *avg* is between min_{th} and $mid_{th}(1)$ (corresponding to the low load state), packets are dropped with probability defined by a nonlinear drop function, when *avg* falls between $mid_{th}(1)$ and $mid_{th}(2)$ (moderate load), packets are dropped with probability which increases linearly and when *avg* falls between $mid_{th}(2)$ and max_{th} (high load), then packet drop probability is defined by another nonlinear function. Although results of the analysis conducted in [13] have shown that TRED improved throughput at low load and maintained low delays at high load, however, parameterization is still a problem in TRED.

Karmeshu et al. in [16] proposed another AQM scheme called Adaptive Queue Management with Random Dropping (AQMRD). In AQMRD, *avg* and its rate of change (*davg*) are used to optimally define the drop function, such that the queue length as well as the delay remains low regardless of how frequently the traffic load changes. However, dropping strategy introduced in AQMRD is so aggressive which leads to poor link utilization and high loss rate.

Another Adaptive AQM was proposed by Tang and Tan called Change Trend Queue Management (CT-AQM) [17]. CT-AQM predicts the change trend in average queue length based on its rate of change as well as the network environment to define packet drop function. AQMRD proposed in [16] is to some extent similar to CT-AQM since both of them use the rate of change in average queue length. Results of the analysis conducted in [17] have shown that CT-AQM was successful in lowering loss-rate and improved throughput for different load situations. However, CT-AQM introduces high delay and allows more packets into queue which may lead to congestion when used in a complex network environment.

In this paper, a Flexible Random Early Detection (FXRED) algorithm is proposed. Unlike the existing enhanced RED schemes, FXRED considers the current traffic load condition and adapts a suitable drop pattern. If the observed traffic load is low, FXRED gently drops packets since in that situation congestion is not a threat, however, when traffic load becomes high, FXRED will switch to aggressive drop pattern in order to avoid congestion and overflow.

3 Flexible Random Early Detection Algorithm

In the proposed Flexible RED (FXRED), the router's queue with finite capacity is divided into four (4) segments (A, B, C and D) via threshold values min_{th}, Δ and max_{th} (Fig. 1), where $\Delta = \frac{1}{2}(min_{th} + max_{th})$.

In FXRED, the data arrival rate from each flow at time t is considered, i.e. $\lambda_n(t)$, $n = 1, ..., N$, where N is the number of active flows at time t. Then, the

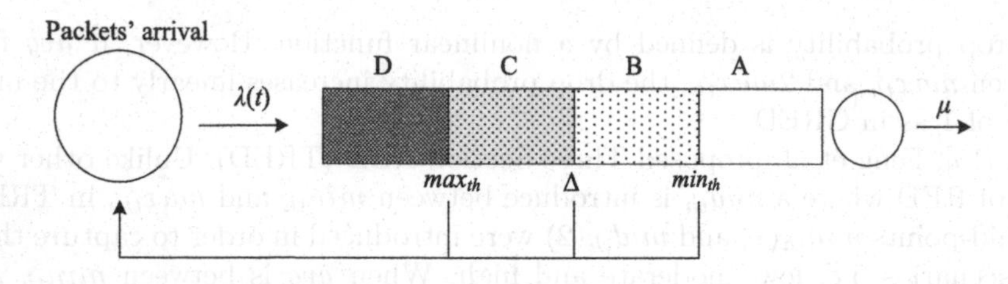

Fig. 1. FXRED's queue

total data arrival rate (from all the active flows) at the router at time t is given by Eq. (1).

$$\lambda(t) = \sum_{n=1}^{N} \lambda_n(t). \tag{1}$$

Let μ be the bandwidth of the bottleneck link, then the traffic load at time t is denoted by $\rho(t)$ and expressed by Eq. (2).

$$\rho(t) = \frac{\lambda(t)}{\mu}. \tag{2}$$

Given (2), at time $t > 0$, the system can be in three (3) possible load states as follows:

State-1: $\rho(t) < 1$,
State-2: $\rho(t) \approx 1$,
State-3: $\rho(t) > 1$.

If the system is in *state*-1 ($\rho(t) < 1$), i.e. for the bottleneck link the traffic load is light, therefore, packets will not accumulate in the queue, if the system remains in this state for a long time, link underutilization will occur. If the system is in *state*-2 ($\rho(t) \approx 1$), the traffic load is moderate for the bottleneck link, hence, performance is optimal. However, if the system is in *state*-3 ($\rho(t) > 1$), the traffic load is high for the bottleneck link, packets will accumulate in the queue and wait to be sent, if this state is maintained for a long time, congestion and overflow may likely occur. Note that with FXRED, network's load states can also be formed based on the *number of sources multiplexed over the bottleneck link.*

FXRED recognizes the current traffic load state and adapts a suitable drop pattern. If the network is in *state*-1, FXRED gently drops packets in order to maximize link utilization and throughput, however, if the network is in *state*-3, FXRED aggressively drops packets in order to avoid congestion and overflow.

Just like in traditional RED and its other variants, in FXRED, average queue length (avg) is computed via Eq. (3).

$$avg = (1 - w)avg' + wq(t), \tag{3}$$

where $q(t)$ is the queue length at time $t > 0$ (instantaneous queue length), avg' is the previously obtained average queue length and w is the predefined weight parameter to compute the avg, $0 < w < 1$.

If $avg \in [0, min_{th})$ (Segment A), then the drop probability is zero, i.e. no packet will be dropped. If $avg \in [min_{th}, \Delta)$ (Segment B), then FXRED proposes a nonlinear drop function expressed by Eq. (4).

$$p_{d(1)} = 2^{\lfloor k \rfloor} \left(\frac{avg - min_{th}}{max_{th} - min_{th}} \right)^{\lfloor k \rfloor} \cdot (1 - \epsilon), \tag{4}$$

where k and ϵ are expressed by Eqs. (5) and (6) respectively,

$$k = c^{\frac{1}{\gamma}}, c \geq 2. \tag{5}$$

$$\epsilon = c^{-\gamma}. \tag{6}$$

In this case, the value of $c \geq 2$ was chosen in order to turn the operation of FXRED in Segment B of the queue into nonlinear mode at low and moderate load states, where γ is the drop mode regulator and is expressed for the defined three network states as follows (if needed, more states can be added in FXRED):

$$\begin{cases} \gamma < 1 & \text{for } \textbf{State-1}, \\ \gamma \approx 1 & \text{for } \textbf{State-2}, \\ \gamma \geq c & \text{for } \textbf{State-3}. \end{cases} \tag{7}$$

If $avg \in [\Delta, max_{th})$ (Segment C), then FXRED proposes a linear drop function expressed by Eq. (8).

$$p_{d(2)} = 2\epsilon \left(\frac{avg - \Delta}{max_{th} - min_{th}} \right) + (1 - \epsilon). \tag{8}$$

Lastly, if $avg \geq max_{th}$ (Segment D), then all incoming packets will be dropped with probability 1. The general drop function of FXRED is presented in Eq. (9) and depicted in Fig. 2.

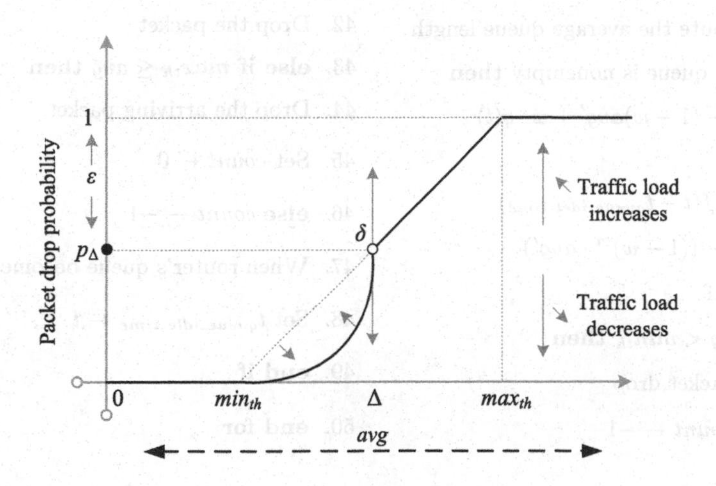

Fig. 2. FXRED's packet dropping probability curve

$$p_d = \begin{cases} 0 & avg < min_{th}, \\ 2^{\lfloor k \rfloor}\left(\frac{avg - min_{th}}{max_{th} - min_{th}}\right)^{\lfloor k \rfloor} \cdot (1-\epsilon) & min_{th} \leq avg < \Delta, \\ 2\epsilon\left(\frac{avg - \Delta}{max_{th} - min_{th}}\right) + (1-\epsilon) & \Delta \leq avg < max_{th}, \\ 1 & avg \geq max_{th}. \end{cases} \tag{9}$$

Algorithm 1: FXRED

1. Input: min_{th}, Δ, max_{th}, w, μ, c
2. Initialization:
3. $avg \leftarrow 0$
4. $count \leftarrow -1$
5. **for each** packet arrival **do**
6. Return $\lambda(t)$ [Mbps]
7. $\rho(t) \leftarrow \frac{\lambda(t)}{\mu}$
8. **if** $\rho(t) < 1$ **then**
9. $\gamma = 0.25$
10. **else if** $\rho(t) \approx 1$ **then**
11. $\gamma = 1$
12. **else if** $\rho(t) > 1$ **then**
13. $\gamma = c$
14. **end if**
15. $k \leftarrow c^{\frac{1}{\gamma}}$
16. $\epsilon \leftarrow c^{-\gamma}$
17. Compute the average queue length
18. **if** the queue is nonempty **then**
19. $avg \leftarrow (1-w)avg' + w \cdot q(t)$
20. **else**
21. $m \leftarrow f(t - t_{queue\_idel\_time})$
22. $avg \leftarrow ((1-w)^m \cdot avg')$
23. **end if**
24. **if** $avg < min_{th}$ **then**
25. No packet drop
26. Set $count \leftarrow -1$
27. **else if** $min_{th} \leq avg < \Delta$ **then**
28. Set $count \leftarrow count + 1$
29. Calculate packet drop probability P_a
30. $P_b \leftarrow 2^{\lfloor k \rfloor}\left(\frac{avg - min_{th}}{max_{th} - min_{th}}\right)^{\lfloor k \rfloor} \cdot (1-\epsilon)$
31. $P_a \leftarrow \frac{P_b}{1 - count \cdot P_b}$
32. Mark the arriving packet with P_a
33. Set $count \leftarrow 0$
34. Drop the packet
35. **else if** $\Delta \leq avg < max_{th}$ **then**
36. Set $count \leftarrow count + 1$
37. Calculate packet drop probability P_a
38. $P_b \leftarrow 2\epsilon\left(\frac{avg - \Delta}{max_{th} - min_{th}}\right) + (1-\epsilon)$
39. $P_a \leftarrow \frac{P_b}{1 - count \cdot P_b}$
40. Mark the arriving packet with P_a
41. Set $count \leftarrow 0$
42. Drop the packet
43. **else if** $max_{th} \leq avg$ **then**
44. Drop the arriving packet
45. Set $count \leftarrow 0$
46. **else** $count \leftarrow -1$
47. When router's queue becomes empty
48. Set $t_{queue\_idle\_time} \leftarrow t$
49. **end if**
50. **end for**

Table 1. FXRED's parameters

Saved variables	Fixed parameters	Other
avg: current average queue length	w: queue weight	t: current time
avg': calculated previous average queue length	min_{th}: queue's minimum threshold	$\lambda(t)$: current total incoming traffic flow (Mbps)
$t_{queue\_idle\_time}$: start of the queue idle time	Δ: queue's middle threshold	$\rho(t)$: current traffic load
count: packets since last marked packets	max_{th}: queue's maximum threshold	$q(t)$: current queue length
k: exponent of the nonlinear drop function	μ: bottleneck link capacity (Mbps)	$f(t)$: a linear function of the time t
γ: drop mode regulator	c: nonlinear index	P_a: current packet-marking probability

4 Simulation Experiments

To analyze the performance of the proposed algorithm, experiments were conducted using NS2 simulator. For the simulation, a network with a double dumbbell topology (Fig. 3) was used, where N TCP flows generated by N sources shared a bottleneck router R_1 (where AQM was implemented), the considered network also comprised of N sinks. The two routers R_1 and R_2 were connected via a bottleneck link with capacity 10Mbps and propagation delay of 20 ms.

Hosts were connected to routers via links with capacity 10 Mbps and propagation delay of 10 ms. The TCP data senders were of the New-Reno type. Different levels of traffic load were created by varying the number of flows across the bottleneck link.

In the experiments, number of flows N was increased from 5 to 95 and each point of the simulation result was obtained after 200s simulation. The queue capacity of router R_1 was 100 packets. The current value of $\rho(t)$ was obtained by applying the current value of $\lambda(t)$ received from queue monitoring object to the bottleneck link capacity. The queue threshold values used were $min_{th} = 25$, $max_{th} = 75$, where $\Delta = \frac{(max_{th}+min_{th})}{2}$, $c = 2$ and $w = 0.002$ (Table 1). The values of γ for three different load states are given by (10).

$$\gamma = \begin{cases} 0.25 & \text{if } \rho(t) < 0.75, \\ 1 & \text{if } \rho(t) \geq 0.75, \\ c & \text{if } \rho(t) \geq 1.25. \end{cases} \tag{10}$$

In this paper, the performance of FXRED is compared with one of the recently proposed enhanced RED versions, i.e. TRED. Since it was discovered in

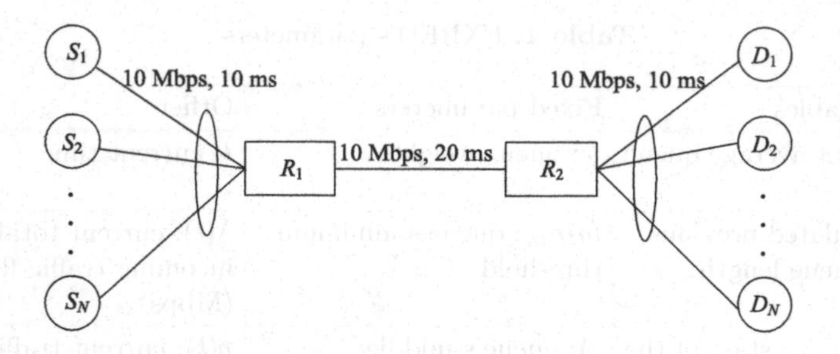

Fig. 3. Simulation topology

[15] that TRED provides better performance than RED, therefore, in this paper RED is not considered. For the analysis of TRED, $max_p = 0.1$ was used. Results obtained from the experiments conducted are presented in Fig. 4, 5, 6 and 7.

Graph presented in Fig. 4 has shown that at low load, with TRED the observed average queue length is slightly above the min_{th} value and as load becomes high the observed average queue length of TRED approaches the max_{th} as exhibited by RED (as observed in [18] and [19]). Although at low load TRED improved link utilization (better than RED [15]), however, at very high load states, TRED cannot cope and will incur forced packet drops and global synchronization [19]. On the other hand, since FXRED adapts its drop pattern based on load, at low load, FXRED adapts a drop pattern that will maintain average queue length well above the min_{th} in order to maximize link utilization and throughput, however, with increase in load, FXRED adapts a drop pattern that will maintain average queue length well below max_{th} in order to maintain stable performance. It can also be observed from graph presented in Fig. 4 that TRED works well at moderate load scenarios.

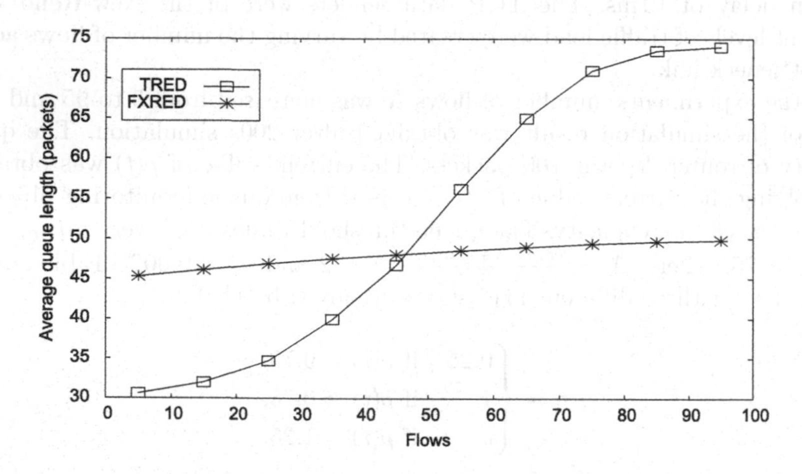

Fig. 4. Average queue length vs. number of flows

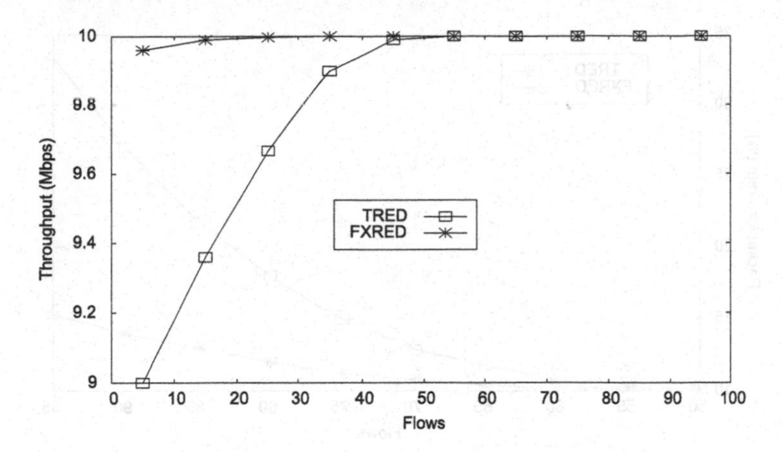

Fig. 5. Throughput vs. number of flows

Results presented in Fig. 5 revealed that FXRED maximizes throughput regardless of load, unlike TRED, at low load the bottleneck link is not fully utilized.

Graph presented in Fig. 6 has shown that delay encountered with FXRED at low load is higher than that of TRED since at low load, FXRED allows more packets to accumulate in the queue than TRED does (Fig. 4), hence, packets will encounter larger queuing delay with FXRED than with TRED, which will finally have impact on the end-to-end delay. However, at high load, the delay observed with FXRED is lower than that of TRED since in high load situation, FXRED has higher drop rate and subsequently sources will send fewer packets (Fig. 4).

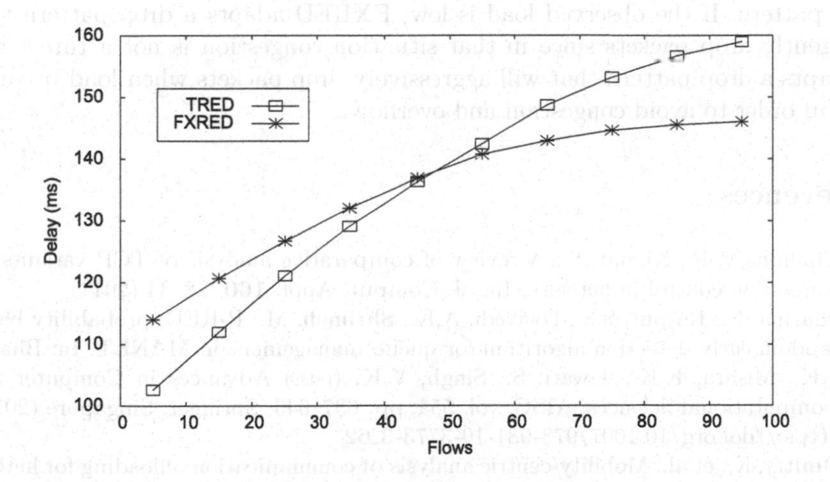

Fig. 6. Delay vs. number of flows

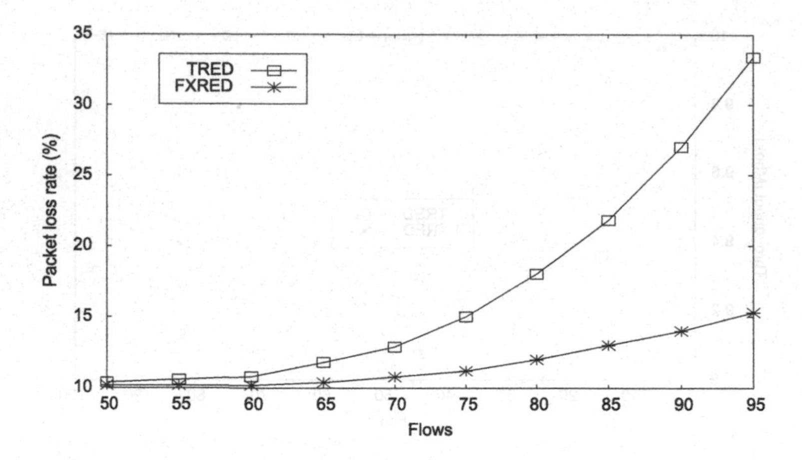

Fig. 7. Packet loss rate vs. number of flows

Finally packet loss rate was analyzed for moderate to high load situations and the results have shown that at high load, TRED has higher packet loss rates than FXRED (Fig. 7), this is because the observed packets losses with FXRED are solely due to its drop policy, while the observed packets losses with TRED are both due to its drop policy and forced packet drops.

5 Conclusion

In this paper, a Flexible Random Early Detection (FXRED) algorithm is proposed. Unlike other enhanced RED schemes with static drop patterns, FXRED considers the state of the current network's traffic load and adapts a suitable drop pattern. If the observed load is low, FXRED adapts a drop pattern that will gently drop packets since in that situation congestion is not a threat and it adapts a drop pattern that will aggressively drop packets when load becomes high in order to avoid congestion and overflow.

References

1. Chaudhary, P., Kumar, S.: A review of comparative analysis of TCP variants for congestion control in network. Int. J. Comput. Appl. **160**, 28–34 (2017)
2. Sharma, N., Rajput, S.S., Dwivedi, A.K., Shrimali, M.: P-RED: probability based random early detection algorithm for queue management in MANET. In: Bhatia, S.K., Mishra, K.K., Tiwari, S., Singh, V.K. (eds.) Advances in Computer and Computational Sciences. AISC, vol. 554, pp. 637–643. Springer, Singapore (2018). https://doi.org/10.1007/978-981-10-3773-3_62
3. Dmitry, K., et al.: Mobility-centric analysis of communication offloading for heterogeneous Internet of Things devices. Green Internet Things (IoT) Enabling Technol. Arch. Perform. Des., 1–10 (2018). https://doi.org/10.1155/2018/3761075

4. Vishnevsky, V., Dudin, A., Kozyrev, D., Larionov, A.: Methods of performance evaluation of broadband wireless networks along the long transport routes. In: Vishnevsky, V., Kozyrev, D. (eds.) DCCN 2015. CCIS, vol. 601, pp. 72–85. Springer, Cham (2016). https://doi.org/10.1007/978-3-319-30843-2_8
5. Samouylov, K.E., Abaev, P.O., Gaidamaka, Y.V., Pechinkin, A.V., Razumchik, R.V.: Analytical modelling and simulation for performance evaluation of sip server with hysteretic overload control. In: 28th European Conference Proceedings on Modelling and Simulation, ECMS, pp. 603–609 (2014)
6. Gaidamaka, Y., Pechinkin, A., Razumchik, R., Samouylov, K., Sopin, E.: Analysis of an $M|G|1|R$ queue with batch arrivals and two hysteretic overload control policies. Int. J. Appl. Math. Comput. Sci. **24**(3), 519–534 (2014)
7. Braden, B., et al.: Recommendations on Queue Management and Congestion Avoidance in the Internet. RFC2309, United States, April 1998
8. Floyd, S., Jacobson, V.: Random early detection gateways for congestion avoidance. IEEE/ACM Trans. Netw. **1**, 397–413 (1993)
9. Cisco Systems, Congestion Avoidance Overview. https://www.cisco.com/c/en/us/td/docs/ios-xml/ios/qos_conavd/configuration/xe-16/qos-conavd-xe-16-book/qos-conavd-oview.html. Accessed 10 Sept 2020
10. Floyd, S.: Recommendation on Using the Gentle Variant of RED. https://www.icir.org/floyd/red/gentle.html. Accessed 10 Sept 2020
11. Cheng, M., Wang, H., Yan, L.: Dynamic RED: a modified random early detection. J. Comput. Inf. Syst. **7**, 5243–5250 (2011)
12. Zheng, B., Atiquzzaman, M.: DSRED: improving performance of active queue management over heterogeneous networks. In: IEEE ICC Proceedings, vol. 8, pp. 2375–2379 (2006)
13. Zhou, K., Yeung, K.L., Li, V.O.K.: Nonlinear RED: a simple yet efficient active queue management scheme. J. Comput. Netw. **50**, 3784–3794 (2006)
14. Zhang, Y., Ma, J., Wang, Y., Xu, C.: MRED: an improved nonlinear RED algorithm. In: International Conference Proceedings on Computer and Automation Engineering (ICCAE 2011), vol. 44, pp. 6–11 (2012)
15. Feng, C.W., Huang, L.F., Xu, C., Chang, Y.C.: Congestion control scheme performance analysis based on nonlinear RED. IEEE Syst. J. **11**, 2247–2254 (2017)
16. Karmeshu, Patel, S., Bhatnagar, S.: Adaptive mean queue size and its rate of change: queue management with random dropping. Telecommun. Syst. **11**, 287 295 (2017)
17. Tang, L., Tan, Y.: Adaptive queue management based on the change trend of queue size. KSII Trans. Internet Inf. Syst. **13**, 1345–1362 (2019)
18. Plasser, E., Ziegler, T., Reichl, P.: On the non-linearity of the RED drop function. In: Proceedings of ICCC, pp. 515–534 (2002)
19. Korolkova, A.V., Kulyabov, D.S., Velieva, T.R., Zaryadov, I.S.: Essay on the study of the self-oscillating regime in the control system. Commun. Eur. Counc. Model. Simul. Caserta Italy, pp. 473–480 (2019). https://doi.org/10.7148/2019-0473
20. Velieva, T.R., Korolkova, A.V., Kulyabov, D.S., Abramov, S.A.: Parametric study of the control system in the TCP network. In: 10th International Congress Proceedings on Ultra Modern Telecommunications and Control Systems, Moscow, Russian Federation, pp. 334–339 (2019). https://doi.org/10.1109/ICUMT.2018.8631267
21. Pavlotsky, O.E., Bobrikova, E.V., Samouylov, K.E.: Comparison of LBOC and RBOC mechanisms for SIP server overload control. In: Galinina, O., Andreev, S., Balandin, S., Koucheryavy, Y. (eds.) NEW2AN/ruSMART -2018. LNCS, vol. 11118, pp. 247–254. Springer, Cham (2018). https://doi.org/10.1007/978-3-030-01168-0_23

22. Kim, C., Dudin, S., Dudin, A., Samouylov, K.: Multi-threshold control by a single-server queuing model with a service rate depending on the amount of harvested energy. Perform. Eval. **127**(128), 1–20 (2018). https://doi.org/10.1016/j.peva.2018.09.001
23. Efrosinin, D., Gudkova, I., Stepanova, N.: Performance analysis and optimal control for queueing system with a reserve unreliable server pool. Commun. Comput. Inf. Sci. 1109, 109–120 (2019). https://doi.org/10.1007/978-3-030-33388-1_10

Architecture and Functionality of the Collective Operations Subnet of the Angara Interconnect

Alexey Simonov[1(✉)] and Oleg Brekhov[2(✉)]

[1] JSC "NICEVT", Varshavskoe Shosse 125, Moscow, Russia
alexey.s.simonov@rambler.ru
[2] National Research University for Aeronautical Engineering, Volokolamkoe Shosse 4, Moscow, Russia
obrekhov@mail.ru
http://www.nicevt.ru

Abstract. The Angara interconnect developed by JSC "NICEVT" is designed to connect the nodes of supercomputers and computing clusters. The paper describes the main architectural solutions, algorithms and functionality of the collective operations subnet of the Angara interconnect and presents the forecast of its characteristics based on the simulation modeling and actual operation. The proposed solutions allow bringing the time complexity of the collective operations execution to the theoretical limit for the kD-torus topology network.

Keywords: Angara interconnect · Multiprocessor computing system · Supercomputer

1 Introduction

Multiprocessor computing systems (hereinafter referred to as MCS) are important for solving application tasks aimed at increasing the scientific and technical potential of the economy and strengthening the country's defense capability. After NEC SX-6 Earth Simulator [1] appeared, it became clear how the characteristics and capabilities of the interconnect are important for ensuring high scalability of the MCS performance in solving computationally complex problems, primarily computer simulation, processing of large data arrays and forecasting.

JSC "NICEVT" started the development of the first-generation Angara high-speed interconnect (hereinafter referred to as Angara interconnect) in 2006. The operation principles and technical appearance of the Angara interconnect were formed basing on the analysis of the world experience in creating custom-made interconnect solutions for the highest performance range supercomputers, primarily the IBM BlueGene series [2–4] and CRAY SeaStar/Gemini [5–7], as well

JSC "NICEVT".

V. M. Vishnevskiy et al. (Eds.): DCCN 2020, LNCS 12563, pp. 209–219, 2020.
https://doi.org/10.1007/978-3-030-66471-8_17

as on the results of a number of studies conducted at JSC "NICEVT" using simulation modeling tools.

Angara interconnect is a Direct Network that supports topologies from 1D-mesh to 4D-torus and allows to create MCS of up 32K nodes. Its first production samples were presented in 2013.

During the development of the Angara interconnect the emphasis was placed on ensuring high scalability of the MCS performance in solving computationally complex problems. Algorithms for solving these problems, as a rule, are based on numerical methods with spatial decomposition of the computational domain, which require the execution of collective operations and boundary condition exchange operations after every iteration between the computational nodes of MCS that are involved in solving the problem.

Collective operations are among the main primitives of the interaction of computational processes in most parallel programming standards oriented to distributed memory computer systems (MPI, SHMEM, PGAS-languages—UPC, X10), and they can make up a significant part of communication exchanges. Such operations primarily include:

- broadcasting a packet from one node of MCS to other nodes allocated to the application (broadcast);
- collecting information from the nodes allocated to the application into one node and performing reduction (reduce).

This paper presents the main architectural solutions, algorithms and functionality of the collective operations subnet of the Angara interconnect and forecast of its characteristics based on the simulation modelling and actual operation.

2 Architecture of the Collective Operations Subnet of the Angara Interconnect

A trivial way to implement collective operations at the hardware level is to adequately replace them with many point-to-point operations, in which each broadcast is replaced with many writes to the memory coming from the root node to all other nodes of MCS allocated to the application, and each operation of collection is replaced by many operations of reading data from the MCS nodes memory to the root node.

This approach has undoubted advantages - simplicity of implementation and predictability of the result, and it can be used to build small MCS up to 100–200 nodes, for which, due to the small number of nodes, broadcast traffic will not have a significant impact on the load on the interconnect, and due to the small diameter of the network and the number of hops, the communication delay will be negligible.

For the medium and large size MCS the situation is different. In this case a large share of duplicate traffic can significantly affect the load of the interconnect, which, together with a large diameter, can negatively affect the latency and bandwidth of the interconnect and will lead to a significant deterioration in MCS performance [8]. That is why the hardware support of collective operations will positively affect the MCS performance scalability [9].

Obviously, a significant increase in the duration of collective operations due to an increase in the number of MCS nodes allocated to the application is associated with two factors - increased traffic due to an increase in the number of packets in the network and an increase in the number of hops due to an increase in the diameter of the network. In this regard, the problem of reducing the number of packets when performing collective operations for medium and large size MCS is very urgent.

The analysis of the trajectories of packets on the network in the case of implementation of the collective operations as a set of simple read and write operations in the memory of a remote node showed that in most cases the trajectories overlap each other. Considering the fact that the proposed solution requires its own routing algorithm that allows duplication and multiplication of packets when passing through transit nodes, it is advisable to implement hardware support for collective operations within a dedicated collective operations virtual subnet based on two virtual channels. A virtual subnet has the topology of a tree built in a torus (developers of the SMPO-10G interconnect came to a similar solution). There are two directions of movement in the tree: from root to leaves and from leaves to root. Each direction has its own virtual channel, VcDown – from root to leaves, and VcUp – from leaves to root. There may be transit nodes in the system; neither injection nor ejection of traffic occurs in these nodes.

When the broadcast operation is performed, each node, when receiving a packet from a node located higher in the tree, sends it to all nodes located lower in the tree. When a packet is injected into the network not from the root node, first a broadcast request is generated, which is sent to the root via the VcUp virtual channel, after which the broadcast operation itself is performed from the root via the VcDown virtual channels.

When the reduce (or all reduce) operation is performed, the node waits for packets from the node's processor if the node isn't transit, and from all the nodes located lower in the tree, performs the commutative associative binary operation indicated in the packet and sends the finished result up to the root. The current implementation supports the operations of maximum, minimum and sum of integers. The reduce operation ends with a hardware sending (without ejection) of the result to the given node using point-to-point operation (broadcast for all reduce). In order to determine direction to and from the root a routing table for collective operations subnet is set at each node. The table form is shown in Table 1.

The routing table in addition to dir-bits indicating the direction of the packet distribution along the subnet down the tree, has isRoot field that determines whether the node is root, toRoot field that indicates the direction to the root node, PE field that determines whether this node is transit or not and dirSum field which stores the number of directions to the leaves of the tree.

Table 1. Form of collective operation subnet routing table

TreeId	isRoot	toRoot	PE	dirSum	dir-bits			
					+X	−X	+Y	−Y
0 × 00	0	+Y	1	3	1	1	0	1
0 × 01	0	+X	0	2	0	0	1	1
...								
0 × 0F								

To set the correct tree, the following criteria must be met:

a) there is only one root;
b) if at some node the direction is set down the tree, then in this direction there should be a node that belongs to the tree in which the direction up the tree is set opposite to the given;
c) directions to nodes down the tree can only be:

- directions next to the direction, specified by the toRoot field, according to the direction order;
- the direction opposite to the direction, specified by the toRoot field.

Since many tasks can run simultaneously on the MCS, the collective operations virtual subnet allows to build up to 16 intersecting trees. At the same time, one task can use several trees. Each tree has its own TreeId identifier. For each TreeId identifier there is a corresponding routing table entry that determines routing direction. A subnet supports up to 16 different reduce packets for each TreeId. Each reduce performing on this tree has its own reduceId identifier (from 0 to 15).

The generalized algorithms of the virtual channels VcDown and VcUp of the collective operation subnet are shown in Fig. 1 and 2. The algorithm presented in Fig. 1 works as follows. From the header fleet received for routing, the TreeId field is selected, according to which the corresponding line is searched in the routing table of the collective operations subnet. If the line is found, the PE field in the routing table is checked. If it is 1, i.e. the node is not transit; the packet is ejected into the node.

Next, the reading and execution of the bitwise conjunction operation of dir-bits of the packet header fleet with dir-bits from the routing table is performed. The resulting bit vector is used to perform a loop search over directions (+X, −X, +Y, −Y . . .) and duplicate the packet into those for which the corresponding bit of the resulting vector is in state 1.

The algorithm presented in Fig. 2 works as follows. The packet type is checked after checking the TreeId identifier. If this is a broadcast packet the field isRoot of the routing table is checked to see if this node is root. If this is a root node the package is ejected into the node and distributed similar to previous algorithm. Otherwise, the toRoot field of the routing table is read and the packet is sent in the direction to the root node of the tree.

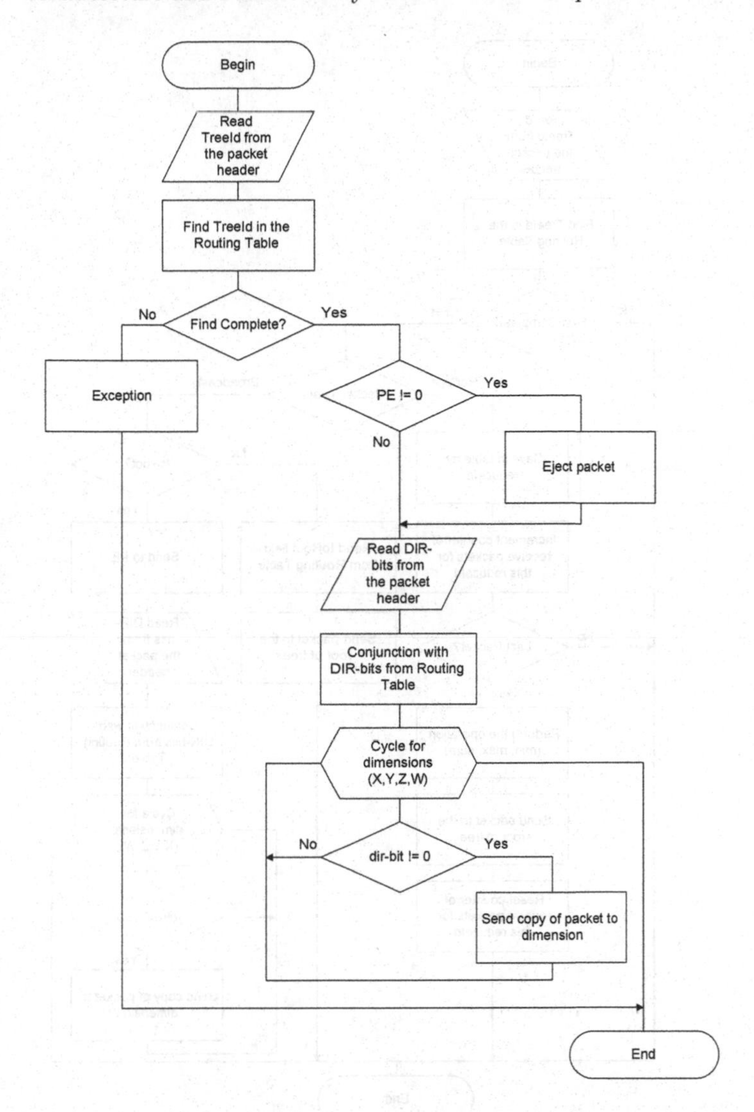

Fig. 1. The algorithm of the virtual channel VcDown of the collective operations subnet of the network with kD-torus topology

Reduce type packets that have got into the virtual channel VcUp are processed as follows. Data is extracted from the package and stored in a special table in the line corresponding to the value of the reduceId field in the package, after which the received packets counter allocated for this reduceId is increased by 1. If packets came from all directions, i.e. the value of received packets counter has become equal to the dirSum value of the routing table, the operation is performed, the packet is sent up the tree to the root node and the received packets counter is reset.

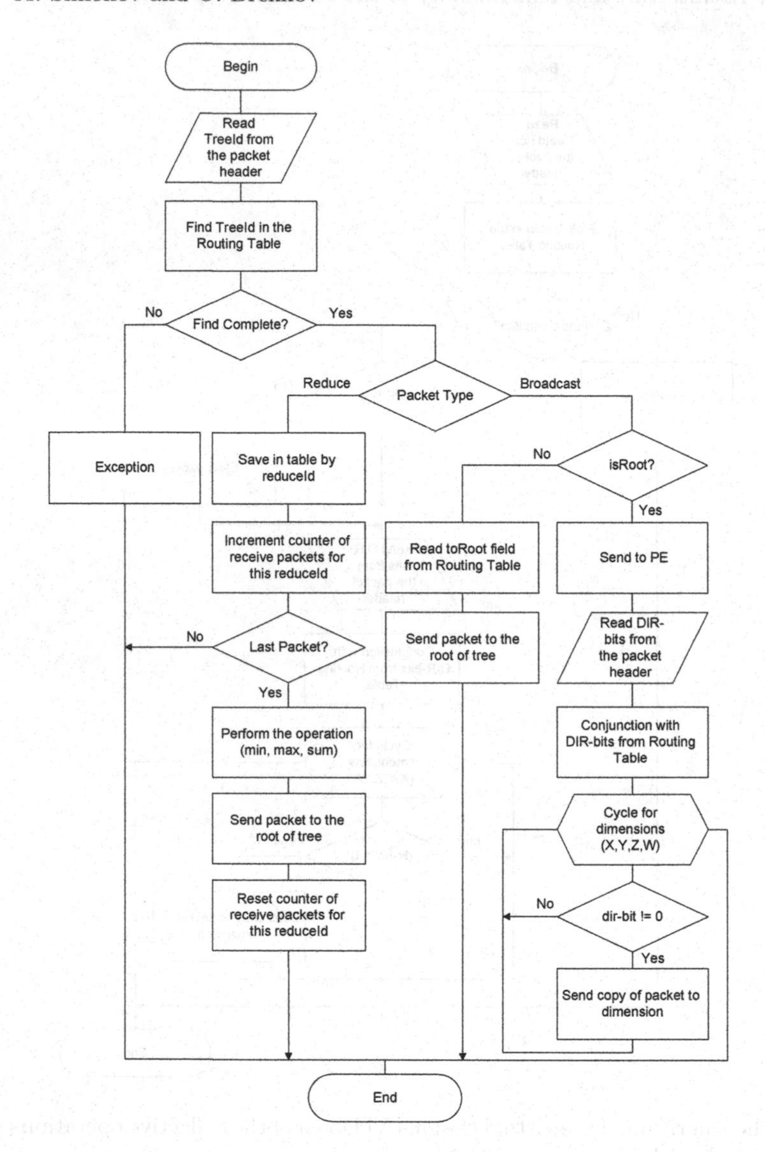

Fig. 2. The algorithm of the virtual channel VcUp of the collective operations subnet of the network with kD-torus topology

From the point of view of the application programmer, the basic versions of collective operations are one-way asynchronous operations. The processor is not blocked after sending the message, and the result is written into memory without the active participation of the receiving party, which allows overlapping computation and communication. Synchronization mechanisms based on collective operations are implemented in order to determine if the collective operation is completed and the result is available to the computational nodes.

Considering the specificity of toroidal networks, the distribution of nodes by tasks is usually carried out on the principle of minimizing the distance between them. As a result, the nodes allocated to the task in the structure of the kD-torus are generally limited to a rectangular region. Since it is advisable to use the principle of minimizing the distances to the most distant nodes of the tree when choosing the root node, it is obvious that it is advisable to choose the node located in the geometric center of the rectangular area of the MCS nodes allocated to the task as the root node.

Two mechanisms are used in collective operations virtual subnet to prevent deadlocks caused by different tasks trees overlapping:

- bubble routing;
- various trees are constructed according to the X, Y, Z, W direction order.

3 Characteristics of the Collective Operations Subnet of the Angara Interconnect

To confirm the efficiency of proposed algorithms time complexity should be estimated. Diameter of a network is the shortest distance between the two most distant nodes in it. In general, for an equilateral kD-torus with even number of nodes diameter equals

$$r = \frac{k}{2}\sqrt[k]{\omega} \qquad (1)$$

where r - is the network diameter;
k - is a number of dimensions in torus;
ω – is the total number of MCS nodes.

For an arbitrary 4D-torus networks r gives the lower-bound estimate.

When using VCT routing method, network message delivery time is determined by the communication delay and the time, required to inject the message of a given size into the communication channel (link) with a certain bandwidth, that is

$$T_d = t^0 + (l-1)t^1 + \frac{Q}{V_L} \qquad (2)$$

where T_d - is the network message delivery time;
t^0 – is the communication latency between two adjacent nodes;
l – is the route length (number of hops);
t^1 – is the communication latency of a hop;
Q – is the message size;
V_L – is the communication channel bandwidth.

If collectives are implemented using point-to-point operations, the time complexity estimation for the broadcast and reduce algorithms will heavily depend on the host interface bandwidth of the root node

$$T(A^1_\omega) = T(A^\omega_1) = (\omega - 1)(t^1 + \frac{Q}{V_P}) + (\frac{k}{2}\sqrt[k]{\omega} - 1)t^1 + t^0 \qquad (3)$$

where A^1_ω – is a broadcast algorithm;

$T(A_\omega^1)$ – is an estimation of the time complexity of the broadcast algorithm;
A_1^ω – reduce algorithm;
$T(A_1^\omega)$ - is an estimation of the time complexity of the reduce algorithm;
V_P – host processor interface bandwidth.

Considering that in the VcDown and VcUp virtual channels algorithms the multiplication of packets during broadcast execution and reduction during reduce execution are performed at all nodes in the tree, the proposed method can significantly reduce both the number of packets and the total execution time. As a result, the estimate of time complexity will be close to the theoretical limit as it is determined by the distance to the most distant nodes. Applied to kD-torus network

$$T(A_\omega^1) = T(A_1^\omega) = t^0 + (\frac{k}{2}\sqrt[k]{\omega} - 1)t^1 + \frac{Q}{V_L} \quad (4)$$

Collective operations execution time were estimated to verify the above relation using simulation model. The tests consisted of sending one packet of 32 flits (256 bytes) using the broadcast operation for a different number of nodes of the simulated network.

Tables 2, 3 present the results obtained under the following initial conditions: $t^0 = 700$ ns; $t^1 = 80$ ns; $Q = 256$ bytes; $V_L = 6$ GB/s; $V_P = 8$ GB/s.

Table 2. Comparison of the estimated execution time of the broadcast operation depending on the number of computational nodes for the 3D-torus network

Parameter	Number of nodes of MCS			
	8	64	512	4096
3D-torus, point-to-point collective operations				
Analytical calculation, s	1,64	8,16	58,81	461,18
Simulation modelling results, s	2,18	6,07	40,16	312,64
Divergence, %	24,5%	25,6%	31,7%	32,2%
3D-torus, proposed method				
Analytical calculation, s	0,9	1,14	1,62	2,58
Simulation modelling results, s	0,96	1,20	1,76	2,80
Divergence, %	5,9%	4,8%	7,8%	7,8%

The above results indicate an acceptable value of the inaccuracy in estimating the time complexity of collective operations execution using expression 4. Expression 3, unfortunately, gives a higher inaccuracy, which, according to the author, is caused by incomplete consideration of certain aspects of the network.

The results obtained using the simulation model made it possible to preliminarily confirm the hypothesis that the proposed method for making a collective operations subnet gives a significant gain in comparison with the use point-to-point collective operations. At the same time, its hardware implementation in comparison with the software implementation also provides a significant gain.

Table 3. Comparison of the estimated execution time of the broadcast operation depending on the number of computational nodes for the 4D-torus network

Parameter	Number of nodes of MCS		
	16	256	4096
4D-torus, point-to-point collective operations			
Analytical calculation, s	2,62	29,82	460,54
Simulation modelling results, s	2,90	21,78	312,65
Divergence, %	9,6%	26,9%	32,1%
4D-torus, proposed method			
Analytical calculation, s	0,98	1,3	1,94
Simulation modelling results, s	1,04	1,44	2,16
Divergence, %	5,5%	9,5%	10,0%

4 Angara Interconnect Design Versions

Currently there are two version of the Angara interconnect:

- switchless version – full-height full-length PCI Express card (see. Fig. 3) that allows to connect up 32K computing nodes with an $8 \times 16 \times 16 \times 16$ 4D-torus topology;
- switch version – 19" 24-port switch and low-profile PCI Express card (half-height full-length) (see. Fig. 4). This version allows to connect up to 2048 computing nodes with 2D-torus topology by connecting up to 256 switches.

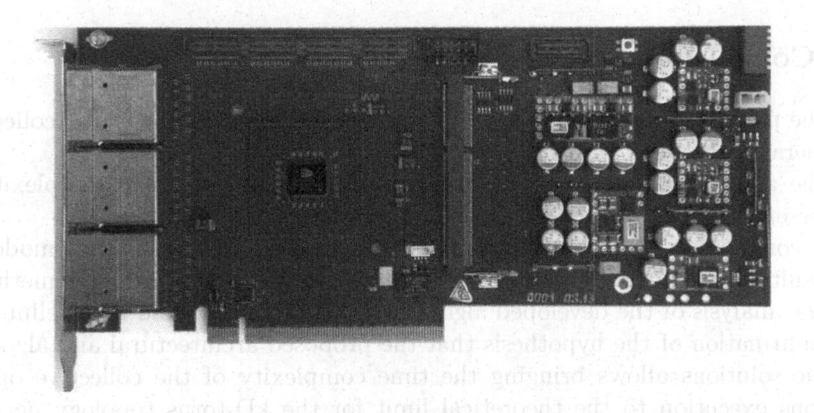

Fig. 3. Switchless version of the Angara interconnect

There are several MCS with high performance scalability based on the Angara interconnect in the Russian Academy of Sciences institutions, research institutes and industrial enterprises.

Fig. 4. The Angara interconnect 24-port switch (a) and low-profile PCI Express card (b)

The Angara interconnect allows achieving good performance scalability of MCS both on evaluation tests and applied computer simulation tasks (see. Fig. 5) [10–16].

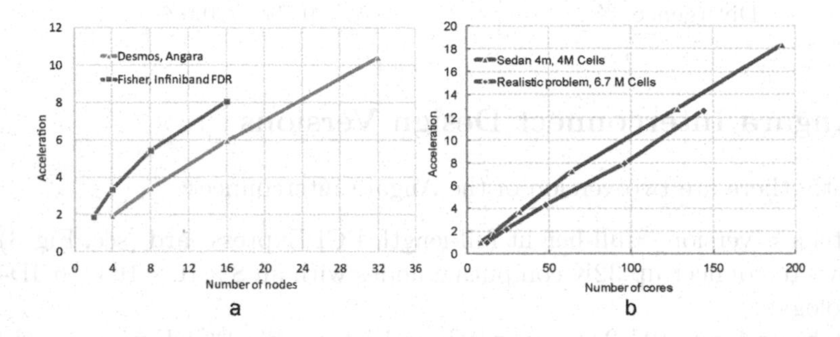

Fig. 5. MCS performance scalability (a – VASP - software package for quantum-chemical calculations, Desmos supercomputer; b – ANSYS FLUENT, Angara-K1 cluster)

5 Conclusion

1. The paper proposes architectural and algorithmic solutions for the collective operations subnet for a kD-torus topology network.
2. The analytical assessment of the developed algorithms time complexity is presented in the paper.
3. A comparison of the analytical assessment with the simulation modeling results was performed, which showed a 10% divergence (for proposed method).
4. The analysis of the developed algorithms was performed and the preliminary confirmation of the hypothesis that the proposed architectural and algorithmic solutions allows bringing the time complexity of the collective operations execution to the theoretical limit for the kD-torus topology network was obtained using the simulation model.

The proposed architectural and algorithmic solutions have a positive effect on the scalability of MVS performance in solving computationally complex problems, primarily computer simulation, processing large data arrays, planning and forecasting.

References

1. Habata, S., et al.: The earth simulator system. NEC Res. Dev. **44**(1), 21–26 (2003)
2. Gara, A.: Overview of the Blue Gene/L system architecture. IBM J. Res. Dev. **49**, 195–212 (2005). http://rsim.cs.illinois.edu/arch/qual_papers/systems/19.pdf
3. Almasi, G., Asaad, S., Bellofatto, R.E., et al.: Overview of the IBM Blue Gene/P project. IBM J. Res. Dev. **52**(1–2), 199–220 (2008). http://scc.acad.bg/ncsa/documentation/team.pdf
4. Chen, D., Eisley, N., Heidelberger, P., et al.: The IBM BlueGene/Q interconnection fabric. IEEE Micro **32**(1), 32–43 (2012). https://www.researchgate.net/publication/220290398_The_IBM_blue_geneQ_interconnection_fabric
5. Abts, D., Storm, C.R.: The Cray XT4 and Seastar 3-D torus interconnect (2010). https://static.googleusercontent.com/media/research.google.com/ru//pubs/archive/36896.pdf
6. Alam, S.R.: Cray XT4: an early evaluation for petascale scientific simulation. In: Proceedings of the 2007 ACM/IEEE Conference on Supercomputing, pp. 1–12. IEEE (2007). https://doi.org/10.1145/1362622.1362675
7. Alverson, R., Roweth, D., Kaplan, L.: The Gemini system interconnect. 2010 IEEE 18th Annual Symposium on High Performance Interconnects (HOTI), pp. 83–87. IEEE (2010). https://doi.org/10.1109/HOTI.2010.23
8. Bala, V., Bruck, J., Cypher, R., et al.: CCL: A Portable and Tunable Collective Communication Library for Scalable Parallel Computers. In: Proceedings of the Parallel Processing Symposium, pp. 835–844 (1994). ISBN 0-8186-5620-6
9. Almási, G., Dózsa, G., Erway, C.C., Steinmacher-Burow, B.: Efficient implementation of allreduce on BlueGene/L collective network. In: Di Martino, B., Kranzlmüller, D., Dongarra, J. (eds.) EuroPVM/MPI 2005. LNCS, vol. 3666, pp. 57–66. Springer, Heidelberg (2005). https://doi.org/10.1007/11557265_12
10. Mukosey, A., Simonov, A., Semenov, A.: Extended routing table generation algorithm for the angara interconnect. In: Voevodin, V., Sobolev, S. (eds.) Extended Routing Table Generation Algorithm for the Angara Interconnect. CCIS, vol. 1129, pp. 573–583. Springer, Cham (2019). https://doi.org/10.1007/978-3-030-36592-9_47
11. Stegailov, V., et al.: Angara interconnect makes CPU based Desmos supercomputer an efficient tool for molecular dynamics calculations. Int. J. High Perform. Comput. Appl. **33**, 507–521 (2019)
12. Stegailov, V., Smirnov, G., Vecher, V.: VASP hits the memory wall: processors efficiency comparison. Concurr. Comput. Pract. Exp., e5136 (2019). https://doi.org/10.1002/cpe.5136
13. Polyakov, S., Podryga, V., Puzyrkov, D.: High performance computing in multiscale problems of gas dynamics. Lobachevskii J. Math. **39**(9), 1239–1250 (2018)
14. Ostroumova, G., Orekhov, N., Stegailov, V.: Reactive molecular-dynamics study of onion-like carbon nanoparticle formation. Diamond Related Mater. **94**, 14–20 (2019)
15. Tolstykh, M., Goyman, G., Fadeev, R., Shashkin, V.: Structure and algorithms of SL-AV atmosphere model parallel program complex. Lobachevskii J. Math. **39**(4), 587–595 (2018). https://doi.org/10.1134/S1995080218040145
16. Akimov, V., Silaev, D., Aksenov, A., Zhluktov, S., Savitskiy, D., Simonov, A.: FlowVision scalability on supercomputers with angara interconnect. Lobachevskii J. Math. **39**(9), 1159–1169 (2018). https://doi.org/10.1134/S1995080218090081

The Model of WBAN Data Acquisition Network Based on UFP

S. Vladimirov[1](✉), V. Vishnevsky[2], A. Larionov[2], and R. Kirichek[1]

[1] St.Petersburg State University of Telecommunications, 22 Bolshevikov Pr., St.Petersburg 193232, Russian Federation
vladimirov.opds@gmail.com, kirichek@sut.ru

[2] V.A. Trapeznikov Institute of Control Sciences of Russian Academy of Sciences, 65 Profsoyuznaya Street, Moscow 117997, Russian Federation
larioandr@gmail.com

Abstract. The paper determines the general model of an unmanned flying platforms based network for collecting information from wearable wireless body area networks. There are considered several possible approaches to its implementation. A list of tasks to be solved and their features in the framework of building a network are given. There are proposed variants of optimal network radio technologies and topologies, antenna devices and types of unmanned flying platforms to create the acquisition network. Variants of implementation of the interaction protocol and methods of organizing secure data transmission, taking into account the peculiarities of the problem being solved, are recommended.

Keywords: WBAN · Data acquisition network · Unmanned flying platform

1 Introduction

In the modern world there is a need for prompt receipt of information. An important part of information collection systems are sensor networks that combine different types of sensors and control units that take readings, process it and send it to the recipient [13,14].

One of the important types of sensor networks are wearable wireless body area networks (WBANs), which are formed from small-sized body sensor units (BSU) located on the human body and taking readings from it, and a central body control unit (BCU), which collects that readings from BSU and processes them [16,19]. The BCU is usually a general purpose user device such as a smartphone or tablet, or a specialized microcontroller-based device. In the first case, the WBAN is built on the basis of widely used radio technology supported by most smartphones, such as Wi-Fi or Bluetooth. In the second case, the most suitable wireless personal area network (WPAN) radio technology is selected, for example, ZigBee or another IEEE 802.15 standard [7,9,16].

Supported by RFBR according to the research project No.20-37-70059.

V. M. Vishnevskiy et al. (Eds.): DCCN 2020, LNCS 12563, pp. 220–231, 2020.
https://doi.org/10.1007/978-3-030-66471-8_18

The main application of WBAN networks is monitoring of human health indicators. Therefore, they are widely used in medical institutions, professional and amateur sports and military sports games. Also, wearable nets are being actively implemented in the army and rescue services. In all these cases, there is a situation when a certain number of WBANs carrying persons are located in a limited area. Information from WBAN controllers needs to be promptly transmitted to some data collecting center (DCC). For example, timely data on the athletes' health obtained by referees during tournaments can prevent accidents and allow doctors to react more quickly to possible injuries and pre-critical conditions [7,12].

2 Approaches to Collecting Data from WBAN

Various approaches are used to transfer data from the WBAN controller to the data collecting center. If the WBAN is geographically located within the coverage area of a larger radio access network, be it a wireless local area network (WLAN), a city-wide network (WMAN), or a global wireless network (WWAN), it makes sense to use this larger network to transfer operating data. However, if the location of the WBAN is not covered by the existing radio access network, or information cannot be transmitted over open networks, it is necessary to deploy a dedicated wireless network, so called data acquisition network (DAN), to collect data from the WBAN.

For most radio networks, the best signal reception conditions are obtained with a clear line of sight between the transmitter and receiver. Therefore, it is logical to place the data acquisition node high above the ground. The higher the node is placed, the more coverage it will provide. Naturally, the height of the acquisition node depends on the maximum operating range of the applied wireless technology, which determines the maximum distance from the acquisition node to the most distant WBAN from which information can be collected.

In order not to depend on the presence of buildings, towers or poles on the territory of the DAN location, it is convenient to implement the acquisition node as an unmanned flying platform (UFP). This method allows to place the acquisition node at the optimal place and height for the provided territory and, if necessary, change its location during operation [8,11].

We represent the organization of a WBAN data acquisition network based on UFPs of various types in the form of Fig. 1.

The parameters that must be taken into account when building an acquisition network based on UFPs, and the issues associated with them, are shown in Fig. 2. In general, their relationship with each other can be represented by the function

$$[z_1, z_2, \ldots, z_n] = F(p_1, p_2, \ldots, p_m), \tag{1}$$

z_i—issues to solve; p_j—network parameters.

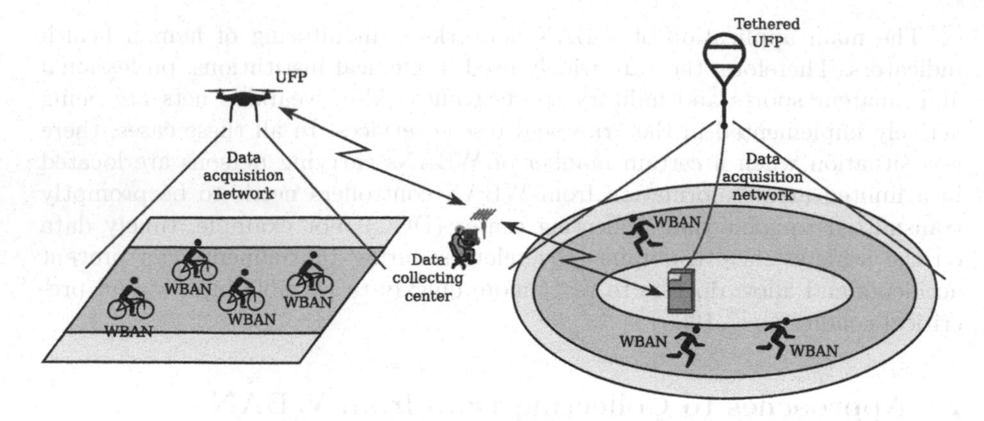

Fig. 1. WBAN data acquisition network based on UFP.

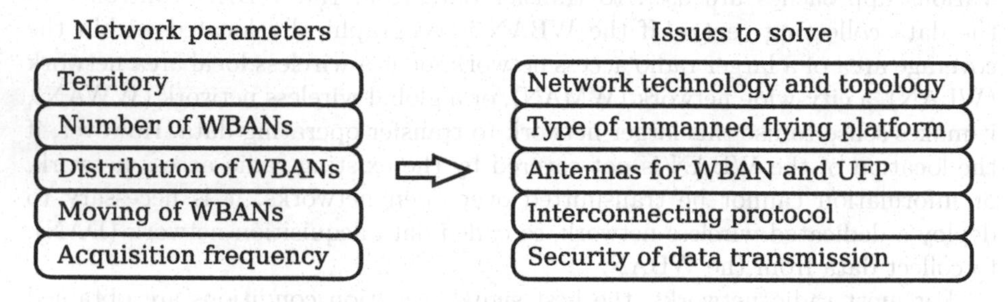

Fig. 2. Network parameters and issues to solve.

3 Issues to Solve

3.1 Network Technology and Topology

The choice of radiotransmitting technology and topology of the deploying DAN primarily depends on such factors as the size and topography of the area where the data acquisition network needs to be deployed. In the case of some sporting events, such as marathon or cycling, or military sports games, the area of the event can be quite large and the use of distributed acquisition wireless networks based on the widely used Wi-Fi technology will require additional deployment costs due to the limited communication range and significant energy consumption of the acquisition nodes. Using cellular technologies based on a single base station can be convenient, but will require a deployment license. Thus, it is advisable to use medium and long range technologies that have low power consumption and operate in unlicensed frequency bands, such as LoRa, DECT-ULE or Bluetooth LE [5,17,26]. Since the very idea of collecting information from a WBAN does not imply the exchange of a large amount of information, the requirements for the data transfer rate in such networks are insignificant.

After determining the technology, you need to select the topology of the data acquisition network. Depending on the network coverage area, either a star-shaped network with one acquisition node or a distributed multinode network should be used (Fig. 3). The height of the UFP with the acquisition node will depend on the maximum working distance of the radio link, which determines the maximum distance from the acquisition node to the most distant WBAN from which information is to be read.

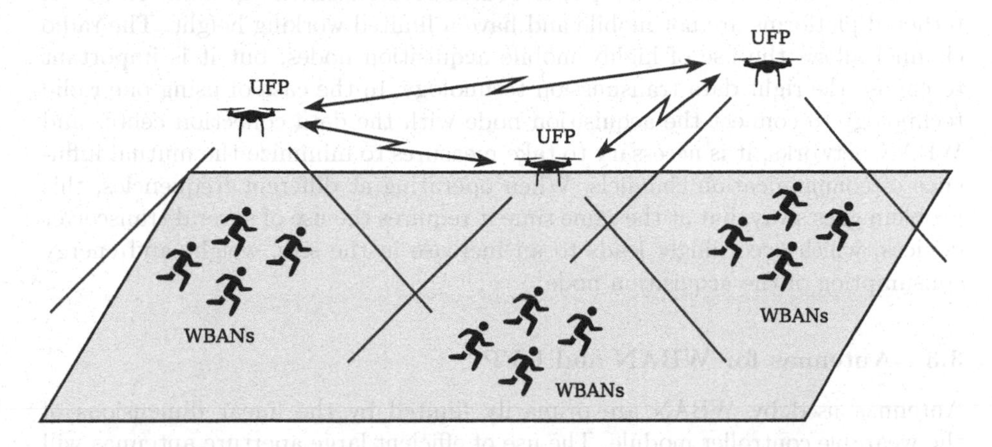

Fig. 3. Distributed multinode data acquisition network

3.2 Type of UFP

Unmanned flying platforms are classified according to several criteria.

By design, there are UFPs heavier than air, which include aircraft and multirotor platforms, and platforms lighter than air—balloon type [2].

To solve the presented issue, it seems efficient to use multi-rotor or balloon UFPs, which can hover for a while at a given point. At the same time, aerostatic platforms will be convenient in situations where a long-term placement of the acquisition node is required without the need for a quick change of position. Multi-rotor UFPs are more versatile and allow moving the acquisition node at a sufficiently high speed, which can be a determining factor in the expected active movement of the served WBANs along a certain route. Such systems have a limited run time and place increased demands on the power supply.

According to the height of placement, two types of flying platforms are usually distinguished: operating at heights of up to a few kilometers, the so-called LAPs—low-altitude platforms, and HAPs—high-altitude platforms, operating in the altitude range from 17 to 21 km [2]. To collect information from WBANs, it is more convenient to use platforms operating at low altitudes up to hundreds of meters.

According to the method of communication with the ground, the platforms are divided into tethered, in which communication is provided via the connecting cable [21,22], and autonomous, using a radio channel.

The choice of communication method depends on the number, location and speed of WBAN's movement. Tethered platforms are useful when there is a lot of information being transferred. In such cases, the UFP, in addition to the functions of the acquisition node, can perform the functions of a video surveillance system. Tethered systems traditionally have greater run time, since they can be powered from a stationary power source located on the ground. However, tethered platforms are not mobile and have a limited working height. The radio channel allows the use of highly mobile acquisition nodes, but it is important to choose the right data transmission technology. In the case of using one radio technology to connect the acquisition node with the data collection center and WBAN networks, it is necessary to take measures to minimize the mutual influence of communication channels. When operating at different frequencies, this problem goes away, but at the same time it requires the use of several transceiver devices, which accordingly leads to an increase in the size, weight and energy consumption of the acquisition node.

3.3 Antennas for WBAN and UFP

Antennas used by WBAN are primarily limited by the linear dimensions of the wearable controller module. The use of efficient large aperture antennas will result in inconvenience for the user to wear the WBAN. It should be additionally taken into account that the location of the antenna relative to the acquisition node is not known in advance, therefore it is advisable to use omnidirectional antennas in a compact design, such as linear and helical monopole antennas and a planar antennas [4,18,25,27,28].

The antennas placed on the UFP have smaller size restrictions, since the linear dimensions of the UFP, which have sufficient carrying capacity and duration of operation to perform the functions of a data acquisition node, already exceed the linear dimensions of the microwave antennas. On UFPs, the best option would be to use efficient antennas with a large aperture. Considering that the UFP is always located above the WBAN (Fig. 1), antenna structure should have a directional pattern towards the ground. Among the effective antennas of this type are dipoles [3,6] and biquadrats [15]. The advantage of these antennas is their flat design, which is convenient for placement on the bottom of the UFP.

It is important to note that the directional pattern of the antenna determines the UFP coverage area. Therefore, when deploying a data acquisition network and designing UFP, this issue must be paid special attention. In particular, it is possible to change the directional pattern using reflector.

3.4 Interconnecting Protocol

For interaction between acquisition nodes and WBAN networks, you can use existing protocols or develop your own exchange protocol. It should be noted that collecting information from WBAN sensor networks involves the transmission of data in a previously known format. Earlier it was noted that the acquisition network usually has a simple structure that does not require the use of complex routing and addressing protocols. Thus, it seems optimal to use a specially developed data transfer protocol that works directly on top of the link layer protocol, which will provide a minimum amount of service information, in contrast to the case with the deployment of a network based on standard TCP/IP protocols.

The WBAN readings acquisition network is generally represented as a dynamic redundant system formed by N WBAN controllers, M mobile UFP nodes and L stationary nodes that form a network from the WBAN to the data collection center. Information transfer is carried out along a chain in three stages: from WBAN to UFP, from UFP to stationary nodes and further from stationary nodes to DCC. At each of these stages, data can be transmitted using various technologies. From the point of view of organizing the data collection, the most interesting is the interaction between WBAN and UFP, which occurs actually at the data collection stage.

The number of WBANs served by one UFP at a collection stage can vary, and the number of UFPs operating simultaneously at a single territory can also change, as shown in Fig. 3. Accordingly, the network must provide a mechanism for registering WBAN to UFP. The entire UFP operation time is divided into cycles, each of which consists of two periods: the W_{reg} time window (registration channel), during which WBANs register with the UFP, and the W_{data} window (data channel), during which the UFP polls the registered WBANs. In the W_{reg} window, the predefined transmission parameters are necessarily used, which are the same for all WBANs and UFPs. Depending on the technology used, it can be a separate frequency channel, CDMA code or chirp spreading factor. In the W_{data} window, work is carried out with parameters that are separate for each UFP. This allows multiple UFPs to operate simultaneously in the same area without interfering with each other.

During the W_{reg} WBANs are registered with the UFP based on channel time contention, treating the registration channel as a common shared medium. As a method of accessing the medium, it is optimal to use the traditional CSMA/CA algorithm based on checking the channel before data transmission. The start of the registration window is determined by the UFP through a Registration Request (RReq) packet containing the UFP identifier and data channel settings. In order to avoid collisions during registration, WBANs that receive an RReq and want to transmit information to the UFP, at a random interval, send a Registration Reply (RRep) packet with the WBAN identifier, the size of the information to be transmitted in bytes and the urgency of the data. After receiving the RRep, the UFP sends the next RReq, which indicates the identifier of the previously registered WBAN. If the RRep packet was not received by the UFP, for example, due to a collision, then the next RReq indicates the identifier of the last registered WBAN. Thus, the WBAN that failed to register is aware of this and can try to register again. In total, the UFP sends N_{req} RReqs during the registration window. Accordingly, during the time W_{reg}, $N_{reg} \leq N_{req}$ WBAN can be registered. WBAN, registered to UFP, switch their transceivers to data channel settings.

The duration of one stage of registration is estimated by the formula

$$T_{reg} = \frac{(L_{RReq} + L_{RRep})}{B} + t_{proc}, \tag{2}$$

L_{RReq}—size of Registration Request from UFP in bits; L_{RRep}—size of Registration Reply from WBAN in bits; B—channel data rate in bps; t_{proc}—time of processing RReq by WBAN and RRep by UFP in seconds.

Accordingly, the duration of the W_{reg} window is

$$T_{W_{reg}} = N_{req} \cdot T_{reg}. \tag{3}$$

An example of the acquisition system operation in the registration window is shown in Fig. 4. The scenario is considered with $N_{req} = 8$. Taking into account three errors of reception and transmission, $N_{reg} = 5$ WBAN out of k possible were registered on the UFP.

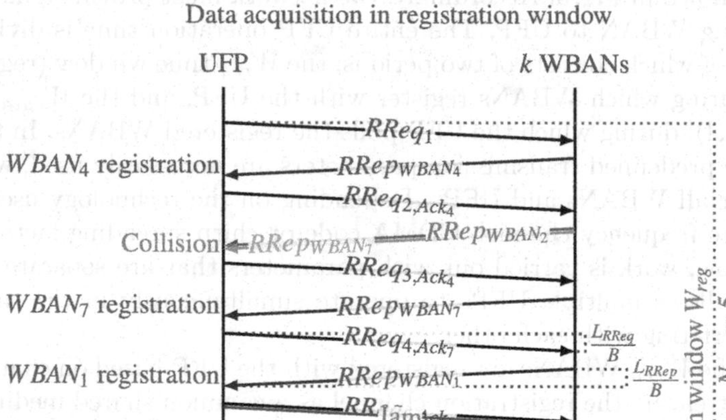

Fig. 4. An example of the registration window W_{reg}

After receiving requests from all registered WBANs, the UFP generates a polling sequence of N_{reg} length, taking into account the urgency indicators received from the WBAN. Further, during the next data channel W_{data} the UFP polls the selected WBANs on a priority basis, sending Data Request (DReq) packets with the identifier of the polled WBAN, for example, $WBAN_1$. The $WBAN_1$, which receives a DReq with its ID, responds with a Data Reply (DRep) packet. The UFP checks the DRep by comparing the IDs and, if necessary, the checksum, and then replies with the next DReq packet, in which it acknowledges data reception and requests readings for the next WBAN in the sequence, for example, $WBAN_2$. If the DRep packet was transmitted with an error or lost, the UFP repeat request data by the DReq with ID of $WBAN_1$. If after 2 repeated requests it was not possible to receive data from $WBAN_1$, it is marked as unavailable and is no longer polled until the next registration request is received from it. The DCC is notified of an unsuccessful attempt to receive data from $WBAN_1$.

The polling time of one WBAN node during the W_{data} window is estimated by the formula

$$T_{WBAN} = \frac{(L_{DReq} + L_{DRep})}{B} + t_{proc}, \tag{4}$$

L_{DReq}—size of Data Request from UFP in bits; L_{DRep}—size of Data Reply from WBAN in bits; B—channel data rate in bps; t_{proc}—time of processing DReq by WBAN and DRep by UFP in seconds.

The duration of the W_{data} window depends on the number of N_{rpt} repeated requests and is equal to

$$T_{W_{data}} = (N_{reg} + N_{rpt}) \cdot T_{WBAN} + \left(\frac{L_{DReq}}{B} + 0.5t_{proc}\right), \tag{5}$$

where the second summand takes into account the confirmation of the final data packet.

An example of the acquisition system operation in a data transmission window is shown in Fig. 5. The scenario with $N_{reg} = 5$ is considered, corresponding to the previously considered registration example in Fig. 4. We will assume that, in accordance with the urgency indicators, the UFP has formed a polling sequence: $WBAN_1$, $WBAN_7$, $WBAN_2$, $WBAN_4$, $WBAN_5$. Due to two transmission errors, $N_{rpt} = 2$ repeated requests were made.

To increase the survivability of the system and prevent the loss of data of a higher category of urgency or importance in the event of a discharge of the $WBAN_i$ power supply, data transmission is provided to the neighboring WBANs and further to the WBAN in the UFP visibility zone for organizing urgent transmission. Transit WBANs do not process the data received from $WBAN_i$, but simply transmit them in the usual order to the UFP, indicating the maximum urgency.

At the end of the data transmission window, UFP and WBAN switch back to registration mode and the cycle repeats.

The suggested protocol packet format is shown in Fig. 6.

At the beginning of the packet there is address information containing the receiver's identifier and the sender's identifier. The content of the data block depends on the direction of transmission. In general, the UFP sends a data transmission command and a list of requested readings to the WBAN nodes, and the WBAN responds with sensor readings. If each WBAN in the acquisition

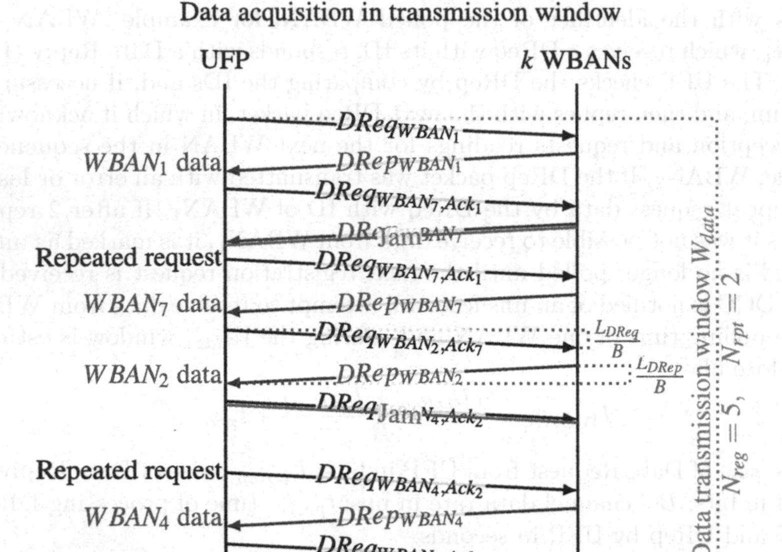

Fig. 5. An example of the data transmission window W_{data}

Receiver Address/ID	Sender Address/ID	Data	Checksum (Hash)

Fig. 6. Protocol packet format for data acquisition network

network contains several sensors, then their readings should be separated from each other. In the case when all WBANs contain the same set of sensors, the order of the readings can be foreseen and only the readings themselves can be transmitted. A more universal option is a separate data block for each readings, consisting of three fields: the header, the block length in bytes and the readings themselves. The sizes of the header and length fields are selected based on the number of possible values. Considering that the number of sensors traditionally used in WBAN is limited, it seems optimal to use 1 byte for the header and field length, as shown in Fig. 7.

This method allows to transmit only the necessary readings and optimize the load on the acquisition network. Also, the specified format of data blocks allows them to be used for negotiating encryption keys in cases where it is necessary.

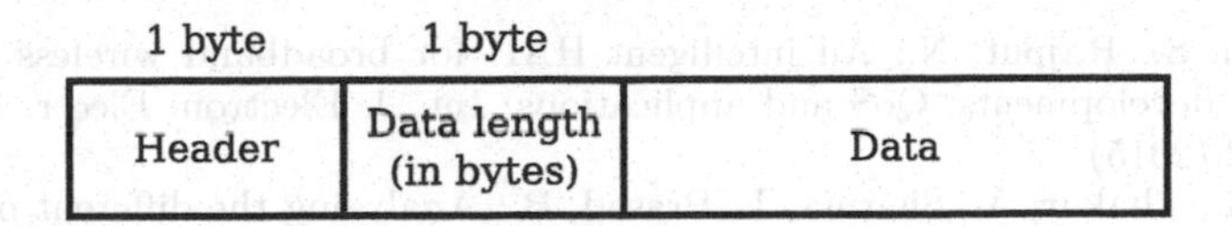

Fig. 7. Data block format for data acquisition network

3.5 Security of Data Transmission

To ensure the secure interaction of the acquisition network nodes and WBANs, it is necessary to solve the problems of mutual identification of the acquisition nodes and the WBAN, as well as to define methods of encryption and integrity checking, which together will provide protection against data interception and substitution by third-party devices.

It is proposed to use various identification methods as device identifiers. The main criterion for choosing a method is the uniqueness of the identifiers used. Ideally, identifiers should be globally unique for maximum flexibility and security. For example, the MAC addresses of the controllers, numeric DOI [1] or the CID and Ecode identification systems [20]. Also there is a developed technique for generating unique hardware identifiers based on degraded NAND [10] or NOR [23,24] flash memory.

For encryption and checking the integrity of the transmitted data, it seems optimal to use the classic combination of symmetric cipher for data exchange, asymmetric encryption for exchange of symmetric encryption keys and hash functions for checking integrity. The choice of specific algorithms depends on the requirements for the acquisition network and the equipment used. For example, when using controllers with low computational capabilities, you should use fast algorithms that do not require productive hardware resources. Also a number of microcontrollers have built-in hardware mechanisms for quickly calculating cryptographic functions, which reduces the time spent on encryption.

4 Conclusion

The article proposes a general model of data acquisition network for collecting information from wireless wearable networks based on unmanned flying platforms. The main conceptual approaches to solving problems of network design and their features are considered. It is planned to continue work within the framework of the tasks considered.

References

1. Albahri, M., Kirichek, R., Muthanna, A., Ateya, A., Borodin, A.: Combating counterfeit for IoT system based on DOA. In: 2018 10th International Congress on Ultra Modern Telecommunications and Control Systems and Workshops (ICUMT), p. 8631257 (2018)

2. Alsamhi, S., Rajput, N.: An intelligent HAP for broadband wireless communications: developments, QoS and applications. Int. J. Electron. Electr. Eng. **3**(2), 134–143 (2015)
3. Bath, A., Thakur, A., Sharma, J., Prasad, B.: Analyzing the different parameters of dipole antenna. Int. J. Electr. Electron. Eng. **1**(1), 11–15 (2014)
4. Chandran, A.R., Conway, G.A., Scanlon, W.G.: Compact slot-loaded patch antenna for 868 MHZ wireless body area networks. In: 2008 Loughborough Antennas and Propagation Conference, LAPC, pp. 433–436 (2008)
5. Das, K., Havinga, P.: Evaluation of DECT-ULE for robust communication in dense wireless sensor networks. In: 2012 3rd IEEE International Conference on the Internet of Things, pp. 183–190. Wuxi (2012)
6. Dw, E.F., Pratama, H., Ihsan, N., Rahmatia, S., Wulandari, P.: Design and performance investigation of dipole antenna using aluminum and iron at 644 MHZ - 736 MHZ. In: International Conference on Engineering, Technologies and Applied Sciences, Kuala Lumpur, Malaysia, January 2017
7. Ghamari, M., Janko, B., Sherratt, R., Harwin, W., Piechockic, R., Soltanpur, C.: A survey on wireless body area networks for eHealthcare systems in residential environments. Sensors **16**, 831 (2016)
8. Ha, I., Cho, Y.: Unmanned aerial vehicles-based health monitoring system for prevention of disaster in activities of the mountain. Int. J. Control Autom. **9**(9), 353–362 (2016)
9. IEEE: IEEE Standard for Local and metropolitan area networks - Part 15.6: Wireless Body Area Networks. IEEE Std 802.15.6-2012 (2012)
10. Jakobsson, M., Johansson, K.A.: Unspoofable device identity using NAND flash memory. SecurityWeek (2010). http://www.securityweek.com/unspoofable-device-identity-using-nand-flash-memory
11. Kirichek, R.: The model of data delivery from the wireless body area network to the cloud server with the use of unmanned aerial vehicles. In: Proceedings 30th European Conference on Modelling and Simulation (ECMS), pp. 603–606 (2016)
12. Kirichek, R., Pirmagomedov, R., Glushakov, R., Koucheryavy, A.: Live substance in cyberspace - Biodriver system. In: 18th International Conference on Advanced Communication Technology (ICACT) 2016, pp. 274–278 (2016)
13. Koucheryavy, A., Salim, A.: Cluster-based perimeter-coverage technique for heterogeneous wireless sensor networks. In: 2009 International Conference on Ultra Modern Telecommunications and Workshops St. Petersburg, p. 5345452 (2009)
14. Koucheryavy, A., Vladyko, A., Kirichek, R.: State of the art and research challenges for public flying ubiquitous sensor networks. Lect. Notes Comput. Sci. **9247**, 299–308 (2015)
15. Li, R., Traille, A., Laskar, J., Tentzeris, M.M.: Bandwidth and gain improvement of a circularly polarized dual-rhombic loop antenna. IEEE Antennas Wirel. Propag. Lett. **5**, 84–87 (2007)
16. Movassaghi, S., Abolhasan, M., Lipman, J., Smith, D., Jamalipour, A.: Wireless body area networks: a survey. IEEE Commun. Surveys Tutorials **16**(3), 1658–1686 (2014)
17. Olatinwo, D.D., Abu-Mahfouz, A., Hancke, G.: A survey on LPWAN technologies in WBAN for remote health-care monitoring. Sensors **19**(23), 5268 (2019)
18. Salim, M., Pourziad, A.: A novel reconfigurable spiral-shaped monopole antenna for biomedical applications. Progress Electromagnet. Res. Lett. **57**, 79–84 (2015)
19. Schmidt, R., Norgall, T., Morsdorf, J., Bernhard, J., Grun, T.: Body area network BAN - a key infrastructure element for patient-centered medical applications. Biomedizinische Technik/Biomed. Eng. **47**(s1a), 365–368 (2002)

20. Soldatos, J., Yuming, G.: Internet of things. EU-China joint white paper on internet-of-things identification. Technical report, European Research Cluster on the Internet of Things (2014)
21. Vishnevskiy, V., Shirvanyan, A., Tumchenok, D.: Mathematical model of the dynamics of operation of the tethered high-altitude telecommunication platform in the turbulent atmosphere. In: 2019 Systems of Signals Generating and Processing in the Field of on Board Communications, SOSG 2019, p. 8706784 (2019)
22. Vishnevsky, V., Efrosinin, D., Krishnamoorthy, A.: Principles of construction of mobile and stationary tethered high-altitude unmanned telecommunication platforms of long-term operation. Commun. Comput. Inf. Sci. **919**, 561–569 (2018)
23. Vladimirov, S., Kirichek, R.: The IoT identification procedure based on the degraded flash memory sector. In: Galinina, O., Andreev, S., Balandin, S., Koucheryavy, Y. (eds.) NEW2AN/ruSMART/NsCC -2017. LNCS, vol. 10531, pp. 66–74. Springer, Cham (2017). https://doi.org/10.1007/978-3-319-67380-6_6
24. Vladimirov, S., Pirmagomedov, R., Kirichek, R., Koucheryavy, A.: Unique degradation of flash memory as an identifier of ICT device. IEEE Access **7**, 107626–107634 (2019)
25. Wang, J.C., Lim, E.G., Leach, M., Wang, Z., Man, K.L., Huang, Y.: Conformal wearable antennas for WBAN applications. In: Proceedings of the International MultiConference of Engineers and Computer Scientists 2016, vol. II, IMECS 2016, 16–18 March 2016, Hong Kong, pp. 651–654 (2016)
26. Wu, F., Wu, T., Yuce, M.R.: An internet-of-things (IoT) network system for connected safety and health monitoring applications. Sensors **19**(1), 21 (2019)
27. Xue, D., Garner, B., Li, Y.: Electrically-small folded cylindrical helix antenna for wireless body area networks. In: 2016 Texas Symposium on Wireless and Microwave Circuits and Systems (WMCS), Waco, TX, pp. 1–4 (2016)
28. Xue, S., Yi, Z., Xie, L., Wan, G., Ding, T.: A displacement sensor based on a normal mode helical antenna. Sensors **19**(17), 3767 (2019)

Development and Investigation of Model Network IMT2020 with the Use of MEC and Voice Assistant Technologies

Maria Makolkina(✉), Nikolay Shypota(✉), and Andrey Koucheryavy(✉)

The Bonch-Bruevich Saint-Petersburg State University of Telecommunications, 22 Prospekt Bolshevikov Russia, Saint Petersburg 193232, Russia
rector@sut.ru

Abstract. Humanity is at a new stage in the development of information and computer technologies. The digitalization of most areas of activity creates a contradiction. On the one hand, to start the operation of the device, you must promptly enter information. Only the information entered by the user, allows the software agent to perform tasks. On the other hand, the technological diversity of modern devices requires new complex skills from the user. Consequently, the search for a solution to the problem of the quick introduction of information by a person without special and long-term preparation becomes relevant. The technology of voice input of information makes it possible to solve this problem. It underlies the work of virtual voice assistants, the use of which is constantly growing. The paper developed a simulation model using voice assistants (VA) based on software-defined (SDN) networks and mobile edge computing (MEC) technology. The analysis of the scope of voice assistants. An experimental study is given of the influence of the dependence of the execution time of processes and the total delay on the method of processing speech information of devices. The dependence of the total delay on the type of traffic and computing device for the final processing of packets'.

Keywords: Voice assistant · Mobile edge computing · Software-defined network · Model network · IMT-2020

1 Introduction

Voice assistants have become a necessary part of the functioning of various online technologies: in distance learning, in online consulting in the field of medicine, in geolocation and GPS navigation, in robotics, in the field of business management as virtual assistants [1–5]. Voice Assistant is now one of the essential components of a software-defined network management system [6].

The publication has been prepared with the support of the "RUDN University Program 5–100".

V. M. Vishnevskiy et al. (Eds.): DCCN 2020, LNCS 12563, pp. 232–243, 2020.
https://doi.org/10.1007/978-3-030-66471-8_19

Voice assistants are also actively used when working with clients to advise and advertise products. They have an impact on the economic results of companies and reduce their costs. This technology has become necessary when interacting with devices operating on the "smart home" platform [7]. Personal voice assistants enable the user to interact comfortably with digital devices.

A large number of diverse VA applications require a certain organization of the communication network structure. Particular requirements are placed on providing performance indicators for the quality of experience of VA services in various areas of human life. The use of VA provides a number of advantages that directly affect the quality and responsiveness of the user experience, and also increases the efficiency of the system itself, within which the technology operates.

Let us consider some of the advantages of such a system: flexibility in network management, significant simplification of network management for the user, automation of network management and administration.

Today, SDN and MEC technologies are most suitable for VA services.

SDN technology is one of the 5G/IMT-2020 core technologies [8]. Its use provides: programmability and flexibility, adaptability of network management, increased reliability, etc. SDN has such advantages as: the ability to dynamically configure traffic flows throughout the network in accordance with changing needs; the administrator can configure, manage, provide protection, quickly optimize network resources using SDN programs, independently develop these algorithms; SDN is based on open standards; network management is provided not by devices and protocols of certain manufacturers, but by software SDN controllers. SDN greatly simplifies network design and operation. The main difference between SDN and traditional networks is the centralized intelligent management and monitoring of the network, which provides verification, control and modification of transmitted data streams. The SDN controller acts as a single centralized control point that interacts with the application layer through an open API, and also monitors and controls physical devices on the network through an open interface - the OpenFlow protocol. The controller consists of several modules or levels, each module is responsible for a number of required functionality.

MEC technology also has several advantages: it reduces the delay in the transmission of a data from sender to receiver and in reverse; increase network bandwidth; realizing the potential of introducing new applications and services based on network structure data; reduced load on the core network. Among the important characteristics are also called the prompt provision of information about the state of the network. This can be used to provide services, focusing attention on information about the network structure in real time. In 5G networks, it is planned to use a multilevel structure MEC, which is based on two types of clouds - Micro- cloud and Mini-cloud. Micro-cloud has little computing power and processes data from various devices and sensors to the base station. Mini-cloud is used at a higher level of the network structure and has greater processing and storage capabilities (compared to Micro-cloud). Paper [9] proposes a modified multilevel MEC structure for augmented reality applications that provides access to computing resources at the edge of the RAN near to AR users. This solution allows you to reduce network latency, as well as increase the efficiency of computing tasks for AR applications.

SDN and MEC are complementary technologies with a common goal: the application of specific control principles to the data plane.

This article developed a model network based on SDN and MEC technologies using VA. Developed various options for organizing the structure 5G/IMT-2020 networks using VA. A comparative analysis of three options for the implementation of the network structure according to the results of an experimental study. An investigation of the work of VA on this model network was carried out, with a different number of simultaneously processed packets and the effect of the number of packets on the delivery delay.

2 Goal of Investigation

The goal of this investigation is to study the technology of voice assistant based on the SDN network using MEC. Including the study of the dependence of the execution time of processes and the total delay in the delivery of data from the method of processing voice information.

3 Related Works

Currently, the future of networks is the transition to the fifth generation 5G/IMT-2020 communication network. Today, a number of works is known dedicated to the study of architecture, requirements for construction, technology and organization of such networks [9,10]. These networks will be able to provide a seamless connection between various devices and numerous applications as shown in work [11]. In article [12], the authors develop and explore approaches to centralized management of IoT devices when implementing the concept of a smart home in fifth-generation communication networks. A number of researchers are working on the study of the network structure and methods of distributing information over the network in order to reduce the load on the network core and thereby increase the quality of service indicators for applications that are especially demanding for delays [13,14]. The advantages of SDN technology and the possibility of its application for implementing applications of the Internet of Things, augmented reality, Tactile Internet, etc. made it main in the implementation of modern services. Many researchers create model networks on the basis of which they study the features of using this technology for various applications and together with other well-proven technologies, for example, MEC [15–18].

4 VA Simulation Model

Let's consider the main processes that take place when Voice Assistant works. As a rule, immediately after calling the Voice Assistant, there is a greeting, code words selected by the user. Interaction with the voice assistant begins with a passphrase, such as "Ok Google" or "Hey Siri". Together with the passphrase, commands are sent to perform any actions. These actions can be requesting the latest news from the Internet, opening a specific application on a smartphone, controlling smart home devices. Next, the voice signal received from the user is transformed. The final processes in the interaction between the user and Voice Assistant will be to analyze the ratio of keywords to the commands being executed and to execute the corresponding command for the keywords.

The types of voice assistants are quite difficult to systematize, due to the fact that even those created to provide assistance in a certain type of activity, designed to perform unique tasks, they can support conversation, report news, read weather information and perform other similar functions, characterized as related not the main ones. At the same time, in our opinion, two large groups of assistants can be distinguished. These are voice assistants with wide functionality and those developed for work in narrow-profile areas.

Voice assistants with a wide range of functions: Amazon Alexa, Microsoft Cortana, Yandex Alice, Google Assistant. Amazon Alexa has extensive functionality: sets an alarm and opens the blinds, turns on the light and sets the air conditioning, reads books and conducts workouts, draws up a schedule and reminds you of visits, calls, necessary purchases, selects and voices news information.

Assistants with limited functionality are trained to assist in activities in a specific area: in business, in banking, in household management, in gps navigation and geolocation, in services for recognizing music and songs.

By systematizing voice assistants according to their functional tasks, we can distinguish services designed for music recognition. As an example, let's take the following: Shazam, SoundHound, MusiXmatch, Midomi, TrackID, BeatFind, MusicID, Spotify, AudioTag, Audiggle.

As you can see, voice assistants have firmly entered various spheres of people's lives and require special attention when organizing a communication network for the timely delivery of information to the user.

In this paper, we will examine the interaction of Voice Assistant and the MEC system. The algorithm of the voice assistant will be as follows. Initially, through a microphone, speech is delivered to a portable device. On this device, the processes of obtaining a voice signal, its filtering and digitization are performed. The presented structure includes a mobile device with a built-in microphone, as well as a voice assistant responsible for determining the actions to be performed on the commands received from the user. The processes for receiving a voice signal, its filtering and digitization in all three versions were carried out on a mobile device.

Three options were developed for organizing the network structure.

The first version of the algorithm for organizing the network structure is VA located on a remote server. The user's mobile device does not extract keywords from the received signal. The steps of highlighting keywords and comparing them with the commands that are performed are performed on the remote server. The second option, the selection of keywords is carried out on a mobile device, and their comparison with the executed commands is performed on a remote server with a voice assistant. In the third option, VA is located on the mobile device and, therefore, all processes are performed on it. Assume that only the remote server can execute the command requested by the user. This means that in the third version of the algorithm, despite the fact that all processes are performed on a mobile device, the need to transfer service data containing information about the command required to be executed is saved to the remote server.

The experimental part of this study was carried out using the AnyLogic 7 simulation software package.

Figure 1 presents a simulation model of the first and second algorithm of networking.

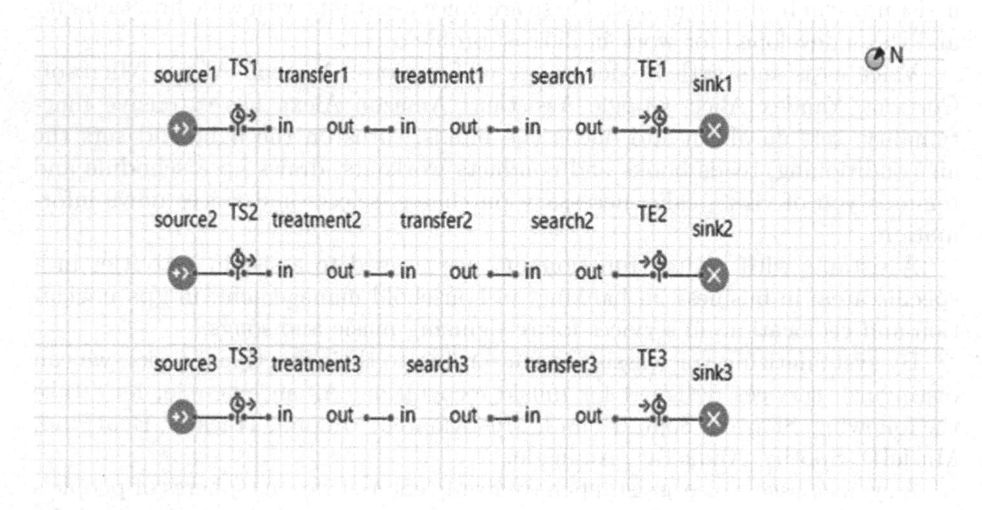

Fig. 1. SDN network simulation model using VA

TS-TimeMeasureStart TE-TimeMeasureEnd CL-cloud

Figure 2 shows the structure of the studied network.

For the experiment, data packets of different types and sizes were prepared. The first data packet consists of 218 bytes and simulates traffic from the voice assistant. The second type of packet transmits data after the database forms key phrases and selects the payload, therefore, it has a shorter length of 138 bytes. The third data packet relates to the transmission of commands for performing certain actions and is 98 bytes.

The first data packet size is incoming traffic from the voice assistant - 218 bytes, of which 160 Bytes related to the payload, and 58 bytes to the headers (HTTP2.0 protocol).

The second size of the data packet occurs after the formation of the database of key phrases and the choice of the payload, and therefore should have a shorter

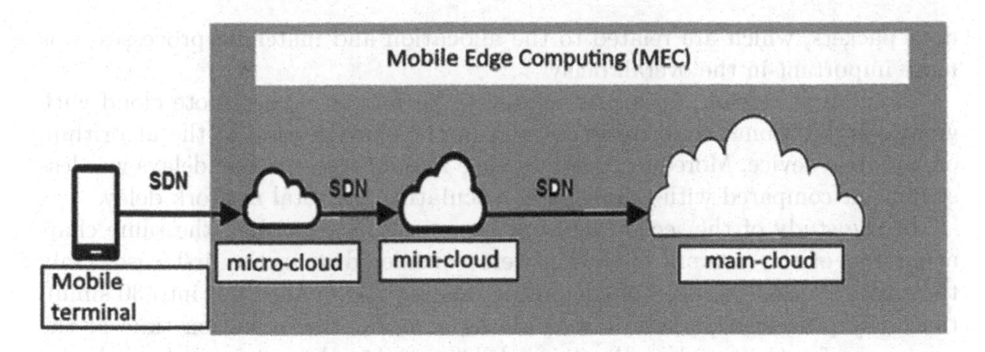

Fig. 2. The structure of the network

data packet length. In this regard, a value equal to a sample of the G.711 codec was determined, that is, equal to 80–, respectively, the packet size is 138 bytes, of which 80 bytes were related to the payload, and 58 bytes were related to headers.

The third data packet relates to the transfer of commands to perform certain actions. The size of this packet is 98 bytes, of which 40 bytes related to the payload, and 58 bytes to the headers.

The data processing speeds on the mobile device, micro-cloud, mini-cloud and main cloud were 2 Mbps, 5 Mbps, 8 Mbps 25 Mbps, respectively [13].

Below are tables with generalized data on the dependence of the execution time of processes and the total delay on the method of processing speech information for three variants of the algorithm implementation.

5 The Results of the Simulation

Table 1 shows the results with generalized data on the dependence of the execution time of processes and the total delay on the method of processing voice information for the first experiment.

Table 1. Results of the first experiment

Network algorithm	Transfer delay	Treatment delay	Search delay	Common delay
1 variant	20 ms	10 ms	7 ms	37 ms
2 variant	13 ms	26 ms	7 ms	46 ms
3 variant	9 ms	26 ms	17 ms	52 ms

In the first version of the algorithm, we considered a scenario when the delay time of data transfer on a mobile device and on a server was comparable. As a result, the delay in the transmission of data, but the delay in the processing of data packets, which are related to the allocation and matching processes, was more important in the overall delay.

In the first version, both processes were performed on a remote cloud with great computational characteristics, and in the third version of the algorithm, on a mobile device. Moreover, the difference in data transmission delays was less significant compared with the delay in calculating the total network delay.

In the study of the second network organization algorithm, the same characteristics of the network operation were considered as in the first case. Only the entire processing speed of computing devices was divided not into 30 simultaneously processed packets, but by 8. As a result, the execution time of the processes of extracting key phrases and determining the correspondence had a lower value, compared to the time of packet transmission, in the formation of the total delay.

Table 2 shows the results of the second experiment.

Table 2. The results of the second experiment

Network algorithm	Transfer delay	Treatment delay	Search delay	Common delay
1 variant	20 ms	3 ms	2 ms	25 ms
2 variant	13 ms	7 ms	2 ms	22 ms
3 variant	9 ms	7 ms	4 ms	20 ms

In the second experiment, the entire processing speed of computing devices was divided not by 30 simultaneously processed packets, but by 8. As a result, when forming the overall delay, the execution time of the processes for extracting key phrases and determining correspondence was less important than the transmission time of packets.

Consequently, the delay in data transmission will have a greater effect on the overall delay than the parameters of the time interval for performing the processes of extracting key phrases in the user's request, their comparison with the database of commands.

6 VA Simulation Model for 5G/IMT-2020 Networks with MEC

Consider the use of a voice assistant in an enterprise environment.

The voice assistant is installed on the user's mobile terminal and is used to control processes on the territory of one enterprise. The multi-layer MEC system consists of a micro-cloud, a mini-cloud and a main computing cloud, which are servers. Computing clouds are presented in ascending order of computing and storage capabilities. The micro-cloud manages the processes of one room, the mini-cloud controls the processes of the entire enterprise. Voice Assistant accepts voice commands, converts them, and determines which actions to take and on which cloud. If the Voice Assistant database does not contain this command, then a request is sequentially sent to the micro-cloud, mini-cloud and the server, whether these computational levels can execute the required command. If one of these levels can execute the command, then the request for the higher-level clouds is not sent. In addition, the Voice Assistant database records which cloud can perform a given action. When using such a search scheme, the database of the mobile device will be significantly smaller compared to the scheme considered in the previous stage of the study. This greatly reduces the memory requirements of the device and brings certain benefits to users. Thus, the process of searching for a match, if it could not be established from the database of the mobile device, will be carried out within the framework of the voice assistant.

Next, a network model was developed, in addition to voice assistant traffic, IoTDM and VLC traffic was present. The network architecture is shown in Fig. 3.

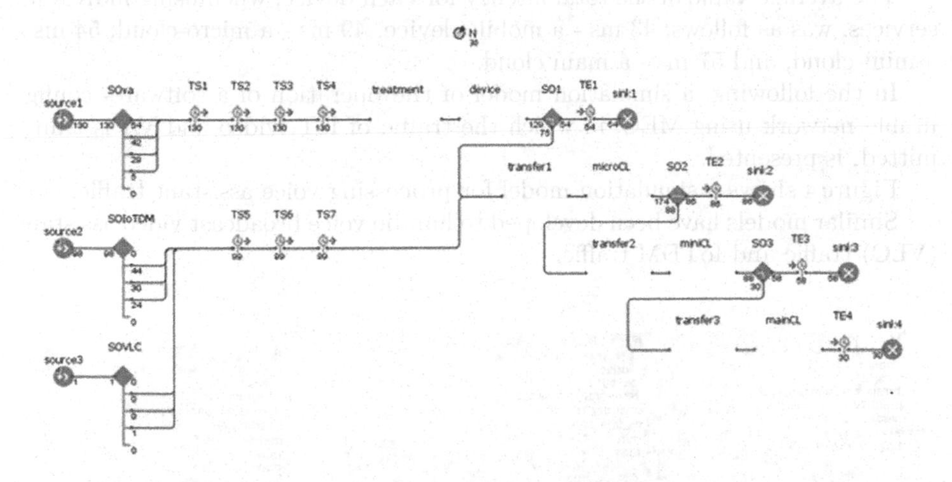

Fig. 3. The structure of the network

TS-TimeMeasureStart SO-SelectOutput TE-TimeMeasureEnd CL-cloud

The intensity of traffic generation is taken as a constant. Its value was calculated by dividing the total number of selected flows over a certain period of time. The packet sampling time was 5 min or 300 s. During this period, 8001 IoTDM traffic packets and 168 VLC traffic were selected.

For the Internet of Things traffic generator (IoTDM), the data packet length was 80 bytes, and for the video broadcast traffic generator (VLC), the data packet length would be 608 bytes

Table 3 shows the dependence of the total delay on the type of traffic and the computing device for the final processing of packets.

Due to the fact that the traffic of the IoT and video broadcasting goes directly to the micro-cloud without additional processing on the mobile device, we do not take into account the traffic processing delay on the user device.

Table 3. The dependence of the total delay on the type of traffic and the computing device for the final processing of packets

Common delay, ms				
Types of traffic generators	Mobile terminal	Micro-cloud	Mini-cloud	Main-cloud
Voice assistant	43	49	54	57
Video translation, VLC	–	62	67	71
Internet of things, IoTDM	–	13	18	22

As a result of the experiment, the average value of the total latency for each processing device was - 43 ms on the mobile device, 33 ms on the micro-cloud, 38 ms on the mini-cloud and 31 ms on the main cloud.

The average value of the total latency for each device, when using individual services, was as follows: 43 ms - a mobile device, 49 ms - a micro-cloud, 54 ms - a mini-cloud, and 57 ms - a main cloud.

In the following, a simulation model of the operation of a software- configurable network using MEC, in which the traffic of IoT, video and VA is transmitted, is presented.

Figure 4 shows a simulation model for processing voice assistant traffic.

Similar models have been developed to handle voice broadcast video assistant (VLC) traffic and IoTDM traffic.

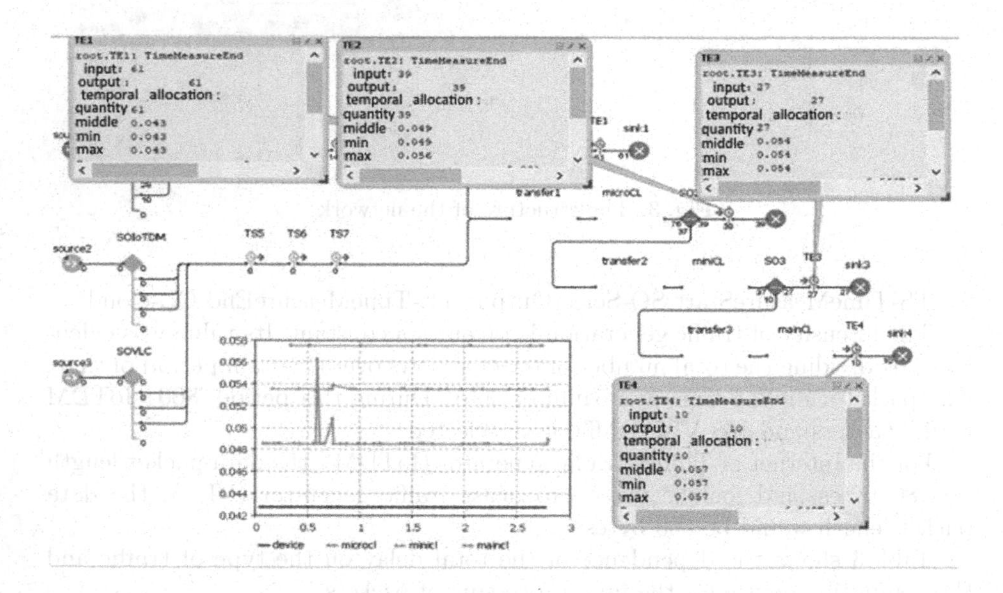

Fig. 4. Simulation model for processing voice assistant traffic

Figure 5 shows the dependence of the total delay on the type of terminal processing device for four different processing levels.

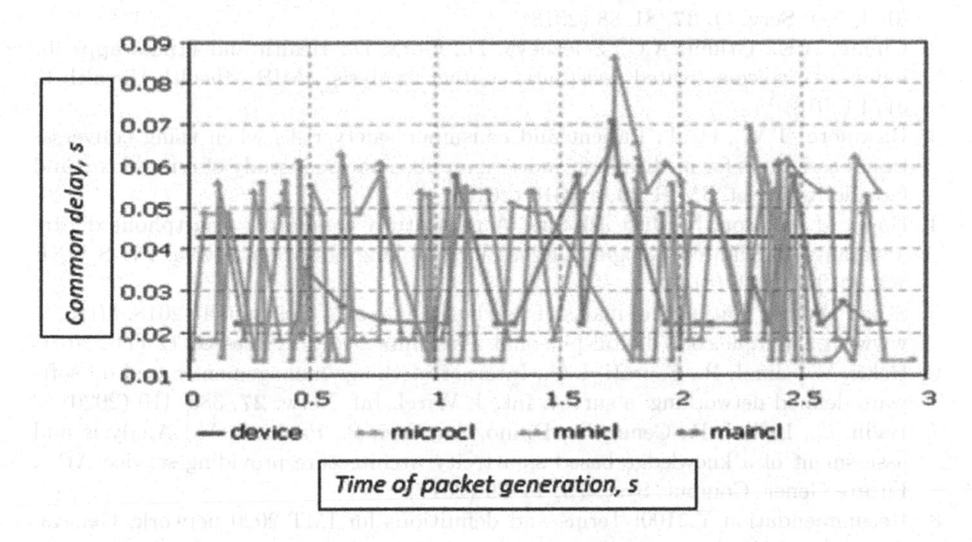

Fig. 5. Dependence of the total delay on the terminal processing device

The abscissa axis takes the time of generation of the packet - at what point in time the model was generated. The y-axis is the total delay. The graph shows a comparison of the delay time for various network units with the processing time. As you can see from the graph, the use of a multi-level architecture for voice assistant traffic processing and SDN technology allows us to fulfill the requirements for ensuring delay in the implementation of various services.

7 Conclusion

The use of SDN and MEC technologies for the implementation of services in modern communication networks improves the quality of service, unloads the core of the network and rational use of resources.

As part of this investigation, options were developed for organizing the network structure for the implementation of various services using a voice assistant. A comparative analysis of the three options for networking based on the results of an experimental study. While processing thirty data packets in the first experiment and eight packets data, respectively, in the second. The dependencies of the execution time of processes and the total delay on the method of processing speech information were investigated. The experimental results showed that, depending on the type of application and its requirements for network characteristics, it is advisable to use various network structures and multilevel traffic offloading architectures.

References

1. Hoy, M.B.: Alexa, Siri, Cortana, and more: an introduction to voice assistants. Med. Ref. Serv. Q. **37**, 81–88 (2018)
2. Chung, A.E., Griffin, A.C., Selezneva, D., Gotz, D.: Health and fitness apps for hands-free voice-activated assistants: content analysis. JMIR Mhealth Uhealth **6**, e174 (2018)
3. Bickmore, T.W., et al.: Patient and consumer safety risks when using conversational assistants for medical information: an observational study of Siri, Alexa, and Google Assistant. JMIR **20**, e11510 (2018)
4. Boyd, M., Wilson, N.: Just ask Siri? A pilot study comparing smartphone digital assistants and laptop Google searches for smoking cessation advice. PLoS ONE **13**, e0194811 (2018)
5. Stone Temple: Rating the smarts of the digital personal assistants in 2018. https://www.stonetemple.com/digital-personal-assistants-study/. Accessed 12 Feb. 2019
6. Bekri, W., Jmal, R., Fourati, L.C.: Internet of things management based on software defined networking: a survey. Int. J. Wirel. Inf. Netw. **27**, 385–410 (2020)
7. Badii, C., Bellini, P., Cenni, D., Difino, A., Nesi, P., Paolucci, M.: Analysis and assessment of a knowledge based smart city architecture providing service APIs. Future Gener. Comput. Syst. **75**, 14–29 (2017)
8. Recommendation Y.3100: Terms and definitions for IMT-2020 network, Geneva, September 2017
9. Yastrebova A., Kirichek R., Ye, K., Borodin, A., Koucheryavy, A.: Future networks 2030: architecture and requirements. In: 2018 10th International Congress on Ultra Modern Telecommunications and Control Systems and Workshops (ICUMT) (2018)
10. Ateya, A.A., Muthanna, A., Koucheryavy, A.: 5G framework based on multi-level edge computing with D2D enabled communication. In: 20th International Conference on Advanced Communication Technology (ICACT) Conference Proceedings, pp. 507–512 (2018)
11. Asadi, A., Wang, Q., Mancuso, V.: A survey on device-to-device communication in cellular networks. IEEE Wireless Commun. 1–19 (2014)
12. Muthanna, A., Gimadinov, R., Kirichek, R., Koucheryavy, A.: Software development for the centralized management of IoT-devices in the smart home systems. In: 2017 IEEE Conference of Russian Young Researchers in Electrical and Electronic Engineering (2017)
13. Ateya, A., Muthanna, A., Gudkova, I., Abuarqoub, A., Vybornova, A., Koucheryavy, A.: Development of intelligent core network for tactile internet and future smart systems. J. Sens. Actuator Netw. **7**(1), 1 (2018)
14. Khakimov, A., Muthanna, A., Kirichek, R., Koucheryavy, A., Ali, M.M.S.: Investigation of methods for remote control IoT- devices based on cloud platforms and different interaction protocols. In: 2017 IEEE Conference of Russian Young Researchers in Electrical and Electronic Engineering (2017)
15. Kirichek, R., Vladyko, A., Zakharov, M., Koucheryavy, A.: Model networks for Internet of Things and SDN. In: Proceedings of the 18th ICACT, pp. 76–79 (2016)
16. Vladyko, A., Muthanna, A., Kirichek, R.: Comprehensive SDN testing based on model network. In: Galinina, O., Balandin, S., Koucheryavy, Y. (eds.) Internet of Things, Smart Spaces, and Next Generation. LNCS, vol. 9870, pp. 539–549. Springer International Publishing (2016)

17. Khan, P.W., Abbas, K.H., Shaiba, H.A., Mutkhanna, A.S.A., Abuarqoub, A., Khayyat, M.: Energy efficient computation offloading mechanism in multi-server mobile edge computing–an integer linear optimization approach. Electronics **9**(6), 1010 (2020)
18. Manariyo, S., Poluektov, D., Khakimov, A., Muthanna, A., Makolkina, M.: Mobile edge computing for video application migration. In: Internet of Things, Smart Spaces, and Next Generation Networks and Systems, pp. 562–571 (2019)
19. Ateya, A.A., Muthanna, A., Koucheryavy, A.: 5G framework based on multi-level edge computing with D2D enabled communication. In: 2018 20th International Conference on Advanced Communication Technology (ICACT), pp. 507–512. IEEE (2018)

On Overall Measure of Non-classicality of N-level Quantum System and Its Universality in the Large N Limit

V. Abgaryan[1,2](✉), A. Khvedelidze[1,3,4], and I. Rogojin[1]

[1] Laboratory of Information Technologies, Joint Institute for Nuclear Research, Dubna, Russia
vahagnab@googlemail.com

[2] Peoples' Friendship University of Russia (RUDN University), 6 Miklukho-Maklaya Street, Moscow 117198, Russian Federation

[3] Andrea Razmadze Mathematical Institute, Iv. Javakhishvili Tbilisi State University, Tbilisi, Georgia

[4] Institute of Quantum Physics and Engineering Technologies, Georgian Technical University, Tbilisi, Georgia

Abstract. In this report we are aiming at introducing a global measure of non-classicality of the state space of N-level quantum systems and estimating it in the limit of large N. For this purpose we employ the Wigner function negativity as a non-classicality criteria. Thus, the specific volume of the support of negative values of Wigner function is treated as a measure of non-classicality of an individual state. Assuming that the states of an N-level quantum system are distributed by Hilbert-Schmidt measure (Hilbert-Schmidt ensemble), we define the global measure as the average non-classicality of the individual states over the Hilbert-Schmidt ensemble. We present the numerical estimate of this quantity as a result of random generation of states, and prove a proposition claiming its exact value in the limit of $N \to \infty$.

Keywords: Wigner function · Phase space formalism · Non-classicality · Hilbert-Schmidt measure

1 Introduction

With the rise of quantum information and computation paradigms alongside the adjacent fields, one time and again encounters the characterization "*non-classical*" when describing the quantum states involved. It must be pointed out that the notion of non-classicality of quantum states is not well defined. Under this label, we usually understand the effects predicted by quantum mechanics which are incomprehensible from the standpoint of classical intuition. These include everything spanning from quantum entanglement and other purely quantum correlations to sub-poissonian statistics and squeezing of electromagnetic fields. Quite often these become resources for new powerful techniques, as is

V. M. Vishnevskiy et al. (Eds.): DCCN 2020, LNCS 12563, pp. 244–255, 2020.
https://doi.org/10.1007/978-3-030-66471-8_20

the case, for example, with quantum entanglement. A question of quantitative description of the degree of non-classicality and hence of the resource itself arises here. Obviously, due to the wideness and vagueness of the question, it would be naive to assume the existence of a universal measure encompassing the intensities of all the quantum effects. However, it seems that the central object of quantum mechanics on the phase space, the Wigner function, somehow encodes the crucial information about the non-classical features of the state through the property of having negative values. Indeed, to name just a few examples: it has been shown that quantum circuits where the initial state together with the quantum operations is representable with positive Wigner functions can be classically efficiently simulated [2]; s-waves are entangled if and only if corresponding Wigner function has negative domains [1]; the negativity volume of the Wigner function is an entanglement indicator for hybrid qubit-bosonic states if certain conditions are met [3], etc.

Elaborating on the property of Wigner function to have negative values, several measures of non-classicality have been introduced (see [4] and references therein). Here, we generalize these well-established ideas from the level of individual states to the whole state space.

The article is organized as follows. In the next section we introduce the necessary basics about the Wigner function. In Sect. 3 we define the main quantities which will be used in the rest of paper. Section 4.1 contains results on the global measure of non-classicality of density matrices from the Hilbert-Schmidt ensemble. Finally, in Sect. 4.2 the analysis of behavior of the introduced measure of non-classicality for large N is given.

2 The Wigner Function of a Density Operator of N-level System

Wigner function [5] was introduced in an attempt of phase space description of quantum mechanics. For quantum states represented by a density operator $\hat{\rho} \in \mathcal{D}(L^2(\mathbb{R}^n))$ acting on the Hilbert space $\mathcal{H} = L^2(\mathbb{R}^n)$ the Wigner function of $\hat{\rho}$ is defined over a phase space $(\mathbb{R}^{2n}, w)$ with the standard symplectic 2-form $w := \sum_j^n dp_j \wedge dq_j$ and is given by the so-called Wigner transform:

$$W_{\hat{\rho}}(\boldsymbol{q}, \boldsymbol{p}) = \left(\frac{1}{2\pi\hbar}\right)^n \int_{\mathbb{R}^n} d\boldsymbol{\eta} \left\langle \boldsymbol{q} - \frac{\boldsymbol{\eta}}{2} \middle| \hat{\rho} \middle| \boldsymbol{q} + \frac{\boldsymbol{\eta}}{2} \right\rangle e^{\frac{i}{\hbar}\boldsymbol{p}\boldsymbol{\eta}}. \tag{1}$$

According to the Weyl-Wigner formalism one can establish an invertible map between the self-adjoint semipositive definite operator $\hat{\rho}$ and its Weyl symbol $(2\pi\hbar)^n W_{\hat{\rho}}(\boldsymbol{q}, \boldsymbol{p})$ in (1)

$$\hat{\varrho} \rightleftarrows W_{\rho}(\boldsymbol{q}, \boldsymbol{p}). \tag{2}$$

Generalizing the Weyl-Wigner mapping (2) to the case of an arbitrary self-adjoint operator,

$$\hat{A} \rightleftarrows W_A(\boldsymbol{q}, \boldsymbol{p}), \tag{3}$$

the quantum mechanical prediction of the operator, i.e., $\mathbb{E}[\hat{A}] = \mathrm{tr}[\hat{\rho}\hat{A}]$ is expressible in the form of conventional ensemble average in classical mechanics defined as the mean of an operator symbol $A(\boldsymbol{q}, \boldsymbol{p})$ over the phase-space with the distribution W_ϱ :

$$\mathbb{E}[\hat{A}] = \overline{A} \qquad \overline{A} := \int_{\mathbb{R}^{2n}} dp\, dq\, W_A(\boldsymbol{q}, \boldsymbol{p})\, W_\rho(\boldsymbol{p}, \boldsymbol{q})\,. \tag{4}$$

However, the similarity between the quantum and classical expressions is somewhat illusive. Though, the marginal distributions of momenta on one side and of coordinates on the other are true probability density functions, due to the limitation of simultaneous measurements of coordinates and momenta in quantum mechanics by Heisenberg uncertainty principle, Wigner function is not free from "faults". Namely, it may be shown that there are states for which Wigner function has negative values. Hence it can't be considered as a true probability density function, and is usually called a quasiprobability density function.

As it was mentioned above the Wigner transform is well adapted to the case of a quantum mechanical system associated to the Hilbert space $\mathcal{H} = L^2(\mathbb{R}^n)$. The natural question arises how to deal with other quantum systems whose Hilbert space $\mathcal{H}$ is different from $L^2(\mathbb{R}^n)$? In 1957, based on the Weyl-Wigner approach, R.L.Stratonovich formulated [6] general principles of constructing the mapping (3), which should be satisfied for any quantum system associated to some Hilbert space. These principles, later on, received the name of Stratonovich-Weyl (SW) correspondence. Since in the present note we are interested in quantification of "quantumness" in systems whose Hilbert space is $\mathcal{H} = \mathbb{C}^N$, below basics of SW correspondence are reproduced in a form adapted to the case of finite-dimensional quantum systems.

The basic idea of realisation of mapping (3) is to use the kernel operator $\Delta(\Omega_N)$ defined over the symplectic manifold Ω_N endowed with some symplectic 2-form. The mapping is given by formulae

$$W_A(\Omega_N) = \mathrm{tr}\,(A\Delta(\Omega_N))\,, \tag{5}$$

$$\hat{A} = \int_{\Omega_N} \mathrm{d}\Omega_N\, \Delta(\Omega_N) W_A(\Omega_N)\,; \tag{6}$$

Here the kernel $\Delta(\Omega_N)$ is self-dual, in sense that the same kernel defines as direct as well an inverse mapping (5), and it is the so-called Staratonovich-Weyl kernel. According to the Stratonovich-Weyl principles in order to have a correct phase-space formulation of quantum theory SW kernel should provide fulfilment of the following compulsory requirements:

- the kernel must be Hermitian, $\Delta(\Omega_N)^\dagger = \Delta(\Omega_N)$ guaranteeing the reality of symbols;
- the kernel must be the trace class operator, i.e., $\int_{\Omega_N} d\Omega_N\, \Delta(\Omega_N) = 1$, ensuring completeness of quantum states as well as classical ones;

– the unitary symmetry of states $\rho' = g\rho g^{\dagger}$, $g \in SU(N)$ induces the adjoint transformation of SW kernel, $\Delta(\Omega_N') = g^{\dagger}\Delta(\Omega_N)g$ where Ω_N' is an image of point Ω_N under the action of g.

It has been shown [7] that for N-level quantum system the Stratonovich-Weyl correspondence clauses admit simple formulation in the form of algebraic equations on spectrum of SW kernel:

$$\operatorname{tr}[\Delta(\Omega_N)] = 1 \quad \text{and} \quad \operatorname{tr}[\Delta(\Omega_N)^2] = N\,. \tag{7}$$

These equations leave $N-2$ parametric freedom of choice of the spectrum of SW kernel. Taking into account this ambiguity we can write the SVD decomposition of SW kernel

$$\Delta(\Omega_N|\boldsymbol{\nu}) = U(\Omega_N)\,P(\boldsymbol{\nu})\,U^{\dagger}(\Omega_N)\,, \tag{8}$$

where $P(\boldsymbol{\nu})$ is a diagonal matrix whose elements are specifically ordered eigenvalues of the SW kernel, $\boldsymbol{spec}(\Delta) = \{\pi_1(\boldsymbol{\nu}), \pi_2(\boldsymbol{\nu}), \dots, \pi_N(\boldsymbol{\nu})\}$. Eigenvalues $\pi(\boldsymbol{\nu})$ are functions of a real $(N-2)$- tuple $\boldsymbol{\nu} = (\nu_1, \cdots, \nu_{N-2})$ parameterising the moduli space of solutions to (7). Hereafter, dealing with the Wigner function of density matrix ρ we will point at this ambiguity by explicitly writing dependence of SW kernel on the moduli space parameters $\boldsymbol{\nu}$:

$$W_{\rho}^{(\boldsymbol{\nu})}(\Omega_N) = \operatorname{tr}[\rho\,\Delta(\Omega_N|\boldsymbol{\nu})]\,. \tag{9}$$

See more on the moduli space of parameters in [8].

Finally, a few remarks on symplectic space Ω_N are in order. From the SVD decomposition (8) it follows that its structure, particularly its dimension depends on the choice of kernel. Now, assuming that its isotropy group $H \in U(N)$ is of the form

$$H_{\boldsymbol{k}} = U(k_1) \times U(k_2) \times U(k_{s+1})\,,$$

then the corresponding phase-space Ω_N can be identified with a complex flag variety $\mathbb{F}^N_{d_1,d_2,\dots,d_s} = U(N)/H$, where $(d_1, d_2, \dots, d_s)$ are positive integers with sum N, such that $k_1 = d_1$ and $k_{i+1} = d_{i+1} - d_i$ with $d_{s+1} = N$. Therefore, each SW kernel is in one-to one correspondence with a point of moduli space (with $\boldsymbol{\nu}$- being the corresponding coordinate) and it is defined over the phase $\Omega_{N,\boldsymbol{k}}$, member of the finite family of flag varieties labeled by an integer $(s+1)$-tuple $\boldsymbol{k} = (k_1, \dots, k_{s+1})$. The volume form on $\Omega_{N,\boldsymbol{k}}$ is determined by the bi-invariant normalised Haar measure $d\mu_{SU(N)}$ on $SU(N)$ group [7]:

$$d\Omega_{N,\boldsymbol{k}} = N\,\mathrm{Vol}(H_{\boldsymbol{k}})\,\frac{d\mu_{SU(N)}}{d\mu_{H_{\boldsymbol{k}}}}\,, \tag{10}$$

where $d\mu_{H_{\boldsymbol{k}}}$ is the bi-invariant measure over the isotropy group $H_{\boldsymbol{k}}$.

3 Measures of Non-classicality of State and Overall Quantum System

Before introducing the main quantity we are interested in, it is worth to remind a few auxiliary notions. We begin with the definition of the state space $\mathfrak{P}_N$ of an N-level quantum system.

Definition 1. *The state space $\mathfrak{P}_N$ is a N^2-1 dimensional subset in the space of $N \times N$ complex matrices $M_N(\mathbb{C})$ given by following conditions:*

$$\mathfrak{P}_N = \{X \in M_N(\mathbb{C}) \mid X = X^\dagger\,, \quad X \geq 0\,, \quad \mathrm{tr}\,(X) = 1\}\,. \tag{11}$$

Let, $\Delta(\Omega_N \,|\, \boldsymbol{\nu})$ be the Stratonovich-Weyl (SW) kernel with moduli parameter $\boldsymbol{\nu}$. Due to possible symmetries of state ρ and SW kernel the corresponding WF function has domain of definition not over the whole Ω_N, but is restricted to its certain subset. Having in mind this fact we introduce two additional definitions.

Definition 2. *$\Omega_N[\rho \,|\, \boldsymbol{\nu}] \in \Omega_N$ represents a support of WF associated to a given state $\rho \in \mathfrak{P}_N$ and SW kernel.*

Definition 3. *We call $\Omega_N^{(-)}[\rho \,|\, \boldsymbol{\nu}]$ the negative support of the Wigner function associated to a given SW kernel and state $\rho \in \mathfrak{P}_N$,*

$$\Omega_N^{(-)}[\rho \,|\, \boldsymbol{\nu}] = \{\omega \in \Omega_N[\rho \,|\, \boldsymbol{\nu}] \mid W_\rho^{\boldsymbol{\nu}}(\omega) < 0\}\,. \tag{12}$$

Associating non classicality with the discrepancy between positivity requirement on classical probability distribution and a property of the Wigner function to attain negative values, we introduce a measure quantifying quantumness of state via a relative volume of the subset of phase space where this discrepancy occurs. The next definitions give formalization of this idea.

Definition 4. *For a state ρ of an N-dimensional quantum system we define its non-classicality measure (or quantumness) $\mathfrak{Q}_N[\rho\,, \boldsymbol{\nu}]$ as*

$$\mathfrak{Q}_N[\rho\,, \boldsymbol{\nu}] = \frac{\mathrm{Vol}(\Omega_N^{(-)}[\rho \,|\, \boldsymbol{\nu}])}{\mathrm{Vol}\,(\Omega_N[\rho \,|\, \boldsymbol{\nu}])}\,. \tag{13}$$

It is necessary to note that in definition (13) it is assumed that the volume is evaluated using the symplectic volume form which is a projection of the corresponding volume form on the phase space Ω_N to the subset $\Omega_N[\rho \,|\, \boldsymbol{\nu}]$.

Definition 5. *We call the following unions,*

$$\Omega_N[\boldsymbol{\nu}] = \bigcup_{\rho \in \mathfrak{P}_N} \Omega_N[\rho \,|\, \boldsymbol{\nu}]\,, \quad \text{and} \quad \Omega_N^{(-)}[\boldsymbol{\nu}] = \bigcup_{\rho \in \mathfrak{P}_N} \Omega_N^{(-)}[\rho \,|\, \boldsymbol{\nu}] \tag{14}$$

as the "symplectic superspace" and the collection of supports of negativity of the Wigner function will be called correspondingly as "negativity supersupport".

This definitions are in given in a sense of the famous Wheeler's superspace notion in General Relativity (see [9]). Basically $\Omega_N[\boldsymbol{\nu}]$ is the collection of the supports of the WF of all possible states of N-level quantum system with fixed SW kernel. Following the same logic as before one can introduce the measure of quantumness on the "symplectic superspace" as well.

Definition 6. *For a given SW kernel, the global non-classicality measure* $\mathfrak{Q}_N[\boldsymbol{\nu}]$ *of N-level quantum system is*

$$\mathfrak{Q}_N[\boldsymbol{\nu}] = \frac{\mathrm{Vol}_{\mathrm{g}}(\Omega^{(-)}_N[\boldsymbol{\nu}])}{\mathrm{Vol}_{\mathrm{g}}\,(\Omega_N[\boldsymbol{\nu}])}\,. \tag{15}$$

In the definition (15) under the volume of the "symplectic superspace" we assume a result of an average of the symplectic volume of $\mathrm{Vol}(\Omega_N[\rho, \boldsymbol{\nu}])$ over all possible states distributed in accordance with the measure $\mathrm{d}\mathfrak{m}_{\mathrm{g}}[\rho]$, associated to a certain Riemannian metric g on $\mathfrak{P}_N$:

$$\mathrm{Vol}_{\mathrm{g}}\,(\Omega_N[\boldsymbol{\nu}]) = \int_{\mathfrak{P}_N} \mathrm{d}\mathfrak{m}_{\mathrm{g}}[\rho]\,\mathrm{Vol}(\Omega_N[\rho, \boldsymbol{\nu}]) \tag{16}$$

Below we introduce notions allowing us to relate the definition of the global indicator of system quantumness $\mathfrak{Q}_N[\boldsymbol{\nu}]$ given in terms of the "symplectic superspace" with the corresponding notion formulated in terms of the state space $\mathfrak{P}_N$.

Definition 7. *For an arbitrary point $\omega \in \Omega_N[\rho\,|\,\boldsymbol{\nu}]$, the subspace $\mathfrak{P}^{(-)}_N[\boldsymbol{\nu}\,|\,\omega] \subset \mathfrak{P}_N$ of state space is defined as*

$$\mathfrak{P}^{(-)}_N[\boldsymbol{\nu}\,|\,\omega] = \{\rho \in \mathfrak{P}_N\,|\,\omega \in \Omega_N[\rho\,|\,\boldsymbol{\nu}]\,,\, W^{\nu}_{\rho}(\omega) < 0\,\}\,. \tag{17}$$

Proposition 1. *The volume of $\mathfrak{P}^{(-)}_N[\boldsymbol{\nu}\,|\,\omega]$ evaluated with respect to the unitary invariant measure on $\mathfrak{P}_N$ is independent of ω,*

$$\frac{\mathrm{d}}{\mathrm{d}\omega}\,\mathrm{Vol}_g\left(\mathfrak{P}^{(-)}_N[\boldsymbol{\nu}\,|\,\omega]\right) = 0\,. \tag{18}$$

Proof. Indeed, let us write down the volume integral (18) over negativity domain via the Heaviside step function $\theta[-W^{\nu}_{\rho}(\omega)]$ and use SVD decomposition (8) for SW kernel $\Delta(\omega|\boldsymbol{\nu})$ of the Wigner function

$$\mathrm{Vol}_g\left(\mathfrak{P}^{(-)}_N[\nu\,|\,\omega]\right) = \int_{\mathfrak{P}_N} \mathrm{d}\mathfrak{m}_{\mathrm{g}}[\rho]\,\theta\left[-\mathrm{tr}(U(\omega)P^{\nu}\,U(\omega)^{\dagger}\rho)\right] = \int_{\mathfrak{P}_N} \mathrm{d}\mathfrak{m}_{\mathrm{g}}[\rho']\,\theta\left[-\mathrm{tr}(P^{\nu}\,\rho')\right], \tag{19}$$

In the last line of (19) we perform transformation $\rho' = U(\omega)^{\dagger}\rho U(\omega)$. Noting that the state space $\mathfrak{P}_N$ is $SU(N)$ invariant space endowed with the invariant measure we get convinced that $\mathrm{Vol}\left(\mathfrak{P}^{(-)}_N[\boldsymbol{\nu}\,|\,\omega]\right)$ is the same for all $\omega \in \Omega_N[\rho\,|\,\boldsymbol{\nu}]$.

Based on this observation, afterwards we choose ω corresponding to the diagonal SW kernels, i.e., $\omega = 0$ and simplify notation of the negativity subset, $\mathfrak{P}^{(-)}_N[\boldsymbol{\nu}]$.

Now we are in position to formulate the Proposition which interrelates two ways of interpretation of the global measure of quantumness.

Proposition 2. *The global non-classicality measure $\mathfrak{Q}_N[\boldsymbol{\nu}]$ can be expressed as the relative volume of the subset $\mathfrak{P}^{(-)}_N[\boldsymbol{\nu}]$ with respect to total volume of state space $\mathfrak{P}_N$:*

$$\mathfrak{Q}_N[\boldsymbol{\nu}] = \frac{\mathrm{Vol_g}(\mathfrak{P}^{(-)}_N[\boldsymbol{\nu}])}{\mathrm{Vol_g}(\mathfrak{P}_N[\boldsymbol{\nu}])} . \tag{20}$$

where the volume of state space is evaluated with respect to the metric g generating the measure $\mathrm{d}\mathfrak{m}_{\mathrm{g}}[\rho]$ *in definition (15).*

Proof. At first let us note that contribution to (20) from components of "symplectic superspace" associated to non-generic states (degenerate and non-maximal rank density matrices) is zero owing to zero integration measure of this states.

Hence, the integration effectively projects only to the components of "symplectic superspace" corresponding to the stratum of the generic states, whose isotropy group is conjugated to subgroup $H = U(1)^N$. Therefore the structure of $\Omega_N[\rho, \boldsymbol{\nu}]$ is solely determined by the SW kernel and does not depend on the ρ, which means that,

$$\mathrm{Vol}_g(\Omega_N[\boldsymbol{\nu}]) = \mathrm{Vol}_g(\mathfrak{P}_N)\mathrm{Vol}\left(\frac{U(N)}{U(1)^N}\right) , \tag{21}$$

The same argumentation lead the relation

$$\mathrm{Vol}_g(\Omega^{(-)}_N[\boldsymbol{\nu}] = \mathrm{Vol}_g(\mathfrak{P}^{(-)}_N[\boldsymbol{\nu}])\mathrm{Vol}\left(\frac{U(N)}{U(1)^N}\right) , \tag{22}$$

thus proving the Proposition.

4 Global Measure of Quantumness as Geometric Probability

In this section we outline an interpretation of the above introduced measure of nonclassicality $\mathfrak{Q}_N[\boldsymbol{\nu}]$ as a certain geometric probability. Indeed, according to the representation (20) the global non-classicality measure $\mathfrak{Q}_N[\boldsymbol{\nu}]$ can be expressed as the relative volume of the subset $\mathfrak{P}^{(-)}_N[\boldsymbol{\nu}] \in \mathfrak{P}_N$, consisting out of states ρ whose Wigner functions $W_\rho(\omega \mid \boldsymbol{\nu})$ evaluated at some fixed point of phase space, say $\omega = 0$ are negative. Therefore, in consent to the Theory of Geometric Probability, this relative volume can be identified with the probability of finding of states with the negative WF among a certain random ensemble of states:

$$\mathcal{P}^{(-)} := \frac{\textit{Number of states with negative WF}}{\textit{Total number of generated states}} \tag{23}$$

Note, that this identification of $\mathfrak{Q}_N[\boldsymbol{\nu}]$ and $\mathcal{P}^{(-)}$ is correct if random states are distributed in ensemble according to the measure $\mathrm{d}\mathfrak{m}_{\mathrm{g}}[\rho]$ in definition (15).

Following this identification of $\mathfrak{Q}_N[\boldsymbol{\nu}]$ and probability $\mathcal{P}^{(-)}$, we will generate random ensemble of the Hilbert-Schmidt states of N-level quantum system and then construct the corresponding Wigner functions with different kernels and test them on negativity.

4.1 Quantumness of Hilbert-Schmidt States for Different SW Kernels

There is an elegant method of generation of random density matrices from the Hilbert-Schmidt ensemble . Its starting point is is the generation of the so-called Ginibre ensemble, i.e., the set of complex matrices whose elements have real and imaginary parts distributed as independent normal random variables. Considering a square $N \times N$ complex random matrix z from the Ginibre ensemble, one can construct the density matrix from the Hilbert-Schmidt ensemble as

$$\rho_{\text{HS}} = \frac{z^\dagger z}{\text{tr}(z^\dagger z)}, \tag{24}$$

We have generated such set of density matrices we evaluate $\mathfrak{Q}_N(\nu)$ according to (23), for the next families of SW kernels:

- Kernels whose isotropy group is $H = SU(N-1)$, i.e., $(N-1)$- fold degenerate eigenvalues $\frac{1+\sqrt{1+N}}{N}$ and one smallest eigenvalue, $\frac{1+(1-N)\sqrt{1+N}}{N}$;
- Kernels whose isotropy is $H = SU(N-2) \times SU(2)$, i.e., $(N-2)$- fold degenerate eigenvalues $\frac{2-N-\sqrt{2}\sqrt{(N-2)(N-1)(1+N)}}{(N-2)N}$ and double degenerate eigenvalues $\frac{2-\sqrt{2}\sqrt{(N-2)(N-1)(1+N)}}{2N}$;
- Kernels whose isotropy group is $H = SU(N-3) \times SU(3)$, i.e., $(N-3)$-fold degenerate eigenvalues $\frac{3-N-\sqrt{3}\sqrt{3-N-3N^2+N^3}}{(N-3)N}$ and triple of degenerate eigenvalues, $\frac{3-\sqrt{3}\sqrt{3-N-3N^2+N^3}}{3N}$;
- Random kernels, which almost always are generic.

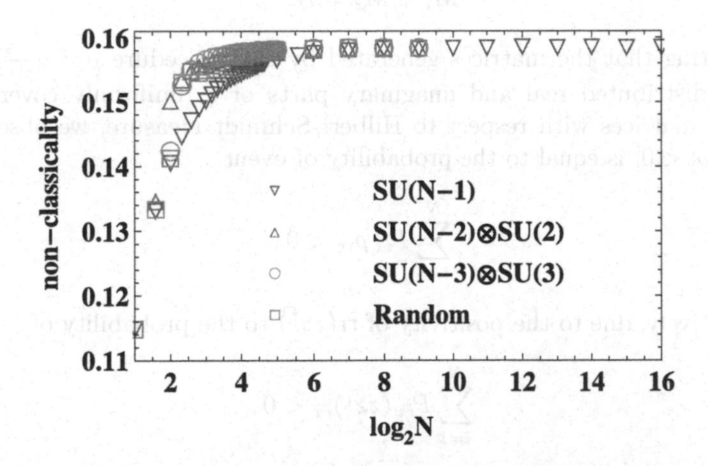

Fig. 1. Dependence of $\mathfrak{Q}_N[\nu]$ on number of levels N for different types SW kernels described in the text. For systems with number of levels greater than $N = 2^8$ the outputs of all but one plots are suppressed, since the difference of values of $\mathfrak{Q}_N[\nu]$ for different kernels are within the statistical error.

In Fig. 1 we have plotted $\mathfrak{Q}_N(\boldsymbol{\nu})$ depending on N, for different SW kernels. Approximately $\sim 10^8$ matrices have been generated and tested on the Wigner function negativity for each N. This plot shows that with growing number of levels the quantumness of system becomes independent of SW kernel and tends to a certain value. In the next section we will give argumentation of this universality of $\mathfrak{Q}_N(\boldsymbol{\nu})$ for the Hilbert-Schmidt ensemble.

4.2 The Large N Limit of Global Non-classicality

Proposition 3. *In the limit $N \to \infty$ the global non-classicality measure $\mathfrak{Q}_N[\boldsymbol{\nu}]$ of the Hilbert-Schmidt ensemble does not depend on the choice of SW kernel. Furthermore, for the infinite level system the quantumness measure is*

$$\lim_{N\to\infty} \mathfrak{Q}_N(\boldsymbol{\nu}) = \operatorname{erfc}\left(\frac{1}{\sqrt{2}}\right) \tag{25}$$

Proof. Suppose that SW kernel $\Delta = UP(\boldsymbol{\nu})U^\dagger$ is given by $P(\boldsymbol{\nu}) = \operatorname{diag}||\pi_1, \pi_2, \cdots, \pi_N||$, where the eigenvalues are presented in a decreasing order and that only m of them are non negative. We assume the following notations

$$Z_1 = \sum_{i=1}^{m} \pi_i, \quad Z_2 = \sum_{i=m+1}^{N} |\pi_i|, \quad M_1 = \sum_{i=1}^{m} \pi_i^2, \quad M_2 = \sum_{i=m+1}^{N} \pi_i^2 . \tag{26}$$

In this notations the equations SW kernel obeys may be rewritten as

$$Z_1 - Z_2 = 1 \tag{27}$$

$$M_1 + M_2 = N. \tag{28}$$

Remembering that the matrices generated by the procedure $\rho = \frac{zz^\dagger}{tr(zz^\dagger)}$, with normally distributed real and imaginary parts of ϱ, uniformly cover the set of density matrices with respect to Hilbert-Schmidt measure, we observe that $\operatorname{Prob}[\operatorname{tr}[P\rho] < 0]$ is equal to the probability of event

$$\sum_{i=1}^{N} P_{ii}\,\rho_{ii} < 0\,, \tag{29}$$

or alternatively, due to the positivity of $\operatorname{tr}(zz^\dagger)$ to the probability of

$$\sum_{i=1}^{N} P_{ii}\,(zz^\dagger)_{ii} < 0\,. \tag{30}$$

Equivalently rewriting this event we get

$$\sum_{i=1}^{m} \pi_i\,(zz^\dagger)_{ii} < \sum_{i=m+1}^{N} |\pi_i|\,(zz^\dagger)_{ii}\,. \tag{31}$$

Further we denote $\xi_j^{(i)} = z_{i,j} z_{i,j}^*$, so that $(zz^\dagger)_{i\,i} = \sum_{j=1}^N \xi_j^{(i)}$, and $\beta_j^{(i)} = |\pi_i|\,\xi_j^{(i)}$. In this terms 31 may be rewritten

$$\sum_{i=1}^{m}\sum_{j=1}^{N}\beta_i^{(j)} < \sum_{i=m+1}^{N}\sum_{j=1}^{N}\beta_i^{(j)}. \tag{32}$$

Since $Re(z_{i,j})$ and $Im(z_{i,j})$ are distributed by normal distribution with zero mean and unit variance, then ξ's are distributed with χ_2^2, distribution while $\beta_j^{(i)}$'s are distributed with mean $\mathbb{E}(\beta_j^{(i)}) = 2\,|\pi_i|$ and variance $\mathrm{var}(\beta_j^{(i)}) = 4\pi_i^2$. Now according to central limit theorem

$$x = \frac{\sum_{i=1}^m \sum_{j=1}^N \beta_j^{(i)} - \sum_{i=1}^m \sum_{j=1}^N \mathbb{E}\left(\beta_j^{(i)}\right)}{\left(\sum_{i=1}^m \sum_{j=1}^N \mathrm{var}\left(\beta_j^{(i)}\right)\right)^{\frac{1}{2}}} = \tag{33}$$

$$\frac{\sum_{i=1}^m \sum_{j=1}^N \beta_j^{(i)} - 2\,N Z_1}{\sqrt{4\,N M_1}} \tag{34}$$

as well as

$$y = \frac{\sum_{i=m+1}^N \sum_{j=1}^N \beta_j^{(i)} - \sum_{i=m+1}^N \sum_{j=1}^N \mathbb{E}\left(\beta_j^{(i)}\right)}{\left(\sum_{i=m+1}^N \sum_{j=1}^N \mathrm{var}\left(\beta_j^{(i)}\right)\right)^{\frac{1}{2}}} = \tag{35}$$

$$\frac{\sum_{i=m+1}^m \sum_{j=1}^N \beta_j^{(i)} - 2\,N Z_2}{\sqrt{4\,N M_2}}, \tag{36}$$

are distributed normally with zero mean and unit variance. Transforming Eq. 32 by subtracting from both sides $2N(Z_1 + Z_2)$ and dividing by $\sqrt{4NM_1M_2}$ we get

$$\frac{x}{\sqrt{M_2}} - \frac{N Z_2}{\sqrt{N M_1 M_2}} < \frac{y}{\sqrt{M_1}} - \frac{N Z_1}{\sqrt{N M_1 M_2}}, \tag{37}$$

or taking into account the equations for the Statonovich-Weyl kernel we get

$$x < y\sqrt{\frac{N - M_1}{M_1}} - \sqrt{\frac{N}{M_1}}. \tag{38}$$

Now, let us denote $t = \sqrt{\frac{N-M_1}{M_1}}$ so that the initial probability is equal to the probability of

$$x < y\,t - \sqrt{t^2 + 1}. \tag{39}$$

Since x and y are distributed normally the probability of the event described by Eq. (39) will be

$$\mathcal{P}(t) = \frac{1}{2\pi}\int_{-\infty}^{\infty} dy \int_{-\infty}^{yt-\sqrt{t^2+1}} dx\, e^{-\frac{y^2}{2}} e^{-\frac{x^2}{2}}. \tag{40}$$

It may be checked that

$$\frac{d\mathcal{P}(t)}{dt} = \frac{1}{2\pi}\int_{-\infty}^{\infty} dy \, \frac{e^{-\frac{y^2}{2}-\frac{1}{2}\left(\sqrt{1+t^2}-ty\right)^2}\left(-t+\sqrt{1+t^2}y\right)}{\sqrt{t^2+1}} = 0 \tag{41}$$

Hence, $\mathcal{P}(t) = \mathcal{P}(0) = \frac{1}{2}\,\mathrm{erfc}\left(\frac{1}{\sqrt{2}}\right) = 0.158655$, where erfc is the complimentary error function. Which proves the preposition.

5 Conclusions

Summarizing, in this work we have introduced a global measure of non-classicality of the state space of N-level quantum systems. By computer simulations the measure was computed for several SW kernel families depending on the number of levels. It was proven that for large N the measure of non-classicality does not depend on the choice of the SW kernel, thus it is an invariant over the moduli space of SW kernels.

However, it must be noted, that it is unreasonable to suppose that a single measure might capture all the non-classical aspects of quantum states, let al.one of the whole state space. Hence, there must be different ways of defining non-classicality measures underlining this or that features of quantum behaviour (see for comparison [10]).

Acknowledgments. The publication has been prepared with the support of the "RUDN University Program 5–100" (recipient V.A.).

References

1. Dahl, J.P., Mack, H., Wolf, Schleich, W.P.: Entanglement versus negative domains of Wigner functions. Phys. Rev. A **74**, 042323 (2006)
2. Mari, A., Eisert, J.: Positive Wigner functions render classical simulation of quantum computation efficient. Phys. Rev. Lett. **109**, 230503 (2012)
3. Arkhipov, I.I., Barasiński, A., Svozilk, J.: Negativity volume of the generalized Wigner function as an entanglement witness for hybrid bipartite states. Sci. Rep. **8**, 16955 (2018)
4. Kenfack, A., Zyczkowski, K.: Negativity of the Wigner function as an indicator of nonclassicality. J. Opt. B: Quantum Semiclass. Opt. **6**, 396–404 (2004)
5. Wigner, E.P.: On the quantum correction for thermodynamic equilibrium. Phys. Rev. **40**, 749 (1932)
6. Stratonovich, R.L.: On distributions in representation space. Sov. Phys. JETP **4**(6), 891 (1957)
7. Abgaryan, V., Khvedelidze, A.: On families of Wigner functions for N-level quantum system (2018). https://arxiv.org/pdf/1708.05981.pdf
8. Abgaryan, V., Khvedelidze, A., Torosyan, A.: On moduli space of the Wigner quasiprobability distributions for N-dimensional quantum systems. J. Math. Sci. **240**, 617 (2019)

9. Wheeler, J.A.: Superspace and the nature of quantum geometrodynamics. In: DeWitt, C.M. Wheeler, J.A. (eds.) Battelle Rencontres, 1967. Lectures in Mathematics and Physics, pp. 242–307 (1968)
10. Abgaryan, V., Khvedelidze, A., Torosyan, A.: The global indicator of classicality of an arbitrary N-level quantum system. Zap. Nauchn. Sem. POMI **485**, 5 (2019)

Analytical Modeling of Distributed Systems

Queues with Markovian Arrivals, Phase Type Services, Breakdowns, and Repairs

Srinivas R. Chakravarthy[1(✉)], Rakesh Kumar Meena[2], and Alka Choudhary[3]

[1] Departments of Industrial and Manufacturing Engineering and Mathematics, Kettering University, Flint, MI 48504, USA
schakrav@kettering.edu

[2] Banaras Hindu University, Varanasi, Uttar Pradesh, India
meena.rk@bhu.ac.in

[3] Central University of Rajasthan, Ajmer, India
alkababal@gmail.com

Abstract. In service sectors, the server interruptions occur naturally and are studied using queueing models with interruptions. Such models have been studied extensively in the literature. The server interruptions may occur due to external events such as shocks and negative customers, or due to internal events which are dictated by a timer or a clock which when expired may result in an interruption. In this paper, we analyze models of $MAP/PH/1$-type with phase type breakdowns and phase type repairs analytically. The breakdowns are modeled using internal clocks. The study of such breakdowns in the context of $MAP/PH/c$-type models is carried out via simulation. Some illustrative examples are discussed.

Keywords: Queueing · Server breakdowns · Server repairs · Markovian arrivals · Phase type distribution · Simulation

1 Introduction

Stochastic modeling plays an important role in all walks of life (see, e.g., [1]). The concept of quality is well-known in queueing, statistics, and quality assurance fields among others. The quality concept, which has many dimensions depending on whether we deal with products (in a manufacturing set up) or services (not only as follow-ups to products needing services but also in service sectors) is playing a key role ever since that has permeated from producers/service providers to consumers on a day-in-day-out basis.

In manufacturing sectors, the machines, the operators, the assembly lines, and other essential things are subject to failures due to a variety of reasons, and need to be fixed. Similar scenarios occur in service sectors. Stochastic modeling, specifically, queueing models help to understand the situations and offer solutions in many cases. Queueing models wherein the servers are subject to failures and repairs are studied as queues with interruptions (see, e.g., [2]). In certain service sectors, the operations (or services) maybe delayed due to service interruptions.

V. M. Vishnevskiy et al. (Eds.): DCCN 2020, LNCS 12563, pp. 259–281, 2020.
https://doi.org/10.1007/978-3-030-66471-8_21

The concept of a server breakdown in queueing theory was first introduced by White and Christie [3]. Since then hundreds of papers have been published in the literature. A comprehensive review of queues with interruptions can be seen in [2]. Some recent works on queues with interruptions include [4–12]. As seen in the literature (see, e.g. [2]) the server interruptions can occur in different forms, some of which could be due to the failure of the server and some could be due to the server seeking help or offering help to fellow servers [4]. Thus, a service interruption is said to occur if the current service has to be put on hold so that the server can tend to other activities not relevant to the on-going service.

To point out the significance of the study of queues with interruptions in different fields, we point out a few recent studies. Some recent work involving unreliable server in the context of retrial queueing models include [13–18]. In the context of priority queues, Sharma and Jain [19] studied an unreliable queues with service interruptions. In manufacturing systems it is common to see machines that are subject to failures and repairs. Using a threshold-based control policy with unreliable workstations, Shekar et al. [20] studied a queueing model with service interruptions. A repairable machine interference problem with unreliable repairman and imperfect standby switch overs was studied in [21]. Kim et al. [8] studied an unreliable queueing system with versatile point processes for both arrivals and for breakdown events. Service interruptions also occur in other notable areas like traffic and communication networks. For example, in the context of a wireless network with licensed shared access, Markova et al. [22] studied unreliable queueing models.

In this paper, our focus is to study a queueing model in the context of service sectors, wherein the server interruptions occur naturally. These interruptions could be in the form of the servers breaking down due to external or internal events. External events such as shocks and negative customers may result in the server getting interrupted and possibly the server breakdowns. Internal events such as a timer/clock expiring result in an interruption. Unless exponential times are assumed for interruptions, there is a difference in modeling the breakdowns with internal clocks or external shocks. For example, in the external one, the events will continue to happen as one doesn't have any control whereas in the internal case, the clock starts and stops depending on whether the server becomes busy or under repair. Further, the applications in real life will dictate what type of events to use to model the breakdowns of the resources such as the servers.

Here, we will concentrate on modeling the server interruptions using an internal clock. The clock will be turned on whenever the server begins to offer a new service or to resume the interrupted service upon returning from the repair facility. The clock will be turned off whenever the server goes to the repair facility. The models studied in this paper are of the type in which the arrivals are governed by correlated processes and the service times are of phase type. Such models have been studied in the context of single and multi-server cases by many authors (see, e.g., [23–26]).

We first investigate analytically a single server queueing system. Secondly, when dealing with multiple-server systems, we resort to simulation. This is mainly due to the complexity of the model as the state space grows exponentially. When dealing with multi-server systems we consider two scenarios. First, we will assume that each server will have its own repairman. Further, all interrupted services will resume from where the interruptions occurred and offered by the same server. For example, if server 1 is interrupted while offering a service, then the customer will have to wait for this server to return from a repair even if there are other servers available to take over the interrupted service. The main reason for this is that there are applications where servers are scattered across various places (offering services remotely like in call centers) and hence it is only reasonable to expect the customer to receive services from the same server. Further, the history and other pertinent information gathered by the server on this customer may not be available to the other servers unless there is a centralized process wherein all the relevant information are entered in which case the other servers can take over. This latter possibility and other scenarios will be pointed out later on in this paper.

The paper is organized as follows. In Sect. 2, we describe the model of interest along with its steady-state analysis. Also, in this section we establish a number of results. In Sect. 3, we present a few illustrative examples, and a simulation study of a multi-server system is presented in Sect. 4. Some concluding remarks are spelled out in Sect. 5.

2 Model Description and Steady-State Analysis

In this section, we describe the model under study and perform its steady-state analysis. We also register a number of interesting results, some of which are serve as useful accuracy checks in numerical computation of the steady-state probability vector. In any numerical implementation, it is of utmost importance to make sure that the algorithm produces correct and accurate (to the level of degree sought) results. Hence, having accuracy checks is crucial in any stochastic modeling involving numerical examples.

2.1 Model Description

Customers arrive to the system according to a MAP with representation matrices D_0 and D_1 of dimension m. These matrices are such that the underlying (irreducible) discrete-state continuous-time Markov chain with generator $D = D_0 + D_1$ governs transitions corresponding to the MAP. Specifically, D_0 governs transitions corresponding to no arrivals to the system. The arrivals of the customers to the system are governed by D_1, and at these epochs, the underlying Markov chain will start in one of the m transient states based on a certain (discrete) probability distribution. The beauty of MAP, as introduced by Neuts [27], lies in the fact that one can model independent as well as correlated inter-arrival times through properly choosing the parameter matrices. For additional details on models with correlated flows, we refer the reader to the recent book by Dudin et al. [24].

The customers are served on a first-come-first-served ($FCFS$) basis. The service times are assumed to follow a phase type ($PH-$) distribution with an irreducible representation of $(\boldsymbol{\beta}, S)$ of order n. Recall that $PH-$distributions, introduced by Neuts [28] and studied extensively in the literature, are defined as the time until absorption in a finite-state Markov chain with all but one transient states, and an absorbing state. These distributions are defined for both discrete and continuous times, and have been used in a variety of practical applications. For full details on MAP and their usefulness in stochastic modeling, as well as on $PH-$ distributions, we refer the reader to [23–35].

Anytime the server starts serving a customer by being idle or returning from a repair, a clock with a lifetime following a $PH-$ distribution with representation $(\boldsymbol{\alpha}, T)$ of order b will be started. The purpose of this clock is to model the breakdown of the server. The clock operates as follows. At the time of the expiry of this clock, if the server is busy serving, then the server is deemed to have failed. The failed server will be attended immediately by a repairman. During this repair time, the service phase will be frozen. Upon completion of a repair, the server will resume service from where the service was interrupted, and the breakdown clock will be turned on with the initial phase determined by the probability vector $\boldsymbol{\alpha}$. Whenever a service completion results in the server becoming idle, the clock will be turned off completely. The clock will be turned on only when the server starts servicing a new arrival. In this case, the initial phase of the clock be chosen according the probability vector $\boldsymbol{\alpha}$.

The repair times of the server are assumed to be of phase type with representation $(\boldsymbol{\gamma}, L)$ of order r. The repair will commence immediately upon the failure of the server.

We assume that all underlying random variables are independent of each other. That is, the inter-arrival times, the service times, the breakdown times, and the repair times are mutually independent of each other. Before we proceed further, we need to set up the following (standard) notation.

- All vectors (columns or rows) will be displayed in bold-faced characters. The noation ($'$) will stand for the transpose operation.
- The column vector with all elements set at 1 is denoted by the symbol $\boldsymbol{e}$.
- The column vector with all element but one set at 0, and the nonzero element of 1 occurring in the jth position s denoted by the symbol $\boldsymbol{e}_j$. Thus, $\boldsymbol{e}_2 = (0, 1, 0, \cdots, 0)$.
- The identity matrix is denoted by I.
- In stochastic modeling Kronecker products (denoted by $\otimes$) and Kronecker sums (denoted by $\oplus$) play a significant role. The matrix $E_m \oplus F_n$ of dimension mn is obtained as the Kronecker sum of matrices E of dimension m and F of dimension n such that $E_m \oplus F_n = E_m \otimes I_n + I_m \otimes F_n$. For definitions and details on these, we refer the reader to the standard books on matrices (see e.g., [36,37]).
- All the vectors and matrices will be displayed in such a way their dimensions should be clear from the context. However, where more clarity is needed, their dimensions will be indicated. For example, $\boldsymbol{e}(m)$, will denote a column vector

of 1's with dimension m. The vector $\boldsymbol{e}_2(m)$ of dimension m is such that the only nonzero entry (which is 1) occurs in the second position.

Let $\boldsymbol{\pi}$ be the invariant vector of irreducible matrix D. That is, $\boldsymbol{\pi}$ satisfies

$$\boldsymbol{\pi} D = 0, \quad \boldsymbol{\pi} \boldsymbol{e} = 1. \tag{1}$$

Letting λ, μ, θ, and ξ, respectively, denote the average rates of the arrival, the service, the breakdown, and the repair, it is easy to verify that

$$\lambda = \boldsymbol{\pi} D_1 \boldsymbol{e}, \quad \mu = \left[\boldsymbol{\beta}(-S)^{-1}\boldsymbol{e}\right]^{-1}, \quad \theta = \left[\boldsymbol{\alpha}(-T)^{-1}\boldsymbol{e}\right]^{-1}, \quad \xi = \left[\boldsymbol{\gamma}(-L)^{-1}\boldsymbol{e}\right]^{-1}. \tag{2}$$

2.2 Generator of the Markov Process

Here, we will illustrate how the model under study can be studied using the well-known QBD-process. First, we need to define a number of random variables. At time t, let

- $N(t)$ represents the number of customer in the system,
- $J_1(t)$ represents the phase of the service, if any,
- $J_2(t)$ represents the phase of the breakdown process,
- $J_3(t)$ represents the phase of the repair process,
- $K(t)$ represents the phase of the MAP.

The process $\{(N(t), J_1(t), J_2(t), J_3(t), K(t)) : t \geq 0\}$ is an irreducible Markov process on the state space Ω given by

$$\begin{aligned}\Omega = \Big\{(0,k) : 1 \leq k \leq m\Big\} \bigcup \Big\{(i, j_1, j_2, k) : 1 \leq j_1 \leq n, 1 \leq j_2 \leq b, 1 \leq k \leq m\Big\} \\ \bigcup \Big\{(i, j_1, j_3, k) : 1 \leq j_1 \leq n, 1 \leq j_3 \leq r, 1 \leq k \leq m\Big\}, \ i \geq 1.\end{aligned}$$

We now define the sets of states, $\mathbf{0}$ and $\boldsymbol{i}$ as

$$\mathbf{0} = \{(0,k) : 1 \leq k \leq m\}, \quad \boldsymbol{i} = \{\boldsymbol{i}_b\} \bigcup \{\boldsymbol{i}_r\}, \ i \geq 1,$$

where

$$\boldsymbol{i}_b = \{(i, j_1, j_2, k) : 1 \leq j_1 \leq n, 1 \leq j_2 \leq b, 1 \leq k \leq m\},$$

$$\boldsymbol{i}_r = \{(i, j_1, j_3, k) : 1 \leq j_1 \leq n, 1 \leq j_3 \leq r, 1 \leq k \leq m\}.$$

Then the generator matrix will be as

$$Q = \begin{pmatrix} D_0 & H_0 & 0 & 0 & \cdots \\ H_1 & A_1 & A_0 & 0 & \cdots \\ 0 & A_2 & A_1 & A_0 & \cdots \\ \vdots & & \ddots & \ddots & \ddots \end{pmatrix}, \tag{3}$$

where the entries of Q are

$$H_0 = (\boldsymbol{\beta} \otimes \boldsymbol{\alpha} \otimes D_1 \;\; 0), \;\; H_1 = \begin{pmatrix} \boldsymbol{S^0} \otimes \boldsymbol{e} \otimes I \\ 0 \end{pmatrix},$$

$$A_0 = I_{n(b+r)} \otimes D_1, \;\; A_2 = \begin{pmatrix} \boldsymbol{S^0}\boldsymbol{\beta} \otimes I_{mb} & 0 \\ 0 & 0 \end{pmatrix}, \;\; A_1 = \begin{pmatrix} S \oplus T \oplus D_0 & I \otimes \boldsymbol{T^0}\boldsymbol{\gamma} \otimes I \\ I \otimes \boldsymbol{L^0}\boldsymbol{\alpha} \otimes I & I \otimes (L \oplus D_0) \end{pmatrix}. \tag{4}$$

2.3 Stability Condition

Here, we will prove the following result related to the stability of the queueing model under study.

Result 1. The model under study is stable if and only if the following condition holds good.

$$\lambda < \frac{\mu \xi}{\theta + \xi}. \tag{5}$$

Proof. Suppose that $A = A_0 + A_1 + A_2$ and that $\hat{\boldsymbol{\pi}} = (\hat{\boldsymbol{\pi}}_1, \hat{\boldsymbol{\pi}}_2)$ is the invariant probability of the matrix A. That is,

$$\hat{\boldsymbol{\pi}} A = 0, \;\; \hat{\boldsymbol{\pi}} \boldsymbol{e} = 1. \tag{6}$$

The above equations can be rewritten as

$$\hat{\boldsymbol{\pi}}_1((S + \boldsymbol{S^0}\boldsymbol{\beta}) \oplus T \oplus D) + \hat{\boldsymbol{\pi}}_2(I \otimes \boldsymbol{L^0}\boldsymbol{\alpha} \otimes I) = 0, \tag{7}$$

$$\hat{\boldsymbol{\pi}}_1(I \otimes \boldsymbol{T^0}\boldsymbol{\gamma} \otimes I) + \hat{\boldsymbol{\pi}}_2(I \otimes (L \oplus D)) = 0 \tag{8}$$

subject to

$$\hat{\boldsymbol{\pi}}_1 \boldsymbol{e} + \hat{\boldsymbol{\pi}}_2 \boldsymbol{e} = 1. \tag{9}$$

Post-multiplying Eq. (7) by $(I_n \otimes \boldsymbol{e}(b) \otimes \boldsymbol{e}(m))$ and Eq. (8) by $(I_n \otimes \boldsymbol{e}(r) \otimes \boldsymbol{e}(m))$ and adding the resulting equations we get

$$\hat{\boldsymbol{\pi}}_1((S + \boldsymbol{S^0}\boldsymbol{\beta}) \otimes \boldsymbol{e}(b) \otimes \boldsymbol{e}(m)) = \mathbf{0}.$$

From which, using the uniqueness of the invariant vector of the irreducible representation $S + \boldsymbol{S^0}\boldsymbol{\beta}$ related to the service times, we get

$$\hat{\boldsymbol{\pi}}_1(I_n \otimes \boldsymbol{e}(b) \otimes \boldsymbol{e}(m)) = d_1 \mu \boldsymbol{\beta}(-S)^{-1}. \tag{10}$$

Similarly, one can show that

$$\hat{\boldsymbol{\pi}}_2(\boldsymbol{e}(n) \otimes I_r \otimes \boldsymbol{e}(m)) = d_2 \xi \boldsymbol{\gamma}(-L)^{-1}. \tag{11}$$

The normalizing condition in (9) with the help of (10) and (11) indicates

$$d_1 + d_2 = 1. \tag{12}$$

Now post-multiplying Eq. (7) by $\boldsymbol{e}(n) \otimes (-T)^{-1} \otimes \boldsymbol{e}(m)$ and simplifying the resulting equation with the help of (10) and (11), we get a second equation involving d_1 and d_2 as

$$\theta d_1 = \xi d_2. \tag{13}$$

Now solving for d_1 and d_2 we get

$$d_1 = \frac{\xi}{\theta + \xi}, \quad d_2 = \frac{\theta}{\theta + \xi}. \tag{14}$$

It is well-known that the $QBD-$process given in (3) is stable if and only if (see, e.g., [33]) the condition $\hat{\boldsymbol{\pi}} A_0 \boldsymbol{e} < \hat{\boldsymbol{\pi}} A_2 \boldsymbol{e}$, holds good. On noting that

$$\hat{\boldsymbol{\pi}}_1(\boldsymbol{e}(n) \otimes \boldsymbol{e}(b) \otimes I_m) + \hat{\boldsymbol{\pi}}_2(\boldsymbol{e}(n) \otimes \boldsymbol{e}(r) \otimes I_m) = \boldsymbol{\pi}, \tag{15}$$

where $\boldsymbol{\pi}$ is as given in (1), and the facts that $\hat{\boldsymbol{\pi}} A_0 \boldsymbol{e} = \lambda$ and $\hat{\boldsymbol{\pi}} A_2 \boldsymbol{e} = \hat{\boldsymbol{\pi}}_1(\boldsymbol{S^0} \otimes \boldsymbol{e} \otimes \boldsymbol{e}) = d_1\mu$, we get the stated result in (5).

2.4 Steady-State Equations

Under the condition that the queueing model under study is stable (i.e., the condition given in (5) holds good), we can discuss the stationary probability vector of the generator Q given in (3). Towards this end, let $\boldsymbol{x} = (\boldsymbol{x}_0, \boldsymbol{x}_1, ...)$ satisfy

$$\boldsymbol{x}Q = \boldsymbol{0}, \quad \boldsymbol{x}\boldsymbol{e} = 1. \tag{16}$$

Using the classical result on $QBD-$ process (see, e.g., [33]), the vector $\boldsymbol{x}$ has modified matrix-geometric structure. That is,

$$\boldsymbol{x}_i = \boldsymbol{x}_1 R^{i-1}, \; i \geq 1, \tag{17}$$

where R, of order $mn(b+r)$, is the minimal nonnegative solution to the matrix quadratic equation $R^2A_2 + RA_1 + A_0 = 0$, and the vectors $\boldsymbol{x}_0$ and $\boldsymbol{x}_1$ are obtained by solving

$$\begin{aligned} \boldsymbol{x}_0 D_0 + \boldsymbol{x}_1 H_1 &= \boldsymbol{0}, \\ \boldsymbol{x}_0 H_0 + \boldsymbol{x}_1 [A_1 + RA_2] &= \boldsymbol{0}, \end{aligned} \tag{18}$$

subject to the normalizing condition

$$\boldsymbol{x}_0 \boldsymbol{e} + \boldsymbol{x}_1 (I - R)^{-1} \boldsymbol{e} = 1. \tag{19}$$

The computation of R matrix can be done using one of the well-known techniques like logarithmic-quadratic algorithm (see, e.g., [38]). However, when m, n, b and r are large, we need to exploit the structure of the coefficient matrices and use iterative methods like (block) Gauss-Seidel (see e.g., [39]). The details are omitted.

The equations given in (18) can further be simplified for computational purposes by exploiting the structure of the coefficient matrices. For example, if R is partitioned as

$$R = \begin{pmatrix} R_1 & R_2 \\ R_3 & R_4 \end{pmatrix}, \tag{20}$$

and if we further partitioning $\boldsymbol{x}_i$ as $\boldsymbol{x}_i = (\boldsymbol{u}_i, \boldsymbol{v}_i),\ \ i \geq 1$, and noting that the vector $\boldsymbol{u}_i$ of dimension mnb giving the steady-state probability vector of the server being busy with i customers in the system and the system is in various phases; the vector $\boldsymbol{v}_i$ of dimension mnr gives the steady-state probability vector that the server is under repair with the system being in various phases, we can rewrite equation in (18) as

$$\begin{aligned} \boldsymbol{x}_0 D_0 + \boldsymbol{u}_1(\boldsymbol{S^0} \otimes \boldsymbol{e} \otimes I) &= \boldsymbol{0}, \\ \boldsymbol{x}_0(\boldsymbol{\beta} \otimes \boldsymbol{\alpha} \otimes D_1) + \boldsymbol{u}_1\left[(S \oplus T \oplus D_0) + R_1(\boldsymbol{S^0}\boldsymbol{\beta} \otimes I)\right] & \\ +\boldsymbol{v}_1\left[(I \otimes \boldsymbol{L^0}\boldsymbol{\alpha} \otimes I) + R_3(\boldsymbol{S^0}\boldsymbol{\beta} \otimes I)\right] &= \boldsymbol{0}, \\ \boldsymbol{x}_0 \boldsymbol{e} + \boldsymbol{x}_1(I - R)^{-1}\boldsymbol{e} &= 1. \end{aligned} \tag{21}$$

Of course, one can further exploit the structure produced by the Kronecker products and sums occurring in the coefficient matrices, and the details are omitted.

2.5 Additional Useful Results

In this section, we will prove some additional interesting results useful in evaluating the system performance measures as well as to serve as accuracy checks in the numerical computation of the vector $\boldsymbol{x}$. Towards this end, we rewrite the steady-state equations in (16) in terms of vectors of smaller dimensions by exploiting the structure of the coefficient matrices as

$$\boldsymbol{x}_0 D_0 + \boldsymbol{u}_1(\boldsymbol{S^0} \otimes \boldsymbol{e} \otimes I) = \mathbf{0}, \tag{22}$$

$$\boldsymbol{x}_0(\boldsymbol{\beta} \otimes \boldsymbol{\alpha} \otimes D_1) + \boldsymbol{u}_1(S \oplus T \oplus D_0) + \boldsymbol{v}_1(I \otimes \boldsymbol{L^0}\boldsymbol{\alpha} \otimes I) + \boldsymbol{u}_2(\boldsymbol{S^0}\boldsymbol{\beta} \otimes I \otimes I) = \mathbf{0}, \tag{23}$$

$$\boldsymbol{u}_1(I \otimes \boldsymbol{T^0}\boldsymbol{\gamma} \otimes I) + \boldsymbol{v}_1(I \otimes (L \oplus D_0)) = \mathbf{0}, \tag{24}$$

$$\boldsymbol{u}_{i-1}(I \otimes I \otimes D_1) + \boldsymbol{u}_i(S \oplus T \oplus D_0) + \boldsymbol{v}_i(I \otimes \boldsymbol{L^0}\boldsymbol{\alpha} \otimes I) + \boldsymbol{u}_{i+1}(\boldsymbol{S^0}\boldsymbol{\beta} \otimes I \otimes I) = \mathbf{0},\ i \geq 2, \tag{25}$$

$$\boldsymbol{v}_{i-1}(I \otimes I \otimes D_1) + \boldsymbol{u}_i(I \otimes \boldsymbol{T^0}\boldsymbol{\gamma} \otimes I) + \boldsymbol{v}_i(I \otimes (L \oplus D_0)) = \mathbf{0},\ i \geq 2. \tag{26}$$

For use in the sequel, we define

$$\boldsymbol{y} = (\boldsymbol{y}^{(1)}, \boldsymbol{y}^{(2)}) = \sum_{i=1}^{\infty} \boldsymbol{x}_i = \boldsymbol{x}_1(I - R)^{-1}. \tag{27}$$

The following result, which is intuitively obvious, gives an expression for the invariant vector of the underlying Markov chain of the MAP in terms of the steady-state probability vector $\boldsymbol{x}$.

Result 2. We have

$$\boldsymbol{x}_0 + \boldsymbol{y}(\boldsymbol{e} \otimes \boldsymbol{e} \otimes I) = \boldsymbol{\pi}. \tag{28}$$

Proof. Post-multiplying Eqs. (23) and (25) by $(\boldsymbol{e}(n) \otimes \boldsymbol{e}(b) \otimes I_m)$ and Eqs. (24) and (26) by $(\boldsymbol{e}(n) \otimes \boldsymbol{e}(r) \otimes I_m)$, and adding them (some over i as well) along with (22), we get

$$[\boldsymbol{x}_0 D + \boldsymbol{y}(\boldsymbol{e} \otimes \boldsymbol{e} \otimes D)] = \boldsymbol{0}.$$

The stated result follows immediately from the uniqueness of the invariant vector, $\boldsymbol{\pi}$, of D.

The following result, which is again intuitively obvious, shows that in steady-state the input and output rates should be equal.

Result 3. We have

$$\sum_{i=1}^{\infty} \boldsymbol{u}_i(\boldsymbol{S}^0 \otimes \boldsymbol{e} \otimes \boldsymbol{e}) = \lambda. \tag{29}$$

Proof. From (22), (23), and (24) we get

$$\boldsymbol{u}_1(\boldsymbol{e} \otimes \boldsymbol{e} \otimes D_1\boldsymbol{e}) + \boldsymbol{v}_1(\boldsymbol{e} \otimes \boldsymbol{e} \otimes D_1\boldsymbol{e}) = \boldsymbol{u}_2(\boldsymbol{S}^0 \otimes \boldsymbol{e} \otimes \boldsymbol{e}), \tag{30}$$

which with the help of (25) and (26) yields

$$\boldsymbol{u}_i(\boldsymbol{e} \otimes \boldsymbol{e} \otimes D_1\boldsymbol{e}) + \boldsymbol{v}_i(\boldsymbol{e} \otimes \boldsymbol{e} \otimes D_1\boldsymbol{e}) = \boldsymbol{u}_{i+1}(\boldsymbol{S}^0 \otimes \boldsymbol{e} \otimes \boldsymbol{e}),\ i \geq 2. \tag{31}$$

It is easy to verify that from Eqs. (22), (30), and (31), we get

$$\sum_{i=1}^{\infty} \boldsymbol{u}_i(\boldsymbol{S}^0 \otimes \boldsymbol{e} \otimes \boldsymbol{e}) = \boldsymbol{x}_0 D_1 \boldsymbol{e} + y(\boldsymbol{e} \otimes \boldsymbol{e} \otimes D_1\boldsymbol{e}), \tag{32}$$

which immediately yields the stated result by applying Result 2.

Result 4. We have

$$\sum_{i=1}^{\infty} \boldsymbol{u}_i(I \otimes \boldsymbol{e} \otimes \boldsymbol{e}) = \lambda \boldsymbol{\beta}(-S)^{-1}. \tag{33}$$

Proof. First verify from (22) that

$$\boldsymbol{x}_0(\boldsymbol{\beta} \otimes D_1\boldsymbol{e}) = \boldsymbol{u}_1(\boldsymbol{S}^0\boldsymbol{\beta} \otimes \boldsymbol{e} \otimes \boldsymbol{e}). \tag{34}$$

Post-multiplying (24) and (26) by $(I \otimes \boldsymbol{e}(r) \otimes \boldsymbol{e}(m))$, and adding the resulting equations over i, we get

$$\sum_{i=1}^{\infty} \boldsymbol{u}_i(I \otimes \boldsymbol{T}^0 \otimes \boldsymbol{e}) = \sum_{i=1}^{\infty} \boldsymbol{v}_i(I \otimes \boldsymbol{L}^0 \otimes \boldsymbol{e}). \tag{35}$$

Now post-multiply (23) and (25) by $(I \otimes \boldsymbol{e}(b) \otimes \boldsymbol{e}(m))$ and add this over i. To this resulting equation when we add (34), and substitute the expression given in (35) we get

$$\sum_{i=1}^{\infty} \boldsymbol{u}_i(I \otimes \boldsymbol{e} \otimes \boldsymbol{e}) = d\mu\boldsymbol{\beta}(-S)^{-1}. \tag{36}$$

The constant d appearing in (36) is obtained as $d = \frac{\lambda}{\mu}$ by applying Result 3. Hence, we have the stated result.

Result 5. We have

$$P(busy) = \frac{\lambda}{\mu}. \tag{37}$$

Proof. Noting that $P(busy) = \sum_{i=1}^{\infty} \boldsymbol{u}_i \boldsymbol{e}$, the stated result follows immediately from Result 4.

Note: Result 5 generalizes the result of the classical $MAP/PH/1$ queue. Also, combining Results 4 and 5, it is obvious that the (conditional) steady-state probability of the service phase given that the server is busy should be equal $\mu\boldsymbol{\beta}(-S)^{-1}$.

Result 6. We have

$$\sum_{i=1}^{\infty} \boldsymbol{v}_i(\boldsymbol{e} \otimes I \otimes \boldsymbol{e}) = P(repair)\xi\boldsymbol{\gamma}(-L)^{-1}. \tag{38}$$

Proof. First, post-multiply Eqs. (24) and (26) by $(\boldsymbol{e} \otimes I \otimes \boldsymbol{e})$. Adding the resulting equations over i, we get

$$\sum_{i=1}^{\infty} \boldsymbol{u}_i(\boldsymbol{e} \otimes \boldsymbol{T^0}\boldsymbol{\gamma} \otimes \boldsymbol{e}) + \sum_{i=1}^{\infty} \boldsymbol{v}_i(\boldsymbol{e} \otimes L \otimes \boldsymbol{e}) = \boldsymbol{0}. \tag{39}$$

Using the above equation along with (35), we notice

$$\sum_{i=1}^{\infty} \boldsymbol{v}_i(\boldsymbol{e} \otimes (L + \boldsymbol{L^0}\boldsymbol{\gamma}) \otimes \boldsymbol{e}) = \boldsymbol{0}, \tag{40}$$

which when using the uniqueness of the invariant vector of the generator $L + \boldsymbol{L^0}\boldsymbol{\gamma}$ yields

$$\sum_{i=1}^{\infty} \boldsymbol{v}_i(\boldsymbol{e} \otimes I \otimes \boldsymbol{e}) = d\xi\boldsymbol{\gamma}(-L)^{-1}, \tag{41}$$

from which the stated result follows on noting that the normalizing constant d is given by $d = \sum_{i=1}^{\infty} \boldsymbol{v}_i \boldsymbol{e}$. Obviously d is the probability of the server being under repair.

Note: (a) Result 6 gives an expression for the phase of the repair process given that the server is under repair. This is again intuitively clear. (b) When the breakdown times are exponential, i.e., $\boldsymbol{\alpha} = 1, T = -\theta$, then it is easy to verify that $P(repair) = \frac{\lambda\theta}{\mu\xi}$, which depends only on the rates of the arrival, the service, the breakdown and the repair processes and not on the nature of the distributions themselves. Further, $P(idle) = 1 - \rho_q$, where $\rho_q = \frac{\lambda(\theta+\xi)}{\mu\,\xi}$ is the traffic intensity of our current model. These could serve as another set of accuracy checks in numerical computation of the steady-state vector.

2.6 Effective Service Time

Due to the possibility of the server breaking down in the middle of offering a service, the service time of a customer consists of a normal service time plus (maybe) an additional random quantity to take into account the chances of having one or more repairs. However, we can model the effective service time, say, Y, of a customer as a $PH-$distribution as shown in the following result.

Result 7. The effective service time, Y, follows a PH distribution with representation $(\boldsymbol{\kappa}, K)$ of order $n(b+r)$ where

$$\boldsymbol{\kappa} = \frac{1}{\lambda}\Big(\boldsymbol{x}_0 D_1 \boldsymbol{e}(\boldsymbol{\beta} \otimes \boldsymbol{\alpha}) + \sum_{i=2}^{\infty} \boldsymbol{u}_i(\boldsymbol{S^0}\boldsymbol{\beta} \otimes I \otimes \boldsymbol{e}), \boldsymbol{0}\Big), \quad K = \begin{pmatrix} S \oplus T & I \otimes \boldsymbol{T^0}\boldsymbol{\gamma} \\ I \otimes \boldsymbol{L^0}\boldsymbol{\alpha} & I \otimes L \end{pmatrix}. \tag{42}$$

Proof. The proof follows immediately on noting that a service is initiated either through a new arrival or upon completion of the current service (with at least one customer waiting in the queue). This results in the choice of the vector $\boldsymbol{\kappa}$. Once a service begins the possible transitions before the customer departs the service facility with a service are governed by the irreducible matrix K.

Note: The mean effective service time, μ_Y', is obtained as

$$\mu_Y' = \boldsymbol{\kappa}(-K)^{-1}\boldsymbol{e}. \tag{43}$$

Result 8. In the case of exponential breakdown times, μ_Y' is explicitly given by

$$\mu'_Y = \frac{\theta + \xi}{\mu}. \tag{44}$$

Proof. In this case, we have $\boldsymbol{\alpha} = 1$ and $T = -\theta$. Verify that, for this special case, the inverse of K (see Eq. (42)) is obtained as

$$(-K)^{-1} = \begin{pmatrix} (-S)^{-1} & \theta[(-S^{-1}) \otimes \boldsymbol{\gamma}(-L)^{-1}] \\ (-S)^{-1} \otimes \boldsymbol{e} & \Big[(I \otimes (-L)^{-1}) + \theta[(-S^{-1}) \otimes \boldsymbol{e}\boldsymbol{\gamma}(-L)^{-1}]\Big] \end{pmatrix}. \tag{45}$$

Using the above form of $(-K)^{-1}$ in (43) along with the equations given in (2), (34) and (37), the stated result follows.

Note: It is interesting to see that the memoryless property of the breakdown times makes the mean effective service time to be dependent only on the rates of the service and the repair, and not on the nature of the distributions governing these random variables. The above result can be used as yet another accuracy check in implementing the numerical computation of the steady-state vector.

2.7 Sojourn Time Distribution

Suppose that W denotes the sojourn time of a tagged customer at an arrival epoch. Following the technique used in [40], we will derive an expression for the tail probability of W. Towards this end, we first need to calculate the steady-state probability vector at arrival epochs. Let

$$M = -(A_1 + RA_2)^{-1}, \quad \tilde{R} = MA_0. \tag{46}$$

Let $\boldsymbol{z} = (\boldsymbol{z}_1, \boldsymbol{z}_2, \cdots)$ denote the steady-state probability vector at arrival epochs. Note that the vector $\boldsymbol{z}_i,\ i \geq 1$, of dimension $mn(b+r)$, gives the steady-state probability vector that soon after an arrival there are $i,\ i \geq 1$, customers in the system, and the service/breakdown clock/repair in various phases depending on whether the server is busy or under repair.

Result 9. The vector $\boldsymbol{z}$ is given by

$$\boldsymbol{z}_i = \frac{1}{\lambda}\boldsymbol{x}_0 H_0 \tilde{R}^{i-1},\ i \geq 1. \tag{47}$$

Proof. Noting $R = A_0 M$ and using Eqs. (17) and (18), the stated result follows after some elementary matrix manipulations.

Defining

$$\boldsymbol{\delta} = \sum_{i=1}^{\infty} \boldsymbol{z}_i = \frac{1}{\lambda}\boldsymbol{x}_0 H_0 (I - \tilde{R})^{-1}, \tag{48}$$

and applying Ozawas's technique (see [40]) along with Result 9, we get the following result expressing the tail probability of W.

Result 10. We have

$$P(W > t) = (\boldsymbol{e}'(mn(b+r)) \otimes \boldsymbol{\delta}) exp\Big(\Big[[(A_0 + A_1)' \otimes I] + [A_2' \otimes \tilde{R}]\Big] t\Big) \boldsymbol{g},\ t \geq 0, \tag{49}$$

where the column vector, $\boldsymbol{g}$, of dimension $[mn(b+r)]^2$ is given by

$$\boldsymbol{g} = \begin{pmatrix} \boldsymbol{e}_1 \\ \boldsymbol{e}_2 \\ \vdots \\ \boldsymbol{e}_{mn(b+r)} \end{pmatrix}. \tag{50}$$

Note: (a) The mean, μ'_W, of W is

$$\mu'_W = (e' \otimes \delta)\Big(- \Big[[(A_0 + A_1)' \otimes I] + [A_2' \otimes \tilde{R}] \Big] \Big)^{-1} g. \tag{51}$$

(b) The classical Little's law holds good here and can be used as another key accuracy check in numerical computation.

2.8 System Performance Measures

Here, we list a number of system performance measures which are needed in the qualitative study of the model under study. While there are a number of measures available, we list only a few for our illustrative examples. Note that some of these measures have been derived in earlier sections. However, we list them for the sake of convenience and ease of reference.

1. P(idle)= $x_0 e$.
2. P(server is busy)= $y^{(1)} e = \frac{\lambda}{\mu}$.
3. P(server is under repair)= $y^{(2)} e$.
4. μ_{NS} = mean number in system = $x_1(I - R)^{-2} e$.
5. μ'_Y = Effective mean service time = $\kappa(-K)^{-1} e$.

3 Illustrative Examples for $MAP/PH/1$-system

In this section, we will illustrate a few representative numerical examples in the context of $MAP/PH/1$ system. Towards this end, we will look at different types of $MAPs$ covering a wide range of scenarios in practice. These scenarios include (a) independent and identically distributed inter-arrival times such that the coefficient of variation (CV) is less than 1 (Erlang); $CV = 1$ (exponential), $CV > 1$ (hyperexponential); (b) negatively correlated arrivals (NCA) such that the 1-lag correlation coefficient of two successive inter-arrival times is negative; (c) positively correlated arrivals (PCA) such that 1-lag correlation coefficient of two successive inter-arrival times is positive. Specifically, we consider the following five $MAPs$. Note that without loss of generality we fix the arrival rate, λ, to be 1.

3.1 MAP for Arrivals

Here we will set up the notation for the $MAPs$ considered.

ERA: This stands for an Erlang distribution with 5 stages with rate 5 in each stage. Note that the standard deviation of this distribution is 0.4472.

EXA: This stands for the classical exponential distribution with parameter 1 and hence the standard deviation is also 1.

HEA: This stands for a hyperexponential distribution with mixing probabilities $(0.5, 0.3, 0.15, 0.05)$ and the corresponding rates given by $(68.5, 6.85, 0.685, 0.0685)$. Here the standard deviation of the distribution is 4.5787.

NCA: This stands for a negatively correlated arrival process. The parameter matrices are given by

$$D_0 = \begin{pmatrix} -1.25 & 1.25 & 0 \\ 0 & -1.25 & 0 \\ 0 & 0 & -2.5 \end{pmatrix}, \quad D_1 = \begin{pmatrix} 0 & 0 & 0 \\ 0.0125 & 0 & 1.2375 \\ 2.4750 & 0 & 0.0250 \end{pmatrix}.$$

The standard deviation of the inter-arrival times and the 1-lag correlation coefficient are given by 1.0392 and −0.3267.

PCA: This stands for a positively correlated arrival process. The parameter matrices are given by

$$D_0 = \begin{pmatrix} -1.25 & 1.25 & 0 \\ 0 & -1.25 & 0 \\ 0 & 0 & -2.5 \end{pmatrix}, \quad D_1 = \begin{pmatrix} 0 & 0 & 0 \\ 1.2375 & 0 & 0.0125 \\ 0.0250 & 0 & 2.4750 \end{pmatrix}.$$

The standard deviation of the inter-arrival times and the 1-lag correlation coefficient are given by 1.0392 and 0.3267.

3.2 PH-distributions for Services, Breakdowns, and Repairs

In order to minimize the number of combinations for our illustrative numerical examples, we will restrict to three type of $PH-$distributions for all the service times, breakdown times, and repair times. In the following we will use the notation (similar to the arrival processes described in Sect. 3.1) ERX, EXX, and HEX, respectively, for Erlang, exponential, and hyperexponential cases dealing with $X-$type distribution, where $X = S, B, R$, depending on whether the services, the breakdowns or the repairs are under consideration. Thus, ERR corresponds to repairs are modeled using Erlang, whereas HES corresponds to hyperexponential services.

ERX: This stands for an Erlang distribution with 5 stages with rate $5a$ in each stage.
EXX: This stands for the classical exponential distribution with parameter a.
HEX: This stands for a hyperexponential distribution with mixing probabilities $(0.7, 0.25, 0.05)$ and the corresponding rates given by $a(8.2, 0.82, 0.082)$.

The value of a is chosen so as to have a given mean in the context where this is used. For example, if we want the service times to have a mean of 2 for all three distributions (so as to compare properly the three different services), then we will take $X = S$ and $a = 0.5$.

As is known these three distributions are qualitatively different and cover a wide range of scenarios for practical usage.

For our illustrative examples, we fix the service rate to be $\mu = 1.3158$, the breakdown rate as $\theta = 0.1$, and the repair rate to be $\xi = 0.4$. A FORTRAN code was written to generate our numerical examples after verifying the correctness with the help of the accuracy checks listed in Results 2 through 6.

Selected key system performance measures, the mean number in the queue, the probability that server is under repair, the probability that the server is idle, and the mean effective service times, are respectively, displayed in Figs. 1 through 4. A few noticeable observations from these figures are summarized below. First, we wanted to point out the labeling in the radar charts. EEE denotes the type of repairs to be Erlang, the type of services to be Erlang, and the type of breakdowns to be Erlang. Similarly, XHX corresponds to exponential repairs, hyperexponential services, and exponential breakdowns.

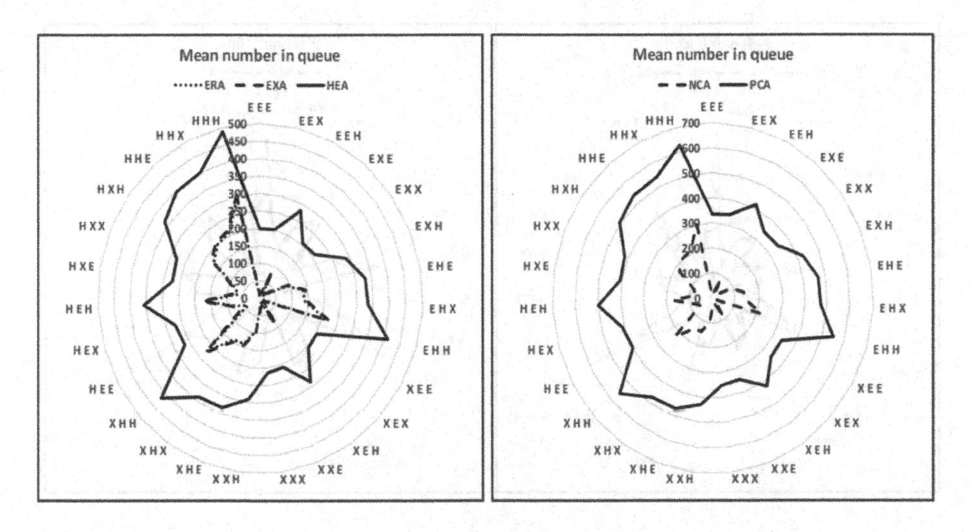

Fig. 1. Mean number in the queue under various scenarios

1. With regard to the mean number of customers in the queue,
   (a) we see that the positively correlated (PCA) arrival process has the largest mean number in queue among all considered.
   (b) It is worth pointing out that the standard deviation of the PCA arrivals is less than that of HEA indicating the role of correlation, especially, the positive ones.

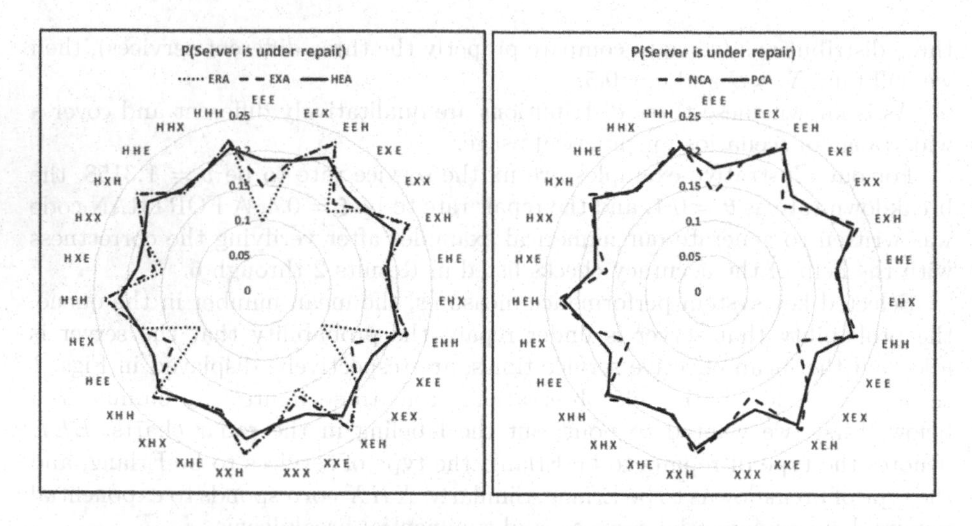

Fig. 2. P (server is under repair) under various scenarios

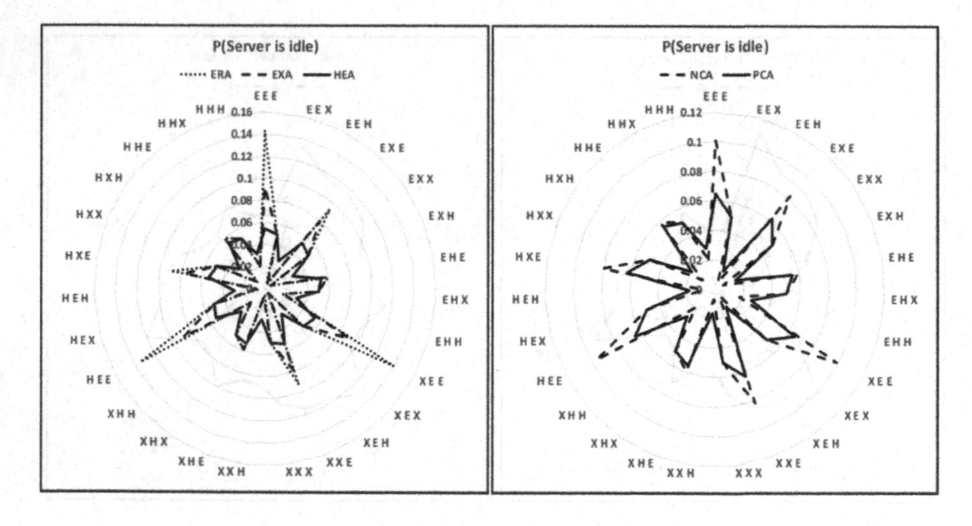

Fig. 3. P (server is idle) under various scenarios

   (c) ERA, EXA and NCA all appear to exhibit a similar pattern.
   (d) Both HEA and PCA processes appear to exhibit patterns that are similar to each other under most scenarios.
2. Looking at the measure, P (server is under repair),
   (a) we see that all five arrival processes exhibit similar patterns when service/repair/breakdown times are varied. First note that the as pointed out earlier (see note (b) following Result 6) that for exponential breakdowns this measure is identical for all scenarios.
   (b) with regard to Erlang and hyperexponential breakdowns, we see some interesting patterns.

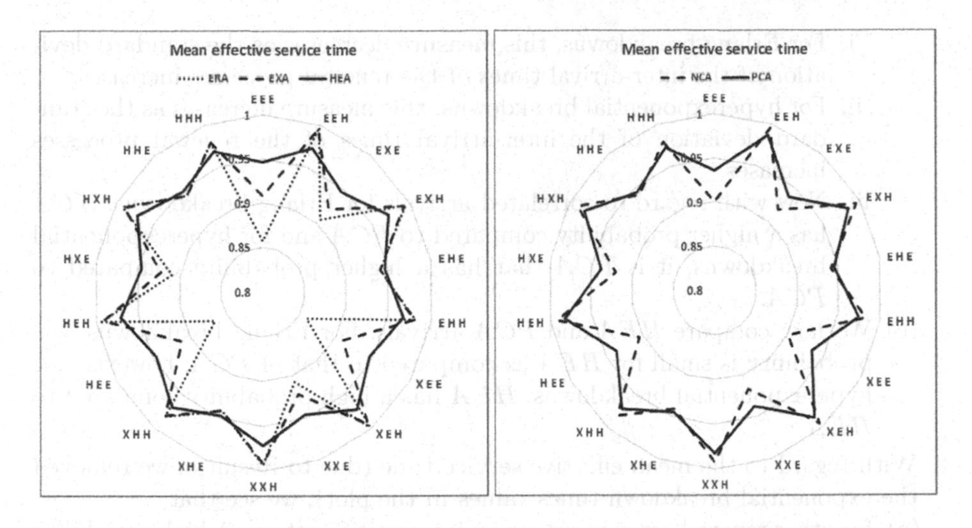

Fig. 4. Mean effective service time under various scenarios

i. For Erlang breakdowns, this measure increases as the standard deviation of the inter-arrival times of the renewal processes increases.
ii. For hyperexponential breakdowns, this measure decreases as the standard deviation of the inter-arrival times of the renewal processes increases.
iii. Now with regard to correlated arrivals for Erlang breakdowns, PCA has a higher probability compared to NCA and for hyperexponential breakdowns, it is NCA that has a higher probability compared to PCA.

(c) We now compare HEA and PCA, mainly due to the fact that HEA has the largest standard deviation and PCA has the largest (positive) correlation among the arrival processes considered. Here, for Erlang breakdowns, the probability is large for HEA as compared to that of PCA; however, for hyperexponential breakdowns, PCA has a high probability compared to HEA. Another interesting observation stressing the importance of the role of (positive) correlation in stochastic modeling.

3. Looking at the measure, P(server is idle),
   (a) we see that all five arrival processes exhibit similar patterns when service/repair/breakdown times are varied. First note that the as pointed out earlier (see note (b) following Result 6) that for exponential breakdowns this measure is identical for all scenarios.
   (b) with regard to Erlang and hyperexponential breakdowns, we see some interesting patterns which are opposite to the ones seen for the measure, P(server is under repair).

i. For Erlang breakdowns, this measure decreases as the standard deviation of the inter-arrival times of the renewal processes increases.
ii. For hyperexponential breakdowns, this measure increases as the standard deviation of the inter-arrival times of the renewal processes increases.
iii. Now with regard to correlated arrivals for Erlang breakdowns, NCA has a higher probability compared to PCA and for hyperexponential breakdowns, it is PCA that has a higher probability compared to PCA.

(c) We now compare HEA and PCA arrivals. For Erlang breakdowns, the probability is small for HEA as compared to that of PCA; however, for hyperexponential breakdowns, HEA has a high probability compared to HEA.

4. With regard to the mean effective service time (due to Result 8, we removed the exponential breakdown times values in the plot), we see that
   (a) for other renewal arrivals, we see an interesting pattern. A high variability breakdown times (like hyperexponential) yield a lower value for HEA (which has the highest variability among all arrivals) whereas a lower variability breakdown times (like Erlang) yield a higher value for HEA arrivals. This appears to be the case for all types of repairs and services.
   (b) When comparing the two correlated arrivals, we see a pattern similar to the ones mentioned for the renewal arrivals. For example, PCA arrivals have a higher mean compared to NCA for the scenario in which we use Erlang breakdown times, and the pattern is reversed when using hyperexponential breakdown times. This is the case for all types repairs and services.

4 Simulation of $MAP/PH/c$ System

The single server system we studied earlier can be generalized to a multi-server system. There are many ways to generalize the single server system studied here to a multi-server system. These correspond to how the repair facility is managed (individual facility versus a common repair facility), how the breakdowns are modeled (each server has its own breakdown clock or a common ones for all busy servers), how the interrupted customers are served (wait for the same server who started the service or another free server at that instant or at a later point in time when a free server is available to continue serving), and where the repaired servers are sent back (either to the same facility from where the server arrived or to the facility that needs to replaced). All possible combinations lead to different models based on the need in practice.

While any such multi-server system can be studied along the lines mentioned here for the single server case, the dimensionality of the problem increases exponentially. Hence, we will resort to simulation so as to get a feel for the system performance measures in going from a single server to multiple servers. Towards this end, we used ARENA (see, e.g., [41]) simulation software to simulate $MAP/PH/c$-system in the context of our model here. The multi-server system that is studied via simulation is as follows.

1. Customers arrive according to a MAP and are served on first-come-first served basis.
2. The service times, breakdown times, and repair times are modeled using (possibly) different phase type distributions.
3. Each server has its own breakdown clock for modeling breakdowns and has its own repairman.
4. An interrupted customer waits for the same server to continue serving until the service gets over.

The above assumptions make sense for applications of server breakdown/repairs in situations where the service facility independently operate.

However, in this paper we will just look at one multi-server system in which the customers are served on a first-come-first-served basis and Before we look at the simulated results, we validate the simulated model against the single server model of the previous section.

All our simulated models were run using 10 replicates with a duration of 200,000 min for each replicate. The error percentages, which are calculated as $\frac{|analytical-simulated|}{analytical} \times 100$, for the mean number in the system and mean effective service time, are generally small except for a few scenarios. For such scenarios we have to run them for even longer duration to bring down the error percentages. This is typical for arrival processes such as HEA and PCA cases.

For input parameters we use the same set of arrival and service processes as in Sect. 3. We vary c, the number of servers, from 2 to 5. In order to compare the models properly when varying c from 2 to 5, we let λ depend on c such that system's arrival rate is proportional to the number of servers in the system. Thus, we take $\lambda = c$, and model the multi-server system with each server's breakdown times and repair times to be identical to those listed in Sect. 3.

We noticed the behavior of all the measures that we discussed in Sect. 3 with regard to a single server case using analytical model holds good for simulated models. Further, we noticed the insensitivity to the type of services as well as repairs when breakdown times are modeled using exponential distribution for the mean effective service and will be interesting to prove this result similar to Result 8 in Sect. 2.6. Some sample graphs (see Fig. 5) are included here to illustrate the similarity to the ones outlined in Sect. 3.

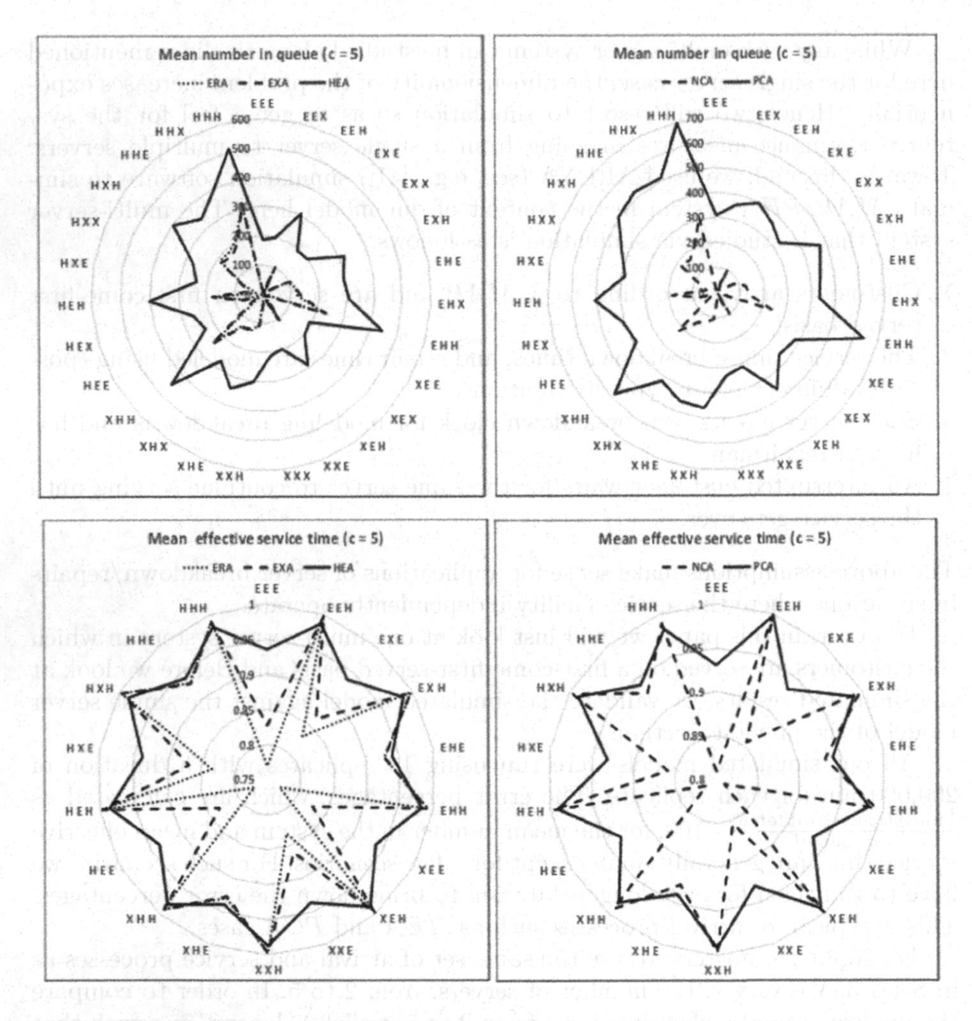

Fig. 5. Selected measures for simulated model for $c = 5$ under various scenarios

5 Concluding Remarks

In this paper we studied both single server (via analytically) and multi-server (via simulation) systems in which the servers are subject to repairs and failures. The arrival of customers are modeled using a versatile Markovian point process and all other underlying random variables to be of phase type. The failures are modeled using internal clocks. Such models have practical use, especially, in service sectors. The model studied in this paper can be extended to cover more multi-server scenarios. For example, we can consider multi-server systems where there is a common repair facility which will be responsible to repair all failed servers. We can relax the restriction that the interrupted customers have to resume their services from the server. Finally, it will be of interest to introduce both internal clocks and external shocks to model the server interruptions. These extensions are currently being explored.

Acknowledgments. The authors deeply appreciate the constructive suggestions of three anonymous reviewers that improved the presentation of the paper.

References

1. Cochran, J.J., Cox, L.A., Keskinocak, P., Kharoufeh, J.P., Smith, C.: Wiley Encyclopedia of Operations Research and Management Science, 15 Jun 2010
2. Krishnamoorthy, A., Pramod, P., Chakravarthy, S.R.: Queues with interruptions: a survey. TOP **22**, 290–320 (2014). https://doi.org/10.1007/s11750-012-0256-6
3. White, H., Christie, L.S.: Queuing with preemptive priorities or with breakdown. Oper. Res. **6**(1), 79–95 (1958)
4. Chakravarthy, S.R.: A multi-server queueing model with server consultations. Eur. J. Oper. Res. **233**(3), 625–639 (2014)
5. Chakravarthy, S.R., Shruti, K.R, Kulshrestha, R.: A queueing model with server breakdowns repairs vacations and backup server. Oper. Res. Perspect. **7**, 100131 (2020)
6. Jain, M., Shekhar, C., Meena, R.K.: Admission control policy of maintenance for unreliable server machining system with working vacation. Arab. J. Sci. Eng. **42**(7), 2993–3005 (2017)
7. Jain, M., Kumar, P., Meena, R.K.: Fuzzy metrics and cost optimization of a fault-tolerant system with vacationing and unreliable server. J. Amb. Intell. Hum. Comput. **11**(11), 5755–5770 (2020). https://doi.org/10.1007/s12652-020-01951-x
8. Kim, C., Klimenok, V.I., Dudin, A.N.: Analysis of unreliable $BMAP/PH/N$ type queue with Markovian flow of breakdowns. Appl. Math. Comput. **314**, 154–172 (2017)
9. Krishnamoorthy, A., Pramod, P.K., Chakravarthy, S.R.: A note on characterizing service interruptions with phase-type distribution. Stochast. Anal. Appl. **31**(4), 671–683 (2013)
10. Krishnamoorthy, A., Divya, V.: (M, MAP)/(PH, PH)/1 queue with non-preemptive priority and working vacation under n-policy. J. Indian Soc. Probab. Stat. **21**(1), 69–122 (2020). https://doi.org/10.1007/s41096-020-00081-z
11. Sethi, R., Jain, M., Meena, R.K., Garg, D.: Cost optimization and ANFIS computing of an unreliable M/M/1 queueing system with customers' impatience under n-policy. Int. J. Appl. Comput. Math. **6**(2), 1–14 (2020)
12. Wu, C.H., Yang, D.Y.: Dynamic control of a machine repair problem with switching failure and unreliable repairmen. Arab. J. Sci. Eng. **45**(3), 2219–2234 (2020)
13. Chang, J., Wang, J.: Unreliable $M/M/1/1$ retrial queues with set-up time. Qual. Technol. Quant. Manage. **15**(5), 589–601 (2018)
14. Chang, F.M., Liu, T.H., Ke, J.C.: On an unreliable-server retrial queue with customer feedback and impatience. Appl. Math. Model. **55**, 171–182 (2018)
15. Choudhury, G., Tadj, L., Deka, M.: An unreliable server retrial queue with two phases of service and general retrial times under Bernoulli vacation schedule. Qual. Technol. Quant. Manag. **12**(4), 437–464 (2015)
16. Jain, M., Bhagat, A., Shekhar, C.: Double orbit finite retrial queues with priority customers and service interruptions. Appl. Math. Comput. **253**, 324–344 (2015)

17. Jain, M., Sanga, S.S., Meena, R.K.: Control F-policy for Markovian retrial queue with server breakdowns. In: IEEE 1st International Conference on Power Electronics, Intelligent Control and Energy Systems, pp. 1–5 (2016)
18. Singh, C.J., Kaur, S.: Unreliable server retrial queue with optional service and multi-phase repair. Int. J. Oper. Res. **14**(2), 35–51 (2017)
19. Sharma, R., Jain, M.: Finite priority queueing system with service interruption. Indian J. Ind. Appl. Math. **8**(1), 90–106 (2017)
20. Shekhar, C., Jain, M., Iqbal, J., Raina, A.A.: Threshold control policy for maintainability of manufacturing system with unreliable workstations. Arab. J. Sci. Eng. **42**(11), 4833–4851 (2017)
21. Ke, J.C., Liu, T.H., Yang, D.Y.: Modeling of machine interference problem with unreliable repairman and standbys imperfect switchover. Reliab. Eng. Syst. Saf. **174**, 12–18 (2018)
22. Markova, E., Satin, Y., Kochetkova, I., Zeifman, A., Sinitcina, A.: Queuing system with unreliable servers and inhomogeneous intensities for analyzing the impact of non-stationarity to performance measures of wireless network under licensed shared access. Mathematics **8**(5), 800 (2020)
23. Chakravarthy, S.R.: Matrix-analytic queueing models, Chapter 8 in Bhat U.N. In: An Introduction to Queueing Theory: Modeling and Analysis in Applications. Birkhäuser (2015)
24. Dudin, A.N., Klimenok, V.I., Vishnevsky, V.M.: The Theory of Queuing Systems with Correlated Flows. Springer, Cham (2020). https://doi.org/10.1007/978-3-030-32072-0
25. He, Q.M.: Fundamentals of Matrix-Analytic Methods. Springer, New York (2014)
26. Lucantoni, D.M.: New results on the single server queue with a batch Markovian arrival process. Stochas. Models **7**, 1–46 (1991)
27. Neuts, M.F.: A versatile Markovian point process. J. Appl. Probab. **16**(4), 764–79 (1979)
28. Neuts, M.F.: Probability distributions of phase type. Liber Amicorum Prof. Emeritus H. Florin (1975)
29. Artalejo, J.R., Gomez-Correl, A., He, Q.M.: Markovian arrivals in stochastic modelling: a survey and some new results. SORT-Stat. Oper. Res. Trans. **34**(2), 101–144 (2010)
30. Bladt, M., Nielsen, B.F.: Matrix-Exponential Distributions in Applied Probability. PTSM, vol. 81. Springer, Boston (2017). https://doi.org/10.1007/978-1-4939-7049-0
31. Chakravarthy, S.R.: The batch Markovian arrival process: a review and future work. Adv. Probab. Theory Stoch. Process. **1**, 21–49 (2001)
32. Chakravarthy, S.R.: Markovian arrival processes. In: Wiley Encyclopedia of Operations Research and Management Science, 15 Jun 2010
33. Neuts, M.F.: Matrix-geometric Solutions in Stochastic Models: an Algorithmic Approach. Johns Hopkins University, Baltimore (1981)
34. Neuts, M.F.: Structured stochastic matrices of $M/G/1$ type and their applications. Marcel Dekker Inc., New York (1989)
35. Neuts, M.F.: Models based on the Markovian arrival process. IEICE Trans. Commun. **75**(12), 1255–1265 (1992)
36. Marcus, M., Minc, H.: Survey of Matrix Theory and Matrix Inequalities. Allyn and Bacon, Boston (1964)
37. Steeb, W.H., Hardy, Y.: Matrix Calculus and Kronecker Product: A Practical Approach to Linear and Multilinear Algebra. World Scientific Publishing, Singapore (2011)

38. Latouche, G., Ramaswami, V.: Introduction to Matrix Analytic Methods in Stochastic Modeling. SIAM, Philadelphia (1999)
39. Stewart, W.J.: Introduction to the Numerical Solution of Markov Chains. Princeton University Press, New Jersey (1994)
40. Ozawa, T.: Sojourn time distributions in the queue defined by a general QBD process. Queueing Syst. **53**, 203–211 (2006)
41. Kelton, W.D., Sadowski, R.P., Swets, N.B.: Simulation with ARENA, 5th edn. McGraw-Hill, New York (2010)

Statistical Analysis of the End-to-End Delay of Packet Transfers in a Peer-to-Peer Network

Natalia M. Markovich[1(✉)] and Udo R. Krieger[2]

[1] V.A. Trapeznikov Institute of Control Sciences Russian Academy of Sciences, Profsoyuznaya Street 65, 117997 Moscow, Russia
markovic@ipu.rssi.ru,nat.markovich@gmail.com

[2] Fakultät WIAI, Otto-Friedrich-Universität, An der Weberei 5, 96047 Bamberg, Germany
udo.krieger@ieee.org

Abstract. The paper is devoted to the statistical analysis of the end-to-end (E2E) delay of packet transfers between source and destination nodes in a peer-to-peer (P2P) overlay network. We focus on the identification of the E2E delay and the longest per-hop delay distributions and the stochastic dependence of the associated random process. The E2E delay is determined by the sum of a random number of dependent per-hop (p-h) delays along the links of a considered overlay path and the longest per-hop delay by their maximum. We propose to use the sum of the p-h delays to get a distribution of the maximum which is motivated by the available statistical data of the E2E delays. Based on recent analytic results derived from extreme-value theory we show that such sums and maxima corresponding to different paths may have the same tail and extremal indexes. These indexes determine the heaviness of the distribution tail and the dependence of extremes. Using the extremal index we identify limit distributions of the maxima of the E2E delays and the maxima of the p-h delays at a path among all source-destination paths. Considering real-time applications with stringent E2E-delay constraints, the distributions are used to identify quality-of-service (QoS) metrics of a P2P model like the packet missing probability and the corresponding playback delay as well as the equivalent capacity of a transport channel.

Keywords: P2P network · End-to-end delay · Per-hop delay · Tail index · Extremal index · Quality-of-service · Packet missing probability · Playback delay · Equivalent capacity

1 Introduction

We consider the delay performance of the packet transfer in a peer-to-peer (P2P) overlay network. The identification of the distribution of the end-to-end (E2E) delays arising between source and destination nodes in a P2P network constitutes an important problem of telecommunication due to live TV and video-on-demand applications. The delay of information transmission through the P2P

V. M. Vishnevskiy et al. (Eds.): DCCN 2020, LNCS 12563, pp. 282–297, 2020.
https://doi.org/10.1007/978-3-030-66471-8_22

network and, hence, the playback delay that is the lag between the generation of a packet and its playout deadline have a big impact on the quality of service and experience. As the E2E delay can be represented as a sum of a random number of the per-hop (p-h) delays, its distribution depends on the distributions of the random length of the overlay path between the source and destination and the p-h delays. The latter are determined by the structure of the P2P overlay network.

In [5], [15] the relation between the distribution of the packet delay and the packet missing probability in a P2P network has been considered. The distribution of the E2E delay of the ith path $D_i(D) = \sum_{j=1}^{L_i(D)} X_{i,j}$ is required. Here, $\{X_{i,j} : 1 \leq j \leq L_i(D)\}$ are the p-h delays of this overlay path i from the source S to the destination node D with a random length $L_i(D)$. The paths between S and D are schematically shown in Fig. 1. Their randomness is caused by the random number of nodes and links of the paths due to the dynamics of the P2P network over the time. The exceedance of the packet delay over the playback deadline b is considered as one of the main reasons to miss a packet. Then this part of the missing probability is the following: $P_m(b) = P\{D_i(D) > b\}$. Considering the E2E delays, we deal here with the sums of a random number of terms which can be heavy-tailed distributed and dependent. These issues constitute a complicated mathematical problem. In [15] the exceedance of the realized packet transmission rate over the equivalent capacity of the transport channel is considered as the second reason to loose packets.

It is one of the objectives of our paper to identify the missing probability under more general assumptions than in [5], [15] in view of the last statistical results obtained in [18]. It was assumed in [5], [15] as well as in [26] that $\{X_{i,j}\}$ are independent and identically distributed (i.i.d.) random variables (r.v.s) with light or heavy tails depending on the P2P overlay structure, and that the number of nodes N in the network and $L_i(D)$ are stationary distributed. The mutual dependence or independence of $X_{i,j}$ and $L_i(D)$ and the assumption which tail of these r.v.s is heavier are essential in order to identify the distribution of the sum, see for instance [8]. Here we assume that the p-h delays $\{X_{i,j}\}$ are now not necessarily i.i.d.. This assumption is realistic since paths may be overlapping as in Fig. 1. We assume that $\{X_{i,j}\}$ are stationary distributed at links located at the same distance with regard to the number of links from S. The random path length $L_i(D)$ is assumed to be stationary distributed, but its mutual independence on the p-h delay is omitted.

Another objective is to find the relation between the local dependence (i.e. cluster) structure and the distributions of the E2E delay and the maximal p-h delay at a path. This allows us to generalize the probability $P_m(b)$ uniformly to all paths of lengths $\{L_i\}$ and to obtain $P\{\max_i D_i(D) > b\}$. Our achievements are based on the results of extreme-value theory obtained in [18].

In [18] it is derived that the tail index (TI) and extremal index (EI) of the asymptotic distributions of sums and maxima of random length sequences are the same subject to some not very restrictive assumptions. One may conclude that the sums and maxima of p-h delays at the paths have the same heaviness

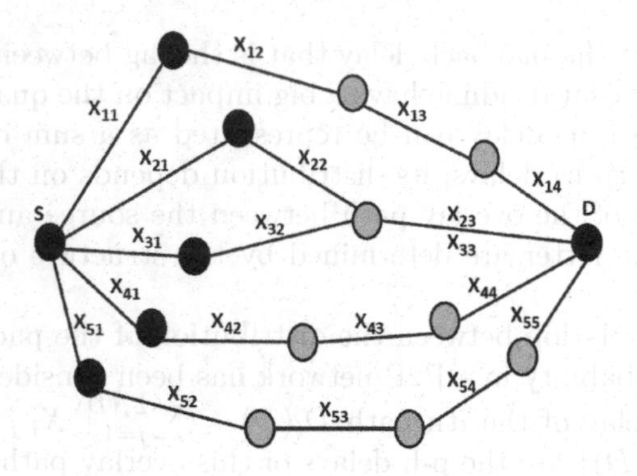

Fig. 1. Paths of random length between source node S and destination node D with the per-hop time delays $\{X_{i,j}\}$ of packet transmissions on the ith path between these nodes; the nodes in black between S and D indicate those ones with a distance of one link from S.

of the distribution tail and the same dependence structure. This feature implies that the distribution of the E2E delay may be approximated asymptotically by the distribution of the maximum of the p-h delays at a source-destination path. As the E2E delays can be made available in practice easier than the p-h delays, this allows us to approximate the distribution of the p-h delays at the most heavy-tailed link using the E2E delay statistics. Then the common TI value and the common EI value can be estimated by a sample of the observed E2E delays. The EI allows us to obtain the common limiting distribution of both the maxima of the E2E delays and the longest p-h delays among all paths. Moreover, one can use the distribution of the maximum to determine the packet missing probability.

The paper is organized as follows. Section 2 contains a survey of related results. In Sect. 3 our main results related to the stochastic model, its nonparametric estimation using a basic statistical algorithm, as well as an illustrative computational example are presented. The exposition is finalized with some conclusions and the discussion of open problems.

2 Related Work

Let the links of a path in the P2P network be enumerated from the source node S (see Fig. 1). We assume that the p-h delay $X_{i,j}$, $i,j \geq 1$, at the link j of path i is regularly varying distributed in a uniform way. This assumption implies that

$$\mathbb{P}\{X_{i,j} > x\} = \ell_j(x)x^{-k_j} \tag{1}$$

holds with the TI k_j and a slowly varying function $\ell_j(x)$, i.e. $\lim_{x\to\infty} \ell_j(tx)/\ell_j(x) = 1$ for any $t > 0$. Positive constants and logarithms provide examples of slowly varying functions $\ell_j(x)$. The links with the same number j are assumed to be stationary distributed and their distributions may be different from the distribution of the links with another number.

The EI θ is sometimes called the local dependence measure having in mind that extremes or consecutive exceedances over a high threshold u occur usually in clusters. Such clusters of exceedances are caused by the dependence in stochastic sequences. The clustering can be intensified by heavy distribution tails.

Definition 1. *[13] The stationary sequence of r.v.s $\{X_n\}_{n\geq 1}$ with cumulative distribution function (cdf) $F(x)$ and $M_n = \max\{X_1, ..., X_n\}$ is said to have the EI $\theta \in [0,1]$ if for each $0 < \tau < \infty$ there is a sequence of real numbers $u_n = u_n(\tau)$ such that it holds*

$$\lim_{n\to\infty} n(1 - F(u_n)) = \tau, \qquad \lim_{n\to\infty} P\{M_n \leq u_n\} = e^{-\tau\theta}. \tag{2}$$

The inverse $1/\theta$ approximates asymptotically the mean cluster size, i.e. the mean number of exceedances per cluster [13]. The cluster structure of a simulated Moving Maxima process [1], for instance, is shown in Fig. 2. The details regarding this process are recalled in Sect. 3.3. A smaller θ corresponds to wider clusters. In this example the values $\theta = 0.3$ and $\theta = 0.8$ imply that the mean cluster may contain approximately 3 and 1 exceedances, respectively.

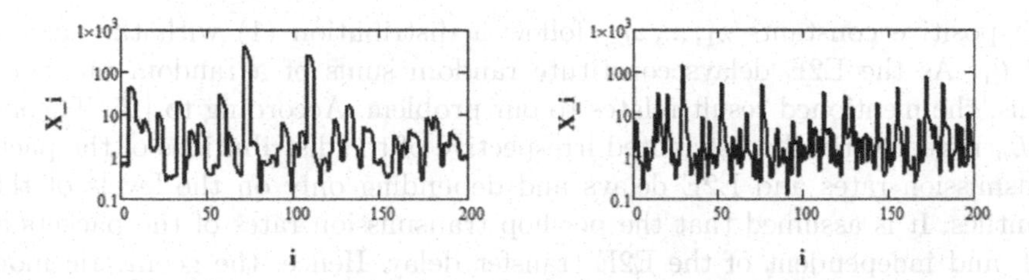

Fig. 2. The Moving Maxima process with larger and smaller clusters of exceedances for the EIs $\theta = 0.3$ (left) and $\theta = 0.8$ (right).

The EI has the following relation to the distribution of the maximum:

$$\mathbb{P}\{M_n \leq u_n\} = \mathbb{P}^{n\theta}\{X_1 \leq u_n\} + o(1) = F^{n\theta}(u_n) + o(1), \quad n \to \infty. \tag{3}$$

It holds $\theta = 1$ if the r.v.s $\{X_n\}$ are i.i.d.. The converse is incorrect. An EI that is close to zero implies a kind of a strong dependence. Stochastic processes with a strong local dependence and $\theta = 0$ exist. A Lindley process that models the waiting times in a G/GI/1 queueing system may provide such an example in case of a sub-exponentially distributed noise term, [2]. Relation (3) implies in the case $\theta = 0$ that the maximum will likely not exceed a sufficiently high threshold u_n, i.e. $\mathbb{P}\{M_n \leq u_n\} \to 1$ holds whenever u_n satisfies the first limit in (2).

In order to use results in [18], we assume that the p-h delays $\{X_{i,j} : i \geq 1\}$ at links with the number $1 \leq j \leq L_i$ of paths with numbers $i \geq 1$ are stationary distributed as in (1) and have their TI $k_j > 0$ and EI $\theta_j \in [0,1]$, and that among all sets of the links there exists a unique set with the minimal TI. Without loss of generality, this can be the set $\{X_{i,1} : i \geq 1\}$ of first links from the source node S with a TI equal to k_1. Such set of links is in a strong-sense stationary distributed with the heaviest distribution tail. Other sets have TIs larger than k_1 and, hence, according to (1) they are not so heavy-tailed distributed. Although some of such $\{k_j\}_{j\geq 2}$ may be equal, the corresponding distributions of the link sets may not be the same if the slowly varying functions $\ell_j(x)$, $j \geq 2$ in (1) are different. An arbitrary dependence between $\{X_{i,j}\}$ and L_i is allowed therein.

In [18] the EI and the TI of sums and maxima of random sequences of random lengths $\{L_n\}$ were considered. One can model the distribution tail of L_n as $\mathbb{P}\{L_n > x\} = \widetilde{\ell}_n(x)x^{-\alpha}$ with the TI $\alpha > 0$. Indeed, the lengths are integer-valued r.v.s. The relevance of such modeling is pointed out in several papers, see [7], [8], [30] among those. Assuming that both the slowly varying functions $\{\ell_j(x)\}$ in (1) and $\{\widetilde{\ell}_n(x)\}$ are bounded uniformly by polynomial functions for sufficiently large x over all sets of links and all path lengths, and that L_n has a lighter tail than the most heavy-tailed distributed p-h delay $X_{n,1}$, i.e. $\alpha > k_1$ holds, it is proved in [18] that the sequences of sums and maxima

$$X_n(z, L_n) = z_1X_{n,1} + z_2X_{n,2} + ... + z_{L_n}X_{n,L_n},$$
$$X_n^*(z, L_n) = \max(z_1X_{n,1}, z_2X_{n,2}, ..., z_{L_n}X_{n,L_n})$$

with positive constants $z_1, ..., z_{L_n}$ follow a distribution (1) with the same k_1 and θ_1. As the E2E delays constitute random sums of a random number of terms, the mentioned result relates to our problem. According to [15, Theorem 1], L_n is geometrically distributed irrespective of the distributions of the packet transmission rates and E2E delays and depending only on the levels of their quantiles. It is assumed that the per-hop transmission rates of the packets are i.i.d. and independent of the E2E transfer delay. Hence, the geometric model meets the result in [18], but L_n is assumed to be regularly varying distributed with a positive TI. The latter assumption is not restrictive since the class of distributions with regularly varying tails is rather wide.

In case that some paths include a node with light-tailed distributed p-h delay and(/or) the distribution of the p-h delays at some link from the source contains a mixture of light- and heavy- tailed distributions, the basic statistical result developed in [18] is still valid. This property follows from the proofs of Theorem 3 and 4 in [18].

3 Statistical Analysis of the End-to-End Delay

3.1 Asymptotic Distribution of the E2E and Maximal P-h Delays

Let us consider a path of random length $L_i(D)$ between the source and destination nodes (S, D). $L_i(D)$ is equal to the number of links between the source S and destination D. Let $n \geq 1$ be the number of possible paths constructed by the nodes of the P2P overlay network. Since the P2P network may be changed dynamically in time, the number of nodes available for the packet transmission is changing and n is random. We can neglect its randomness considering the approach as a conditional one, since n is proportional to the number of nodes N in the network and the latter can be large. The theoretical result in [18] assumes that n is deterministic and tends to infinity.

Let us consider the double-indexed array of the p-h delays $\mathcal{X} = (X_{i,j} : i, j \geq 1)$. The "row index" i corresponds to the p-h delays belonging to the same path i between the source S and destination D, and the "column index" j corresponds to the p-h delays arising at the jth link enumerated from the source node. All p-h delays relate to the same source-destination pair (S, D). We consider the corresponding matrix

$$\mathcal{X} = \begin{pmatrix} X_{1,1} & X_{1,2} & 0 & ...0 & X_{1,L_1} \\ X_{2,1} & X_{2,2} & X_{2,3} & ...0 & X_{2,L_2} \\ ... & ... & ... & ... & ... \\ X_{n,1} & X_{n,2} & X_{n,3} & ...X_{n,L_n-1} & X_{n,L_n} \end{pmatrix} \tag{4}$$

$$\begin{pmatrix} k_1, & k_2, & k_3, & ..., & k_{L_n-1}, & k_{L_n} \\ \theta_1, & \theta_2, & \theta_3, & ..., & \theta_{L_n-1}, & \theta_{L_n} \end{pmatrix}$$

where the first and last columns corresponding to the one-hop links to the source and destination nodes are full and the internal columns are completed by zeros up to the maximal dimension $L_{max} = \max\{L_1, ..., L_n\}$, let's say $L_n = L_{max}$. We assume the most general case: the columns can be dependent, and each column may consist of dependent p-h delays, and the distribution of each column is stationary with the positive TI value k_j. Its local dependence structure is described by the EI value θ_j.

For any location of zeros in the matrix $\mathcal{X}$, the minimal TI (and the corresponding EI) of the internal columns taken together is determined by the distribution of the most heavy-tailed distributed element. This property follows from the proof of Theorem 3 in [18]. The sum $D_i = \sum_{j=1}^{L_i} z_j X_{i,j}$ and maximum $M_i = \max_{j=1,...,L_i}\{z_j X_{i,j}\}$ of weighted elements of the ith string set determine the weighted E2E delay between the source and destination nodes of the ith path and the longest weighted p-h delay at the ith path, respectively. The weights $\{z_i\}$ may reflect a priority which can be proportional to the capacities of links or impact on the scheduling of the peer selection process determining the path. In the simplest case, $\{z_i\}$ are all equal to one.

We suppose, without loss of generality, that the minimal TI value k_1 belongs to the first column and $k_1 < k$, with $k = \lim_{n\to\infty} \inf_{2\leq j \leq l_n} k_j$, $\quad l_n = \lceil n^{\chi} \rceil$, $0 < \chi < (k - k_1)/(k_1(k+1))$ holds. The value k_1 corresponds to the heaviest distribution tail among the columns. According to Theorem 4 in [18] it follows

$$\mathbb{P}\{M_i > x\} = \mathbb{P}\{D_i > x\}(1 + o(1)) = \ell_1(x)x^{-k_1}(1 + o(1)) \tag{5}$$

as $x \to \infty$. This result means that the most heavy-tailed distributed column of the p-h delays determines the distributions of the E2E delay and the maximal p-h delay at the ith path. Instead of the E2E delays, one can consider the maximal p-h delays at each path (or vice versa) since they have the same heaviness of tail, i.e. the same distribution up to the slowly varying functions. This allows us to model the distribution of the p-h delays since the E2E delays can be easily gathered as statistics in practice, rather than the p-h delays.

By Theorem 4 in [18] the EI of M_i and D_i is equal to the value θ_1 corresponding to k_1. Then the maxima of the sequences $\{D_i\}$ and $\{M_i\}$, $i = 1, ..., n$, have the same limiting distributions. More exactly, it holds

$$\lim_{n\to\infty} \mathbb{P}\{M_n^s \le u_n\} = \lim_{n\to\infty} \mathbb{P}\{M_n^m \le u_n\} = e^{-\tau\theta_1} \tag{6}$$

by (2) with

$$\lim_{n\to\infty} n\mathbb{P}\{M_n > u_n\} = \lim_{n\to\infty} n\mathbb{P}\{D_n > u_n\} = \tau, \tag{7}$$

where we denote

$$M_n^s = \max\{D_1, ..., D_n\}, \quad M_n^m = \max\{M_1, ..., M_n\},$$

and $\{u_n\}$ is an increasing sequence of thresholds. In [18] u_n is selected by (5) and (7) in such a way that $\tau = (z_1/y)^{k_1}$ with a constant $y > 0$ holds, namely, $u_n = yn^{1/k_1}\ell_1^\sharp(n)$, where $\ell_1^\sharp(n)$ is a slowly varying function.

Regarding the transmission rates of the packet flows we can argue in the same way. Following [15], each node is a bottleneck and it may upload an own superimposed flow coming from other nodes. Then a transported packet is associated with the sequence of transmission rates $\{R_{i,1}, R_{i,2}, ..., R_{i,L_i}\}$ corresponding to the links of the ith path. We approximate these transmission rates as ratios $R_{i,j} = Y_i/Z_{i,j}$, where Y_i is the packet length and $Z_{i,j}$ is the inter-arrival time between the considered packet and the previous (or next) one arriving at the jth node. Clearly, the rates $\{R_{i,j}\}$, $j = 1, 2, \ldots$ are all dependent for a fixed i.

Considering the matrix $\mathcal{X}$ in (4) one can substitute $X_{i,j}$ by $R_{i,j}$ assuming that the columns of the transmission rates have the TIs $\{k_i^*\}$ and EIs $\{\theta_i^*\}$ and that a unique minimal TI k_1^* exists as for the p-h delays:

$$\mathcal{R} = \begin{pmatrix} R_{1,1} & R_{1,2} & 0 & ...0 & R_{1,L_1} \\ R_{2,1} & R_{2,2} & R_{2,3} & ...0 & R_{2,L_2} \\ \cdots & \cdots & \cdots & \cdots & \cdots \\ R_{n,1} & R_{n,2} & R_{n,3} & ...R_{n,L_n-1} & R_{n,L_n} \end{pmatrix} \tag{8}$$

$$\begin{pmatrix} k_1^*, & k_2^*, & k_3^*, & ..., & k_{L_n-1}^*, & k_{L_n}^* \\ \theta_1^*, & \theta_2^*, & \theta_3^*, & ..., & \theta_{L_n-1}^*, & \theta_{L_n}^* \end{pmatrix}.$$

Then we obtain (6) with corresponding replacements. As the result stated in [18] concerns weighted sums and maxima, one can think that some links may have a priority which can be proportional to their capacities or that the weights $\{z_i\}$ can impact on the scheduling of the peer selection process determining the path.

The probability of the successful transmission P_{st} of n packets over their n paths is determined by

$$P_{st} = \mathbb{P}\{M_n^{m*} \le u_n^*\} + \mathbb{P}\{M_n^m \le b_n\}, \tag{9}$$

where

$$M_n^{m*} = \max\{M_1^*, ..., M_n^*\}, \quad M_i^* = \max_{j=1,...,L_i} \{z_j R_{i,j}\}$$

are the maximal transmission rates of the packets over n paths and over the path i, respectively. The excess of the rate over the equivalent channel capacity u_n^* may cause the miss of packets [15]. In (9) $\mathbb{P}\{M_n^m \leq b_n\}$ is the probability that the longest (weighted) p-h delay M_n over n paths is less than the playback delay b_n. The sequences $\{u_n^*\}$ and $\{b_n\}$ are determined to be increasing as $n \to \infty$ in the same way as $\{u_n\}$ in [18], i.e.

$$u_n^* = yn^{1/k_1^*}, \qquad b_n = yn^{1/k_1} \tag{10}$$

omitting the slowly varying functions for simplicity. Such sequences correspond to high quantiles of the rates and p-h delays. Then it holds

$$P_{st}(y) \approx e^{-\tau^* \theta_1^*} + e^{-\tau \theta_1} = e^{-(z_1/y)^{k_1^*} \theta_1^*} + e^{-(z_1/y)^{k_1} \theta_1}, \; y > 0 \tag{11}$$

for sufficiently large n, where y is selected in such a way to keep $P_{st}(y) < 1$. Hence, the approximate probability to loose at least one packet during the transmission over n paths is given by

$$P_m(y) = 1 - P_{st}(y). \tag{12}$$

Taking $P_m(y) = \eta$, where $\eta \in (0,1)$ is small, one can find a corresponding y. In Fig. 3 an example is shown where $y = 0.755$ provides the solution to $P_m(y) = 0.05$.

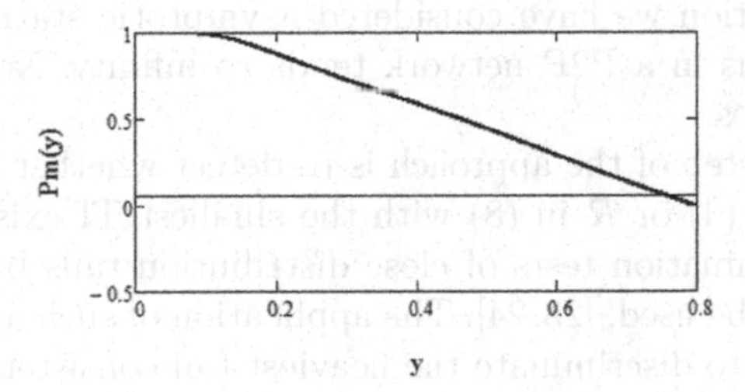

Fig. 3. P_m against y for $\alpha_1 = 1.2$, $\alpha_1^* = 2$, $\theta_1 = 0.3$, $\theta_1^* = 0.7$ and $z_1 = 1$ (thick solid line), $\eta = 0.05$ (thin solid line).

One may also consider a simple example of such calculation. Let us suppose that $e^{-\tau^* \theta_1^*} = e^{-\theta_1}$ holds. Then we get $\tau^* = \theta_1/\theta_1^*$, $1 - e^{-\theta_1} - e^{-\tau\theta_1} = \eta > 0$ and $\eta < 1 - e^{-\theta_1}$. Taking $\tau = (z_1/y)^{\alpha_1}$, we obtain

$$y = z_1 \exp\left(-\frac{1}{\alpha_1} \ln\left(-\frac{1}{\theta_1} \ln(1 - e^{-\theta_1} - \eta)\right)\right). \tag{13}$$

For example, if $z_1 = 1$, $\theta_1 = 0.3$, $\theta_1^* = 0.7$ holds, we may take $\eta < 0.259$, $\tau^* = 0.429$. Given $\alpha_1 = 1.2$ and $\eta = 0.01$, we get $y = 0.221$. However, such y does not depend on the TI and EI of the transmission rates.

To prepare our further statistical estimation techniques, let us finally summarize our general assumptions about the proposed statistical model of the E2E and p-h delays regarding the packet transfer in a P2P network. We assume that

(i) the P2P overlay network may be dynamic;
(ii) the p-h packet delays at the different links of the same path may be arbitrary dependent;
(iii) the p-h packet delays at the different links located on the same distance from the source node and belonging to different source-destination paths are stationary distributed, but not necessarily independent;
(iv) the length of a path and the p-h delays at its links may be dependent;
(v) the distribution of the path length has a lighter tail than the p-h delays with the heaviest tail;
(vi) there exists a unique set of links located on the same distance from the source that has the heaviest distribution tail compared to other sets of links among the overlay paths.

We recall that the E2E delays are regularly varying (heavy-tailed) distributed which follows from (5). The assumption (v) is fulfilled since the normalized path length is geometrically distributed irrespective of the distributions of the transmission rates and E2E delays and depending only on the levels of their quantiles [15].

3.2 Nonparametric Statistical Estimation

In the previous section we have considered asymptotic statistical results when the number of paths in a P2P network tends to infinity. Now we consider the case of finite samples.

The important step of the approach is to detect whether the unique column of the matrix $\mathcal{X}$ in (4) or $\mathcal{R}$ in (8) with the smallest TI exists or not. For this purpose the discrimination tests of close distribution tails built by only higher order statistics can be used, [23,24]. The application of such a test to each pair of columns of $\mathcal{X}$ or $\mathcal{R}$ to discriminate the heaviest tail consistently may constitute a calculation problem that is out of scope of this paper. Here, this problem can be solved from another perspective.

Many proposed network architectures place nodes with large upload capacities close to the source [5]. Thus, one may expect the smallest capacities and transmission rates at the last link before the destination node. This property may lead to the heaviest distribution tail of the p-h delays or the transmission rates and the smallest TI at the last link. Thus, one can estimate and compare the TIs and EIs of the p-h delays or the rates at the internal part and the last column of the matrix $\mathcal{X}$ or $\mathcal{R}$, respectively, and find the minimal k_1 or k_1^*, respectively.

Estimation of the TI. Let $X^n = \{X_1, ..., X_n\}$ be a sample of r.v.s with cumulative distribution function (cdf) $F(x)$. These r.v.s could be the transmission rates $R_{i,j}$ or the p-h delays $X_{i,j}$.

The Hill's estimator is well known and the simplest one to estimate the TI, but it requires an i.i.d. sample, [9], [14], [20], [29]. Regarding dependent data one can recommend estimators based on sums and maxima of non-intersecting data blocks, [19], [22]. The reduced bias estimator of the extreme value index that is the reciprocal of the TI is proposed in [4].

Several nonparametric estimators of the TI can be written by means of the statistic proposed in [21]

$$G_n(k,r,v) = \frac{1}{k}\sum_{i=0}^{k-1} g_{r,v}\left(\frac{X_{n-i,n}}{X_{n-k,n}}\right), \quad g_{r,v}(x) := x^r \ln^v(x),$$

where $r \in R$, $v > -1$. For instance, this estimator includes the Hill's estimator

$$\gamma_n^{(H)}(k) = G_n(k,0,1) = \frac{1}{k}\sum_{i=0}^{k-1} \ln\left(\frac{X_{n-i,n}}{X_{n-k,n}}\right), \tag{14}$$

or the moment-ratio estimator

$$\hat{\gamma}_n^{(mr)}(k) = G_n(k,0,2)\left(2G_n(k,0,1)\right)^{-1} \tag{15}$$

proposed in [6] to estimate the extreme value index $\gamma = 1/\alpha$ which is the reciprocal of the TI α. Here, $1 \le k \le n-1$ is the number of the largest order statistics

$$X_{n-k,n} \le X_{n-k+1,n} \le \cdots \le X_{n,n}$$

of the sample $\{X_1, \ldots, X_n\}$ used for the estimation, and r is a tuning parameter. The statistics $G_n(k,r,v)$ are special cases of the statistics introduced in [25].

The choice of k constitutes another problem. The simplest visual method is given by the Hill plot $\{(k, \gamma_n^{(H)}(k)) : k = 1, \ldots, n-1\}$. Then the estimate of k is selected from the interval $[k_-, k_+]$ of stability of the function $\gamma_n^{(H)}(k)$, [14]. Alternatives could be the exceedance plot or a bootstrap method as well as an exact calculation of k and r as in [20].

Estimation of the EI. Among the nonparametric estimators of the EI, the blocks, runs and intervals estimator are the most popular ones, [3]. As the reciprocal of the EI approximates the mean cluster size, the estimators differ by the definition of the cluster of exceedances. Particularly, the cluster of the blocks estimator is a data block with at least one exceedance over a threshold u. The blocks and runs estimators require a tuning parameter and the threshold u whereas the intervals estimator needs only u, [11].

The intervals estimator is calculated by a specific sample $\{T_1(u)\}_{i=1}^{L}$ of the length $L = L(u) < n$ generated by the initial sample $X^n = \{X_1, ..., X_n\}$. Namely,

$$T_1(u) = \min\{j \geq 1 : M_{1,j} \leq u, X_{j+1} > u | X_1 > u\}$$

denotes the number of consecutive non-exceedances between two consecutive clusters of exceedances, where $M_{1,j} = \max\{X_2, ..., X_j\}$, $M_{1,1} = -\infty$ holds. Here the cluster of exceedances determines a set of consecutive exceedances of the underlying stochastic sequence over the threshold u between two consecutive non-exceedances. Then the intervals estimator is defined as

$$\hat{\theta}_n(u) = \begin{cases} \min(1, \hat{\theta}_n^1(u)), \text{ if } \max\{(T_1(u))_i : 1 \leq i \leq L-1\} \leq 2, \\ \min(1, \hat{\theta}_n^2(u)), \text{ if } \max\{(T_1(u))_i : 1 \leq i \leq L-1\} > 2, \end{cases} \tag{16}$$

where

$$\hat{\theta}_n^1(u) = \frac{2(\sum_{i=1}^{L-1}(T_1(u))_i)^2}{(L-1)\sum_{i=1}^{L-1}(T_1(u))_i^2}, \tag{17}$$

$$\hat{\theta}_n^2(u) = \frac{2(\sum_{i=1}^{L-1}((T_1(u))_i - 1))^2}{(L-1)\sum_{i=1}^{L-1}((T_1(u))_i - 1)((T_1(u))_i - 2)} \tag{18}$$

holds. Among the last achievements, one can mention the K-gaps estimator that improves the intervals estimator, [28]. In [12] one can find the IMT method to calculate an optimal pair (u, K) for the K-gaps estimator.

Usually, u is chosen among those quantiles that are higher than 95% of an underlying sequence. u can be selected visually as corresponding to the stability interval of the plot $\{(u, \widehat{\theta}(u))\}$ in the same way as the Hill plot. One can apply a bootstrap method [17] or the discrepancy method [16] for its automatic selection.

Then we can determine the basic nonparametric estimation algorithm by these statistical means.

Estimation Algorithm. Let us consider the last columns of the matrices $\mathcal{X}$ in (4) and $\mathcal{R}$ in (8), namely, $\{X_{i,L_i}\}$ and $\{R_{i,L_i}\}$, $i = 1, 2, ..., n$, as initial data. Here n is the number of possible paths between the source and destination nodes (S, D) of the P2P overlay network.

1. Estimate the TIs α_{L_n} and $\alpha_{L_n}^*$ by $\{X_{i,L_i}\}$ and $\{R_{i,L_i}\}$ using one of the nonparametric estimators, e.g. (14) or (15).
2. Estimate the EIs θ_{L_n} and $\theta_{L_n}^*$ by $\{X_{i,L_i}\}$ and $\{R_{i,L_i}\}$ using one of the nonparametric estimators, e.g. (16)–(18).
3. Calculate y as $y = \arg\{t : P_m(t) = \eta\}$ or by (13) for a predefined $0 < \eta < 1$.
4. Calculate the probabilities $P_{st}(y)$ and $P_m(y)$ by (11) and (12).
5. Calculate the equivalent capacity u_n^* and the playback delay b_n by (10).

3.3 An Illustrative Example

In this section our aim is to demonstrate the sketched methodology using simulated examples of sequences that are arising from regularly varying distributed r.v.s. We simulate samples of the transmission rates $\{R_{i,L_i}, i \in \{1, 2, \ldots, n\}\}$ as Moving Maxima (MM) process and of the p-h delays $\{X_{i,L_i}, i \in \{1, 2, \ldots, n\}\}$ as MA(2) process. Considering a P2P network in practice, indeed, the real processes could be different. However, our methodology is a pure nonparametric approach and can be applied to any process model.

The mth order MM process is determined by

$$X_t = \max_{i=0,\ldots,m} \{\beta_i Z_{t-i}\}, \qquad t \in \mathbb{Z},$$

where $\{\beta_i\}$ are constants with $\beta_i \geq 0$, $\sum_{i=0}^{m} \beta_i = 1$, and Z_t are i.i.d. standard Fréchet distributed r.v.s with the cdf $F(x) = \exp(-1/x)$ for $x > 0$. The EI of the process is equal to $\theta = \max_i\{\beta_i\}$ [1]. The distribution of $\{X_t\}_{t\geq 1}$ is standard Fréchet. Its TI is equal to one. In our study the values $m = 3$ and $\theta = 0.5$ corresponding to $\beta \in \{0.5, 0.3, 0.15, 0.05\}$ are selected.

The MA(2) process is determined by

$$X_i = pZ_{i-2} + qZ_{i-1} + Z_i, \qquad i \geq 1, \tag{19}$$

with $p > 0$, $q < 1$, and i.i.d. Pareto random variables $Z_{-1}, Z_0, Z_1, \ldots$ with $\mathbb{P}\{Z_0 > x\} = 1$ if $x < 1$, and $\mathbb{P}\{Z_0 > x\} = x^{-\alpha}$ if $x \geq 1$ hold for some $\alpha > 0$ [27]. The EI of the process is given by $\theta = (1 + p^\alpha + q^\alpha)^{-1}$. The case $\alpha = 2$, $(p, q) = (1/\sqrt{2}, 1/\sqrt{2})$ with a corresponding value $\theta = 0.5$ is considered. Since the distribution of the sum of weighted i.i.d. Pareto r.v.s behaves like a Pareto distribution in the tail, namely,

$$\mathbb{P}\{\sum_{i=1}^{n} Z_i > x\} \sim n(1 + x/\beta)^{-\alpha} \cdot L(x), \quad x \to \infty, \tag{20}$$

where $L(x)$ is a slowly varying function at infinity, $\beta > 0$ is a scale parameter, and $\alpha > 0$ is the TI, (see [10, Ch. 8, pp. 268–272]), then X_i is also Pareto distributed with the TI α.

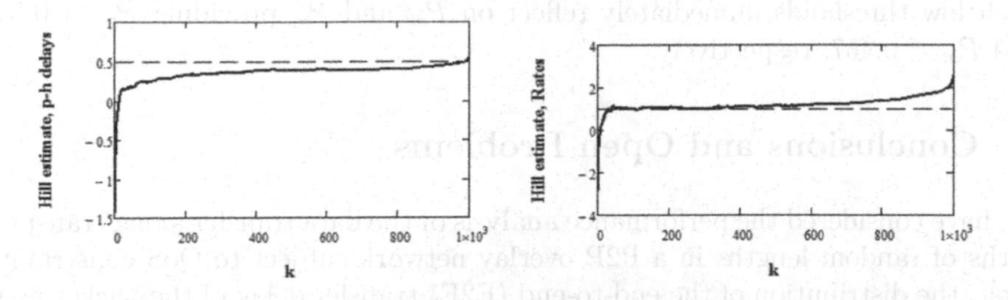

Fig. 4. The Hill's estimate of the p-h delays modeled as MA(2) process with the TI $\alpha = 2$ (and the EVI $\gamma = 0.5$) (lhs) and the transmission rates modeled as MM process with the TI $\alpha = 1$ (and $\gamma = 1$) (rhs); the sample size is given by $n = 1000$.

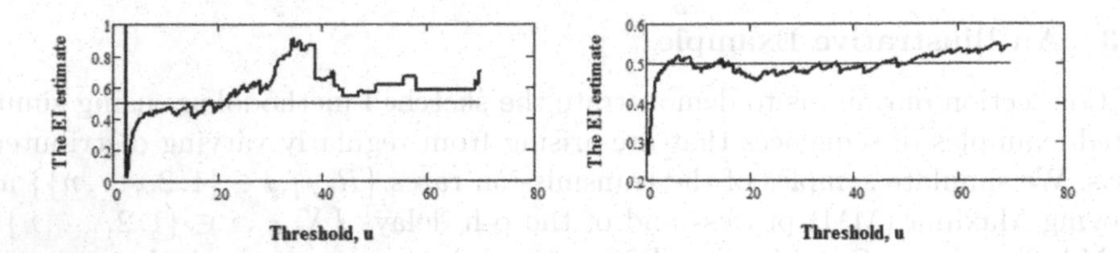

Fig. 5. The intervals estimate of the p-h delays modeled as MA(2) process with the EI $\theta = 0.5$ (lhs) and the transmission rates modeled as MM process with the EI $\theta = 0.5$ (rhs); the sample size is given by $n = 20000$.

Our objective is to estimate the parameters and to calculate all relevant metrics according to the proposed estimation algorithm using these simulated samples.

The Hill's estimator is very sensitive to the presence of a slowly varying function in the distribution tail. Thus, the estimate on the left-hand side of Fig. 4 corresponding to (19) and (20) is rather biased. In practice it is therefore reasonable to use several estimators of the TI.

Now we consider the EI estimation by the intervals estimator (16)–(18) that is applied to the same processes MA(2) and MM, see Fig. 5. Here we have to generate larger samples with $n = 20000$. The intervals estimator requires a large sample size of $\{X^n\}$ to get a better estimation since it is based on the sample of the inter-exceedance times $\{T_1(u)\}_i$, $i = 1, 2, ..., L(u)$, generated from the underlying sample X^n. The size of $\{T_1(u)\}_i$ can be much smaller than n depending on the threshold u, the higher u the smaller $L(u)$.

We can obtain $y \in \{0.768, 0.732\}$ for a given $\eta \in \{0.05, 0.1\}$, respectively, and for given $\alpha_1^* = 1$, $\alpha_1 = 2$, $\theta_1^* = \theta_1 = 0.5$ in the same way as in Fig. 3. By formulae (10) we then obtain for $n = 1000$ $u_n^* \in \{768, 732\}$ and $b_n \in \{24.286, 23.148\}$, respectively. Regarding such y the probabilities $P_{st}(y)$ and $P_m(y)$ calculated by (11) and (12) are equal to $\{0.95, 0.898\}$ and $\{0.05, 0.102\}$, respectively. We note that the maximal values of the generated random sequences $\max_{1 \leq i \leq n} X_i$ are equal to 26.946 w.r.t. the p-h delays and 743.439 w.r.t. the transmission rates. It implies that u_n^* and b_n exceed these maxima, and $P_{st} = 0.95$ is not realistic for these models. A calculation of y by (13) provides $u_n^* = 522$ and $b_n = 16.507$. Such low thresholds immediately reflect on P_{st} and P_m providing $P_{st} = 0.543$ and $P_m = 0.457$, respectively.

4 Conclusions and Open Problems

We have considered the performance analysis of the data transfer along transport paths of random lengths in a P2P overlay network subject to QoS constraints. First, the distribution of the end-to-end (E2E) transfer delay of the packet flows between the source and destination nodes is modeled. The E2E transfer delay is determined by the sum of a random number of p-h delays along the links of an overlay path. Based on recent statistical results in [18] and assuming that the per hop (p-h) delays and the lengths of the paths are regularly varying distributed, it is shown that the sums and maxima of the p-h delays corresponding to different paths of random lengths may have the same tail and extremal indexes TI and EI, respectively. These indexes determine the heaviness of the tail of the delay distribution and the dependence indicator that measures the cluster tendency (i.e., how extreme values arise by groups of observations). Using the EI, then the limit distributions of the maxima of the E2E and p-h delays over all source-destination paths are identified. Considering real-time applications with stringent E2E delay constraints, the latter distributions are used to identify important QoS metrics of a P2P-model like the packet missing probability, the corresponding playback delay, and the required equivalent capacity to transfer the packet flows of the data.

The proposed approach requires the verification and comparison of the TIs of the p-h delays to find the set of links whose delays have the heaviest tail. Regarding modern network architectures one can expect that the last link before the destination node has the heaviest distribution tail. Then known statistical tests allow us to compare pairs of samples in the columns of the matrix $\mathcal{X}$ (or $\mathcal{R}$) regarding the similarity of their distributions. We note that the lengths of the overlay paths of packet flows in a P2P network can be observed if the packet header is providing a counter of the visited nodes along the path. Then the TI of the lengths can be estimated by these means.

The described asymptotic results are valid for sufficiently high thresholds that are in our context the playback delay and the equivalent capacity of the transport channel. Our statistical results provide the basis for an improved control scheme regarding the optimal selection of transport paths in a P2P overlay network subject to QoS constraints on the E2E delay and packet loss metrics.

Regarding the application of a P2P overlay concept in 5G networks, we may look at the deployment of a blockchain functionality on top of an underlying network of mining peers that are validating transactions of IoT data processing or the use of P2P video streaming as important examples. In the case of such real-time applications, we are looking for short playback delays, but they may lead to a large packet missing probability. In this respect the derived asymptotic performance analysis models of the E2E transfer delay provide a tendency with an increasing probability of successful packet transmission as both the playback delay and the equivalent capacity increase. But these performance analysis models require an adjustment for short playback delays and not high, realistic capacities.

Our future studies will focus on these analysis and design issues of modern teletraffic theory.

Acknowledgments. The first author was partly supported by Russian Foundation for Basic Research (grant 19-01-00090).

References

1. Ancona-Navarrete, M.A., Tawn, J.A.: A comparison of methods for estimating the extremal index. Extremes **3**(1), 5–38 (2000). https://doi.org/10.1023/A:1009993419559
2. Asmussen, S.: Subexponential asymptotics for stochastic processes: extremal behavior, stationary distributions and first passage probabilities. Ann. Appl. Probab. **8**, 354–374 (1998)
3. Beirlant, J., Goegebeur, Y., Teugels, J., Segers, J.: Statistics of Extremes: Theory and Applications. Wiley, Chichester (2004)
4. Caeiro, F., Gomes, M.I., Beirlant, J., de Wet, T.: Mean-of-order p reduced-bias extreme value index estimation under a third-order framework. Extremes **19**, 561–589 (2016). https://doi.org/10.1007/s10687-016-0261-5
5. Dán, G., Fodor, V.: Delay asymptotics and scalability for peer-to-peer live streaming. IEEE Trans. Parallel Distrib. **20**(10), 1499–1511 (2009)
6. Danielsson, J., Jansen, D.W., de Vries, C.G.: The method of moments ratio estimator for the tail shape parameter. Commun. Stat. Theory. **25**, 711–720 (1986)
7. Jelenkovic, P.R., Olvera-Cravioto, M.: Information ranking and power laws on trees. Adv. Appl. Prob. **42**(4), 1057–1093 (2010)
8. Jessen, A.H., Mikosch, T.: Regularly varying functions. Publ. Inst. Math. (Beograd) (N.S.) **80**, 171–192 (2006)
9. Hill, B.M.: A simple general approach to inference about the tail of a distribution. Ann. Stat. **3**, 1163–1174 (1975)
10. Feller, W.: An Introduction to Probability and Its Application, 2nd edn. Wiley, New York (1971)
11. Ferro, C.A.T., Segers, J.: Inference for clusters of extreme values. J. R. Stat. Soc. B. **65**, 545–556 (2003)
12. Fukutome, S., Liniger, M.A., Süveges, M.: Automatic threshold and run parameter selection: a climatology for extreme hourly precipitation in Switzerland. Theoret. Appl. Climatol. **120**, 403–416 (2015)
13. Leadbetter, M.R., Lingren, G., Rootzen, H.: Extremes and Related Properties of Random Sequence and Processes. Chap. 3. Springer, New York. https://doi.org/10.1007/978-1-4612-5449-2 (1983)
14. Markovich, N.M.: Nonparametric Estimation of Univariate Heavy-Tailed Data. Wiley, Chichester (2007)
15. Markovich, N.M.: Quality assessment of the packet transport of peer-to-peer video traffic in high-speed networks. Perform. Eval. **70**, 28–44 (2013)
16. Markovich, N. M.: Nonparametric estimation of extremal index using discrepancy method. In: Proceedings of the X International Conference "System Identification and Control Problems" SICPRO-2015, Moscow, V.A. Trapeznikov Institute of Control Sciences, 26–29 January, pp. 160–168 (2015)
17. Markovich, N.M., Ryzhov, M.S., Krieger, U.R.: Statistical clustering of a random network by extremal properties. In: Vishnevskiy, V.M., Kozyrev, D.V. (eds.) DCCN 2018. CCIS, vol. 919, pp. 71–82. Springer, Cham (2018). https://doi.org/10.1007/978-3-319-99447-5_7
18. Markovich, N.M., Rodionov, I.V.: Maxima and sums of non-stationary random length sequences. Extremes **23**(3), 451–464 (2020). https://doi.org/10.1007/s10687-020-00372-5

19. Markovich, N., Vaičiulis, M.: Modification of moment-based tail index estimator: sums versus maxima. In: Bertail, P., Blanke, D., Cornillon, P.-A., Matzner-Løber, E. (eds.) ISNPS 2016. SPMS, vol. 250, pp. 85–101. Springer, Cham (2018). https://doi.org/10.1007/978-3-319-96941-1_6
20. Vaičiulis, M., Markovich, N.M.: A class of semiparametric tail index estimators and its applications. Autom. Remote Control **80**(10), 1803–1816 (2019). https://doi.org/10.1134/S0005117919100035
21. Paulauskas, V., Vaičiulis, M.: Several new tail index estimators. Ann. Inst. Stat. Math. **69**, 461–487 (2017)
22. McElroy, T., Politis, D.N.: Moment-based tail index estimation. J. Statist. Plan. Infer. **137**, 1389–1406 (2007)
23. Rodionov, I.V.: On discrimination between classes of distribution tails. Probl. Inform. Transm. **54**(2), 124–138 (2018)
24. Rodionov, I.V.: Discrimination of close hypotheses about the distribution tails using higher order statistics. Theory Probab. Appl. **63**(3), 364–380 (2019)
25. Segers, J.: Residual estimators. J. Stat. Plan. Inf. **98**, 15–27 (2001)
26. Shih, M.F., Hero, A.O.: Unicast-based inference of network link delay distributions using mixed finite mixture models. IEEE Trans. Signal Process. **51**(8), 2219–2228 (2003)
27. Sun, J., Samorodnitsky, G.: Multiple thresholds in extremal parameter estimation. Extremes **22**, 317–341 (2019). https://doi.org/10.1007/s10687-018-0337-5
28. Süveges, M., Davison, A.C.: Model misspecification in peaks over threshold analysis. Ann. Appl. Statist. **4**(1), 203–221 (2010)
29. Vaičiulis, M.: Local-maximum-based tail index estimator. Lith. Math. J. **54**(4), 503–526 (2014)
30. Volkovich, Y.V., Litvak, N.: Asymptotic analysis for personalized web search. Adv. Appl. Prob. **42**(2), 577–604 (2010)

Multidimensional Central Limit Theorem of the Multiclass M/M/1/1 Retrial Queue

Anatoly Nazarov[1], Tuan Phung-Duc[2], and Yana Izmailova[1(✉)]

[1] Institute of Applied Mathematics and Computer Science, National Research Tomsk State University, 36 Lenina ave., 634050 Tomsk, Russian Federation
nazarov.tsu@gmail.com, evgenevna.92@mail.ru

[2] Faculty of Engineering Information and Systems, University of Tsukuba, 1-1-1 Tennodai, Tsukuba, Ibaraki 305-8573, Japan

Abstract. In this paper, we consider the multiclass M/M/1/1 retrial queueing system. Customers of each class arrive from outside the system according to a Poisson process. The service times of customers are assumed to be exponentially distributed with the parameter corresponding to the type of the customer. If the server is busy incoming customers join the orbit according to their type and make a delay for an exponentially distributed time. Equations for the characteristic function of the multi-dimensional probability distribution of the numbers of customers in the orbits are obtained. These equations are investigated by method of asymptotic analysis under the long delay condition of customers in the orbits. It is shown that the probability distribution can be approximated by a multi-dimensional Gaussian distribution. Equations are obtained for finding the parameters of this probability distribution.

Keywords: Retrial queueing system · A multiclass system · Asymptotic analysis

1 Introduction

Retrial queues have become popular in the queueing research due to the challenging in the analysis as well as the needs of modelling retrial phenomenon in real world systems, e.g., telecommunication systems [10,15], call center and other service systems [6]. Retrial queues reflect the situations that customers who arrive at a service system when the system is fully occupied, do not wait but retry to access the system in a later time. For example, customers of a call center may make a phone call again if all the operators are busy [6]. Retrial queues with single class of customers have been extensively studied in the literature [1,2]. For a survey on advances of retrial queues, we refer to [14]. The main difficulty in the analysis of retrial queues arises from the fact that customers retry independently leading to inhomogeneous transition structures of the underlying Markov

The reported study was funded by RFBR according to the research project No. 18-01-00277.

V. M. Vishnevskiy et al. (Eds.): DCCN 2020, LNCS 12563, pp. 298–310, 2020.
https://doi.org/10.1007/978-3-030-66471-8_23

chains. As a result, even for the pure Markovian model (i.e. Poisson arrivals and exponential service times), explicit results are found in only some special cases only. Single server model with pure Markovian assumptions is explicitly analyzed [2]. For the case of more than one server, generating functions for the number of customers in the orbit is represented in terms of hypergeometric functions for the case of two servers [2,4], while the distribution of the number of customers in the orbit is expressed in terms of continued fraction for the case of three and four servers [11,12] and matrix continued fraction for the case of arbitrary number of servers [5,13]. The main difficulty of the analysis is that the generating functions of the joint queue length distribution are solutions of a system of differential equations whose solution cannot be explicitly obtained in general.

For multiclass retrial queues, the analysis is even more difficult and analytical solution for the joint stationary distribution has not been obtained even for single server case. To the best of our knowledge, only the stability conditions [7,9] and moments of the number of customers in the orbit have been obtained [2,8]. The difficulty is the fact that the joint generating functions of the numbers of customers in the orbits are the solution of a system of partial differential equations. In this paper, we consider the system under an asymptotic regime of slow retrials. First, we consider obtain the first order asymptotic result that the scaled numbers of customers in the orbits converge to the constants having clear physical meaning. Next, we obtain the second order asymptotic result which states that the joint distribution of the centered numbers of customers in the orbits converges to a Gaussian distribution with explicit mean and covariance matrix.

The rest of our paper is organized as follows. Section 2 presents the model and problem formulation. Section 3 show the detailed analysis of the first and second order asymptotics. Section 4 presents some numerical examples while concluding remarks are presented in Sect. 5.

2 Model Description and Problem Statement

We consider a multiclass retrial queueing system. Let N be the number of classes of incoming customers. Customers of each class arrive from outside the system according to a Poisson process with a rates $\lambda_n, n = \overline{1, N}$. If an arriving customer finds the server free, the customer occupies the server and gets a service. The service times for each class of customers are assumed to be exponentially distributed with service rates $\mu_n, n = \overline{1, N}$ depending on the class. If the server is busy incoming customers join the orbit according to their type and make a delay for an exponentially distributed time with rate $\sigma_n, n = \overline{1, N}$ then repeat their request for service.

Let $i_n(t), n = \overline{1, N}$ be the random processes of the numbers of customers in the orbits. We denote in vector notation as $\mathbf{i}(t) = [i_1(t) \dots i_N(t)]$. The aim of the current research is to derive the stationary probability distribution of this vector process. Let $k(t)$ be the random process that defines the server states: 0

if the server is free, n if the server is busy serving an incoming call of n-th type, $n = \overline{1, N}$.

The process $\mathbf{i}(t)$ is not Markovian, therefore we consider the $(N+1)$-dimensional continuous time Markov chain $\{k(t), \mathbf{i}(t)\}$.

Denoting $P_k(\mathbf{i}, t) = P\{k(t) = k, i_1(t) = i_1, \ldots, i_N(t) = i_N\}, k = \overline{0, N}$ it is possible to write down the following equalities

$$P_0(\mathbf{i}, t + \Delta t) = P_0(\mathbf{i}, t) \prod_{m=1}^{N} (1 - \lambda_m \Delta t)(1 - i_m \sigma_m \Delta t) + \sum_{m=1}^{N} P_m(\mathbf{i}, t) \mu_m \Delta t + o(\Delta t),$$

$$P_n(\mathbf{i}, t + \Delta t) = P_n(\mathbf{i}, t)(1 - \mu_n \Delta t) \prod_{m=1}^{N} (1 - \lambda_m \Delta t) + P_0(\mathbf{i}, t) \lambda_n \Delta t$$

$$+ P_0(\mathbf{i} + \mathbf{e}_n, t)(i_n + 1)\sigma_n \Delta t + \sum_{\nu=1}^{N} P_n(\mathbf{i} - \mathbf{e}_\nu, t) \lambda_\nu \Delta t + o(\Delta t), n = \overline{1, N}.$$

Here $\mathbf{e}_n$ is the vector whose n-th component is equal to unity, and the rest are zero.

We will consider the system in a steady state regime under the stability condition [9]:

$$\sum_{m=1}^{N} \frac{\lambda_m}{\mu_m} < 1.$$

We denote $P_k(\mathbf{i}) = \lim_{t \to \infty} P_k(\mathbf{i}, t)$ the stationary probability distribution of the system states $\{k(t), \mathbf{i}(t)\}$.

Let us write the system of equations for the probability distribution

$$\{P_0(\mathbf{i}), P_1(\mathbf{i}), \ldots, P_N(\mathbf{i})\}, \mathbf{i} \geq 0,$$

using equalities the above:

$$P_0(\mathbf{i}) \sum_{m=1}^{N} (-\lambda_m - i_m \sigma_m) + \sum_{m=1}^{N} P_m(\mathbf{i}) \mu_m = 0,$$

$$- P_n(\mathbf{i}) \left(\mu_n + \sum_{m=1}^{N} \lambda_m \right) + P_0(\mathbf{i}) \lambda_n + P_0(\mathbf{i} + \mathbf{e}_n)(i_n + 1)\sigma_n \tag{1}$$

$$+ \sum_{\nu=1}^{N} P_n(\mathbf{i} - \mathbf{e}_\nu) \lambda_\nu = 0, n = \overline{1, N}.$$

Here it is assumed that $P_k(\mathbf{i}) = 0, k = \overline{0, N}$, if at least one component of the vector $\mathbf{i}$ is negative.

Let us introduce the multidimensional partial characteristic functions

$$H_k(\mathbf{u}) = \sum_{i_1=0}^{\infty} \cdots \sum_{i_N=0}^{\infty} P_k(i_1, \ldots, i_N) \exp\left\{ j \sum_{m=1}^{N} u_m i_m \right\}$$

$$= \sum_{\mathbf{i}=0}^{\infty} e^{j\mathbf{u}^T \mathbf{i}} P_k(\mathbf{i}), k = \overline{0, N}, \tag{2}$$

where $j = \sqrt{-1}$ is an imaginary unit, $\mathbf{u}$ – vector with components $u_n, n = \overline{1,N}$.
Substituting functions (2) into (1), the following system of equations is obtained.

$$-H_0(\mathbf{u})\sum_{m=1}^{N}\lambda_m + j\sum_{m=1}^{N}\frac{\partial H_0(\mathbf{u})}{\partial u_m}\sigma_m + \sum_{m=1}^{N}H_m(\mathbf{u})\mu_m = 0,$$
$$-H_n(\mathbf{u})\left(\mu_n + \sum_{m=1}^{N}\lambda_m\right) + H_0(\mathbf{u})\lambda_n - j\sigma_n e^{-ju_n}\frac{\partial H_0(\mathbf{u})}{\partial u_n} \tag{3}$$
$$+\sum_{m=1}^{N}H_n(\mathbf{u})\lambda_m e^{ju_m} = 0, n = \overline{1,N}.$$

3 Asymptotic Analysis Under the Long Delay Condition

Denote

$$\sigma_n = \sigma\gamma_n, n = \overline{1,N}.$$

The main idea of this paper is to find the solution of system (3) by using an asymptotic analysis method under the limit condition of the long delay customers in the orbits, i.e., when $\sigma \to 0$.

3.1 Asymptotic of the First-Order

We make the following substitutions in the system (3):

$$\sigma = \epsilon, \mathbf{u} = \epsilon\mathbf{w}, H_k(\mathbf{u}) = F_k(\mathbf{w},\epsilon), k = \overline{0,N}.$$

As the result, we get the following equations:

$$-F_0(\mathbf{w},\epsilon)\sum_{m=1}^{N}\lambda_m + j\sum_{m=1}^{N}\frac{\partial F_0(\mathbf{w},\epsilon)}{\partial w_m}\gamma_m + \sum_{m=1}^{N}F_m(\mathbf{w},\epsilon)\mu_m = 0,$$
$$-F_n(\mathbf{w},\epsilon)\left(\mu_n + \sum_{m=1}^{N}\lambda_m\right) + F_0(\mathbf{w},\epsilon)\lambda_n - j\gamma_n e^{-j\epsilon w_n}\frac{\partial F_0(\mathbf{w},\epsilon)}{\partial w_n} \tag{4}$$
$$+\sum_{m=1}^{N}F_n(\mathbf{w},\epsilon)\lambda_m e^{j\epsilon w_m} = 0, n = \overline{1,N}.$$

Denoting the asymptotic solution of the system of Eqs. (4) in the form $F_k(\mathbf{w}) = \lim_{\epsilon\to 0} F_k(\mathbf{w},\epsilon), k = \overline{0,N}$, we obtain solution named as "first-order asymptotic". We prove the following theorem.

Theorem 1. *The first-order asymptotic characteristic function of the probability distribution of the numbers of customers in the orbits has the form:*

$$F_k(\mathbf{w}) = R_k \exp\left\{\sum_{m=1}^{N} jw_m x_m\right\}, k = \overline{0,N},$$

where parameter

$$R_n = \frac{\lambda_n}{\mu_n}, n = \overline{1,N}, R_0 = 1 - \sum_{m=1}^{N} \frac{\lambda_m}{\mu_m} \tag{5}$$

is the stationary probability distribution of the state server ($\mathbf{R} = \{R_k\}, k = \overline{0,N}$ *in matrix form),*

$$x_n = \frac{\lambda_n}{\gamma_n} \frac{1 - R_0}{R_0}, n = \overline{1,N}. \tag{6}$$

Proof. In system (4), we take the limit as $\epsilon \to 0$. Then, we get the system of equations:

$$\begin{aligned} -F_0(\mathbf{w})\sum_{m=1}^{N} \lambda_m + j\sum_{m=1}^{N} \frac{\partial F_0(\mathbf{w})}{\partial w_m}\gamma_m + \sum_{m=1}^{N} F_m(\mathbf{w})\mu_m = 0, \\ -F_n(\mathbf{w})\mu_n + F_0(\mathbf{w})\lambda_n - j\gamma_n\frac{\partial F_0(\mathbf{w})}{\partial w_n} = 0, n = \overline{1,N}. \end{aligned} \tag{7}$$

We will look for a solution the above system of equations in the following form

$$F_k(\mathbf{w}) = R_k\Phi(\mathbf{w}), k = \overline{0,N}. \tag{8}$$

Substituting (8) into (7) and multiplying the equations of the system by $\frac{1}{\Phi(\mathbf{w})}$, we derive equations:

$$\begin{aligned} -R_0\sum_{m=1}^{N} \lambda_m + jR_0\sum_{m=1}^{N} \frac{\partial\Phi(\mathbf{w})/\partial w_m}{\Phi(\mathbf{w})}\gamma_m + \sum_{m=1}^{N} R_m\mu_m = 0, \\ -R_n\mu_n + R_0\lambda_n - j\gamma_n R_0\frac{\partial\Phi(\mathbf{w})/\partial w_n}{\Phi(\mathbf{w})} = 0, n = \overline{1,N}. \end{aligned} \tag{9}$$

The solution of Eq. (9) is as follows:

$$\Phi(\mathbf{w}) = \exp\left\{\sum_{m=1}^{N} jw_m x_m\right\}. \tag{10}$$

Substituting this expression into the system (9) yields

$$\begin{aligned} -R_0\sum_{m=1}^{N} \lambda_m - R_0\sum_{m=1}^{N} \gamma_m x_m + \sum_{m=1}^{N} R_m\mu_m = 0, \\ -R_n\mu_n + R_0\lambda_n + \gamma_n x_n R_0 = 0, n = \overline{1,N}. \end{aligned}$$

We express R_n from the second equation of system and get relation

$$R_n = \frac{\lambda_n + \gamma_n x_n}{\mu_n} R_0. \tag{11}$$

We sum the equations of the system (4) in order to get the equation

$$\sum_{m=1}^{N}\lambda_m(e^{j\epsilon w_m}-1)\sum_{m=1}^{N}F_m(\mathbf{w},\epsilon)+j\sum_{m=1}^{N}\frac{\partial F_0(\mathbf{w},\epsilon)}{\partial w_m}(1-e^{-j\epsilon w_m})\gamma_m=0. \quad (12)$$

Let us use expansion

$$e^{j\epsilon w_m}=1+j\epsilon w_m+o(\epsilon)$$

where $o(\epsilon)$ is an infinitesimal of the order greater than ϵ, in Eq. (12). Dividing these equations by $j\epsilon$ and making the transition $\epsilon\to 0$, one obtains the following equation:

$$\sum_{m=1}^{N}\lambda_m w_m\sum_{m=1}^{N}F_m(\mathbf{w})+j\sum_{m=1}^{N}\frac{\partial F_0(\mathbf{w})}{\partial w_m}w_m\gamma_m=0. \quad (13)$$

Substituting (8) and (10) to (13), we have

$$\sum_{m=1}^{N}\lambda_m w_m\sum_{m=1}^{N}R_m-R_0\sum_{m=1}^{N}w_m\gamma_m x_m=0.$$

Using condition of standardization $\sum_{m=0}^{N}R_m=1$, we obtain

$$\sum_{m=1}^{N}\lambda_m w_m-R_0\sum_{m=1}^{N}(\lambda_m+\gamma_m x_m)w_m=0. \quad (14)$$

After some transformations, one obtains the following equation:

$$\lambda_m-R_0(\lambda_m+\gamma_m x_m)=0.$$

We obtain an expressions for $x_n, n=\overline{1,N}$, which coincide with (6).

Using this expression, we can write

$$R_0=\frac{\lambda_m}{\lambda_m+\gamma_m x_m}.$$

Substituting this expression in formula (11), one obtains expression (5).

The values x_n represent the average values of the numbers of customers in the orbits normalized by the value σ.

3.2 Asymptotic of the Second-Order

In the system (3) let us denote

$$H_k(\mathbf{u})=H_k^{(2)}(\mathbf{u})\exp\left\{\sum_{m=1}^{N}j\frac{u_m}{\sigma_m}\gamma_m x_m\right\}, k=\overline{0,N}. \quad (15)$$

The functions $H_k^{(2)}(\mathbf{u})$ are the partial characteristic functions of the values of centered random processes $i_m(t) - \frac{x_m}{\sqrt{\sigma}}$. Substituting

$$\sigma_n = \sigma\gamma_n, \sigma = \epsilon^2, \mathbf{u} = \epsilon\mathbf{w}, H_k^{(2)}(\mathbf{u}) = F_k^{(2)}(\mathbf{w}, \epsilon), k = \overline{0, N}. \tag{16}$$

and expression (15) into the system (3) we get:

$$-F_0^{(2)}(\mathbf{w}, \epsilon)\sum_{m=1}^{N}(\lambda_m + \gamma_m x_m) + j\epsilon\sum_{m=1}^{N}\frac{\partial F_0^{(2)}(\mathbf{w}, \epsilon)}{\partial w_m}\gamma_m + \sum_{m=1}^{N}F_m^{(2)}(\mathbf{w}, \epsilon)\mu_m = 0,$$
$$-F_n^{(2)}(\mathbf{w}, \epsilon)\left(\mu_n + \sum_{m=1}^{N}\lambda_m(1 - e^{j\epsilon w_m})\right) + F_0^{(2)}(\mathbf{w}, \epsilon)(\lambda_n + \gamma_n x_n e^{-j\epsilon w_n})$$
$$-j\epsilon\gamma_n e^{-j\epsilon w_n}\frac{\partial F_0^{(2)}(\mathbf{w}, \epsilon)}{\partial w_n} = 0, n = \overline{1, N}. \tag{17}$$

Denoting the asymptotic solution of the system of Eqs. (17) in the form $F_k^{(2)}(\mathbf{w}) = \lim_{\epsilon\to 0} F_k^{(2)}(\mathbf{w}, \epsilon), k = \overline{0, N}$, we obtain this solution, named as "second-order asymptotic". We prove the following theorem.

Theorem 2. *The second-order asymptotic characteristic function of the probability distribution of the number of customers in the orbits has the form:*

$$F_k^{(2)}(\mathbf{w}) = R_k \exp\left\{-\frac{1}{2}\sum_{\nu=1}^{N}\sum_{m=1}^{N}w_\nu K_{\nu m}w_m\right\}, k = \overline{0, N}, \tag{18}$$

where parameters $K_{\nu m}$ are the solution of the following system:

$$\gamma_m R_0 K_{mm} - \lambda_m R_0\sum_{l=1}^{N}\frac{\gamma_l}{\mu_l}K_{lm} = \lambda_m(1 - R_0)(1 - R_m) + \lambda_m^2\sum_{l=1}^{N}\frac{R_l}{\mu_l}, \nu = m,$$
$$\gamma_m R_0 K_{m\nu} + \gamma_\nu R_0 K_{\nu m} - \lambda_m R_0\sum_{l=1}^{N}\frac{\gamma_l}{\mu_l}K_{l\nu} - \lambda_\nu R_0\sum_{l=1}^{N}\frac{\gamma_l}{\mu_l}K_{lm}$$
$$= 2\lambda_m\lambda_\nu\sum_{l=1}^{N}\frac{R_l}{\mu_l} - (R_m\lambda_\nu + R_\nu\lambda_m)(1 - R_0), \nu \neq m. \tag{19}$$

Proof. We will look for a solution of (17) in the following form:

$$F_k^{(2)}(\mathbf{w}, \epsilon) = \Phi_2(\mathbf{w})\left(R_k + \sum_{m=1}^{N}j\epsilon w_m f_{km} + o(\epsilon)\right). \tag{20}$$

Substituting (20) and the expansion $e^{j\epsilon w_m} = 1 + j\epsilon w_m + o(\epsilon)$ into (17), we obtain

$$-\Phi_2(\mathbf{w})\left(R_0 + \sum_{m=1}^{N} j\epsilon w_m f_{0m}\right)\sum_{m=1}^{N}(\lambda_m + \gamma_m x_m) + \Phi_2(\mathbf{w})\sum_{m=1}^{N}\mu_m\sum_{\nu=1}^{N} j\epsilon w_\nu f_{m\nu}$$
$$+\Phi_2(\mathbf{w})\sum_{m=1}^{N}\mu_m R_m + j\epsilon\sum_{m=1}^{N}\gamma_m(R_0 + j\epsilon\sum_{\nu=1}^{N} w_\nu f_{0\nu})\frac{\partial\Phi_2(\mathbf{w})}{\partial w_m} + o(\epsilon) = 0,$$
$$\Phi_2(\mathbf{w})\left(R_n + \sum_{\nu=1}^{N} j\epsilon w_\nu f_{n\nu}\right)\left(j\epsilon\sum_{m=1}^{N}\lambda_m w_m - \mu_n\right) + \Phi_2(\mathbf{w})R_0(\lambda_n + \gamma_n x_n)$$
$$-j\epsilon\Phi_2(\mathbf{w})R_0 w_n\gamma_n x_n + \Phi_2(\mathbf{w})(\lambda_n + \gamma_n x_n)\sum_{\nu=1}^{N} w_\nu f_{0\nu}$$
$$-j\epsilon\gamma_n R_0\frac{\partial\Phi_2(\mathbf{w})}{\partial w_n} + o(\epsilon) = 0, n = \overline{1,N}.$$

Using equality (11) and multiplying the above equation by $\dfrac{1}{j\epsilon\Phi_2(\mathbf{w})}$, we obtain

$$-\sum_{m=1}^{N}(\lambda_m + \gamma_m x_m)\sum_{\nu=1}^{N} w_\nu f_{0\nu} + \sum_{m=1}^{N}\mu_m\sum_{\nu=1}^{N} w_\nu f_{m\nu} + \sum_{m=1}^{N}\gamma_m R_0\frac{\partial\Phi_2(\mathbf{w})/\partial w_n}{\Phi_2(\mathbf{w})} = 0,$$
$$R_n\sum_{\nu=1}^{N}\lambda_\nu w_\nu - \mu_n\sum_{\nu=1}^{N} w_\nu f_{n\nu} - R_0\gamma_n x_n w_n \tag{21}$$
$$+(\lambda_n + \gamma_n x_n)\sum_{\nu=1}^{N} w_\nu f_{0\nu} - \gamma_n R_0\frac{\partial\Phi_2(\mathbf{w})/\partial w_n}{\Phi_2(\mathbf{w})} = 0, n = \overline{1,N}.$$

The solution of Eq. (21) is as follows:

$$\Phi_2(\mathbf{w}) = \exp\left\{-\frac{1}{2}\sum_{m=1}^{N}\sum_{\nu=1}^{N} w_m K_{m\nu} w_\nu\right\}, \tag{22}$$

where the quantities $K_{m\nu}$ are elements of the covariance matrix $\mathbf{K} = \{K_{m\nu}\}$.
Substituting (22) into (21), we rewrite (21) as follows:

$$\sum_{m=1}^{N}(\lambda_m + \gamma_m x_m)\sum_{\nu=1}^{N} w_\nu f_{0\nu} - \sum_{m=1}^{N}\mu_m\sum_{\nu=1}^{N} w_\nu f_{m\nu} + \sum_{m=1}^{N}\gamma_m R_0\sum_{\nu=1}^{N} w_\nu K_{m\nu} = 0,$$
$$R_n\sum_{\nu=1}^{N}\lambda_\nu w_\nu - \mu_n\sum_{\nu=1}^{N} w_\nu f_{n\nu} - R_0\gamma_n x_n w_n + (\lambda_n + \gamma_n x_n)\sum_{\nu=1}^{N} w_\nu f_{0\nu} \tag{23}$$
$$+\gamma_n R_0\sum_{\nu=1}^{N} w_\nu K_{n\nu} = 0, n = \overline{1,N}.$$

From the system (23) it follows that we can write the following equalities:

$$\begin{aligned} R_n\lambda_n - R_0\gamma_n x_n + R_0\gamma_n K_{nn} - \mu_n f_{nn} + (\lambda_n + \gamma_n x_n) f_{0n} = 0, \nu = n, \\ R_n\lambda_\nu + R_0\gamma_n K_{n\nu} - \mu_n f_{n\nu} + (\lambda_n + \gamma_n x_n) f_{0\nu} = 0, \nu \neq n. \end{aligned} \tag{24}$$

We then rewrite (24) as:

$$\begin{aligned} f_{nn} = R_n^2 - R_n(1 - R_0) + \frac{\gamma_n R_0}{\mu_n} K_{nn} + \frac{R_n}{R_0} f_{0n}, \nu = n, \\ f_{n\nu} = R_n \frac{\lambda_\nu}{\mu_n} + \frac{\gamma_n R_0}{\mu_n} K_{n\nu} + \frac{R_n}{R_0} f_{0\nu}, \nu \neq n. \end{aligned} \tag{25}$$

Summing equations of the system (17) and using expansion $e^{j\epsilon w_m} = 1 + j\epsilon w_m + \frac{(j\epsilon w_m)^2}{2} + o(\epsilon^2)$, we get the relation

$$j\epsilon \sum_{m=1}^{N} \lambda_m \left(w_m + \frac{j\epsilon w_m^2}{2} \right) \sum_{m=1}^{N} F_m(\mathbf{w}, \epsilon) + (j\epsilon)^2 \sum_{m=1}^{N} \gamma_m w_m \frac{\partial F_0(\mathbf{w}, \epsilon)}{\partial w_m}$$
$$- j\epsilon \sum_{m=1}^{N} \gamma_m x_m \left(w_m - \frac{j\epsilon w_m^2}{2} \right) F_0(\mathbf{w}, \epsilon) + o(\epsilon^2) = 0.$$

Substituting in the above equation expansion (20) and dividing each part of this equation by $j\epsilon$, we obtain the following equation for the function $\Phi_2(\mathbf{w})$

$$\sum_{m=1}^{N} \lambda_m \left(w_m + \frac{j\epsilon w_m^2}{2} \right) \sum_{m=1}^{N} \left(R_m + j\epsilon \sum_{\nu=1}^{N} w_\nu f_{m\nu} \right) - \sum_{m=1}^{N} \gamma_m x_m \left(w_m - \frac{j\epsilon w_m^2}{2} \right) R_0$$
$$- \sum_{m=1}^{N} \gamma_m x_m \left(w_m - \frac{j\epsilon w_m^2}{2} \right) j\epsilon \sum_{\nu=1}^{N} w_\nu f_{0\nu}$$
$$+ j\epsilon \sum_{m=1}^{N} \gamma_m w_m R_0 \frac{\partial \Phi_2(\mathbf{w}) / \partial w_m}{\Phi_2(\mathbf{w})} + o(\epsilon) = 0.$$

Using Eq. (14), we divide both sides of it by $j\epsilon$ and take the limit $\epsilon \to 0$. We then have:

$$\sum_{m=1}^{N} \lambda_m (1 - R_0) w_m^2 + \sum_{m=1}^{N} \gamma_m R_0 w_m \frac{\partial \Phi_2(\mathbf{w}) / \partial w_m}{\Phi_2(\mathbf{w})}$$
$$= \sum_{m=1}^{N} \gamma_m x_m w_m \sum_{\nu=1}^{N} w_\nu f_{0\nu} - \sum_{m=1}^{N} \lambda_m w_m \sum_{l=1}^{N} \sum_{\nu=1}^{N} w_\nu f_{l\nu}.$$

Substituting (22), we have

$$\sum_{m=1}^{N}\lambda_m(1-R_0)w_m^2-\sum_{m=1}^{N}\gamma_m R_0 w_m\sum_{\nu=1}^{N}K_{m\nu}w_\nu$$
$$=\sum_{m=1}^{N}\gamma_m x_m w_m\sum_{\nu=1}^{N}w_\nu f_{0\nu}-\sum_{m=1}^{N}\lambda_m w_m\sum_{l=1}^{N}\sum_{\nu=1}^{N}w_\nu f_{l\nu}.$$

We get the system of equations for $K_{m\nu}, m=\overline{1,N}, \nu=\overline{1,N}$. Considering the above equation

$$\lambda_m(1-R_0)-\gamma_m R_0 K_{mm}=\gamma_m x_m f_{0m}-\lambda_m\sum_{l=1}^{N}f_{lm}, \nu=m,$$

$$-R_0(\gamma_m K_{m\nu}+\gamma_\nu K_{\nu m})=-\lambda_m\sum_{l=1}^{N}f_{l\nu}-\lambda_\nu\sum_{l=1}^{N}f_{lm}+\gamma_m x_m f_{0\nu}+\gamma_\nu x_\nu f_{0m}, \nu\neq m$$

and then, substituting (25), it can be rewritten as

$$\gamma_m R_0 K_{mm}-\lambda_m R_0\sum_{l=1}^{N}\frac{\gamma_l}{\mu_l}K_{lm}=\lambda_m(1-R_0)(1-R_m)+\lambda_m^2\sum_{l=1}^{N}\frac{R_l}{\mu_l}, \nu=m,$$

$$\gamma_m R_0 K_{m\nu}+\gamma_\nu R_0 K_{\nu m}-\lambda_m R_0\sum_{l=1}^{N}\frac{\gamma_l}{\mu_l}K_{l\nu}-\lambda_\nu R_0\sum_{l=1}^{N}\frac{\gamma_l}{\mu_l}K_{lm}$$
$$=2\lambda_m\lambda_\nu\sum_{l=1}^{N}\frac{R_l}{\mu_l}-(R_m\lambda_\nu+R_\nu\lambda_m)(1-R_0), \nu\neq m.$$

Thus, the proof is completed.

Replacing (18) to (16) and (15), we can write expression for approximation the partial characteristic function at small values of σ:

$$H_k(\mathbf{u})\approx R_k\exp\left\{j\sum_{m=1}^{N}\frac{u_m}{\sigma}x_m-\frac{1}{2}\sum_{m=1}^{N}\sum_{\nu=1}^{N}\frac{u_m}{\sqrt{\sigma}}K_{m\nu}\frac{u_\nu}{\sqrt{\sigma}}\right\}, k=\overline{0,N}.$$

Summing up all values $k=\overline{0,N}$, we obtain an approximation of the characteristic function of probability distribution of number customers in the orbits

$$H(\mathbf{u})\approx\exp\left\{j\sum_{m=1}^{N}\frac{u_m}{\sigma}x_m-\frac{1}{2}\sum_{m=1}^{N}\sum_{\nu=1}^{N}\frac{u_m}{\sqrt{\sigma}}K_{m\nu}\frac{u_\nu}{\sqrt{\sigma}}\right\}.$$

Thus, the number of customers in the orbits in the multiclass retrial queueing system is asymptotically Gaussian.

Table 1. The model of parameters

The rate of arrival flow, λ_n	The service rate, μ_n	The rate of a delay in the orbit, $\sigma_n = \sigma\gamma_n$
$\lambda_1 = 0.7$	$\mu_1 = 2$	$\sigma_1 = 0.01$
$\lambda_2 = 0.6$	$\mu_2 = 3$	$\sigma_2 = 0.02$
$\lambda_3 = 0.5$	$\mu_3 = 4$	$\sigma_3 = 0.03$

4 Example

We consider particular: $n = 3$ (Table 1).

Given these values row-vector

$$\mathbf{x} = \begin{bmatrix} x_1 \; x_2 \; x_3 \end{bmatrix}$$

and row-vector

$$\mathbf{R} = \begin{bmatrix} R_0 \; R_1 \; R_2 \; R_3 \end{bmatrix}$$

are defined as:

$$\mathbf{x} = \begin{bmatrix} 1.454 \; 0.623 \; 0.346 \end{bmatrix}, \mathbf{R} = \begin{bmatrix} 0.325 \; 0.35 \; 0.2 \; 0.125 \end{bmatrix}.$$

Finally, we specify the matrix covariance

$$\mathbf{K} = \begin{bmatrix} 2.963 \; 0.739 \; 0.428 \\ 0.739 \; 1.018 \; 0.241 \\ 0.428 \; 0.241 \; 0.497 \end{bmatrix}.$$

Using these parameters, we find mean κ_1 and variance κ_2 of total number of customers in the orbits:

$$\kappa_1 = \frac{x_1 + x_2 + x_3}{\sigma}, \; \kappa_2 = \frac{K_{11} + K_{22} + K_{33} + 2K_{12} + 2K_{13} + 2K_{23}}{\sigma}$$

for $\sigma = 0.01$. Let us denote normal distribution function with moments κ_1 and κ_2 by $F(x)$, $P(i)$ be discrete distribution of nonnegative quantity which is defined by

$$P(i) = (F(i + 0.5) - F(i - 0.5))(1 - F(-0.5))^{-1}, i \geq 0.$$

The graph of asymptotic probability distribution $P(i)$ of total number of customers in the orbits is given in Fig. 1.

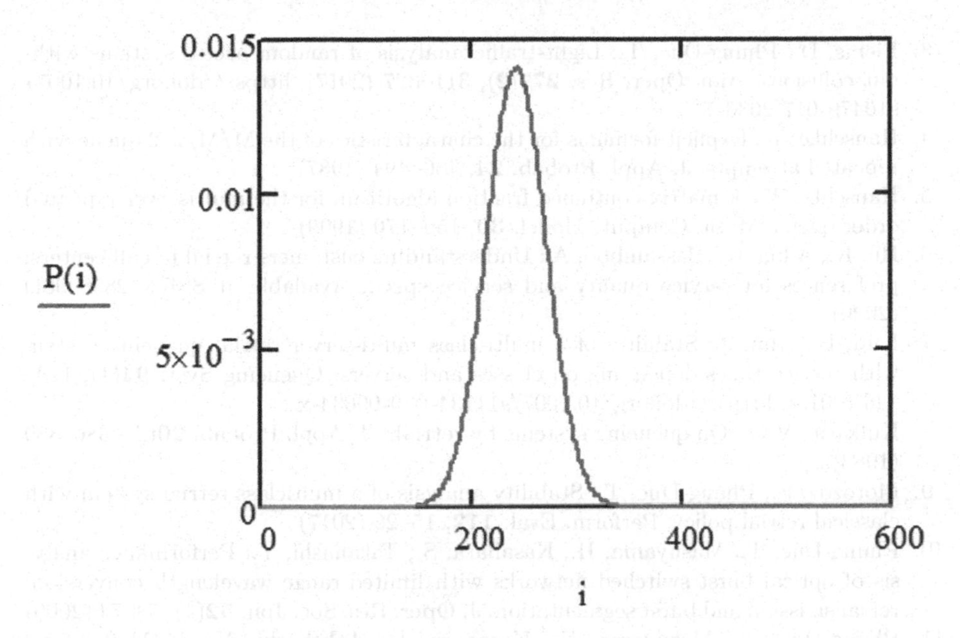

Fig. 1. Graph of distribution $P(i)$

5 Conclusion

In this paper, we considered a multiclass retrial queueing system. Equations for characteristic functions of the multi-dimensional probability distribution of the numbers of customers in the orbits are obtained. We then used the method of asymptotic analysis under condition of a long delay customers in the orbits to find the limiting probability distribution of the number of the customers in the orbits. This probability distribution turned out to be Gaussian. We as well derived the expressions for mean and stationary probability distribution of state server. Equations are obtained for finding the elements of the covariance matrix. In particular, we considered case $n = 3$. We obtained the values of mean and variance of total number of customers in the orbits. Graph for the probability distribution of the total number of customers in the orbits is given.

In the future it is planned to research a multiclass retrial queueing system in which service times of customers in a class follow an arbitrary distribution.

References

1. Artalejo, J.R., Gómez-Corral, A.: Retrial Queueing Systems. Springer, Heidelberg (2008). https://doi.org/10.1007/978-3-540-78725-9
2. Falin, G., Templeton, J.G.: Retrial Queues. CRC Press, Boca Raton (1997)

3. Fiems, D., Phung-Duc, T.: Light-traffic analysis of random access systems without collisions. Ann. Oper. Res. **277**(2), 311–327 (2017). https://doi.org/10.1007/s10479-017-2636-7
4. Hanschke, T.: Explicit formulas for the characteristics of the M/M/2/2 queue with repeated attempts. J. Appl. Probab. **24**, 486–494 (1987)
5. Hanschke, T.: A matrix continued fraction algorithm for the multiserver repeated order queue. Math. Comput. Model. **30**, 159–170 (1999)
6. Hu, K., Allon, G., Bassamboo, A.: Understanding customers retrial in call centers: preferences for service quality and service speed. Available at SSRN 2838998D (2020)
7. Kim, B., Kim, J.: Stability of a multi-class multi-server retrial queueing system with service times depending on classes and servers. Queueing Syst. **94**(1), 129–146 (2019). https://doi.org/10.1007/s11134-019-09634-x
8. Kulkarni, V.G.: On queueing systems by retrials. J. Appl. Probab. **20**(2), 380–389 (1983)
9. Morozov, E., Phung-Duc, T.: Stability analysis of a multiclass retrial system with classical retrial policy. Perform. Eval. **112**, 15–26 (2017)
10. Phung-Duc, T., Masuyama, H., Kasahara, S., Takahashi, Y.: Performance analysis of optical burst switched networks with limited-range wavelength conversion, retransmission and burst segmentation. J. Oper. Res. Soc. Jpn. **52**(1), 58–74 (2009)
11. Phung-Duc, T., Masuyama, H., Kasahara, S., Takahashi, Y.: M/M/3/3 and M/M/4/4 retrial queues. J. Ind. Manag. Optim. **5**(3), 431–451 (2009)
12. Phung-Duc, T., Masuyama, H., Kasahara, S., Takahashi, Y.: State-dependent M/M/c/c+ r retrial queues with Bernoulli abandonment. J. Ind. Manage. Optim. **6**(3), 517–540 (2010)
13. Phung-Duc, T., Masuyama, H., Kasahara, S., Takahashi, Y.: A matrix continued fraction approach to multiserver retrial queues. Ann. Oper. Res. **202**(1), 161–183 (2013)
14. Phung-Duc, T.: Retrial Queueing Models: A Survey on Theory and Applications, to appear in by Dohi, T., Ano, K., Kasahara, S. (eds.) Stochastic Operations Research in Business and Industry. World Scientific Publisher, Singapore (2017)
15. Tran-Gia, P., Mandjes, M.: Modeling of customer retrial phenomenon in cellular mobile networks. IEEE J. Sel. Areas Commun. **15**(8), 1406–1414 (1997)

A Retrial Queueing System in Which Server Searches to Accumulate Customers for Optimal Bulk Serving

Varghese C. Joshua, Ambily P. Mathew(✉), and Achyutha Krishnamoorthy

Department of Mathematics, CMS College Kottayam, Kottayam, Kerala, India
{vcjoshua,ambilypm,krishnamoorthy}@cmscollege.ac.in

Abstract. In this paper, we introduce the concept of orbital search as a means for accumulating customers in a bulk service retrial queueing system. In this model, customers enter the service facility from a finite buffer. If the buffer is full at the time of arrival of a customer, it enters an orbit from where retrials for entering the buffer has been made. Here search is done as a means for accumulating customers in the buffer so that optimum level of bulk service can be provided. A service policy considered here is (a, b) bulk service policy with search. Under this policy, search will be initiated when the number of customers in the buffer reaches a and it has been continued until either the search has been done for a random duration of time or the number of customers in the buffer increases to b where $a < b$. Steady state analysis of the model has been done and some important measures of performance has been evaluated. We analyze the model numerically as well as graphically.

Keywords: Retrial queue · Orbit · Bulk service · Buffer · Search

1 Introduction

It is quite common that we encounter many queueing situations in which the service is provided in batches of varying size or in bulk. Optimum level of bulk service will surely improve the efficiency as well as the cost effectiveness of the queueing models arising out of such situations. In this paper, we consider a retrial queueing system a finite buffer, an infinite orbit, and a server providing bulk service under (a, b) bulk service policy with search. Under this policy search has been done as a means for accumulating customers for optimum bulk service. Even if the server can start service with a minimum of a number of customers, it will not do so. It looks for an optimal level of bulk service and a is only a minimum threshold to start service. Fast accumulation of customers in the buffer occurs as a result of search. This will surely improve the cost effectiveness of the model. A detailed description about (a, b) bulk service policy will be given the next section.

There are a large number of literature found in connection with bulk service retrial queues with orbital search. The concept of search by server introduced

V. M. Vishnevskiy et al. (Eds.): DCCN 2020, LNCS 12563, pp. 311–321, 2020.
https://doi.org/10.1007/978-3-030-66471-8_24

by Neuts and Ramalhoto in the paper [13] has been extended to more general cases. In the case of M/G/1 queues with retrials, search for orbital customers was introduced by Artalejo et al. in [1]. Chakravarthy et al. analyzed multi server queues with search of customers from the orbit in [4]. Krishnamoorthy et al. in [9] incorporated non persistency of customers in $M/G/1$ retrial queues with orbital search. Analysis of multi server queues with orbital search was done by Chakravarthy et al. in [4]. More literature related to orbital search can be found in [6–8] and [10]. Two types of searches and different types of services to primary/orbital customers (retrial/type I/type II searches) retaining the assumption that the arrival process is Markovian Arrival Process or batch Markovian process (BMAP) is considered in [7]. Chakravarthy[3], provides a review of queueing models with batch Markovian arrivals.

Bulk service may be given in batches of fixed size or varying size and with varying rates. The (L, K) policy of bulk service was first introduced by Marcel F. Neuts in [12]. Under this policy a minimum of L customers is needed to start each service. In (L, K) policy, the maximum size to be served at a time is restricted to K where $L \leq K$. A generalization of this (L, K) policy with L and K are random variables on lattice points (a, b) with $a \leq b$ is considered in [12]. Here the service times of individual customers are assumed to be independent and identically distributed random variables and the server accepts only as many customers as to satisfy the condition that their total service time lies in between given upper and lower bounds. In [2] Chakravarthy analyzed a single server finite queue with group service. A survey of queueing models with bulk service is given in [14]. In retrial queues some literature dealing with bulk service can be found. In [5], Chakravarthy et al. consider a retrial queue with two types of customers, type-1 and type-2. Among the two servers considered in [5], one is dedicated to give group service to type-1 customers and the other servers both type-1 and type-2 but one at a time.

The motivation for the present model comes from a common business strategy which can be adopted by the travel agencies and tour operators. Requests for operating travels to some specific locations has been arriving at random intervals of time. A journey is planned when a threshold number of requests are obtained. The operator goes in search of more customers even if the schedule is fixed. This is for accumulating more customers through search before the service starts. The operators or the agencies get more benefits from such accumulated customers. For search, they may adopt various strategies like providing some additional packages, discounts etc. But addition of such accumulated customers may provide huge rewards when compared to the search costs. In such situations the optimum level of bulk service can be determined.

2 Model Description

We consider a single server retrial queueing system with an infinite orbit and a finite buffer. The finite buffer can hold maximum of N customers. Customers arrive according to a Poisson process with parameter λ. Any customer who upon arrival finding the buffer full, enters an orbit. The orbit is considered to be of infinite capacity. Retrials from the orbit to enter the buffer occur at exponentially distributed time intervals and let μ be the rate at which these retrials occur. The service policy considered in this model is (a, b) bulk service policy with search. It may be described as follows:

We assume that a and b are any integers such that $a < b$ and $b < N$. Under this policy the server can provide bulk service to any number of customers in the buffer if the number lies in between a and b including both a and b. If there are b or more customers waiting in the system at the time of a service completion, the service of the next batch starts immediately. If there are less than a customers waiting in the buffer at the time of a departure, the server waits until the buffer size reaches a. Buffer size increases as a result of primary arrivals as well as retrials from the orbit. If the buffer accumulates a customers, the server is ready for service, but wait for some random time and this time is utilized for search to accumulate more customers. As a result of search, customers enter the buffer from the orbit at exponentially distributed time intervals. A search clock starts at the moment the search begins. The search clock operates for a time interval which is exponentially distributed. Let ζ be the exponential rate at which the random clock expires. The server will go in search of customers until the buffer size reaches b or the search clock expires whichever occurs first. The service starts at the moment the search stops. If the number of customers waiting in the buffer is less than b but greater than or equal to a, search along with the exponential clock starts immediately after each service completion. The search continues up to a stage when the buffer size reaches b or the exponential clock expires, whichever occurs first. The clock is in 'on' condition if search is going on and it is in 'off' position if no search is going on. The clock status represents the status of the search. The customers are served in groups of varying size k where $a \leq k \leq b$. The service times of the one batch of customers are assumed to follow exponential distribution with rate ν. It is assumed that the duration of the time the search clock works is less than the service time. Let θ be the rate at which customers enter the buffer from the orbit as a result of search. At the departure epoch of the batches being served, let i and j be the number of customers present in the orbit and the buffer respectively.

Depending on i and j any of the following can happen.

- If $j > b$, the first b customers are taken for service while the others wait in the buffer.
- If $i \neq 0$ and $a \leq j < b$, the search clock starts working and the server searches until the buffer size b is reached or the clock expires whichever occurs first. If b customers are accumulated before the expiry of the clock, it encounters a force stop and service starts immediately. Otherwise the service starts on expiry of the search clock and all the customers available in the buffer are taken for bulk service.
- If $j < a$, then the server waits until the number of customers in the buffer reaches a and initiate the search along with the simultaneous starting of the

clock. Search will go on until the clock expires or the number of customers becomes b. Immediate service is given when the search stops.

3 Mathematical Formulation

Let
$N_1(t)$ denote the number of customers in the orbit at time t;
$N_2(t)$ be the number of customers in the finite buffer at time t;
$S(t)$ be the server status at time t;

$$S(t) = \begin{cases} 0, & \text{if the server is idle} \\ 1, & \text{if the server is busy} \end{cases}$$

$C(t)$ be the clock status at time t;

$$C(t) = \begin{cases} 0, & \text{if the clock is off} \\ 1, & \text{if the clock is on} \end{cases}$$

Then $\phi(t) = \{(N_1(t), N_2(t), S(t), C(t))\}$ is a Markov process and it describes the process under consideration. This model can be considered as a Level Independent Quasi-Birth -Death (LIQBD) process. We define the state space of the QBD under consideration and analyze the structure of its infinitesimal generator.

The state space is

$\Omega = \Omega_1 \bigcup \Omega_2 \bigcup \Omega_3$
$\Omega_1 = \{(i, j, 0, 0)\} \bigcup \{(i, j, 1, 0)\}$ where $i \geq 0; j = 0, 1, \ldots, a-1$
$\Omega_2 = \{(i, j, 1, 0)\} \bigcup \{(i, j, 0, 1)\}$ where $i \geq 0; j = a, a+1, \ldots, b-1$
$\Omega_3 = \{(i, j, 1, 0)\}$ where $i \geq 0; j = b, b+1, \ldots, N$

We analyze the transitions of the Markov chain $\phi(t)$ during an interval having an infinitesimal duration. We can form the matrices defining the transition rates of this chain. The infinitesimal generator Q of the Markov chain $\phi(t)$ is of the form

$$Q = \begin{pmatrix} B & A_0 & & \\ A_2 & A_1 & A_0 & \\ & A_2 & A_1 & A_0 \\ & & \cdots\cdots\cdots & \end{pmatrix} \tag{1}$$

where A_0 represents the rate matrix corresponding to the arrival of a customer to the orbit, that is transition from level $i \to i+1$ where $i \geq 0$. A_2 represents the rate matrix corresponding to the entry of a customer from the orbit to the finite buffer, that is transitions from level $i \to i-1$; for $i = 1, 2, \ldots,$. and A_1 describes all transitions in which the level does not change (transitions within levels i).

The structure of the matrices B, A_0, A_1, and A_2 can all be defined in terms of transitions of the states given in the Table 1. The first column defines the state from which a transition can occur, the second column defines a state to which a transition can occur, the third column describes the condition when this transition occurs and the last column contains the rate of the corresponding transition.

Table 1. Intensities of Transitions.

From	To	Description	Transition rate
$(i, j, k, 0)$	$(i, j+1, k, 0)$	Arrival of a customer to the buffer when $j < a-1$ where $k = 0, 1$	λ
$(i, j, 0, 1)$	$(i, j+1, 0, 1)$	Direct arrival of a customer when search is going on and for $a < j < b$	λ
$(i, a-1, 0, 0)$	$(i, a, 0, 1)$	Start of search when the buffer size becomes a by means of a direct arrival when the server is idle	λ
$(i, a-1, 0, 0)$	$(i-1, a, 0, 1)$	Start of search when the buffer size become a by a successful retrial from the orbit when the server is idle	μ
$(i, j, k, 0)$	$(i-1, j+1, k, 0)$	A successful retrial from the orbit to the buffer when $i \neq 0$ and $j < a-1$ where $k = 0, 1$	μ
$(i, j, 0, 1)$	$(i-1, j+1, 0, 1)$	Entry of customers to the buffer when search is on when $i \neq 0$ and $a \leq j < b$	$\mu + \theta$
$(i, j, 0, 1)$	$(i, 0, 1, 0)$	Start of service on expiry of the search clock when $b-1 > j \geq a$	ζ
$(i, b-1, 0, 1)$	$(i, 0, 1, 0)$	Start of service upon a direct arrival or on expiry of the search clock	$\lambda + \zeta$
$(i, b-1, 0, 1)$	$(i-1, 0, 1, 0)$	Start of service upon a successful retrial or search when $i > 0$	$\mu + \theta$
$(i, j, 1, 0)$	$(i, j-b, 1, 0)$	Continuous service when the buffer size j is such that $b \leq j \leq N$	ν
$(i, N, 1, 0)$	$(i+1, N, 1, 0)$	Entry of customers in to orbit when the buffer is is full with N customers	λ

4 Steady-State Analysis

4.1 Stability Condition

Theorem 1. *The Markov chain with generator Q is stable if and only if*

$$\lambda < (N+b+1)\mu + (b-a)\theta \tag{2}$$

Proof. Let $A = A_0 + A_1 + A_2$. We can see that A is an irreducible infinitesimal generator matrix and so there exists the stationary vector $\boldsymbol{\pi}$ of A such that

$$\boldsymbol{\pi} A = 0 \tag{3}$$

$$\boldsymbol{\pi}\mathbf{e} = 1 \tag{4}$$

The Markov chain with generator Q is stable if and only if

$$\pi A_0 e < \pi A_2 \mathbf{e} \tag{5}$$

i.e, the system is stable if and only if

$$\lambda < (N + b + 1)\mu + (b - a)\theta \tag{6}$$

4.2 Stationary Distribution

The bulk service retrial queue with search considered here can be studied as a level independent Quasi-Birth-Death process and can be solved by Matrix Analytic Method [11].

Under steady state conditions, the stationary distribution of the Markov process under consideration is obtained by solving the set of equations

$$\mathbf{x}Q = \mathbf{0}; \mathbf{xe} = \mathbf{1}. \tag{7}$$

Let $\mathbf{x}$ be decomposed as follows:
$\mathbf{x} = (\mathbf{x_0}, \mathbf{x_1}, \mathbf{x_2}, \dots)$ where $\mathbf{x_i} = (\mathbf{x_{i0}}, \mathbf{x_{i1}}, \dots \mathbf{x_{iN}})$
$\mathbf{x_{ij}} = (\mathbf{x_{ij00}}, \mathbf{x_{ij10}})$for $j = 0, 1, \dots, a-1$
$\mathbf{x_{ij}} = (\mathbf{x_{ij01}}, \mathbf{x_{ij10}})$for $j = a, a+1, \dots, b-1$ and
$\mathbf{x_{ij}} = \mathbf{x_{ij10}}$ for $j = b, b+1, \dots, N$

From $\mathbf{x}Q = 0$, it may be shown that there exists a constant matrix R such that $\mathbf{x_i} = \mathbf{x_{i-1}}R$. The sub vectors $\mathbf{x_i}$ are geometrically related by the equation

$$\mathbf{x_i} = \mathbf{x_0}R^i \tag{8}$$

R can be obtained from the matrix quadratic equation

$$R^2 A_2 + RA_1 + A_0 = O \tag{9}$$

R can be obtained by successive substitution procedure

$$R_0 = 0 \tag{10}$$

and

$$R_{k+1} = -V - R_k^2 W \tag{11}$$

where

$$V = A_2 A_1^{-1} \tag{12}$$

$$W = A_0 A_1^{-1} \tag{13}$$

x_0 can be evaluated using $\mathbf{xe} = 1$

5 Performance Measures

In this section we evaluate some performance measures of the system.

1. Expected number of customers in the orbit

$$E[N] = \sum_{i=0}^{\infty} i\mathbf{x_i e} \tag{14}$$

2. Expected number of customers in the finite buffer

$$E[N_b] = \sum_{i=0}^{\infty} \sum_{j=0}^{N} j\mathbf{x_{ij} e} \tag{15}$$

3. Probability that the buffer is full

$$P[F_b] = \sum_{i=0}^{\infty} \mathbf{x_{iN} e} \tag{16}$$

4. Expected rate at which customers enter the orbit

$$E[R_o] = \lambda \sum_{i=0}^{\infty} \mathbf{x_{iN} e} \tag{17}$$

5. Probability that the search clock is on

$$P[S_1] = \sum_{i=0}^{\infty} \sum_{j=a}^{b} x_{ij01} \tag{18}$$

6. Probability that the server is idle

$$a_0 = \sum_{i=0}^{\infty} \sum_{j=0}^{(a-1)} x_{ij00} + \sum_{i=0}^{\infty} \sum_{j=a}^{(b-1)} x_{ij01} \tag{19}$$

7. Probability that the server is busy

$$a_1 = \sum_{i=0}^{\infty} \sum_{j=0}^{(a-1)} x_{ij10} + \sum_{i=0}^{\infty} \sum_{j=a}^{(b-1)} x_{ij11} \tag{20}$$

8. Expected rate at which customers enter the buffer from the orbit

$$E[R_b] = \mu \sum_{i=1}^{\infty} \sum_{j=0}^{(N-1)} \mathbf{x_{ij} e} + \theta \sum_{i=1}^{\infty} \sum_{j=a}^{(b-1)} x_{ij01} \tag{21}$$

9. The rate of successful retrials from the orbit

$$\mu^* = \sum_{i=1}^{\infty} \sum_{j=0}^{(N-1)} x_{ij}\mathbf{e} \tag{22}$$

10. The overall rate of retrials from the orbit

$$\mu^{**} = \sum_{i=1}^{\infty} \sum_{j=0}^{N} x_{ij}\mathbf{e} \tag{23}$$

11. The fraction of successful rate of retrials

$$F[S_R] = \mu^*/\mu^{**} \tag{24}$$

6 Numerical Example

In this example, we study the effect of ζ on the expected number of customers in the buffer, the idle time of the server and the probability that the search clock is on. For the results in this example, we fix the following values:

$$N = 26; a = 4; b = 8; \theta = 5; \mu = 5; \nu = 6; \lambda = 9$$

Table 2. Effect of ζ on some performance measures

ζ	$E[N_b]$	a_0	$P[S_1]$
8	2.305	0.711	0.033
12	2.1249	0.6932	0.035
16	2.0223	0.6824	0.0363
20	1.9574	0.6752	0.0371
24	1.9133	0.6702	0.0377
28	1.8814	0.6664	0.0381
32	1.8575	0.6635	0.0384

Following are some observations that can be made from the data given in the Table 2, the figures Fig. 1, Fig. 2 and Fig. 3.

- Figure 1 shows that $P[S_1]$, the probability that the search is going on increases with an increase in the rate ζ.
- As ζ increases, $P[S_1]$ increases and as a result more customers are accumulated in the buffer. As a result more batch of customers can enter the service facility without waiting for long. So the idle time of the server a_0 decreases as the mean rate ζ at which the random clock expires increases. It can be seen from Fig. 2

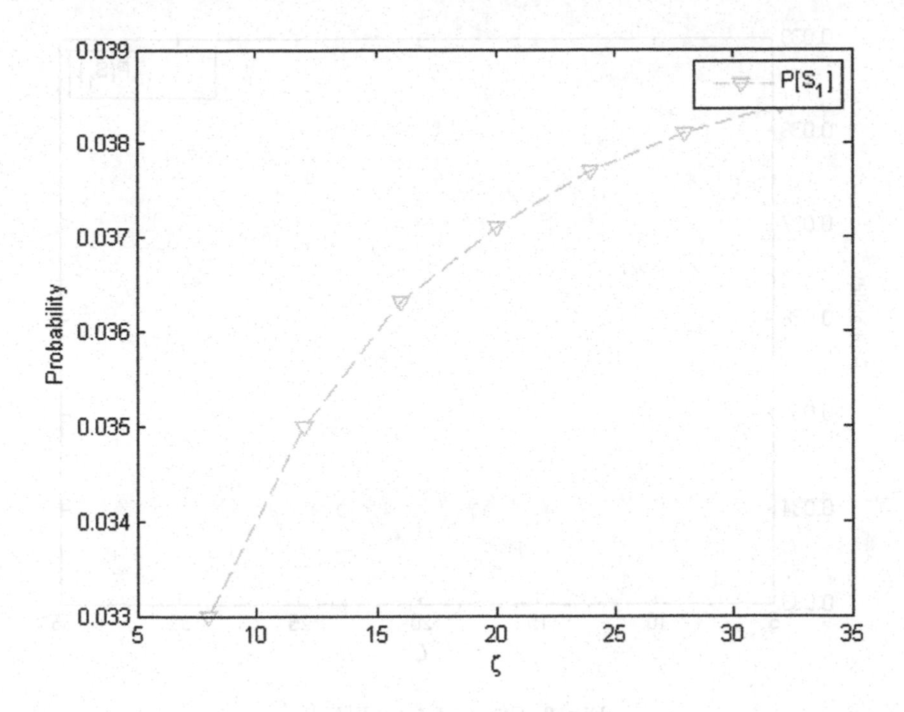

Fig. 1. Effect of ζ on the $P[S_1]$

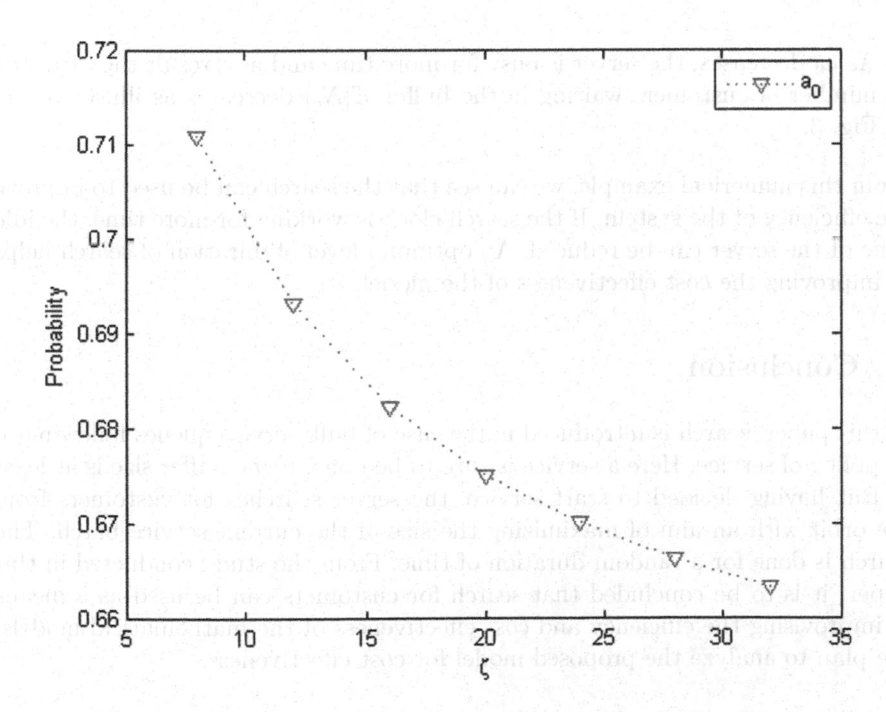

Fig. 2. Effect of ζ on the idle time of the server a_0

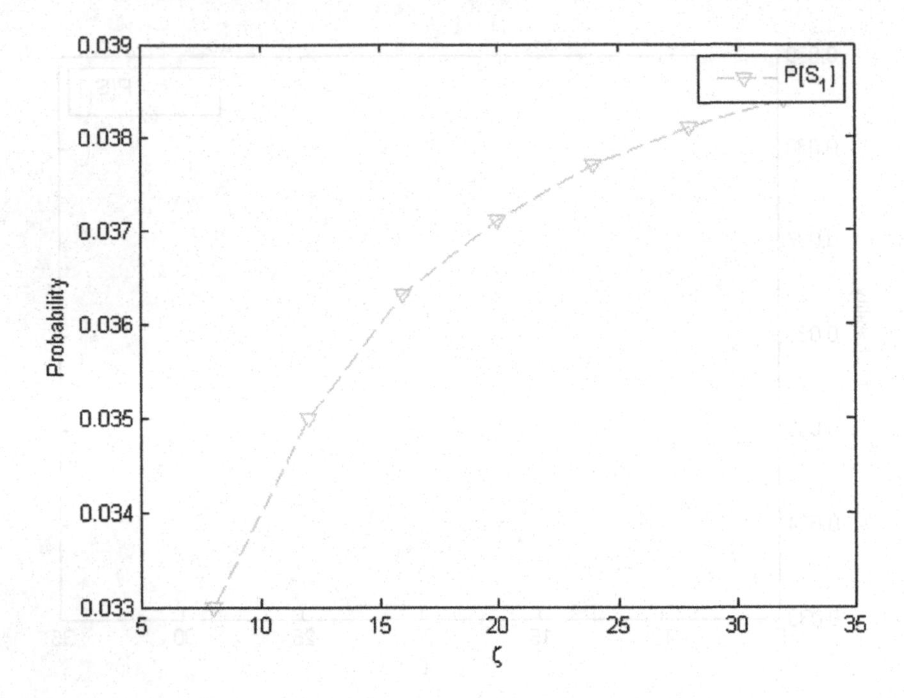

Fig. 3. Effect of ζ on $E[N_b]$

- As a_0 decreases, the server is busy for more time and as a result the expected number of customers waiting in the buffer $E[N_b]$ decreases as illustrated in Fig. 3.

From this numerical example, we can see that the search can be used to improve the efficiency of the system. If the search clock is working for more time, the idle time of the server can be reduced. An optimum level of duration of search helps in improving the cost effectiveness of the model.

7 Conclusion

In this paper, search is introduced in the case of bulk service queues following(a, b) policy of service. Here a service is sure to be done, if the buffer size is at least a. But having decided to start service, the server searches for customers from the orbit with an aim of maximizing the size of the current service batch. The search is done for a random duration of time. From the study conducted in this paper, it is to be concluded that search for customers can be used as a means of improvising the efficiency and cost effectiveness of the mathematical models. We plan to analyze the proposed model for cost effectiveness.

References

1. Artalejo, J.R., Joshua, V.C., Krishnamoorthy, A.: An M/G/1 retrial queue with orbital search by server. In: Advances in Stochastic Modelling, Notable Publications, New Jersey, pp. 41–54 (2002)
2. Chakravarthy, S.R.: A finite capacity $GI/PH/1$ queue with group services. Naval Res. Logist. (NRL) **39**, 345–357 (1992)
3. Chakravarthy, S. R.: The batch Markovian arrival process: a review and future work. In: Advances in Probability Theory and Stochastic Processes, Notable Publications, NJ, pp. 21–49 (2001)
4. Chakravarthy, S.R., Krishnamoorthy, A., Joshua, V.C.: Analysis of a multi-server retrial queue with search of customers from the orbit. Perform. Eval. **63**, 776–798 (2006)
5. Chakravarthy, S.R., Dudin, A.: Analysis of a retrial queueing model with MAP arrivals and two types of customers. Math. Comput. Model. **37**, 343–363 (2003)
6. Babu, D., Krishnamoorthy, A., Joshua, V.C.: MAP/PH/1 retrial queue with abandonment, flush out and search of customers. In: Vishnevskiy, V.M., Kozyrev, D.V. (eds.) DCCN 2018. CCIS, vol. 919, pp. 144–156. Springer, Cham (2018). https://doi.org/10.1007/978-3-319-99447-5_13
7. Dudin, A.N., Krishnamoorthy, A., Joshua, V.C., Tsarenkov, G.: Analysis of $BMAP/G/1$ retrial system with search of customers from the orbit. Eur. J. Oper. Res. **157**, 169–179 (2004)
8. Dudin, A., Deepak, T.G., Joshua, V.C., Krishnamoorthy, A., Vishnevsky, V.: On a $BMAP/G/1$ retrial system with two types of search of customers from the orbit. In: Dudin, A., Nazarov, A., Kirpichnikov, A. (eds.) ITMM 2017. CCIS, vol. 800, pp. 1–12. Springer, Cham (2017). https://doi.org/10.1007/978-3-319-68069-9_1
9. Krishnamoorthy, A., Deepak, T.G., Joshua, V.C.: An $M/G/1$ retrial queue with non persistent customers and orbital search. Stochast. Anal. Appl. **23**, 975–997 (2005)
10. Krishnamoorthy, A., Joshua, V.C., Mathew, A.P.: A retrial queueing system with abandonment and search for priority customers. In: Vishnevskiy, V.M., Samouylov, K.E., Kozyrev, D.V. (eds.) DCCN 2017. CCIS, vol. 700, pp. 98–107. Springer, Cham (2017). https://doi.org/10.1007/978-3-319-66836-9_9
11. Latouche, G., Ramaswami, V.: Introduction to Matrix Analytic Methods in Stochastic Modeling, vol. 5. ASA/SIAM Series on Statistics and Applied Probability (1999)
12. Neuts, M.F.: A general class of bulk queues with poisson input. Ann. Math. Stat. **38**, 759–770 (1967)
13. Neuts, M.F., Ramalhoto, M.F.: A service model in which the server is required to search for customers. J. Appl. Probab. **21**, 57–166 (1984)
14. Sasikala, K., Indira, S.: Bulk service queueing models-a survey. Int. J. Pure Appl. Math. **106**(6), 43–56 (2016)

Approximate Analysis of the Queuing System with Heterogeneous Servers and N-Policy

Agassi Melikov[1(✉)], Sevinc Aliyeva[2], and Mammed Shahmaliyev[3]

[1] Institute of Control Systems, National Academy of Science of Azerbaijan, Baku, Azerbaijan
agassi.melikov@gmail.com
[2] Baku State University, Baku, Azerbaijan
s@aliyeva.info
[3] National Aviation Academy of Azerbaijan, Mardakan, Azerbaijan
s@aliyeva.info

Abstract. In this paper, we propose an approximate method to investigate the Markovian queuing system with two separate pools of heterogeneous servers (HS) under N-policy. It is assumed that fast servers (F-servers) remain awake all the time while slow servers (S-servers) will go to sleep independently when number of calls in the buffer less than some threshold. At the end epoch of a sleep period, if the number of the calls gathered in the system buffer reaches or exceeds a given threshold, the corresponding S-server will wake up independently; otherwise, the S-server will begin another sleep period. An approximate method is applied under the condition that the sleep rates is essentially less than both arrival intensity of calls and their service intensity. The joint probability distribution of the number of calls in system and number of busy S-servers is determined by simple computational procedures. Illustrative numerical examples show the high accuracy of the proposed approximate method.

Keywords: Queuing system · Heterogeneous servers · N-policy · Space merging · Calculation algorithm

1 Introduction

Because of the intensive development of computer technology, cloud data centers use servers of various capacities. Therefore, for their correct mathematical analysis models of queuing systems with heterogeneous servers (QSwHS) are used. Moreover, such kind of models might be used to study systems where arrived calls are processed by people rather than by machines.

It seems that first serious paper devoted to QSwHS is [1]. In [1], a Markovian system with infinite queue and randomized call admission control (CAC) scheme was investigated. Algrotihm to calculation of steady-state probabilities as well

V. M. Vishnevskiy et al. (Eds.): DCCN 2020, LNCS 12563, pp. 322–334, 2020.
https://doi.org/10.1007/978-3-030-66471-8_25

as formulas to determine the mean number of calls in system are proposed. Since the seventies of the last centure, QSwHS have been intensively studied. At the begining, many works were devoted to the generalization of classical results for systems with homogeneous servers [2–5].

In QSwHS, problems of determining optimal CAC schemes are important. In the early stages of the study of QSwHS, models with heuristc CAC such as FSF (Fast Server First), SSF (Slow Server First), ordered entry or randomized assigning the servers was used [6–15]. However, for the first time in paper [16] it is proven that these CAC are not optimal even suboptimal. More spesific, in the indicate paper the optimality of threshold-type CAC is proved. It means that for QSwHS with two servers the following CAC is optimal to minimize mean number of calls in the system: the fast server (F-server) always works if there is at least one call in the system, and slow server (S-server) is only involved when queue length reachs a cetain threshold value (N). This CAC usually called N-policy as well. Later this result is stated for models with unreliable servers [17]; generalization of N-policy to models with more than two servers has been proposed in [18–21].

Recently in [22] Markovian QSwHS is applied to study energy consumption problem in cloud data center. In the indicated paper authors propose a clustered Virtual Machine (VM) allocation strategy based on a sleep-mode with a wake-up threshold. The VMs are clustered into two pools, namely, pool of F-servers and pool of S-servers; it is assumed that F-servers remain awake at all times, while S-servers asynchronously go to sleep under a light workload. After a sleep time expires, the S-server will resume processing calls only if the number of waiting calls reaches the wake-up threshold. In other words, in [22], QSwHS with two separate pools of servers and N-policy is investigated. To study constucted two-dimensional Markov chain (2D MC) matrix-geometric method by Neuts [23] is used.

Note that in literature models of QSwHS in case of identical calls are studied in detail. However, models of QSwHS with calls of different kinds represents some practical and scientific interests. Recently models of buffer-less QSwHS and QSwHS with buffers and jump priorities were investigated in papers [24] and [25], respectively. Note that the indicated papers contain review of available papers devoted to QSwHS.

This paper in spirit is close to [22] and here, to improve readability, we keep its notation. In this paper, we consider alternate (approximate) method to investigate the model that proposed in [22]. An approximate method is applied under the condition that the sleep rates is less than both arrival intensity of calls and their service intensity. As the result of the applying an approximate method it is shown that the joint probability distribution of the number of calls in system and number of busy S-servers is determined by simple computational procedures. Several illustrative examples show the high accuracy and range of applicability of the proposed approximate method. Moreover, results of numerical experiments demonstrate that the proposed formulas are asymptotic exact for the models where sleep intensity is close to zero.

Note that proposed here approach based on the idea of theory of space merging of Markov chains [26]. For the first time, space-merging algorithm (SMA) was developed to calculation of the steady-state probabilities of 2D MC in [27] (recently hierarchical SMA to study 3D MC is developed [28]). Many researchers develop similar SMAs (see [29–34]). In these papers, authors note high efficiency of SMA (in both sense of accuracy and complexity) in comparison with other algorithms.

The paper is organized as follows. In Sect. 2 the model of the QSwHS under study is described. Algorithm to calculate the elements of generating matrix (Q-matrix) and explicit formulas for performance measures are given in Sect. 3. In Sect. 4 SMA to approximately calculation of steady-state probabilities and performance measures is developed. Results of numerical experiments are demonstrated in Sect. 5. Concluding remarks are given in Sect. 6.

2 The Model

The investigated system contains two pools of heterogeneous servers. Pool I contains F-servers and their number is equal to c; pool II contains S-servers and their number is equal to d. The input flow is Poisson one with rate λ. The service rates of calls in Pool I and in Pool II are assumed to be exponentially distributed with parameters μ_1 and μ_2, respectively and $\mu_1 > \mu_2$.

In the system, the following service mechanism is used. The F-servers remain awake all the time while S-servers will go to sleep independently when number of calls in the buffer less than a given wake-up threshold N. At the end epoch of a sleep period, if the number of the calls gathered in the system buffer reaches or exceeds wake-up threshold, the corresponding S-server will wake up independently. Otherwise, the S-server will begin another sleep period. The sleep time length is assumed to follow an exponential distribution with parameter θ ($\theta > 0$).

Rules for transitions from awake state to sleep state and from sleep state to awake state are defined as follows.

· For a busy S-server, the state transition from awake state to sleep state occurs only at the instant when a call either in Pool I or Pool II is completely processed. When a call is completely processed in F-server, if the number of call in system is zero and there is at least one call being processed in S-server, one of the calls being processed in Pool II will migrate Pool I, and then the evacuated S-server will go to sleep. When a call is completely processed in S-server, if the number of call in system is zero, the evacuated S-server will go to sleep directly.

· For a sleeping S-server, the state transition from sleep state to awake state occurs only at the end epoch of a sleep period. When a sleep timer expires, if the number of call in system is equal to or greater than the wake-up threshold N, the corresponding S-server will wake up to process the first call waiting in the system buffer; otherwise, a new sleep timer will be started and the S-server will begin another sleep period.

The problem is finding the joint probability distribution of the number of calls in system and number of busy S-servers. Determination of the indicated probability distribution allows calculate the desired QoS metrics as well.

3 Exact Method

State of the system is defined by the two-dimensional (2D) vector (i, j), where i is the total number of calls in the system, $i = 0, 1, ...,$ and j indicates the number of busy S-channels, $j = 0, 1, ..., d$. Based on the distribution function of the random variables involved in the formation of the model, we determine that the given system is described by the two-dimensional Markov chain (2D MC).

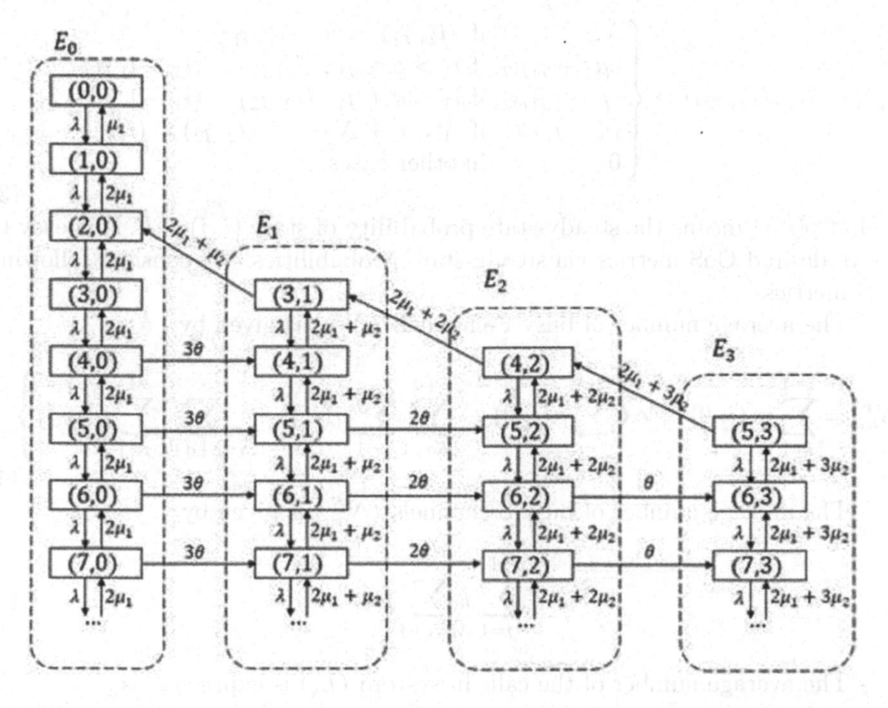

Fig. 1. The state transition graph for case $c = 2, d = 3, N = 2$ (the graph is borrowed from paper [22] and modified).

The state space of this 2D MC is defined as following (see Fig. 1)

$$E = \bigcup_{i=0}^{d} E_i \ , \ E_{i_1} \bigcap E_{i_2} = \emptyset \ , \text{ if } i_1 \neq i_2 \ , \tag{1}$$

where $E_0 = \{(i, 0) : i \geq 0\}$, $E_k = \{(i, k) : i \geq c + k\}$, $k = \overline{1, d}$.

The transition rate from the initial state $(i_1, j_1) \in E$ to the final state $(i_2, j_2) \in E$ is denoted as $q((i_1, j_1), (i_2, j_2))$. The combination of these quantities forms Q-matrix of given 2D MC and are determined from the following relations.

For the initial state of the type $(i_1, 0)$:

$$q((i_1, 0), (i_2, j_2)) = \begin{cases} \lambda, & \text{if } (i_2, j_2) = (i_1 + 1, 0), \\ \min(i_1, c)\mu_1, & \text{if } (i_2, j_2) = (i_1 - 1, 0), \\ d\theta, & \text{if } i_1 \geq c + N, (i_2, j_2) = (i_1, 1), \\ 0 & \text{in other cases.} \end{cases} \tag{2}$$

For the initial state of the type (i_1, j_1) , $i_1 \geq c + j_1$:

$$q((i_1, j_1), (i_2, j_2)) = \begin{cases} \lambda, & \text{if } (i_2, j_2) = (i_1 + 1, j_1), \\ c\mu_1 + j_1\mu_2, & \text{if } i_1 > c + j_1, (i_2, j_2) = (i_1 - 1, j_1), \\ c\mu_1 + j_1\mu_2, & \text{if } i_1 = c + j_1, (i_2, j_2) = (i_1 - 1, j_1 - 1), \\ (d - j_1)\theta, & \text{if } i_1 \geq c + N + j_1, (i_2, j_2) = (i_1, j_1 + 1), \\ 0 & \text{in other cases.} \end{cases} \tag{3}$$

Let $p(i, j)$ means the steady-state probability of state $(i, j) \in E$. It is easy to derive desired QoS metrics via steady-state probabilities. We consider following QoS metrics.

· The average number of busy F-channels (N_{av}^F) is given by

$$N_{av}^F = \sum_{i=1}^{c-1} ip(i, 0) + c\left\{\sum_{i=c}^{\infty} p(i, 0) + \sum_{i=c+1}^{c+d}\sum_{j=1}^{i-c} p(i, j) + \sum_{i=c+d+1}^{\infty}\sum_{j=1}^{d} p(i, j)\right\}. \tag{4}$$

· The average number of busy S-channels (N_{av}^S) is given by

$$N_{av}^S = \sum_{j=1}^{d} j \sum_{i=c+j}^{\infty} p(i, j). \tag{5}$$

· The average number of the calls in system (L_s) is expressed as

$$L_s = \sum_{i=1}^{c} ip(i, 0) + \sum_{i=c+1}^{c+d} i \sum_{j=0}^{i-c} p(i, j) + \sum_{i=c+d+1}^{\infty} i \sum_{j=0}^{d} p(i, j). \tag{6}$$

· The average sojourn time of the calls in system (W_s) is calculated by Little' formula, i.e.

$$W_s = \frac{1}{\lambda} L_s. \tag{7}$$

4 Approximate Method

Here we propose an effective and simple numerical method for approximate analysis of the investigated system. It based on space merging approach to calculate the stationary distribution of 2D MC. For the correct use of this method, we assume that $\theta << min(\lambda, c\mu_1 + d\mu_2)$.

In accordance to our assumption the transition intensities between states in the classes E_i , $i = \overline{0,d}$, are too large than intensities between states from different classes. By using this fact consider the following merge function on the state space E is determined as follows:

$$U((i,j)) =< j > \text{ if } (i,j) \in E_j ,$$

where $< j >$ is a merge state, which includes all the states of the E_j , $j = \overline{0,d}$. Let $\Omega = \{< j >: j = \overline{0,d}\}$.

The approximate values of steady-state probabilities of the initial model are defined as follows:

$$\tilde{p}(i,j) \approx \rho_j(i) \pi(< j >) , \tag{8}$$

where $\rho_j(i)$ denotes the state probability of state (i,j) within the splitting model with state spaceE_j, and $\pi(< j >)$ is the probability of the merge state $< j > \in \Omega$.

From splitting scheme (1) it is clear that all the splitting models are one-dimensional birth and death processes (1D BDP), so that in the class of states E_j the second component is constant. Therefore, in the splitting model with state space E_j microstate $(i,j) \in E$ can be represent by scalar i.

From (2) we get that probabilities $\rho_0(i)$ coincide with the steady-state probabilities of the classical $M/M/c/\infty$ system with individual server utilization, $\nu_0 = \lambda/c\mu_1$, i.e. if $\nu_0 < 1$ we have

$$\rho_0(i) = \begin{cases} \frac{(c\nu_0)^i}{i!}\rho_0(0) , 0 \le i \le c, \\ \frac{\nu_0^i c^c}{c!}\rho_0(0) , \quad i > c, \end{cases} \tag{9}$$

where $\rho_0(0) = \left(\sum_{i=0}^{c-1} \frac{(c\nu_0)^i}{i!} + \frac{(c\nu_0)^c}{c!} \cdot \frac{1}{1-\nu_0}\right)^{-1}$.

From (3) we get that probabilities $\rho_j(i)$, $j = \overline{1,d}$, coincide with steady-state probabilities of the classical $M/M/1/\infty$ with load ν_j where $\nu_j = \lambda/(c\mu_1 + j\mu_2)$, i.e. $\nu_j < 1$ we have

$$\rho_j(i) = \nu_j^{i-c-j}(1-\nu_j) , \; i = c+j, c+j+1, ... \tag{10}$$

Note. Since conditions $\nu_j < 1$ should be satisfied for each $j = \overline{0,d}$, so we obtain ergodicity condition of the initial model, i.e. $\nu_0 < 1$.

The transition intensity from the merge state $< j_1 >$ to other merge state $< j_2 >$ is denoted as $q(< j_1 >, < j_2 >)$, $< j_1 >, < j_2 > \in \Omega$. These quantities are calculated as follows:

$$q(< j_1 >, < j_2 >) = \sum_{\substack{(i_1,j_1) \in E_{j_1} \\ (i_2,j_2) \in E_{j_2}}} q((i_1,j_1), (i_2,j_2)) \rho_{j_1}(i_1) . \tag{11}$$

After certain algebras on the bases of (2), (3) and (9)-(11) we obtain:

$$q(< j_1 >, < j_2 >) = \begin{cases} \alpha(j_1) , j_2 = j_1 + 1 , \\ \beta(j_1) , j_2 = j_1 - 1 , \\ 0 \qquad \text{in other cases,} \end{cases} \tag{12}$$

where $\alpha(j_1) = (d - j_1)\theta\left(1 - \sum_{i=\chi(j_1)}^{c+N+j_1-1} \rho_{j_1}(i)\right)$,
$\chi(j_1) = \begin{cases} c + j_1, j_1 > 0, \\ 0, \quad j_1 = 0, \end{cases} j_1 = \overline{0, d-1}$;

$$\beta(j_1) = (c\mu_1 + j_1\mu_2)\rho_{j_1}(c + j_1), \ j_1 = \overline{1, d}.$$

From (12) we conclude that the probabilities of the merging states $\pi(<j>), <j> \in \Omega$, are calculated as the state probabilities of 1-D BDP. In other words,

$$\pi(<j>) = \prod_{i=1}^{j} \frac{\alpha(i-1)}{\beta(i)} \pi(<0>), \ j = 1, ..., d, \tag{13}$$

where $\pi(<0>)$ is derived from normalizing condition, i.e. $\sum_{j=0}^{d} \pi(<j>) = 1$.

Finally, taking into account the relations (9), (10), (13) from (8) we calculate the steady-state probabilities of the initial 2D MC. After certain algebras, we obtain the following approximate formulas for calculating the desired performance measures of the system:

$$N_{av}^F \approx c(1 - \pi(<0>)) + \left\{ \sum_{i=1}^{c-1} i\rho_0(i) + c\left(1 - \sum_{i=0}^{c-1} \rho_0(i)\right) \right\} \pi(<0>) \ ; \tag{14}$$

$$N_{av}^S \approx \sum_{j=1}^{d} j\pi(<j>) \ ; \tag{15}$$

$$L_s \approx \pi(<0>) \sum_{i=1}^{c} i\rho_0(i) + \sum_{i=1}^{d} (c+i) \sum_{j=0}^{i} \rho_j(c+i)\pi(<j>)$$
$$+ \pi(<0>) \left(L_s^{(0)} - \sum_{i=1}^{c+d} i\rho_0(i)\right) + \sum_{j=1}^{d} \pi(<j>)(1 - \nu_j)\nu_j^{-(c+j)} \left(L_s^{(j)} - \sum_{i=1}^{c+d} i\nu_j^i\right). \tag{16}$$

In formula (16) we use the following notations: $L_s^{(0)}$ is average number of calls in the system $M/M/c/\infty$ system with load ν_0 and $L_s^{(j)}$ is average number of calls in the system $M/M/1/\infty$ with load ν_j.

The approximate value of average sojourn time of the calls in system is calculated from (16) by using (7).

5 Numerical Results

Numerical experiments have two goals: firstly, to analyze the accuracy of the state probabilities and performance measures calculated by the developed approximate formulas; secondly, to perform some numerical experiments to study behavior of the accuracy measures (4)–(7) with respect to both sleep parameter and wake-up threshold.

The accuracy of calculating the approximate values of stationary probabilities is estimated using the following norms:

Cosine similarity:

$$\|\mathrm{N}\|_1 = \frac{\sum_{(i,j)\in E} p\left(i,j\right) \tilde{p}\left(i,j\right)}{\left(\sum_{(i,j)\in E} \left(p\left(i,j\right)\right)^2\right)^{\frac{1}{2}} \left(\sum_{(i,j)\in E} \left(\tilde{p}\left(\left(i,j\right)\right)\right)^2\right)^{\frac{1}{2}}}; \tag{17}$$

Jaccard coefficient [35]:

$$\|\mathrm{N}\|_2 = \frac{\sum_{(i,j)\in E} \min\left\{p\left(i,j\right), \tilde{p}\left(i,j\right)\right\}}{\sum_{(i,j)\in E} \max\left\{p\left(i,j\right), \tilde{p}\left(i,j\right)\right\}}; \tag{18}$$

Euclidean distance:

$$\|\mathrm{N}\|_3 = \left(\sum_{(i,j)\in E} \left(p\left(i,j\right) - \tilde{p}\left(i,j\right)\right)^2\right)^{\frac{1}{2}}. \tag{19}$$

In numerical experiments, the values of system parameters are same as in [22] (the purpose of choosing exactly such initial data is to compare our results with the results of paper [22]). In other words, we assume that $c+d = 50$, $\lambda = 7$; other values of parameters are shown in Table 1. Note that these values of system parameters fulfill the conditions that accepted to correct application of SMA, i.e. $\theta << min(\lambda,\ c\mu_1 + d\mu_2)$.

Some results of the comparative analysis of the both steady-state probabilities and performance measures calculations for the exact and approximate approaches are given in Table 1.

The balance equations is solved by using MATLAB where maximal size of queue is restricted by 1000. Solving time of balance equations depends on its dimension, and for our model, it takes several hours. It is worthy to note that, in the same PC approximately one second are needed for the calculation of performance measures by developed formulas.

From the Table 1 we conclude that the developed approximate formulas to calculate the steady-state probabilities has high accuracy, because the values of both norms (17) and (18) are very close to 1 while Euclidean distance (19) is negligible. It is important to note that our results completely coincide with the results obtained in [22].

For the given initial data the analysis of accuracy of the calculation of system performance measures (4)–(7) was performed as well. It should be noted that the performance measures are almost the same when using exact and approximate approaches (see Table 1). At the same time, the complexity of the proposed algorithm for calculating the steady-state probabilities and performance measures is significantly lower than the algorithm proposed in [22].

It is clear that the accuracy of proposed algorithms essentially depends on ratio of system parameters; in particular, it depends on ratio of load quantities

Table 1. Estimation of calculation accuracy for steady-state probabilities and performance measures; EV – exact value, AV – approximate value.

(c,d)	(N,θ)	(μ_1,μ_2)	Norms			N_{av}^F		N_{av}^S		W_s	
			$\Vert N \Vert_1$	$\Vert N \Vert_2$	$\Vert N \Vert_3$	EV	AV	EV	AV	EV	AV
(45, 5)	(10, 0.5)	(0.2, 0.1)	1	0.97	0.004	34.989	35.073	0.022	0.008	5.032	5.050
		(0.3, 0.2)	1	1	4E-08	23.333	23.333	1.7E-07	2E-08	3.333	3.333
		(0.4, 0.3)	1	1	3E-08	17.500	17.500	4.2E-10	6E-09	2.500	2.500
	(11, 1.0)	(0.2, 0.1)	1	0.97	0.005	34.989	35.117	0.023	0.013	5.032	5.059
		(0.3, 0.2)	1	1	3E-08	23.333	23.333	1.4E-07	3E-08	3.333	3.333
		(0.4, 0.3)	1	1	3E-08	17.500	17.500	2.9E-10	1E-08	2.500	2.500
	(12, 1.5)	(0.2, 0.1)	1	0.96	0.006	34.990	35.138	0.021	0.016	5.032	5.064
		(0.3, 0.2)	1	1	2E-08	23.333	23.333	9.8E-08	2E-08	3.333	3.333
		(0.4, 0.3)	1	1	3E-08	17.500	17.500	1.6E-10	2E-08	2.500	2.500
(42, 8)	(10, 0.5)	(0.2, 0.1)	0.99	0.80	0.031	34.931	35.688	0.138	0.133	5.097	5.298
		(0.3, 0.2)	1	1	8E-07	23.333	23.333	3.6E-06	7E-07	3.333	3.333
		(0.4, 0.3)	1	1	3E-08	17.500	17.500	2.2E-10	1E-08	2.500	2.500
	(11, 1.0)	(0.2, 0.1)	0.98	0.68	0.054	34.932	36.306	0.137	0.301	5.097	5.455
		(0.3, 0.2)	1	1	6E-07	23.333	23.333	2.9E-06	7E-07	3.333	3.333
		(0.4, 0.3)	1	1	3E-08	17.500	17.500	1.5E-10	2E-08	2.500	2.500
	(12, 1.5)	(0.2, 0.1)	0.96	0.61	0.070	34.937	36.738	0.326	0.439	5.099	5.566
		(0.3, 0.2)	1	1	5E-07	23.333	23.333	2.1E-06	6E-07	3.333	3.333
		(0.4, 0.3)	1	1	3E-08	17.500	17.500	8.2E-11	3E-08	2.500	2.500
(44, 6)	(10, 0.5)	(0.2, 0.1)	1	0.92	0.012	34.973	35.284	0.053	0.039	5.045	5.116
		(0.3, 0.2)	1	1	2E-07	23.333	23.333	6.6E-07	1E-07	3.333	3.333
		(0.4, 0.3)	1	1	3E-08	17.500	17.500	2.4E-11	1E-08	2.500	2.500
	(11, 1.0)	(0.2, 0.1)	1	0.89	0.019	34.974	35.503	0.051	0.077	5.045	5.163
		(0.3, 0.2)	1	1	1E-07	23.333	23.333	5.2E-07	1E-07	3.333	3.333
		(0.4, 0.3)	1	1	3E-08	17.500	17.500	1.5E-11	2E-08	2.500	2.500
	(12, 1.5)	(0.2, 0.1)	1	0.86	0.0228	34.977	35.631	0.046	0.101	5.046	5.192
		(0.3, 0.2)	1	1	8E-08	23.333	23.333	3.4E-07	1E-07	3.333	3.333
		(0.4, 0.3)	1	1	4E-08	17.500	17.500	7.9E-10	3E-08	2.500	2.500

and sleep parameter. It is possible to study the dependence of accuracy on any system parameters.

For definiteness, here we study the dependency of norms (17), (18) and (19) on sleep parameter θ (see Fig. 2) and on wake-up threshold N(see Fig. 3). As it was expected, accuracy of developed formulas is decreasing with respect to increasing of sleep parameter while it is increasing with respect to increasing of wake-up threshold. These facts are explained by general observation: with a decrease in the intensity of transitions between classes E_i, $i = \overline{0,d}$, in splitting (1), the accuracy of the proposed formulas increases.

It is clear that when sleep parameter is decreased and/or wake threshold is increased then the intensity of transitions between classes E_i, $i = \overline{0,d}$, in splitting (1) is decreased as well. Note that an increase in the number of F-servers also leads to an increase in the accuracy of proposed approximate formulas (see Figs. 2 and 3).

We study behavior of the performance measures (4)–(6) with respect to both sleep parameter and to number of F-servers. Due to size limitation these results does not demonstrated here.

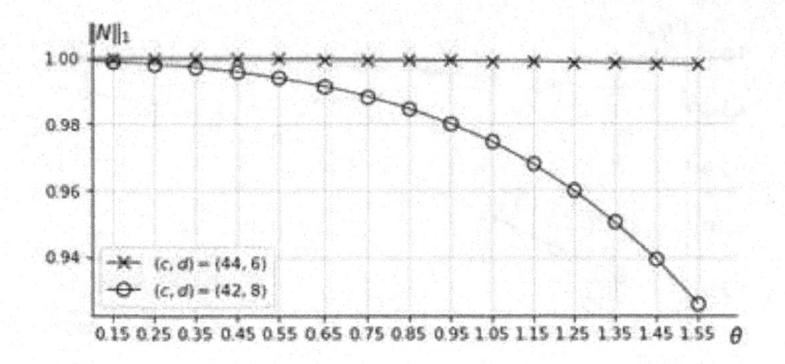

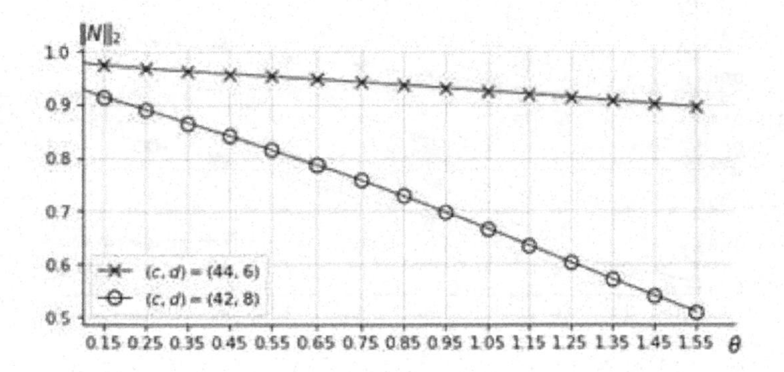

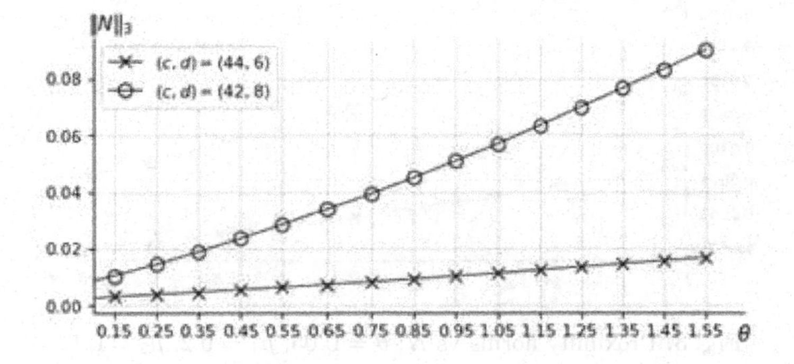

Fig. 2. Proximity norms vs θ; $N = 11$, $\mu_1 = 0.2$, $\mu_2 = 0.1$.

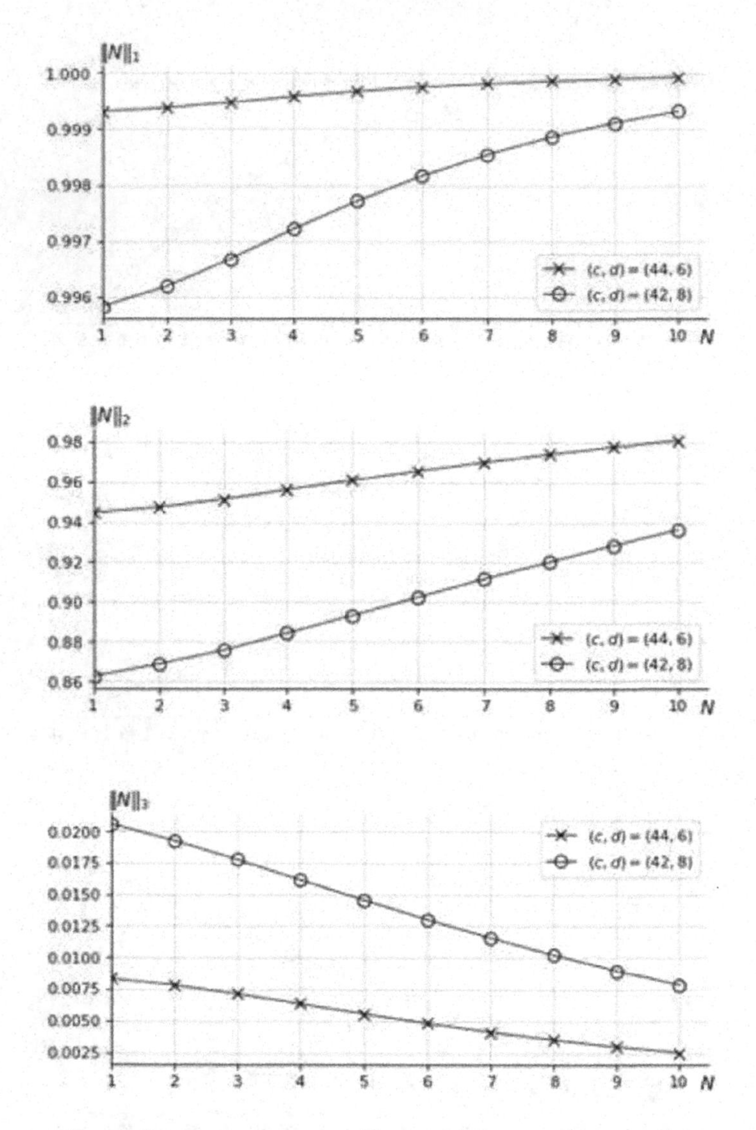

Fig. 3. Proximity norms vs N; $\theta = 0.05$, $\mu_1 = 0.2$, $\mu_2 = 0.1$.

6 Conclusion

In this paper, we have analyzed a queuing system with heterogeneous servers in which N-policy is used. Fast servers are active for all the time while slow servers will wake-up if the number of the calls in the buffer reaches or exceeds a given threshold; S-servers go to sleep asynchronously when number of calls in the buffer less than indicated threshold. An approximate method based on space merging approach is applied to calculate the steady-state probabilities of constructed two-dimensional Markov chain with infinite state space as well as performance measures of investigated system. Illustrative numerical examples show the high accuracy of the proposed approximate method.

The proposed approach might be applied to study similar model with finite size of queue. This issue is subject of separate investigation.

References

1. Gumbel, H.: Waiting lines with heterogeneous servers. Oper. Res. **8**(4), 504–511 (1960)
2. Singh, V.S.: Two-server Markovian queues with balking: heterogeneous vs. homogeneous servers. Oper. Res. **18**, 145–159 (1970)
3. Singh, V.S.: Markovian queues with three servers. IIE Trans. **3**, 45–48 (1971)
4. Fakinos, D.: The M/G/k blocking system with heterogeneous servers. J. Oper. Res. Soc. **31**, 919–927 (1980)
5. Fakinos, D.: The generalized M/G/k blocking system with heterogeneous servers. J. Oper. Res. Soc. **33**, 801–809 (1982)
6. Nath, G., Enns, E.: Optimal service rates in the multi-server loss system with heterogeneous servers. J. Appl. Probab. **18**, 776–781 (1981)
7. Lin, B.W., Elsayed, E.A.: A general solution for multichannel queuing systems with ordered entry. Comput. Oper. Res. **5**, 219–225 (1978)
8. Elsayed, E.A.: Multichannel queuing systems with ordered entry and finite source. Comput. Oper. Res. **10**, 213–222 (1983)
9. Yao, D.D.: The arrangement of servers in an ordered entry system. Oper. Res. **35**, 759–763 (1987)
10. Pourbabai, B., Sonderman, D.: Server utilization factors in queuing loss systems with ordered entry and heterogeneous servers. J. Appl. Probab. **23**, 236–242 (1986)
11. Pourbabai, B.: Markovian queuing systems with retrials and heterogeneous servers. Comput. Math. Appl. **13**(12), 917–923 (1987)
12. Nawijn, W.M.: On a two-server finite queuing system with ordered entry and deterministic arrivals. Eur. J. Oper. Res. **18**, 388–395 (1984)
13. Nawijn, W.M.: A note on many-server queuing systems with ordered entry with an application to conveyor theory. J. Appl. Probab. **20**, 144–152 (1983)
14. Yao, D.D.: Convexity properties of the overflow in an ordered entry system with heterogeneous servers. Oper. Res. Lett. **5**, 145–147 (1986)
15. Isguder, H.O., Kocer, U.U.: Analysis of GI/M/n/n queuing system with ordered entry and no waiting line. Appl. Math. Model. **38**, 1024–1032 (2014)
16. Larsen, R.L., Agrawala, A.K.: Control of heterogeneous two-server exponential queuing system. IEEE Trans. Softw. Eng. **9**(4), 522–526 (1983)
17. Efrosinin, D., Sztrik, J., Farkhadov, M., Stepanova, N.: Reliability analysis of a two-server heterogeneous unreliable queueing system with a threshold control policy. In: Dudin, A., Nazarov, A., Kirpichnikov, A. (eds.) ITMM 2017. CCIS, vol. 800, pp. 13–27. Springer, Cham (2017). https://doi.org/10.1007/978-3-319-68069-9_2
18. Viniotis, I., Ephremides, A.: Extension of the optimality of a threshold policy in heterogeneous multi-server queuing systems. IEEE Trans. Autom. Control **33**, 104–109 (1988)
19. Rosberg, Z., Makowski, A.M.: Optimal routing to parallel heterogeneous servers - small arrival rates. IEEE Trans. Autom. Control **35**(7), 789–796 (1990)
20. Rykov, V.V.: Monotone control of queuing systems with heterogeneous servers. Queuing Syst. **37**, 391–403 (2001)

21. Rykov, V.V., Efrosinin, D.: On the slow server problem. Autom. Remote Control **70**(12), 2013–2023 (2009)
22. Jin, S., Qie, X., Zhao, W., Yue, W., Takahash, Y.: A clustered virtual machine allocation strategy based on a sleep-mode with wake-up threshold in a cloud environment. Ann. Oper. Res. https://doi.org/10.1007/s10479-019-03339-3
23. Neuts, M.F.: Matrix-Geometric Solutions in Stochastic Models: An Algorithmic Approach, p. 332. John Hopkins University Press, Baltimore (1981)
24. Melikov, A.Z., Ponomarenko, L.A., Mekhbaliyeva, E.V.: Analysis of models of systems with heterogeneous servers. Cybern. Syst. Anal. **56**(1), 89–99 (2020)
25. Melikov, A.Z., Mekhbaliyeva, E.V.: Analysis and optimization of system with heterogeneous servers and jump priorities. J. Comput. Syst. Sci. Int. **58**(5), 718–735 (2019)
26. Korolyuk, V.S., Korolyuk, V.V.: Stochast. Models Syst., p. 185. Kluwer Academic Publishers, Boston (1999)
27. Melikov, A.Z., Fattakhova, M.I.: Computational algorithms to optimization of buffer allocation strategies in a packet switching networks. Appl. Comput. Math. **1**(1), 51–58 (2002)
28. Korolyuk, V.S., Melikov, A.Z., Ponomarenko, L.A., Rustamov, A.M.: Asymptotic analysis of the system with server vacations and perishable inventory. Cybern. Syst. Anal. **53**(4), 543–553 (2017). https://doi.org/10.1007/s10559-017-9956-0
29. Brian, P.G.: Approximate analysis of an unreliable M/M/2 retrial queue. Thesis presented for the Degree of Master of Science in Operations Research. Air Force Institute of Technology, p. 84, March 2007
30. Liang, C.C., Luh, H.: Cost estimation queuing model for large-scale file delivery service. Int. J. Electron. Commer. Stud. **2**(1), 19–34 (2011)
31. Liang, C.C., Luh, H.: Optimal services for content delivery based on business priority. J. Chin. Inst. Eng. **36**(4), 422–440 (2013)
32. Liang, C.C., Luh, H.: Solving two-dimensional Markov chain model for call center. Ind. Manage. Data Syst. **115**(5), 901–922 (2015)
33. Raiah, L., Oukid, N.: An M/M/2 retrial queue with breakdowns and repairs. Rom. J. Math. Comput. Sci. **17**(1), 11–20 (2017)
34. Elhaddad, M., Belarbi, F.: On the analysis of unreliable Markovian multi-server queue with retrials and impatience. Math. Sci. Appl. E-notes **7**(2), 205–217 (2019)
35. Jaccard, P.: Étude Comparative de la Distribution Florale dans une Portion des Alpes et des Jura. Bulletin del la Sociétè Vaudoise des Sciences Naturelles **37**, 547–579 (1901)

Analysis of a Resource-Based Queue with the Parallel Service and Renewal Arrivals

Ekaterina Lisovskaya[1(✉)], Ekaterina Pankratova[2], Svetlana Moiseeva[3], and Michele Pagano[4]

[1] Peoples' Friendship University of Russia (RUDN University), 6 Miklukho-Maklaya Street, Moscow 117198, Russian Federation
lisovskaya-eyu@rudn.ru

[2] V. A. Trapeznikov Institute of Control Sciences of Russian Academy of Sciences, 65 Profsoyuznaya Street, Moscow 117997, Russian Federation
pankate@sibmail.com

[3] National Research Tomsk State University, 36 Lenina Avenue, Tomsk, Russian Federation
smoiseeva@mail.ru

[4] Department of Information Engineering, University of Pisa, Via Caruso 16, 56122 Pisa, Italy
michele.pagano@iet.unipi.it

Abstract. Classical queueing theory is often not suitable to model modern computer and communication systems, in which the service itself can require random amounts of multiple resources. For instance, this is true for distributed computation and wireless devices connected through different access technologies.

To model such systems we propose a resource queueing system with customer duplication, in which the service time and the amount of requested resources in each block are independent random variables. In more detail, we assume that customers arrive according to a general renewal process and, taking advantage of the dynamic screening and the asymptotic analysis methods, we derive a Gaussian approximation for the stationary probability distribution of the occupied resources in the system blocks. Finally, simulation experiments point out the applicability region (in terms of arrival rate) of the proposed approximation.

Keywords: Resource queueing system · Total resource amount · Dynamic screening method · Asymptotic analysis method

---

The publication has been prepared with the support of the "RUDN University Program 5-100" (recipient E. Lisovskaya, Conceptualization and Investigation) and of the University of Pisa under the PRA 2018–2019 Research Project "CONCEPT – COmmunication and Networking for vehicular CybEr-Physical sysTems" (recipient M. Pagano, Simulation).

V. M. Vishnevskiy et al. (Eds.): DCCN 2020, LNCS 12563, pp. 335–349, 2020.
https://doi.org/10.1007/978-3-030-66471-8_26

1 Introduction

In classical queueing systems (QSs), servers play the role of discrete resources required by the customers. In resource QSs (RQSs), in addition to servers and waiting areas, customers may require various resources [12,13]. It can be a random resource amount occupied for the waiting time to the start of service, or for the service time only, or for the entire time the customer is in the system. For example, this can represent the memory capacity of a device or radio frequencies of wireless networks [17]. The interest to RQSs is explained by the possibility of their application for modeling a fairly wide area of technical devices and information and computing systems, in general.

An example of tasks with a set of resources are wireless network technologies such as Long Term Evolution (LTE), or New Radio (NR), or Wi-Fi [1,2,7]. In such networks, each active session takes up a certain amount of radio resources such as the bandwidth of the frequency spectrum and the transmission power of the radio frequency amplifier. It should be noted, that a large number of scientific papers is devoted to modeling wireless communication systems using RQSs [6, 8,23]. The growth of their popularity makes it necessary to create effective tools for assessing the radio interfaces operation telecommunications tenants [3,5,22].

1.1 Related Works

The first publication devoted to the analysis of the QS with the allocation to each arrival of some random amount of a resource in addition to a server is the paper [21]. In this paper, mathematical relations that allow calculating the main stationary characteristics of a QS operating in discrete time are obtained and simple algorithms are presented for numerical calculations. These results are extended in the works [25,26] etc.; to finite QSs with arbitrary service time distribution and resource volumes, including systems, in which service time and resource amount are dependent random variables. A part of the papers deals with models of resource systems with the processor sharing discipline (Egalitarian Processor Sharing, EPS) and active queue management (AQM). The paper [26] investigated single-line systems for servicing customers of random size with the Poisson arrival process under the assumption that the customer volume and service time are independent and the systems implement the AQM mechanism. Similar results were obtained in [25] for multi-line systems.

Despite the undoubted importance of studying RQSs, there are very few works on the development of analytical methods for analyzing such systems, due to the fact that for the correct construction of a Markov process it is necessary to take into account the volumes of all customers received in the system. Therefore, we highlight the set of works on the study of multi-line QS with multiple resources, a recurrent incoming flow, and arbitrary service times distributions and resource volumes. In paper [15], it was proposed to investigate a simplified version of the system, in which the resources volumes released upon service completion may differ from those that were allocated to the customer at the beginning of its service. The stochastic process describing the behavior of such simplified system is easier to analyze since there is no need to memorize the resource amounts occupied by each served customer. Further, in [14], using simulation it was shown that the stationary probability distributions of the total volumes of occupied resources are close. Then, in [16], it was proved that the analysis of the simplified system does not change the stationary distribution of the total volumes of occupied resources in the case of the Poisson arrival process and exponential service time.

1.2 Our Contributions

This paper continues the cycle of the authors' work on the study of heterogeneous RQSs with non-Poisson arrival process, non-exponential service times, an unlimited server number, and an unlimited resource amount.

In one of the first papers [19], the multidimensional dynamic screening method for non-Markov heterogeneous queuing systems was used to study systems with a renewal arrival process and an arbitrary distribution function of the service time. Later, this approach was developed for the study of tandem queues and queuing networks. For example, the paper [10] originally presents a modification of the dynamic screening method for a two-phase resource queuing system.

The paper [9,20] proposes the modification of the asymptotic analysis method for the case of multi-resource systems with the splitting of the heterogeneous customers. Such models and the results of its investigation may be applied to analyze the wireless networks, working with different access technologies, where each active session occupies a random amount of radio resources.

Note, the main advantage of the dynamic screening method is that the resource amount released upon service completion coincides with the resource amount allocated upon customer arrival. The studies carried out allowed us to conclude that in all cases the approximate distributions are Gaussian and their parameters depends in an explicit way on the stochastic process that describes the behavior of the system under study.

In this paper, we will carry out the study of a RQS with parallel servicing [24] of double customers, the arrival process of which is renewal. As a possible application of the model, we can consider a computing system that distributes requests to two processors; or some device (a smartphone/a laptop/a tablet) that can be connected to a wireless network for audio-calls and video-conference through different communication channels using different access technologies (see Fig. 1a). Moreover, in this formulation of the problem, the service duration does not depend in any way on the allocated resource amounts (bits or frequencies) to each session, as in streaming traffic [4].

The paper is organized as follows. Section 2 describes in detail the mathematical model and formalizes the stochastic process that will be explored later. In Sect. 3, the modification of the dynamic sieving method is described and it is applied to the analysis of the stochastic process, deriving the corresponding Kolmogorov equations. We obtain an approximation of the stationary probability distribution of the process under study using the asymptotic analysis method in Sect. 4. Then, in Sect. 5 numerical experiments to determine the obtained approximation accuracy are carried out. Finally, Sect. 6 draws the main conclusions about the paper.

2 Mathematical Model

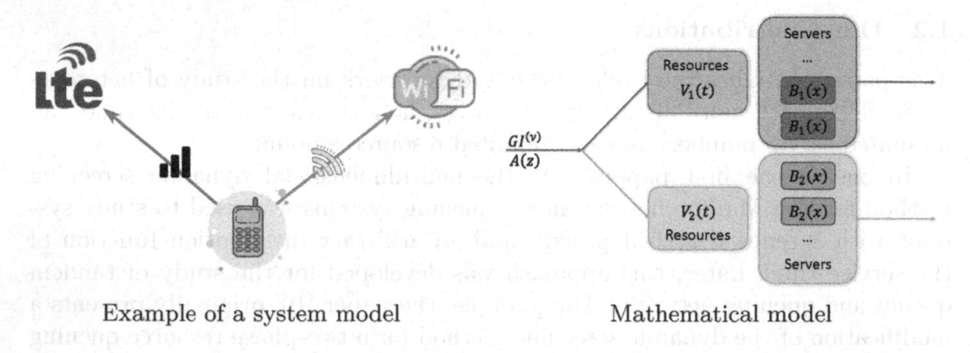

Fig. 1. The visualization of the parallel servicing

Consider the two-block RQS with an unlimited number of servers and resources shown in Fig. 1b. Customers arrive in the system according to a renewal arrival process, given by distribution function $A(z) = P\{\zeta < z\}$ of the inter-arrival time ζ. Each arrival customer generates a simultaneous request for both system resources, the resulting two customers instantly occupy two free servers in different blocks for a random times ξ_i with distribution function $B_i(x) = P\{\xi_i < x\}$, $i = 1, 2$, respectively, and also taking random resources amount ν_i with distribution function $G_i(y) = P\{\nu_i < y\}$, $i = 1, 2$. Upon completion of service, each customer leaves the system, frees the server, and all used resources. The occupied resources amount and the service time are independent of each other.

The goal of this paper is to obtain the probability distribution of the total volumes of occupied resources in the system. We study the system using the dynamic screening and asymptotic analysis methods, originally proposed in papers [11,18].

3 Dynamic Screening Method

Let denote by $V_1(t)$ and $V_2(t)$ the total amount of occupied resource on the first and second block at time t, respectively. The goal is to find the stationary probability distribution of the two-dimensional random process $\{V_1(t), V_2(t)\}$. However, this process is non-Markovian, therefore, we will use the dynamic screening method for its investigation.

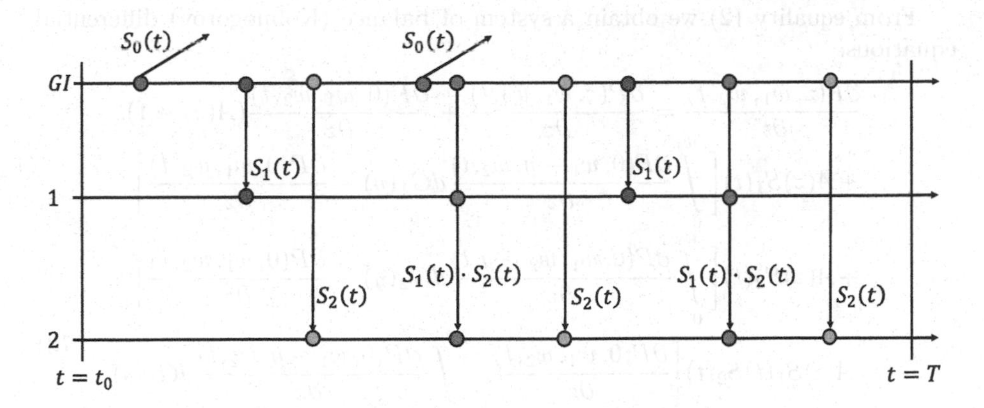

Fig. 2. Screening of arrived of customers

Let the system be empty at moment t_0, and let us fix a certain time moment $T > t_0$. The GI axis shows all arrival customers (see Fig. 2). We generate the points of the screened processes (axes 1 and 2) from the moments of arrivals. Consider the probability that a customer arriving at time t will not finish its service until the moment T. Let us denote the probability of screened arrivals on axis i as $S_i(t) = 1 - B_i(T - t)$, $(i = 1, 2)$ and on both axes as $S_1(t) \cdot S_2(t) = (1 - B_1(T - t)) \cdot (1 - B_2(T - t))$.

Let us denote the total resource amounts occupied by screened arrivals in the interval $[t_0, t)$ by $W_1(t)$ and $W_2(t)$ on the first and second axes, respectively. At the time moment $t = T$, the probability distributions of random variables $\{V_1(t), V_2(t)\}$ and $\{W_1(t), W_2(t)\}$ coincide:

$$P\{V_1(T) < y_1, V_2(T) < y_2\} = P\{W_1(T) < y_1, W_2(T) < y_2\},\\ y_1 > 0, y_2 > 0. \tag{1}$$

Let us denote the remaining time from the moment t to the moment of the next customer arrival by $z(t)$. For the probability distribution of the three-dimensional Markovian process $\{z(t), W_1(t), W_2(t)\}$ $P\{z(t) < z, W_1(t) < w_1, W_2(t) < w_2\} = P(z, w_1, w_2, t)$, $z, w_1, w_2 > 0$ we can write the equality:

$$\begin{aligned} P(z, w_1, w_2, t + \Delta t) &= [P(z + \Delta t, w_1, w_2, t) - P(\Delta t, w_1, w_2, t)] \\ &+ P(\Delta t, w_1, w_2, t)(1 - S_1(t))(1 - S_2(t))A(z) \\ &+ S_1(t)(1 - S_2(t))A(z) \int_0^{w_1} P(\Delta t, w_1 - y, w_2, t)dG_1(y) \\ &+ S_2(t)(1 - S_1(t))A(z) \int_0^{w_2} P(\Delta t, w_1, w_2 - y, t)dG_2(y) \end{aligned} \tag{2}$$

$$S_1(t)S_2(t)A(z) \int_0^{w_2} \int_0^{w_1} P(\Delta t, w_1 - y_1, w_2 - y_2, t)dG_1(y_1)dG_2(y_2) + o(\Delta t).$$

From equality (2) we obtain a system of balance (Kolmogorov) differential equations:

$$\frac{\partial P(z,w_1,w_2,t)}{\partial t} = \frac{\partial P(z,w_1,w_2,t)}{\partial z} + \frac{\partial P(0,w_1,w_2,t)}{\partial z}(A(z)-1)$$
$$+A(z)S_1(t)\left[\int_0^{w_1} \frac{\partial P(0,w_1-y,w_2,t)}{\partial z} dG_1(y) - \frac{\partial P(0,w_1,w_2,t)}{\partial z}\right]$$
$$+A(z)S_2(t)\left[\int_0^{w_2} \frac{\partial P(0,w_1,w_2-y,t)}{\partial z} dG_2(y) - \frac{\partial P(0,w_1,w_2,t)}{\partial z}\right]$$
$$+A(z)S_1(t)S_2(t)\left[\frac{\partial P(0,w_1,w_2,t)}{\partial z} - \int_0^{w_1} \frac{\partial P(0,w_1-y,w_2,t)}{\partial z} dG_1(y)\right.$$
$$-\int_0^{w_2} \frac{\partial P(0,w_1,w_2-y,t)}{\partial z} dG_2(y)$$
$$\left.+\int_0^{w_2}\int_0^{w_1} \frac{\partial P(0,w_1-y_1,w_2-y_2,t)}{\partial z} dG_1(y_1)dG_2(y_2)\right], \tag{3}$$

with the initial condition

$$P(z,w_1,w_2,t_0) = \begin{cases} R(z), \text{ if } w_1 = w_2 = 0, \\ 0, \text{ otherwise,} \end{cases}$$

where $R(z)$ is the stationary probability distribution of the stochastic process $z(t)$:

$$R(z) = \lambda \int_0^z (1-A(x))dx, \quad \lambda = \left[\int_0^\infty (1-A(x))\right]^{-1}.$$

Let us introduce the partial characteristic function

$$h(z,u_1,u_2,t) = \int_0^\infty e^{ju_1w_1} \int_0^\infty e^{ju_2w_2} P(z,dw_1,dw_2,t), z > 0,$$

where $j = \sqrt{-1}$ is the imaginary unit.

Then from (3) we obtain the differential equation:

$$\frac{\partial h(z,u_1,u_2,t)}{\partial t} = \frac{\partial h(z,u_1,u_2,t)}{\partial z} + \frac{\partial h(0,u_1,u_2,t)}{\partial z}\Big[A(z)-1$$
$$+A(z)S_1(t)(G_1^*(u_1)-1) + A(z)S_2(t)(G_2^*(u_2)-1) \tag{4}$$
$$+A(z)S_1(t)S_2(t)(G_1^*(u_1)-1)(G_2^*(u_2)-1)\Big],$$

where

$$G_i^*(u_i) = \int_0^\infty e^{ju_iy} dG_i(y), \; i = 1,2,$$

with the initial condition

$$h(z, u_1, u_2, t_0) = R(z).$$

4 Asymptotic Analysis

We will seek the solution of Eq. (4) by the asymptotic analysis method under the condition of increasing the intensity of arrivals.

Let $A_1(Nz) = A(z)$ be the distribution function of interarrival times, where N is a high-intensity parameter (theoretically, $N \to \infty$). Substituting this into (4), we obtain:

$$\begin{aligned} \tfrac{1}{N}\tfrac{\partial h(z,u_1,u_2,t)}{\partial t} &= \tfrac{\partial h(z,u_1,u_2,t)}{\partial z} + \tfrac{\partial h(0,u_1,u_2,t)}{\partial z}\Big[A(z) - 1 \\ &+ A(z)S_1(t)(G_1^*(u_1) - 1) + A(z)S_2(t)(G_2^*(u_2) - 1) \\ &+ A(z)S_1(t)S_2(t)(G_1^*(u_1) - 1)(G_2^*(u_2) - 1)\Big]. \end{aligned} \tag{5}$$

Theorem 1. *Under the conditions of increasing intensity of arrivals, the first-order asymptotic characteristic function of the stochastic process $\{z(t), W_1(t), W_2(t)\}$ has the form:*

$$h(z, u_1, u_2, t) \approx R(z) \exp\left\{ ju_1 N\lambda a_1 \int_{t_0}^{t} S_1(\tau)d\tau + ju_2 N\lambda a_2 \int_{t_0}^{t} S_2(\tau)d\tau \right\},$$

where a_1 and a_2 are the means of the occupied resource amounts.

Proof. Let us make the following substitutions in Eq. (5):

$$\varepsilon = \frac{1}{N}, \quad u_1 = \varepsilon x_1, \quad u_2 = \varepsilon x_2, \quad h(z, u_1, u_2, t) = f_1(z, x_1, x_2, t, \varepsilon).$$

We obtain

$$\begin{aligned} \varepsilon\tfrac{\partial f_1(z,x_1,x_2,t,\varepsilon)}{\partial t} &= \tfrac{\partial f_1(z,x_1,x_2,t,\varepsilon)}{\partial z} + \tfrac{\partial f_1(0,x_1,x_2,t,\varepsilon)}{\partial z}\Big[A(z) - 1 \\ &+ A(z)S_1(t)(G_1^*(\varepsilon x_1) - 1) + A(z)S_2(t)(G_2^*(\varepsilon x_2) - 1) \\ &+ A(z)S_1(t)S_2(t)(G_1^*(\varepsilon x_1) - 1)(G_2^*(\varepsilon x_2) - 1)\Big]. \end{aligned} \tag{6}$$

Step 1. Let us find the asymptotic solution $f_1(z, x_1, x_2, t) = \lim_{\varepsilon \to 0} f_1(z, x_1, x_2, t, \varepsilon)$. Let $\varepsilon \to 0$; Eq. (6) becomes:

$$\frac{\partial f_1(z, x_1, x_2, t)}{\partial z} + \frac{\partial f_1(0, x_1, x_2, t)}{\partial z}[A(z) - 1] = 0$$

and assume that:

$$f_1(z, x_1, x_2, t) = R(z)\Phi_1(x_1, x_2, t), \tag{7}$$

where $\Phi_1(x_1, x_2, t)$ is a scalar differentiable function satisfying the condition $\Phi_1(x_1, x_2, t_0) = 1$.

Step 2. In Eq. (6), we make the transition to the limit as $z \to \infty$:

$$\begin{aligned} \varepsilon\frac{\partial f_1(\infty,x_1,x_2,t,\varepsilon)}{\partial t} = \frac{\partial f_1(0,x_1,x_2,t,\varepsilon)}{\partial z}\Big[S_1(t)(G_1^*(\varepsilon x_1) - 1) \\ + S_2(t)(G_2^*(\varepsilon x_2) - 1) + S_1(t)S_2(t)(G_1^*(\varepsilon x_1) - 1)(G_2^*(\varepsilon x_2) - 1)\Big]. \end{aligned} \tag{8}$$

We substitute (7) into (8). Then we use the exponent expansion for functions in the form:

$$G_i^*(\varepsilon x) = \int_0^\infty e^{j\varepsilon x_i y} dG_i(y) = \int_0^\infty (1 + j\varepsilon x_i y + O(\varepsilon^2)) dG_i(y) = 1 + j\varepsilon x_i a_i + O(\varepsilon^2),$$

further we divide everything by ε and perform the passage to the limit as $\varepsilon \to 0$. Notice that $R'(0) = \lambda$, we obtain equation

$$\frac{\partial \Phi_1(x_1, x_2, t)}{\partial t} = \Phi_1(x_1, x_2, t)[jx_1\lambda a_1 S_1(t) + jx_2\lambda a_2 S_2(t)]. \tag{9}$$

The solution of the differential Eq. (9) is

$$\Phi_1(x_1, x_2, t) = \exp\left\{ jx_1\lambda a_1 \int_{t_0}^{t} S_1(\tau)d\tau + jx_2\lambda a_2 \int_{t_0}^{t} S_2(\tau)d\tau \right\},$$

that leads to the following asymptotic approximate inequality for $\varepsilon \to 0$:

$$h(z, u_1, u_2, t) \approx R(z) \exp\left\{ ju_1 N\lambda a_1 \int_{t_0}^{t} S_1(\tau)d\tau + ju_2 N\lambda a_2 \int_{t_0}^{t} S_2(\tau)d\tau \right\}.$$

The theorem is proved.

Theorem 2. *Under the conditions of increasing intensity of arrivals, the second-order asymptotic characteristic function of the stochastic process*

$\{z(t), W_1(t), W_2(t)\}$ has the form:

$$h(z,u_1,u_2,t) \approx R(z)\exp\Bigg\{ju_1N\lambda a_1\int_{t_0}^{t}S_1(\tau)d\tau + ju_2N\lambda a_2\int_{t_0}^{t}S_2(\tau)d\tau$$
$$+\frac{(ju_1)^2}{2}\left(N\lambda\alpha_1\int_{t_0}^{t}S_1(\tau)d\tau + N\kappa a_1^2\int_{t_0}^{t}S_1^2(\tau)d\tau\right)$$
$$+\frac{(ju_2)^2}{2}\left(N\lambda\alpha_2\int_{t_0}^{t}S_2(\tau)d\tau + N\kappa a_2^2\int_{t_0}^{t}S_2^2(\tau)d\tau\right)$$
$$+ju_1ju_2N(\lambda+\kappa)a_1a_2\int_{t_0}^{t}S_1(\tau)S_2(\tau)d\tau\Bigg\},$$

where α_1 and α_2 are the second raw moments of the occupied resources.

Proof. Let us introduce the function

$$h(z,u_1,u_2,t) = h_2(z,u_1,u_2,t)\cdot$$
$$\exp\left\{ju_1N\lambda a_1\int_{t_0}^{t}S_1(\tau)d\tau + ju_2N\lambda a_2\int_{t_0}^{t}S_2(\tau)d\tau\right\}.$$

Then from (5) we obtain the equation regarding $h_2(z,u_1,u_2,t)$:

$$\begin{aligned}\tfrac{1}{N}\tfrac{\partial h_2(z,u_1,u_2,t)}{\partial t} &+ h_2(z,u_1,u_2,t)(ju_1\lambda a_1S_1(t)+ju_2\lambda a_2S_2(t))\\ &= \frac{\partial h_2(z,u_1,u_2,t)}{\partial z}+\frac{\partial h_2(0,u_1,u_2,t)}{\partial z}\Big[A(z)-1\\ &+A(z)S_1(t)(G_1^*(u_1)-1)+A(z)S_2(t)(G_2^*(u_2)-1)\\ &+A(z)S_1(t)S_2(t)(G_1^*(u_1)-1)(G_2^*(u_2)-1)\Big].\end{aligned}\tag{10}$$

Let us make the following substitutions in Eq. (10)

$$\varepsilon^2=\frac{1}{N},\quad u_1=\varepsilon x_1,\quad u_2=\varepsilon x_2,\quad h_2(z,u_1,u_2,t)=f_2(z,x_1,x_2,t,\varepsilon).$$

We obtain

$$\begin{aligned}\varepsilon^2\tfrac{\partial f_2(z,x_1,x_2,t,\varepsilon)}{\partial t} &+ f_2(z,x_1,x_2,t,\varepsilon)(j\varepsilon x_1\lambda a_1S_1(t)+j\varepsilon x_2\lambda a_2S_2(t))\\ &= \frac{\partial f_2(z,x_1,x_2,t,\varepsilon)}{\partial z}+\frac{\partial f_2(0,x_1,x_2,t,\varepsilon)}{\partial z}\Big[A(z)-1\\ &+A(z)S_1(t)(G_1^*(\varepsilon x_1)-1)+A(z)S_2(t)(G_2^*(\varepsilon x_2)-1)\\ &+A(z)S_1(t)S_2(t)(G_1^*(\varepsilon x_1)-1)(G_2^*(\varepsilon x_2)-1)\Big].\end{aligned}\tag{11}$$

Step 1. Let us find asymptotic solution $f_2(z,x_1,x_2,t) = \lim\limits_{\varepsilon\to 0} f_2(z,x_1,x_2,t,\varepsilon)$. Let $\varepsilon\to 0$; Eq. (11) becomes:

$$\frac{\partial f_2(z,x_1,x_2,t)}{\partial z}+\frac{\partial f_2(0,x_1,x_2,t)}{\partial z}[A(z)-1]=0.$$

We will find the function $f_2(z, x_1, x_2, t)$ as:

$$f_2(z, x_1, x_2, t) = R(z)\Phi_2(x_1, x_2, t), \tag{12}$$

where $\Phi_2(x_1, x_2, t)$ is a scalar differentiable function satisfying the condition $\Phi_2(x_1, x_2, t_0) = 1$.

Step 2. We write the solution of Eq. (11) as power expansion

$$\begin{aligned} f_2(z, x_1, x_2, t, \varepsilon) = \Phi_2(x_1, x_2, t)\big[R(z) \\ + (j\varepsilon x_1 \lambda a_1 S_1(t) + j\varepsilon x_2 \lambda a_2 S_2(t))f(z) + O(\varepsilon^2)\big], \end{aligned} \tag{13}$$

where $f(z)$ is some differentiable function.

We substitute (13) into (11) and get a differential equation for the unknown function $f(z)$:

$$f(z) = f'(0)\int_0^z (1 - A(x))dx + \int_0^z (R(x) - A(x))dx.$$

Step 3. In Eq. (11), we make the transition to the limit as $z \to \infty$. The function $f_2(z, x_1, x_2, t)$ is monotonically increasing and bounded above on z, then:

$$\lim_{z\to\infty} \frac{\partial f_2(z, x_1, x_2, t, \varepsilon)}{\partial z} = 0.$$

We use the exponent expansion for functions in the form:

$$\begin{aligned} G_i^*(\varepsilon x_i) = \int_0^\infty e^{j\varepsilon x_i y} dG(y) = \int_0^\infty \left(1 + j\varepsilon x_i y + \tfrac{(j\varepsilon x_i y)^2}{2} + O(\varepsilon^3)\right) dG_i(y) \\ = 1 + j\varepsilon x_i a_i + \frac{(j\varepsilon x_i)^2}{2}\alpha_i + O(\varepsilon^3). \end{aligned} \tag{14}$$

We substitute (13) into (11), using (14), dividing everything by ε^2, for $z \to \infty$ and $\varepsilon \to 0$, we obtain the equation for $\Phi_2(x_1, x_2, t)$:

$$\begin{aligned} \tfrac{\partial \Phi_2(x_1, x_2, t)}{\partial t} = \Phi_2(x_1, x_2, t)\Bigg[\tfrac{(jx_1)^2}{2}(\lambda\alpha_1 S_1(t) + \kappa a_1^2 S_1^2(t)) \\ + \frac{(jx_2)^2}{2}(\lambda\alpha_2 S_2(t) + \kappa a_2^2 S_2^2(t)) + jx_1 jx_2 a_1 a_2 (\lambda + \kappa) S_1(t) S_2(t)\Bigg], \end{aligned}$$

where $\kappa = 2f'(0) - 2f(\infty)$. Let $f(\infty) = const$, then putting $f(\infty) = 0$ we get that $\kappa = 2f'(0)$. The solution of the differential equation is

$$\begin{aligned} \Phi_2(x_1, x_2, t) = \exp\Bigg\{ \tfrac{(jx_1)^2}{2}\left(\lambda\alpha_1 \int_{t_0}^t S_1(\tau)d\tau + \kappa a_1^2 \int_{t_0}^t S_1^2(\tau)d\tau\right) \\ + \frac{(jx_2)^2}{2}\left(\lambda\alpha_2 \int_{t_0}^t S_2(\tau)d\tau + \kappa a_2^2 \int_{t_0}^t S_2^2(\tau)d\tau\right) \\ + jx_1 jx_2(\lambda + \kappa)a_1 a_2 \int_{t_0}^t S_1(\tau)S_2(\tau)d\tau\Bigg\}. \end{aligned} \tag{15}$$

We substitute (15) in (12), then following the reverse substitutions, we write the approximate asymptotic equality regarding $h(z, u_1, u_2, t)$.

The theorem is proved.

Corollary 1. *The asymptotic characteristic function of the stationary probability distribution of the stochastic process* $\{V_1(t), V_2(t)\}$ *has the form:*

$$h(u_1, u_2) \approx \exp\left\{ju_1 N\lambda a_1 b_1 + ju_2 N\lambda a_2 b_2 + \frac{(ju_1)^2}{2}(N\lambda \alpha_1 b_1 + N\kappa a_1^2 \beta_1) + \frac{(ju_2)^2}{2}(N\lambda \alpha_2 b_2 + N\kappa a_2^2 \beta_2) + ju_1 ju_2 N a_1 a_2 (\lambda + \kappa) b_{12}\right\},$$

where

$$b_i = \int_0^\infty (1 - B_i(\tau))d\tau, \quad \beta_i = \int_0^\infty (1 - B_i(\tau))^2 d\tau, \quad i = 1, 2,$$

$$b_{12} = \int_0^\infty (1 - B_1(\tau))(1 - B_2(\tau))d\tau.$$

Corollary 2. *The vector of mathematical expectations and the covariance matrix of the resulting Gaussian approximation have the form:*

$$\mathbf{a} = N\lambda \left[a_1 b_1 \quad a_2 b_2\right],$$

$$\mathbf{K} = N \begin{bmatrix} \lambda \alpha_1 b_1 + \kappa a_1^2 \beta_1 & a_1 a_2 (\lambda + \kappa) b_{12} \\ a_1 a_2 (\lambda + \kappa) b_{12} & \lambda \alpha_2 b_2 + \kappa a_2^2 \beta_2 \end{bmatrix}.$$

5 Simulation and Numerical Examples

Let the system parameters be as follows:

- the renewal arrival process is defined by a uniform distribution function on the interval $\left[\frac{0.5}{N}; \frac{1.5}{N}\right]$. Note, that N is a high-intensity asymptotic parameter. Thus, intensity of arrivals $\lambda = N$.
- the service times have gamma distributions with parameters $\alpha_1 = \beta_1 = 0.5$ for the first block and $\alpha_2 = \beta_2 = 1.5$ for the second block;
- the resource requirements have uniform distributions on the intervals $[1; 2]$ and $[0; 1]$, respectively for the first and the second blocks.

For such initial data the parameters of Gaussian approximation are shown in Table 1.

Table 1. The approximation parameters

	Mean	Variance
First block	$1.5 \cdot N$	$1.584 \cdot N$
Second block	$0.5 \cdot N$	$0.201 \cdot N$

To determine the applicability area of the obtained approximate result, we perform simulation experiments for various values of the arrivals intensity N and use the Kolmogorov distance:

$$\Delta_{12} = \sup_{x,y} |F_{sim}(x, y) - F_{as}(x, y)|,$$

where $F_{sim}(x, y)$ is the empirical and $F_{as}(x, y)$ is the asymptotic 2D distribution function of the total resource amounts in the both blocks. In addition, we can find the Kolmogorov distances for one-dimensional probability distributions as

$$\Delta_i = \sup_x \left|F^i_{sim}(x) - F^i_{as}(x)\right|,$$

where $F^i_{sim}(x)$ is empirical and $F^i_{as}(x)$ is asymptotic distribution functions of the total resource amounts in the i-th block.

The first row of the Table 2 shows the Kolmogorov distance between the 2D distributions while the second and the third rows refer to the marginal distributions on each block for various values of the parameter N. We can conclude that the accuracy of the approximation increases with increasing arrivals intensity (with increasing the asymptotic parameter N), and on the Fig. 3 we can see confirmation of this. In this figure, the empirical probability mass function and the Gaussian probability density function with the parameters from Table 1 are shown. Here, we can see a faint resemblance between functions for $N = 5$ and the most accurate results for $N = 50$. If, for example, we suppose that the Kolmogorov distance less than 0.03 is enough for our purposes, we can conclude that the obtained asymptotic approximations are applicable when $N \geq 5$.

Table 2. Kolmogorov distances

N	1	3	5	7	10	20	50	100
Δ_{12}	0.240	0.041	0.021	0.015	0.011	0.008	0.005	0.003
Δ_1	0.300	0.049	0.024	0.021	0.018	0.013	0.009	0.007
Δ_2	0.300	0.048	0.023	0.016	0.013	0.009	0.006	0.004

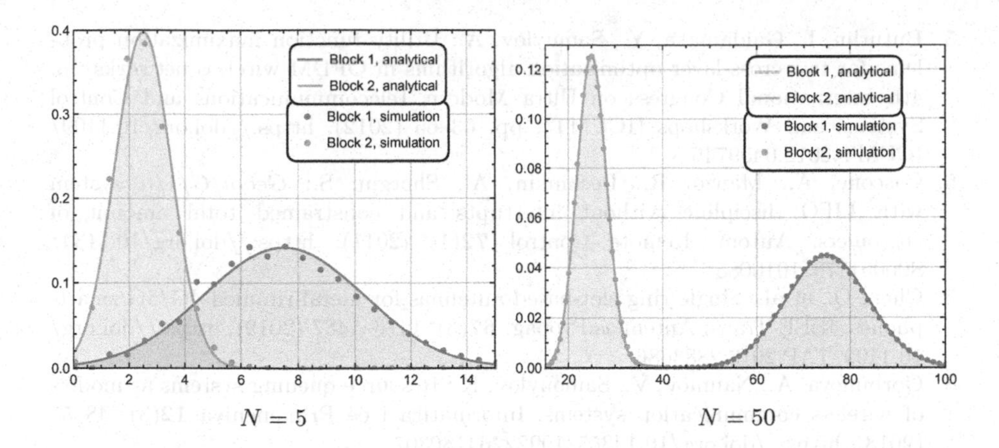

Fig. 3. Distribution of the total resource amounts for each blocks of the system

6 Conclusion

This paper analyzed a two-blocks resource queueing system with parallel servicing and a renewal arrival process. In the considered system, each customer takes up random resource amounts that do not affect the time of their service. By applying the dynamic screening and the asymptotic analysis methods the approximation of the stationary probability distribution of the total occupied resources amount in the system blocks is obtained. In more detail, it is shown that such probability distribution is 2D Gaussian, its parameters (means vector and covariance matrix) are obtained, and the correctness of the asymptotic is verified using discrete-event simulation.

References

1. Andrews, J.G., Buzzi, S., Choi, W., Hanly, S.V., Lozano, A., Soong, A.C.K., Zhang, J.C.: What will 5G be? IEEE J. Sel. Areas Commun. **32**(6), 1065–1082 (2014). https://doi.org/10.1109/JSAC.2014.2328098
2. Ayoub, W., Samhat, A.E., Nouvel, F., Mroue, M., Prévotet, J.: Internet of mobile things: overview of LoRaWAN, DASH7, and NB-IoT in LPWANs standards and supported mobility. IEEE Commun. Surv. Tutor. **21**(2),1561–1581 (2019). https://doi.org/10.1109/COMST.2018.2877382
3. Begishev, V., et al.: Quantifying the impact of guard capacity on session continuity in 3GPP new radio systems. IEEE Trans. Veh. Technol. **68**(12), 12345–12359 (2019). https://doi.org/10.1109/TVT.2019.2948702
4. Begishev, V., et al.: Resource allocation and sharing for heterogeneous data collection over conventional 3GPP LTE and emerging NB-IoT technologies. Comput. Commun. **120**, 93–101 (2018). https://doi.org/10.1016/j.comcom.2018.01.009

5. Buturlin, I., Gaidamaka, Y., Samuylov, A.: Utility function maximization problems for two cross-layer optimization algorithms in OFDM wireless networks. In: 4th International Congress on Ultra Modern Telecommunications and Control Systems and Workshops (ICUMT), pp. 63–65 (2012). https://doi.org/10.1109/ICUMT.2012.6459745
6. Cascone, A., Manzo, R., Pechinkin, A., Shorgin, S.: $Geom/G/1/n$ system with LIFO discipline without interrupts and constrained total amount of customers. Autom. Remote Control **72**(1) (2011). https://doi.org/10.1134/S0005117911010085
7. Chen, Q., et al.: Single ring slot-based antennas for metal-rimmed 4G/5G smartphones. IEEE Trans. Antennas Propag. **67**(3), 1476–1487 (2019). https://doi.org/10.1109/TAP.2018.2883686
8. Gorbunova, A., Naumov, V., Samouylov, K.: Resource queuing systems as models of wireless communication systems. Informatika i ee Primeneniya **12**(3), 48–55 (2018). https://doi.org/10.14357/19922264180307
9. Lisovskaya, E., Pankratova, E., Gaidamaka, Y., Moiseeva, S., Pagano, M.: Heterogeneous queueing system MAP/GI$^{(n)}$/∞ with random customers' capacities. In: Vishnevskiy, V.M., Samouylov, K.E., Kozyrev, D.V. (eds.) DCCN 2019. LNCS, vol. 11965, pp. 315–329. Springer, Cham (2019). https://doi.org/10.1007/978-3-030-36614-8_24
10. Lisovskaya, E.Y., Moiseev, A.N., Moiseeva, S.P., Pagano, M.: Modeling of mathematical processing of physics experimental data in the form of a non-Markovian multi-resource queuing system. Russ. Phys. J. **61**(12), 2188–2196 (2019). https://doi.org/10.1007/s11182-019-01655-6
11. Moiseev, A., Nazarov, A.: Queueing network $MAP - (GI/\infty)^K$ with high-rate arrivals. Euro. J. Oper. Res. **254**(1), 161–168 (2016). https://doi.org/10.1016/j.ejor.2016.04.011
12. Naumov, V.A., Samuilov, K.E.: Analysis of networks of the resource queuing systems. Autom. Remote Control **79**(5), 822–829 (2018). https://doi.org/10.1134/S0005117918050041
13. Naumov, V., Samouylov, K.: Conditions for the product form of the stationary probability distribution of Markovian resource loss systems. Tomsk State University Journal of Control and Computer Science (46), 64–72 (2019). https://doi.org/10.17223/19988605/46/8. (in Russian)
14. Naumov, V., Samouylov, K., Sopin, E., Andreev, S.: Two approaches to analyzing dynamic cellular networks with limited resources. In: 6th International Congress on Ultra Modern Telecommunications and Control Systems and Workshops (ICUMT), pp. 485–488 (2014)
15. Naumov, V., Samouylov, K.: On the modeling of queueing systems with multiple resources. Discrete Continuous Models Appl. Comput. Sci. **3**, 60–64 (2014)
16. Naumov, V.A., Samuilov, K.E., Samuilov, A.K.: On the total amount of resources occupied by serviced customers. Autom. Remote Control **77**(8), 1419–1427 (2016). https://doi.org/10.1134/S0005117916080087
17. Naumov, V., Beschastnyi, V., Ostrikova, D., Gaidamaka, Y.: 5G new radio system performance analysis using limited resource queuing systems with varying requirements. In: Vishnevskiy, V.M., Samouylov, K.E., Kozyrev, D.V. (eds.) DCCN 2019. LNCS, vol. 11965, pp. 3–14. Springer, Cham (2019). https://doi.org/10.1007/978-3-030-36614-8_1
18. Nazarov, A.A., Moiseev, A.N.: Analysis of the GI/PH/∞ system with high-rate arrivals. Autom. Control Comput. Sci. **49**(6), 328–339 (2015). https://doi.org/10.3103/S0146411615060085

19. Pankratova, E., Moiseeva, S.: Queueing system $GI|GI|\infty$ with n types of customers. In: Dudin, A., Nazarov, A., Yakupov, R. (eds.) ITMM 2015. CCIS, vol. 564, pp. 216–225. Springer, Cham (2015). https://doi.org/10.1007/978-3-319-25861-4_19
20. Pankratova, E., Moiseeva, S., Farkhadov, M., Moiseev, A.: Heterogeneous system $MMPP/GI(2)/\infty$ with random customers capacities. J. Siberian Fed. Univ. Math. Phys. **12**(2), 231–239 (2019). https://doi.org/10.17516/1997-1397-2019-12-2-231-239
21. Romm, E., Skitovich, V.: On certain generalization of problem of erlang. Autom. Remote Control **32**(6), 1000–1003 (1971)
22. Samuylov, A., et al.: Characterizing resource allocation trade-offs in 5G NR serving multicast and unicast traffic. IEEE Trans. Wirel. Commun. **19**(5), 3421–3434 (2020). https://doi.org/10.1109/TWC.2020.2973375
23. Samuylov, A., Beschastnyi, V., Moltchanov, D., Ostrikova, D., Gaidamaka, Y., Shorgin, V.: Modeling coexistence of unicast and multicast communications in 5G new radio systems. In: 30th Annual International Symposium on Personal, Indoor and Mobile Radio Communications (PIMRC), pp. 1–6 (2019). https://doi.org/10.1109/PIMRC.2019.8904350
24. Sinyakova, I., Moiseeva, S.: Investigation of output flows in the system with parallel service of multiple requests. In: IV International Conference "Problems of Cybernetics and Informatics" (PCI'2012), pp. 180–181 (2012)
25. Tikhonenko, O.: Queuing system with processor sharing and limited resources. Autom. Remote Control **71**(5), 803–815 (2010). https://doi.org/10.1134/S0005117910050073
26. Tikhonenko, O., Kempa, W.: On the queue-size distribution in the multi-server system with bounded capacity and packet dropping. Kybernetika **49**(6), 855–867 (2013)

Two-Phase Resource Queueing System with Requests Duplication and Renewal Arrival Process

Anastasia Galileyskaya[1], Ekaterina Lisovskaya[2(✉)], Michele Pagano[3], and Svetlana Moiseeva[1]

[1] National Research Tomsk State University, 36 Lenina Ave., Tomsk, Russian Federation
n.galileyskaya@bk.ru, smoiseeva@mail.ru

[2] Peoples' Friendship University of Russia (RUDN University), 6 Miklukho-Maklaya St., Moscow 117198, Russian Federation
lisovskaya-eyu@rudn.ru

[3] Department of Information Engineering, University of Pisa, Via Caruso 16, 56122 Pisa, Italy
michele.pagano@iet.unipi.it

Abstract. In this paper, we analyze a two-phase resource queueing system with duplication at the second phase under the assumption that customers enter the system according to a renewal process and take up random resource amounts that do not affect the time of their service (for video-conference or streaming traffic). We apply the dynamic screening method and the asymptotic analysis method to obtain an approximation for the stationary probability distribution of the total amount of occupied resources in the system under increasing arrival rate. In more detail, we show that the three-dimensional probability distribution of the total resource amounts on the system blocks is three-dimensional Gaussian, obtain its parameters (means vector and covariance matrix) and verify the correctness of the asymptotic using discrete-event simulation.

Keywords: Queueing system · Copying of requirement · Asymptotic analysis method · Arbitrary service time

1 Introduction

With the increasing load of multimedia traffic, wireless networks are constantly evolving, with higher data rates, lower latency, and better quality of service for

The publication has been prepared with the support of the "RUDN University Program 5-100" (recipient E. Lisovskaya, Conceptualization and Methodology), of the University of Pisa under the PRA 2018–2019 Research Project "CONCEPT – COmmunication and Networking for vehicular CybEr-Physical sysTems" (recipient M. Pagano, Software), and the reported study was funded by RFBR and Tomsk region according to the research project 19-41-703002 (recipient A. Galileyskaya, Investigation).

V. M. Vishnevskiy et al. (Eds.): DCCN 2020, LNCS 12563, pp. 350–364, 2020.
https://doi.org/10.1007/978-3-030-66471-8_27

end-users [6]. An example of the deployment of an urban mobile network is shown in Fig. 1. However, user demands are growing even faster than the capabilities of the network [13]. Consequently, network operators need powerful productivity tools that address the critical features of today's cellular networks.

Fig. 1. The schematic base stations deployment in the city

To transmit data over a wireless network, each session requests some amount of radio resources (frequencies) at a target base station [4]. Resource allocation and transmission mode selection are automatically determined by the signal-to-noise ratio of a singular session [2]. Therefore, at the same required data rates, each user session takes up different and usually random non-negative amounts of radio resources. This feature of modern wireless networks is of particular interest and cannot be ignored, since the required number of radio resources varies depending on the users geographic location [10,11,16].

The random variables associated with resource requirements can be either discrete or continuous, and depend on the specific signal propagation conditions, etc. [9,14,15].

Performance analysis of such networks, as a rule, can be carried out using queueing systems with limited resources. However, given the ongoing network densification, the load on future networks is expected to change significantly and unpredictably [1]. This requires new methods to model dense heterogeneous networks of the next generation [5,8].

We will consider a scenario where each session is preprocessed on the user equipment and then split into two parts, which are served independently on two base stations of different radio technologies. Moreover, we will be interested in services, for example, video conference or video streaming, for which the duration of the sessions does not depend on the number of allocated radio frequencies, but this number will affect the video quality [3].

We will consider a two-phase resource queueing system with copying (or splitting) customers in the second phase. The arrival process is assumed to be renewal with an arbitrary distribution of the interarrival time; the service time will also have an arbitrary probability distribution function.

This paper is aimed at demonstrating an approach to the study of resource queueing system in order to determine the required resource amount at a base station at known intensities of connection requests and service parameters, and, accordingly, the density of base station deployment in the target area [7].

The article is organized as follows. Section 2 describes the mathematical model, and the application of the dynamic screening method to the considered two-phase resource queueing system, while in Sect. 3 the corresponding Kolmogorov balance equations are derived. Section 4 presents the approximation of the total resource amount probability distribution and the derivation of the approximation parameters and their applicability is verified in Sect. 5 by means of discrete-event simulation. Finally, Sect. 6 concludes the paper with some final remarks.

2 Mathematical Model

Consider a queueing system with unlimited servers number and arbitrary service time. The renewal arrival process describing arrivals is determined by the arbitrary distribution function $A(z)$ of the interarrival time, its have a finite the first and second raw moments. We will assume that each customer is characterized by some random request to the resource.

Each arriving customer immediately occupies the first free server in the first phase and requires resources. We denote by $B_0(\tau)$ and $G_0(y)$ the distributions of the service time and of the requested resources, respectively After service in the first phase, the customer proceeds to the next phase and is duplicated there. The amounts of requested resources and the service times in the two blocks at the second phase are independent and characterized by the corresponding arbitrary distribution functions $G_1(y)$, $B_1(\tau)$ and $G_2(y)$, $B_2(\tau)$, respectively. When the service is complete in the second phase (independently of each other), the customer leaves the system. Resource amount and service times are mutually independent and do not depend on the epochs of customer arrivals. Figure 2 shows the structure of the system.

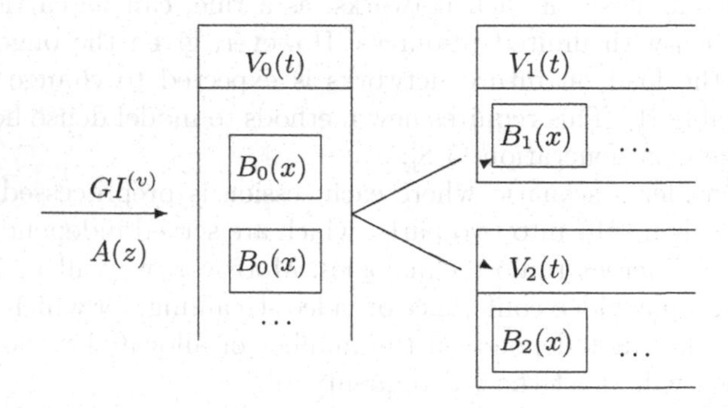

Fig. 2. Queueing system with the customers copying at the second phase and renewal arrival process

Denote by $V_i(t)$ the total resource amounts in the i-th block at time t, $(i = 0, 1, 2)$. Our goal is to derive the probabilistic characterization of the three-dimensional process $\{V_0(t), V_1(t), V_2(t)\}$. This process, in general, is not Markovian and, therefore, we use the dynamic screening method for its investigation.

Consider the four time axes that are presented in the Fig. 3. Let axis *GI* show the epochs of customers arrivals, axis 0, 1 and 2 will correspond to the first, second and third screened processes, respectively.

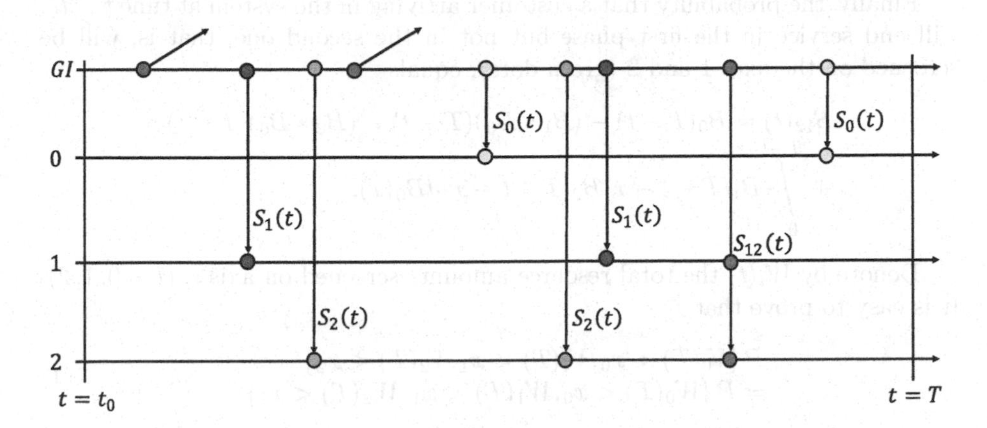

Fig. 3. Screening of the customers arrivals (Color figure online)

We introduce the functions (dynamic probabilities) $S_0(t), S_1(t), S_2(t), S_{12}(t)$, the values of which lie in the range $[0, 1]$ and satisfy the property $S_0(t) + S_1(t) + S_2(t) + S_{12}(t) \leq 1$.

The arrival process event can be screened only on one of the axes 0, 1, 2, or on axes 1 and 2 simultaneously. Let the system be empty at moment t_0, and let us fix some arbitrary moment T in the future.

In more detail, $S_0(t)$ represents the probability that a customer arriving at the time t will be still serviced in the first phase by moment T, that is, will be screened on the axis 0 (yellow dots). It is easy to show that

$$S_0(t) = 1 - B_0(T - t) \text{ for } t_0 \leq t \leq T.$$

The probability that the customer that entered the system at time $t > t_0$ by time T will finish service in the first phase and first block of the second phase, but not in the second block of the second phase, that is, will be screened on the axis 1 (purple dots), is equal to

$$S_1(t) = (B_2 * B_0)(T - t) - \int_0^{T-t} B_1(T - t - x)B_2(T - t - x)dB_0(x),$$

where * denotes the convolution.

The probability that the customer that entered the system at time $t > t_0$ by time T will finish service in the first phase and second block of the second phase, but not in the first block of the second phase, that is, will be screened on the axis 2 (pink dots), is equal to

$$S_2(t) = (B_1 * B_0)(T-t) - \int_0^{T-t} B_1(T-t-x)B_2(T-t-x)dB_0(x).$$

Finally, the probability that a customer arriving in the system at time $t > t_0$, will end service in the first phase but not in the second one, that is, will be screened on the axes 1 and 2 (green dots), equals

$$S_{12}(t) = B_0(T-t) - (B_1 * B_0)(T-t) - (B_2 * B_0)(T-t) + \int_0^{T-t} B_1(T-t-x)B_2(T-t-x)dB_0(x).$$

Denote by $W_i(t)$ the total resource amounts screened on axis i, $(i = 0, 1, 2)$. It is easy to prove that

$$\begin{aligned} &P\{V_0(T) < x_0, V_1(T) < x_1, V_2(T) < x_2\} \\ &= P\{W_0(T) < x_0, W_1(T) < x_1, W_2(T) < x_2\}, \end{aligned} \tag{1}$$

for $x_i > 0, (i = 0, 1, 2)$ [12]. We use the equality (1) to investigate the process $\{V_0(t), V_1(t), V_2(t)\}$ via the analysis of the process $\{W_0(t), W_1(t), W_2(t)\}$.

3 Integro-Differential Equations

Let us consider the four-dimensional Markovian process $\{z(t), W_0(t), W_1(t), W_2(t)\}$, where $z(t)$ is the residual time from t to the next arrival. Denoting the probability distribution of this process by

$$P\{z(t) < z, W_0(t) < x_0, W_1(t) < x_1, W_2(t) < x_2\} = P(z, x_0, x_1, x_2, t).$$

and taking into account the total probability formula, we can write the following system of Kolmogorov integro-differential equations

$$\begin{aligned} &\frac{\partial P(z,x_0,x_1,x_2,t)}{\partial t} = \frac{\partial P(z,x_0,x_1,x_2,t)}{\partial z} + \frac{\partial P(0,x_0,x_1,x_2,t)}{\partial z}(A(z)-1) \\ &+ A(z)\Bigg[S_0(t)\left(\int_0^{x_0} \frac{\partial P(0,x_0-y,x_1,x_2,t)}{\partial z} dG_0(y) - \frac{\partial P(0,x_0,x_1,x_2,t)}{\partial z}\right) \\ &+ S_1(t)\left(\int_0^{x_1} \frac{\partial P(0,x_0,x_1-y,x_2,t)}{\partial z} dG_1(y) - \frac{\partial P(0,x_0,x_1,x_2,t)}{\partial z}\right) \\ &+ S_2(t)\left(\int_0^{x_2} \frac{\partial P(0,x_0,x_1,x_2-y,t)}{\partial z} dG_2(y) - \frac{\partial P(0,x_0,x_1,x_2,t)}{\partial z}\right) \\ &+ S_{12}(t)\left(\int_0^{x_1}\int_0^{x_2} \frac{\partial P(0,x_0,x_1-y_1,x_2-y_2,t)}{\partial z} dG_2(y_2)dG_1(y_1) \right. \\ &\left. - \frac{\partial P(0,x_0,x_1,x_2,t)}{\partial z}\right)\Bigg] \end{aligned}$$

with the initial condition

$$P(z, x_0, x_1, x_2, t_0) = \begin{cases} R(z), x_0 = x_1 = x_2 = 0, \\ 0, \text{ otherwise}, \end{cases}$$

where $R(z)$ denotes the stationary probability distribution of the random variable, which is determined by equality

$$R(z) = \lambda \int_0^z (1 - A(x))dx, \quad \lambda = \left[\int_0^\infty (1 - A(x))dx \right]^{-1}.$$

We introduce the partial characteristic function

$$h(z, v_0, v_1, v_2, t) = \int_0^\infty e^{jv_0 x_0} \int_0^\infty e^{jv_1 x_1} \int_0^\infty e^{jv_2 x_2} P(z, dx_0, dx_1, dx_2, t), \quad j = \sqrt{-1}.$$

Then, we can rewrite

$$\begin{aligned} \frac{\partial h(z, v_0, v_1, v_2, t)}{\partial t} &= \frac{\partial h(z, v_0, v_1, v_2, t)}{\partial z} \\ &+ \frac{\partial h(0, v_0, v_1, v_2, t)}{\partial z} \{A(z) - 1 + A(z) [S_0(t) (G_0^*(v_0) - 1) \\ &+ S_1(t) (G_1^*(v_1) - 1) + S_2(t) (G_2^*(v_2) - 1) + S_{12}(t) (G_1^*(v_1) G_2^*(v_2) - 1)]\}, \end{aligned} \tag{2}$$

with the initial condition

$$h(z, v_0, v_1, v_2, t_0) = R(z), \tag{3}$$

where $G^*(v) = \int_0^\infty e^{jvy} dG(y)$.

4 Gaussian Approximation

A direct solution of equation (2) is impossible to find. Therefore, to solve the problem (2)–(3), we use the asymptotic analysis method under the condition of infinitely growing arrival rate. We write the distribution function of the interarrival times in the form $A(Nz)$, where $N \to \infty$ is a parameter of high flow intensity

Then, the Eq. (2) takes the form

$$\begin{aligned} \frac{1}{N} \frac{\partial h(z, v_0, v_1, v_2, t)}{\partial t} &= \frac{\partial h(z, v_0, v_1, v_2, t)}{\partial z} \\ &+ \frac{\partial h(0, v_0, v_1, v_2, t)}{\partial z} \{A(z) - 1 + A(z) [S_0(t) (G_0^*(v_0) - 1) \\ &+ S_1(t) (G_1^*(v_1) - 1) + S_2(t) (G_2^*(v_2) - 1) + S_{12}(t) (G_1^*(v_1) G_2^*(v_2) - 1)]\}. \end{aligned} \tag{4}$$

The result of the asymptotic analysis under the condition of the growing intensity of the arrivals is the approximation of the stationary probability distribution of the three-dimensional process describing the total amounts of the occupied resource on each system unit. Let us formulate the result as the following theorem, and give the proof below.

Theorem 1. *The stationary joint three-dimensional probability distribution of the total resource amount in the two-phase resource system with customers copying is asymptotically three-dimensional Gaussian with means vector:*

$$\mathbf{a} = N\lambda \left[a_1^{(0)} b_0 \; a_1^{(1)} b_1 \; a_1^{(2)} b_2\right],$$

where $a_1^{(i)}, i = 0, 1, 2$ are the means of resource requirements for a single customer and

$$b_0 = \int_0^\infty (1 - B_0(\tau))d\tau, \quad b_1 = \int_0^\infty (B_0(\tau) - (B_1 * B_0)(\tau))d\tau,$$

$$b_2 = \int_0^\infty (B_0(\tau) - (B_2 * B_0)(\tau))d\tau,$$

and covariance matrix:

$$\mathbf{K} = N \begin{bmatrix} \lambda a_2^{(0)} b_0 + \kappa \left(a_1^{(0)}\right)^2 \beta_0 & \kappa a_1^{(0)} a_1^{(1)} \beta_{01} & \kappa a_1^{(0)} a_1^{(2)} \beta_{02} \\ \kappa a_1^{(0)} a_1^{(1)} \beta_{01} & \lambda a_2^{(1)} b_1 + \kappa \left(a_1^{(1)}\right)^2 \beta_1 & \lambda a_1^{(1)} a_1^{(2)} b_{12} + \kappa a_1^{(1)} a_1^{(2)} \beta_{12} \\ \kappa a_1^{(0)} a_1^{(2)} \beta_{02} & \lambda a_1^{(1)} a_1^{(2)} b_{12} + \kappa a_1^{(1)} a_1^{(2)} \beta_{12} & \lambda a_2^{(2)} b_2 + \kappa \left(a_1^{(2)}\right)^2 \beta_2 \end{bmatrix},$$

where $a_2^{(i)}, i = 0, 1, 2$ are the second raw moments of resource requirements for a single customer, $\kappa = \lambda^3 \left(\sigma^2 - a^2\right)$, a and σ^2 are the mean and variance of the random variable with distribution function $A(z)$, and

$$\beta_0 = \int_0^\infty (1 - B_0(\tau))^2 d\tau, \quad \beta_{01} = \int_0^\infty (1 - B_0(\tau))(B_0(\tau) - (B_1 * B_0)(\tau))d\tau,$$

$$\beta_1 = \int_0^\infty (B_0(\tau) - (B_1 * B_0)(\tau))^2 d\tau, \quad \beta_2 = \int_0^\infty (B_0(\tau) - (B_2 * B_0)(\tau))^2 d\tau,$$

$$\beta_{02} = \int_0^\infty (1 - B_0(\tau))(B_0(\tau) - (B_2 * B_0)(\tau))d\tau,$$

$$\beta_{12} = \int_0^\infty (B_0(\tau) - (B_1 * B_0)(\tau))(B_0(\tau) - (B_2 * B_0)(\tau))d\tau,$$

$$b_{12} = \int_0^\infty (B_0(\tau) - (B_1 * B_0)(\tau) - (B_2 * B_0)(\tau) + \int_0^\tau B_1(\tau - x)B_2(\tau - x)dB_0(x))d\tau.$$

Proof. First of all, we carry out the asymptotic analysis of the first-order; we formulate the obtained result in the form of the following intermediate lemma.

Lemma 1. *The first-order asymptotic characteristic function of the four-dimensional stochastic process $\{z(t), W_0(t), W_1(t), W_2(t)\}$ has the form:*

$$h^{(1)}(z, v_0, v_1, v_2, t) = R(z)\exp\left\{N\lambda\left[jv_0 a_1^{(0)}\int_{t_0}^{t} S_0(\tau)d\tau + jv_1 a_1^{(1)}\int_{t_0}^{t} S_1(\tau)d\tau + jv_2 a_1^{(2)}\int_{t_0}^{t} S_2(\tau)d\tau + \left(jv_1 a_1^{(1)} + jv_2 a_1^{(2)}\right)\int_{t_0}^{t} S_{12}(\tau)d\tau\right]\right\}.$$

We skip the proof for sake of space, since it is not difficult and similar to the one used in [5].

At the next step, we carry out the asymptotic analysis of the second-order; we formulate the obtained result in the form of the following intermediate lemma.

Lemma 2. *The second-order asymptotic characteristic function of the four-dimensional stochastic process $\{z(t), W_0(t), W_1(t), W_2(t)\}$ has the form:*

$$\begin{aligned}
h^{(2)}(z, v_0, v_1, v_2, t) = R(z)\exp\Bigg\{ & N\lambda\left[jv_0 a_1^{(0)}\int_{t_0}^{t} S_0(\tau)d\tau + jv_1 a_1^{(1)}\int_{t_0}^{t} S_1(\tau)d\tau + jv_2 a_1^{(2)}\int_{t_0}^{t} S_2(\tau)d\tau + \left(jv_1 a_1^{(1)} + jv_2 a_1^{(2)}\right)\int_{t_0}^{t} S_{12}(\tau)d\tau\right] \\
& + \frac{(jv_0)^2}{2}N\left[\lambda a_2^{(0)}\int_{t_0}^{t} S_0(\tau)d\tau + \kappa\left(a_1^{(0)}\right)^2\int_{t_0}^{t} S_0(\tau)d\tau\right] \\
& + \frac{(jv_1)^2}{2}N\left[\lambda a_2^{(1)}\int_{t_0}^{t}(S_1(\tau)+S_{12}(\tau))d\tau + \kappa\left(a_1^{(1)}\right)^2\int_{t_0}^{t}(S_1(\tau)+S_{12}(\tau))^2 d\tau\right] \\
& + \frac{(jv_2)^2}{2}N\left[\lambda a_2^{(2)}\int_{t_0}^{t}(S_2(\tau)+S_{12}(\tau))d\tau + \kappa\left(a_1^{(2)}\right)^2\int_{t_0}^{t}(S_2(\tau)+S_{12}(\tau))^2 d\tau\right] \\
& + jv_0 jv_1 a_1^{(0)} a_1^{(1)} N\kappa\int_{t_0}^{t} S_0(\tau)(S_1(\tau)+S_{12}(\tau))d\tau \\
& + jv_0 jv_2 a_1^{(0)} a_1^{(2)} N\kappa\int_{t_0}^{t} S_0(\tau)(S_2(\tau)+S_{12}(\tau))d\tau \\
& + jv_1 jv_2 N a_1^{(1)} a_1^{(2)}\left[\lambda\int_{t_0}^{t} S_{12}(\tau)d\tau + \kappa\int_{t_0}^{t}(S_1(\tau)+S_{12}(\tau))(S_2(\tau)+S_{12}(\tau))d\tau\right]\Bigg\}.
\end{aligned}$$

Proof. In Eq. (2), *by substituting*

$$h(z, v_0, v_1, v_2, t) = h_2(z, v_0, v_1, v_2, t) \exp \left\{ N\lambda \left[jv_0 a_1^{(0)} \int_{t_0}^{t} S_0(\tau)d\tau \right.\right.$$
$$+ jv_1 a_1^{(1)} \int_{t_0}^{t} S_1(\tau)d\tau + jv_2 a_1^{(2)} \int_{t_0}^{t} S_2(\tau)d\tau \tag{5}$$
$$\left.\left. + \left(jv_1 a_1^{(1)} + jv_2 a_1^{(2)}\right) \int_{t_0}^{t} S_{12}(\tau)d\tau \right]\right\},$$

we get the equation for the function $h_2(z, v_0, v_1, v_2, t)$

$$\frac{1}{N}\frac{\partial h_2(z, v_0, v_1, v_2, t)}{\partial t} + h_2(z, v_0, v_1, v_2, t)\lambda \left[jv_0 a_1^{(0)} S_0(t) + jv_1 a_1^{(1)} S_1(t) \right.$$
$$\left. + jv_2 a_1^{(2)} S_2(t) + \left(jv_1 a_1^{(1)} + jv_2 a_1^{(2)}\right) S_{12}(t) \right] = \frac{\partial h_2(z, v_0, v_1, v_2, t)}{\partial z}$$
$$+ \frac{\partial h_2(0, v_0, v_1, v_2, t)}{\partial z} \left\{ A(z) - 1 + A(z) \left[S_0(t)(G_0^*(v_0) - 1) \right.\right. \tag{6}$$
$$\left.\left. + S_1(t)(G_1^*(v_1) - 1) + S_2(t)(G_2^*(v_2) - 1) + S_{12}(t)(G_1^*(v_1)G_2^*(v_2) - 1) \right]\right\},$$

with the initial condition

$$h_2(z, v_0, v_1, v_2, t_0) = R(z). \tag{7}$$

Let us make the following substitution here

$$\begin{gathered} \varepsilon^2 = \frac{1}{N}, \quad v_0 = \varepsilon y_0, \quad v_1 = \varepsilon y_1, \quad v_2 = \varepsilon y_2, \\ h_2(z, v_0, v_1, v_2, t) = f_2(z, y_0, y_1, y_2, t, \varepsilon). \end{gathered} \tag{8}$$

Using the notation (5), *the problem* (6)–(7) *can be rewrite as*

$$\varepsilon^2 \frac{\partial f_2(z, y_0, y_1, y_2, t, \varepsilon)}{\partial t} + f_2(z, y_0, y_1, y_2, t, \varepsilon)\lambda \left[j\varepsilon y_0 a_1^{(0)} S_0(t) \right.$$
$$\left. + j\varepsilon y_1 a_1^{(1)} S_1(t) + j\varepsilon y_2 a_1^{(2)} S_2(t) + \left(j\varepsilon y_1 a_1^{(1)} + j\varepsilon y_2 a_1^{(2)}\right) S_{12}(t) \right]$$
$$= \frac{\partial f_2(z, y_0, y_1, y_2, t, \varepsilon)}{\partial z} + \frac{\partial f_2(0, y_0, y_1, y_2, t, \varepsilon)}{\partial z} [A(z) - 1 \tag{9}$$
$$+ A(z)((G_0^*(\varepsilon y_0) - 1)S_0(t) + (G_1^*(\varepsilon y_1) - 1)S_1(t) + (G_2^*(\varepsilon y_2) - 1)S_2(t)$$
$$+ (G_1^*(\varepsilon y_1)G_2^*(\varepsilon y_2) - 1)S_{12}(t))],$$

with the initial condition $f_2(z, y_0, y_1, y_2, t_0, \varepsilon) = R(z)$.

Let us find the asymptotic solution of this problem (when $\varepsilon \to 0$), *that is*

$$f_2(z, y_0, y_1, y_2, t) = \lim_{\varepsilon \to 0} f_2(z, y_0, y_1, y_2, t, \varepsilon).$$

Step 1. We carry out the passage to the limit as $\varepsilon \to 0$ *in* (9), *obtaining*

$$\frac{\partial f_2(z,y_0,y_1,y_2,t,\varepsilon)}{\partial z} + \frac{\partial f_2(0,y_0,y_1,y_2,t,\varepsilon)}{\partial z}(A(z)-1) = 0.$$

Then, we represent the function $f_2(z,y_0,y_1,y_2,t)$ *in the form*

$$f_2(z,y_0,y_1,y_2,t) = R(z)\Phi_2(y_0,y_1,y_2,t), \tag{10}$$

where $\Phi_2(y_0,y_1,y_2,t)$ *is some scalar differentiable function satisfying the initial condition* $\Phi_2(y_0,y_1,y_2,t_0) = 1$.

Step 2. We write the solution of Eq. (9) *in the form*

$$\begin{aligned} f_2(z,y_0,y_1,y_2,t,\varepsilon) &= \Phi_2(y_0,y_1,y_2,t)\Big\{R(z) + g(z)\Big(j\varepsilon y_0 a_1^{(0)} S_0(t) \\ &+ j\varepsilon y_1 a_1^{(1)} S_1(t) + \Big(j\varepsilon y_1 a_1^{(1)} + j\varepsilon y_2 a_1^{(2)}\Big) S_{12}(t)\Big)\Big\} + o(\varepsilon^2), \end{aligned} \tag{11}$$

where $g(z)$ *is some scalar function. We substitute this expression in* (9) *using the decomposition* $e^{j\varepsilon x} = 1 + j\varepsilon x + o(\varepsilon^2)$, *and also given that* $R'(z) = \lambda(1 - A(z))$, *we get a differential equation for the function* $g(z)$

$$\lambda R(z) = g'(z) + \lambda A(z) + g'(0)(A(z)-1),$$

and the solution of this equation gives the following result

$$g(z) = \lambda \int_0^z (A(u) - R(u))du + g'(0)(A(z)-1).$$

It is easy to show that

$$g'(0) = \lambda g(\infty) + \frac{\kappa}{2}. \tag{12}$$

Step 3. In (9), *we perform the limit transition for* $z \to \infty$. *Due to the way the function* $f_2(z,y_0,y_1,y_2,t,\varepsilon)$ *is constructed, it is a monotonically increasing and top-bounded function in* z. *Therefore,* $\lim\limits_{\varepsilon\to 0} \frac{\partial f_2(\infty,y_0,y_1,y_2,t,\varepsilon)}{\partial z} = 0$. *Given this equality and using the decomposition* $e^{j\varepsilon x} = 1 + j\varepsilon x + \frac{(j\varepsilon x)^2}{2} + o(\varepsilon^3)$, *as a result of transformations we obtain*

$$\begin{aligned} &\varepsilon^2 \frac{\partial f_2(\infty,y_0,y_1,y_2,t,\varepsilon)}{\partial t} + f_2(\infty,y_0,y_1,y_2,t,\varepsilon)\lambda\Big(j\varepsilon y_0 a_1^{(0)} S_0(t) + j\varepsilon y_1 a_1^{(1)} S_1(t) \\ &+ j\varepsilon y_2 a_1^{(2)} S_2(t) + \Big(j\varepsilon y_1 a_1^{(1)} + j\varepsilon y_2 a_1^{(2)}\Big) S_{12}(t)\Big) = \frac{\partial f_2(0,y_0,y_1,y_2,t,\varepsilon)}{\partial z} \\ &\times \Bigg[\Big(j\varepsilon y_0 a_1^{(0)} + \frac{(j\varepsilon y_0)^2}{2} a_1^{(0)}\Big) S_0(t) + \Big(j\varepsilon y_1 a_1^{(1)} + \frac{(j\varepsilon y_1)^2}{2} a_1^{(1)}\Big) S_1(t) \\ &+ \Big(j\varepsilon y_2 a_1^{(2)} + \frac{(j\varepsilon y_2)^2}{2} a_1^{(2)}\Big) S_2(t) + \Big(j\varepsilon y_1 a_1^{(1)} + j\varepsilon y_2 a_1^{(2)} \\ &+ \frac{(j\varepsilon y_1)^2}{2} a_2^{(1)} + \frac{(j\varepsilon y_2)^2}{2} a_2^{(2)} + (j\varepsilon)^2 y_1 y_2 a_1^{(1)} a_1^{(2)}\Big) S_{12}(t)\Bigg] + o(\varepsilon^3). \end{aligned}$$

Substituting the expansion (11), *we write*

$$\varepsilon^2\frac{\partial\Phi_2(y_0,y_1,y_2,t)}{\partial t}+\Phi_2(y_0,y_1,y_2,t)\lambda\Big(j\varepsilon y_0a_1^{(0)}S_0(t)+j\varepsilon y_1a_1^{(1)}S_1(t)$$
$$+\ j\varepsilon y_2a_1^{(2)}S_2(t)+\Big(j\varepsilon y_1a_1^{(1)}+j\varepsilon y_2a_1^{(2)}\Big)S_{12}(t)\Big)$$
$$+\Phi_2(y_0,y_1,y_2,t)g(\infty)\lambda\Big(j\varepsilon y_0a_1^{(0)}S_0(t)+j\varepsilon y_1a_1^{(1)}S_1(t)+j\varepsilon y_2a_1^{(2)}S_2(t)$$
$$+\Big(j\varepsilon y_1a_1^{(1)}+j\varepsilon y_2a_1^{(2)}\Big)S_{12}(t)\Big)^2=\Phi_2(y_0,y_1,y_2,t)\lambda\Bigg[\Big(j\varepsilon y_0a_1^{(0)}$$
$$+\ \frac{(j\varepsilon y_0)^2}{2}a_2^{(0)}\Big)S_0(t)+\Big(j\varepsilon y_1a_1^{(1)}+\frac{(j\varepsilon y_1)^2}{2}a_2^{(1)}\Big)S_1(t)$$
$$+\ \Big(j\varepsilon y_2a_1^{(2)}+\frac{(j\varepsilon y_2)^2}{2}a_2^{(2)}\Big)S_2(t)+\Big(j\varepsilon y_1a_1^{(1)}+j\varepsilon y_2a_1^{(2)}+(j\varepsilon)^2y_1y_2a_1^{(1)}a_1^{(2)}$$
$$+\ \frac{(j\varepsilon y_1)^2}{2}a_2^{(1)}+\frac{(j\varepsilon y_2)^2}{2}a_2^{(2)}\Big)S_{12}(t)\Bigg]+\Phi_2(y_0,y_1,y_2,t)g'(0)\Big(j\varepsilon y_0a_1^{(0)}S_0(t)$$
$$+\ j\varepsilon y_1a_1^{(1)}S_1(t)+j\varepsilon y_2a_1^{(2)}S_2(t)+\Big(j\varepsilon y_1a_1^{(1)}+j\varepsilon y_2a_1^{(2)}\Big)S_{12}(t)\Big)^2+o(\varepsilon^3).$$

Applying the appropriate transformations, we divide by ε^2, *substitute* (12) *and pass to the limit at* $\varepsilon\to 0$. *We obtain the following differential equation*

$$\frac{\partial\Phi_2(y_0,y_1,y_2,t)}{\partial t}=\Phi_2(y_0,y_1,y_2,t)\Bigg\{\frac{(jy_0)^2}{2}\Big(\lambda a_2^{(0)}S_0(t)+\kappa\Big(a_1^{(0)}S_0(t)\Big)^2\Big)$$
$$+\ \frac{(jy_1)^2}{2}\Big[\lambda a_2^{(1)}(S_1(t)+S_{12}(t))+\kappa\Big(a_1^{(1)}\Big)^2(S_1(t)+S_{12}(t))^2\Big]$$
$$+\ \frac{(jy_2)^2}{2}\Big[\lambda a_2^{(2)}(S_2(t)+S_{12}(t))+\kappa\Big(a_1^{(2)}\Big)^2(S_2(t)+S_{12}(t))^2\Big]$$
$$+j^2y_0y_1a_1^{(0)}a_1^{(1)}\kappa S_0(t)(S_1(t)+S_{12}(t))+j^2y_0y_2a_1^{(0)}a_1^{(2)}\kappa S_0(t)(S_2(t)+S_{12}(t))$$
$$+j^2y_1y_2a_1^{(1)}a_1^{(2)}\Big[\lambda S_{12}(t)+\kappa\Big(S_1(t)+S_{12}(t))(S_2(t)+S_{12}(t)\Big)\Big]\Bigg\}.$$

Next, we write the solution to this equation, taking into account the initial condition, as an exponential. Then, we substitute this solution into (10), *make the inverse changes to* (8), *we get a second-order asymptotic characteristic function of the four-dimensional stochastic process* $\{z(t),W_0(t),W_1(t),W_2(t)\}$, *as in the statement of the Lemma 2.*

In the equality in the formulation of Lemma 2, we put $z\to\infty, t=T, t_0\to-\infty$, thus we obtain the characteristic function of the stationary probability distribution of the three-dimensional process $\{V_0(t),V_1(t),V_2(t)\}$, which has the form:

$$h(v_0, v_1, v_2) = \exp\Big\{N\lambda\left[jv_0 a_1^{(0)} b_0 + jv_1 a_1^{(1)} b_1 + jv_2 a_1^{(2)} b_2\right]$$
$$+ \frac{(jv_0)^2}{2} N\left[\lambda a_2^{(0)} b_0 + \kappa\left(a_1^{(0)}\right)^2 \beta_0\right] + \frac{(jv_1)^2}{2} N\left[\lambda a_2^{(1)} b_1 + \kappa\left(a_1^{(1)}\right)^2 \beta_1\right]$$
$$+ \frac{(jv_2)^2}{2}\left[\lambda a_2^{(2)} b_2 + \kappa\left(a_1^{(2)}\right)^2 \beta_2\right] + jv_0 jv_1 \kappa a_1^{(0)} a_1^{(1)} \beta_{01}$$
$$+ jv_0 jv_2 \kappa a_1^{(0)} a_1^{(2)} \beta_{02} + jv_1 jv_2 a_1^{(1)} a_1^{(2)} \left[\lambda b_{12} + \kappa \beta_{12}\right]\Big\}.$$

and coincides with the characteristic function of the three-dimensional Gaussian distribution, the parameters of which are presented in the formulation of Theorem 1.

5 Numerical Example

We assume that the renewal arrival process is characterized by the following distribution function

$$A(Nz) = \begin{cases} 0, & z < \frac{0.5}{N}, \\ Nz - 0.5, & z \in \left[\frac{0.5}{N}; \frac{1.5}{N}\right], \\ 1, & z > \frac{1.5}{N}, \end{cases}$$

and service time has gamma distribution with parameters

$$\alpha_0 = \beta_0 = 0.5; \quad \alpha_1 = \beta_1 = 1.5; \quad \alpha_2 = \beta_2 = 2.5.$$

Resource requests have a uniform distribution in the range: [0; 3] for the first phase, [0; 2] for the first block of the second phase and [0; 1] for the second block of the second phase. Our goal is to find a lower bound of parameter N for the applicability of the proposed approximation. To this aim, we carried out series of simulation experiments (in each of them 10^{10} arrivals are generated) for increasing values of N and compared the asymptotic distributions with the empiric ones by using the Kolmogorov distance $\Delta = \sup_x |F(x) - G(x)|$ as an accuracy measure. Here $F(x)$ is the cumulative distribution function built on the basis of simulation results, and $G(x)$ is the Gaussian approximation based on Theorem 1; the corresponding parameters for the three classes are summarized in Table 1.

Table 2 presents the Kolmogorov distances between the asymptotic and empirical distribution functions of total amount of resources occupied in three system blocks. The approximation accuracy increases with incoming process intensity N, which is also illustrated by Fig. 4.

Table 1. Gaussian approximation parameters

System block	Mean	Variance
First phase	$1.5 \cdot N$	$2.251 \cdot N$
First block of the second phase	$1 \cdot N$	$1.088 \cdot N$
Second block of the second phase	$0.5 \cdot N$	$0.266 \cdot N$

Table 2. Kolmogorov distances

N	1	5	10	50
Δ_0	0.303	0.036	0.021	0.009
Δ_1	0.312	0.038	0.021	0.009
Δ_2	0.303	0.035	0.019	0.008

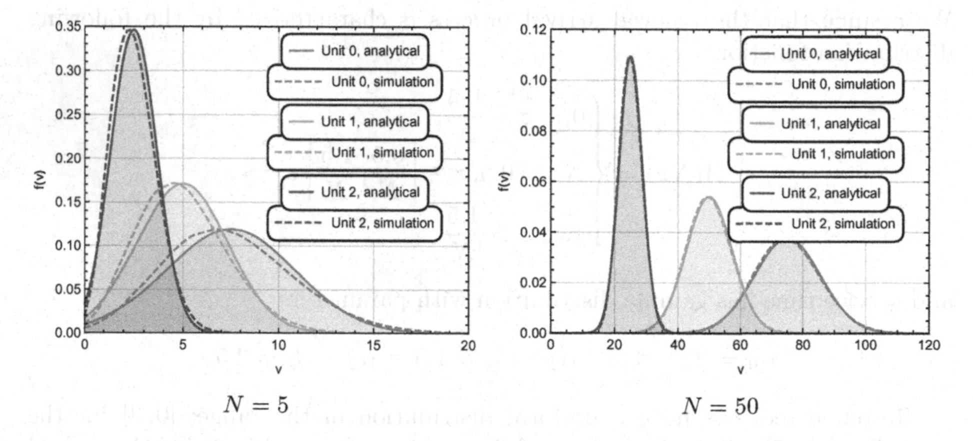

Fig. 4. Distribution of the total resource amounts for each blocks of the system

6 Conclusion

The paper investigated a tandem resource queueing system with unlimited server number and resource, with customers copying at the second phase and with a renewal arrival process. Using the asymptotic analysis method, it is shown that the joint asymptotic probabilities distribution of the total amount of occupied resources in each block converges to a three-dimensional Gaussian distribution in the asymptotic condition of increasing arrival intensity.

References

1. Cisco VNI Global IP Traffic Forecast, 2017–2022. https://www.cisco.com/c/dam/m/en_us/network-intelligence/service-provider/digital-transformation/knowledge-network-webinars/pdfs/1213-business-services-ckn.pdf. Accessed 21 June 2020

2. IEEE Standard for Terminology and Test Methods of Digital-to-Analog Converter Devices. IEEE Std 1658–2011, pp. 1–126 (2012). https://doi.org/10.1109/IEEESTD.2012.6152113
3. IEEE Standard for Information technology-Telecommunications and information exchange between systems Local and metropolitan area networks-Specific requirements - Part 11: Wireless LAN Medium Access Control (MAC) and Physical Layer (PHY) Specifications. IEEE Std 802.11-2016 (Revision of IEEE Std 802.11-2012) (2016). https://doi.org/0.1109/IEEESTD.2016.7786995
4. ETSI TR 138 900: Study on channel model for frequency spectrum above 6 GHz. https://www.etsi.org/deliver/etsi_tr/138900_138999/138900/14.02.00_60/tr_138900v140200p.pdf. Accessed 15 Aug 2020
5. Galileyskaya, A., Lisovskaya, E., Pagano, M.: On the total amount of the occupied resources in the multi-resource QS with renewal arrival process. Commun.Comput. Inf. Sci. **1109**, 257–269 (2019). https://doi.org/10.1007/978-3-030-33388-1_21
6. GSMA: The State of Mobile Internet Connectivity 2019. https://www.gsma.com/mobilefordevelopment/wp-content/uploads/2019/07/GSMA-State-of-Mobile-Internet-Connectivity-Report-2019.pdf. Accessed 10 Aug 2020
7. Li, W., Tian, L., Zhang, J., Cheng, Y.: Analysis of base station deployment impact on LOS probability model for 5G indoor scenario. In: 2017 IEEE/CIC International Conference on Communications in China (ICCC), pp. 1–5 (2017). https://doi.org/10.1109/ICCChina.2017.8330328
8. Lisovskaya, E., Pankratova, E., Gaidamaka, Y., Moiseeva, S., Pagano, M.: Heterogeneous queueing system $MAP/\mathrm{GI}^{(n)}/\infty$ with random customers' capacities. In: Vishnevskiy, V.M., Samouylov, K.E., Kozyrev, D.V. (eds.) DCCN 2019. LNCS, vol. 11965, pp. 315–329. Springer, Cham (2019). https://doi.org/10.1007/978-3-030-36614-8_24
9. Naumov, V., Samouylov, K.: Analysis of networks of the resource queuing systems. Autom. Remote Control **79**, 822–829 (2018). https://doi.org/10.1134/S0005117918050041
10. Naumov, V., Samouylov, K.: Conditions for the product form of the stationary probability distribution of Markovian resource loss systems. Tomsk State Uni. J. Control Comput. Sci. **46**, 64–72 (2019). https://doi.org/10.17223/19988605/46/8. ((in Russian)
11. Naumov, V., Beschastnyi, V., Ostrikova, D., Gaidamaka, Y.: 5G new radio system performance analysis using limited resource queuing systems with varying requirements. In: Vishnevskiy, V.M., Samouylov, K.E., Kozyrev, D.V. (eds.) DCCN 2019. LNCS, vol. 11965, pp. 3–14. Springer, Cham (2019). https://doi.org/10.1007/978-3-030-36614-8_1
12. Nazarov, A., Moiseev, A.: Analysis of the $GI/PH/\infty$ system with high-rate arrivals. Autom. Control Comput. Sci. **49**, 328–339 (2015). https://doi.org/10.3103/S0146411615060085
13. Stahlbuhk, T., Shrader, B., Modiano, E.: Throughput maximization in uncooperative spectrum sharing networks. IEEE/ACM Trans. Netw. 1–14 (2020). https://doi.org/10.1109/TNET.2020.3012273
14. Tikhonenko, O., Ziółkowski, M.: Queueing models of systems with Non-homogeneous customers and their applications in computer ience. In: 2019 IEEE 15th International Scientific Conference on Informatics, pp. 427–432 (2019). https://doi.org/10.1109/Informatics47936.2019.9119317

15. Tikhonenko, O., Ziółkowski, M.: Unreliable single-server queueing system with customers of random capacity. In: Gaj, P., Gumiński, W., Kwiecień, A. (eds.) CN 2020. CCIS, vol. 1231, pp. 153–170. Springer, Cham (2020). https://doi.org/10.1007/978-3-030-50719-0_12
16. Tzanakaki, A., Anastasopoulos, M.P., Simeonidou, D.: Converged optical, wireless, and data center network infrastructures for 5G services. IEEE/OSA J. Opt. Commun. Netw. **11**(2), A111–A122 (2019). https://doi.org/10.1364/JOCN.11.00A111

Deep Neural Networks for Emotion Recognition

Eugene Yu. Shchetinin[1], Leonid A. Sevastianov[2,3], Dmitry S. Kulyabov[2,3](✉), Edik A. Ayrjan[3,4], and Anastasia V. Demidova[2]

[1] Government of the Russian Federation, Financial University, Moscow, Russian Federation
riviera-molto@mail.ru

[2] Peoples' Friendship University of Russia (RUDN University), Moscow, Russian Federation
{sevastianov-la,kulyabov-ds,demidova-av}@rudn.ru

[3] Joint Institute for Nuclear Research, Dubna, Russian Federation
ayrjan@jinr.ru

[4] Dubna State University, Dubna, Russian Federation

Abstract. The paper investigates the problem of recognizing human emotions by voice using deep learning methods. Deep convolutional neural networks and recurrent neural networks with bidirectional LSTM memory cell were used as models of deep neural networks. On their basis, an ensemble of neural networks is proposed. We carried out computer experiments on using the constructed neural networks and popular machine learning algorithms for recognizing emotions in human speech contained in the RAVDESS audio record database. The computational results showed a higher efficiency of neural network models compared to machine learning algorithms. Accuracy estimates for individual emotions obtained using neural networks were 80%. The directions of further research in the field of recognition of human emotions are proposed.

Keywords: Paralinguistic model · Emotion recognition · Deep learning · Convolutional neural networks · Recurrent neural networks · BLSTM model

1 Introduction

Paralinguistics is a field of linguistics that studies various nonverbal aspects of speech, such as emotions, intonation, pronunciation, and other characteristics of the human voice [13]. Recognition of human emotions is one of the most urgent and dynamically developing areas of modern speech technologies, and recognition of emotions in human speech (REHS) is the most demanded part of them. Computer paralinguistics is one of the most relevant and dynamically developing areas of modern speech technologies, and the recognition of emotions in human

V. M. Vishnevskiy et al. (Eds.): DCCN 2020, LNCS 12563, pp. 365–379, 2020.
https://doi.org/10.1007/978-3-030-66471-8_28

speech is the most popular part of them [14,15]. Computer recognition of emotions deals with the problem of identifying the signs of a person's emotional speech on the basis of audio recordings, video recordings of speaking people, and other modalities. Computer classification of emotions sets the task of identifying signs of emotional speech of a person based on audio recordings, video recordings of people who uttered this statement, and other modalities. To do this, various paralinguistic models are used that evaluate the physical parameters of the voice, such as pitch, intensity, formants, and harmonics, to determine emotions. Such classifiers are used in the development of emotional intelligence systems, security systems, biometric research, telemedicine, mobile assistants, and others.

The complexity of the task is the need to determine such features that are sufficiently resistant to emissions and noise, while maintaining all the main characteristics and features of the voice. Also, the model used must take into account the dynamics of features over time for effective analysis of changes in the voice. Most often, the method of feature extraction based on a sliding window is used to solve these problems. This method solves the problem of data normalization and prevents the model from being retrained. Another problem in the way of solving the problem is the lack of an explicit standard for naming human emotions. A possible solution is to adopt the MPEG4 standard, which divides human emotions into 6 groups: aggression, discontent, joy, sadness and surprise. The additional 6 group is called "neutral", and serves to classify the signal in which any emotions are absent.

Creating an automatic classifier consists of several stages: collecting information to form training and test samples, selecting features from the information that the model will be trained on, selecting the model and its architecture, configuring hyperparameters, training the model, and validation – checking the finished model against new data. For example, if we want to teach the model to recognize two types of emotions – anger and benevolence, then our sample must contain audio recordings with labels corresponding to these two classes. Once trained, the model will be able to make its own forecast when it receives new data on which one of the two initial types of emotions is expressed.

In practice, creating a paralinguistic model looks much more complicated. People express their emotions through a variety of meta-channels-both visual, gestures, facial expressions and body positions, and audio – through changes in voice, the presence of various non-verbal markers in speech. Moreover, our body itself can give away the emotions we experience – a person can blush from embarrassment or shame, sweat under severe stress, and so on. When using the voice as a channel for transmitting information, a person forms two information streams at once. The first and main one is directly the speech that a person expresses, in its linguistic meaning-words, phrases and sentences. The second information stream consists of paralinguistic means that can be transmitted separately or mixed with linguistic information. These paralinguistic tools include the tone, timbre of the voice, the types of vocalizations it uses, various sounds that fill in gaps in speech, and so on. By analyzing the information that makes up the first stream, we get only a distant idea of the meaning of what the person wanted to convey through his speech, and by analyzing the second channel with him, we can determine the true meaning of the words spoken – whether the person spoke with irony, sarcasm, or simply lied. Because of the great potential that comes from having this kind of information, determining emotions by voice has recently become not only a popular topic for scientific research, but also a base for creating startups and industrial technologies that solve the problem of determining emotions.

The most common REHS modeling and classification methods are Gaussian mixture models (GMM), hidden Markov models (HMM), support vector machines (SVM), and artificial neural networks (ANN) [5,18]. With the advent of deep learning methods and the creation of deep neural networks (DNN), research in the field of computer analysis of emotions has taken a qualitatively new direction.

In this paper, we propose a computer paralinguistic model of emotion recognition based on an ensemble of a bidirectional recurrent neural network with an LSTM memory cell and a deep convolutional neural network ResNet18. Using the RAVDESS database, which contains recordings of human emotional speech, computer experiments were carried out to recognize emotions using the proposed model and a comparative analysis of the results obtained with other models of neural networks, as well as the most popular machine learning algorithms was performed.

2 Development of a Computer System for Paralinguistic Analysis of Emotions

The paper proposes to develop a computer system for the paralinguistic analysis of human emotions in the following steps:

- *Selecting the appropriate database.* The database combines sets of audio recordings intended for training and testing the model, sets of tags depending on the task being solved, as well as accompanying documentation and meta-information.
- *Data pre-processing.* The purpose of preprocessing is to eliminate as much as possible the influence of external factors on the audio signal – recording quality, external noise, differences in the sensitivity of recording equipment, and so on. Typical types of preprocessing are filtering noise by frequency, cropping audio recordings by purity, reverberation, and normalization by audio recording by volume.
- *Allocation of low-level descriptors (LLD).* At this stage, you can directly select features from the audio recording. This happens using a sliding window algorithm, usually 10–30 ms wide. Window functions of various types are used: rectangular ones for selecting features based on the time distribution of the signal, and smooth ones for selecting spectral and frequency features. The main types of acoustic LLD include: intonation (tone, frequency, etc.), intensity (energy, Teagerfunction), linear cepstral coefficients

(LPCC), mel-spectrograms and mel-cepstral coefficients (MFCCs), formants-amplitudes and so on, harmonic signs (the ratio of harmonics to noise, noise to harmonics, and so on).

- *Dimension reduction.* At this stage, the feature space is transformed so as to reduce covariances outside the main diagonal of the covariance matrix, often this is achieved by shifting and rotating the source space. Dimension compression techniques include principal component analysis (PCA) and linear discriminant analysis (LDA).
- *Feature selection.* After compressing the space using PCA or LDA methods, it is further compressed by eliminating unnecessary features. To select features, you need a criterion – often using criteria based on entropy and information growth – the Ginny criterion, the Shannon entropy, the Akaike information criterion, and so on. You can also use the value of the error functional of the trained model. In addition, you should choose an algorithm for searching for the optimal data set, since a complete search of all combinations in a large feature space is a very complex and resource-consuming computational task. One of the options may be to start with a full set of features and then exclude them, or vice versa – to start with one of the features and then add all the others. Algorithms for random search of feature combinations are often used – Random Search, Grid Search, and so on.
- *The choice of the model parameters.* At this stage, the parameters of the trained paralinguistic model are fine-tuned. For neural networks, this process will consist in choosing the correct network topology – the number and organization of layers, the presence of network normalization mechanisms-Dropout, and so on, the number of neurons in hidden layers, the network initialization option, choosing an optimization algorithm and its configuration – the learning speed, choosing the value of annealing, and so on. Most often, these values are selected manually, based on experience with a specific algorithm, but there are also options for machine search for a suitable set of model parameters – the already mentioned Grid and Random Search.
- *Model training.* At this stage, the model is trained to solve the problem-classification, regression, or something else. The model uses the labels of the training set in order to find dependencies in the data and to learn how to extrapolate them to new data.

3 Basic Models of Deep Neural Networks Used in Emotion Recognition

Recurrent neural networks (RNN) are a group of neural network models used in sequence processing. This allowed determination of flexible long-term data dependencies, which is especially important in the context of human speech analysis. For this purpose, the computational graph RNN contains cycles reflecting the influence of the previous information from the sequence of events on the current information. However, it was found that, despite the ability to model long-term dependencies, in practice, recurrent neural network models do not implement the requirements and suffer from problems with gradient descent [3]. To preserve the context for long time intervals and to solve the problem of gradient damping, a special neural network architecture known as long short-term memory (LSTM) was developed, which was proposed in [16].

An LSTM module is a memory cell that has multiple inputs and outputs that allow adding or removing information about the cell state. The addition or removal of information is controlled by the gates. The LSTM contains three such gates to control the cell status. These are sigmoid layers (rectangles inside an RNN cell, see Fig. 1) that output numbers between zero and one, describing how much information should be skipped. A zero value means that we are not missing anything, while a value of one means that we are missing all information. Thus, we have the following architecture of the recurrent neural network shown in Fig. 1.

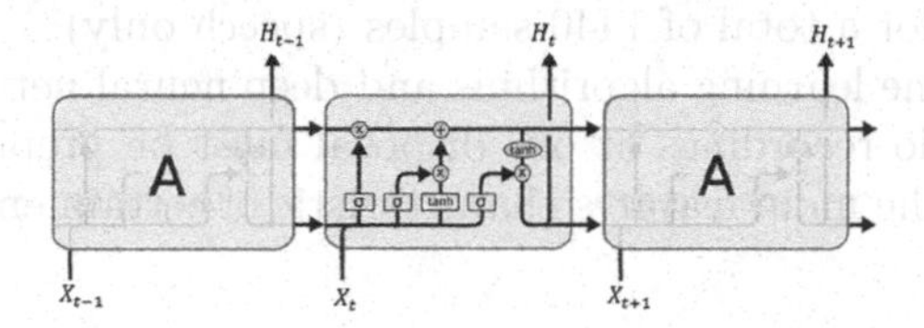

Fig. 1. Recurrent neural network with LSTM memory cells.

In this form, the model only retains past information, since it processes the sequence in one direction. To eliminate this drawback, a model of a bidirectional recurrent neural network with an LSTM memory cell was proposed in [16]. Bidirectional LSTM networks work in both directions, combining the output of two hidden LSTM layers that transmit information in opposite directions - one in the course of time, the other against it, and, thus, simultaneously receiving data from the past and future states.

A wide class of convolutional neural networks was also considered, from which the ResNet18 model was chosen [2]. ResNet is a class of convolutional neural networks consisting of residual blocks: multiple convolutional layers and skip-connections. Its distinctive feature lies in the fact that convolutional units are trained to correct the previous charts of characteristics. In other words, each convolutional block can be interpreted as some increment that needs to be added to the performance maps to improve accuracy (see Fig. 2). This allows for fast and stable training of a deep convolutional network with a large number of convolutional layers.

4 Data Description and Preliminary Processing

We carried out computer studies of the RAVDESS database containing the emotional human speech [11]. RAVDESS contains 7356 files (total size: 24.8 GB).

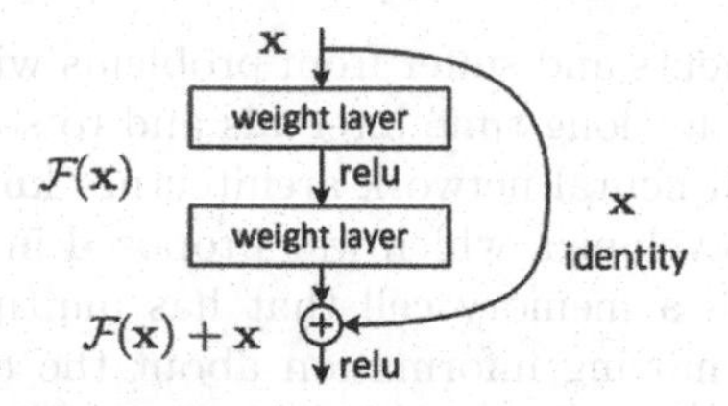

Fig. 2. Residual block of RNN network model

For each of the 24 actors (male and female), three modality formats have been created: audio only (16 bit, 48 kHz .wav), audio-video (720p H. 264, AAC 48 kHz, mp4) and video only (no sound). The records contain the following emotions: 0 – neutral, 1 – calm, 2 – happiness, 3 – sadness, 4 – anger, 5 – fear, 6 – disgust, 7 – surprise. In total, the database contains 16 classes (8 emotions, divided into male and female) for a total of 1440 samples (speech only).

To train machine learning algorithms and deep neural networks to recognize emotions, the audio recordings at our disposal must be preprocessed in such a way as to extract the main features characteristic of certain emotions. Let's look at the main ones:

- Tone, volume, and frequency are the main characteristics of an audio signal. Tone refers to the pitch of the sound inside the window, and directly depends on the frequency of the signal [6]. Sound volume is defined as the sound pressure level. The frequency of the signal expresses the number of vibrations of the sound wave per second;
- Zero-crossing rate (ZR). ZR is a measure that expresses the number of times the audio signal graph crosses zero within a given window. ZR, as a feature, allows you to classify different types of content on audio recordings well – the values of this indicator for human conversation and, for example, for music differ significantly. However, this feature is subject to a strong influence of noise on audio recordings, so it requires preliminary data cleaning to use it [4];
- Linear predictive coefficients (LPC). LPC is a method that allows you to predict the value of the next window based on the value of previous Windows. It can be expressed by the following formula:

$$s^{\sim}(t) = \sum_{i=1}^{n} w_i s(t-i), \tag{1}$$

  where wi is the weight of previous window values, $s(t-i)$ means the window value during $t-i$.
- Sound energy. This indicator expresses the vibrations of the particles of the medium that carries sound waves. Sound energy is defined as the sum of potential and kinetic energy densities integrated by volume:

$$W = W_{potential} + W_{kinetic} = \int_V \frac{p^2}{2p_0c^2}dV + \int_V \frac{pv^2}{2}dV \tag{2}$$

The following features are calculated based on the sound wave spectrogram 2. Spectrogram shows the dependence of the signal power density on time, and allows you to evaluate the signal in terms of different frequencies (see Fig. 3). The original signal is decomposed into a spectrum using the Fourier transform.

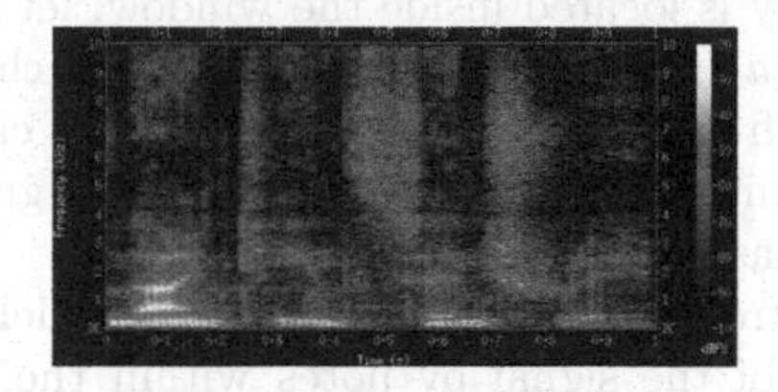

Fig. 3. An example of a spectrogram of the audio

Spectral Centroid. This feature of the spectrum shows where his center of mass, a different interpretation of the median values of the spectrum (Fig. 2). Spectral is a good description of a certain tone – brightness [10] (Fig. 4).

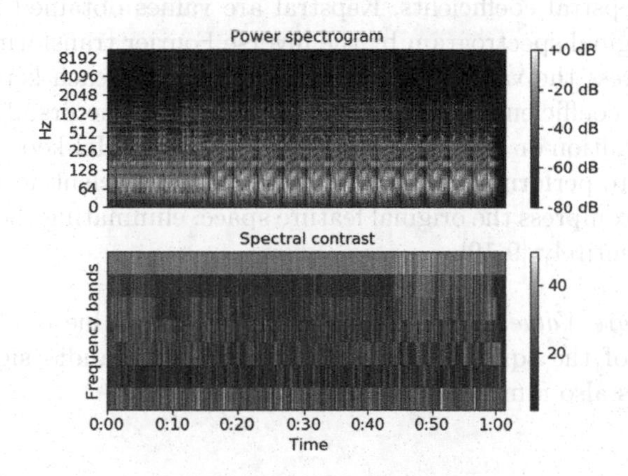

Fig. 4. Examples of spectral centroid and spectral contrast graphs

Spectral Contrast. This feature shows the differences between spectral peaks and spectral troughs at a specific time, in the context of each frequency of the spectrum (Fig. 2). It is a good measure of the range of the spectrum at a given time.

Spectral Flatness. This characteristic is also called the tonality coefficient or Wiener entropy. This indicator characterizes the content of pure tone in the audio signal, or, if the values are too high, the noise content. Thus, a high spectral flatness coefficient, close to one, is observed in white noise. A low value of this indicator means that the sound is mostly concentrated in two or three specific frequencies – such a sound can be made up of several pure sine wave signals, and its spectrum will consist of only a few peaks [7].

Spectral Rolloff. This indicator expresses the frequency below which a certain amount of sound energy is located inside the window, let's say 85%.

Spectralflux, or Delta of spectral values, characterizes changes in the spectrum of an audio recording from window to window. In the case of speech analysis tasks, such as determining emotions, this feature is of great importance, since each type of emotion has its own form of spectrogram.

In addition to spectrograms, chromatograms are widely used. *Chromagrams* show the distribution of the signal by notes within the height class (Fig. 3). Chromagrams are often used in problems of classifying an audio signal by musical genres, but they can also be used in problems of determining gender and age from an audio recording [9].

Mel-Features. Mel are units of measurement of audibility of a sound signal, calculated from the physiological features of the structure of the human ear. A mel spectrogram is a spectrogram obtained by decomposing the signal using the Fourier transform, translated into mel (see Fig. 5). On this basis it is possible to calculate Mel-cepstral coefficients. Kepstral are values obtained from the logarithm of the original spectrogram by the inverse Fourier transform. Mel-cepstral coefficients express the value of sound energy falling on each kepstr. A total of 24 Mel-kepstral coefficients are calculated, and usually the first 13 are sufficient for speech recognition or classification tasks. The use of Mel-kepstral coefficients allows not only to perform speech classification with sufficient accuracy, but also to significantly compress the original feature space, eliminating the use of spectra or mel-spectra entirely [9,19].

Root-Mean-Square Value (RMS). RMS expresses the value of the square root of the average of the squares of the amplitude of the audio signal inside the window. RMS is also a measure of signal volume.

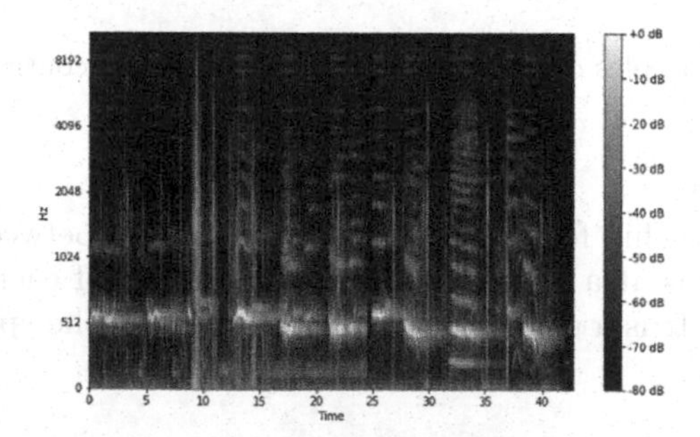

Fig. 5. An example of a mel-spectrogram

For computer experiments, only the audio part of the RAVDESS set was taken, containing 1440 3-second audio recordings made by 24 actors. The audio recordings are equally divided into 8 classes according to the emotions expressed in them: neutral, calm, upset, joy, irritation, fear, disgust and surprise. Each emotion was recorded with two levels of intensity - medium and high. Intensity of expression of emotion is a very important quality and has a serious significance in many paralinguistic studies. The use of intense emotions enables you to train the algorithms, which should provide accurate, vivid emotions. Using medium-intensity emotions in training allows you to create a less precise, but more general algorithm that allows you to determine emotions in an expression that is similar to the real one. The original audio recordings were normalized in terms of loudness and cleaned of noise outside the range from 300 to 3400 Hz. Next, using a fast Fourier transform with a window width of 93 ms, overlapping windows of 46.5 ms, the audio recordings were decomposed into a frequency spectrum. From the resulting spectrum, the following features were extracted:

1. Mel-frequency cepstral coefficients (Cepstral Mel-coefficients 1–24, Delta Cepstral Mel-coefficient, Second-order Delta Cepstral Mel-coefficient, Cepstral Mel-coefficient mean, Cepstral Mel-coefficient standard deviation);
2. Chroma features (Chromagram, Energy Normalized Chromagram);
3. Spectral features (Tonal centroid features, Spectral Contrast, Zero Crossing Rate, Spectral Centroid, Spectral Bandwidth, Spectral Flatness, Spectral Rolloff, RMS).

The pre-processing of the data depended on the model used. For the BLSTM model, using a sliding window the data were divided into sections of the required width with a step of one sample. In order to apply convolutional neural networks, the sound file was presented in the form of spectrograms in a linear scale or mel-scale, after which the resulting spectrograms are operated as with ordinary 2D images. For the ResNet18 neural network, only features extracted from the spectrogram were used. The spectrogram was converted into an image of dimension (224, 224) and normalized so that the average for each image channel had the following values: [0.485; 0.456; 0.406], and the standard deviation was [0.229: 0.224; 0.225]. For linear models-linear regression and a classifier based on the support vector machine, data was normalized by subtracting its average from each column of the matrix and dividing it by the standard deviation. For models based on decision trees – a single tree, boosting over trees, and a decision forest - no additional data preprocessing was performed.

Due to the specifics of the task and the neural networks used, the selected models solved slightly different problems. Neural networks have the ability to work at the input with multidimensional data, so they solved the problem of assigning a single class to a set of samples-a matrix, or in other words, the problem of Many-to-One classification. Other algorithms do not have the ability to work at the input with multidimensional data, so they were trained to classify each sample separately – One-to-One classification, and the received responses of the algorithms were then grouped by the record ID, and the final response of the algorithm was obtained from the voting of tags of all samples included in this audio recording. The data was divided into three parts – training and test samples. The size of the training sample was 1010 audio records, or 218160 samples, the size of the test sample was 215 audio records, or 46440 samples in total.

5 Computer Studies of Emotion Recognition Models

We carried out computer studies of various models of neural networks for recognizing emotions using the example of the data described above. In addition, machine learning algorithms were used for comparative analysis. Thus, during the experiments, the following models were trained: logistic regression (LR), support vector machine (SVM) classifier, decision tree (DT), random forest (RF), gradient boosted trees (XGBoost), convolutional neural network (CNN), recurrent neural network (RNN) (ResNet18), and an ensemble of convolutional and recurrent networks Stacked CNN-RNN. The optimization algorithm for logistic regression was LBFGS. The regularization parameter for the support vector machine algorithm is chosen equal to 1 and the kernel is radial basis function (RBF).

The DT algorithm was trained with the following parameters: without limiting the depth of the tree, the minimum number of samples for splitting was 5000. The RF algorithm was trained with the following parameters: the number of trees was 1000, no depth limit, the minimum number of samples for splitting was 5000. The XGBoost algorithm was trained with the following settings: the number of trees was 500, the maximum depth of each tree was 3, minimum number of samples for splitting was 5000. For all algorithms based on decision trees, the tree structure was optimized based on the Gini impurity.

The architecture of the ResNet18 convolutional network consisted of 18 convolutional layers, for each of these layers a Batch Normalization layer was also added. ReLU functions served as functions of activation for all layers. At the output of the last convolutional layer, the Average Pooling layer is installed, after which a fully connected linear layer is installed, which transforms the output of the last convolutional layer of dimension (512, 1, 1) into a vector of dimension (8), in which each value means the probability of a sample being assigned to one of the classes. The total number of the trained parameters of the model was 11 180 616, the weight of the model was 43 Mbytes. Its architecture is shown in Fig. 6.

The Cross-Entropy Loss function is used as the loss function for training the convolutional neural network. The SGD with momentum algorithm was used as an optimizer with the following training settings: learning rate = 0.001, momentum = 0.9. The optimal value of the number of epochs for training was chosen equal to 35. The number of batches (the number of sample elements with which we work within one iteration before changing the weights) during training was 64.

As an RNN model, the model of the following architecture was trained: a bidirectional LSTM model with two hidden layers, with 128 hidden states stored in the memory of the neural network. ReLU functions were used as activation

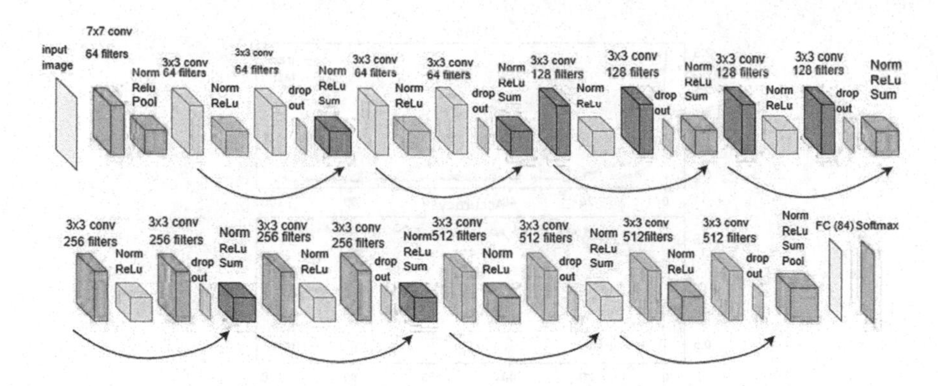

Fig. 6. Architecture of the ResNet18 convolutional neural network.

functions. The output of the last recurrent layer is processed by the Softmax function, after which it is converted using a fully connected linear layer to a vector of dimension 8, the values of which reflect the probability of assigning a sample to any of the classes. To improve the generalizing ability of the model, the Dropout regularization algorithm was also applied, with a regularization value of 0.3. The RMS Prop method with the following settings was chosen as an optimizer: alpha = 0.99, moment = 0.9, learning rate = 0.0001. The optimal number of epochs for training was chosen equal to 100.

Also, an ensemble of the two described above models of the convolutional and recurrent neural network ResNet18 was trained. The architecture of the final ensemble consists of the combined architectures of the already described models, with the exception that the last fully connected layer was removed from the convolutional neural network, which performed the transformation of the values obtained after convolution into a vector of final labels. Thus, these values were immediately input into the reconfigured recurrent neural network. The Focal Loss function was also chosen as the loss function for this model, and the RMS Prop function was used as the optimizer. The learning process of the constructed ensemble model is shown in Fig. 7.

All described neural network models were developed using the PyTorch module for the Python programming language [1]. Google Colab was used as the development environment, and model training was performed on the NVIDIA Tesla V100 GPU. The program code of the described neural network models is presented in the Appendix of this article. The final performance of all described models is shown in Table 1. It shows the values of the emotion classification accuracy metrics (average accuracy, F1-measure, average AUC) obtained after applying the trained models on the test sample. As can be seen from the results, neural network models showed a much higher accuracy of recognition and classification of emotions than the used machine learning algorithms. Of the three presented neural network models, the ResNet18+BLSTM ensemble showed a higher accuracy, which turned out to be possible due to the use of long-term memory modules in this architecture, which allow two-way work with the

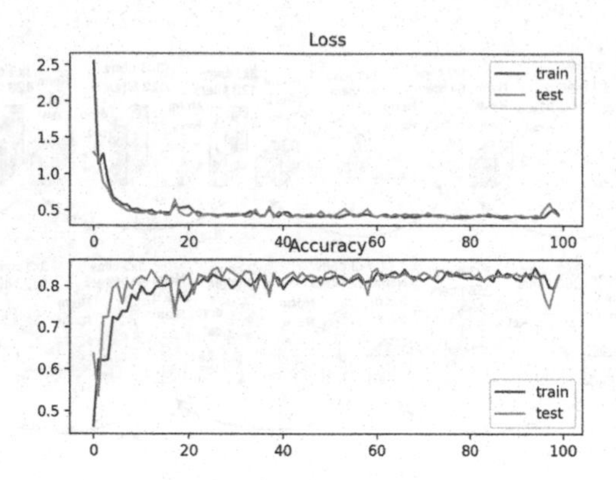

Fig. 7. Graphs of training a model of an ensemble of neural networks Stack-CNN + BLSTM. Upper plot: loss function – number of epochs; bottom graph: precision-number of epochs

context of the information being processed. Figure 8 shows a plot of the ROC analysis function and the AUC value for all classes of emotions, produced by the ResNet18+BLSTM on the test data. Also we presented the confusion matrix for this model on Fig. 9.

In further experiments, records containing the emotions of disgust, surprise, and neutrality for both sexes of the actors were removed from the analyzed data, resulting in 10 classes of emotions. This increased the classification accuracy up to 77%. Similar studies of this data using deep neural networks, for example, [12,20], report the achieved recognition accuracy of 58.6% and 64%, respectively. Also, for this database, computer experiments were carried out to classify the gender of the actor, and, in addition, the positivity (negativity) of the expressed emotions. In these cases, the classification accuracy was 97.4% and 98.7%, respectively. It is obvious that the reduction of the classes of emotions

Table 1. Results of computer experiments on emotion recognition using machine learning algorithms and neural networks.

Model	Test sample accuracy	F1-measure	Avg. AUC
LogReg	0.1723	0.1245	0.521952
SVC	0.1278	0.1137	0.4938
DT	0.3872	0.3884	0.6477
RF	0.5571	0.5555	0.7425
XGBoost	0.3065	0.3073	0.5947
CNN	0.6984	0.6879	0.7266
BLSTM	0.714603	0.6953	0.8324
ResNet18+BLSTM	0.7487	0.7235	0.8612

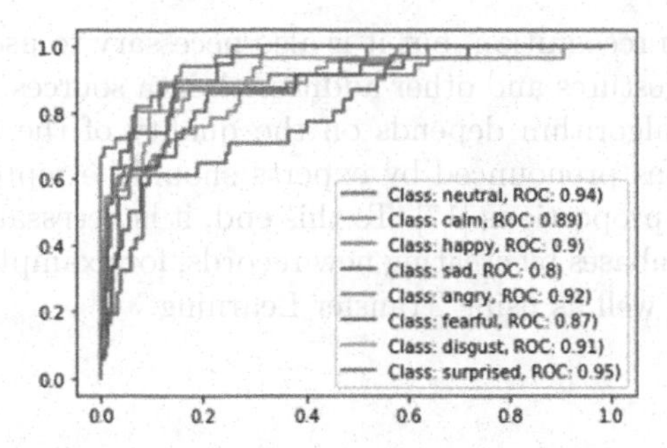

Fig. 8. Plots of AUC accuracy indicators for emotion classes obtained using the CNN + BLSTM ensemble model

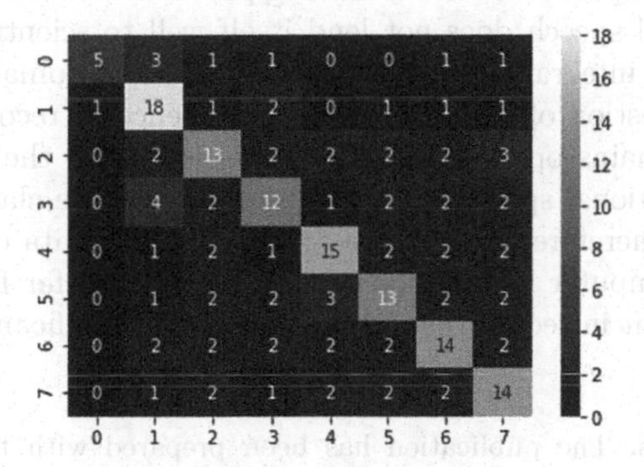

Fig. 9. Confusion matrix graph for emotions classification, produced by ensemble of ResNet18+BLSTM model

or their binarisation leads to the expected significant increase in the classification accuracy. In similar studies, for example, [8,12], the obtained accuracy is reported to be 58% and 60%, respectively.

6 Discussion of Results and Conclusions

In this paper, we have investigated methods for recognizing emotions in human speech using deep neural networks and machine learning algorithms. Audio recordings contained in the RAVDESS database were used as the analyzed data. The paper presents a model based on an ensemble of a BLSTM neural network and a convolutional neural network ResNet18, and also developed an algorithm for fine tuning its parameters. A comparative analysis of the results of using various models of neural networks and machine learning algorithms has shown the advantage and high accuracy of the proposed architecture of an ensemble of neural networks. This became possible due to the use of LSTM, which allows two-way processing of the information context.

Based on the results of the studies carried out, the following conclusions were drawn in the work. It is obvious that speech alone is not enough for high accuracy of emotion recognition, but it is also necessary to use video recordings, facial expressions, gestures and other additional data sources. To a large extent, the success of the algorithm depends on the quality of the training database. All types of emotions pronounced by experts should be represented in it, and, preferably, in equal proportions [17]. To this end, it is necessary to replenish and expand existing databases by creating new records, for example, using generative neural networks, as well as using Transfer Learning.

7 Conclusion

Emotions play an important role in human communications, are complex and have a significant impact on decision-making processes in various areas of human activity. Emotional speech does not lend itself well to scientific understanding and is difficult to integrate into the procedures for automating technological processes. The question of using artificial intelligence to recognize emotions in the real world remains open. First of all, this is due to the ambiguity in the assessment of emotional speech. Some statements may be classified in different ways by experts, therefore, ambiguously marked in the data corpus. In general, the problem of computer recognition of emotions is still far from being solved, despite the fact that in recent years there have been significant advances in this area.

Acknowledgments. The publication has been prepared with the support of the. "RUDN University Program 5-100" and funded by Russian Foundation for Basic Research (RFBR) according to the research project No 19-01-00645.

References

1. Deep learning library (2020). https://pytorch.org/. Accessed 4 Oct 2020
2. He, K., Zhang, X., Ren, S., Sun, J.: Deep residual learning for image recognition. CoRR abs/1512.03385 (2015). http://arxiv.org/abs/1512.03385
3. Hochreiter, S., Bengio, Y., Frasconi, P., Schmidhuber, J.: Gradient flow in recurrent nets: the difficulty of learning long-term dependencies. In: Kremer, S.C., Kolen, J.F. (eds.) A Field Guide to Dynamical Recurrent Neural Networks. IEEE Press (2001)
4. Ishi, C., Ishiguro, H., Hagita, N.: Using prosodic and voice quality features for paralinguistic information extraction. In: Proceedings of the Speech Prosody 2006, pp. 883–886, Dresden (2006)
5. Karpov, A.A., Kaya, H., Salakh, A.A.: Actual problems and achievements of paralinguistic speech analysis. Nauchno-tekhnicheskiy vestnik informatsionnykh tekhnologiy, mekhaniki i optiki **16**(4), 581–592 (2016). (in Russian)
6. Kennedy, L., Ellis, D.: Pitch-based emphasis detection for characterization of meeting recordings. In: Proceedings of the ASRU, pp. 243–248, Virgin Islands (2003)
7. Kockmann, M., Burget, L., Cernock, J.: Brno university of technology system for interspeech 2010 paralinguistic challenge, pp. 2822–2825 (2010)
8. Kurkov, N.A., Shchetinin, E.Y.: Emotion classification by voice using the blstm neural network. In: Information and Telecommunication Technologies and Mathematical Modeling of High-Tech Systems, pp. 461–464 (2019)

9. Lee, C., Narayanan, S., Pieraccini, R.: Recognition of negative emotions from the speech signal, pp. 240–243 (2001)
10. Liu, J., Chen, C., Bu, J., You, M., Tao, J.: Speech Emotion Recognition Using an Enhanced Co-training Algorithm, pp. 999–1002. Springer, Heidelberg (2007). https://doi.org/10.1109/ICME.2007.4284821
11. Livingstone, S.R., Russo, F.A.: The Ryerson audio-visual database of emotional speech and song (Ravdess): a dynamic, multimodal set of facial and vocal expressions in North American English. PLoS ONE **13**(5), 1–35 (2018). https://doi.org/10.1371/journal.pone.0196391
12. Popova, A.S., Rassadin, A.G., Ponomarenko, A.A.: Emotion recognition in sound. In: Kryzhanovsky, B., Dunin-Barkowski, W., Redko, V. (eds.) NEUROINFORMATICS 2017. SCI, vol. 736, pp. 117–124. Springer, Cham (2018). https://doi.org/10.1007/978-3-319-66604-4_18
13. Rabiner, L., Juang, B.: Fundamental of Speech Recognition. Prentice-Hall, Englewood Cliffs (1993)
14. Schuller, B.: The computational paralinguistics challenge. IEEE Signal Process. Mag. **29**(4), 1264–1281 (2012)
15. Schuller, B., Batliner, A.: Computational Paralinguistics: Emotion Affect and Personality in Speech and Language Processing. Wiley, New York (2013)
16. Schuster, M., Paliwal, K.: Bidirectional recurrent neural networks. IEEE Trans. Signal Process. **45**, 2673–2681 (1997). https://doi.org/10.1109/78.650093
17. Sevastyanov, L.A., Shchetinin, E.Y.: On methods of increasing the accuracy of multiclass classification based on unbalanced data. Inf. Appl. **14**(1), 67–74 (2020). https://doi.org/10.14357/19922264200109
18. Singh, N., Agrawal, A., Khan, R.A.: Automatic speaker recognition: current approaches and progress in last six decades. Global J. Enterp. Inf. Syst. **9**(3), 45–52 (2017). https://doi.org/10.18311/gjeis/2017/15973
19. Steidl, S.: Automatic Classification of Emotion-Related User States in Spontaneous Children's Speech. Logos Verlag, Berlin (2009)
20. Sterling, G., Prikhodko, P.: Deep learning in the problem of recognizing emotions from speech. In: Proceedings of the Conference Information Technologies and Systems 2016 IITP RAS. pp. 451–456 (2016). (in Russian)

A Simulation Approach to Reliability Assessment of a Redundant System with Arbitrary Input Distributions

H. G. K. Houankpo[1], D. V. Kozyrev[1,2](✉), E. Nibasumba[1], M. N. B. Moualé[1], and I. A. Sergeeva[1]

[1] Peoples' Friendship University of Russia (RUDN University), 6 Miklukho-Maklaya Street, Moscow 117198, Russia
gibsonhouankpo@yahoo.fr, kozyrev-dv@rudn.ru, ema.patiri2015@yandex.ru, bmouale@mail.ru, sergeevair@mail.ru

[2] V. A. Trapeznikov Institute of Control Sciences of RAS, 65 Profsoyuznaya Street, Moscow 117997, Russia

Abstract. With the rapid development and spread of computer networks and information technologies, researchers are faced with new complex challenges of both applied and theoretical nature in investigating the reliability and availability of communication networks and data transmission systems. In the current paper, we perform the system-level reliability analysis for a redundant system with arbitrary distributions of uptime and repair time of its elements using a simulation approach. Also, we obtained the values of the relative recovery speed at which the desired level of reliability is achieved, presented dependency plots of the probability of system uptime and plots of the uniform difference of the obtained simulation results against the relative speed of recovery; also plots of the empirical distribution function $F^*(x)$ and reliability function $R^*(x)$ relative to the reliability assessment. Software implementation of simulation algorithms was carried out on the basis of the R language.

Keywords: Simulation · Stochastic modeling · Reliability of redundant systems · Redundant communications · Relative repair rate · Probability of the failure-free operation · Sensitivity analysis

1 Introduction

With the rapid development of computer networks and information technologies, researchers are faced with new complex applied and theoretical problems on studying the reliability and availability of networks and data transmission systems [1].

Currently, simulation is effectively utilized in modeling of info-communication network systems, validating mathematical methods, testing information technologies, elaborating new computational models for analysis of functioning of computer networks, modeling teletraffic, etc. Previously in [2], it was shown

V. M. Vishnevskiy et al. (Eds.): DCCN 2020, LNCS 12563, pp. 380–392, 2020.
https://doi.org/10.1007/978-3-030-66471-8_29

that explicit analytical expressions for the stationary distribution of the system under consideration cannot always be obtained. The simulation model developed in this work allowed to investigate the reliability of the system, defined as the stationary probability of its failure-free operation, as well as to assess the reliability measures of the system; also numerical research and graphical analysis have shown that this dependence becomes vanishingly small under a "fast" recovery, that is, with the growth of the relative repair rate ρ.

Recently, the functioning of various aspects of modern society has become critically dependent on communication networks [3,4]. With the migration of critical communications tools, it has become vital to ensure the reliability and accessibility of data networks and systems.

A number of previous studies [5–9] have focused on analyzing the reliability of various complex telecommunications systems. In particular, a study was conducted on the reliability of cold-standby data transmission systems. Paper [10] focused on reliability analysis of a combined power plant running on a gas turbine engine. In a series of works by Enrico Zio et al. [11–13] the Monte Carlo simulation method was applied to reliability assessment and risk analysis of multi-state physics systems. The aim of [14] was to develop a model for studying system reliability and analyzing the sensitivity of system availability. In stochastic systems stability often means insensitivity or low sensitivity of their output characteristics to the shapes of some input distributions. The proof of the insensitivity can significantly simplify the model of the system under study by using more convenient distributions (from the exponential family). This urges the importance of the sensitivity analysis. Actually, the term "sensitivity analysis" can be understood differently in civil engineering than in basic sciences [15]. In operations research, sensitivity analysis is developed as a method of critical assessment of decisional variables, and is capable to identify those sensitive variables that influence the final desired result [16]. Another suitable complement to probabilistic reliability analysis is structural sensitivity analysis [17,18].

In [19], a simulation method was considered to simulate the reliability of a task by a complex system by modeling a task cyclogram, modeling a run-time profile and a method of dynamic reliability modeling. In [20], modeling and estimation methods were presented that allow temperature optimization of the reliability of a multiprocessor system on a chip for specific applications.

The current paper summarizes the results of previous studies of the authors in the case of cold standby of the system $\langle GI_N/GI/1\rangle$ with an arbitrary distribution function (DF) of uptime and an arbitrary DF of repair time of its elements. The aim of the work is to conduct simulation to find the value of the coefficient ρ (relative repair rate), at which a given level of reliability is achieved and to graph the dependence of the probability of failure-free operation of the system on the relative repair rate. The results of calculating the reliability estimate for different input distributions are presented.

2 Problem Statement and Model Description

As a simulation model of a redundant data transmission system consisting of N different types of data transmission channels, we consider a repairable multiple cold standby system $\langle GI_N/GI/1\rangle$ with one repair device, with an arbitrary distribution function (DF) of uptime and an arbitrary DF of repair time of its elements.

In this paper, we consider the dependence of the probability of failure-free operation of the system $\langle GI_N/GI/1\rangle$ on the relative repair rate. The task is to develop a simulation model for calculating the steady-state probabilities of the system, to find the stationary probability of failure-free operation of the system for some special cases of distributions and assess the reliability of the system, for $N = 3$.

2.1 Simulation Model for Calculating Steady-State Probabilities of $\langle GI_N/GI/1\rangle$ System

Let's define the following states of the simulated system:

- State 0: One (main) element works, $N-1$ are in a cold standby;
- State 1: One element failed and is being repaired, one – works, $N-2$ are in a cold standby;
- State 2: Two elements have failed, one is being repaired, the other is waiting for its turn for repair, one – works, $N-3$ are in a cold standby;
- State N: All the items have failed, one is being repaired, the rest are waiting their turn for repair.

To describe the reliability modeling algorithm for the $\langle GI_N/GI/1\rangle$ system we introduce the following variables:

- double t - simulation clock; changes in case of failure or repair of the system's elements;
- int i, j - system state variables; when an event occurs, the transition from i to j takes place;
- double $t_{nextfail}$ – service variable, which stores the time until the next element failure;
- double $t_{nextrepair}$ – service variable, which stores the time until the next repair of the failed element;
- int k - counter of iterations of the main loop.

For clarity, the simulation model is presented graphically in Fig. 1 in the form of a flowchart. The criterion for stopping the main cycle of the simulation model is to achieve the maximum model execution time T.

For a better understanding and reproducibility of the simulation model, in addition to the flowchart, the algorithm of the discrete-event process of simulation modeling is also provided in the form of pseudo-code with comments (Algorithm 1).

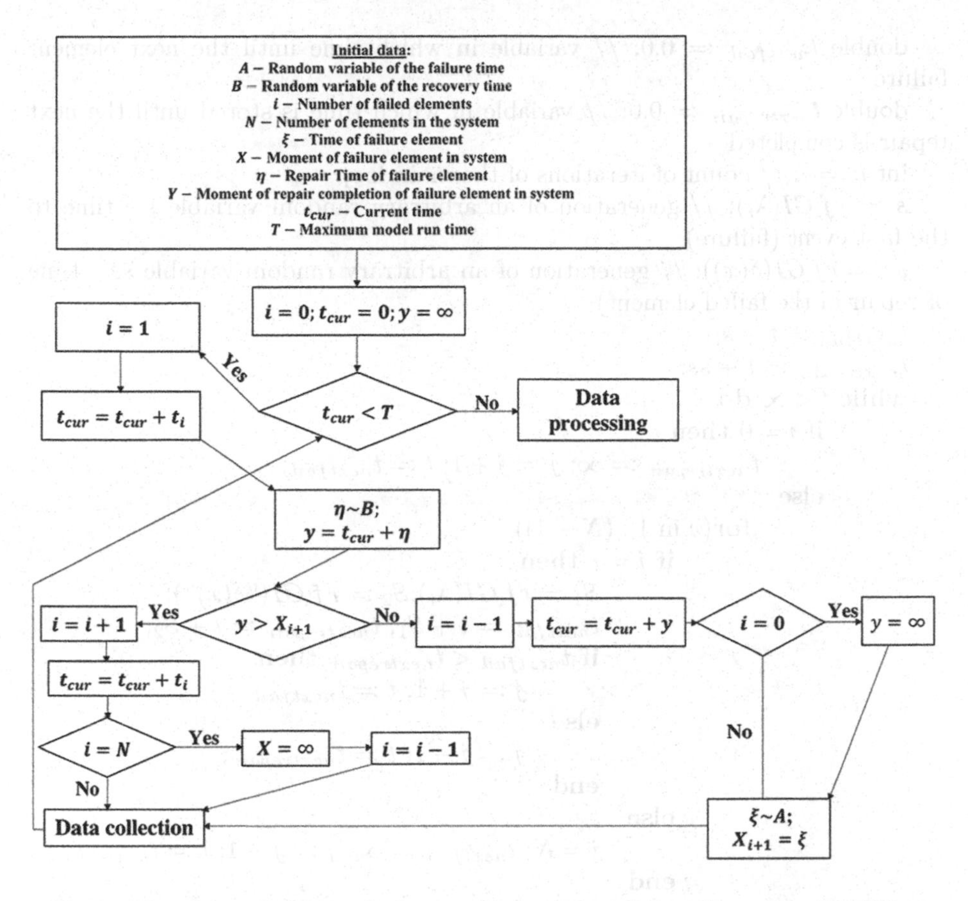

Fig. 1. Flowchart of the simulation model for estimating stationary probabilities.

Algorithm 1. Pseudocode of the simulation process of the system $\langle GI_N/GI/1 \rangle$.

Input: a1, b1, N, T, NG, GI.

a1 - Average time between element failures,
b1 - Average repair time,
N - Number of elements in the system,
T - Maximum model run time,
NG - Number of Trajectory (Path) Graphs,
"GI" - Arbitrary Distribution function.

Output: steady-state probabilities $P_0, P_1, P_2, \ldots, P_N$.

Begin

array $r[] := [0, 0, 0]$; // multi-dimensional array containing results, k-step of the main cycle (loop)

double $t := 0.0$; // time clock initialization

int $i := 0$; $j := 0$; // system state variables

double $t_{nextfail}$:= 0.0; // variable in which time until the next element failure

double $t_{nextrepair}$:= 0.0; // variable in which time is stored until the next repair is completed

int $k := 1$; // count of iterations of the main loop

$s := rf\_GI(\lambda_i)$; // generation of an arbitrary random variable $s-$ time to the first event (failure)

$ss := rf\_GI(\delta(x))$; // generation of an arbitrary random variable $ss-$ time of repair of the failed element)

```
t_nextfail := t + s;
t_nextrepair := t + ss;
while t < ∞ do
        if i = 0 then
                t_nextrepair := ∞; j := j + 1; t := t_nextfail;
        else
                for(v in 1 : (N − 1))
                        if i = v then
                                S1 := rf_GI(λi);S2 := rf_GI("δ(x)");
                                t_nextfail := t + S1; t_nextrepair := t + S2;
                                if t_nextfail < t_nextrepair then
                                        j := j + 1; t := t_nextfail;
                                else
                                        j := j − 1; t := t_nextrepair;
                                end
                        else
                                i = N; t_nextfail := ∞; j := j − 1; t := t_nextrepair ;
                        end
                        if t > Tthen
                                t = T
                        end
                r[, , k] := [t, i, j]; i := j; k := k + 1;
        end do
```

Calculate the duration of stay in each state $i, i = 0, 1, 2, \ldots N$. The formula for calculating stationary probabilities is:

$$\widehat{P}_i = \frac{1}{NG}\sum_{j=1}^{NG}(\text{duration of stay in state } i/T)_j$$

end

Table 1 shows the values of the coefficient $\rho = \frac{a_1}{b_1}$ — the relative repair rate (i.e. the ratio of the average uptime of the main element to the average repair time of the failed element), at which the specified level of stationary reliability $1 - \pi_3 = 0.9; 0.99; 0.999$. To analyze and compare the results, the following distributions were chosen: Exponential (M), Weibull-Gnedenko (WB), Lognormal (LN).

We consider particular cases of the model at $\rho = 25$; $N = 3$; $NG = 100$; $T = 1000$; where $b_1 = 1$; T_1 - system uptime; T_2 - repair time of a failed element.

Table 1. Values of the relative repair rate, at which a given level of the system's stationary reliability is achieved.

T_1	T_2								
	$M(1/b_1)$			$WB(W)$			$LN(sig)$		
	0.9	0.99	0.999	0.9	0.99	0.999	0.9	0.99	0.999
$M(1/a_1)$	1.6	4.2	9.1	1.5	4.7	11.3	1.6	4.4	9.7
$WB(W)$	3.2	12.2	25	2.9	11.6	25	3.3	11.9	25
$LN(sig)$	1.6	3.9	7.6	1.6	4.5	9.7	1.6	4	7.2

A sufficiently high level of system reliability is achieved with a relatively small excess of the average values of the uptime by the repair time, except when the uptime of the system elements is distributed according to the Weibull-Gnedenko distribution.

Figure 2 presents graphs of the probability of system uptime; and Fig. 3 shows the uniform difference of the results of the simulation model between the exponential and the non-exponential cases.

The obtained results demonstrate a high asymptotic insensitivity of the stationary reliability of the system. It can be seen that the differences between the curves during "fast" recovery become vanishingly small for all the considered distributions of the repair time of the system elements. For example, already starting from the value $\rho = 10$, all the curves are almost indistinguishable.

Graphical results from Fig. 3 show that the uniform difference between the models $\langle M_3/M/1\rangle$ and $\langle LN_3/M/1\rangle$; $\langle M_3/M/1\rangle$ and $\langle LN_3/LN/1\rangle$ tends to zero with a small increase in ρ.

2.2 Simulation Model for Assessment of the $\langle GI_N/GI/1\rangle$ System Reliability

In this case, the system stops functioning after all N elements have failed, and the maximum model run time T is equal to ∞. For clarity, the simulation model is presented graphically in Fig. 4.

For a better understanding and reproducibility of the simulation model, in addition to the flowchart, the algorithm of the discrete-event process of simulation modeling is also provided in the form of pseudo-code with comments (Algorithm 1).

Algorithm 2. Pseudocode of the simulation process of the system $\langle GI_N/GI/1\rangle$.

Input: a1, b1, N, NG, GI.

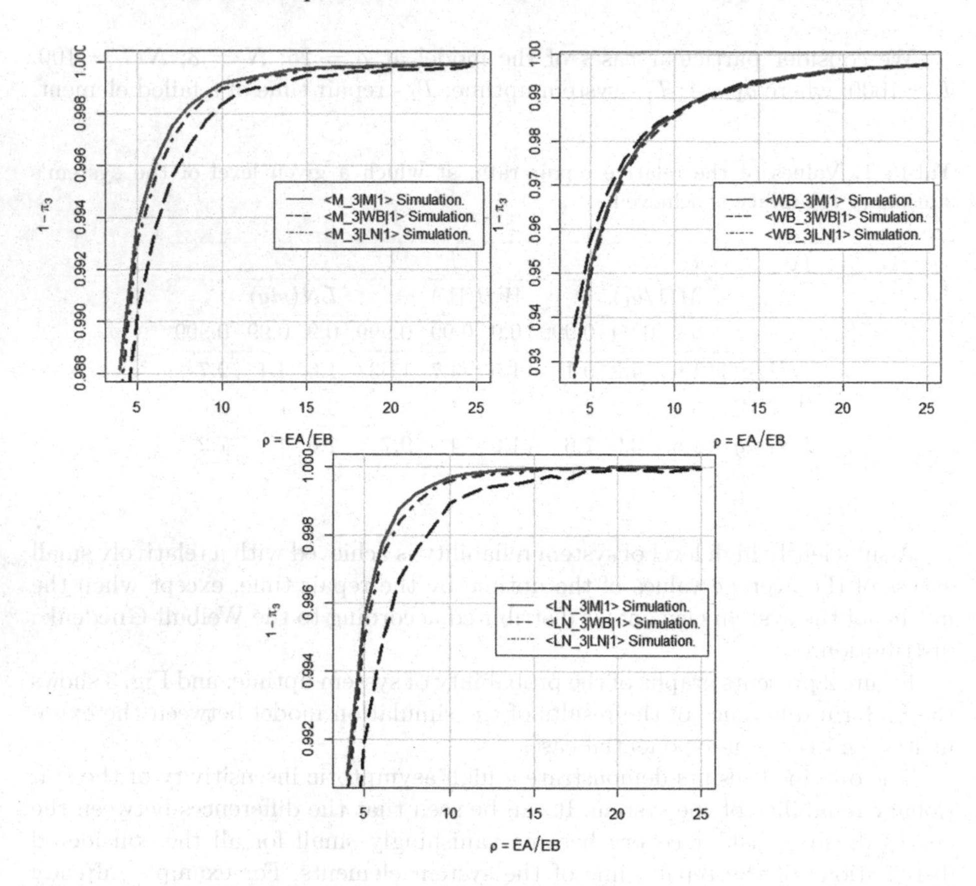

Fig. 2. Graphs of the probability of system uptime versus relative recovery rate for the systems $\langle M_3/GI/1\rangle$, $\langle WB_3/GI/1\rangle$ and $\langle LN_3/GI/1\rangle$.

Output: Reliability assessment $\widehat{ET}$.
Begin
array $r[] := [0, 0, 0]$; // multi-dimensional array containing results, k-step of the main cycle (loop)
double $t := 0.0$; // time clock initialization
int $i := 0$; $j := 0$; // system state variables
double $t_{nextfail} := 0.0$; // variable in which time until the next element failure
double $t_{nextrepair} := 0.0$; // variable in which time is stored until the next repair is completed
int $k := 1$; // count of iterations of the main loop
$s := rf\_GI(\lambda_i)$; // generation of an arbitrary random variable $s-$ time to the first event (failure)
$ss := rf\_GI(\delta(x))$; // generation of an arbitrary random variable $ss-$ time of repair of the failed element)

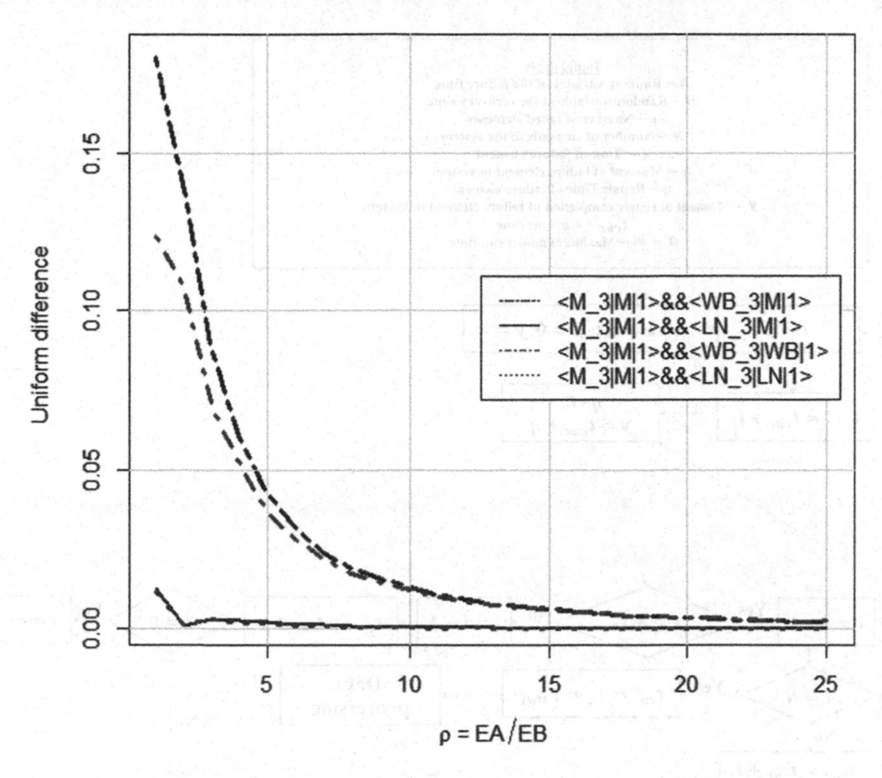

Fig. 3. Graphs of the uniform difference in the results of the simulation model as a function of ρ.

$t_{nextfail} := t + s;$
$t_{nextrepair} := t + ss;$
while $t < \infty$ **do**
  if $i = 0$ **then**
    $t_{nextrepair} := \infty;\ j := j + 1;\ t := t_{nextfail};$
  else
    for$(v$ in $1 : (N - 1))$
      if $i = v$ **then**
        $S_1 := rf\_GI(\lambda_i); S_2 := rf\_GI(''\delta(x)'');$
        $t_{nextfail} := t + S_1;\ t_{nextrepair} := t + S_2;$
        if $t_{nextfail} < t_{nextrepair}$ **then**
          $j := j + 1;\ t := t_{nextfail};$
        else
          $j := j - 1;\ t := t_{nextrepair};$
        end
      else
        $i = N$; **then break** ;
      end
    end

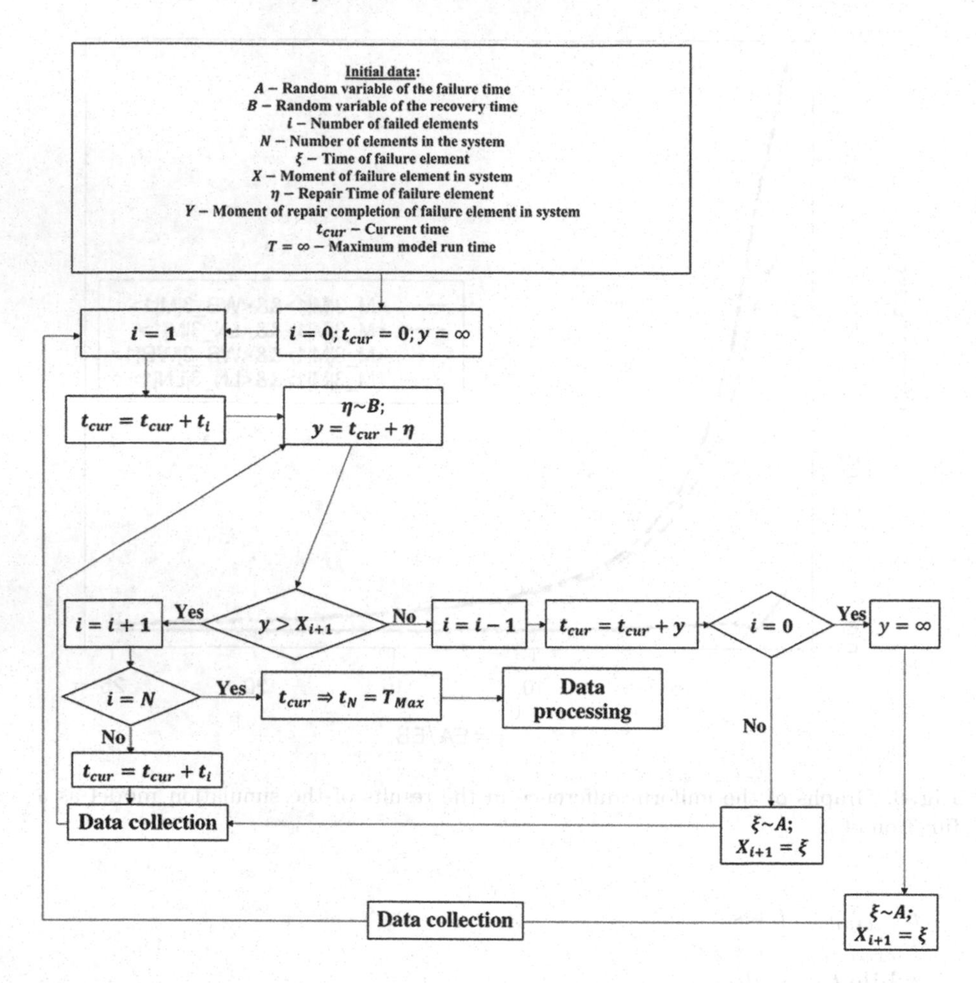

Fig. 4. Flowchart of the simulation model for evaluating system reliability.

$r[,,k] := [t, i, j];\ i := j;\ k := k + 1;$

end do

Calculate the duration of stay in state N. The formula for calculating the reliability measure is:

$$\widehat{ET} = \frac{1}{NG} \sum_{i=1}^{NG} (\text{duration of stay in state } N)_i$$

end

Table 2 shows the values of the reliability estimates of the system (estimates of the mean time to failure of the system) with the time spent on modeling. The same distributions were chosen: Exponential, Weibull-Gnedenko, Lognormal.

We consider particular cases of the model at $\rho = 25$; $N = 3$; $NG = 10000$; where $b_1 = 1$; T_1 - system uptime; T_2 - repair time of a failed element.

Table 2. Values of the estimates of the mean time to failure of the $\langle GI_3/GI/1\rangle$ system.

T_1	T_2		
	$M(1/b_1)$	$WB(W)$	$LN(sig)$
$M(1/a_1)$	16530.34	19566.77	25.18033
$WB(W)$	28.57675	927.8087	564.099
$LN(sig)$	249458.5	71212.42	190780.8

As it can be seen from Table 2, the most reliable model is a model with a lognormal distribution of uptime and an exponential distribution of the repair time of a failed element.

Figure 5 presents graphs of the empirical distribution function $F^*(t)$ and the empirical reliability function $R^*(t)$.

The results also show the high asymptotic insensitivity of the empirical distribution function and the corresponding empirical reliability function of the system to the shapes of the uptime and repair time distributions of the system's elements.

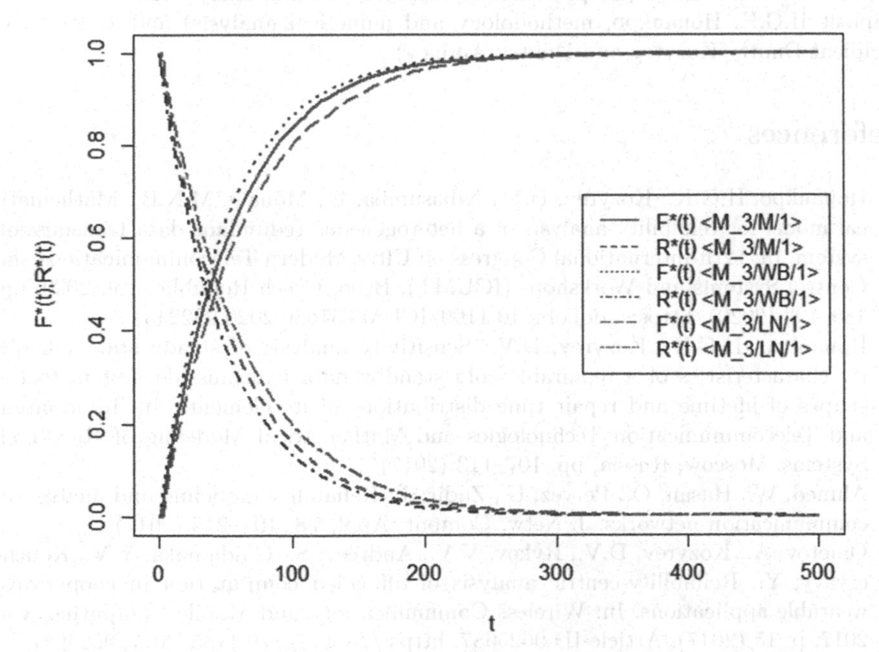

Fig. 5. Graphs of the empirical distribution function $F^*(t)$ and the empirical reliability function $R^*(t)$

3 Conclusion

In practice, redundancy is a common approach to enhance the reliability of communication systems, which may be designed to avoid communication failure by including redundant components that are active upon the failure of a primary component. To address these practical issues, in the current work we considered a repairable multiple cold standby system $\langle GI_N/GI/1\rangle$ with one repair device, with an arbitrary distribution function of uptime and an arbitrary distribution function of repair time of its elements. This paper is a continuation of the previous studies in this area that were focused on analytical models. For the considered system we applied the discrete-event simulation approach to perform the assessment of the system-level reliability and obtained the values of the relative repair rate at which the given level of the system's stationary reliability is achieved. Graphic and numerical results show a high asymptotic insensitivity of the stationary system reliability to the input distributions. The differences between the curves under "fast" recovery become vanishingly small for all the studied special cases of distributions. It was shown that the most reliable case is the model with a lognormal distribution of uptime and an exponential distribution of the repair time of a failed element. The graphic results also show a high asymptotic insensitivity of the empirical distribution function and the empirical reliability function of the system.

Acknowledgments. The publication has been prepared with the support of the "RUDN University Program 5-100" and funded by RFBR according to the research projects No. 20-37-90137 (recipient Dmitry Kozyrev, formal analysis, validation, and recipient H.G.K. Houankpo, methodology and numerical analysis) and 19-29-06043 (recipient Dmitry Kozyrev and Dmitry Aminev).

References

1. Houankpo, H.G.K., Kozyrev, D.V., Nibasumba, E., Mouale, M.N.B.: Mathematical model for reliability analysis of a heterogeneous redundant data transmission system. In: 12th International Congress on Ultra Modern Telecommunications and Control Systems and Workshops (ICUMT), Brno, Czech Republic, vol. 2020, pp. 189–194 (2020). https://doi.org/10.1109/ICUMT51630.2020.9222431
2. Houankpo, H.G.K., Kozyrev, D.V.: Sensitivity analysis of steady state reliability characteristics of a repairable cold standby data transmission system to the shapes of lifetime and repair time distributions of its elements. In: Information and Telecommunication Technologies and Mathematical Modeling of High-Tech Systems, Moscow, Russia, pp. 107–113 (2017)
3. Ahmed, W., Hasan, O., Pervez, U., Zadir, J.: Reliability modeling and analysis of communication networks. J. Netw. Comput. Appl. **78**, 191–215 (2017)
4. Ometov, A., Kozyrev, D.V., Rykov, V.V., Andreev, S., Gaidamaka, Y.V., Koucheryavy, Y.: Reliability-centric analysis of offloaded computation in cooperative wearable applications. In: Wireless Communications and Mobile Computing, vol. 2017, p. 15 (2017). Article ID 9625687. https://doi.org/10.1155/2017/9625687

5. Rykov, V., Kozyrev, D., Zaripova, E.: Modeling and simulation of reliability function of a homogeneous hot double redundant repairable system. In: Proceedings of the 31st European Conference on Modelling and Simulation, ECMS2017, pp. 701–705 (2017). https://doi.org/10.7148/2017-0701
6. Efrosinin, D., Rykov, V.: Sensitivity analysis of reliability characteristics to the shape of the life and repair time distributions. In: Dudin, A., Nazarov, A., Yakupov, R., Gortsev, A. (eds.) ITMM 2014. CCIS, vol. 487, pp. 101–112. Springer, Cham (2014). https://doi.org/10.1007/978-3-319-13671-4_13
7. Efrosinin, D., Rykov, V.V., Vishnevskiy, V.: Sensitivity of reliability models to the shape of life and repair time distributions. In: 9th International Conference on Availability, Reliability and Security (ARES 2014), pp. 430–437. IEEE (2014). Published in CD: 978-I-4799-4223-7/14. https://doi.org/10.1109/ARES.40
8. Rykov, V.V., Kozyrev, D.V.: Analysis of renewable reliability systems by markovization method. In: Rykov, V.V., Singpurwalla, N.D., Zubkov, A.M. (eds.) ACMPT 2017. LNCS, vol. 10684, pp. 210–220. Springer, Cham (2017). https://doi.org/10.1007/978-3-319-71504-9_19
9. Rykov, V., Kozyrev, D.: On sensitivity of steady-state probabilities of a cold redundant system to the shapes of life and repair time distributions of its elements. In: Pilz, J., Rasch, D., Melas, V.B., Moder, K. (eds.) IWS 2015. SPMS, vol. 231, pp. 391–402. Springer, Cham (2018). https://doi.org/10.1007/978-3-319-76035-3_28
10. Lisnianski, A., Laredo, D., Haim, H.B.: Multi-state markov model for reliability analysis of a combined cycle gas turbine power plant, Published. In: Second International Symposium on Stochastic Models in Reliability Engineering. Life Science and Operations Management (SMRLO) (2016). https://doi.org/10.1109/SMRLO.2016.31
11. Wang, W., Di Maio, F., Zio, E.: Three-loop Monte Carlo simulation approach to multi-state physics modeling for system reliability assessment. Reliab. Eng. Syst. Saf. **167**, 276–289 (2017)
12. Li, X.-Y., Huang, H.-Z., Li, Y.-F., Zio, E.: Reliability assessment of multi-state phased mission system with non-repairable multi-state components. Appl. Math. Model. **61**, 181–199 (2018)
13. Lin, Y.-H., Li, Y.-F., Zio, E.: A comparison between Monte Carlo simulation and finite-volume scheme for reliability assessment of multi-state physics systems. Reliab. Eng. Syst. Saf. **174**, 1–11 (2018)
14. Tourgoutian, B., Yanushkevich, A., Marshall, R.: Reliability and availability model of offshore and onshore VSC-HVDC transmission systems. In: 11th IET International Conference on AC and DC Power Transmission, 13 July 2015. https://doi.org/10.1049/cp.2015.0101
15. Kala, Z.: Sensitivity analysis in probabilistic structural design: a comparison of selected techniques. Sustainability **12**(11), 19 (2020). https://doi.org/10.3390/su12114788
16. Kala, Z.: Quantile-oriented global sensitivity analysis of design resistance. J. Civil Eng. Manage. **25**(4), 297–305 (2019). https://doi.org/10.3846/jcem.2019.9627. ISSN 1392–3730. E-ISSN 1822–3605
17. Kala, Z.: Estimating probability of fatigue failure of steel structures. Acta et Commentationes Universitatis Tartuensis de Mathematica **23**(2), 245–254 (2019). https://doi.org/10.12697/ACUTM.2019.23.21. ISSN 1406–2283. E-ISSN 2228–4699
18. Kala, Z.: Global sensitivity analysis of reliability of structural bridge system. Eng. Struct. **194**, 36–45 (2019). https://doi.org/10.1016/j.engstruct.2019.05.045. ISSN 1644–9665

19. Cao, J., Wang, Z., Shen, Y.: Research on modeling method of complex system mission reliability simulation. In: 2012 International Conference on Quality, Reliability, Risk, Maintenance, and Safety Engineering (2012). https://doi.org/10.1109/IC'R2MSE.2012.6246242
20. Gu, J., Zhu, C., Shang, L., Dick, R.: Application-specific multiprocessor system-on-chip reliability optimization. IEEE Trans. Very Large Scale Integr. (VLSI) Syst. **16**(5) (2008). https://doi.org/10.1109/TVLSI.2008.917574

Problem of Overbooking for a Case of a Random Environment Existence

Alexander Andronov[1,2], Iakov Dalinger[2], and Diana Santalova[3](✉)

[1] Transport and Telecommunication Institute, Lomonosova 1, Riga 1019, Latvia
aleksander.andronov1@gmail.com

[2] Saint-Petersburg State University of Civil Aviation, Pilotov 38, 196210 Saint-Petersburg, Russia
iakovdalinger@gmail.com

[3] Møreforsking AS, Britvegen 4, 6411 Molde, Norway
dsantalova@gmail.com

Abstract. An overbooking supposes that a booking of some product or service exceed given possibilities. It takes into consideration that a part of the booking will be cancelled. This situation is considered following to example of aviation ticket booking. It is supposed that an external random environment exists. The environment is described as a continuous-time finite irreducible Markov chain. A demand on the booking depends on the state of the random environment. An using of economical criterion supposes a consideration of such indices as costs of engaged seats of the aircraft and the penalty for the refusal of passenger with sold tickets. This criterion is optimized by means of dynamic programming method. A numerical example is considered.

Keywords: Continuous-time Markov chain · Dynamic programming · Overbooking's problem

1 Introduction

An overbooking policy assumes that a sale and a booking of some product or service exceed given possibilities. It takes into consideration that a part of the sale or the booking will be cancelled. The overbooking is used in different spheres of a transport, hotel's businesses etc. Numerous publications are devoted to this problem [1–8].

In this paper we consider the problem of the overbooking in the case of an external random environment existence. Airline overbooking will be considered for concreteness. Notably we use one word "to buy" both as for "to buy" and for "to book".

At the beginning the following positing of the problem will be considered following to the text [8].

It is considered one aircraft trip, having the capacity n^* passengers. The preddeparture time, when passengers buy tickets, is a random variable with the

V. M. Vishnevskiy et al. (Eds.): DCCN 2020, LNCS 12563, pp. 393–405, 2020.
https://doi.org/10.1007/978-3-030-66471-8_30

density $f(x), x \geq 0$. The passengers buy tickets independently of each other. There is the probability q that the passenger with the ticket, doesn't come for the trip.

Then an external random environment exists, having k states with numbers $1, \ldots, k$. The environment is described as a continuous-time finite irreducible Markov chain $J(t)$ with the matrix $\lambda = (\lambda_{i,j})_{k \times k}$ of transition probabilities between states.

An average demand for a considered trip depends on the state of the random environment and equals d_i for the i-state, $i = 1, \ldots, k$. Therefore the intensity of customer's arrivals at time t till a departure, if the i-th state occurs, is calculated as follows:

$$\tilde{d}_i(t) = d_i f(t), t \geq 0.$$

At every moment of time t, the number of the sold tickets n and the state i of the environment J are known. A decision on overbooking is adopted with intervals Δ, at the instants $s\Delta$, $s = 1, 2, \ldots$, until departure. The i-th stage is called the time interval $(s\Delta, (s-1)\Delta)$. The average number of the passengers, whose buy tickets on the s-th stage, if $J((s^* - s)\Delta) = i$, is

$$\alpha_i(s) = \int_{t=\Delta(s-1)}^{\Delta s} \tilde{d}_i(t) dt, s = s^*, s^* - 1, \ldots, 1. \quad (1)$$

Additionally we are guided by the maximal value of the overbooking. Let it be $m_{n,i}(s)$: the maximal value of overbooking at instant $t = s\Delta$, if n place are busy and $J(t) = i$. Here $m_{n,i}(\Delta) \leq m^*$, where m^* is given.

From the beginning we suppose that the values $m_{n,i}(s)$ are given. Then we look for the following functions $m_{n,i}(s)$, that optimizes some effectiveness criteria. For example, the average aircraft load. The application of economical criteria implies considering of indices such as the costs of engaged seats of the aircraft and the penalty for refusing a passenger with sold tickets. Let c and r be the cost of one engaged seat and the penalty for one refusal respectively.

At the beginning we consider the first case, then the second case.

2 Main Results

We will consider the described process with the step $\Delta > 0$. Let τ be the time, when all seats are occupied, $s^* = \tau/\Delta$ be an integer, so the step number s of the step belongs to the set $\{s^*, s^* - 1, \ldots, 0\}$. The s-th step corresponds to the time interval $(\Delta s, \Delta(s-1))$.

The random environment is represented by means of the continuous-time irreducible finite Markov chain $J(t)$ with k states and matrix $\lambda(t) = (\lambda_{i,j}(t))$ of transition intensities between states. Let $P_{i,j}(t)$ be the probability that chain $J(t)$ will be in state j at instant t if the initial state is i, $P(t) = (P_{i,j}(t))$ be the corresponding matrix. This matrix is calculated as follows [9,10].

Let $(1 \ldots 1)^T_{k\times 1}$ be the column-vector from the units, $\Lambda = \lambda\,(1 \ldots 1)^T_{k\times 1}$ be the column-vector, $diag(\Lambda)$ be the diagonal matrix with vector Λ on the main diagonal. The $k \times k$-matrix $A = \lambda - diag(\Lambda)$ is called generator of the Markov chain. We denote eigenvalues and eigenvectors of this matrix by $\chi_1, \chi_2, \dots, \chi_k$ and $\beta_1, \beta_2, \dots, \beta_k$ correspondingly. It is supposed that all values $\chi_1, \chi_2, \dots, \chi_k$ are different.

Let $B = (\beta_1, \dots, \beta_k)$ be the matrix, whose columns are eigenvectors $\beta_1, \dots, \beta_k$ of the generator A, $\tilde{\beta}_1, \dots, \tilde{\beta}_k$ be rows of the inverse matrix B^{-1}, so that $B^{-1} = (\tilde{\beta}_1^T, \dots, \tilde{\beta}_k^T)^T$, $diag(\exp(t\chi))$ be the diagonal matrix with the vector $\exp(t\chi) = (\exp(t\chi_1), \dots, \exp(t\chi_k))$ on the main diagonal. Then

$$P(t) = \sum_{i=1}^{k} \exp(\chi_i t)\beta_i\tilde{\beta}_i = Bdiag(\exp(t\chi))B^{-1}, t \geq 0. \tag{2}$$

Now we can calculate the average value $ET_{i,\nu,j}(\Delta)$ of sojourn time in the state ν on interval $(0, \Delta)$ jointly with probability $P\{J(\Delta) = j\}$, if $J(0) = i$:

$$ET_{i,\nu,j}(\Delta) = \int_0^{\Delta} P_{i,\nu}(u)P_{\nu,j}(\Delta - u)du, \quad i, \nu, j \in \{1, \dots, k\}. \tag{3}$$

Further let us calculating the probability $\tilde{P}r_{\eta,i,j}(s)$, that η new requests on tickets are received during stage s and the final state $J(\Delta(s-1))$ equals j, if the i–th state take place at instant Δs. With respect to the paper [8] we use the following formula:

$$\tilde{P}r_{\eta,i,j}(s) = \frac{1}{n!}\left(\sum_{\nu=1}^{k} \alpha_\nu(s)ET_{i,\nu,j}(\Delta)\right)^{\eta} \exp\left(-\sum_{\nu=1}^{k} \alpha_\nu(s)ET_{i,\nu,j}(\Delta)\right), \eta = 0, 1, \dots. \tag{4}$$

Let $Pr_{n,i}(s)$ be the probability that at beginning of the s–th stage the following situation occurs: n claims are gotten, the state of MC $J(\Delta s)$ equals i. We will consider the following values of n: $n \in \{0, 1, \dots, n^* + m^* + 1\}$. If values $n \leq n^*$ then $Pr_{n,i}(s)$ means the probability that n places are busy. If value n belong to interval $[n^* + 1, n^* + m^*]$ then $Pr_{n,i}(s)$ means the probability of corresponding overbooking. The probability $Pr_{n^*+m^*+1,i}(s)$ means the probability that the number of claims exceeds $n^* + m^*$.

We know values of $n = n^*$ and $i = i_0$ for the initial stage with number s^*, therefore

$$Pr_{n,i}(s^*) = \begin{cases} 1 & \text{if } n = n^*, i = i_0, \\ 0 & \text{otherwise.} \end{cases} \tag{5}$$

Further for $s = s^* - 1, s^* - 2, \dots, 0$, $n \geq n^*$,

$$Pr_{n,j}(s) = \begin{cases} \displaystyle\sum_{i=1}^{k}\sum_{\eta=n^*}^{n} Pr_{\eta,i}(s+1)\tilde{P}r_{n-\eta,i,j}(s+1), & \text{if } n^* \leq n \leq n^* + m^*(s+1); \\ \displaystyle P_{i_0,j}((s^* - s)\Delta) - \sum_{n=n^*}^{n^*+m^*(s+1)} Pr_{n,j}(s), & \text{if } n = n^* + m^*(s+1) + 1. \end{cases} \tag{6}$$

The part of the last formula, corresponding to the case $n = n^* + m^*(s+1)$, follows from the equality

$$P_{i_0,j}((s^* - s)\Delta) = \sum_{n=n^*}^{\infty} Pr_{n,j}(s).$$

The zero stage $s = 0$ corresponds to the instant of the trip beginning. Now n means the number of all booked or sold tickets. Each of the corresponding passengers can not arrive to the trip with probability q, independently on the other passengers. Therefore, the probability that n passengers are in front of the take off

$$PP_n = \sum_{\eta=n}^{n^*+m^*} \frac{\eta!}{n!(\eta - n)!}(1-q)^n q^{\eta-n} \sum_{j=1}^{k} Pr_{\eta,j}(0), n \le n^* + m^*. \tag{7}$$

Finally the probability that n passengers have flown away is as follows:

$$PFA_n = \begin{cases} PP_n, & \text{if } n < n^*, \\ \sum_{n=n^*}^{n^*+m^*} PP_n, & \text{if } n = n^*. \end{cases} \tag{8}$$

The represented formulas allow to calculate various efficiency indices. Firstly, the average number of engaged seats:

$$Avr = \sum_{n=1}^{n^*} n \times PFA_n. \tag{9}$$

Secondly, the probability that n passengers with bought or booked tickets meet with refusal equals $P_{n+n^*}, n = 1, \ldots, m^*$. Average number of those passengers AvrR is calculated as follows:

$$AvrR = \sum_{n=1}^{m^*} nP_{n+n^*}. \tag{10}$$

Finally, the probability that a customer encounters a refusal to purchase the ticket equals $\sum_{i=1}^{k} \sum_{s=1}^{s^*} Pr_{n^*+m^*+1,i}(s)$.

3 Numerical Example

Our example has the following input data. The Markov chain has three states ($k = 3$) and the transition intensities matrix

$$\lambda = \begin{pmatrix} 0 & 0.3 & 0.4 \\ 0.5 & 0 & 0.4 \\ 0.5 & 0.6 & 0 \end{pmatrix}.$$

The initial state of Markov chain is known and fixed: $J(0) = i_0 = 1$. The capacity of the aircraft n^* equals 20. A time before departure, when passengers buy tickets, has Erlang distribution with parameters $\mu = 0.5$ and $\theta = 3$, and the density

$$f(x) = \frac{1}{4}(0.5x)^2 \exp(-0.5x), x \geq 0. \tag{11}$$

Further we assume, that the average demand for given trip depends on the state of the random environment and equals $d_1 = 22.5$, $d_2 = 18$, $d_3 = 13.5$ for the first, second and third states. Now the arrivals intensity of passengers at time t until departure, if the i-th state occurs, is calculated by formula (1).

Let $\tau = 3$ be the time, when all 20 seats are occupied. We consider the selling and the booking process with the step $\Delta = 1$. Therefore we have $s^* = \tau/\Delta = 3$ stages. At last we suppose that the probability q, that a passenger doesn't come to a trip, equals 0.15.

The results for these initial data are given below. Firstly, the case $m_{n,i}(s) = m^* = 3$ for all n and s is considered, when the maximal number of additional overbooking equals 3 and doesn't depend on the number s of the step and the number n of booked and sold tickets. Figure 1 contains graph of density (11).

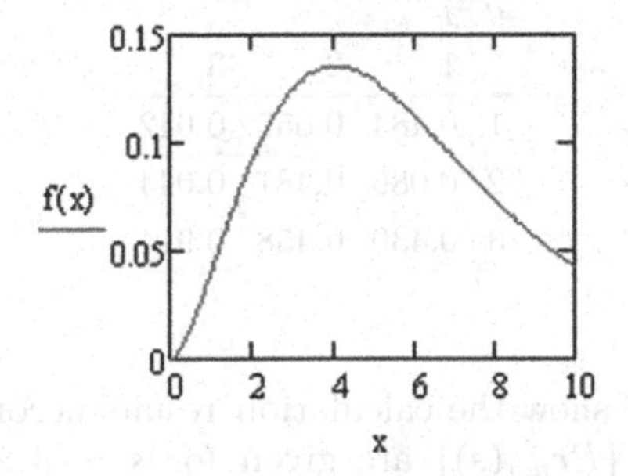

Fig. 1. Graph of density (11)

The expression (2) for the transition probabilities between states $J(t)$ has the following form:

$$P(t) = \begin{pmatrix} -0.704 & 0.577 & -0.323 \\ 0.503 & 0.577 & -0.323 \\ 0.503 & 0.577 & 0.889 \end{pmatrix} \begin{pmatrix} e^{-1.2t} & 0 & 0 \\ 0 & 1 & 0 \\ 0 & 0 & e^{-1.5t} \end{pmatrix} \begin{pmatrix} -0.829 & 0.829 & 0 \\ 0.722 & 0.549 & 0.462 \\ 0 & -0.825 & 0.825 \end{pmatrix}.$$

Tables 1, 2 and 3 contain the conditional average times $\tilde{ET}_{i,\nu,j}(\Delta) = ET_{i,\nu,j}(\Delta)/P_{i,j}(\Delta)$ of the sojourn of process $J(t)$ in the state ν on interval $(0, \Delta)$, if $J(0) = i, J(t) = j$. The columns of the tables correspond to the final states $j = 0, 1, 2$, the rows correspond to the intermediate states ν. Note that for each column the sum of all its elements equals 1.

Table 1. Conditional average sojourn times $ET_{0,\nu,j}(\Delta)$

ν	j		
	1	2	3
1	0.944	0.466	0.500
2	0.026	0.440	0.056
3	0.030	0.094	0.444

Table 2. Conditional average sojourn times $ET_{1,\nu,j}(\Delta)$

ν	j		
	1	2	3
1	0.484	0.031	0.086
2	0.455	0.931	0.470
3	0.061	0.038	0.444

Table 3. Conditional average sojourn times $ET_{2,\nu,j}(\Delta)$

ν	j		
	1	2	3
1	0.484	0.055	0.042
2	0.086	0.487	0.044
3	0.430	0.458	0.914

Tables 4, 5, 6 and 7 show the calculation results according to formulas (5)–(6). The probabilities $\{Pr_{n,j}(s)\}$ are given for $s = 3, 2, 1, 0$, $j = 0, 1, 2$ and $n = 20, \ldots, 24$. Let us remind, that: 1) the value for $n = 24$ means the probability, that the number of claims exceeds $n^* + m^* = 23$; 2) the initial state $J(s^*) = i0 = 1$.

Table 4. Probabilities $Pr_{n,j}(3)$

j	n				
	20	21	22	23	24
1	1	0	0	0	0
2	0	0	0	0	0
3	0	0	0	0	0

Table 5. Probabilities $Pr_{n,j}(2)$

j	n				
	20	21	22	23	24
1	0.051	0.125	0.153	0.125	0.015
2	0.023	0.049	0.954	0.039	0.112
3	0.027	0.055	0.056	0.038	0.077

Table 6. Probabilities $Pr_{n,j}(1)$

j	n				
	20	21	22	23	24
1	0.011	0.041	0.077	0.096	0.207
2	0.009	0.030	0.052	0.060	0.153
3	0.009	0.031	0.050	0.055	0.119

Table 7. Probabilities $Pr_{n,j}(0)$

j	n				
	20	21	22	23	24
1	0.009	0.034	0.066	0.085	0.228
2	0.007	0.027	0.049	0.062	0.168
3	0.007	0.024	0.044	0.054	0.138

The probabilities (7), that n passengers are in front of the take-off, are presented in Table 8.

Table 8. Probabilities PP_n

n	12	13	14	15	16	17	18	19	20	21	22	23
PP_n	0.001	0.002	0.006	0.019	0.047	0.095	0.158	0.209	0.213	0.158	0.075	0.017

The probabilities (8) that n passengers have flown away are presented in Table 9.

Table 9. Probabilities PFA_n

n	13	14	15	16	17	18	19	20
PFA_n	0.002	0.006	0.019	0.047	0.095	0.158	0.209	0.464

These tables allow the calculating of various efficiency indices. The average number of engaged seats Avr, calculated by the formula (9), equals 18.853. The probability PP_{20+n} that n passengers with bought or booked tickets encounter a refusal equals 0.158, 0.075, and 0.017 for $n = 1, 2$, and 3. There the average number of such passengers

$$1 \times 0.158 + 2 \times 0.075 + 3 \times 0.017 = 0.359.$$

It is noteworthy to compare these results with those whose will be without overbooking, when $m_{n,i}(s) = m^* = 0$. Instead of Table 9, Table 10 takes place.

Table 10. Probabilities PFA_n for the case $m_{n,i}(s) = m^* = 0$

n	11	12	13	14	15	16	17	18	19	20
PFA_n	0.001	0.005	0.016	0.045	0.103	0.182	0.243	0.229	0.137	0.039

The average number of engaged places Avr equals 17 instead of 18.853. We see that the difference is significant.

4 Optimization of an Economical Criterion

We need to determine values $m_{n,i}(s) \in (0, \ldots, m^*)$ for each stage $s = s^*, s^* - 1, \ldots, 1$, and values n and i so, that to maximize the average reward of engaged seats without penalty for refusal of the passenger with sold or booked tickets. Let $m(s) = (m_{n,i}(s))_{(m^*+1)\times k}$ be the corresponding matrix, $\boldsymbol{m} = (m(s), m(s), \ldots, m(s))^T$ be the block-matrix.

Let c be the cost of one ticket, r is the penalty for one refusal. We take into account an additional index: a company derives income from earlier selling or booking of tickets. So we introduce the function of the treble $\varepsilon(s)$: it is the additional income for one given ticket per one day. The function $\varepsilon(s)$ has the following properties: 1) it is non-negative and decreasing; 2) $\varepsilon(0) = 0$.

The average number of the of engaged seats $Avr(\boldsymbol{m})$ is calculated by formula (9). The average number the refusals $AvrR(\boldsymbol{m})$ is calculated by formula (10).

Let $\Sigma\varepsilon$ be the total additional income owing to the treble. Therefore average reward is calculated as follows:

$$AvReward(\boldsymbol{m}) = c \times Avr(\boldsymbol{m}) - r \times AvrR(\boldsymbol{m}) + \Sigma\varepsilon. \tag{12}$$

We have the typical problem of the dynamic programming [11,12]. In our case the decision variable on stage s for fixed n and i is $m_{n,i}(s), s = s^*, s^* - 1, \ldots, 1$.

Let $\tilde{c}_{n,i}(s)$ be the maximal average reward of the best overall policy for the remaining stages $s, s-1, \ldots, 1$, given that n tickets are sold and the random environment has state i. Given n and i, let $\tilde{m}_{n,i}(s)$ denote any value of the decision variable $m_{n,i}(s)$ that maximizes $\tilde{c}_{n,i}(s)$.

We begin with zero stage $s = 0$ and end by the last stage s^*. The zero stage gives the number n of the passengers, whose have bought or booked tickets. As before each of corresponding passengers could come not come to the trip with probability q, independently of the other passengers. The probability that η passengers from n are in front of the take-off is calculated with respect to the binomial distribution analogously to formula (7). Therefore

$$\tilde{c}_n(0) = \sum_{\eta=0}^{n} \frac{n!}{\eta!(n-\eta)!}(1-q)^{\eta} q^{n-\eta}(c \times min(\eta, n^*) - r \times max(\eta - n^*, 0))$$
$$0 \leq n \leq n^* + m^*, \forall i. \tag{13}$$

Values $\tilde{c}_{n,i}(s)$ for the other stages s are calculated recurrently for $s = 1, 2, \ldots, s^*$; $0 \leq n \leq n^* + m^*$; $\forall i$:

$$\tilde{c}_{n,i}(s) = \varepsilon(s)n + \max_{0 \leq m_{n,i}(s) \leq m^*} \left\{ \begin{array}{c} \displaystyle\sum_{\eta=0}^{n^* + m_{n,i}(s) - n} \sum_{j=1}^{k} \tilde{Pr}_{\eta,i,j}(s)\tilde{c}_{n+\eta,j}(s-1) \\ \displaystyle + \sum_{j=1}^{k} \Big(\sum_{\eta=n^*+m_{n,i}(s)-n+1}^{\infty} \tilde{Pr}_{\eta,i,j}(s) \Big) \tilde{c}_{n^*+m_{n,i}(s),j}(s-1). \end{array} \right\} \tag{14}$$

The ultimate solution is reached at the end of stage s^* as $\tilde{c}_{0,i}(s^*), i = 1, \ldots, k$.

We see that dynamic programming finds it by successively calculating $\tilde{c}_n(0)$, $\tilde{c}_{n,i}(1)$, ..., $\tilde{c}_{n,i}(s^*-1)$, $\tilde{c}_{n,i}(s^*)$ for all $0 \leq n \leq n^* + m^*$, $\forall i$. Because this procedure involves moving backward stage by stage, some authors also call s the number of remaining stages to the problem's solution.

The described procedure supposes using matrices $(\tilde{c}_{n,i}(s))$ for every stage s. Analogous matrices are necessary to keep up optimal decisions values $\tilde{m}_{n,i}(s)$. It is possible to act otherwise, using so called forward procedure. Namely, the optimal values $\tilde{m}_{n,i}(s)$ are found after the computing of all $\tilde{c}_{n,i}(\eta)$, $\eta = s^*, s^* - 1, \ldots, 0$.

5 Numerical Example (Continue)

Additionally to previous conditions we set $m^* = 6$, $c = 100$, $r = 200$. At beginning we set $\varepsilon(s) = 0$ for all s: Table 11 shows results of calculations by formula (13).

Table 11. Values of $\tilde{c}_n(0)$

n	1	2	3	4	5	6	7	8	9	10	11	12	13
$\tilde{c}_n(0)$	85	170	255	340	425	510	595	680	765	850	935	1020	1105
n	14	15	16	17	18	19	20	21	22	23	24	25	26
$\tilde{c}_n(0)$	1190	1275	1360	1445	1530	1615	1700	1775	1821	1819	1767	1670	1543

Tables 12, 13 and 14 contain the calculation results according to formula (14) for $s = 1, 2, 3$.

Table 12. Values of $\tilde{c}_{n,i}(1)$

i	n						
	20	21	22	23	24	25	26
1	1721	1787	1821	1819	1767	1670	1541
2	1719	1786	1821	1819	1767	1670	1541
3	1716	1784	1821	1819	1767	1670	1541

Table 13. Values of $\tilde{c}_{n,i}(2)$

i	n						
	20	21	22	23	24	25	26
1	1783	1812	1821	1819	1767	1670	1541
2	1777	1810	1821	1819	1767	1670	1541
3	1772	1808	1821	1819	1767	1670	1541

Table 14. Values of $\tilde{c}_{n,i}(3)$

i	n						
	20	21	22	23	24	25	26
1	1814	1820	1821	1819	1767	1670	1541
2	1812	1819	1821	1819	1767	1670	1541
3	1809	1819	1821	1819	1767	1670	1541

We see that the state of the random environment influences on the results weakly. The optimal values of the decision are the same for various states. Moreover, they are equally at different stages in our case. Theses optimal decisions are presented in the Table 15.

Table 15. Optimal decisions $m_{n,\cdot}(\cdot)$

n	20	21	22	23	24	25	26
$m_{n,\cdot}(\cdot)$	2	1	0	0	0	0	0

From the given tables it follows that optimal value of the overbooking $m_{n,i}(s)$ for all stages s and states i is 2 or 1. This corresponds to $n = 20 + 2 = 22$, and from given tables we see, that the optimal value of the rewards equals 1821.

Now compare this result with the previous case, where it is supposed, that $m_{n,i}(s) = m^* = 3$ for all $n < 21$, i and s. In this case the average number of engaged seats $Avr = 18.853$, the average number of passengers with bought and booked tickets, whose meet with refusal equals 0.359. It gives the following reward, with respect to formula (12):

$$AvReward(\boldsymbol{m}) = c \times Avr(\boldsymbol{m}) - r \times AvrR(\boldsymbol{m}) = 100 \times 18.853 - 200 \times 0.359 = 1813.5.$$

We see that the increment is less.

Now we take into account the function of the treble $\varepsilon(s)$. Let it be the following form:

$$\varepsilon(s) = \delta(\exp(\alpha(s^* - s)), s \geq 0, \tag{15}$$

where $\delta, \alpha \geq 0$.

Graph of this function for $\delta = 15$ and $\alpha = 2$ is presented on Fig. 2. Now the Tables 12, 13 and 14 look as follows (see Tables 16, 17 and 18).

Table 16. Values of $\tilde{c}_{n,i}(1)$ for the expression (15)

i	n					
	20	21	22	23	24	25
1	1727	1793	1827	1826	1773	1548
2	1724	1791	1182	1826	1773	1677
3	1722	1790	1827	1826	1773	1677

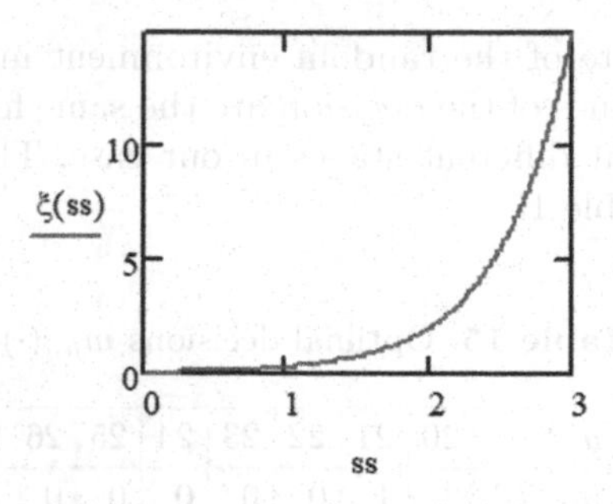

Fig. 2. Graph of the function (15)

Table 17. Values of $\tilde{c}_{n,i}(2)$ for the expression (15)

i	n						
	20	21	22	23	24	25	26
1	1815	1852	1865	1866	1815	1721	1594
2	1808	1849	1865	1866	1815	1721	1594
3	1802	1847	1865	1866	1815	1721	1594

Table 18. Values of $\tilde{c}_{n,i}(3)$ for the expression (15)

i	n						
	20	21	22	23	24	25	26
1	2108	2138	2157	2171	2133	2052	1938
2	2103	2137	2157	2171	2133	2052	1938
3	2098	2135	2157	2171	2133	2052	1938

As earlier the random environment influences on the results weakly, but now the optimal values of the decision depends on different stages. These optimal decisions are presented in the Table 19.

Table 19. Optimal decisions $m_{n,\cdot}(s)$ for the expression (15)

n	20	21	22	23	24	25	26
$m_{n,\cdot}(1)$	2	1	0	0	0	0	0
$m_{n,\cdot}(2)$	3	2	1	0	0	0	0
$m_{n,\cdot}(3)$	3	2	1	0	0	0	0

6 Conclusion

The considered model can be generalized in many ways. Firstly to discriminate between sold and booked tickets. Secondly it takes into account a possibility of a cancellation of booked tickets during a period of our consideration. Our future researches will be connected with the realization of these possibilities.

References

1. Shlifer, R., Vardi, Y.: An airline overbooking policy. Transp. Sci. **9**(2), 101–114 (1975)
2. Liberman, V., Yechiali, U.: On the hotel overbooking problem - an inventory system with stochastic cancellations. Manage. Sci. **24**(11), 1117–1126 (1978)
3. Chatwin, R.E.: Optimal control of continuous-time terminal-value birth-and-death processes and airline overbooking. Naval Res. Logist. **43**(2), 159–168 (1996)
4. Chatwin, R.E.: Multi-period airline overbooking with a single fare class. Opns. Res. **46**(6), 805–819 (1998)
5. Chatwin, R.E.: Continuous-time airline overbooking with time-dependent fares and refunds. Transp. Sci. **33**(2), 182–191 (1999)
6. Fard, F.A., Sy, M., Ivanov, D.: Optimal overbooking strategies in the airlines using dynamic programming approach in continuous time. Transp. Res. Part E Logistics Transp. Rev. **128**, 384–399 (2019)
7. Phillips, R.: Pricing and Revenue Optimization. Stanford University Press, Stanford (2005)
8. Sulima, N.: Probabilistic model of overbooking for an airline. Autom. Control Comput. Sci. **46**(1), 68–78 (2012)
9. Kijima, M.: Total positivity. Markov Processes for Stochastic Modeling, pp. 313–318. Springer, Boston (1997). https://doi.org/10.1007/978-1-4899-3132-0_8
10. Pacheco, A., Tang, L.C., Prabhu, N.U.: Markov-Modulated Processes & Semiregenerative Phenomena. World Scientific, Hoboken, New York (2009)
11. Bellman, R.: Dynamic Programming. Princeton University Press, Princeton (1957)
12. Bellman, R.E., Dreyfus, S.E.: Applied Dynamic Programming. Princeton University Press, Princeton (1962)

Optimization of Signals Processing in Nodes of Sensor Network with Energy Harvesting and Expenditure for Admission and Transmission

Sergey Dudin[1], Olga Dudina[1], Alexander Dudin[1,2], and Chesoong Kim[3](✉)

[1] Department of Applied Mathematics and Computer Science, Belarusian State University, Minsk 220030, Belarus
dudin85@mail.ru, dudina@bsu.by, dudin-alexander@mail.ru
[2] Peoples' Friendship University of Russia (RUDN University), 6 Miklukho-Maklaya Street, Moscow 117198, Russia
[3] Sangji University, Wonju, Kangwon 26339, Republic of Korea
dowoo@sangji.ac.kr

Abstract. Operation of a sensor node of a wireless sensor network with energy harvesting is described by the single-server queue. Customers and energy units arrive according to the Markov arrival processes (MAP) and are stored in the corresponding buffers. Service of a customer is possible only in presence of an energy unit. In contrast to previously investigated in the literature models, we assume that, besides the use of one energy unit for service of any customer, one more unit is expended at the moment of a customer arrival if the customer is accepted to the system. To optimize operation of the system, a parametric strategy of admission control is used. The goal of control is to minimize the risk of the server starvation in case of too strict control and the risk of wasting the energy due to acceptance of too many customers that finally will not receive a service (due to the lack of energy or impatience) in case of too liberal control. Under the fixed value of control parameter, the behavior of the system is described by the six-dimensional Markov chain. The generator of this Markov chain is obtained. Expressions for computation of the key performance indicators of the system are presented. Numerical results illustrating the effectiveness of the proposed control strategy are presented.

Keywords: Energy harvesting and consumption · Admission control · Markov arrival process · Impatience · Phase-type distribution

This work was supported by the Basic Science Research Program through the National Research Foundation of Korea (NRF) funded by the Ministry of Education (NRF-2020R1A2C1006999) and by the RUDN University Program 5-100.

V. M. Vishnevskiy et al. (Eds.): DCCN 2020, LNCS 12563, pp. 406–421, 2020.
https://doi.org/10.1007/978-3-030-66471-8_31

1 Introduction

Wireless sensor networks have a huge number of applications including environmental applications (forest fire and flood detection, monitoring the pesticides level in the drinking water, the level of soil erosion, and the level of air pollution in realtime), health applications (telemonitoring and tracking the human physiological data, location of doctors, patients and drugs inside a hospital), military applications (including C4ISRT – command, control, communications, computing, intelligence, surveillance, reconnaissance and targeting – systems), applications for home automation, control in office buildings, managing inventory control, vehicle tracking and detection, etc. The relevant surveys are given, e.g., in [1–4].

Nodes of sensor networks have small batteries with limited power and storage space. When the battery of a node is exhausted, sometimes it is not possible to replace it and the node dies. When a certain number of nodes will die, the network may not be able to perform its designated task. Recent advances in energy harvesting technology have resulted in the design of new types of sensor nodes which are able to extract energy from the surrounding environment. The major sources of energy harvesting include solar, wind, sound, vibration, thermal, and electromagnetic power. Energy harvesting sensor nodes have the potential to perpetuate the life time of a battery through continuous energy harvesting. Literature about the wireless sensor networks with energy harvesting includes, in particular, [5–10].

Since a signal detection occurs at the random instants and the process of energy harvesting can be also interpreted as the stochastic process, it is reasonable to apply the theory of queues for description of operation and optimization of sensor nodes with energy harvesting. As examples of papers where such application is made, we can refer to [11–18]. As more recent papers, we can mention [19] and [20] as well as several papers, which are published in 2019, cited in [20]. It is worth noting that queueing models with energy harvesting are quite close to previously investigated in the literature queueing models with additional items like so-called queueing/inventory models, queueing models with paired customers, assembly-like queues, passenger-taxi or double-ended queues, coupled queues, etc. More details about the mentioned queues can be found, e.g., in the paper [21]. In that paper, the role of the additional items is played by the lag between the critical and current value of the server's temperature. Among the papers devoted to queueing models with additional items we can mention recent papers [22–24].

Importance of queueing models with energy harvesting among more wide class of other queueing models with additional items stems from their evident practical motivation and the necessity of choosing the required source of energy harvesting, the capacity of the battery and of control by the model operation to maximally reduce energy consumption. This allows to optimize performance of the sensors. Various variants of the optimization problem's formulation naturally arise. E.g., in [18,25] the $MAP/PH/1$ type queues were considered. Flow of energy is also described by the MAP. It is assumed that a unit of energy is spent for implementation of each phase of service. If, at a some phase completion epoch, the next phase of service of a customer should be implemented but the energy is absent, this customer is lost. After the server becomes idle, vacation period starts. This period finishes when the number of customers accumulated in the buffer reaches the preassigned threshold, say, K. The problem of choosing the optimal value of the threshold K is not trivial. If K is too large, the idle time of the server is too long and performance of the system is low. If K is too small, there is an essential risk that the number of phases of an arbitrary customer service is large and service will not be completed due to the absence of the energy. Therefore, this customer will be lost and all energy used for its (incomplete) service is wasted. This also leads to poor performance of the system. The influence of the threshold K on the performance measures of the system was analysed in [18] for the system with a buffer for customers and in [25] for the system with customers retrials. Importance of account of correlation in the arrival processes of customers and energy is demonstrated.

In [26], also the $MAP/PH/1$ type queues were considered. Flow of energy is also described by the MAP. It is assumed that a unit of energy is spent for providing service to one customer. Server takes a vacation if at least of one of the queues of customers or energy is empty. Waking up the server requires certain number of energy units. Decision about waking the server up is made by means of comparison of the number of customers and energy units in the system with some thresholds. The problem of clarification of influence of these thresholds on various performance measures is solved. In [22], similar problem was considered for the system with two types of customers requiring different number of energy units and limited preemptive priority of one type of customers. Arrival of the priority customer interrupts service of non-priority customer only if the energy required for its service is available. If not, the priority customer is lost. In [27], an analog of this system with customers retrials is analysed. In [28], model with energy harvesting was analysed in assumption that the processes of arrival of customers and energy and service of customers depend on the state of some external Markov process called as random environment. In [19,20,28], quite general models were considered. The system can operate in several modes. Modes are distinguished by the required number of energy units, service rate and probability of service without error. Thresholds in state space of the energy are fixed and decision about the service mode for the next customer is based on the relation of the available number of energy units with these thresholds. Possibility of optimal choice of the thresholds is shown.

In all queueing models considered in the cited papers, it is assumed that energy units are spent only for service of a customer (transmission of a registered signal to the neighboring or gateway node). In our model, we account the fact that indeed the energy may be required not only for transmission of the signal but also for receiving and registering the signal from a sensor. Therefore, all harvested energy has to be split by the node between its activity related to the signal admission and buffering and the activity related to the signal transmission. Due to the possible deficit of the energy, this sharing has to be organized in a proper way. From one hand, it does not make sense to accept too many signals (customers) if the chance to further provide service to them is small (due to the possible lack of energy or obsolescence of information delivered by these customers). From another hand, if the acceptance discipline is too strict, many signals are lost and the node badly implements its task. In addition, the risk of the future server starvation even in presence of energy may be high. To optimize the splitting of the harvested energy for customers admission and service in the system, in this paper, we introduce a parametric strategy of admission control and analyse influence of the parameter defining this strategy on the performance of the system.

The outline of the presentation of the results is as follows. In Sect. 2, the mathematical model is formulated and control strategy is fixed. The process of system states is defined in Sect. 3. This process is the six-dimensional continuous-time Markov chain. Due to the chosen strategy of control, this Markov chain has a finite state space and always has a stationary distribution of the states. Generator of this Markov chain is presented and the problem of computation of the stationary distribution is briefly discussed in this section. In Sect. 4, formulas for computation of the key performance indicators of the system are given. Numerical results are presented in Sect. 5. Section 6 concludes the paper.

2 Mathematical Model

We consider a single-server queuing system with an infinite buffer for customers and a finite buffer of capacity R for energy. The structure of this system is presented in Fig. 1.

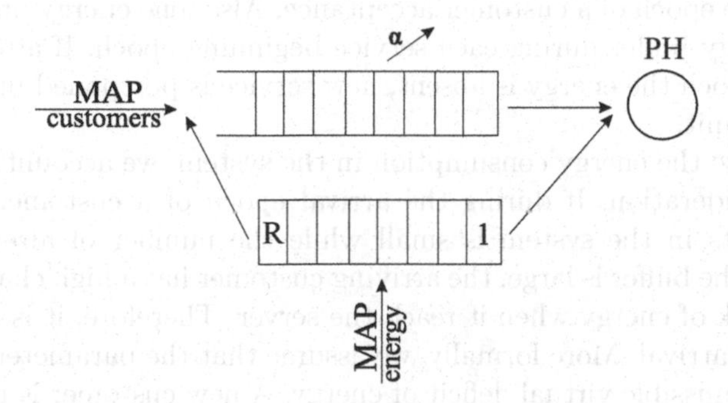

Fig. 1. System under study.

Arrival of customers and energy units is defined by the two independent MAP arrival processes. Arrivals of customers in the MAP are directed by an irreducible continuous-time Markov chain ν_t, $t \geq 0$, with the finite state space $\{1, ..., V\}$. The sojourn time of the Markov chain ν_t, $t \geq 0$, in the state ν has an exponential distribution with the parameter λ_ν, $\nu = \overline{0, V}$. After this sojourn time expires, with probability $p_1(\nu, \nu')$ the process ν_t transits to the state ν' and a customer arrives to the system. With probability $p_0(\nu, \nu')$ this transition of the process ν_t does not cause arrival of a customer.

The intensities of transitions of the process ν_t from one state to another one, that are accompanied by the arrival of k customers, $k = 0, 1$, are combined to the square matrices D_k, $k = 0, 1$, of size $V + 1$. The matrix generating function of these matrices is $D(z) = D_0 + D_1 z$, $|z| \leq 1$. The matrix $D(1)$ is an infinitesimal generator of the process ν_t, $t \geq 0$. The stationary distribution vector $\boldsymbol{\theta}$ of this process satisfies the system of equations $\boldsymbol{\theta} D(1) = \mathbf{0}$, $\boldsymbol{\theta}\mathbf{e} = 1$. Here, $\mathbf{e}$ is a column vector consisting of 1's and $\mathbf{0}$ is a row vector consisting of 0's.

The average intensity λ_c of customers arrival is defined by $\lambda_c = \boldsymbol{\theta} D_1 \mathbf{e}$. The squared coefficient of the variation of inter-arrival times is calculated as $c_{var}^2 = 2\lambda_c \boldsymbol{\theta}(-D_0)^{-1}\mathbf{e} - 1$, and the coefficient of correlation of the intervals between the successive arrivals is given as $c_{cor} = (\lambda_c \boldsymbol{\theta}(-D_0)^{-1} D_1 (-D_0)^{-1}\mathbf{e} - 1)/c_{var}^2$.

We assume that the customers staying in the buffer are impatient. If any customer is not picked up for service during a period of time that is exponentially distributed with the parameter α, $0 \leq \alpha < \infty$, then the customer leaves the buffer and the system (is lost), independently of other customers.

Arrivals of energy units in the corresponding MAP are defined by an irreducible continuous-time Markov chain w_t, $t \geq 0$, with the finite state space $\{1, ..., W\}$ and the matrices H_0 and H_1. The average intensity of energy units arrival λ_e is defined by the formula $\lambda_e = \boldsymbol{\zeta} H_1 \mathbf{e}$ where $\boldsymbol{\zeta}$ is the row vector of the stationary probabilities of the Markov chain w_t. This vector is the unique solution to the system $\boldsymbol{\zeta}(H_0 + H_1) = \mathbf{0}$, $\boldsymbol{\zeta}\mathbf{e} = 1$. An arriving unit of energy joins the buffer for energy if it is not full and is lost otherwise.

We assume that energy units are spent both to receive customers and provide service to them. Namely, we assume that the number of energy units decreases by one during the epoch of a customer acceptance. Also, one energy unit disappears from the energy buffer during each service beginning epoch. If after the service completion epoch the energy is absent, new service is postponed until an arrival of an energy unit.

To optimize the energy consumption in the system, we account the following intuitive consideration. If during the arrival epoch of a customer the number of energy units in the system is small while the number of already accepted customers in the buffer is large, the arriving customer has a high chance to be lost due to the lack of energy when it reach the server. Therefore, it is reasonable to reject it upon arrival. More formally, we assume that the parameter R_1, $R_1 \geq 0$, defines the admissible virtual deficit of energy. A new customer is rejected if the number of available energy units is less or equal to $\max\{0, i - R_1\}$, where i is the number of customers in the buffer at the arrival epoch. The control parameter R_1 defines the tolerance of the admission strategy. If $R_1 = 0$, a new customer is rejected if availability of energy unit for its service is not guaranteed even if all accumulated energy will be spent for service of customers (without acceptance of future arrivals). If $R_1 \geq 1$, some temporal deficit of energy is admissible (in anticipation of arrival of new energy units during service of customers staying in the queue). It is clear that the system operation can be optimised via the proper choice of the threshold R_1. This choice is not trivial, moreover we consider the system with MAP processes of customers and energy which may exhibit correlation that may cause an alternation of periods of the frequent and rare arrivals. In case when the parameter R_1 is chosen too small, many customers can be rejected during the periods of frequent arrivals and further this may cause the starvation of the server and the loss of possible profit that may be gained via the service of customers. In case when the parameter R_1 is chosen too large, it may cause the waste of energy because: (i) some accepted customers will not be served due to unavailability of energy, (ii) due to the large number of customers in the buffer some of them will be lost due to their impatience, while the energy was already spent for their acceptance.

If a customer arrives when the number of energy units in the energy buffer is more than $\max\{0, i - R_1\}$, then the customer is accepted for service. Two scenarios are possible:

1) If the server is idle and the number of energy units is not less than two, the customer immediately starts service and the number of energy units decreases by two. One unit of energy is spent on receiving this customer and one more unit is consumed for its service. If the server is idle and the number of energy units is equal to one, the unit of energy decreases to zero and the customer joins the buffer to wait until energy arrival (if it will not depart earlier due to impatience);
2) If the server is busy, the number of energy units decreases by one and the customer joins the buffer.

Customers are picked up for service from the buffer according to the First-In-First-Out discipline.

The service time of a customer by a server has the PH distribution with the irreducible representation $(\boldsymbol{\beta}, S)$. This service time can be interpreted as the time until the underlying Markov process $m_t, t \geq 0$, with a finite state space $\{1, \ldots, M, M+1\}$ reaches the single absorbing state $M+1$ conditional on the fact that the initial state of this process is selected among the states $\{1, \ldots, M\}$ according to the probabilistic row vector $\boldsymbol{\beta} = (\beta_1, \ldots, \beta_M)$. The transition rates of the process m_t within the set $\{1, \ldots, M\}$ are defined by the sub-generator S, and the transition rates into the absorbing state (what leads to service completion) are given by the entries of the column vector $\mathbf{S}_0 = -S\mathbf{e}$. The mean service time is calculated as $b_1 = \boldsymbol{\beta}(-S)^{-1}\mathbf{e}$.

Our goal is to analyse the behavior of the described queueing model.

3 Process of System States and Its Stationary Distribution

Let, during the epoch t, $t \geq 0$,

- i_t, $i_t = \overline{0, R + R_1}$, be the number of customers in the infinite buffer,
- n_t, be the state of the server: if $n_t = 0$, the server is idle, if $n_t = 1$, the server is busy,
- r_t, $r_t = \overline{0, R}$, be the number of available energy units,

- ν_t, $\nu_t = \overline{1,V}$, be the state of the underlying process of the MAP of customers,
- w_t, $w_t = \overline{1,W}$, be the state of the underlying process of the MAP of units of energy,
- m_t, $m_t = \overline{1,M}$, be the state of PH service process.

The Markov chain $\xi_t = \{i_t, n_t, r_t, \nu_t, w_t, m_t\}$, $t \geq 0$, is the regular irreducible continuous-time Markov chain.

The Markov chain ξ_t, $t \geq 0$, has the following state space:

$$\left(\{0,0,r,\nu,w\}\right)\bigcup\left(\{i,0,0,\nu,w\},\ i=\overline{1,R+R_1}\right)\bigcup\left(\{i,1,r,\nu,w,m\},\ i=\overline{0,R+R_1}\right),$$
$$r=\overline{0,R},\ \nu=\overline{1,V},\ w=\overline{1,W},\ m=\overline{1,M}.$$

Let us introduce the following notations:

- I is the identity matrix and O is a zero matrix of an appropriate dimension. If it is necessary, dimension of the matrix is indicated by the suffix;
- E^- is the square matrix of size $R+1$ with all zero entries except the entries $(E^-)_{i,i-1}$, $i=\overline{1,R}$, which are equal to 1;
- E^+ is the square matrix of size $R+1$ with all zero entries except the entries $(E^+)_{i,i+1}$, $i=\overline{0,R-1}$, and $(E^+)_{R,R}$ which are equal to 1;
- $\hat{I}_i$ is the square matrix of size $R+1$ defined as $\hat{I}_i = \text{diag}\{\underbrace{1,1,\ldots,1}_{a_i},0,0,\ldots,0\}$ where $a_i = \min\{\max\{0, i-R_1\}, R\}+1$, $i=\overline{0,R+R_1}$;
- $\hat{\mathbf{e}}$ is the column vector of size $R+1$ defined as $\hat{\mathbf{e}} = (1,0,0,\ldots,0)$;
- $\tilde{\mathbf{e}}$ is the column vector of size $R+1$ defined as $\tilde{\mathbf{e}} = (0,1,0,\ldots,0)$;
- $\otimes$ and $\oplus$ are the symbols of the Kronecker product and sum of matrices, see, e.g., [29].

Let us enumerate the states of the Markov chain ξ_t in the lexicographic order and refer to the set of states of the chain having value i of the first components of the Markov chain as *level* i, $i \geq 0$.

Let Q be the generator of the Markov chain ξ_t, $t \geq 0$.

Lemma 1. *The generator Q has the following block-tridiagonal structure:*

$$Q=\begin{pmatrix}
Q_{0,0} & Q_{0,1} & O & O & \cdots & O & O & O\\
Q_{1,0} & Q_{1,1} & Q_{1,2} & O & \cdots & O & O & O\\
O & Q_{2,1} & Q_{2,2} & Q_{2,3} & \cdots & O & O & O\\
\vdots & \vdots & \vdots & \vdots & \ddots & \vdots & \vdots & \vdots\\
O & O & O & O & \cdots & Q_{R+R_1-1,R+R_1-2} & Q_{R+R_1-1,R+R_1-1} & Q_{R+R_1-1,R+R_1}\\
O & O & O & O & \cdots & O & Q_{R+R_1,R+R_1-1} & Q_{R+R_1,R+R_1}
\end{pmatrix}.$$

The non-zero blocks $Q_{i,j}$, $i,j \geq 0$, containing the intensities of the transitions from level i to level j have the following form:

$$Q_{i,j}=\begin{pmatrix} Q_{i,j}^{(0,0)} & Q_{i,j}^{(0,1)}\\ Q_{i,j}^{(1,0)} & Q_{i,j}^{(1,1)}\end{pmatrix},$$

where the sub-block $Q_{i,j}^{(n,n')}$ *contains the intensities of transition from the states of level* i *with the value* n *of the component* n_t *of the Markov chain* ξ_t *to the states of level* j *with the value* n' *of the component* n_t, $n, n' = 0, 1$.

The non-zero sub-blocks $Q_{i,j}^{(n,n')}$ *are given by:*

$$Q_{0,0}^{(0,0)} = I_{R+1} \otimes (D_0 \oplus H_0) + \hat{I}_0 \otimes D_1 \otimes I_W + E^+ \otimes I_V \otimes H_1,$$

$$Q_{0,0}^{(0,1)} = (I - \hat{I}_0)(E^-)^2 \otimes D_1 \otimes I_W \otimes \boldsymbol{\beta},\ Q_{0,0}^{(1,0)} = I_{(R+1)VW} \otimes \boldsymbol{S}_0,$$

$$Q_{0,0}^{(1,1)} = I_{R+1} \otimes (D_0 \oplus H_0 \oplus S) + \hat{I}_0 \otimes D_1 \otimes I_{WM} + E^+ \otimes I_V \otimes H_1 \otimes I_M,$$

for $i = \overline{1, R + R_1}$:

$$Q_{i,i}^{(0,0)} = D_0 \oplus H_0 - i\alpha I_{VW} + D_1 \otimes I_W,\ Q_{i,i}^{(1,0)} = \hat{\mathbf{e}}^T \otimes I_{VW} \otimes \boldsymbol{S}_0,$$

$$Q_{i,i}^{(1,1)} = I_{R+1} \otimes (D_0 \oplus H_0 \oplus S) + \hat{I}_i \otimes D_1 \otimes I_{WM} + E^+ \otimes I_V \otimes H_1 \otimes I_M - i\alpha I_{(R+1)VWM},$$

$$Q_{0,1}^{(0,0)} = (I - \hat{I}_0)\tilde{\mathbf{e}}^T \otimes D_1 \otimes I_W,\ Q_{0,1}^{(1,1)} = (I - \hat{I}_0)E^- \otimes D_1 \otimes I_{WM},$$

for $i = \overline{1, R + R_1 - 1}$:

$$Q_{i,i+1}^{(1,1)} = (I - \hat{I}_i)E^- \otimes D_1 \otimes I_{WM},\ Q_{1,0}^{(0,0)} = \alpha\hat{\mathbf{e}} \otimes I_{VW},$$

$$Q_{1,0}^{(0,1)} = \hat{\mathbf{e}} \otimes I_V \otimes H_1 \otimes \boldsymbol{\beta},\ Q_{1,0}^{(1,1)} = \alpha I_{(R+1)VWM} + E^- \otimes I_{VW} \otimes \boldsymbol{S}_0\boldsymbol{\beta},$$

for $i = \overline{2, R + R_1}$:

$$Q_{i,i-1}^{(0,0)} = i\alpha I_{VW},\ Q_{i,i-1}^{(0,1)} = \hat{\mathbf{e}} \otimes I_V \otimes H_1 \otimes \boldsymbol{\beta},\ Q_{i,i-1}^{(1,1)} = i\alpha I_{(R+1)VWM} + E^- \otimes I_{VW} \otimes \boldsymbol{S}_0\boldsymbol{\beta}.$$

Proof of the lemma is performed by means of analysis of the intensities of all possible transitions of the Markov chain ξ_t during the time interval having an infinitesimal length.

The Markov chain ξ_t, $t \geq 0$, is an irreducible and has a finite state space. Therefore, the stationary probabilities $\pi(i, 1, r, \nu, w, m)$, $i = \overline{0, R + R_1}$, $r = \overline{0, R}$, $\nu = \overline{1, V}$, $w = \overline{1, W}$, $m = \overline{1, M}$, and $\pi(0, 0, r, \nu, w)$, $r = \overline{0, R}$, $\nu = \overline{1, V}$, $w = \overline{1, W}$, and $\pi(i, 0, 0, \nu, w)$, $i = \overline{1, R + R_1}$, $\nu = \overline{1, V}$, $w = \overline{1, W}$, of the system states exist. Let us form the row vectors $\boldsymbol{\pi}_i$ of these probabilities as follows

$\boldsymbol{\pi}_i = (\boldsymbol{\pi}(i, 0), \boldsymbol{\pi}(i, 1))$, $i = \overline{0, R + R_1}$, where
$\boldsymbol{\pi}(0, 0) = (\boldsymbol{\pi}(0, 0, 0), \boldsymbol{\pi}(0, 0, 1), \dots, \boldsymbol{\pi}(0, 0, R))$,
$\boldsymbol{\pi}(0, 0, r) = (\boldsymbol{\pi}(0, 0, r, 1), \boldsymbol{\pi}(0, 0, r, 2), \dots, \boldsymbol{\pi}(0, 0, r, V))$,
$\boldsymbol{\pi}(0, 0, r, \nu) = (\pi(0, 0, r, \nu, 1), \pi(0, 0, r, \nu, 2), \dots, \pi(0, 0, r, \nu, W))$, $r = \overline{0, R}$, $\nu = \overline{1, V}$,
$\boldsymbol{\pi}(i, 0) = \boldsymbol{\pi}(i, 0, 0) = (\boldsymbol{\pi}(i, 0, 0, 1), \boldsymbol{\pi}(i, 0, 0, 2), \dots, \boldsymbol{\pi}(i, 0, 0, V))$,
$\boldsymbol{\pi}(i, 0, 0, \nu) = (\pi(i, 0, 0, \nu, 1), \pi(i, 0, 0, \nu, 2), \dots, \pi(i, 0, 0, \nu, W))$, $i = \overline{1, R + R_1}$, $\nu = \overline{1, V}$, and
$\boldsymbol{\pi}(i, 1) = (\boldsymbol{\pi}(i, 1, 0), \boldsymbol{\pi}(i, 1, 1), \dots, \boldsymbol{\pi}(i, 1, R))$,
$\boldsymbol{\pi}(i, 1, r) = (\boldsymbol{\pi}(i, 1, r, 1), \boldsymbol{\pi}(i, 1, r, 2), \dots, \boldsymbol{\pi}(i, 1, r, V))$,
$\boldsymbol{\pi}(i, 1, r, \nu) = (\boldsymbol{\pi}(i, 1, r, \nu, 1), \boldsymbol{\pi}(i, 1, r, \nu, 2), \dots, \boldsymbol{\pi}(i, 1, r, \nu, W))$,
$\boldsymbol{\pi}(i, 1, r, \nu, w) = (\pi(i, 1, r, \nu, w, 1), \pi(i, 1, r, \nu, w, 2), \dots, \pi(i, 1, r, \nu, w, M))$,
$i = \overline{0, R + R_1}$, $r = \overline{0, R}$, $\nu = \overline{1, V}$, $w = \overline{1, W}$.

It is well known that the probability vectors $\boldsymbol{\pi}_i$, $i = \overline{0, R+R_1}$, satisfy the following system of linear algebraic equations (equilibrium or Chapman-Kolmogorov equations):

$(\boldsymbol{\pi}_0, \boldsymbol{\pi}_1, \ldots, \boldsymbol{\pi}_{R+R_1})Q = \mathbf{0}$, $(\boldsymbol{\pi}_0, \boldsymbol{\pi}_1, \ldots, \boldsymbol{\pi}_{R+R_1})\mathbf{e} = 1$ where Q is the infinitesimal generator of the Markov chain ξ_t, $t \geq 0$. This system is the finite one and there are several numerically stable methods for its solving that effectively use the sparse structure of the generator, see, e.g., [30–34].

4 Computation of Performance Measures

As soon as the vectors $\boldsymbol{\pi}_i$, $i = \overline{0, R+R_1}$, have been calculated, we are able to find various performance measures of the system.

The average number N_c of customers in the buffer is computed by

$N_c = \sum\limits_{i=1}^{R+R_1} i\boldsymbol{\pi}_i\mathbf{e}$.

The average number N_e of energy units in the buffer is computed by

$N_e = \sum\limits_{r=1}^{R} r\boldsymbol{\pi}(0,0,r)\mathbf{e} + \sum\limits_{i=0}^{R+R_1}\sum\limits_{r=1}^{R} r\boldsymbol{\pi}(i,1,r)\mathbf{e}$.

The probability P_{busy} that at an arbitrary moment the server is busy is computed by $P_{busy} = \sum\limits_{i=0}^{R+R_1} \boldsymbol{\pi}(i,1)\mathbf{e}$.

The probability $P_c^{imp-loss}$ that an arbitrary customer is lost due to impatience is computed by $P_c^{imp-loss} = \frac{1}{\lambda_c}\sum\limits_{i=1}^{R+R_1} i\alpha\boldsymbol{\pi}_i\mathbf{e} = \frac{\alpha N_c}{\lambda_c}$.

The probability $P_c^{ent-loss}$ that an arbitrary customer is lost at the entrance to the system due to lack of energy is computed by

$P_c^{ent-loss} = \frac{1}{\lambda_c}\sum\limits_{i=0}^{R+R_1}\Big[\boldsymbol{\pi}(i,0,0)(D_1 \otimes I_W)\mathbf{e} + \sum\limits_{r=0}^{\max\{0,i-R_1\}} \boldsymbol{\pi}(i,1,r)(D_1 \otimes I_{WM})\mathbf{e}\Big]$.

The intensity λ_{out} of the output flow of successfully served customers from the system is computed by $\lambda_{out} = \sum\limits_{i=0}^{R+R_1} \boldsymbol{\pi}(i,1)(\mathbf{e}_{(R+1)VW} \otimes \mathbf{S}_0)\mathbf{e}$.

The loss probability P_c^{loss} of an arbitrary customer is computed by $P_c^{loss} = P_c^{imp-loss} + P_c^{ent-loss} = 1 - \frac{\lambda_{out}}{\lambda_c}$.

The probability P_e^{loss} of the loss of an arbitrary unit of energy (due to the buffer overflow) is computed by

$P_e^{loss} = \frac{1}{\lambda_e}\Big[\boldsymbol{\pi}(0,0,R)(I_V \otimes H_1)\mathbf{e} + \sum\limits_{i=0}^{R+R_1} \boldsymbol{\pi}(i,1,R)(I_V \otimes H_1 \otimes I_M)\mathbf{e}\Big]$.

The probability $P_{idle}^{lack-of-energy}$ that the server is idle at an arbitrary moment due to lack of energy is computed by

$P_{idle}^{lack-of-energy} = \sum\limits_{i=1}^{R+R_1} \boldsymbol{\pi}(i,0,0)\mathbf{e}$.

The probability P_{imm} that an arbitrary customer occupies the server immediately upon arrival is computed by

$$P_{imm} = \frac{1}{\lambda_c}\sum_{r=2}^{R}\boldsymbol{\pi}(0,0,r)(D_1 \otimes I_W)\mathbf{e}.$$

The probability $P_{ent-loss}^{no-energy}$ that an arbitrary customer will be lost at the entrance to the system due to absence of energy is computed by

$$P_{ent-loss}^{no-energy} = \frac{1}{\lambda_c}\sum_{i=0}^{R+R_1}[\boldsymbol{\pi}(i,0,0)(D_1 \otimes I_W)\mathbf{e} + \boldsymbol{\pi}(i,1,0)(D_1 \otimes I_{WM})\mathbf{e}].$$

The probability $P_{imp-loss}^{no-energy}$ that an arbitrary customer will be lost due to impatience at the moment of absence of energy is computed by

$$P_{imp-loss}^{no-energy} = \frac{1}{\lambda_c}\sum_{i=1}^{R+R_1} i\alpha\boldsymbol{\pi}(i,0)\mathbf{e}.$$

The probability $P_{imp-loss}^{busy-server}$ that an arbitrary customer will be lost due to impatience when the server is busy is computed by

$$P_{imp-loss}^{busy-server} = \frac{1}{\lambda_c}\sum_{i=1}^{R+R_1} i\alpha\boldsymbol{\pi}(i,1)\mathbf{e}.$$

5 Numerical Examples

To illustrate the influence of the parameter R_1, in this section we present dependencies of the main performance measures on this parameter. It is discovered during the numerical experiments that the shape of these dependencies essentially depends on correlation in arrival processes. To show the importance of account of correlation in the arrival processes of customers and energy units, we consider the following four sets (combinations) of arrival processes of customers and energy. Combination denoted $M+M$ supposes the stationary Poisson arrivals of customers and energy; combination $M+MAP$ supposes the stationary Poisson arrivals of customers and the Markovian arrival process of energy; combination $MAP+M$ suggests the Markovian arrival process of customers and the stationary Poisson arrivals of energy, and combination $MAP+MAP$ supposes the Markovian arrival processes of customers and energy. In all these combinations, the flows have the same average arrival rate of customers $\lambda_c = 0.65$ and energy $\lambda_e = 1.2$, but different coefficients of correlation of sequential inter-arrival times.

In the first set $M+M$, the arrival processes of customers and energy are stationary Poisson and defined as follows: $D_0 = (-0.65)$, $D_1 = (0.65)$, $V = 1$; $H_0 = -1.2$, $H_1 = 1.2, W = 1$. The coefficient of correlation of both processes is equal to zero.

In the second set $M+MAP$, the arrival process of customers is stationary Poisson and defined as the previous one ($D_0 = (-0.65)$, $D_1 = (0.65)$, $V = 1$), while the arrival flow of energy is Markovian with coefficient of correlation of 0.4 and defined as follows:

$$H_0 = \begin{pmatrix} -2.02746 & 0. \\ 0. & -0.0658053 \end{pmatrix},\ H_1 = \begin{pmatrix} 2.01398 & 0.0134798 \\ 0.0366523 & 0.0291531 \end{pmatrix},\ W = 2.$$

In the third set $MAP + M$, the Markovian arrival process of customers is defined by the matrices

$$D_0 = \begin{pmatrix} -0.878566 & 0 \\ 0 & -0.0285157 \end{pmatrix},\ D_1 = \begin{pmatrix} 0.872725 & 0.00584124 \\ 0.0158826 & 0.012633 \end{pmatrix},\ V = 2$$

and has the coefficient of correlation equal 0.4. The arrival process of energy is stationary Poisson and is defined by $H_0 = (-1.2)$, $H_1 = (1.2)$, $W = 1$.

In the fourth set $MAP + MAP$, the arrival process of customers coincides with the MAP in the set $MAP + M$, and the arrival flow of energy is the same as the MAP in the set $M + MAP$.

PH service process of customers is characterized by the vector $\beta = \{0.1, 0.9\}$ and the matrix $S = \begin{pmatrix} -0.4 & 0.2 \\ 0.3 & -2 \end{pmatrix}$. The mean service time is $b_1 = 1.14865$.

The capacity R of the buffer for energy is equal to 15. The customers staying in the buffer are impatient and leave the system after an exponentially distributed time with the parameter $\alpha = 0.005$. Let us vary the parameter R_1 in the interval $R_1 \in [0; 40]$.

Figures 2, 3, 4, 5, 6, 7, 8, 9, 10, 11, 12, 13, 14 show the dependencies of various performance measures of the system on the parameter R_1 for different sets of arrival processes of customers and energy units. Some of these measures (e.g., N_c, N_e, loss probabilities with various reasons of the loss) essentially depend on value of R_1. Also, these figures demonstrate the necessity of account of correlation in arrival processes of customers and energy because values of some performance measures drastically differ for various combinations. Summarizing, we conclude that, generally speaking, the best values of performance measures are achieved for combination $M + M$, in which arrival rates of customers and energy units are constant. The worst values are achieved for combination $MAP+MAP$ in which both arrival processes exhibit correlation and, as a consequence, more irregular arrivals. Correlation in the arrival process of energy has more significant negative effect than the correlation in the arrival process of customers. However, certain exceptions take place. E.g., the loss probability $P_{imp-loss}^{no-energy}$ is the highest in combination $MAP + M$. These observations make clear why it is necessary to consider the MAP arrival flows, not to restrict ourselves to the much more simple stationary Poisson processes. Assumptions that arrival flows are the stationary Poisson flows lead to too optimistic evaluation of performance measures of the system. Let us assume that the quality of the system operation is evaluated by the cost criterion in the form

$$J(R_1) = c_1\lambda_c P_{ent-loss} + c_2\lambda_c P_{imp-loss}^{busy-server} + c_3\lambda_c P_{imp-loss}^{no-energy}.$$

This criterion includes the probabilities of several kinds of customers loss with different cost coefficients (charges paid due to a customer loss caused by the corresponding reasons). The cost coefficient are fixed as $c_1 = 10$, $c_2 = 20$, $c_3 = 40$.

Figure 15 illustrates the dependencies of cost criterion $J(R_1)$ on the threshold R_1 for various combinations of arrival flows. The optimal values of the threshold and cost criterion for various combinations of arrival processes are given in the Table 1.

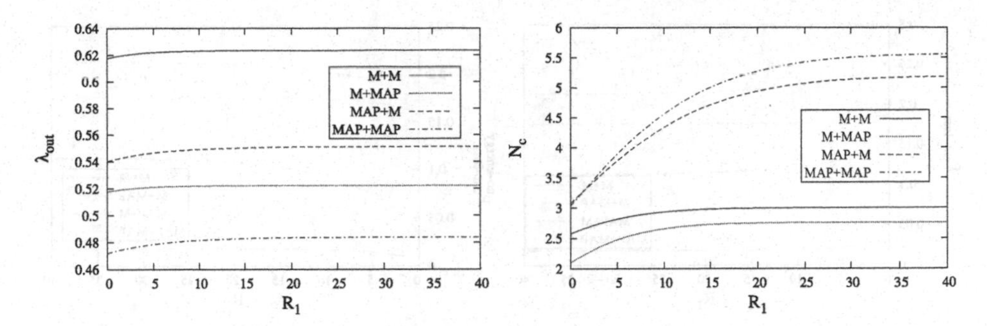

Fig. 2. Dependence of the intensity λ_{out} of the output flow of successfully served customers from the system on R_1.

Fig. 3. Dependence of the average number N_c of customers in the buffer on R_1.

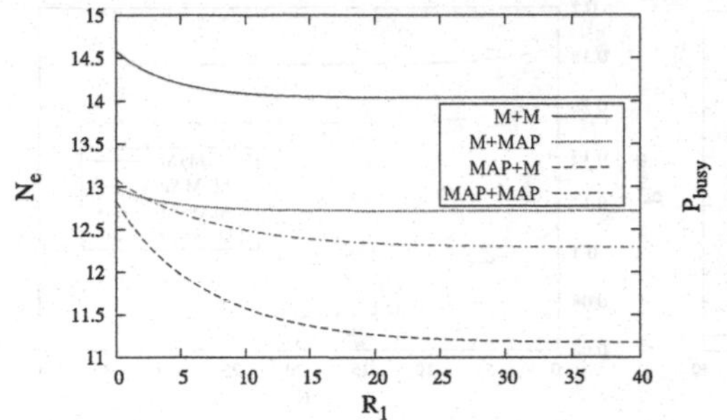

Fig. 4. Dependence of the average number N_e of units of energy in the buffer on R_1.

Fig. 5. Dependence of the probability P_{busy} that the server is busy at an arbitrary moment on R_1.

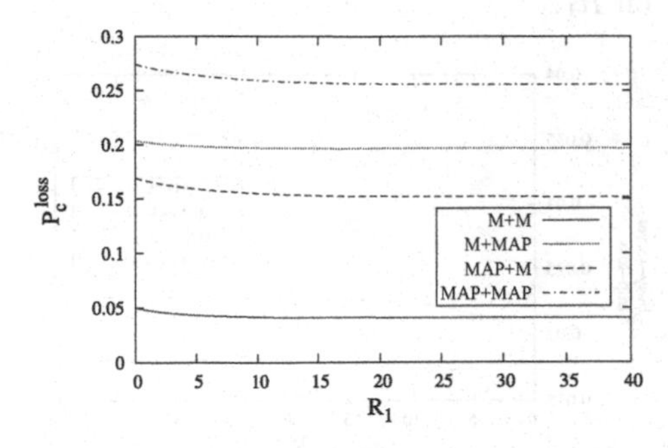

Fig. 6. Dependence of the loss probability P_c^{loss} of an arbitrary customer on R_1.

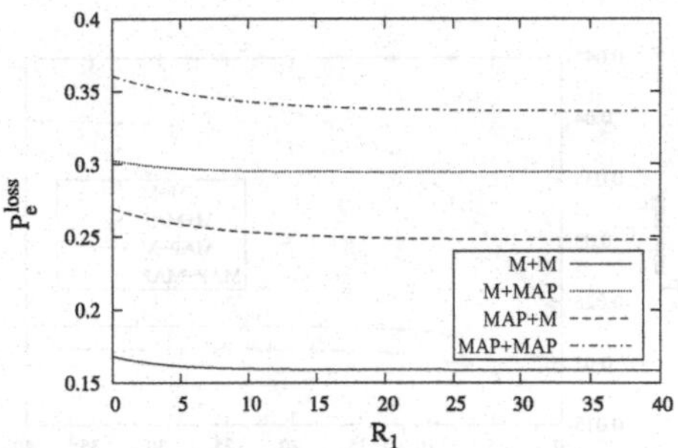

Fig. 7. Dependence of the probability P_e^{loss} of the loss of an arbitrary unit of energy on R_1.

It is evidently seen from this table that the optimal value of parameter R_1 as well as the corresponding value of cost criterion essentially depend on correlation of arrival flows of customers and energy units.

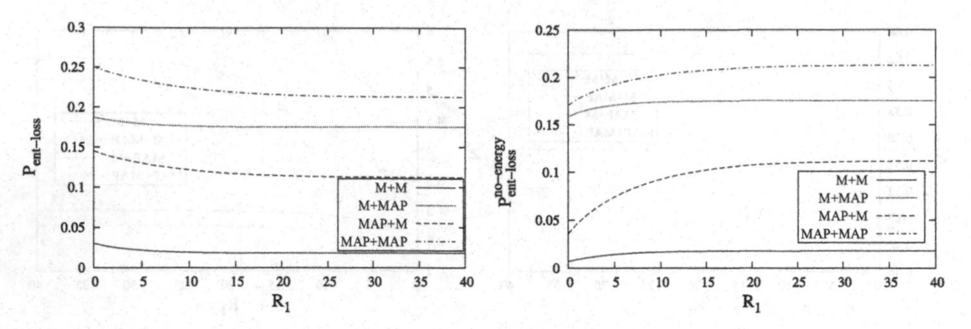

Fig. 8. Dependence of the probability $P_{ent-loss}$ that an arbitrary customer will be lost at the entrance to the system due to lack of energy on R_1.

Fig. 9. Dependence of the probability $P_{ent-loss}^{no-energy}$ that an arbitrary customer will be lost at the entrance to the system due to absence of energy on R_1.

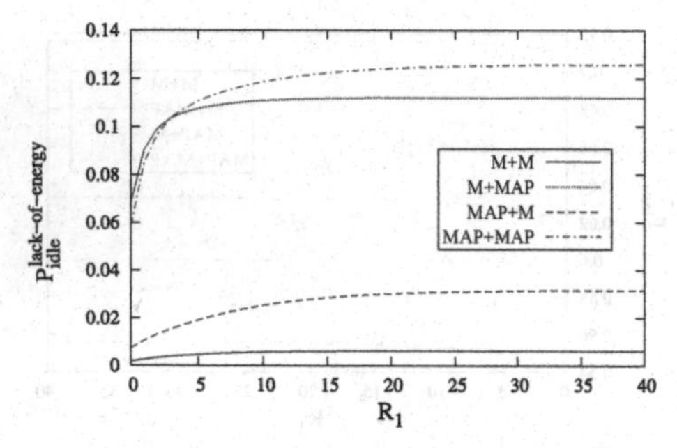

Fig. 10. Dependence of the probability $P_{idle}^{lack-of-energy}$ that the server is idle at an arbitrary moment due to lack of energy on R_1.

Fig. 11. Dependence of the probability P_{imm} that an arbitrary customer occupies the server immediately upon arrival on R_1.

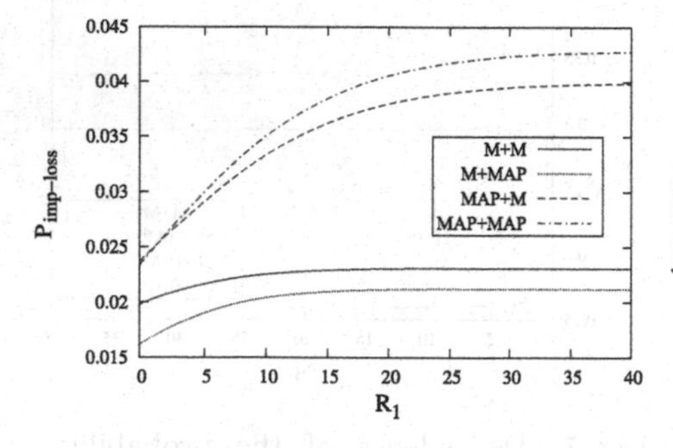

Fig. 12. Dependence of the probability $P_{imp-loss}$ that an arbitrary customer will be lost due to impatience on R_1.

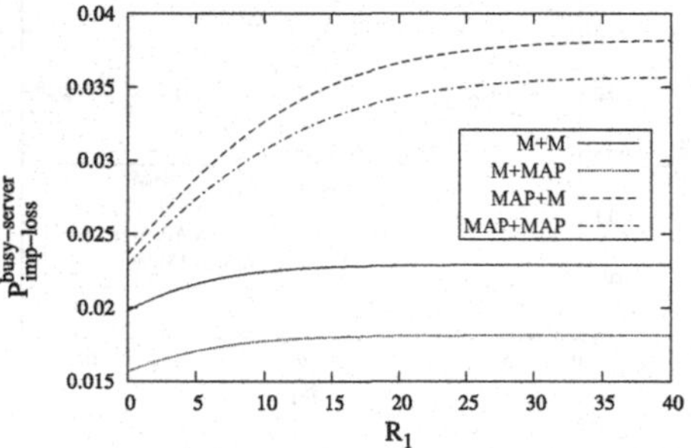

Fig. 13. Dependence of the probability $P_{imp-loss}^{busy-server}$ that an arbitrary customer will be lost due to the busy server on R_1.

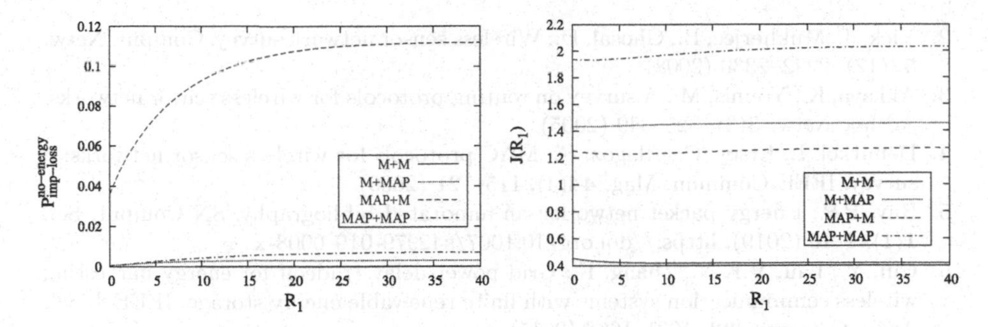

Fig. 14. Dependence of the probability $P_{imp-loss}^{no-energy}$ that an arbitrary customer will be lost due to impatience at the moment of absence of energy on R_1.

Fig. 15. Dependence of the cost criterion $J(R_1)$ on R_1.

Table 1. The optimal values of the threshold and cost criterion for various combinations of arrival processes.

Combination of flows	$J_{min}(R_1^{min})$	R_1^{min}
$M+M$	0.418242	13
$M+MAP$	1.432350	1
$MAP+M$	1.227877	5
$MAP+MAP$	1.933670	2

6 Conclusion

In this paper, we considered a novel queueing model describing operation of the node of a wireless sensor network. The node is autonomous and harvests energy for its operation. The units of energy are spent both for customers admission to the system and customers service. Aiming to optimize performance of the node via the careful spending of the accumulated energy, we introduce the parametric strategy of customers admission. Under the fixed value of the parameter, operation of the system is described by six-dimensional continuous-time Markov chain. The problem of computation of the stationary distribution of this chain is solved what gives an opportunity to compute the variety of performance measures of the system and numerically analyse the impact of the parameter. Importance of account of correlation in arrival processes of both customers and energy units is shown. Example of formulation and solution of optimization problem is presented.

References

1. Akyildiz, I.F., Su, W., Sankarasubramaniam, Y., Cayirci, E.: Wireless sensor networks: a survey. Comput. Netw, **38**(4), 393–422 (2002)

2. Yick, J., Mukherjee, B., Ghosal, D.: Wireless sensor network survey. Comput. Netw. **52**(12), 2292–2330 (2008)
3. Akkaya, K., Younis, M.: A survey on routing protocols for wireless sensor networks. Ad hoc Netw. **3**(3), 325–349 (2005)
4. Demirkol, I., Ersoy, C., Alagoz, F.: MAC protocols for wireless sensor networks: a survey. IEEE Commun. Mag. **44**(4), 115–121 (2006)
5. Ray, P.P.: Energy packet networks: an annotated bibliography. SN Comput. Sci. **1**(1), 1–10 (2019). https://doi.org/10.1007/s42979-019-0008-x
6. Cui, Y., Lau, V.K.N., Zhang, F.: Grid power-delay tradeoff for energy harvesting wireless communication systems with finite renewable energy storage. IEEE J. Sel. Areas Commun. **33**, 1651–1666 (2015)
7. Lu, X., Wang, P., Niyato, D., Kim, D.I., Han, Z.: Wireless networks with RF energy harvesting: a contemporary survey. IEEE Commun. Surv. Tutor. **17**, 757–789 (2015)
8. Zhang, F., Lau, V.K.N.: Delay-sensitive dynamic resource control for energy harvesting wireless systems with finite energy storage. IEEE Commun. Mag. **53**, 106–113 (2015)
9. Ulukus, S., et al.: Energy harvesting wireless communications: a review of recent advances. IEEE J. Sel. Areas Commun. **33**, 360–381 (2015)
10. Kanoun, O.: (Ed.) Energy Harvesting for Wireless Sensor Networks: Technology, Components and System Design. Walter de Gruyter GmbH & Co KG. (2018)
11. Sharma, V., Mukherji, U., Joseph, V., Gupta, S.: Optimal energy management policies for energy harvesting sensor nodes. IEEE Trans. Wirel. Commun. **9**(4), 1326–1336 (2010)
12. Tutuncuoglu, K., Yener, A.: Optimum transmission policies for battery limited energy harvesting nodes. IEEE Trans. Wirel. Commun. **11**(3), 1180–1189 (2012)
13. Yang, J., Ulukus, S.: Optimal packet scheduling in an energy harvesting communication system. IEEE Trans. Commun. **60**(1), 220–230 (2012)
14. Yang, J., Ulukus, S.: Optimal packet scheduling in a multiple access channel with energy harvesting transmitters. J. Commun. Netw. **14**, 140–150 (2012)
15. Gelenbe, E.: Synchronising energy harvesting and data packets in a wireless sensor. Energies **8**(1), 356–369 (2015)
16. Gelenbe, E.: A sensor node with energy harvesting. ACM SIGMETRICS Perform. Eval. Rev. **42**(2), 37–39 (2014)
17. Patil, K., De Turck, K., Fiems, D.: A two-queue model for optimising the value of information in energy-harvesting sensor networks. Perform. Eval. **119**, 27–42 (2018)
18. Dudin, S.A., Lee, M.H.: Analysis of single-server queue with phase-type service and energy harvesting. Math. Prob. Eng. **2016**, 1–16 (2016). ID592794
19. Kim, C.S., Dudin, S., Dudin, A., Samouylov, K.: Multi-threshold control by a single-server queuing model with a service rate depending on the amount of harvested energy. Perform. Eval. **127–128**, 1–20 (2018)
20. Dudin, A., Kim, C., Dudin, S.: Optimal control by the queue with rate and quality of service depending on the amount of harvested energy as a model of the node of wireless sensor network. In: Vishnevskiy, V.M., Samouylov, K.E., Kozyrev, D.V. (eds.) DCCN 2019. LNCS, vol. 11965, pp. 165–178. Springer, Cham (2019). https://doi.org/10.1007/978-3-030-36614-8_13
21. Dudina, O., Dudin, A.: Optimization of queueing model with server heating and cooling. Mathematics **7**(9), 1–19 (2019)

22. Baek, J.H., Dudina, O., Kim, C.S.: Queueing system with heterogeneous impatient customers and consumable additional items. Appl. Math. Comput. Sci. **27**(2), 367–384 (2017)
23. Sun, B., Dudin, A., Dudin, S.: Queueing system with impatient customers, visible queue and replenishable inventory. Appl. Comput. Math. **17**(2), 161–174 (2018)
24. Shajin, D., Krishnamoorthy, A., Dudin, A.N., Joshua, V.C., Jacob, V.: On a queueing-inventory system with advanced reservation and cancellation for the next K time units ahead: the case of overbooking. Queueing Syst. **94**(1–2), 3–37 (2020)
25. Dudin, A., Dudina, O.: Analysis of the $MAP/PH/1$ service system with repeat calls and energy audit. Autom. Control Comput. Sci. **45**(5), 277–285 (2015)
26. Dudin, A.N., Lee, M.H., Dudin, S.A.: Optimization of service strategy in queueing system with energy harvesting and customers impatience. Appl. Math. Comput. Sci. **26**(2), 367–378 (2016)
27. Shajin, D., Dudin, A., Dudina, O., Krishnamoorthy, A.: A two-priority single server retrial queue with additional items. J. Ind. Manage. Optim. https://doi.org/10.3934/jimo.2019085
28. Dudin, A., Dudin, S., Dudina, O., Kim, C.: Analysis of a wireless sensor node with varying rates of energy harvesting and consumption. In: Rykov, V.V., Singpurwalla, N.D., Zubkov, A.M. (eds.) ACMPT 2017. LNCS, vol. 10684, pp. 172–182. Springer, Cham (2017). https://doi.org/10.1007/978-3-319-71504-9_16
29. Graham, A.: Kronecker Products and Matrix Calculus with Applications. Ellis Horwood, Cichester (1981)
30. Klimenok, V.I., Dudin, A.N.: Multi-dimensional asymptotically quasi-Toeplitz Markov chains and their application in queueing theory. Queueing Syst. **54**, 245–259 (2006)
31. Neuts, M.F.: Matrix-Geometric Solutions in Stochastic Models. The Johns Hopkins University Press, Baltimore (1981)
32. Klimenok, V.I., Kim, C.S., Orlovsky, D.S., Dudin, A.N.: Lack of invariant property of Erlang $BMAP/PH/N/0$ model. Queueing Syst. **49**, 187–213 (2005)
33. Baumann, H., Sandmann, W.: Multi-server tandem queue with Markovian arrival process, phase-type service times, and finite buffers. Eur. J. Oper. Res. **256**, 187–195 (2017)
34. Kim, C.S., Dudin, S., Taramin, O., Baek, J.: Queueing system $MMAP/PH/N/N+R$ with impatient heterogeneous customers as a model of call center. Appl. Math. Model. **37**, 958–976 (2013)

The Analysis of Resource Sharing for Heterogenous Traffic Streams over 3GPP LTE with NB-IoT Functionality

Sergey N. Stepanov[1](✉), Mikhail S. Stepanov[1], Umer Andrabi[2], and Juvent Ndayikunda[1]

[1] Department of Communication Networks and Commutation Systems, Moscow Technical University of Communication and Informatics, 8A, Aviamotornaya str., Moscow 111024, Russia

stpnvsrg@gmail.com, mihstep@yandex.ru, juvndayi@mail.ru

[2] Moscow Institute of Physics and Technology (State University), 9 Institutskiy per., Dolgoprudny, Moscow Region 141701, Russia

umer.andrabi@rediffmail.com

Abstract. The main feature in the development of Internet of Things (IoT) applications is the necessity of conjoint servicing of heterogenous data streams over existent network infrastructure. This trend has been recognized and supported by 3GPP with introducing of NarrowBand IoT (NB-IoT) technology, which allows to use the same resource by 3GPP LTE high-end equipment and NB-IoT low-end devices. The need of sharing the limited amount of available resource efficiently emphasizes the importance of theoretical study of formulated problem. The model of resource allocation and sharing for conjoint servicing of real time video traffic of surveillance cameras and NB-IoT data traffic of smart meters and actuators over LTE cell facilities is constructed. In the model the access control is used to create the conditions for differentiated servicing of coming sessions. All random variables used in the model have exponential distribution with corresponding mean values but the obtained results are valid for models with arbitrary distribution of service times. Using the model the main performance measures of interest are given with help of values of probabilities of model's stationary states. The recursive algorithm of performance measures estimation is suggested. The model and derived algorithms can be used for study the scenarios of resource sharing between heterogenous data streams over 3GPP LTE with NB-IoT functionality.

Keywords: NB-IoT technology · Resource allocation and sharing · System of state equations · Recursive algorithm

1 Introduction

The essential trend in the development of telecommunications is growth of the volumes and the diversity of Internet of Things (IoT) applications. The IoT can

V. M. Vishnevskiy et al. (Eds.): DCCN 2020, LNCS 12563, pp. 422–435, 2020.
https://doi.org/10.1007/978-3-030-66471-8_32

be typically defined as a network of multiple physical smart objects (vehicles, actuators, sensors, etc.) which have ability to produce, process and exchange data without involvement of human beings [1–6]. These digital devices have low storage capabilities and processing capacities and aim more at reliability of information delivery to the data centers for collecting and proceeding. Together with usage of low-traffic smart meters we see the growing impact of multimedia traffic, for example collected by video surveillance systems deployed for security and safety reasons [7]. This trend has been recognized and supported by 3GPP with developing of NarrowBand IoT (NB-IoT) technology, which allows to use the same spectrum by 3GPP LTE high-end equipment and NB-IoT low-end devices [2,3]. By providing the technical instruments that can be used for radio resources sharing between LTE and NB-IoT technologies, 3GPP does not formulate the concrete solutions on how these resources should be shared. This problem can be solved by mathematical modeling with taking into account the features of traffic streams forming and accepting for servicing [7–10].

In this paper we address the above mentioned challenges by constructing an analytical framework for modeling the process of resource sharing for an operator planning to create and exploit surveillance system. The system consists of numerous video cameras to perform video monitoring and a large number of smart meters. Both network segments collecting and transfer heterogenous data streams to analytical centers over existent infrastructure of LTE network (see Fig. 1).

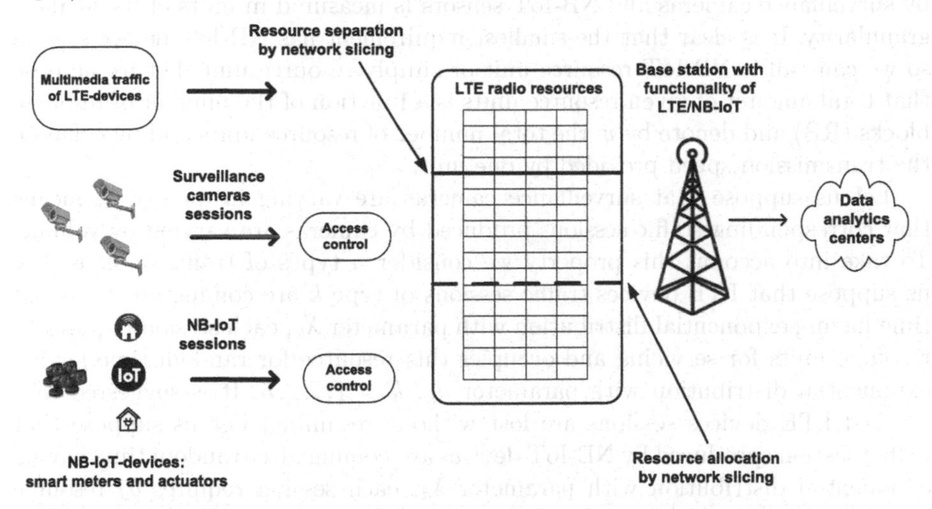

Fig. 1. The functional model of resource sharing between LTE surveillance cameras and NB-IoT sensors.

The proposed model generalizes the results of [7–10] by considering arbitrary number of traffic streams created by video cameras (LTE-devices) and one traffic stream originated from NB-IoT devices. In the model the access control is used to create the conditions for differentiated servicing of coming sessions. All random variables used in the model have exponential distribution with corresponding mean values but the obtained results are valid for models with arbitrary distribution of service times. Three scenarios of resource sharing by coming traffic streams are considered: Slicing when resources are strictly divided among LTE and NB-IoT devices traffic streams; Fully shared, when resources are fully shared and Access controlled, when the access to resource is restricted depending on the amount of resource occupied by corresponding traffic stream.

The rest of the paper is organized as follows. In Sect. 2 the mathematical description of the model will be presented. Here the system of state equations that relates the model's stationary probabilities is outlined and main performance measures are defined. In Sect. 3 the recursive algorithm of characteristics calculation is formulated. Next section is devoted to the discussion of approachers that alow to increase the effectiveness of recursive algorithm. Numerical assessment of the suggested scenarios of resource sharing is performed in Sect. 5. Conclusions are drawn in the last section.

2 Model Description

We consider an LTE cell with a base station placed in its center and formalize the process of resources sharing. The volume of available radio resources of LTE cell in uplink direction given by network slicing for serving traffic streams originated by surveillance cameras and NB-IoT sensors is measured in units of its smallest granularity. It is clear that the smallest requirement has NB-IoT device session so we can call it NB-IoT resource unit or simply resource unit. Let us suppose that total amount of given resource units is a function of the number of resource blocks (RB) and denote by v, the total number of resource units and by c denote the transmission speed provided by one unit.

Let us suppose that surveillance cameras are varying in quality. It means that corresponding traffic sessions produced by cameras are varying by volume. To take into account this property we consider n types of traffic sessions. Let us suppose that LTE devices traffic sessions of type k are coming after random time having exponential distribution with parameter λ_k, each session requires b_k resource units for servicing and occupies this resource for random time having exponential distribution with parameter μ_k, $k = 1, \ldots, n$. It is suggested that blocked LTE devices sessions are lost without resuming. Let us suppose that traffic sessions produced by NB-IoT devices are coming after random time having exponential distribution with parameter λ_d, each session requires b_d resource units for transmitting of files having exponential distribution with mean F. The service time of NB-IoT session has exponential distribution with mean value $\frac{F}{b_d}$ and parameter $\frac{b_d}{F}$. It is suggested that blocked LTE devices sessions are lost without resuming.

Let us formalize scenarios of resource sharing by coming traffic streams. The simplest scenario corresponds to the case when all v resource units are strictly divided among LTE devices sessions and NB-IoT devices sessions. Let us denote by v_ℓ the number of resource units that is given for exclusive usage to LTE devices sessions and by $v_b = v - v_\ell$ we denote the number of resource units given for exclusive usage to NB-IoT devices sessions correspondingly. By varying the values of v_ℓ and v_b we can give the priority in resource usage to the chosen traffic type but as we show later this way of resource sharing greatly decreases the usage of resource unit.

Next scenario is related with access control. Let us denote for k-th flow of LTE devices sessions by c_k the maximum allowed number of traffic sessions that can be on service at the same time. In a similar way let us denote for NB-IoT devices sessions by c_d maximum allowed number of traffic sessions that can be on service at the same time. For this type of resource usage the traffic session of k-th flow can be blocked for two reasons: (1) if $v_k = c_k b_k$ resource units have already been occupied by sessions from the k-th flow or (2) if total number of busy resource units is greater than $v - b_k$. The same is true for NB-IoT devices sessions. The coming session of this type can be blocked for two reasons: (1) if $v_d = c_d b_d$ resource units have already been occupied by NB-IoT devices sessions or (2) if total number of busy resource units is greater than $v - b_d$. We show later that by using the access control (by choosing the values of v_k, $k = 1, \ldots, n$ and v_d) we can give the priority in resource usage to chosen traffic type and increase the usage of resource unit compare to static scenario.

The last scenario corresponds to the case when resources are fully shared without giving priority to some traffic streams. In this case we usually increase the usage of resource unit compare to formulated above scenarios but we are not able to reach the same level of sessions losses for all type of traffic streams considered in the model. All three formulated scenarios can be modeled by proper choosing of v and access boundaries v_k, $k = 1, \ldots, n$ and v_d so further we will study only model of resource sharing based on access control.

Let us denote by $i_k(t)$ the number of LTE devices sessions of k-th flow being on servicing at time t, and by $d(t)$ we denote the number of sessions of NB-IoT devices being on servicing at time t. The dynamic of a model states changing is described by Markov process

$$r(t) = (i_1(t), \ldots, i_n(t), d(t)),$$

defined on the finite set of model's states S. Let us denote by $(i_1, \ldots, i_n, d)$ the state of $r(t)$. The vector $(i_1, \ldots, i_n, d)$ belongs to S when i_k, $k = 1, \ldots, n$, d varies as follows

$$0 \le i_k \le c_k,\ k = 1, \ldots, n;\ \ 0 \le d \le c_d;\ \ i_1 b_1 + \ldots + i_n b_n + d b_d \le v. \quad (1)$$

Let us denote by i for state $(i_1, \ldots, i_n, d) \in S$ the number of occupied resource units $i = i_1 b_1 + \ldots + i_n b_n + d b_d$.

Let us denote by $p(i_1, \ldots, i_n, d)$ the value of stationary probability of state $(i_1, \ldots, i_n, d) \in S$. It can be interpreted as portion of time the model stays in the state $(i_1, \ldots, i_n, d)$. This interpretation gives the possibility to use the values of $p(i_1, \ldots, i_n, d)$ for estimation of model's main performance measures. Let us define for k-th flow of LTE devices traffic by π_k the portion of lost sessions and by m_k the mean number of occupied resource units. Their formal definitions are looking as follows

$$\pi_k = \sum_{(i_1,\ldots,i_n,d)\in U_k} p(i_1,\ldots,i_n,d); \quad m_k = \sum_{(i_1,\ldots,i_n,d)\in S} p(i_1,\ldots,i_n,d) i_k b_k.$$

Here U_k is defined as subset of S having property $(i_1,\ldots,i_n,d) \in U_k$, if $i_k+1 > c_k$ or $i + b_k > v$. In the same way we define the performance measures of NB-IoT devices traffic servicing. They are π_d the portion of lost sessions and m_d the mean number of occupied resource units

$$\pi_d = \sum_{(i_1,\ldots,i_n,d)\in U_d} p(i_1,\ldots,i_n,d); \quad m_d = \sum_{(i_1,\ldots,i_n,d)\in S} p(i_1,\ldots,i_n,d) d b_d.$$

Here U_d is defined as subset of S having property $(i_1,\ldots,i_n,d) \in U_d$, if $d+1 > c_d$ or $i + b_d > v$.

System of state equations is obtained after equating the intensity of transition $r(t)$ out of the arbitrary model's state $(i_1,\ldots,i_n,d) \in S$ to the intensity of transition $r(t)$ into the state $(i_1,\ldots,i_n,d)$

$$P(i_1,\ldots,i_n,d) \times \Big(\sum_{k=1}^{n}(\lambda_k I(i+b_k \le v, i_k+1 \le c_k) + i_k\mu_k) + \tag{2}$$

$$+\lambda I(i+b_d \le v, d+1 \le c_d) + d\mu_d\Big) =$$

$$= \sum_{k=1}^{n} P(i_1,\ldots,i_k-1,\ldots,i_n,d)\lambda_k I(i_k > 0) +$$

$$+P(i_1,\ldots,i_n,d-1)\lambda_d I(d > 0) +$$

$$+\sum_{k=1}^{n} P(i_1,\ldots,i_k+1,\ldots,i_n,d)(i_k+1)\mu_k I(i+b_k \le v, i_k+1 \le c_k) +$$

$$+P(i_1,\ldots,i_n,d+1)(d+1)\mu_d I(i+b_d \le v, d+1 \le c_d).$$

Here $I(\cdot)$ — is indicator function. Values $P(i_1,\ldots,i_n)$ should be normalized.

It can be proved that $r(t)$ is reversible Marcov process. From relations of detailed balance follows that values of $P(i_1,\ldots,i_n,d)$ can be found as a unique solution of the system of state equation that has a product form [11–13].

$$p(i_1,\ldots,i_n,d) = \frac{1}{N}\frac{a_1^{i_1}}{i_1!}\cdots\frac{a_n^{i_n}}{i_n!}\frac{a_d^d}{d!}. \tag{3}$$

Here $a_k = \lambda_k/\mu_k$ and $a_d = \lambda_d/\mu_d$ are offered traffic expressed in Erlangs and N is a normalizing constant

$$N = \sum_{(i_1,\ldots,i_n,d)\in S} \frac{a_1^{i_1}}{i_1!}\cdots\frac{a_n^{i_n}}{i_n!}\frac{a_d^d}{d!}.$$

3 Recursive Algorithm

The values of introduced performance measures can be found by convolution algorithm [11,14] based on the product form (3). Let us introduce two auxiliary notions. For two vectors $x = (x(0), x(1), \dots, x(a_x))$ and $y = (y(0), y(1), \dots, y(a_y))$ we define the convolution operator that being applied to x, y gives vector z with components found in accordance with formula

$$z(s) = \sum_{u=0}^{s} x(u)y(s-u)I(u \le a_x, s-u \le a_y), \quad s = 0, 1, \dots, a_z. \tag{4}$$

Now let us take the k-th flow of LTE devices traffic and find the state probabilities of resource occupancy as if it was the only traffic stream offered to v_k resource units. We denote by $P_k(i)$ the value of unnormalized probability that i resource units are occupied. We shall call $P_k(i)$, $i = 0, 1, \dots, v_k$, as the individual distribution of the k-th flow. Let $P_k(0) = 1$ than

$$P_k(i) = \begin{cases} \frac{a_k^{i_k}}{i_k!}, & i = i_k b_k, \quad i_k = 0, 1, \dots, c_k, \\ 0, & \text{in opposite case.} \end{cases} \tag{5}$$

In the same way we can find the individual distribution for NB-IoT devices traffic

$$P_d(i) = \begin{cases} \frac{a_d^d}{d!}, & i = db_d, \quad d = 0, 1, \dots, c_d, \\ 0, & \text{in opposite case.} \end{cases} \tag{6}$$

Recursive algorithm of estimation of the π_d and m_d consists of making the following three steps.

1. By using (5),(6) we find $P_k(i)$, $i = 0, 1, \dots, v_k$, $k = 1, \dots, n$ and $P_d(i)$, $i = 0, 1, \dots, v_d$.
2. In order of numbering the LTE devices traffic streams we do the successive convolution of all n individual state distributions. Let us denote by $P^{(n)}(\cdot)$ the vector obtained after convolving the all n individual distributions.
3. Finally we convolve vectors $P^{(n)}(\cdot)$ and $P_d(\cdot)$ and obtain after normalization the system state distribution

$$P(i) = \sum_{u=0}^{i} P_d(u)\, P^{(n)}(i-u)\, I\big(u \le c_d\, b_d,\, i-u \le v\big),\; i = 0, 1, \dots, v, \tag{7}$$

and values of π_d and m_d

$$\pi_d = \sum_{i=v-b_d+1}^{v} P(i) + P_d(c_d\, b_d) \sum_{i=c_d\, b_d}^{v-b_d} P^{(n)}(i - c_d\, b_d); \tag{8}$$

$$m_d = a_d(1 - \pi_d).$$

The performance measures for all flows of sessions can be found after performing the above mentioned steps for each flow by putting it at the end of the convolution procedure. This action needs to perform $n(n+1)$ convolutions.

4 Enhancements of Recursive Algorithm

Let us consider two enhancements of the used approach that can decrease the amount of computational works when calculating the performance measures for all traffic streams and improve the stability of algorithm when calculating the performance measures for large values of v. We start from the procedure that decreases the total number of convolutions by remembering the results of intermediate convolutions [15]. Let us represent the number of flows $n+1$ in a binary form $n+1 = 2^{l_1} + 2^{l_2} + \cdots + 2^{l_s}$ and relate with each of binary terms 2^{l_r}, $r = 1, \ldots, s$ a binary tree consisting of $(l_r + 1)$ levels where level 0 has 2^{l_r} leaves and level l_r is the root of the binary tree. We accept further that all trees are numbered in decreasing order of corresponding binary groups and that levels of the same number are located on the same line. An example of such a group of trees for $n+1 = 7$ is shown in Fig. 2.

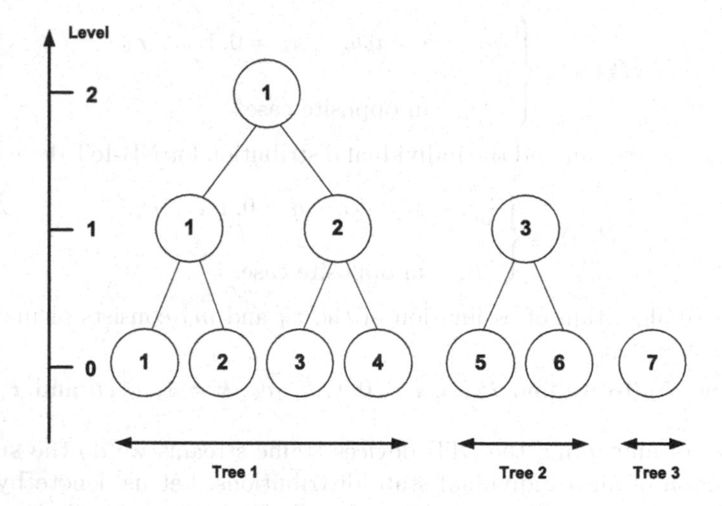

Fig. 2. The distribution of binary trees for $n+1 = 7$.

Let us construct an algorithm for determining performance measures of all streams by making only $4n - 2$ convolutions.

1. Number leaves and nodes of each level in the order used in Fig. 2 and relate with leaf number m, $m = 1, \ldots, n+1$ the vector of individual state distribution of the corresponding flow of sessions (we suppose that the flow of NB-IoT devices sessions has number $(n+1)$).
2. Relate with each non-leaf node of all trees the result of convolution of vectors associated with two sons of the node considered. It can be done by $w_1 = n + 1 - s - I(s = 1)$ convolutions (here $I(\cdot)$—is an indicator function).
3. For the root node of the tree number r, $r = 1, \ldots, s$, convolute all vectors associated with root nodes of the trees number $f = 1, \ldots, s, f \neq r$ and

associate the resultant vector with root node of the tree number r. It gives $w_2 = 3(s-2+I(s=1))$ convolutions.

4. For tree number r, $r = 1, \dots, s$, associate with each node located on level j consequently for $j = l_r - 1, l_r - 2, \dots, 0$, the result of convolution of the vector associated with this node and the vector associated with his father if $j = l_r - 1$ or brother of his father if $j < l_r - 1$. This step is made if $l_r > 0$. To do this we need to perform $w_3 = 2(n+1-s-I(s=1))$ convolutions.
5. Finally, convolve each vector associated with leaf node with the vector formerly associated with his brother or with him if the considered binary group consists of one element and calculate the individual performance measures by using (8). Last step takes $w_4 = n+1$ convolutions.

It is easy to check that totally we need to make $4n-2$ convolutions to find the performance measures for all streams. This is less than $n(n+1)$ if we try to find characteristics for all flows without keeping the results of intermediate convolutions.

Next refinement of recursive algorithm devoted to analysis of cases with big values of v. In the process of construction of the model (see Sect. 2) it was supposed that the transmission speed provided by one resource unit corresponds to the requirement of NB-IoT session. Because this requirement is quite small compare to transmission speed provided by all available resource it increases the values of v up to several hundreds of resource units or even more. This leads to numerical instability of the implementation of recursive algorithm. To overcome this difficulties we construct modified version of convolution algorithm by introducing the truncation levels for individual distributions into account. Let us suppose that we a priory know the truncation levels for vectors of individual and convoluted distributions. The modified version of recursive algorithm is looking as follows.

As was done in Sect. 3 we calculate the state probabilities for each stream as if it was the only traffic stream offered to all available resource. We start the calculation procedure from the state having the maximum relative value of probability. Let us show the details on the example of the k-th flow of LTE devices traffic. We suppose that for k-th flow this state is i_k^*. It is easy to verify that $i_k^* = min(\lfloor a_k \rfloor, c_k)$. Next we take $P_k(i_k^*) = 1$ and make recurrence in both or one (depending on the relation between a_k and c_k) directions of decreasing $P_k(i)$ by using relations

$$P_k(i) = \frac{P_k(i-1)(i-1)}{a_k}, \quad i = i_k^*+1, i_k^*+2, \dots, b_k^u,$$

$$P_k(i) = \frac{P_k(i+1)a_k}{i}, \quad i = i_k^*-1, i_k^*-2, \dots, b_k^l.$$

Here b_k^u, b_k^l are correspondingly upper and lower truncation levels for the individual state distribution of the k–th stream. This calculation is made for all streams. The convolution operator that being applied for two truncated vectors x, y where $x = (x(t_x^l), x(t_x^l+1), \cdots, x(t_x^u))$, $y = (y(t_y^l), y(t_y^l+1), \cdots, y(t_y^u))$ gives vector z with components $z(i) = \sum_{j=l(i)}^{u(i)} x(i-j) \cdot y(j)$, $i = t_z^l, t_z^l + 1, \cdots, t_z^u$, with functions $u(i), l(i)$

$$u(i) = \begin{cases} i - t_x^l, & t_z^l \leq i < t_x^l + t_y^u \\ t_y^u, & t_x^l + t_y^u \leq i \leq t_z^u, \end{cases} \qquad l(i) = \begin{cases} t_y^l, & t_z^l \leq i < t_x^u + t_y^l \\ i - t_x^u, & t_x^u + t_y^l \leq i \leq t_z^u. \end{cases}$$

The consequence of making convolutions remains the same as in the previous case. The choice of truncation levels and estimation of the error caused by truncations can be done similar to solving the analogous problems in [15]. All random variables used in the model have exponential distribution with corresponding mean values but the obtained results are valid for models with arbitrary distribution of service times [13].

5 Numerical Assessment

By using the elaborated mathematical model and algorithms of it's performance measures estimation we can analyze the effectiveness of suggested scenarios of resource allocation. The level of traffic load can be characterized by ρ the offered load per one resource unit. To define ρ it is necessary to find the offered load of each traffic stream considered in the model. Let us denote by A_k the offered load expressed in resource units for k-th flow of LTE devices traffic $A_k = \frac{\lambda_k}{\mu_k} b_k = a_k b_k$. Let us denote by A_d the offered load expressed in resource units for flow of NB-IoT devices sessions $A_d = \frac{\lambda_d}{\mu_d} b_d = a_d b_d = \frac{\lambda_d F}{b_d}$. The value of ρ can be defined from relation $\rho = \frac{A_1 + \ldots + A_n + A_d}{v}$.

Let us consider the model with following values of input parameters: $v = 200$ resource units (r.u.); transmission rate that is provided by one resource unit is $c = 100$ kbit/c; $n = 1$; $b_1 = 10$ r.u.; $b_d = 1$ r.u.; $F = 100$ kbit; $1/\mu_1 = 10$ c; $1/\mu_d = 1$ c. We begin the model's numerical assessment with Fig. 3 that presents the values of π_1 and π_d and Fig. 4 with mean values of unit usage by LTE devices traffic — δ_1 and NB-IoT devices traffic—δ_d and the their sum $\delta = \delta_1 + \delta_d$ vs the value of ρ the offered load of one resource unit. The values of performance measures are obtained by recursive algorithm presented in Section 3 with enhancements discussed in Sect. 4. Let us suppose that both traffic flows considered in the model generate the same offered load $A_1 = A_d = \frac{v\rho}{2}$. It allows to find the intensities λ_1, λ_d of sessions coming for each flow considered in the model from known values of ρ. The results presented in Fig. 3 and Fig. 4 show that despite equality of offered traffic NB-IoT-devices sessions obtain priority in occupying the transmission resource that is clearly seen in overload conditions, when $\rho > 1$, (see Fig. 4).

The only way to overcome mentioned difficulties is to create the conditions for differentiated servicing of coming sessions. Three scenarios of resource sharing are compared: Slicing when resources are strictly divided among LTE devices and NB-IoT devices traffic streams, Fully shared, when resources are fully shared and Access controlled, when the access to resource is restricted depending of

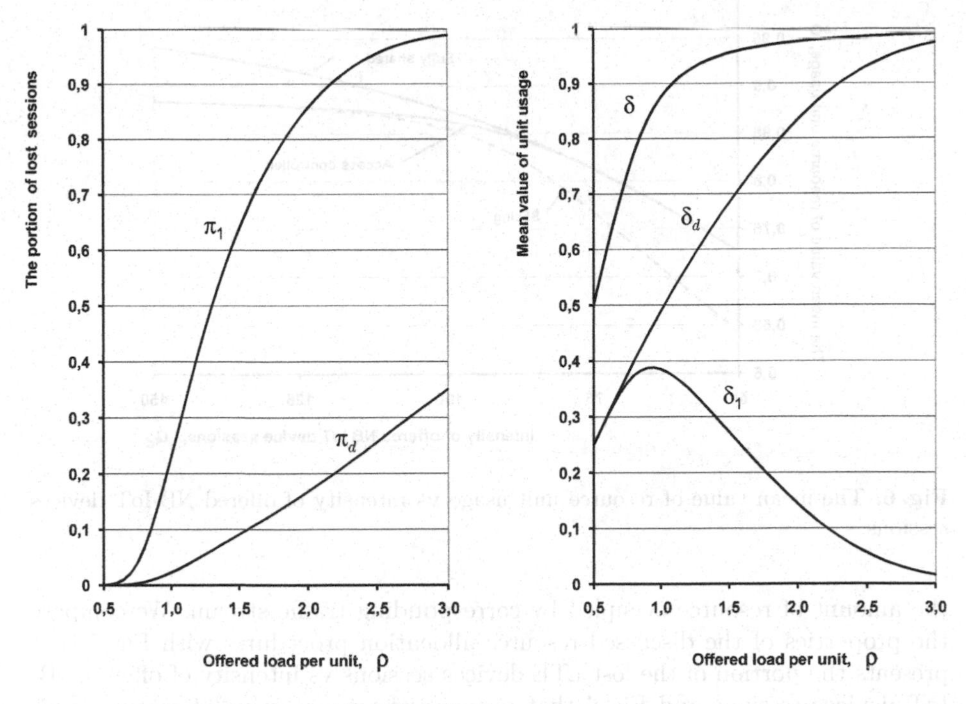

Fig. 3. The portions of lost sessions for LTE and NB-IoT devices.

Fig. 4. The values of unit usage by LTE and NB-IoT devices.

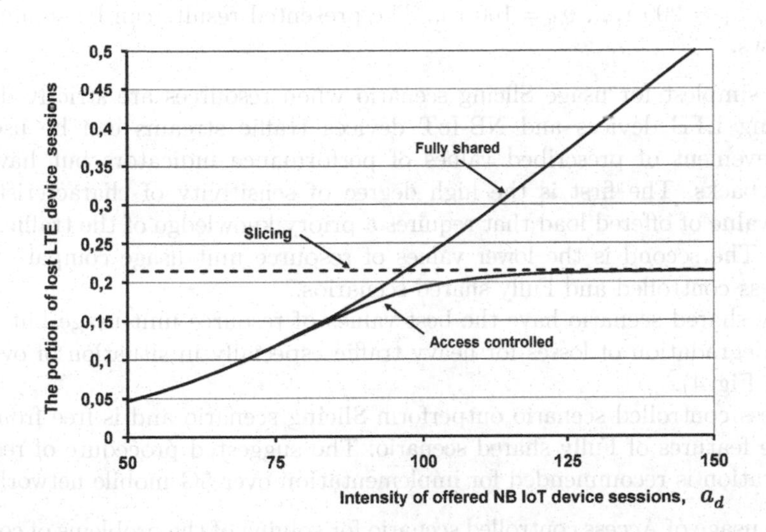

Fig. 5. The portion of the lost LTE devices sessions vs intensity of offered NB-IoT devices sessions.

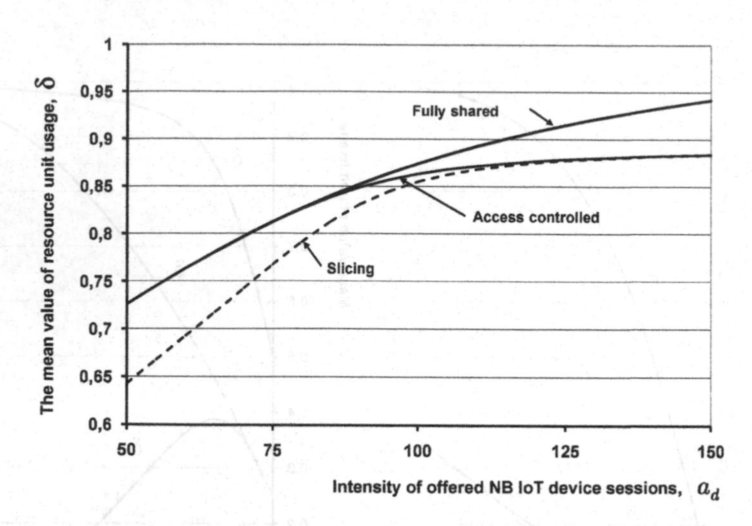

Fig. 6. The mean value of resource unit usage vs intensity of offered NB-IoT devices sessions.

the amount of resource occupied by corresponding traffic stream. We compare the properties of the discussed resource allocation procedures with Fig. 5 that presents the portion of the lost LTE devices sessions vs intensity of offered NB-IoT devices sessions and Fig. 6 that presents the mean value of resource unit usage vs intensity of offered NB-IoT devices sessions.

The model input parameters are the same as was used in Fig. 3 and Fig. 4 except $a_1 = 10$ Erl. For Slicing scenario $v_\ell = v_b = 100$ r.u. For Access controlled scenario $v_1 = 200$ r.u., $v_d = 100$ r.u. The presented results can be summarized as follows.

1. The simplest for usage Slicing scenario when resources are strictly divided among LTE devices and NB-IoT devices traffic streams can be used for achievement of prescribed values of performance indicators but have two drawbacks. The first is the high degree of sensitivity of characteristics to the value of offered load that requires a priory knowledge of the traffic intensity. The second is the lower values of resource unit usage compare to the Access controlled and Fully shared scenarios.
2. Fully shared scenario have the best values of resource unit usage but allows the degradation of losses for heavy traffic especially in situation of overload (see, Fig. 4).
3. Access controlled scenario outperform Slicing scenario and is free from negative features of Fully shared scenario. The suggested procedure of resource allocation is recommended for implementation over 5G mobile networks.

The usage of Access controlled scenario for solving of the problems of conjoint servicing of NB-IoT devices sessions and LTE devices sessions that was discussed above in Fig. 3 and Fig. 4 is presented in Fig. 7 and Fig. 8 for the same values of model input parameters and $v_1 = 200$ r.u., $v_d = 100$ r.u.

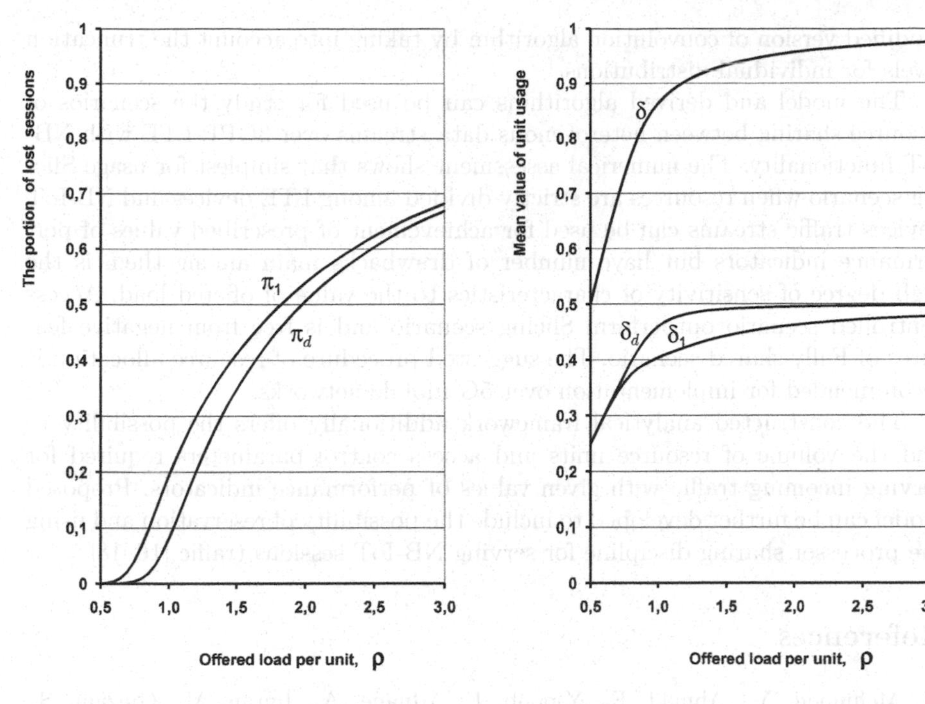

Fig. 7. The portions of lost sessions for LTE and NB-IoT devices

Fig. 8. The values of unit usage by LTE and NB-IoT devices

6 Conclusion

The model of resource allocation and sharing for conjoint servicing of real time video traffic of surveillance cameras and NB-IoT data traffic of smart meters and actuators over LTE cell facilities is constructed. In the model the access control is used to create the conditions for differentiated servicing of coming sessions. All random variables used in the model have exponential distribution with corresponding mean values but the obtained results are valid for models with arbitrary distribution of service times. Using the model the main performance measures of interest are given with help of values of probabilities of model's stationary states. The recursive algorithm of performance measures estimation is suggested.

Two enhancements of the used approach that can decrease the amount of computational works when calculating the performance measures for all traffic streams and improve the stability of algorithm when calculating the performance measures for large values of v are considered. The first enhancement is based on the remembering the results of intermediate convolutions that allows to decrease the total number of convolutions. Next refinement of recursive algorithm devoted to analysis of cases with big values of v. This leads to numerical instability of the implementation of recursive algorithm. To overcome this difficulties we construct modified version of convolution algorithm by taking into account the truncation levels for individual distributions.

The model and derived algorithms can be used for study the scenarios of resource sharing between heterogenous data streams over 3GPP LTE with NB-IoT functionality. The numerical assessment shows that simplest for usage Slicing scenario when resources are strictly divided among LTE devices and NB-IoT devices traffic streams can be used for achievement of prescribed values of performance indicators but have number of drawbacks main among them is the high degree of sensitivity of characteristics to the value of offered load. Access controlled scenario outperform Slicing scenario and is free from negative features of Fully shared scenario. The suggested procedure of resource allocation is recommended for implementation over 5G mobile networks.

The constructed analytical framework additionally offers the possibility to find the volume of resource units and access control parameters required for serving incoming traffic with given values of performance indicators. Proposed model can be further developed to include the possibility of reservation and using the processor sharing discipline for serving NB-IoT sessions traffic [16–18].

References

1. Mehmood, Y., Ahmad, F., Yaqoob, I., Adnane, A., Imran, M., Guizani, S.: Internet-of- things-based smart cities: recent advances and challenges. IEEE Commun. Mag. **55**(9), 16–24 (2017)
2. Rico-Alvarino, A., et al.: An overview of 3GPP enhancements on machine to machine communications. IEEE Commun. Mag. **54**(6), 14–21 (2016)
3. 3GPP. Standardization of NB-IOT completed (2016). http://www.3gpp.org/news-events/3gpp-news/1785-nb_iot_complete
4. Nokia. Dynamic end-to-end network slicing for 5G. White Paper (2017)
5. ElHalawany, B.M., Hashad, O., Wu, K., Tag Eldien, A.S.: Uplink resource allocation for multi-cluster internet-of-things deployment underlaying cellular networks. Mob. Netw. Appl. **25**(1), 300–313 (2019). https://doi.org/10.1007/s11036-019-01288-6
6. Malik, H., Pervaiz, H., Alam, M.M., et al.: Radio resource management scheme in NB-IoT systems. IEEE Access. **6**, 15051–15064 (2018)
7. Begishev, V., Begishev, V., et al.: Resource allocation and sharing for heterogeneous data collection over conventional 3GPP LTE and emerging NB-IoT technologies. Comput. Commun. **120**(2), 93–101 (2018)
8. Gudkova, I., et al.: Analyzing impacts of coexistence between M2M and H2H communication on 3GPP LTE system. In: Mellouk, A., Fowler, S., Hoceini, S., Daachi, B. (eds.) WWIC 2014. LNCS, vol. 8458, pp. 162–174. Springer, Cham (2014). https://doi.org/10.1007/978-3-319-13174-0_13
9. Stepanov, S., Stepanov, M., Tsogbadrakh, A., Ndayikunda, J., Andrabi, U.: Resource allocation and sharing for transmission of batched NB-IoT traffic over 3GPP LTE. In: The Proceedings of the 24th Conference of Open Innovations Association (FRUCT), pp. 422–429. Moscow Technical University of Communications and Informatics. Moscow (2019)

10. Stepanov, S.N., Stepanov, M.S.: Efficient algorithm for evaluating the required volume of resource in wireless communication systems under joint servicing of heterogeneous traffic for the internet of things. Autom. Remote Control **80**(11), 2017–2032 (2019). https://doi.org/10.1134/S0005117919110067
11. Iversen, V.B.: Teletraffic Engineering and Network Planning. Technical University of Denmark (2010)
12. Ross, K.W.: Multiservice Loss Models for Broadband Telecommunications Networks. Springer, Heidelberg (1995). https://doi.org/10.1007/978-1-4471-2126-8
13. Kelly, F.P.: Blocking probabilities in large circuit-switched networks. Adv. Appl. Prob. **18**, 473–505 (1986)
14. Iversen, V.B.: The exact evaluation of multi-service loss system with access control. Teleteknik **31**(2), 56–61 (1987)
15. Iversen, V.B., Stepanov, S.N.: The usage of convolution algorithm with truncation for estimation of individual blocking probabilities in circuit-switched telecommunication networks. In: Ramaswami, V., Wirth, P.E. (eds.) Proceedings ITC 15, pp. 1327–1336. Elservier, Amsterdam (1997)
16. Stepanov, S.N., Stepanov, M.S.: Planning transmission resource at joint servicing of the multiservice real time and elastic data traffics. Autom. Remote Control. **78**(11), 2004–2015 (2017)
17. Stepanov, S.N., Stepanov, M.S.: Planning the resource of information transmission for connection lines of multiservice hierarchical access networks. Autom. Remote Control **79**(8), 1422–1433 (2018). https://doi.org/10.1134/S0005117918080052
18. Stepanov, S.N., Stepanov, M.S.: The model and algorithms for estimation the performance measures of access node serving the mixture of real time and elastic data. In: Vishnevskiy, V.M., Kozyrev, D.V. (eds.) DCCN 2018. CCIS, vol. 919, pp. 264–275. Springer, Cham (2018). https://doi.org/10.1007/978-3-319-99447-5_23

Performance Measures of Emergency Services in Case of Overload

Sergey N. Stepanov(✉), Mikhail S. Stepanov, and Maxim O. Shishkin

Department of Communication Networks and Commutation Systems,
Moscow Technical University of Communication and Informatics,
8A, Aviamotornaya str., Moscow 111024, Russia
stpnvsrg@gmail.com, mihstep@yandex.ru, mackschischkin1@yandex.ru

Abstract. The mathematical model of public-safety answering points (PSAP) functioning is constructed and analyzed. In the model the usage of interactive voice response (IVR) and the possibility of call repetition in case of blocking or unsuccessful waiting are taken into account. Algorithm of characteristics estimation based on truncation of used infinite space of states and solving the system of state equations is suggested. Relative error of characteristics calculation caused by truncation is found. The first two or three terms in the asymptotic expansion of basic stationary performance measures into a power series of the intensity of primary calls as it tends to infinity are derived. Approximate algorithm of performance measures estimation is constructed. The usage of the model for elimination of negative effects of PSAP overload is considered.

Keywords: Public-safety answering points · System of state equations · Approximate algorithm · Asymptotic for large load · Repeated attempts

1 Introduction

Emergency services play an important role in people's lives. Modern technical facilities allow citizens to ask for help at any time and in fact from anywhere. Technically the possibility to reach primary emergency services such as police, firefighting, ambulance is organized through public-safety answering points (PSAP) [1]. PSAP is the foundation stone of the whole system because it handles inbound requests from the citizens and dispatches an intervention resources if necessary. In some countries, one number is used for all the emergency services (e.g. 112 in continental Europe including Russia, 911 in the USA, 999 in the UK). Nowadays, the most common channel for connecting PSAP is through phone calls. Because of this, the availability of mobile and fixed telephone networks plays an essential role. Most such systems have technical ability to get the location of the caller. The functional model of the emergency service in network is presented on Fig. 1.

The emergency services are typically connected to public mobile or fixed telephone networks by access lines. This resource of the PSAP should be carefully

V. M. Vishnevskiy et al. (Eds.): DCCN 2020, LNCS 12563, pp. 436–449, 2020.
https://doi.org/10.1007/978-3-030-66471-8_33

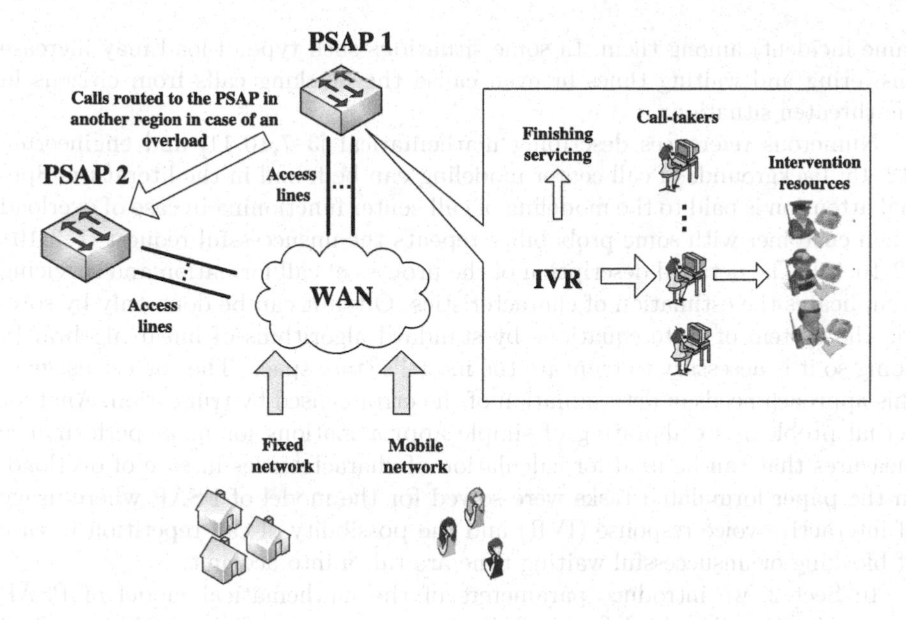

Fig. 1. The functional model of the PSAP place in the network.

estimated because it's necessary to have more available lines than call-takers. In the case of an overload in PSAP1, the call can be routed to the PSAP2 responsible for another geographical region (see Fig. 1). After entering to the emergency service call should be distributed to one of available call-taker. After processing the call it's necessary to respond it. Such aspects as availability of intervention resources (police, firefighters, ambulance etc.) and correct location information are critical for the quick response. This type of bottleneck isn't connected with technical aspects of overload directly but it may increase the call volume in case of unavailability of intervention resources.

The key performance indicators of PSAP are the ratios of calls that were blocked due to occupation of access lines and the lack of free call-takers [1,2]. Besides, mean time of waiting for the beginning of service should be estimated. Using these characteristics and the intensities of coming calls which can be taken from the data collected by emergency service, we can calculate the number of call-takers sufficient for calls servicing with given QoS requirements [2–15]. Under normal circumstances it's easy to predict for emergency services the volume of arriving calls for each period of the day and the number of call-takers to handle it. However, all emergency services can be affected by an overload of calls in case of different unforeseen situations. Among them there are both natural (earthquakes, floods) and man-made (fires, terrorist attacks, nuclear accidents) disasters [1,2]. A large number of citizens seeing or suffering the same incident may contact emergency services at the same time. Nowadays it's very easy to do with the mobile phone. The statistics of incoming calls have to be thoroughly analyzed because there can be a large number of repeated calls (or calls concerning the same incident) among them. In some situations such type of load may increase answering and waiting times or even cause the blocking calls from citizens in life-threaten situations.

Numerous references describing mathematical [3–7,10,11] and engineering [12–15] backgrounds of call center modeling can be found in the literature. Special attention is paid to the modeling of call center functioning in case of overload when customer with some probability repeats the unsuccessful request [1–3,10–12,16,17]. The detailed description of the process of call formation and servicing complicates the estimation of characteristics. Often it can be done only by solving the system of state equations by standard algorithms of linear algebra. In doing so it is necessary to truncate the model's state space. The correct usage of this approach needs in determination of the error caused by truncation. Another actual problem is elaborating of simple approximations for main performance measures that can be used for calculation of characteristics in case of overload. In the paper formulated tasks were solved for the model of PSAP where usage of interactive voice response (IVR) and the possibility of call repetition in case of blocking or unsuccessful waiting time are taken into account.

In Sect. 2, we introduce parameters of the mathematical model of PSAP functioning and give the definitions of main performance measures. Algorithm of characteristics estimation is suggested based on truncation of used infinite state space and solving the system of state equations. Relative error of characteristics calculation caused by truncation is studied in Sect. 3. The first two or three terms in the asymptotic expansion of basic stationary performance measures into a power series of the intensity of primary calls as it tends to infinity are derived in Sect. 4. Approximate algorithm of performance measures estimation is constructed in Sect. 5. The usage of the model for elimination of negative effects of PSAP overload is considered in Sect. 6.

2 Model Description

Calls for getting emergency service are entering the PSAP through telephone access lines. After occupying an access line a call can be served by IVR and if it is necessarily by PSAP call-takers. Let us denote by v the overall number of call-takers and by $w + v$ we denote the overall number of access lines. It is supposed that w access lines are used for waiting the beginning of service and v access lines are used in the process of call servicing by call-taker.

The PSAP functioning is considered in case of overload. It means that apart from primary calls that arrive for servicing according to the Poisson model with intensity λ the emergency center serves the flow repeated calls caused by insufficient number of free call-takers and access lines or by unsuccessful finishing the time of waiting the beginning of service. In both situations, a calling citizen with probability H repeats the request for servicing after random time having exponential distribution with parameter ν and with additional probability $1 - H$ the blocked citizen stops his attempts to find free call-taker and leaves the system unserved. It is supposed that maximum allowed time of waiting the beginning of servicing at PSAP has exponential distribution with parameter σ.

The process of call servicing at PSAP includes two stages. The first stage consists in getting a recorded message from the IVR and second consists in exchanging of information with call-taker. It is supposed that duration of call-taker's service has exponential distribution with parameter μ. The transition to call-taker's servicing is depending on the type of the call: primary or repeated. With probability q_p for primary call and with probability q_r for repeated attempt after getting service from IVR a citizen is trying to get servicing from PSAP call-taker. With additional probabilities $1 - q_p$ and $1 - q_r$ correspondingly a citizen leaves the system satisfying by the servicing at IVR.

The administration of PSAP has a few scenarios of using the process of call servicing at IVR. In one scenario IVR can be considered as robotized call-taker that provides a citizen with asked information and gives a cheap way to decrease the required number of costly call-takers in case of overload. In another scenario the IVR-message can be considered as ruling directives to citizen concerning his behavior in case of overload. In particular it can contain the urgent advice do not repeat the call in case of overload. By doing this we get a way to filter the input flow and as result to decrease the required number of call-takers in case of overload. Examples of realizing the suggested approachers will be considered in Sect. 6.

The structure of mathematical model used for description of PSAP functioning is shown on Fig. 2.

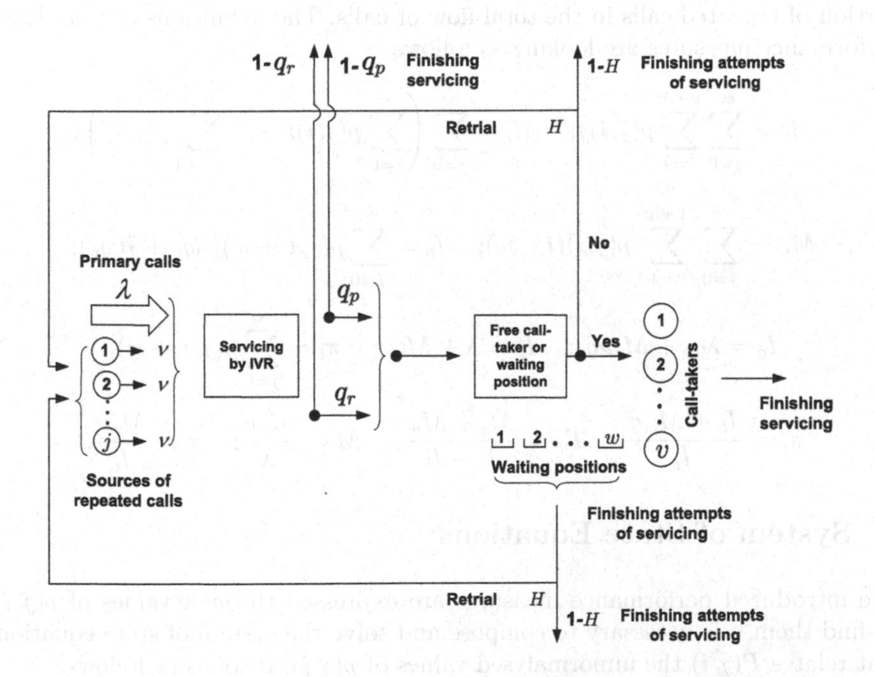

Fig. 2. The structure of mathematical model used for description of PSAP functioning.

By taking into account the call repetitions the constructed model allows to investigate the avalanche effects of calls occurs in case of overload. Let us denote the state of the PSAP model by vector (j, i) where j is the number of citizens repeating a call and i is the number of occupied call-takers and access lines. The values of j, i varies as follows $j = 0, 1, \ldots,$; $i = 0, 1, \ldots, v + w$ and forms the model's space of states S. Let us denote by $j(t)$ the number of citizens repeating a call at time t and by $i(t)$ we denote the number of busy call-takers and occupied waiting places at time t. The dynamic of changing the model's states is described by Markov process $r(t) = (j(t), i(t))$, defined on the infinite space of states S.

Let us denote by $p(j, i)$ the probability of stationary state $(j, i) \in S$ of the considered PSAP model and define main performance measures of the process of calls serving. The first group of characteristics are mean values of components of the model state. Let us denote by M_r, M_i, M_w mean numbers of, respectively, citizens repeating a call, occupied call-takers and occupied waiting positions. Next group are intensities of arrived and blocked calls. We denote by I_b, I_o and I_t correspondingly the intensity of calls lost in attempt to get service from call-takers, arrived to get service from call-takers and arrived to get service at PSAP. Key performance measures of PSAP functioning are defined as follows. Let us denote by π_t the portion of time when all call-takers and waiting positions are occupied, by π_c we denote the ratio of lost calls arrived to get service at PSAP, by T_w we denote the mean time for call to be on waiting or servicing, by M we denote the mean number of retrials per one primary call, by τ we denote the portion of repeated calls in the total flow of calls. The definitions of introduced performance measures are looking as follows

$$M_r = \sum_{j=0}^{\infty}\sum_{i=0}^{v+w} p(j,i)j; \quad M_i = \sum_{j=0}^{\infty}\left(\sum_{i=0}^{v} p(j,i)i + v\sum_{i=v+1}^{v+w} p(j,i)\right); \tag{1}$$

$$M_w = \sum_{j=0}^{\infty}\sum_{i=v+1}^{v+w} p(j,i)(i-w); \quad I_b = \sum_{j=0}^{\infty} p(j, v+w)(\lambda q_p + j\nu q_r);$$

$$I_o = \lambda q_p + M_r \nu q_r; \quad I_t = \lambda + M_r \nu; \quad \pi_t = \sum_{j=0}^{\infty} p(j, v+w);$$

$$\pi_c = \frac{I_b + M_w \sigma}{I_t}; \quad T_w = \frac{M_i + M_w}{I_o - I_b}; \quad M = \frac{M_r \nu}{\lambda}; \quad \tau = \frac{M_r \nu}{I_t}.$$

3 System of State Equations

The introduced performance measures are expressed through values of $p(j, i)$. To find them it is necessary to compose and solve the system of state equations that relates $P(j, i)$ the unnormalysed values of $p(j, i)$. It looks as follows:

$$P(j,v+w)\Big\{\lambda q_p H + j\nu(1-q_r H) + v\mu + w\sigma\Big\} = \tag{2}$$
$$= P(j,v+w-1)\lambda q_p + P(j+1,v+w-1)(j+1)\nu q_r +$$
$$+ P(j-1,v+w)\lambda q_p H I(j>0) + P(j+1,v+w)(j+1)\nu(1-q_r H);$$
$$j=0,1,\ldots$$

$$P(j,i)\Big\{\lambda q_p + j\nu + i\mu I(i\le v) + (v\mu + (i-w)\sigma)I(i>v)\Big\} =$$
$$= P(j,i-1)\lambda q_p I(i>0) + P(j+1,i-1)(j+1)\nu q_r I(i>0) +$$
$$+ P(j+1,i)(j+1)\nu(1-q_r) + P(j,i+1)(i+1)\mu I(i+1\le v) +$$
$$+ P(j,i+1)(v\mu + (i+1-v)\sigma(1-H))I(i+1>v) +$$
$$+ P(j-1,i+1)(i+1-v)\sigma H I(j>0, i+1>v);$$
$$j=0,1,\ldots;\quad i=0,1,\ldots,v+w-1.$$

By $I(\cdot)$ in (2) the indicator function is defined. Values of $P(j,i,)$ should be normalized. Let us derive some auxiliary relations that help us to estimate the error caused by truncation of the infinite state space S. We begin with summing up (2) over i from 0 to $v+w$ with j fixed and over j from 0 to ∞ with i fixed. The summation gives the following two sets of relations

$$p(j,v+w)\lambda q_p H + \sum_{i=v+1}^{v+w} p(j,i)(i-v)\sigma H = \tag{3}$$
$$= p(j+1,v+w)(j+1)\nu(1-q_r H) + \sum_{i=0}^{v+w-1} p(j+1,i)(j+1)\nu,$$
$$j=0,1,\ldots$$

$$\sum_{j=0}^{\infty} p(j,i)(\lambda q_p + j\nu q_r) = \sum_{j=0}^{\infty} p(j,i+1)\Big((i+1)\mu I(i+1\le v) + \tag{4}$$
$$+(v\mu + (i+1-v)\sigma)I(i+1>v)\Big),$$
$$i=0,1,\ldots,v+w-1.$$

Next we sum up (3) over j from 0 to ∞ and (4) over i from 0 to $v+w-1$. The summation gives the following two relations

$$M_r\nu = (I_b + M_w\sigma)H; \tag{5}$$

$$I_t = \lambda(1-q_p) + M_r\nu(1-q_r) + I_b + M_w\sigma + M_i\mu. \tag{6}$$

To solve (2) it is necessarily to truncate the number of repeating citizens by applying inequality $j \le j_m$, where j_m is some integer number, and find the values of $p(j,i)$ by ordinary algorithms of linear algebra. Let us denote performance measures of truncated model by the same symbols that used for initial model only with superscript $*$ and find the error caused by truncation. The analog of (5) for truncated model is looking as follows

$$M_r^* \nu = (I_b^* + M_w^* \sigma)H - \gamma, \tag{7}$$

where γ defined as

$$\gamma = p^*(j_m, v + w)\lambda H q_p + \sum_{i=v+1}^{v+w} p^*(j_m, i)(i - v)\sigma H.$$

Let us denote by Δ the difference between exact value of characteristic and their estimate obtained with help of truncated model, for example, $\Delta M_r = M_r - M_r^*$. By using the basic property of exponentially distributed variables and ideas used in [16] it can be proved that the following inequalities are true

$$\Delta M_r \geq 0; \quad \Delta I_b + \Delta M_w \sigma \geq 0; \quad \Delta M_i \geq 0. \tag{8}$$

For main performance measures from (5)–(7) follows upper estimates of error caused by truncations as function of γ

$$\gamma \leq \Delta M_r \nu \leq \frac{\gamma}{1 - q_r H}; \quad 0 \leq \Delta I_b + \Delta M_w \sigma \leq \frac{\gamma q_r}{1 - q_r H}; \tag{9}$$

$$0 \leq \Delta M_i \mu \leq \gamma q_r.$$

For other model's characteristics that can be expressed as function of M_r, M_i, M_w, I_t and model's input parameters the estimation of relative error can be obtained with help of (9). For example, for π_c the following inequality is true

$$\delta \pi_c = \left| \frac{\Delta \pi_c}{\pi_c} \right| \leq \frac{\gamma}{1 - q_r H} \left(\frac{q_r}{I_b + M_w \sigma} + \frac{1}{\lambda + M_r \nu} \right). \tag{10}$$

From (9) follows that error of estimation of M_r is proportional to γ. Let us denote by $\Delta^b M_r$ and by $\Delta^a M_r$ correspondingly the lower and upper estimates of ΔM_r presented at (9) and consider a numerical example that illustrates their accuracy. Model input parameters are as follows: $\lambda = 30$; $q_p = 0{,}5$; $q_r = 0{,}9$; $H = 0{,}9$; $\nu = 5$; $\mu = 1$; $\sigma = 0{,}5$; $v = 10$; $w = 5$. The value of j_m varies from 2 to 40. The values of characteristics found for $j_m = 40$ are considered as found for unlimited interval of varying j. As a time unit was chosen the mean time of servicing a request by call-taker. In the Table 1 are presented the values of j_m and depending on j_m the values of π_c, M_r, ΔM_r, the lower $\Delta^b M_r$ and upper $\Delta^a M_r$ estimation of ΔM_r found from (9) and value of γ.

From the content of the table it is seen that upper estimate of ΔM_r has very good accuracy. As result after making calculation of performance measures with help of truncated model we can find the error caused by truncation in terms of characteristics of truncated model. More details about using the concept of truncation can be found in [16].

Table 1. The dependence of error caused by truncation on j_m.

j_m	π_c	M_r	$\Delta^b M_r$	ΔM_r	$\Delta^a M_r$	γ
2	0,280257	1,07070753	$7{,}127 \cdot 10^{-1}$	$3{,}694 \cdot 10^{0}$	$3{,}751 \cdot 10^{0}$	$3{,}564 \cdot 10^{0}$
4	0,360144	2,12947699	$5{,}055 \cdot 10^{-1}$	$2{,}636 \cdot 10^{0}$	$2{,}661 \cdot 10^{0}$	$2{,}528 \cdot 10^{0}$
6	0,415571	3,06574402	$3{,}250 \cdot 10^{-1}$	$1{,}699 \cdot 10^{0}$	$1{,}710 \cdot 10^{0}$	$1{,}625 \cdot 10^{0}$
8	0,451382	3,79243499	$1{,}857 \cdot 10^{-1}$	$9{,}727 \cdot 10^{-1}$	$9{,}773 \cdot 10^{-1}$	$9{,}284 \cdot 10^{-1}$
10	0,472605	4,28008825	$9{,}249 \cdot 10^{-2}$	$4{,}850 \cdot 10^{-1}$	$4{,}868 \cdot 10^{-1}$	$4{,}625 \cdot 10^{-1}$
12	0,483811	4,55746731	$3{,}957 \cdot 10^{-2}$	$2{,}076 \cdot 10^{-1}$	$2{,}082 \cdot 10^{-1}$	$1{,}978 \cdot 10^{-1}$
14	0,488935	4,68929331	$1{,}444 \cdot 10^{-2}$	$7{,}580 \cdot 10^{-2}$	$7{,}599 \cdot 10^{-2}$	$7{,}219 \cdot 10^{-2}$
16	0,490928	4,74145936	$4{,}500 \cdot 10^{-3}$	$2{,}363 \cdot 10^{-2}$	$2{,}368 \cdot 10^{-2}$	$2{,}250 \cdot 10^{-2}$
18	0,491585	4,75875655	$1{,}206 \cdot 10^{-3}$	$6{,}335 \cdot 10^{-3}$	$6{,}347 \cdot 10^{-3}$	$6{,}030 \cdot 10^{-3}$
20	0,491770	4,76361833	$2{,}804 \cdot 10^{-4}$	$1{,}473 \cdot 10^{-3}$	$1{,}476 \cdot 10^{-3}$	$1{,}402 \cdot 10^{-3}$
22	0,491814	4,76479159	$5{,}706 \cdot 10^{-5}$	$2{,}999 \cdot 10^{-4}$	$3{,}003 \cdot 10^{-4}$	$2{,}853 \cdot 10^{-4}$
24	0,491823	4,76503758	$1{,}025 \cdot 10^{-5}$	$5{,}388 \cdot 10^{-5}$	$5{,}395 \cdot 10^{-5}$	$5{,}126 \cdot 10^{-5}$
26	0,491825	4,76508285	$1{,}638 \cdot 10^{-6}$	$8{,}609 \cdot 10^{-6}$	$8{,}620 \cdot 10^{-6}$	$8{,}189 \cdot 10^{-6}$
28	0,491825	4,76509023	$2{,}343 \cdot 10^{-7}$	$1{,}232 \cdot 10^{-6}$	$1{,}233 \cdot 10^{-6}$	$1{,}171 \cdot 10^{-6}$
30	0,491825	4,76509130	$3{,}017 \cdot 10^{-8}$	$1{,}586 \cdot 10^{-7}$	$1{,}588 \cdot 10^{-7}$	$1{,}509 \cdot 10^{-7}$
40	0,491825	4,76509146				

4 Estimation of Characteristics for Large Load

Let us find in explicit form the first terms of expansion of introduced characteristics into a power series of λ the intensity of primary calls as it tends to infinity. In the considered case a citizen with probability tends to one gets a refusal in servicing in primary or repeated attempts and further behave according to rules given by probabilities q_p, q_r and H. The mean number of retrials per one primary call tends to the value

$$q_pH(1-q_rH)\big(1+q_rH\cdot 2+(q_rH)^2\cdot 3+\ldots+\big)=\frac{q_pH}{1-q_rH}. \tag{11}$$

Because primary calls are coming according to Poisson model we can anticipate that as $\lambda\to\infty$ the flow of repeated calls is asymptotically Poissonian with intensity $\frac{\lambda q_pH}{1-q_rH}$. This property will be used in asymptotic analysis of the model.

From (4) follows that as $\lambda\to\infty$

$$\sum_{j=0}^{\infty}p(j,i)=o\left(\frac{1}{\lambda^{v+w-1-i}}\right),\quad i=0,1,\ldots,v+w-1. \tag{12}$$

By using (5), (6) we can find alternative formula for calculation of M_r

$$M_r=\frac{\lambda q_pH}{\nu\,(1-q_rH)}-\frac{M_i\mu H}{\nu\,(1-q_rH)}. \tag{13}$$

From (12), (13) follows that as $\lambda \to \infty$

$$M_i = v + o\left(\frac{1}{\lambda^w}\right); \quad M_r = \frac{\lambda q_p H}{\nu(1-q_r H)} - \frac{v\mu H}{\nu(1-q_r H)} + o\left(\frac{1}{\lambda^w}\right). \tag{14}$$

By using (12), (13) it is possible to derive more terms of expansion for main characteristics as $\lambda \to \infty$

$$\pi_t = 1 - \frac{(v\mu + w\sigma)(1-q_r H)}{\lambda q_p} + o\left(\frac{1}{\lambda}\right); \tag{15}$$

$$M_w = w - \frac{(v\mu + w\sigma)(1-q_r H)}{\lambda q_p} + o\left(\frac{1}{\lambda}\right);$$

$$M = \frac{q_p H}{1-q_r H} - \frac{v\mu H}{\lambda(1-q_r H)} + o\left(\frac{1}{\lambda^{w+1}}\right);$$

$$\pi_c = \frac{q_p}{1-H(q_r-q_p)} - \frac{v\mu(1-q_r H)}{\lambda(1-H(q_r-q_p))^2} - \frac{(v\mu)^2 H(1-q_r H)}{\lambda^2(1-H(q_r-q_p))^3} + o\left(\frac{1}{\lambda^2}\right).$$

The number of terms for π_c depends on the value of w. In (15) the first and second terms are valid for $w \geq 0$, the first, second and third terms are valid for $w \geq 1$. Asymptotic expansions for other characteristics can be found with help of (12)-(15), conservation laws (5), (6) and definitions of characteristics. From (14) follows good accuracy of estimation M_r as $\lambda \to \infty$ and $w > 1$. It means that for calculation of some characteristics it is better to use their expressions through M_r instead of found expansions into power of λ. For example more accurate result in calculation of π_c gives the formula $\pi_c = \frac{M_r \nu}{H(\lambda + M_r \nu)}$ than expansion (15). The same is true for calculation of τ with help of formula $\tau = \frac{M}{1+M}$.

Let us denote by M_r^s and π_c^s the expansions of M_r and π_c defined by relations (14) and (15) correspondingly and by π_c^f we define the result of calculation $\pi_c^f = \frac{M_r^s \nu}{H(\lambda + M_r^s \nu)}$. Let us consider a numerical example that illustrates the accuracy of obtained asymptotic formulae. Model input parameters are as follows: $q_p = 0{,}5$; $q_r = 0{,}9$; $H = 0{,}9$; $\nu = 5$; $\mu = 1$; $\sigma = 0{,}5$; $v = 10$; $w = 5$. The value of λ varies from 22,5 to 60. As a time unit was chosen the mean time of servicing a request by call-taker. In the Table 2 are presented the values of λ and depending on λ the values of M_r, M_r^s, π_c, π_c^s and π_c^f. From the content of the table it is seen that asymptotic expressions found in explicit form have quite good accuracy especially in the case of overload. More terms of expansion can be found if we use the ideas formulated in [17].

5 Approximate Calculation of Performance Measures

Let us derive the approximate algorithm of performance measures calculation and show that obtained results are asymptotically correct when $\lambda \to \infty$. In doing this we construct simplifying model of PSAP functioning by supposing that the flow of retrials in the considered model is poissonian with some intensity $x - \lambda$,

Table 2. The dependence of asymptotic expressions of performance measures on λ.

λ	M_r	M_r^s	π_c	π_c^s	π_c^f
22,5	1,574889	1,184211	0,288051	0,446235	0,231481
25,0	2,538859	2,368421	0,374188	0,491333	0,357143
27,5	3,623002	3,552632	0,441254	0,526315	0,436047
30,0	4,765091	4,736842	0,491825	0,554148	0,490196
32,5	5,932285	5,921053	0,530186	0,576764	0,529661
35,0	7,109745	7,105263	0,559877	0,595466	0,559701
37,5	8,291286	8,289474	0,583394	0,611165	0,583333
40,0	9,474434	9,473684	0,602431	0,624514	0,602410
45,0	11,842247	11,842105	0,631316	0,645955	0,631313
50,0	14,210558	14,210526	0,652175	0,662384	0,652174
60,0	18,947372	18,947368	0,680272	0,685819	0,680272

where x is unknown intensity of total poissonian flow of primary and repeated calls. Let us indicate the obtained estimates by the same symbols that was used before for corresponding characteristics of the initial model only with superscript $*$ and suppose that for obtained in this way estimates the relations (5)–(6) are true. It gives

$$x = (I_b^* + M_w^* \sigma)H + \lambda; \tag{16}$$

$$x = \lambda(1 - q_p) + (x - \lambda)(1 - q_r) + (I_b^* + M_w^* \sigma) + M_i^* \mu,$$

where characteristics

$$I_b^*(x) = \lambda q_p \pi_t^*(x) + (x - \lambda)\pi_t^*(x) q_r; \quad \pi_t^*(x) = p(v + w);$$

$$M_w^*(x) = \sum_{i=v+1}^{v+w} p(i)(i - v); \quad M_i^*(x) = \sum_{i=1}^{v} p(i)i + v \sum_{i=v+1}^{v+w} p(i)$$

are functions of x. Values of $p(i)$, $i = 0, \ldots, v + w$ are calculated from relations of detailed balance

$$p(i)\Lambda = p(i+1)(i+1)\mu, \quad i = 0, 1, \ldots, v - 1; \tag{17}$$

$$p(i)\Lambda = p(i+1)(v\mu + (i + 1 - v)\sigma), \quad i = v, v + 1, \ldots, v + w - 1$$

after subsequent normalization. Parameter $\Lambda = \lambda q_p + (x - \lambda) q_r$.

From (16) we obtain equation for determination of x. It looks in the following way

$$x = \frac{\lambda(1 + \pi_t^*(x)H(q_p - q_r)) + M_w^*(x)\sigma H}{1 - \pi_t^*(x) q_r H}. \tag{18}$$

It is easy to prove that (18) has solution, this solution is unique and can be obtained by successive substitutions. By implementing the approach used in

Section 4 it is possible to prove that suggested estimates are satisfying the asymptotic relations (14) and (15) as $\lambda \to \infty$. Let us consider a numerical example that illustrates the accuracy of obtained estimates.

Model input parameters are as follows: $q_p = 0{,}5$; $q_r = 0{,}9$; $H = 0{,}9$; $\nu = 0{,}5$; $\mu = 1$; $\sigma = 0{,}5$; $v = 10$; $w = 5$. The value of λ varies from 10 to 40. As a time unit was chosen the mean time of servicing a request by call-taker. In the Table 3 are presented the values of λ and depending on λ the values of M_r, M_r^*, π_c, π_c^* M, M^*, τ and τ^*. From the content of the table it is seen that found estimates have quite good accuracy especially in the case of overload.

Table 3. The dependence of approximate values of characteristics on λ.

λ	M_r	M_r^*	π_c	π_c^*	M	M^*	τ	$\tau*$
10,0	0,028148	0,026270	0,001562	0,001458	0,001407	0,001313	0,001405	0,001312
12,5	0,160472	0,141170	0,007087	0,006239	0,006419	0,005647	0,006378	0,005615
15,0	0,653758	0,535829	0,023697	0,019497	0,021792	0,017861	0,021327	0,017548
17,5	2,161588	1,685996	0,064630	0,051064	0,061760	0,048171	0,058167	0,045957
20,0	6,017536	4,916144	0,145296	0,121613	0,150438	0,122904	0,130766	0,109452
22,5	13,547018	12,719305	0,257097	0,244850	0,301045	0,282651	0,231387	0,220365
25,0	24,025644	23,824177	0,360620	0,358572	0,480513	0,476484	0,324558	0,322715
27,5	35,592975	35,557679	0,436543	0,436280	0,647145	0,646503	0,392889	0,392652
30,0	47,384792	47,377748	0,490291	0,490250	0,789747	0,789629	0,441262	0,441225
35,0	71,054619	71,054053	0,559709	0,559707	1,015066	1,015058	0,503738	0,503736
40,0	94,737269	94,737181	0,602411	0,602411	1,184216	1,184215	0,542170	0,542170

6 The Usage of the Model for Elimination of PSAP Overload

The process of normal functioning of PSAP can be disturbed by increasing the intensity of coming requests. The overload can be caused by many reasons that are discussed in Sect. 1. To decrease the negative consequences of input flow fluctuations we can use the procedures of the filtering the input flows of primary calls or repeated attempts. Another possibility is to redirect part of the input flow of primary calls to other PSAPs (see Fig. 1). The consequences of usage these and other procedures aimed to elimination of overload and estimation of necessary volumes of access lines and call-takers can be studied with help of constructed model. Because shortage of place let us consider only one example.

In case of overload part of primary flow can be redirected to other PSAP with similar service facilities. The exact proportion can be found with help of constructed model. We illustrate the procedure of redirecting by numerical example. Model's input parameters are are as follows: $\lambda = 24$; $v = 10$; $w = 5$; $H = 0{,}9$; $\nu = 5$; $\jmath_m = 50$; $\mu = 1$; $\sigma = 0{,}1$. As a time unit was chosen the mean time of servicing a request by call-taker. The portion r of redirected primary calls defined as $r = \frac{24-\lambda_c}{24}$ and varies from 0 to 0,4. Here λ_c is current value of intensity of primary calls that in the considered case consequently decreases from 24. The required level of service should satisfy the inequality $\pi_c < 0{,}05$. The Fig. 3 shows the dependence of π_c on r. The results of calculations show that by redirecting primary calls we can decrease the value of losses in efficient way. The constructed model allows to study the process of forming and servicing requests in case of overload and choose the right values of parameters that can be used for control the PSAP functioning in case of overload.

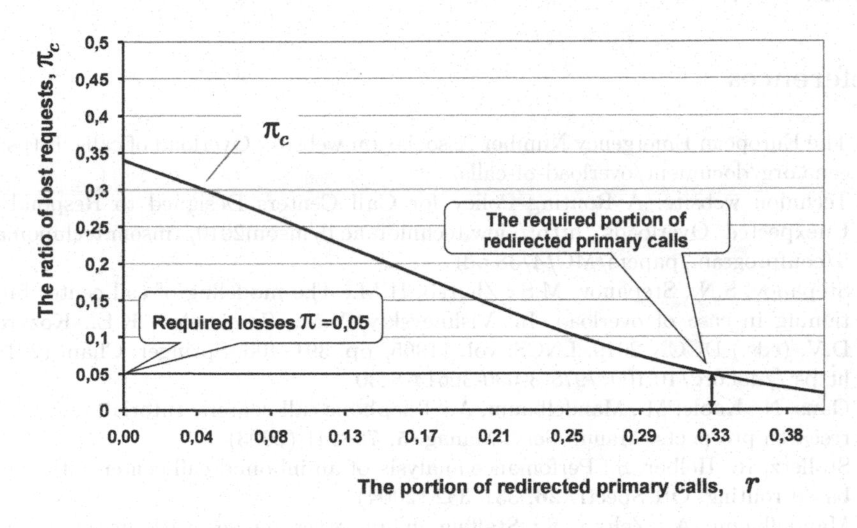

Fig. 3. The dependence of π_c on the portion of primary calls redirected to other PSAP.

7 Conclusion

In this paper the mathematical model of PSAP is constructed and analyzed. In the model the usage of interactive voice response and the possibility of call repetition in case of blocking or unsuccessful waiting time are taken into account. Primary and repeated calls are coming after exponentially distributed time intervals. It is supposed that service time has exponential distribution. Markov process that describes model functioning is constructed. In the framework of the proposed model the definitions of main performance measures are formulated through values of probabilities of model's stationary states. Algorithm of characteristics estimation is suggested based on truncation of the used infinite state space and solving the system of state equations. Relative error of characteristics calculation caused by truncation is found. The first two or three terms in the asymptotic expansion of basic stationary performance measures into a power series of the intensity of primary calls as it tends to infinity are derived. Approximate algorithm of performance measures estimation is constructed. The usage of the model for elimination of negative effects of PSAP overload is considered.

The model and derived algorithms of performance measures estimation can be used to produce the quantitative and qualitative analysis of the dependence of model's performance measures on the values of input parameters with taking into account the customer behavior in case of overload. The constructed analytical framework additionally offers the possibility to find the numbers of call-takers and waiting positions (access lines) required for serving coming calls with given values of performance indicators. Proposed model can be further developed to include the possibility of non exponential distribution of time interval between retrials.

References

1. The European Emergency Number Association website. Overload of calls. https://eena.org/document/overload-of-calls/
2. Technion website. A Routing Policy for Call Centers Designed to Respond to Unexpected Overloads. http://iew.technion.ac.il/msom2010//msom.technion.ac.il/confprogram/papers/MC/4/38.pdf
3. Stepanov, S.N., Stepanov, M.S., Zhurko, H.M.: The modeling of call center functioning in case of overload. In: Vishnevskiy, V.M., Samouylov, K.E., Kozyrev, D.V. (eds.) DCCN 2019. LNCS, vol. 11965, pp. 391–406. Springer, Cham (2019). https://doi.org/10.1007/978-3-030-36614-8_30
4. Gans, N., Koole, M., Mandelbaum, A.: Telephone call-centers: tutorial, review and research prospects. Manuf. Serv. Manag. **5**, 79–141 (2003)
5. Stolletz, R., Helber, S.: Perfomance analysis of an inbound call-center with skills-based routing. OR Spectr. **26**, 331–352 (2004)
6. Mandelbaum, A., Zeltyn, S.: Staffing many-server queues with impatient customers: constraint satisfaction in call centers. Oper. Res. **57**(5), 1189–1205 (2009)
7. Bhulai, S., Koole, G.: A queueing model for call blending in call centers. IEEE Trans. Autom. Control **48**(8), 1434–1438 (2003)
8. Ferrari, S.C., Morabito, R.: Application of queueing models with abandonment for call center congestion analysis. Gestao Producao **27**(1), e3765 (2020)
9. Colin, M.: Call center service level: a customer experience model from benchmarking and multivariate analysis. Esic Mark. Econ. Bus. J. **51**(3), 467–496 (2020)
10. Stepanov, S.N., Stepanov, M.S.: Construction and analysis of a generalized contact center model. Autom. Remote Control **75**(11), 1936–1947 (2014)
11. Stepanov, S.N., Stepanov, M.S.: Algorithms for estimating throughput characteristics in a generalized call center model. Autom. Remote Control **77**(7), 1195–1207 (2016)
12. Aguir, S., Karaesmen, F., Aksin, O.Z., Chauvet, F.: The impact of retrials on call center performance. OR Spectr. **26**(3), 353–376 (2004)
13. Aksin, Z., Armony, M., Mehrotra, A.: The modern call center: a multi- disciplinary perspective on operations management research. Prod. Oper. Manag. **16**(6), 665–688 (2007)
14. Whitt, W.: Engineering solution of a basic call-center model. Manag. Sci. **51**(2), 221–235 (2005)
15. Whitt, W.: Staffing a call center with uncertain arrival rate and absenteeism. Prod. Oper. Manag. **15**(1), 88–102 (2006)

16. Stepanov, S.N.: Markov models with retrials: the calculation of stationary performance measures based on the concept of truncation. Math. Comput. Model. **30**, 207–228 (1999)
17. Stepanov, S.N.: Generalized model with retrials in case of extreme load. Queueing Syst. **27**, 131–151 (1998)

Evaluation and Prediction of an Optimal Control in a Processor Sharing Queueing System with Heterogeneous Servers

Dmitry Efrosinin[1,2](✉), Vladimir Rykov[2,4], and Natalia Stepanova[3]

[1] Johannes Kepler University Linz, Altenbergerstrasse 69, 4040 Linz, Austria
dmitry.efrosinin@jku.at

[2] Peoples' Friendship University of Russia (RUDN University), Miklukho-Maklaya Street 6, 117198 Moscow, Russia
vladimir_rykov@mail.ru

[3] V.A. Trapeznikov Institute of Control Sciences of RAS, Profsoyuznaya Street, 65, 117997 Moscow, Russia
natalia0410@rambler.ru

[4] Institute for Information Transmission Problems, Bolshoy Karetny per. 19, build.1, 127051 Moscow, Russia

http://www.jku.at, http://www.eng.rudn.ru, http://www.ipu.ru/en, http://www.iitp.ru/en

Abstract. In this paper we study the problem of optimal controlling in a processor sharing (PS) $M/M/2$ queueing system with heterogeneous servers. The servers differ in the service intensities, operating and usage costs. The objective is to find the optimal policy to allocate the customers either to an idle or partially loaded server, or to the queue at each arrival and service completion epoch to minimize the long-run average cost per unit of time. We handle this optimization problem as Markov decision problem and study numerically structural properties of the optimal control policy. Using a policy-iteration algorithm we show that this policy for the current model is of threshold type. In this case the faster server handles customers with maximum capacity, while the number of simultaneously serviced customers at the slower server can be increased only when the number of waiting customers exceeds a certain threshold level. The data-sets generated by classic methodology of analyzing the controlled queues are used to explore predictions for optimal thresholds through artificial neural networks. The presented theoretical results are accompanied by heuristic solution and numerical examples.

Keywords: Processor sharing · Heterogeneous servers · Policy-iteration algorithm · Heuristic solution · Artificial neural network

The publication has been prepared with the support of the "RUDN University Program 5–100". (recipients D. Efrosinin and V. Rykov, mathematical model development). The reported study was funded by RFBR, project number 19-29-06043 mk.

V. M. Vishnevskiy et al. (Eds.): DCCN 2020, LNCS 12563, pp. 450–462, 2020.
https://doi.org/10.1007/978-3-030-66471-8_34

1 Introduction

Queues with processor shared service have attracted considerable interest over the past few decades as models for sharing the resources in a service area. If the capacity of a server with processor sharing (PS) is N it means that N customers presented in the system receives service at this server simultaneously at an equal fraction of $1/N$ of the total service rate. A lot of research papers [2,3,7,8,14] and surveys [1,21] has been devoted to processor sharing models specially in the context of time-shared computer systems, bandwidth sharing protocols in telecommunication network, task sharing in a multi-system software and other real applications. While the first steps in the analysis of single-server queueing systems with PS have already been developed, a missing link to an applicability of these models is the study of multi-server PS queues especially when servers are heterogeneous and controllable.

The proposed heterogeneous queueing system is appropriate for stochastic modeling of a group of servers with different processor types as a consequence of system updates, nodes in telecommunication networks where links have different capacities and availability, nodes in wireless systems serving different mobile users, peer-to-peer services like streaming, file sharing and storage, multi-processor systems with processors of different throughput and power consumption. For the system with heterogeneous service attributes it is necessary to fix some decision rule or policy to allocate the customers between servers. As it was shown in [4,16], the optimal control policy, which minimizes the average cost, in case of an ordinary heterogeneous queueing system is of threshold type. Threshold control policy prescribes the usage of the fastest server whenever it is idle. The slower server k must be used only when all $(k-1)$ faster servers are busy and the queue length exceeds some threshold level $q_k > 1$. The techniques to prove such results are based on monotonicity properties of the dynamic programming relative value function (see, e.g., [10,17,20]).

The contribution of this paper is twofold. First, we consider a two-server heterogeneous queueing system with PS. In many cases a server can be shared by unlimited number of users. Similarly to [3], we assume that the capacities of servers $N_1 < \infty$ and $N_2 < \infty$ to guarantee minimal acceptable quality of service to a customer. The servers are differentiated by their service rates and operation costs. The objective is to allocate customers either to an available server, which can be idle or partially loaded, or to the queue, at certain event epochs in order to minimize the long-run average cost per unit of time. We calculate the optimal control policy for the allocation of customers between two servers with PS and show that it exhibits a threshold structure. According to this policy the faster server accept for simultaneous service a maximal possible number of customers N_1. A slower server, on the contrary, is used only if the number of waiting customers reaches a threshold level q_k depending on the number of customers $k < N_2$ being served at the moment on the slower server. Although for this particular model it is sufficient to use the policy-iteration algorithm [9,12] to calculate the optimal thresholds, searching for optimal policy in more complicated systems using this algorithm can be time-consuming, and in some cases, for example in heavy traffic, has problems with a convergence. It motivates us for the second fold of the paper. We provide some simple heuristic solution (HS) based on a simple deterministic fluid approximation that performs well and can be accepted as sub-optimal policy in order to avoid the evaluation of the optimal one. Our study investigates also the possibility to use the data-sets from queueing system generated by existing classical methods as training data to provide prediction of the optimal thresholds with one of a supervised machine learning such as artificial neural networks (NN) [6,13,15]. The results from the NN are positive with a good accuracy and indicate that this approach could be used for predicting the optimal thresholds if the amount of data provided for the training phase is quite large.

The paper is organized as follows. In Sect. 2 we formulate the mathematical model and provide the MDP model formulation. Section 3 introduces a heuristic choice of thresholds for the usage of the slower server and demonstrates the combination of the policy-iteration algorithm with a neural network to predict the optimal thresholds. Finally, some concluding remarks are given in Sect. 4.

2 Mathematical Model and MDP Formulation

We propose an infinite-capacity $M/M/2$ queueing system with heterogeneous servers which serve the customers according to a processor sharing discipline, see Fig. 1. The customers arrive to the system according to a homogeneous Poisson process with a rate λ. The jth server can serve simultaneously up to N_j customers, each of them is served in exponentially distributed time with a service rate $\frac{\mu_j}{N_j}$, where $\mu_1 \geq \mu_2$. The server j is called an available server if its capacity does not exceed the admissible value N_j. The service of customers is assumed to be without preemption, i.e. a customer being served on a server can not change it. The inter-arrival and service times are mutually independent. The system costs include an operating cost $c_{j,1}, j = 1, 2$, per unit of time for busy server j, usage cost $c_{j,2}, j = 1, 2$, per unit of time for each customer being served on the jth server, and holding cost c_0 per unit of time for any customer waiting in the queue.

The controller or decision maker, which has a full information about system states, dispatchers customers either to one of available servers or to queue at a new arrival and service completion epoch if it occurs with a nonempty queue. The main goal is to minimize the long-run average cost per unit of time. The system dynamics is common for the systems with one queue and heterogeneous servers. At each arrival epoch the customer joins the queue and the controller can allocate the customer staying at the head of the queue to an available server j. At service completion epochs the controller may decide to allocate the customer from the head of nonempty queue to an available server or leave the customer in the queue.

We formulate the above optimization problem as a Markov decision problem and use dynamic programming approach to calculate the average cost under optimal control policy. The main idea consists in calculating of optimal policy by

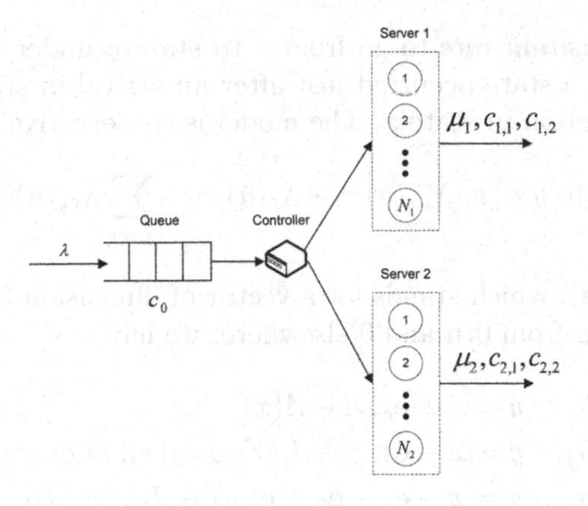

Fig. 1. Controllable two-server queueing system with processor sharing discipline.

constructing a sequence of improved policies that converges to the optimal one. It can be performed by solution of the system of Bellman's optimality equation [12,17]. Let $Q(t) \in E_Q \equiv \mathbb{N}_0$ denote the number of customers in the queue, $D_j(t) \in E_{D_j} = \{0, 1, \ldots, N_j\}, j \in \{1, 2\}$, the number of customers at server j. The system states at time t are described by a multi-dimensional continuous-time Markov chain

$$\{X(t)\}_{t \geq 0} = \{Q(t), D_1(t), D_2(t)\}_{t \geq 0}. \tag{1}$$

with a state space E_X defined as

$$E_X = E_Q \times E_{D_1} \times E_{D_2}. \tag{2}$$

A state $x \in E_X$ is a triplet of the form $x = (q(x), d_1(x), d_2(x))$, where the notations $q(x)$ and $d_j(x)$ will be used further in the paper to specify the components of the vector state $x \in E_X$. The following notations are used to specify respectively a set of idle, busy available and busy unavailable servers in state $x \in E_X$,

$$J_0(x) = \{j : d_j(x) = 0\},\ J_1(x) = \{j : d_j(x) \in E_{D_j} \setminus \{0, N_j\}\},\ J_2(x) = \{j : d_j(x) = N_j\}.$$

The controllable model associated with a Markov chain (1) is a five-tuple

$$\{E_X, A, A(x), \lambda_{xy}(a), c(x, a)\}, \tag{3}$$

where

- $A = \{0, 1, 2\}$ is an *action space* with elements $a \in A$, where $a \neq 0$ means "to allocate a customer to a server a" and $a = 0$ means "to allocate a customer to the queue".
- $A(x) = J_0(x) \cup J_1(x) \cup \{0\} \subseteq A$ denotes a *subset of admissible actions* in state $x \in E_X$.

- $\lambda_{xy}(a)$ is a *transition rate* to go from x to state y under a control action a associated with a state occurred just after an arrival in state x or just after a service completion in state x. The model is conservative, i.e.

$$\lambda_{xy}(a) \geq 0,\ y \neq x,\ \lambda_{xx}(a) = -\lambda_x(a) = -\sum_{y \neq x} \lambda_{xy}(a), \lambda_x(a) < \infty.$$

For $y \neq x$ and $\mathbf{e}_j$, which stands for a vector of dimension 3 with 1 in the jth position starting from 0th and 0 elsewhere, we have

$$\lambda_{xy}(a) = \begin{cases} \lambda, & y = x + \mathbf{e}_a,\ a \in A(x) \\ \mu_j, & y = x - \mathbf{e}_j,\ j \in J_1(x) \cup J_2(x),\ q(x) = 0, \\ \mu_j, & y = x - \mathbf{e}_j - \mathbf{e}_0 + \mathbf{e}_a,\ j \in J_1(x) \cup J_2(x),\ q(x) > 0, \\ & a \in A(x - \mathbf{e}_j - \mathbf{e}_0), \\ 0, & \text{otherwise}. \end{cases} \quad (4)$$

- $c(x,a)$ is an *immediate cost* in state $x \in E_X$ under a control action $a \in A(x)$

$$c(x,a) = c(x) = c_0 q(x) + \sum_{j \in J_1(x) \cup J_2(x)} c_{1,j} + \sum_{j=1}^{2} c_{2,j} d_j(x), \quad (5)$$

which is independent of a. The setting $c_0 = 1, c_{1,j} = 0, c_{2,j} = 1, j = 1, 2$, leads to the number of customers in state x.

A controller chooses an action according to the following decision rule which will refer to as *stationary policy.*

Definition 1. *A stationary policy is a function $f : E_X \to A(x)$ which prescribes a selection of a control action $a = f(x) \in A(x)$ whenever the process $\{X(t)\}_{t \geq 0}$ is in state $x \in E_X$ just after a new arrival or just after a service completion at busy server if $q(x) > 0$. Any other transition cannot impute a selection of some action.*

Remark 1. Since the maximum throughput rate of the system ist $\mu_1 + \mu_2$, the stability condition coincides with a condition for the uncontrollable ordinary $M/M/2$ queueing system, i.e.

$$\rho = \frac{\lambda}{\sum_{j=1}^{2} \mu_j} < 1. \quad (6)$$

For the infinite state Markov decision process with unbounded costs the existence of an optimal stationary policy and convergence of the sequence of improved policies to the optimal one under the average cost criterion and stability condition (6) must be verified. To do it we employ the main theorem in [19], by consecutive checking whether our model satisfies Assumptions 1,2,3, and 3* of [19].

For any fixed stationary policy f we wish to guarantee that the process $\{X(t)\}_{t\geq 0}$ is an irreducible, positive recurrent Markov chain with a state space E_X and infinitesimal generator Λ_X. As it is known, for ergodic Markov chains with costs the long-run average cost per unit of time for the policy f coincides with corresponding assemble average,

$$g^f = \limsup_{t\to\infty} \frac{1}{t} V^f(x,t) = \sum_{y\in E_X} c(y)\pi_y^f, \tag{7}$$

where

$$V^f(x,t) = \mathbb{E}^f\Big[\int_0^t \Big(c_0 Q(t) + \sum_{j=1}^{2}[c_{1,j}1_{\{D_j(t)>0\}} + c_{2,j}D_j(t)]\Big)dt|X(0)=x\Big]$$

denotes the *total average cost* up to time t given initial state is x and $\pi_y^f = \mathbb{P}^f[X(t)=y]$ denotes a stationary state probability of the process under given policy f. The policy f^* is said to be optimal when

$$g^* = \inf_f g^f. \tag{8}$$

One fruitful approach to finding optimal policy f^* is through solving the Bellman's average cost optimality equation

$$Bv(x) = v(x) + g, \tag{9}$$

where B is a *dynamic programming operator* acting on a relative value function $v: E_X \to \mathbb{R}$ which indicates a transient effect of an initial state x to the total average cost and

$$V^f(x,t) = g^f t + v^f(x) + o(1),\ x\in E,\ t\to\infty. \tag{10}$$

The functions v^f and g^f further in the paper will be denoted by v and g without upper index f.

Theorem 1. *The Bellman's optimality equation in uniformized model is defined as follows*

$$Bv(x) = c(x) + \lambda T_0 v(x) + \sum_{j\in J_1(x)\cup J_2(x)} \mu_j T_j v(x) + \sum_{j\in J_0(x)} \mu_j v(x), \tag{11}$$

where $T_j, j=0,1,2,$ are event operators defined as

$$T_0 v(x) = \min_{a\in A(x)} v(x+\mathbf{e}_a),\ T_j v(x) = \begin{cases} v(x-\mathbf{e}_j), & q(x)=0,\\ T_0 v(x-\mathbf{e}_j-\mathbf{e}_0), & q(x)>0. \end{cases} \tag{12}$$

Proof. According to [18] in general case we have

$$v(x) = \min_a \Big\{\frac{1}{\lambda_x(a)}\Big[c(x) + \sum_{y\neq x}\lambda_{xy}(a)v(y) - g\Big]\Big\}.$$

Evaluating these equations for analyzed queueing system and taking into account the transition rates of the specified Markov decision model we get

$$v(x) = \frac{1}{\lambda + \sum_{j \in J_1(x) \cup J_2(x)} \mu_j} \Big[c(x) + \lambda T_0 v(x) + \sum_{j \in J_1(x) \cup J_2(x)} \mu_j T_j v(x) - g \Big],$$

where the term $c(x)$ is an immediate cost in state $x \in E_X$, the second term represents the changing of the state accompanying with a new arrival which occurs with a rate λ. The third term represents transitions due to service completions at server j with a rate μ_j. Due to the transition rates (4) together with dummy transitions μ_j in states x with $j \in J_0(x)$, uniformization constant $\lambda + \sum_{j=1}^{2} \mu_j = 1$ and definition of the event operators (12), the last equation implies (11).

Algorithm 1. Policy-iteration algorithm

1: **procedure** PIA($\lambda, W, N_j, \mu_j, c_{j,k}, j, k = 1, 2, c_0$)
2: $f^{(0)}(x) = \text{argmax}_{j \in J_0(x) \cup J_1(x)} \{\mu_j\}$ ▷ Initial policy
3: $n \leftarrow 0$
4: $g^{(n)} = \lambda v^{(n)}(\mathbf{e}_1)$ ▷ Evaluation of value function
5: **for** $x = (0, 1, 0)$ **to** (B, N_1, N_2) **do**
6:

$$\begin{aligned} v^{(n)}(x) = & \frac{1}{\lambda + \sum_{j \in J_1(x) \cup J_2(x)} \mu_j} \Big[c(x) - g^{(n)} + \lambda v^{(n)}(x + \mathbf{e}_{f^{(n)}(x)}) \\ & + \sum_{j \in J_1(x) \cup J_2(x)} \mu_j v^{(n)}(x - \mathbf{e}_j) 1_{\{q(x)=0\}} \\ & + \sum_{j \in J_1(x) \cup J_2(x)} \mu_j v^{(n)}(x - \mathbf{e}_j - \mathbf{e}_0 + \mathbf{e}_{f^{(n)}(x - \mathbf{e}_j - \mathbf{e}_0)}) 1_{\{q(x)>0\}} \Big] \end{aligned}$$

7: **end for**
8: ▷ Policy improvement

$$f^{(n+1)}(x) = \text{argmin}_{a \in A(x)} v^{(n)}(x + \mathbf{e}_a)$$

9: **if** $f^{(n+1)}(x) = f^{(n)}(x)$, $x \in E_X$ **then return** $f^{(n+1)}(x), v^{(n)}(x), g^{(n)}$
10: **else** $n \leftarrow n + 1$, **go to step 4**
11: **end if**
12: **end procedure**

Remark 2. Optimization problem (8) is solved numerically using policy-iteration algorithm 1. The infinite buffer queueing system is approximated by a finite buffer equivalent system. For the bounded puffer size B the number of states is

$$|E_X| = (N_1 + 1)(N_2 + 1)(W + 1).$$

If the queue length $q \geq q_{N_2}$, the system behaves like a $M/M/1$ queueing system with a service rate $\mu_1 + \mu_2$ and the stationary state probabilities have geometric solution $\pi_q = \pi_{q_{N_2}} \rho^{q-q_{N_2}}, q \geq q_{N_2}$. For details and theoretical substantiation see e.g. [5]. Threshold level q_{N_2} can be estimated using HS (14). The buffer size B is chosen in such a way that it satisfies the condition for the loss probability

$$\sum_{q(y)=W}^{\infty} \pi_y = \pi_{q_{N_2}} \sum_{q(y)=W}^{\infty} \rho^{q(y)-q_{N_2}} \leq \sum_{q(y)=W}^{\infty} \rho^{q(y)-q_{N_2}} = \frac{\rho^{W-q_{N_2}}}{1-\rho} < \varepsilon$$

which after simple algebra implies

$$W > \frac{\log \varepsilon(1-\rho)}{\log(\rho)} + q_{N_2}.$$

Remark 3. We convert the three-dimensional state space E_X of the Markov decision process ordered in a certain way to a one-dimensional equivalent state space $\mathbb{N}_0$, $\Delta : E_X \to \mathbb{N}_0$, for state $x = (q(x), d_1(x), d_2(x)) \in E_X$,

$$\Delta(x) = q(x) \prod_{j=1}^{2} (N_j + 1) + d_1(x) + d_2(x)(N_1 + 1). \tag{13}$$

Therefore, in one-dimensional case the changing of the state x can be represented in the form,

$$\begin{aligned} \Delta(x \pm \mathbf{e}_0) &= \Delta(x) \pm \prod_{j=1}^{2} (N_i + 1), \\ \Delta(x \pm \mathbf{e}_1) &= \Delta(x) \pm 1, \\ \Delta(x \pm \mathbf{e}_2) &= \Delta(x) \pm (N_1 + 1). \end{aligned}$$

The policy-iteration algorithm 1 allows to construct a sequence of improved policies until the average cost optimal is reached. In this algorithm the vector states are transformed by a function (13), d.h. $x \mapsto \Delta(x), x \in E_X$. The policy-iteration algorithm consists of two main parts. In first part, for the given policy $f^{(n)}(x)$ of the nth iteration the relative value function $v^{(n)}(x), x \in E_X$, and the gain $g^{(n)}$ must be calculated as solution of the system of linear equations (9) assuming $v^{(n)}(0) = 0$. In second part, the improved policy $f^{(n+1)}(x)$ must be evaluated by minimizing the function $v^{(n)}(x + \mathbf{e}_a)$ in any state $x \in E_X$ over the control action $a \in A(x)$.

Example 1. Consider a queueing system with $\lambda = 20$, $N_1 = N_2 = 10$ and

j, k	0	1,1	1,2	2,1	2,2
$c_0, c_{j,k}$	3	10	0.2	5	0.1
μ_j	-	50		5	

The buffer size is $W = 80$ which for $\varepsilon = 0.0001$ guaranties that

$$W > \frac{\log 0.0001(1 - 20/55)}{\log(20/55)} + 9 = 18.$$

The table of evaluated control actions for selected system states is of the form:

States x	Queue length $q(x)$													
(d_1, d_2)	0	1	2	3	4	5	6	7	8	9	10	11	12	...
$(N_1, 0)$	0	0	0	0	0	0	0	$\underline{2}$	2	2	2	2	2	2
$(N_1, 2)$	0	0	0	0	0	0	0	0	$\underline{2}$	2	2	2	2	2
$(N_1, 8)$	0	0	0	0	0	0	0	0	0	$\underline{2}$	2	2	2	2

Threshold levels q_k, $k = 1, \ldots, 10$, can be evaluated by comparing the optimal actions $f(q, N_1, k-1) < f(q+1, N_1, k-1)$ for $q = 0, \ldots, W-1$. This example confirms the fact that the optimal allocation policy is of threshold type. In this example the optimal policy f^* is defined here through a sequence of threshold levels $(7, 7, 8, 8, 8, 8, 8, 8, 9, 9)$ and $g^* = 4.13$.

3 Estimation and Prediction for Optimal Thresholds

In this section we derive a heuristic solution to estimate threshold level q_k using a fluid approximation and combine policy-iteration algorithm with a multilayer neural networks to predict optimal thresholds for any values of system parameters. The fluid approximation is illustrated in Fig. 2. Assume that q_k is an optimal threshold to allocate the customer to server 2 in state $(q_k - 1, N_1, k-1)$, $k = 1, \ldots, N_2$. Now we compare the queues of the system

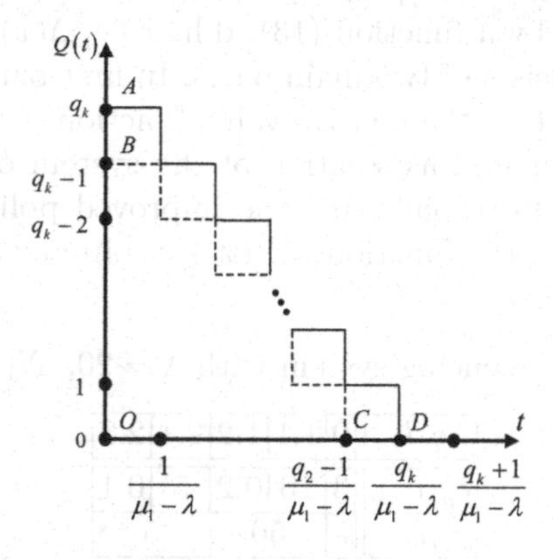

Fig. 2. Fluid approximation.

given initial states are $(q_k, N_1, k-1)$, where the second server is not used for a new customer, and $(q_k - 1, N_1, k)$, where the number of customers served at server 2 has increased by one. In Fig. 2, the queue lengths are labeled by $A = q_k$ and $B = q_k - 1$. If the queue dynamics corresponded to the deterministic fluid, it would decrease at the rate $\mu_1 - \lambda$. When this rate is maintained until the queue is empty, it occurs respectively at points $D = \frac{q_k}{\mu_1 - \lambda}$ and $C = \frac{q_k - 1}{\mu_1 - \lambda}$. The total holding times of customers in a queue with lengths q_k and $q_k - 1$ are equal obviously to the areas

$$F_{AOD} = \frac{q_k(q_k+1)}{2(\mu_1 - \lambda)} \quad \text{and} \quad F_{BOC} = \frac{q_k(q_k-1)}{2(\mu_1 - \lambda)}$$

of triangles AOD and BOC. The operating and usage costs at server 1 until the queue is empty starting from state $(q_k, N_1, k-1)$ are equal to $\frac{q_k(c_{1,1}+N_1 c_{1,2})}{\mu_1}$ and starting from state $(q_k - 1, N_1, k)$ – are equal to $\frac{(q_k-1)(c_{1,1}+N_1 c_{1,2})}{\mu_1}$. According to a deterministic fluid schema we formulate

Proposition 1. *The optimal thresholds $q_k, k = 1, \ldots, N_2$, are defined by*

$$q_k \approx \hat{q}_k = \min\left\{1, \left\lfloor \frac{\mu_1 - \lambda}{c_0}\left[\frac{c_{2,1} + k\, c_{2,2}}{\mu_2} - \frac{c_{1,1} + N_1 c_{1,2}}{\mu_1}\right]\right\rfloor + 1\right\}. \quad (14)$$

Proof. Denote by $V(x)$ the overall average system cost until the system is empty given initial state is $x \in E$. The decision to perform the allocation to server 2 by arriving a new customer to state $(q_k - 1, N_1, k-1)$ must lead to a reduction of the overall system costs under deterministic fluid schema, i.e.

$$V(q_k, N_1, k-1) - V(q_k - 1, N_1, k) > 0. \quad (15)$$

where

$$V(q_k, N_1, k-1) = c_0 F_{AOD} + \frac{q_k(c_{1,1} + N_1 c_{1,2})}{\mu_1} + V(0, N_1, k-1), \quad (16)$$

$$\begin{aligned} V(q_k - 1, N_1, k) &= \frac{c_{2,1} + k c_{2,2}}{\mu_2} + V(q_k - 1, N_1, k-1) \\ &= \frac{c_{2,1} + k c_{2,2}}{\mu_2} + c_0 F_{BOC} + \frac{(q_k - 1)(c_{1,1} + N_1 c_{1,2})}{\mu_1} \\ &\quad + V(0, N_1, k-1). \end{aligned}$$

After substitution of (16) into (15) and some simple manipulations we get that the heuristic solution for the optimal threshold q_k is defined then as the integer larger as 1 and the smallest integer (14) satisfying the inequality (15).

Example 2. Consider the queueing system with $N_1 = N_2 = 10$. We generate a data-set using the policy-iteration algorithm in form of the following list

$$\begin{aligned} S =& \{(\lambda, \mu_j, c_0, c_{j,k}) \to (q_1, q_5, q_{10}) : \lambda \in [1, 45], \mu_1 \in [\lambda, 50], \mu_2 \in [1, \mu_1], \\ & c_0 \in [0.5, 1.5], c_{1,1}, c_{2,1} \in [1, 10], c_{1,2}, c_{2,2} \in [0.1, 0.5]\}. \end{aligned} \quad (17)$$

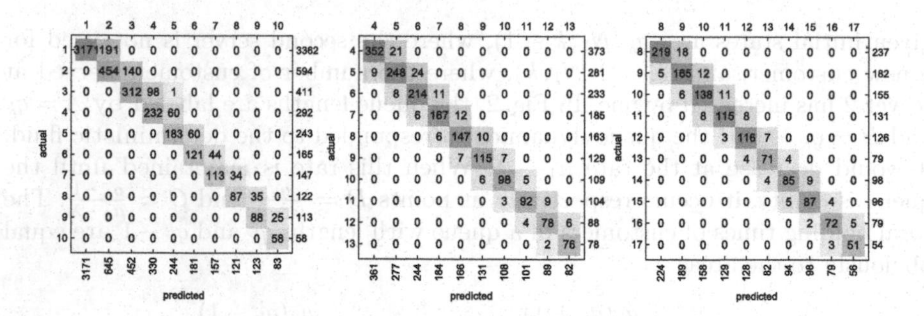

Fig. 3. Confusion matrices for prediction using HS

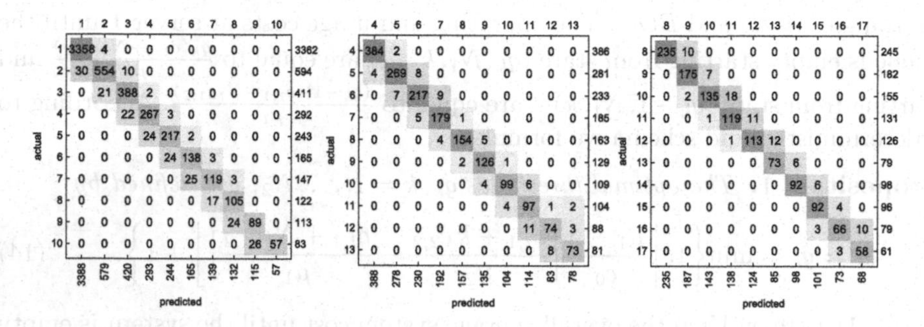

Fig. 4. Confusion matrices for prediction using NN

The parameters λ, μ_j and $c_{j,1}, j = 1,2$, are varied with a lag 1, c_0 and $c_{j,2}, j = 1,2$, – with a lag 0.1. To accelerate the data generation using the policy-iteration algorithm, a heuristic solution (14) can be used as an initial control policy. We select randomly from the list (17) the system parameters and evaluate with HS the corresponding thresholds. Confusion matrices in Fig. 3 visualize the performance of proposed heuristics respectively for the threshold levels (q_1, q_5, q_{10}). Each row of these matrices represents the instances in a predicted value while each column represents the instances in an actual value. The overall accuracies for each threshold q_k, which represent closeness of the measurements to a specific value and are calculated through the ratio of correct predictions to total predictions, are summarized respectively for $k = 1, \ldots, 5$ in Table 1 and for $k = 6, \ldots, 10$ in Table 2.

Table 1. Accuracy of classification for threshold levels $q_k, k = 1, \ldots, 5$.

Method	q_1	q_2	q_3	q_4	q_5
HS	0.8711	0.8912	0.8945	0.9001	0.9022
NN	0.9566	0.9565	0.9554	0.9516	0.9505

Table 2. Accuracy of classification for threshold levels $q_k, k = 6, \ldots, 10$.

Method	q_6	q_7	q_8	q_9	q_{10}
HS	0.9137	0.9015	0.9124	0.9003	0.8938
NN	0.9501	0.9413	0.9401	0.9336	0.9249

The data-set S generated in Example 2 can be used to explore predictions for the optimal threshold levels through a supervised learning such as artificial neural networks. 70% of S which was not used for HS is referred to as training data and the rest of S – as validation data. We train a multilayer (6-layer) NN using an adaptive moment estimation method [11] and NN-toolbox integrated in *Mathematica*© program of the Wolfram Research. Then we verify the approximated function $\hat{q}_k := \hat{q}_k(\lambda, \mu_j, c_0, c_{j,k})$ which should be accurate enough to be used to predict new output from verification data.

Example 3. The results of predictions are summarized in confusion matrices shown in Fig. 4. The overall accuracy of classification is given in Table 1. We can see that the NN methodology exhibits more accurate predictions for the optimal thresholds. Therefore we may conclude that classical system analysis can and must be supplemented and extended by more active use of machine learning technologies.

4 Conclusion

In this paper we have confirmed that the processor sharing in heterogeneous queueing system preserves threshold structure of the optimal allocation policy, but now threshold level for the usage of slower server depends on a state of this server. When analyzing and comparing the results obtained by algorithms of the Markov decision theory and by supervised learning we may conclude that these methodologies can be seen as complementary rather than competitive.

Acknowledgements. The publication has been prepared with the support of the "RUDN University Program 5–100". (recipients D. Efrosinin and V. Rykov, mathematical model development). The reported study was funded by RFBR, project number 19-29-06043 mk.

References

1. Aalto S., et al.: Beyond processor sharing. In: ACM SIGMETRICS Performance Evaluation Review, vol. 34 (2007)
2. Brandt, A., Brandt, M.: Waiting times for M/M systems under state-dependent processor sharing. Queueing Syst. **59**, 297–319 (2008)
3. Dudin, A.N., et al.: Analysis of queueing model with processor sharing discipline and customers impatience. Oper. Res. Perspect. **5**, 245–255 (2018)

4. Efrosinin, D.: Controlled queueing systems with heterogeneous servers: Dynamic optimization and monotonicity properties of optimal control policies in multiserver heterogeneous queues. VDM Verlag (2008)
5. Efrosinin, D., Sztrik, J.: An algorithmic approach to analyzing the reliability of a controllable unreliable queue with two heterogeneous servers. Eur. J. Oper. Res. **271**, 934–952 (2018)
6. Gershenson, C.: Artificial neural networks for beginners (2003). http://arxiv.org/abs/cs/0308031
7. Guillemin, F., Boyer, J.: Analysis of the $M/M/1$ queue with processor sharing via spectral theory. Queueing Syst. **39**, 377–397 (2001)
8. Guillemin, F., Robert, P., Zwart, B.: Tail asymptotics for processor-sharing queues. Adv. Appl. Prob. **36**, 525–543 (2004)
9. Howard, R.: Dynamic Programming and Markov Processes. Wiley, Hoboken (1960)
10. Özkan, E., Kharoufeh, J.P.: Optimal control of a two-server queueing system with failures. Prob. Eng. Inf. Sci. **28**, 489–527 (2014)
11. Kingma, D.P., Ba, J.L.: Adam: A Method for stochastic optimization (2015). https://arxiv.org/abs/1412.6980
12. Puterman M.L.: Markov Decision Process. Wiley series in Probability and Mathematical Statistics (1994)
13. Rätsch, G.: A brief introduction into machine learning. Friedrich Miescher Laboratory of the Max Planck Society (2004)
14. Roberts, J., Massoulie, L.: Bandwidth sharing and admission control for elastic traffic. In: Proceedings of Infocom 1999 (1999)
15. Russel, S.J., Norvig, P.: Artificial Intelligence: A Modern Approach. Prentice-Hall Inc., Upper Saddle River (1995)
16. Rykov, V., Efrosinin, D.: On the slow server problem. Autom. Remote Control **70**, 2013–2023 (2010)
17. Rykov, V.: Monotone control of queueing systems with heterogeneous servers. Queueing Syst. **37**, 391–403 (2001)
18. Rykov, V.V.: Controlled queueing systems (in Russian). Itogi Nauki i Techniki. Serie Probability Theory. Math. Stat. Theor. Cybern. **12**, 43–153 (1975)
19. Sennott, L.I.: Average cost optimal stationary policies in infinite state Markov decision processes with unbounded costs. Oper. Res. **37**, 626–633 (1989)
20. Yang, R., Bhulai, S., van der Mei, R.: Structural properties of the optimal resource allocation policy for single-queue systems. Ann. Oper. Res. **202**, 211–233 (2013)
21. Yashkov, S.F., Yashkova, A.S.: Processor sharing: a survey of the mathematical theory. Autom. Remote Control **68**, 1662–1731 (2007)

On Exponential Convergence of Dynamic Queueing Network and Its Applications

Elmira Yu. Kalimulina(✉)

Institute of Control Sciences of Russian Academy of Sciences, Moscow, Russia
elmira.yu.k@gmail.com
https://www.ipu.ru/staff/elmira

Abstract. This paper is a continuation of previous research in ergodicity of some models for unreliable networks. The set of random graphs and the sequence of matrixes describing the failure and recovery process has been used instead of the fixed graph for network structure. The main results about an ergodicity and bounds for rate of convergence to stationary distribution are formulated under more general assumptions on intensity rates.

Keywords: Ergodicity · Stochastic networks · Convergence rate · Spectral gap

1 Introduction

Let's remind what the standard queueing network (Jackson's type) is. The standard queueing network is the network with following parameters (see Fig. 1) [2,9]:

- the network consists of m nodes, $M = \{1, 2, \dots, m\}$;
- each node is a multi-server system with an infinite waiting room;
- the algorithm of service is FCFS (First Come First Served);
- all customers are supposed to be indistinguishable;
- there is an external Poisson arrival flow with intensity Λ (only the open queueing network is considered in this research);
- denote the routing matrix as $R = (r_{ij}), i, j = 0, 1, \dots, m$; without loss of generality R is supposed to be regular;
- denote the traffic vector as $\lambda = (\lambda_0, \lambda_1, \dots, \lambda_m)$;
- denote service rates as $\mu = (\mu_1(n_1), \dots, \mu_m(n_m))$;
- the number of customers in the system is denoted as $\mathbf{n} = (n_1, \dots, n_m)$.

The state space for stochastic process describing this system is following:

$$\mathbf{n} = (n_1, , n_2, \cdots, n_m) \in \mathbf{Z}_+^m = \mathbf{E}, \tag{1}$$

The publication has been prepared with the support of the Russian Foundation for Basic Research according to the research project No.20-01-00575 A.

V. M. Vishnevskiy et al. (Eds.): DCCN 2020, LNCS 12563, pp. 463–474, 2020.
https://doi.org/10.1007/978-3-030-66471-8_35

and with the following transitions:

$$T_{ij}\mathbf{n} = (n_1, \cdots, n_i - 1, \cdots, n_j + 1, \cdots, n_m),$$
$$T_{0j}\mathbf{n} = (n_1, \cdots, n_j + 1, \cdots, n_m),$$
$$T_{i0}\mathbf{n} = (n_1, \cdots, n_i - 1, \cdots, n_m).$$

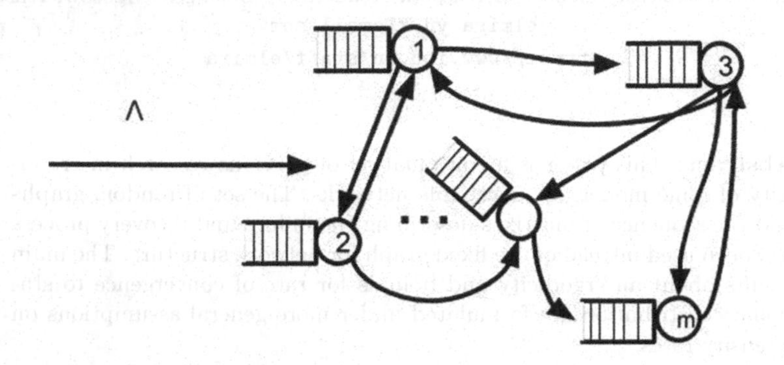

Fig. 1. Standard queueing network.

The important problem for such a network is the existence of a limit distribution and the rate of convergence to it. These problem are well studied by many researchers. One of the well-known results can be found in [9–11].

We are interested to study some modification of this standard model. The general motivation for our research is the real systems modelling such as transport networks, computer networks, telecommunication traffic models and etc. One of the key feature of these real systems is a changing structure. These systems are well described (in some approach) by queuing systems and networks models. But the changing structure (due unreliable nodes or part-time regime of operation) demands some modification of the standard approach. The problem with the standard approach is that the classical models don't include parameters specific for real systems. They are more complicated than standard queueing networks models.

So the following modification of a standard model described above (unreliable network) is considered here [3,4]:

- each node may switch on/off (ex. break down and repair) with intensities α_i, β_i, $i = 1, ..., m$;
- a dynamic routing is being applied as a failure management mechanisms.

The principle of "dynamic routing" is in selecting the alternative node if the target node is under failure. The alternative node is selected from the nearest to the failed one. This modification make this model different from another similar ones.

There are several alternative failure management mechanisms: one of them is "blocking" (before service and after service), for details see [7,8]. The approach suggested here is more specious for real systems, but it demands the more complicated random process to be considered.

2 Dynamic Routing

The "Dynamic routing" failure management mechanisms results the extended state space by adding some component to standard state space of the process. The Fig. 2 shows the initial structure of the network. The standard approach implies this graph to be fixed. Recoveries and failures form a new process that describes the transformation of this graph to another in the suggested model.

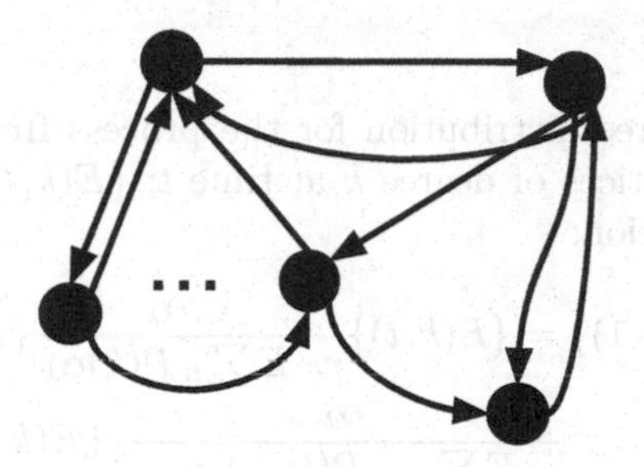

Fig. 2. Initial structure of the network.

The number of nodes is fixed, new nodes don't appear in contrast with growing networks (see [15,16]).

In our model nodes can be blocked (by deleting/adding edges to it). This way of transformation is shown on Fig. 3.

The nodes marked with red color are under failure, the transition forward and back from one graph (with working node i) to another (when the node is under failure) occurs with intensities α_i (failure rate) and β_i (recovery rate). This way of graph transformation generates the Markov process with a finite state space (because the number of nodes is finite).

We denote the state space of the graph transformation process as the set G: the node i is "removed" with some intensity α_i (failure rate for this node) or it can be restored with some intensity β_i.

So, the standard state space (1) for our network process is extended by adding the component G and is following:

$$\tilde{\mathbf{n}} = (G, n_1, n_2, ..., n_m) \in |G| \times \mathbf{Z}_+^m =: \mathbf{E},$$

where G is a component describing the graph (or transition matrix) transformation.

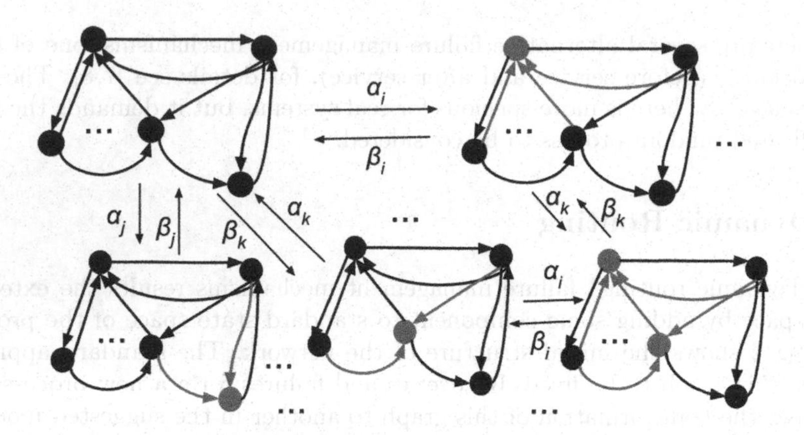

Fig. 3. The network graph evolution process. (Color figure online)

We can find a degree distribution for the process from state space G. The average number of vertices of degree k at time t: $\{E(k,t)\} = E\ P(k,t)$ can be described by the equation:

$$\{E(k,t+1)\} = \{E(k,t)\} - \frac{\alpha_k}{E\sum_k P(k)\alpha_k}\{E(k,t)\} + \frac{\alpha_{k-1}}{E\sum_{k-1} P(k-1)\alpha_{k-1}}\{E(k-1,t)\} + \frac{\alpha_{k+1}}{E\sum_{k+1} P(k+1)\alpha_{k+1}}\{E(k+1,t)\}.$$

It describes the evolution of graph of our network structure in time and for the continuous time takes the form:

$$E\frac{\partial P(k,t)}{\partial t} = -\alpha_k P(k,t) + \alpha_{k-1}P(k-1,t) + P(k+1,t) + \alpha_{k+1}P(k+1,t). \tag{2}$$

Is easy to see for this equation that (2) is linear homogeneous equation (under assumption of constant failure and recovery rates) and has a stationary solution:

$$P(k) = \lim_{t\to\infty} P(k,t). \tag{3}$$

3 Main Results

3.1 Convergence of Process $X_S(t)$

The state for this network process is described by the following vector

$$\boldsymbol{n} = ((n_1,s_1),(n_2,s_2),...,(n_m,s_m)),$$

where n_i – the number of customers at the i-th node and

$$s_i = \begin{cases} 0, \text{if the } i\text{th node works}, 1, \text{otherwise.} \end{cases}$$

The behaviour of $\boldsymbol{n}$ is a Markov chain in continuous time. It includes an embedded homogeneous Markov chain with positive probabilities for transitions:

$$\begin{aligned} s_i &\longrightarrow (1 - s_i), \\ n_i &\longrightarrow (n_i \pm 1). \end{aligned} \tag{4}$$

Exponential convergence of reliability process $\boldsymbol{S} = (s_1, \ldots, s_m)$ converges to stationary distribution with exponential rate.

Let's consider the reliability process $X_S(t)$ of our model separately.

$$\{X_{S_i}(t+1) = X_{S_i}(t)\} = \frac{\sum\limits_{j=1}^{m} \gamma_j - \gamma_i}{\sum\limits_{j=1}^{m} \gamma_j},$$

$$\{X_{S_i}(t+1) = 1 - X_{S_i}(t)\} = \frac{\gamma_i}{\sum\limits_{j=1}^{m} \gamma_j},$$

where

$$\gamma_i = \alpha_i \mathbf{1}\{s_i = 0\} + \beta_i \mathbf{1}\{s_i = 1\}.$$

3.2 Convergence of Process $X_R(t)$

The behaviour of the process $X_R(t)$ is defined by the process $X_S(t)$ with the same transition probabilities. It takes values from the finite set $(R = \|r_{ij}(t)\|)$, so $X_R(t)$ has the stationary distribution and converges to it exponentially. The sequence of $R = \|r_{ij}(t)\|$ has a limit $\tilde{R} = \|\tilde{r}_{ij}\|$, where $\tilde{r}_{ij}$ are dependent random variables.

Processes $X_S(t)$ and $X_R(t)$ describe only reliability of our network. At this moment we still haven't took into consideration the service process and an input flow, that are our main interest of studying.

But they are ergodic and don't depend on the input flow and service process (in further we will apply these facts).

Definition of the Main Network Process. The process has the following state space:

$$\tilde{\mathbf{n}} = (G, n_1, n_2, \ldots, n_m) \in G \times \mathbf{Z}_+^m =: \mathbf{E}$$

The following transitions in a network are possible:

$$\begin{aligned} T_{ij}\tilde{\mathbf{n}} &:= (G, n_1, \ldots, n_i - 1, \ldots, n_j + 1 \ldots, n_m), \\ T_{0j}\tilde{\mathbf{n}} &:= (G, n_1, \ldots, n_j + 1, \ldots, n_m), \\ T_{i0}\tilde{\mathbf{n}} &:= (G, n_1, \ldots, n_i - 1, \ldots, n_m), \\ T_f\tilde{\mathbf{n}} &:= (G^+, n_1, \ldots, n_m), \\ T_r\tilde{\mathbf{n}} &:= (G^-, n_1, \ldots, n_m). \end{aligned}$$

Definition. We will call a "dynamic routing network" the process

$$\mathbf{X} = (X(t), t \geq 0)$$

defined by the following infinitesimal generator:

$$\mathbf{Q}f(\mathbf{n}) = \sum_{i=1}^{m}\sum_{j=1}^{m}(f(T_{0j}\mathbf{n}) - f(\mathbf{n}))\lambda_i r_{ij} + \sum_{i=1}^{m}\sum_{j=1}^{m}(f(T_{ij}\mathbf{n}) - f(\mathbf{n}))\mu_i(n_i) r_{ij} + \sum_{k\in G^+}(f(T_k\mathbf{n}) - f(\mathbf{n}))\alpha_k + \sum_{k\in G\setminus G^+}(f(T_k\mathbf{n}) - f(\mathbf{n}))\beta_k + \sum_{i=1}^{m}(f(T_{i0}\mathbf{n}) - f(\mathbf{n}))\mu_i(n_i) r_{i0}, \quad (5)$$

where $\lambda = (\lambda_0, \lambda_1, \ldots, \lambda_m)$ satisfies the balance equations.

Suppose the infinitesimal generator (5) satisfies the following **assumptions**:

1. $$\inf_{\mathbf{n},i}\sum_{i=1}^{m}\frac{\alpha_i\mu_i(\mathbf{n})}{\alpha_i + \beta_i} > \Lambda;$$
2. $\tilde{R} = \|\tilde{r}_{ij}\|$ is irreducible, so the expectation of steps visited by one customer within the network is finite.

The second condition may be checked for $R(t)$ under large t. The convergence rate of $R(t)$ may be estimated from the Markov-Doeblin condition (see, e.g. Doeblin, 1938 [14]).

The second condition guarantees the existence on non-zero values for the traffic vector $\lambda = (\lambda_0, \lambda_1, \ldots, \lambda_m)$. It leads every customer to leave the system with non-zero probability. So the number of nodes each customer visited within the network is less than some geometrically distributed random variable and has a finite expectation.

Some Notations for Network Process. If $\mathbf{X} = (X_t, t \geq 0)$ is a Markov process, the following notations will be used:
$Q = [q(\mathbf{e}, \mathbf{e}')]_{\mathbf{e},\mathbf{e}' \in \mathbf{E}}$ – transition intensities;
π – stationary distribution;
infinitesimal generator:

$$\mathbf{Q}f(\mathbf{e}) = \sum_{\mathbf{e}' \in \mathbf{E}} (f(\mathbf{e}') - f(\mathbf{e}))q(\mathbf{e}, \mathbf{e}');$$

the scalar product for some functions f and g on $L_2(\mathbf{E}, \pi)$:

$$\langle f, g \rangle_{pi} = \sum_{\mathbf{e} \in \mathbf{E}} f(\mathbf{e})g(\mathbf{e})\pi(\mathbf{e}). \tag{6}$$

Spectral gap for $\mathbf{X}$ [1,6]:

$$Gap(\mathbf{Q}) = \inf\{-\langle f, \mathbf{Q}f \rangle_\pi : \|f\|_2 = 1, \langle f, \mathbf{1} \rangle_\pi = 0\} \tag{7}$$

Theorem 1. *If $\mathbf{X}$ - the "dynamic routing network" process, with $\mathbf{Q}$ - infinitesimal generator (suppose bounded), minimal service and recovery intensities $\mu > 0$ and $\beta > 0$, and assumptions satisfy (1–2), then*

$$Gap(\mathbf{Q}) > 0$$

iff for each $i = 1, \ldots, m$, the birth and death process with λ_i, $\mu_i(n_i)$, α_i, β_i, has $Gap_i(\mathbf{Q}_i) > 0$.

Theorem 2. *If $\mathbf{X}$ - the "dynamic routing network" process with infinitesimal generator $\mathbf{Q}$ (suppose bounded), minimal service and recovery intensities $\mu > 0$ and $\beta > 0$, $X(t)$ satisfies the assumptions (1–2), then*

$$Gap(\mathbf{Q}) > 0$$

iff for each $i = 1, \ldots, m$, distribution $\pi = (\pi_i), i \geq 0$ is strongly light-tailed, i.e.

$$\inf_k \frac{\pi_i(k)}{\sum_{j>k} \pi_i(j)} > 0.$$

Theorem 3. *Let $\mathbf{X}$ - the "dynamic routing network" process with generator $\mathbf{Q}$ (given above) and the corresponding transition semigroup P_t, with minimal service and recovery intensities $\mu > 0$ and $\beta > 0$, and $X(t)$ satisfies the assumptions (1–2). Suppose that G satisfies the condition (2). If π_i is strongly light-tailed, for each $i = 1, \cdots, m$, then following statements are equivalent*

– *for all $f \in L_2(\mathbf{E}, \pi)$*

$$\|P_t f - \pi(f)\|_2 \leq e^{-Gap(\mathbf{Q})t} \|f - \pi(f)\|_2, t > 0, \tag{8}$$

– *for each* $\mathbf{e} \in \mathbf{E}$ *there exists* $C(\mathbf{e}) > 0$ *such that*

$$\|\delta_{\mathbf{e}} - \pi(f)\|_{TV} \leq C(\mathbf{e})e^{-Gap(\mathbf{Q})t}, t > 0. \quad (9)$$

Proof. The proofs of these results are based on the standard techniques developed by T. Ligget and extended for queueing systems by other researchers [6,12,13]. There are two main results from Liggett [6]:

– Assume that $\mathbf{Z}$ is a birth and death process on $\mathbf{Z}_+$ with state independent birth rates $\mu > 0$, and possibly state dependent death rates $\mu(n) > 0$, and for all $i \geq 0$, and for some $b, c > 0$, we have

$$\sum_{j>i} \pi(i) \leq c\pi(i)\lambda$$

and

$$\sum_{j>i} \pi(i) \leq b\pi(i).$$

Then for the corresponding generator $\mathbf{Q}$

$$Gap(\mathbf{Q}) \geq \frac{(\sqrt{b+1} + \sqrt{b})^2}{c} \geq \frac{1}{2c(1+2b)}.$$

– Suppose that $\mathbf{X}$ is a Markov process with generator $\mathbf{Q}$ and stationary distribution π evolves on the product state space

$$\mathbf{E} = \mathbf{E}_1 \times \mathbf{E}_2 \times \ldots \mathbf{E}_m, \, m \geq 1,$$

having coordinates which are independent Markov processes such that i-th coordinate has generator $\mathbf{Q}_i$, denumerable state space $\mathbf{E}_i$ and invariant probability measure π_i.
Then π is the product measure of π_i and

$$Gap(\mathbf{Q}) = \inf_i Gap(\mathbf{Q}_i).$$

Consider generators $\hat{\mathbf{Q}}, \hat{\mathbf{Q}}_i, \hat{\mathbf{Q}}_0, i = 1, \ldots, m$ associated with independent processes $(\hat{\mathbf{X}}_{0t}, \hat{\mathbf{X}}_t), \hat{\mathbf{X}}_{0t}, \hat{\mathbf{X}}_t$ describing the evolution of each node of our "dynamic routing network" separately. The process $\hat{\mathbf{X}}_{0t}$ defined on the finite state space, the stationary distribution π_0 and $Gap(\hat{\mathbf{Q}}_0) > 0$. From Liggett's results we may conclude that

$$Gap(\hat{\mathbf{Q}}) = \min_i Gap(\hat{\mathbf{Q}}_i)$$

and so

$$Gap(\hat{\mathbf{Q}}) > 0.$$

3.3 Rate of Convergence

The next important result relates to the convergence rate of the process $\mathbf{X}$ and is a consequence of Mu-Fa Chen results (see [5]).

Theorem 4. *Let* $\mathbf{X}$ *- the "dynamic routing network" process with generator* $\mathbf{Q}$ *(given above) and the corresponding transition semigroup* P_t*, then the classical variations formula holds*

$$Gap(Q) = \inf\{-\langle f, \mathbf{Q}\rangle_\pi : \pi(f) = 0, \|f\|_2 = 1\}$$

where

$$\langle f, g\rangle = \int f(x)g(x)\pi(dx),$$

$$\pi(f) = \int f(x)\pi(dx), Var_\pi(f) = \pi(f^2) - (\pi(f))^2,$$

and π *is invariant for* (P_t)*.*

Let $f \in L_2(\mathbf{E}, \pi)$*, then*

$$C = Gap(\mathbf{Q})^{-1}$$

is optimal in Poincare inequality:

$$Var_\pi(f) \leq C - \langle f, \mathbf{Q}\rangle_\pi.$$

Proof. The above result for our network is a consequence from two theorems for Markov process from [5]. The first one is a Poincare inequality, the second one is the theorem about constant C existence:

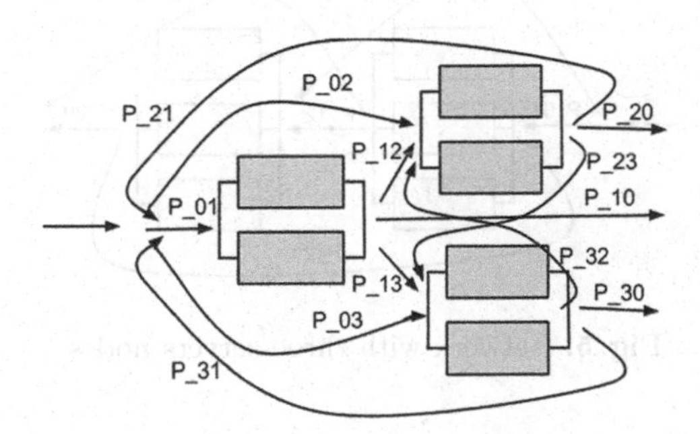

Fig. 4. Network with two-servers nodes

- Poincare inequality holds if and only if Markov process converges according to $L_2(\mathbf{E}, \pi)$-exponential convergence: for all

$$f \in L_2(\mathbf{E}, \pi)$$

$$\|P_t f - f\|_2^2 = Var(P_t f) \leq Var(f)exp(-2Gap(\mathbf{Q})t), \quad t > 0;$$

- Suppose that $\mathbf{E}$ is countable and P_t reversible.
  Then for all $f \in L_2(\mathbf{E}, \pi)$

$$\|P_t f - \pi(f)\|_2^2 = Var(P_t f) \leq Var(f)\exp(-2Gap(\mathbf{Q})t), \quad t > 0,$$

  iff for each $\mathbf{e} \in \mathbf{E}$ there exists $C(\mathbf{e}) > 0$ such that

$$\|\delta_e P_t - \pi\|_{tv} \leq C(\mathbf{e})\exp(-2Gap(\mathbf{Q})t), t > 0.$$

  So we can show that

$$C = Gap(\mathbf{Q})^{-1}$$

and from Poincare inequality for general Markov process:

$$Var_\pi(f) \leq C - \langle f, \mathbf{Q} \rangle_\pi.$$

4 The Numerical Example

We consider two numerical examples of network state probabilities calculation from [4]:

Example 1: The network consists of three nodes, each node is a system with two servers (see Fig. 4).

Example 2: The network consists of two nodes, each node is a system with three servers (see Fig. 5).

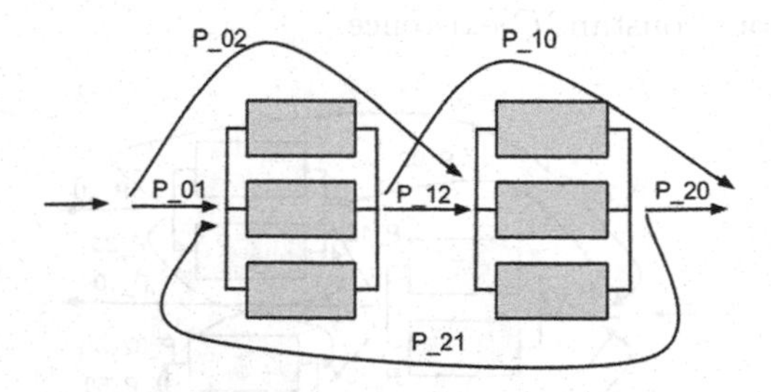

Fig. 5. Network with three-servers nodes

For the transition probabilities matrix (the same from [4]):

$$P_{ij} = \begin{pmatrix} 0.03 & 0.57 & 0.35 & 0.05 \\ 0.1 & 0.002 & 0.398 & 0.5 \\ 0.35 & 0.25 & 0.15 & 0.25 \\ 0.2 & 0.25 & 0.3 & 0.25, \end{pmatrix}$$

we suppose following failure rates α_i:

$$\alpha_0 = \alpha_1 = \alpha_2 = 3.0,$$

and recovery rates β_i:

$$\beta_0 = \beta_1 = \beta_2 = 6.0.$$

The obtained results for characteristics of this network are: - The probability of denial of service (the probability that all sites are occupied) = 0.0054; - Availability factor of the system (the system is completely free) = 0.29.

4.1 Further Work and Conclusion

The bounds derived above are valid only for light-tailed distribution. The convergence rates estimations for heavy-tailed distribution of service may be received via more complicated technique such as coupling method [17] and the generalized Lorden's inequality [18,19], but only under large t. The future plan is to find polynomial bounds via this approach. This bounds will be valid only for large $t > T$, where T is computable.

References

1. Liggett, T.H. Interacting Particle Systems. Springer, New York (1999). https://doi.org/10.1007/978-1-4613-8542-4
2. van Doorn, E.A.: Representations for the rate of convergence of birth-death processes. Memorandum / Department of Applied Mathematics; No. 1584. University of Twente, Department of Applied Mathematics (2001)
3. Kalimulina, E.Y.: Analysis of unreliable open queueing network with dynamic routing. In: Vishnevskiy, V.M., Samouylov, K.E., Kozyrev, D.V. (eds.) DCCN 2017. CCIS, vol. 700, pp. 355–367. Springer, Cham (2017). https://doi.org/10.1007/978-3-319-66836-9_30
4. Kalimulina, E.Y.: Rate of convergence to stationary distribution for unreliable Jackson-type queueing network with dynamic routing. In: Vishnevskiy, V.M., Samouylov, K.E., Kozyrev, D.V. (eds.) DCCN 2016. CCIS, vol. 678, pp. 253–265. Springer, Cham (2016). https://doi.org/10.1007/978-3-319-51917-3_23
5. Chen, M.-F.: Eigenvalues, Inequalities, and Ergodic Theory. Springer, Heidelberg (2005)
6. Liggett, T.H.: Exponential convergence of attractive reversible nearest particle systems. Ann. Prob. **17**, 403–432 (1989)
7. Sauer, C., Daduna, H.: Availability, formulae and performance measures for separable degradable networks. Econ. Qual. Control **18**(2), 165–194 (2003)
8. Lawler, G.F., Sokal, A.D.: Bounds on the L_2 spectrum for Markov chains and Markov processes: a generalization of Cheeger's inequality. Trans. Am. Math. Soc. **309**(2), 557–580 (1988)
9. van Doorn, E.: Representations for the rate of convergence of birth-death processes. Theory Probab. Math. Stat. **65**, 37–43 (2002)
10. Zeifman, A.I.: Some estimates of the rate of convergence for birth and death processes. J. Appl. Probab. **28**, 268–277 (1991)
11. Van, Z.A.I., Doorn, E.A., Zeifman, A.I.: On the speed of convergence to stationarity of the Erlang loss system. Queueing Syst. **63**, 241 (2009). https://doi.org/10.1007/s11134-009-9134-9

12. Lorek, P., Szekli, R.: Computable bounds on the spectral gap for unreliable Jackson networks. Adv. Appl. Probab. **47**(2), 402–424 (2015)
13. Lorek, P.: The exact asymptotic for the stationary distribution of some unreliable systems. arXiv:1102.4707 (2011)
14. Doeblin, W.: Mathématique de l'Union Interbalkanique (1938)
15. Albert, R., Barabasi, A.: Statistical mechanics of complex networks. Rev. Mod. Phys. **74**(1), 47–97. arXiv:cond-mat/0106096
16. Dorogovtsev, S.N., Mendes, J.F.F.: Evolution of Networks: From Biological Nets to the Internet and WWW (Physics). Oxford University Press Inc., New York (2003)
17. Zverkina, G.: Stationary coupling method for renewal process in continuous time (application to strong bounds for the convergence rate of the distribution of the regenerative process). arXiv preprint. arXiv:1704.04808 (2017)
18. Zverkina, G.: Lorden's inequality and coupling method for backward renewal process. arXiv preprint. arXiv:1706.00922 (2017)
19. Kalimulina, E.Y., Zverkina, G.A.: On some generalization of Lorden's inequality for renewal processes. arXiv preprint. arXiv:1910.03381

Leader Nodes in Communities for Information Spreading

Natalia M. Markovich(✉) and Maxim S. Ryzhov(✉)

V.A. Trapeznikov Institute of Control Sciences Russian Academy of Sciences, Profsoyuznaya Street 65, 117997 Moscow, Russia
markovic@ipu.rssi.ru, nat.markovich@gmail.com, maksim.ryzhov@frtk.ru

Abstract. The paper is devoted to the effective information spreading in random complex networks. Our objective is to elect leader nodes or communities of the network, which may spread the content among all nodes faster. We consider a well-known SPREAD algorithm by Mosk-Aoyama and Shah (2006), which provides the spreading and the growth of the node set possessing the information. Assuming that all nodes have asynchronous clocks, the next node is chosen uniformly among nodes of the network by the global clock tick according to a Poisson process. The extremal index measures the clustering tendency of high threshold exceedances. The node extremal index shows the ability to attract highly ranked nodes in the node orbit. Considering a closeness centrality as a measure of a node's leadership, we find the relation between its extremal index and the minimal spreading time.

Keywords: Complex network · Random graph · Community · Extremal index · Information spreading

1 Introduction

Analysis and modeling of a fast content spreading is an important issue in distributed computing [2,14], and social networks. Proposed solutions for this issue do not only help to observe the information diffusion but also serve as a valuable resource to predict the characteristics of the network. The spreading time of infection within a human contact network [9] can dramatically reflect on the life of humanity.

In [12,13], a statistical clustering of the random network by the node extremal index (EI) is proposed. The EI plays a key role in the extreme value analysis since it allows to obtain a limit distribution of maximum of a sample of random variables when the latter are dependent.

The reported study was partly funded by RFBR, project number 19-01-00090 (recipient N. M. Markovich, conceptualization, mathematical model development, methodology development; recipient M. S. Ryzhov, numerical analysis, validation.

V. M. Vishnevskiy et al. (Eds.): DCCN 2020, LNCS 12563, pp. 475–484, 2020.
https://doi.org/10.1007/978-3-030-66471-8_36

Definition 1. *A stationary sequence* $\{Y_n\}_{n\geq 1}$ *with distribution function (df)* $F(x)$ *and* $M_n = \max_{1\leq j\leq n} Y_j$ *is said to have EI* $\theta \in [0,1]$ *if for each* $0 < \tau < \infty$ *there is a sequence of real numbers* $u_n = u_n(\tau)$ *such that*

$$\lim_{n\to\infty} n(1 - F(u_n)) = \tau \qquad \text{and} \quad \lim_{n\to\infty} P\{M_n \leq u_n\} = e^{-\tau\theta},$$

hold ([10], p.63).

For independent random variables, the EI is equal to one. The converse is incorrect. As closer θ to zero, as stronger the dependence. The EI measures the local clustering tendency of high threshold exceedances. Its reciprocal $1/\theta$ approximates the mean number of exceedances per cluster (the mean cluster size).

In classical settings, the cluster of exceedances over threshold is defined as a block of data with at least one exceedance over threshold u. In [6] the cluster is a set of consecutive exceedances between two consecutive non-exceedances. We follow this definition and modify it with respect to graphs. There are several problems here. First, the EI of the graph may not exist in a sense of Definition 1 particularly since the real network may be non-stationary with regard to any characteristic of the nodes. The node degree (i.e. the number of its links), PageRank, and a centrality index may show the node leading in the network. The network may be partition into communities by interests, that are sets of nodes with a large number of internal links. One can expect that the communities can be rather homogeneous and even stationary distributed. Communities can be selected by applying such measures like the conductance, clustering coefficient and modularity [3,15]. Next, nodes in the graph are not numerated as in a random sequence, but clusters require an order. Thus, in [12,13], generations of followers of a node are used as blocks and potential clusters. The hypothesis in [12,13] states that the node EI shows the ability to attract highly ranked nodes in the node orbit and to spread information faster. This means that a coupled tree-like graph of the node taken as a root may contain influential nodes. Since the EI relates to the stationary sequence, we use a stationary community to which the node belongs instead of the sequence.

We aim to study the dependence between the EI of some influence characteristic of nodes and the minimum time needed to spread the information to all nodes in a undirected graph starting from a node. To this end, we use a well known SPREAD algorithm by [14] which provides the spreading and the growth of the set possessing the node information. As the leading measures we consider the node degree and closeness centrality [17]. We partition the graph into communities to find leaders. The question arises, what EI the community must have to be a leader with regard to the minimum spreading time?

The paper is organized as follows. In Sect. 2, related works regarding the community and leading nodes identification as well as the information spreading algorithm, the tail index (TI) estimation and a stationarity test are recalled. In Sect. 3, our main result concerning the EI estimation in random graphs based on the intervals estimator by [6] is presented. Moreover, a simulation study is given in Sect. 4. The exposition is finalized by conclusions in Sect. 5.

2 Related Works

2.1 Graph Community Characteristics

For graph $G = (V, E)$ with number of vertices $\|V\| = n$ and the number of edges $\|E\|$, the conductance measures the minimum relative connection strength between "isolated" subsets $\{S\}$ and the rest of the network [2,11,14]. The conductance is defined by

$$\Phi(G) = \min_{S \subseteq V, |S| \leq n/2} \phi(S, V), \quad \phi(S, V) = \frac{\sum_{i \in S, j \in V \setminus S} P_{i,j}}{|S|},$$

where P is the stochastic probability matrix associated with the communication of nodes [14]. The conductance satisfies $0 < \Phi(G) \leq 1$. In [2] the node i chooses a neighbor j with probability $P_{i,j} = 1/D_i$, where D_i is the degree of node i. In [14] it is proposed to use $P_{i,j} = 1/D_{max}$ if $(i, j) \in E$ and $P_{i,j} = 1 - D_i/D_{max}$ if $i = j$, but it requires to know the maximum degree in the graph $D_{max} = \max_{i \in V} D_i$.

The modularity Q is a measure to partition the network into communities [3]. It shows how many edges exist within communities and between them:

$$Q = \frac{1}{2m} \sum_{vw} \left[A_{vw} - \frac{D_w D_v}{2m} \right] \delta(c_w, c_v).$$

Here, $m = \frac{1}{2} \sum_{vw} A_{vw} = \|E\|/2$ is a number of edges in the undirected graph, A is an adjacency matrix, $\delta(c_w, c_v)$ is equal to 1 when nodes w and v belong to the same community. The main result in [3] is that the modularity reaches a minimum value when it is possible to bipartite a graph. A Greedy Modularity Maximization Algorithm (GMMA) [3] is used to detect the community structure fast.

2.2 Information Spreading Time

The information spreading time is determined in [14] for global broadcast problem, i.e. when one aims to disseminate a content to all nodes in the network. Suppose node $i \in V$ has a message m_i. $S_i(t)$ denotes the set of nodes that have the message m_i at time t. For $\delta \in (0, 1)$ the δ-information-spreading time of the algorithm P is determined as

$$T_P^{spr}(\delta) = inf\{t \geq 1 : Pr(\cup_{i=1}^n \{S_i(t) \neq V\}) \leq \delta\}.$$

In [14] it is derived that there exists an information dissemination algorithm P such that, for any $\delta \in (0, 1)$, $T_P^{spr}(\delta) = O((log(n) + log(\delta^{-1}))/\Phi(G))$, where n is the number of nodes in the network. Spreading Algorithm 1 has been proposed in [14].

Algorithm 1. Algorithm SPREAD(P)

When a node i initiates communication at round t:

1: Node i chooses a node u at random, and contacts u. The choice of the communication partner u is made independently of all other random choices, and the probability that node i chooses any node j is $P_{i,j}$.
2: Node u sends all of the messages it has to node i, so that $M_i(t+1) = M_i(t) \cup M_u(t)$, where $M_i(t)$ is a set of received messages on round t.

2.3 Leader Node Identification

Let $G = (V, E)$ be undirected connected graph of order n. A standard graph index used for the leader election is the closeness centrality C_x [17]:

$$C_x = \frac{n-1}{\sum_{y,y\neq x} d(x,y)}, \qquad 0 < C_x \leq 1, \tag{1}$$

where d(x, y) is the shortest path $(x, \ldots, y)$ between nodes x and y. When a node is closer to other nodes, then its value C_x is closer to 1. The node degree D_x may also be used as a measure of the leadership, since a node with large D_x is connected with a large number of nodes.

2.4 Tail Index Identification

Let $X_1, \ldots, X_n$ be a stationary sequence of i.i.d r.v.s. A TI $\alpha = 1/\gamma$ is reciprocal of the extreme value index γ. γ may be estimated by the moment estimator $\widehat{\gamma}_M(k)$ [4] and the Hill's estimator $\widehat{\gamma}_H(k)$:

$$\widehat{\gamma}_M(k) = \widehat{\gamma}_H(k) + 1 - 0.5\left(1 - \frac{\widehat{\gamma}_H^2(k)}{S_{n,k}}\right)^{-1}, \quad \widehat{\gamma}_H(k) = \frac{1}{k}\sum_{i=1}^{k} log(\frac{X_{(n-i+1)}}{X_{(n-k)}}),$$

by order statistics $X_{(1)} \leq X_{(2)} \leq \ldots \leq X_{(n)}$, where $S_{n,k} = \frac{1}{k}\sum_{i=1}^{k} \left(log(\frac{X_{(n-i+1)}}{X_{(n-k)}})\right)^2$. k is a number of the largest order statistics. Its optimal value is further chosen by the bootstrap method [8]. Bootstrap confidence intervals are calculated by [1].

2.5 Test of Stationarity

In our further analysis we have to test communities of nodes on stationarity. We use the stationarity test statistic

$$V/S = V_n/\widehat{s}^2_{n,q}, \quad V_n = \frac{1}{n^2}\left[\sum_{k=1}^{n}(S_k^*)^2 - \frac{1}{n}\left(\sum_{k=1}^{n} S_k^*\right)^2\right], \quad \widehat{s}^2_{n,q} = q^{-1}\sum_{i,j=1}^{q} \widehat{\gamma}_{i-j}, \tag{2}$$

$$S_k^* = \sum_{j=1}^{k}(X_j - \overline{X}_n), \ \widehat{\gamma}_j = n^{-1}\sum_{i=1}^{n-j}(X_i - \overline{X}_n)(X_{i+j} - \overline{X}_n), \ \ 0 \le j < n,$$

proposed in [7]. The null hypothesis of stationarity is rejected, if $V/S > c_\alpha$, where c_α is a quantile of the asymptotic df of the Kolmogorov statistic $F_K(\pi\sqrt{x})$. $c_\alpha \in \{0.190, 0.153, 0.1, 0.069\}$ holds for significant level $\alpha \in \{5, 10, 30, 50\}\%$, respectively.

3 Extremal Index Estimation

Let $G = (V, E)$ be undirected graph of the order n and $(X_1, \dots, X_n)$ be a sample of node characteristics with a marginal df $F(u)$. For stochastic sequence, a cluster is determined as consecutive exceedances over a predefined threshold u between two consecutive non-exceedances. An inter-cluster size is [6]

$$T(u) = \min\{t \ge 1 : X_{j+t} > u\} \text{ given } X_j > u. \tag{3}$$

According to Theorem 1 in [6] $P\{\overline{F(u_n)}T(u_n) > t\} \to \theta\exp(-\theta t)$ for $t > 0$ holds as $n \to \infty$ assuming that $\{X_n\}_{n\ge1}$ is a strictly stationary sequence of r.v.s under some specific mixing condition. The intervals estimator of the EI is [6]

$$\widehat{\theta}(u) = \min(1, \theta^*), \ \theta^* = \begin{cases} \frac{2(\sum_{i=1}^{N-1} T(u)_i - 1)^2}{(N-1)\sum_{i=1}^{N-1}(T(u)_i - 1)(T(u)_i - 2)}, & max\{T(u)_i\} > 2 \\ \frac{2(\sum_{i=1}^{N-1} T(u)_i)^2}{(N-1)\sum_{i=1}^{N-1}(T^2(u)_i)}, & \text{otherwise}, \end{cases} \tag{4}$$

where $N = N(u) = \sum_{i=1}^{n} 1\!\!\mathrm{I}(X_i > u)$ is the number of exceedances. To introduce the intervals estimator for graphs, we propose to define $T(u)$ as the number of edges of the shortest path between nodes which characteristics are larger than u.

3.1 Extremal Index of the Community

Let us choose a node x with characteristic X_x and a community of nodes to which it may belong. To determine the EI of the community S (assuming that its EI exists), we take a high quantile of $F(x)$ as the threshold u^*. Let a community S be a strict-sense stationary set with EI θ. The stationarity implies that the distribution of all nodes of S remains the same irrespective of the numeration of the nodes.

To estimate the EI of the community $\theta(S)$ one has to determine $\{T(u^*)_i\}$ for all possible pairs $(x, y) \in S$. The event $\{T(u^*)_i = m\}$ means that the characteristics $X_{i_1}, X_{i_2}, \dots, X_{i_m}$ in a sequence X_{xy} are all less than u^*, but X_x and X_y exceed u^*. The shortest path between nodes x and y with such restriction is calculated based on algorithm given in [5].

Algorithm 2. Community EI estimation

1: Set sequences $X_{xy} = \{X_x, X_{i_1}, X_{i_2}, \ldots, X_{i_m}, X_y\}$, $m \geq 1$ corresponding to shortest paths $(x, \ldots, y)$ from a fixed node x to each node y of the community S.
2: Define $\{T(u^*)_i\}$, $i = 1, 2, \ldots N$ by (3) for all sequences, where N is a total number of inter-cluster times over all possible shortest paths X_{xy}.
3: Estimate the EI $\widehat{\theta} = \widehat{\theta}(u^*)$ of community by (4).

4 Simulation Study

4.1 Community and Leader Election

For the simulated undirected connected graph $G = (V, E)$, the spreading time is modeled as the time needed to send the message m_x from node x to other nodes by Algorithm 1. A new node is uniformly selected among all nodes of the network by the clock tick. The clock ticks $\{I_j\}$ are modeled as Poisson process with the rate $n = \|V\|$, where the inter-arrival times between ticks $\{I_{j+1} - I_j\}$ are i.i.d. exponential random variables of rate n.

The spreading time is calculated by the sum of such inter-arrivals $T_x^{spr} = \sum_{j=1}^{k-1}(I_{j+1} - I_j)$, where k is the number of ticks required to deliver the message m_x to all other nodes. When the newly selected node has no direct links to 'the informed set', i.e. the set of nodes which received m_x earlier, then the message m_x cannot be delivered to the node.

The spreading time is a random variable that depends both on the Poisson clock ticks and random communications between nodes which impacts the random value k. The k is thus determined by the conductance of the graph. Indeed, the well-connected graph with a high conductance requires a smaller k and vice versa. For a fixed k a part of nodes of the network may not receive m_x, if k is not large enough. It seems, the Poisson clock used to select a next node is a realistic model of user behavior. However, a renewal process with heavy-tailed inter-arrival time intervals may also be used instead of Poisson model that may impact the spreading time.

We partition the graph into communities by algorithm GMMA [3]. Using the closeness centrality C_x as a node characteristic, the TIs and EIs of communities are estimated as in Sect. 2.4 and 3.1. To compare the node and community leadership we analyze relations between D_x, C_x, T^{spr}, the TI and the EI.

We simulate a geometric graph with the number of nodes $n = 200$ and the radius $r = 0.11$ [16]. This is the undirected graph constructed by n nodes uniformly placing in a unit square. An edge connects two nodes, if the distance between them is less than r. We partition the graph into communities in Fig. 1 (left) and state the stationarity of all communities by test (2) in Fig. 1 (right).

To apply the intervals estimator (4) and the moment estimator we have to check the stationarity of closeness centrality in communities. Figure 1 (right) shows that communities with values of V/S-statistic which do not exceed the critical level $c_\alpha = 0.19$ are stationary with the probability more then 95%. Nodes with high C_x spread information faster, but the tendency is weaker preserved

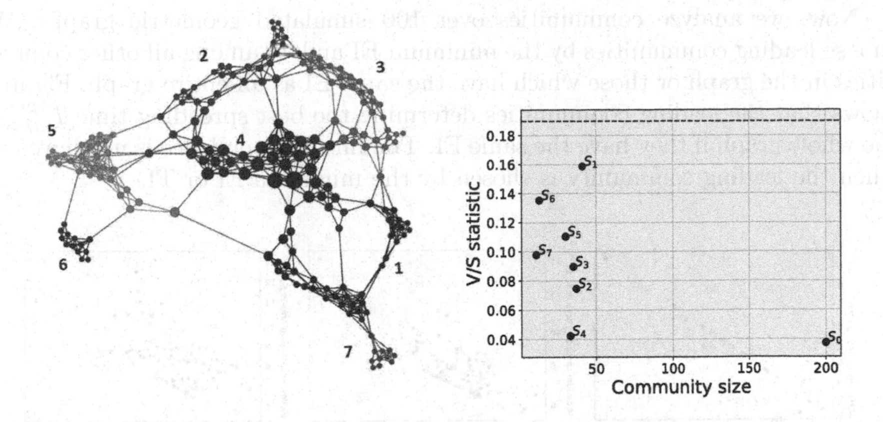

Fig. 1. Geometric graph with communities $\{S_i\}_{i=1}^{7}$ and circle sizes of nodes marked proportional to C_x (left); the V/S-statistics of communities against the community sizes and with the 5% critical value 0.19 (dotted line) (right).

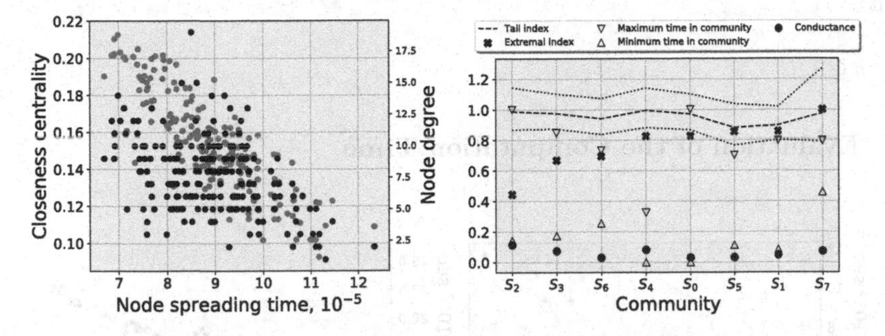

Fig. 2. C_x (grey dotes) and D_x (black dotes) against the T^{spr} (left); Minimum and maximum T^{spr}s over nodes in communities $\{S_i\}$ normalized to $[0, 1]$, the EIs $\theta(S_i)$, the conductances $\phi(S_i, V)$, the TIs with 95% bootstrap confidence intervals (dotted lines) (right).

with regard to node degrees D_x, Fig. 2 (left). This may indicate that C_x is better index of the leadership than D_x with regard to the spreading.

In Fig. 2 (right), we compare communities of the graph in Fig. 1 regarding the spreading time, the TI, the EI and the conductance. S_0 denotes the entire graph. The EI and the TI of S_4 are close to ones of S_0. In this respect, S_4 is a leading community. Its T^{spr} has the smallest variation and it contains a leader node with the smallest spreading time. S_2 has the smallest EI value and the highest conductance, but its minimum T^{spr} is not the best. S_7 is the worse spreader despite its conductance is middle since it contains weak connected independent nodes, and its EI is equal to 1. The TIs of all communities are close, that indicates their similar heaviness of tails.

Now, we analyze communities over 100 simulated geometric graphs. We choose leading communities by the minimum EI and TI among all other communities in the graph or those which have the same EI as an entire graph. Figure 3 shows that the leading communities determine the best spreading time T^{spr}_{min} in the whole graph if they have the same EI. The linear dependence is much weaker when the leading community is chosen by the minimum EI or TI.

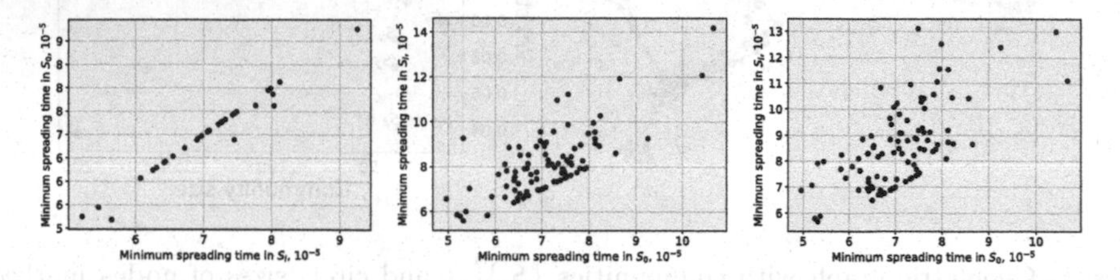

Fig. 3. T^{spr}_{min}s in leading communities and entire graphs when they have the same EIs (left), when the communities have the minimum TI (middle) and the minimum EI (right).

4.2 Evaluation of the Computation Time

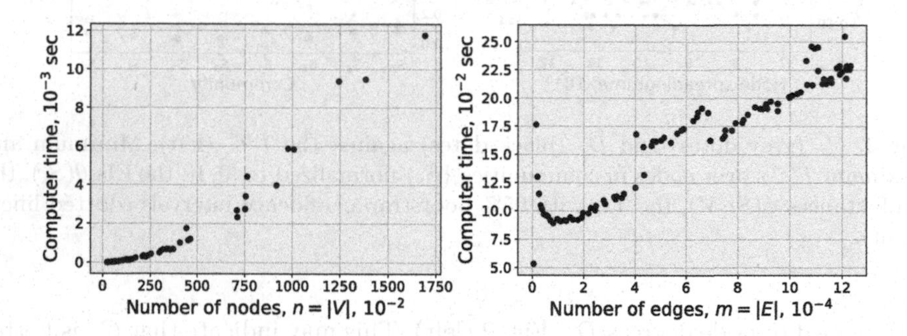

Fig. 4. The minimal computer time t of the spreading algorithm within a geometric graph over 200 retrials versus the number of vertices $n = \|V\|$ (left) and for the graph with $n = 500$ versus the number of edges $m = \|E\|$.

For calculations we used CPU Intel(R) Core(TM) i5-5200, 2.20 GHz, and 8 GB operative memory. The programm was written in Python 3.8. The computer time to realize the spreading Algorithm 1 is shown in Fig. 4 versus the number of nodes $n = \|V\|$ and the number of edges $m = \|E\|$. First, for each node of the geometric graph we calculate the minimal spreading time to all nodes among 200 retrials of this procedure. Increasing the number of nodes in the graph we observe a polynomial increasing of the computer time in Fig. 4, (left). Moreover, we consider geometric graphs all with 500 nodes and increase their number of edges by changing of the radius r, the parameter of the geometric graph. We provided 200 retrials for each simulated geometric graph and obtained the increasing computer time against the number of edges close to polynomial one in Fig. 4, (right).

5 Conclusion

We study the leadership of nodes and communities regarding their minimum spreading time. The modification of the intervals estimator of the EI for graphs is proposed. Taking the closeness centrality as a node characteristic and estimating its EI, we found that a community with the same EI as its entire graph identifies the best spreading time of the entire graph. Our future research will concern to analytical relations between the EI and the minimum spreading time as well as the fast algorithm of the intervals estimator of the EI on graphs.

References

1. Caers, J., Beirlant, J., Vynckier, P.: Bootstrap confidence intervals for tail indices. Comput. Stat. Data Anal. **26**(3), 259–277 (1998). https://doi.org/10.1016/S0167-9473(97)00033-9
2. Censor-Hillel, K., Shachnai, H.: Partial information spreading with application to distributed maximum coverage. In: Proceedings of the 29th ACM SIGACT-SIGOPS symposium on Principles of distributed computing (PODC 2010), pp. 161–170. ACM, New York (2010). https://doi.org/10.1145/1835698.1835739
3. Clauset, A., Newman, M.E., Moore, C.: Finding community structure in very large networks. Phys. Rev. E. **70**(6), 066111 (2004). https://doi.org/10.1103/PhysRevE.70.066111
4. Dekkers, A.L.M., Einmahl, J.H.J., De Haan, L.: A moment estimator for the index of an extreme-value distribution. Ann. Stat. **17**(4), 1833–1855 (1989). https://doi.org/10.1214/aos/1176347397
5. Dijkstra, E.W.: A note on two problems in connexion with graphs. Numer. Math. **1**, 269–271 (1959). https://doi.org/10.1007/BF01386390
6. Ferro, C., Segers, J.: Inference for clusters of extreme values. J. R. Stat. Soc. B **65**(2), 545–556 (2003). https://doi.org/10.1111/1467-9868.00401
7. Giraitis, L., Leipus, R., Philippe, A.: A test for stationarity versus trend and unit root for a wide class of dependent errors. Econ. Theory **22**, 989–1029 (2006). https://doi.org/10.1017/S026646660606049X
8. Hall, P.: Using the bootstrap to estimate mean squared error and select smoothing parameter in nonparametric problems. J. Multivar. Anal. **32**, 177–203 (1990). https://doi.org/10.1016/0047-259X(90)90080-2
9. Holme, P., Litvak, N.: Cost-efficient vaccination protocols for network epidemiology. PLoS Comput. Biol. **13**(9) (2017). https://doi.org/10.1371/journal.pcbi.1005696
10. Leadbetter, M.R.: Extremes and local dependence in stationary sequences. Zeitschrift für Wahrscheinlichkeitstheorie und Verwandte Gebiete **65**(2), 291–306 (1983). https://doi.org/10.1007/BF00532484

11. Leskovec, J., Lang, K.J., Dasgupta, A., Mahoney, M.W.: Community structure in large networks: natural cluster sizes and the absence of large well-defined clusters. Internet Math. **6**(1) (2008). https://doi.org/10.1080/15427951.2009.10129177
12. Markovich, N.M., Ryzhov, M.S., Krieger, U.R.: Nonparametric analysis of extremes on web graphs: PageRank versus Max-Linear model. CCIS **700**, 13–26 (2017). https://doi.org/10.1007/978-3-319-66836-9_2
13. Markovich, N.M., Ryzhov, M.S., Krieger, U.R.: Statistical clustering of a random network by extremal properties. CCIS **919**, 71–82 (2018). https://doi.org/10.1007/978-3-319-99447-5_7
14. Mosk-Aoyama, D., Shah, D.: Computing separable functions via gossip. In: Proceedings of the Twenty-Fifth Annual ACM Symposium on Principles of Distributed Computing (PODC 2006), pp. 113–122. ACM, New York (2006). https://doi.org/10.1145/1146381.1146401
15. Newman, M.E.J.: Networks: An Introduction, 2nd edn. Oxford University Press, Oxford (2018). ISBN-10:0198805098
16. Penrose, M.: Random Geometric Graphs. Oxford Studies in Probability (2003). https://doi.org/10.1093/acprof:oso/9780198506263.001.0001
17. Stephenson, K., Zelen, M.: Rethinking centrality: methods and examples. Soc. Netw. **11**(1), 1–37 (1989). https://doi.org/10.1016/0378-8733(89)90016-6

Sensitivity Analysis of a k-out-of-n:F System Characteristics to Shapes of Input Distribution

V. V. Rykov[1,2,4](✉), N. M. Ivanova[1,3], and D. V. Kozyrev[1,3]

[1] Peoples' Friendship University of Russia (RUDN University), 6 Miklukho-Maklaya Street, Moscow 117198, Russian Federation
vladimir_rykov@mail.ru, nm_ivanova@bk.ru, kozyrev-dv@rudn.ru

[2] Gubkin Russian State Oil and Gas University, 65 Leninsky Prospekt, Moscow 119991, Russia

[3] V.A.Trapeznikov Institute of Control Sciences of Russian Academy of Sciences, 65 Profsoyuznaya Street, Moscow 117997, Russia

[4] Kharkevich Institute for Information Transmission Problems of Russian Academy of Sciences (IITP RAS), 19, Building 1, B. Karetny Per., Moscow 127051, Russia

Abstract. The problem of sensitivity of a redundant system's probability characteristics to shapes of the input distributions is considered. In some previous works, closed-form representations have been obtained for stationary characteristics of hot-standby redundant systems with exponential lifetime distribution of their elements and general distribution of their repair time. In the current paper we carry out the sensitivity analysis of a k-out-of-n:F system with the help of a simulation approach. Comparison of analytic and simulation results is presented.

Keywords: k-out-of-n:F system · Steady state probabilities · Sensitivity analysis · Mathematical modeling and simulation · AnyLogic environment

1 Introduction

Any system that can be considered as a functioning unit has many useful properties, among which stability, reliability, and flexibility can be especially highlighted. These properties are of key importance from both a practical and research points of view.

The term "sensitivity analysis" can be understood differently in civil engineering than in basic sciences [1]. In operations research, sensitivity analysis is

The publication has been prepared with the support of the "RUDN University Program 5–100" (problem setting and simulation model development) and funded by RFBR according to the research projects No. 20-01-00575 (recipient Vladimir Rykov, review and analytic results) and No. 19-29-06043 (recipient Dmitry Kozyrev, formal analysis, validation).

V. M. Vishnevskiy et al. (Eds.): DCCN 2020, LNCS 12563, pp. 485–496, 2020.
https://doi.org/10.1007/978-3-030-66471-8_37

developed as a method of critical assessment of decisional variables, and is capable to identify those sensitive variables that influence the final desired result [2]. Another suitable complement to probabilistic reliability analysis is structural sensitivity analysis [3,4]. In stochastic systems stability often means insensitivity or low sensitivity of their output characteristics to the shapes of some input distributions.

The first results of studies on the insensitivity of systems' characteristics to output parameters dealt with systems with Poisson arrivals and fixed mean value of the service time. Thus, in 1957, Sevastyanov in [5] proved the insensitivity of Erlang formulas to the shape of service time distribution for systems with losses. Kovalenko continued the study of the insensitivity of the stationary characteristics of a renewable system. In [6] he found a necessary and sufficient condition for the reliability characteristics of such a system in the case of an exponential distribution of the lifetime and the general distribution of the recovery time. The sufficiency of this condition for the case of general lifetime and repair time distributions has been found by Rykov [7] with the help of multi-dimensional alternating processes theory.

In papers of Gnedenko and Soloviev [8–10] the convergence of the reliability function of a reparable double redundant system with general distributions of their components' life and repair time to an exponential curve under quick repair has been shown. These works gave an impetus to investigation of the asymptotic insensitivity of the systems' reliability characteristics to the shape distributions of their input parameters. The asymptotic insensitivity of the characteristics of systems has also been the subject of several recent studies that investigated the problem of the stability of stationary characteristics of systems in the case when one of the input distributions (lifetime or repair time) is exponential [11,12]. Moreover, the problem of asymptotic insensitivity of a heterogeneous double redundant hot standby renewable system with Marshall-Olkin failure model for both life- and repair time general distributions with the help of simulation method was studied in [13].

Among investigations on topics of reliability the k-out-of-n:F systems are very popular. These systems consist of n components, and fail, when at least k components fail [14]. The k-out-of-n:F systems are widely used in practice, for example, in telecommunications, the oil and gas industry, data transmission, production management [15,16]. Therefore, it is very important to investigate various characteristics of such models.

Currently, many different scenarios of the k-out-of-n system have already been considered. Kuo and Zuo [17] introduced many of the system reliability models such as parallel, series, standby, multi-state, maintainable systems, etc. In [18] a general closed form equation was developed for system reliability of a k-out-of-n warm standby system. Zhang et al. [19] analyzed the availability and reliability of k-out-of-$(m + n)$ standby system with two types of failure. Yuge et al. [20] introduced the reliability of a k-out-of-n system with common-cause failures using multivariate exponential distribution. In [21] the authors provided a reliability analysis of k-out-of-n system with two independent exponentially distributed types of failure.

The results of the study of reliability function for a k-out-of-n:F system with exponential distribution of the lifetime and general distribution of the repair time were presented at the MMR-2019 conference and published in [22]. This research was carried out using the so-called markovization method, which consists in introduction of supplementary variables [23] (see also [24]) that allow to describe the system behavior by a two-dimensional Markov process. This approach also makes it possible to obtain analytical formulas for the system steady states probabilities (s.s.p.) that will be used in the current paper.

The main idea of the paper consists in investigation of sensitivity properties of system s.s.p.'s in the case when both components life- and repair time are generally distributed with the help of the simulation method. Simulation and analytical results are then compared and numerical analysis is conducted for the case when components' lifetimes have exponential distributions.

In the current paper a modified Kendall's notation $\langle GI|GI|1\rangle$ is used, where the symbol GI in first position means general distributions for independent lifetime of the components, this symbol in second position means general distributions of independent repair times. Number 1 in the last position means the amount of repair facilities. Symbols "$\langle\ \rangle$" mean that the considered system is a closed one.

The paper is organized as follows. In the next section, the problem setting will be done. Section 3 deals with analytical results for stationary distribution for a special case when $k = 3$, $n = 6$. And the next 4-th section is devoted to the sensitivity analysis of the model in general case with the help of the simulation method. The paper ends with the conclusion and some future research directions.

2 Problem Setting and Notation

Consider a hot standby k-out-of-n ($k < n$) repairable system. The system fails when at least k of its components fail and such a system is called as a k-out-of-n:F system. In the model the lifetime of different components and the same component after their repair are supposed to be independent identically distributed (i.i.d.) random variables (r.v.'s) A_i, $(i = 1, 2, \ldots)$ and their common cumulative distribution function (c.d.f.) are denoted by $A(x) = \mathbf{P}\{A_i \le x\}$ $(i = 1, 2, \ldots)$. The system is repairable which means that failed components are repaired with the help of a single repair facility, and upon repair completion they are as good as new. When dealing with the repairable model we need to consider some procedures of the system's restoration after failure. There are at least two possibilities:

- Partial repair, when after any component's failure as well as after the system's failure the failed unit is repaired during a random time B_i $(i = 1, 2, \ldots)$, where the r.v.'s B_i are supposed to be i.i.d. r.v.'s with common c.d.f. $B(x) = \mathbf{P}\{B_i \le x\}$.
- Full repair, when after each component's failure it is repaired according to the same mechanism as above, but after the system failure the renewal of the

whole system begins that requires another random time unit, say F, for the system to be repaired and become as good as new, i.e. return to state 0.

In this paper we deal with the second case, furthermore, we consider a special case when the random time F has the same distribution as partial repair time. For simplicity it is supposed that all distributions are absolute continuous and their probability density functions (p.d.f.) are denoted by $a(x)$, $b(x)$ correspondingly.

The system state space can be represented as $\mathbf{E} = \{0, 1, 2, ..., k-1, k\}$, where

- 0 means that all n components operate,
- i means that i components out of n $(1 \leq i \leq k-1)$ have failed, one of them is being repaired, and the other $(n-i)$ operate,
- k means that k components have failed, thus, the system has failed and is being repaired.

To describe the system's behavior we introduce a stochastic process $J = \{J(t), t \geq 0\}$ on a phase space $\mathbf{E}$:

$$J(t) = j \text{ if at time } t \text{ the system is in state } j \in \mathbf{E}.$$

The current paper deals with the system's s.s.p.

$$\pi_j = \lim_{t\to\infty} \mathbf{P}\{J(t) = j\}$$

and properties of their asymptotic insensitivity to the shapes of life- and repair times of its components.

3 Steady State Probabilities. Analytical Results for $\langle M_{3<6}|GI|1\rangle$ System

In this section analytical results for the k-out-of-n : F system will be represented for the case when the lifetime of its components has an exponential distribution. To calculate the s.s.p. the method of supplementary variables introduction is used [24]. For our case as a supplementary variable we use the elapsed repair time of the failed component and the whole system. Thus, we consider a two-dimensional process $Z = \{Z(t), t \geq 0\}$, with $Z(t) = \{J(t), X(t)\}$ where $J(t)$ is the system's state at time t, and $X(t)$ represents the elapsed repair time of the failed component or the whole system. Due to the supplementary variable the process Z is a Markov one with a states space $\mathbf{E} = \{0, (1, x), (2, x), ..., (k-1, x), (k, x)\}$, and a transition graph, represented in Fig. 1.

Hereinafter we will use the following notations:

- α is the failure rate of the system's components;
- $a = \alpha^{-1}$ is the expectation of components lifetime;
- $\lambda_i = (n-i)\alpha$, $(i = \overline{0, k-1})$ is the system's partial failure rate when i of n components fail;

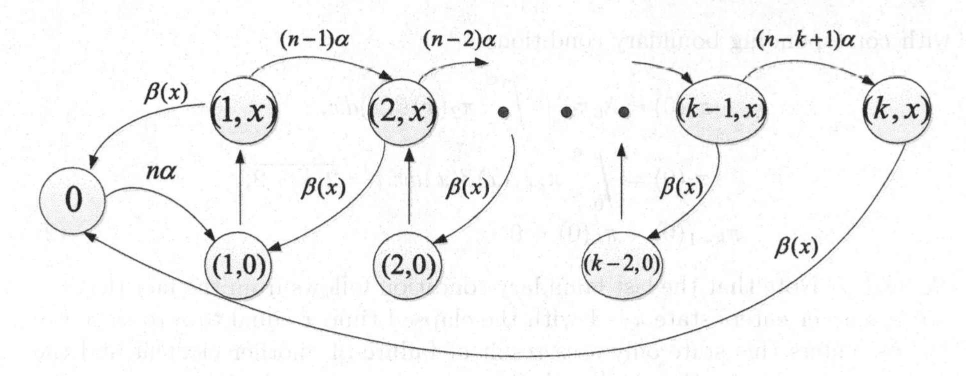

Fig. 1. Transition graph of the k-out-of-n:F system with full repair.

- $b = \int\limits_0^\infty (1 - B(x))dx$ is the mean time to repair;
- $\beta(x) = (1 - B(x))^{-1}b(x)$ is the conditional partial and full repair rate, given the elapsed repair time is x;
- $\tilde{b}(s) = \int\limits_0^\infty e^{-sx}b(x)dx$ is the moment generation function (m.g.f.) of the repair time or Laplace transform (LT) of its p.d.f.

The state probabilities of the process are denoted by

$$\pi_0(t) = \mathbf{P}\{N(t) = 0\},$$
$$\pi_i(t,x)dx = \mathbf{P}\{N(t) = i,\ x < X(t) \le x + dx\} \quad (i = \overline{1,k})$$

and corresponding s.s.p.'s are

$$\pi_0 = \lim_{t\to\infty} \pi_0(t), \quad \pi_i(x) = \lim_{t\to\infty} \pi_i(t;x) \quad (i = \overline{1,k}).$$

Using the method of comparison of the input and the output flows of failures and repair, we obtain the following system of balance equations for the system's s.s.p.:

$$\begin{aligned}
\lambda_0\pi_0 &= \int\limits_0^\infty \pi_1(x)\beta(x)dx + \int\limits_0^\infty \pi_k(x)\beta(x)dx,\\
\frac{d\pi_1(x)}{dx} &= -(\lambda_1 + \beta(x))\pi_1(x),\\
\frac{d\pi_i(x)}{dx} &= -(\lambda_i + \beta(x))\pi_i(x) + \lambda_{i-1}\pi_{i-1}(x), \ (i = \overline{2, k-1}),\\
\frac{d\pi_k(x)}{dx} &= -\beta(x)\pi_k(x) + \lambda_{k-1}\pi_{k-1}(x)
\end{aligned} \tag{1}$$

with corresponding boundary conditions

$$\pi_1(0) = \lambda_0\pi_0 + \int_0^\infty \pi_2(x)\beta(x)dx,$$
$$\pi_i(0) = \int_0^\infty \pi_{i+1}(x)\beta(x)dx \; i = \overline{2, k-2},$$
$$\pi_{k-1}(0) = \pi_k(0) = 0. \tag{2}$$

Remark 1. Note that the last boundary condition follows from the fact that the process never enters state $k-1$ with the elapsed time x equal to zero since the process enters this state only as a result of failure of another element and the transition from state $(k-2, x)$ with the same elapsed repair time.

Theorem 1. *The macro-states s.s.p. of the* 3*-out-of-*6 : *F system in terms of LT in case of full repair have the following form:*

$$\pi_0 = \frac{1 + 5\tilde{b}(5\alpha) - 5\tilde{b}(4\alpha)}{1 + 6\alpha b + 5\tilde{b}(5\alpha) - 5\tilde{b}(4\alpha)},$$
$$\pi_1 = \frac{6}{5} \cdot \frac{1 - \tilde{b}(5\alpha)}{1 + 5\tilde{b}(5\alpha) - 5\tilde{b}(4\alpha)}\pi_0,$$
$$\pi_2 = \frac{3}{2} \cdot \frac{1 + 4\tilde{b}(5\alpha) - 5\tilde{b}(4\alpha)}{1 + 5\tilde{b}(5\alpha) - 5\tilde{b}(4\alpha)}\pi_0,$$
$$\pi_3 = \frac{3}{10} \cdot \frac{20\alpha b - 16\tilde{b}(5\alpha) + 25\tilde{b}(4\alpha) - 9}{1 + 5\tilde{b}(5\alpha) - 5\tilde{b}(4\alpha)}\pi_0. \tag{3}$$

Proof. For the case when $n = 6$, $k = 3$ the system of balance Eqs. (1) takes the form

$$\lambda_0\pi_0 = \int_0^\infty \pi_1(x)\beta(x)dx + \int_0^\infty \pi_3(x)\beta(x)dx,$$
$$\frac{d\pi_1(x)}{dx} = -(\lambda_1 + \beta(x))\pi_1(x),$$
$$\frac{d\pi_2(x)}{dx} = -(\lambda_2 + \beta(x))\pi_2(x) + \lambda_1\pi_1(x),$$
$$\frac{d\pi_3(x)}{dx} = -\beta(x)\pi_3(x) + \lambda_2\pi_2(x) \tag{4}$$

with the following boundary conditions

$$\pi_1(0) = \lambda_0\pi_0 + \int_0^\infty \pi_2(x)\beta(x)dx,$$
$$\pi_2(0) = \pi_3(0) = 0. \tag{5}$$

The solution of the second of Eqs. (4) gives the probability $\pi_1(x)$ in the form

$$\pi_1(x) = C_1 e^{-\lambda_1 x}(1 - B(x)).$$

The solution of the homogeneous part of the third equation gives

$$\pi_2(x) = C_2 e^{-\lambda_2 x}(1 - B(x)).$$

Using the method of constants variation allows to obtain the following solution for $\pi_2(x)$

$$\pi_2(x) = e^{-\lambda_2 x}(1 - B(x)) \left[C_2 - C_1 \frac{\lambda_1}{\lambda_1 - \lambda_2} e^{-(\lambda_1 - \lambda_2)x} \right].$$

The solution of homogeneous part of the fourth equation of (4) has the form

$$\pi_3(x) = C_3(1 - B(x)),$$

and also the constants' variation method gives the solution for $\pi_3(x)$ in the form

$$\pi_3(x) = (1 - B(x)) \left[C_3 b - C_2 e^{-\lambda_2 x} + C_1 \frac{\lambda_2}{\lambda_1 - \lambda_2} e^{-\lambda_1 x} \right].$$

The boundary conditions (5) allow to find constants C_i, $(i = 1, 2, 3)$. The last condition gives the representation of C_2 and C_3 via C_1

$$C_2 = C_1 \frac{\lambda_1}{\lambda_1 - \lambda_2},$$
$$C_3 = C_2 - C_1 \frac{\lambda_2}{\lambda_1 - \lambda_2} = C_1.$$

Substitution of these expressions into the expressions for probabilities $\pi_i(x)$ $(i = 1, 2, 3)$ allows to represent them in terms of C_1:

$$\pi_1(x) = C_1 e^{-\lambda_1 x}(1 - B(x)),$$
$$\pi_2(x) = C_1 e^{-\lambda_2 x}(1 - B(x)) \frac{\lambda_1}{\lambda_1 - \lambda_2} \left[1 - e^{-(\lambda_1 - \lambda_2)x} \right],$$
$$\pi_3(x) = C_1(1 - B(x)) \left[b - \frac{\lambda_1}{\lambda_1 - \lambda_2} e^{-\lambda_2 x} + \frac{\lambda_2}{\lambda_1 - \lambda_2} e^{-\lambda_1 x} \right]$$

Finally, the first of boundary conditions (5) gives the representation of constant C_1 via π_0

$$C_1 = \lambda_0 \pi_0 \left(1 + \frac{\lambda_1}{\lambda_1 - \lambda_2} \left(\tilde{b}(\lambda_1) - \tilde{b}(\lambda_2) \right) \right)^{-1}.$$

The last probability π_0 is found from the normalizing equation

$$\pi_0 + \pi_1 + \pi_2 + \pi_3 = 1.$$

The reverse replacement $\lambda_i = (n - i)\alpha$ and the simple calculation end the proof of the theorem.

4 Numerical Results

In this section we use a simulation approach to show the asymptotic insensitivity of the system's s.s.p. to the shapes of its components' life- and repair time distributions for a special case of a 3-out-of-6 : F system. For the simulation of the system we use the multi-method modeling environment AnyLogic and the following notations (all parameters are considered in the same time scale):

- $\mathbf{E}A$ is the mean lifetime of all components,
- $\mathbf{E}B$ is the mean time to repair of all components and the whole system,
- $T = 10^6$ is the total simulation time,
- the parameters of all distributions are chosen in such a way that the coefficient of variation (the ratio of the standard deviation $\sigma = \sqrt{\mathbf{D}B}$ to the mean $\mathbf{E}B$) $c = \sigma/\mathbf{E}B$ takes a fixed value and the mean lifetime a increases,
- $\rho = \mathbf{E}A/\mathbf{E}B$ is the relative recovery rate of the system's components.

It is shown that as $\rho \to \infty$ the sensitivity of the system s.s.p. to the shapes of its components' life- and repair time distributions becomes negligible. In our experiments the following distributions are used for the repair time:

- Gamma ($\Gamma = \Gamma(k,\theta)$)

$$\mathbf{E}B(\mathbf{E}A) = k \cdot \theta, \quad c = \sqrt{k}/k;$$

- Gnedenko-Weibull ($GW = GW(k,\lambda)$)

$$\mathbf{E}B = \lambda \cdot \Gamma\left(1 + \frac{1}{k}\right), \quad c = \frac{\sqrt{\lambda^2 \cdot \Gamma(1 + 2/k) - \mathbf{E}B^2}}{\mathbf{E}B};$$

- Pareto ($P = P(k,x_m)$)

$$\mathbf{E}B = \frac{k \cdot x_m}{k-1}, \quad c = \frac{1}{k} \cdot \sqrt{\frac{k}{k-2}};$$

- Rayleigh ($R = R(\sigma)$)

$$\mathbf{E}B = \sqrt{\pi/2} \cdot \sigma, \quad c = \sqrt{\frac{2 - \pi/2}{\pi/2}};$$

For the lifetime we used Exponential and Gamma distributions.

In the first numerical example we compare analytical (using formula (3)) and simulation results for the availability of the system $1 - \pi_3$ with the following distributions: Exponential for lifetime and Γ, GW and R for the repair time. In this case the mean lifetime $\mathbf{E}A = 1$, the coefficient of variation $c = 0.5$ equals 0.5 and the relative recovery rate of the system's components runs from 0.00001 to 10, thus the mean full and partial repair time $\mathbf{E}B$ lies within the appropriate limits of the definition ρ.

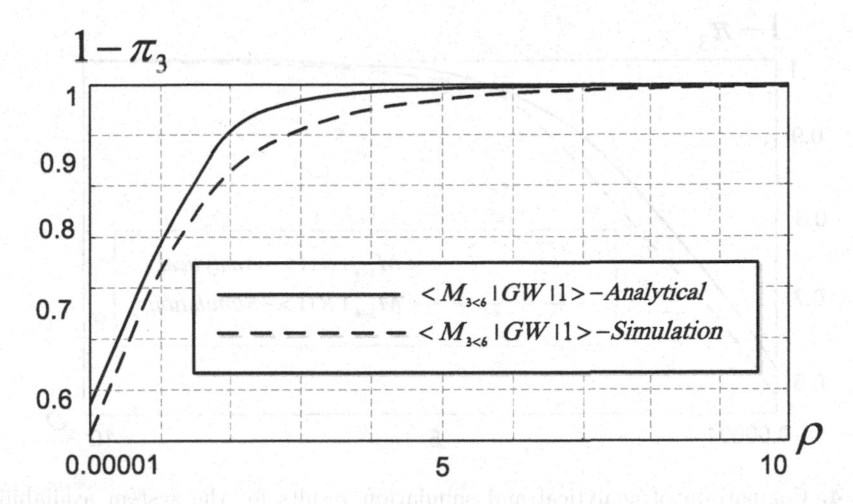

Fig. 2. Comparison of analytical and simulation results for the system availability $1 - \pi_3$, when $B(x) \sim GW$.

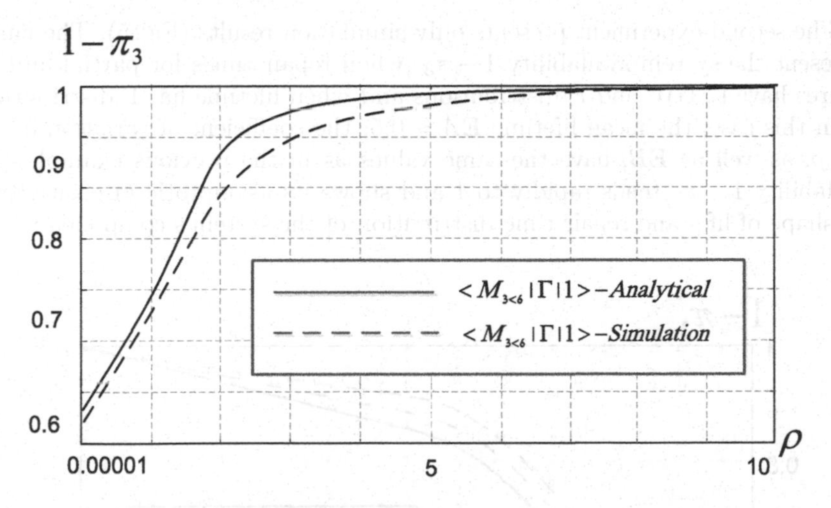

Fig. 3. Comparison of analytical and simulation results for the system availability $1 - \pi_3$, when $B(x) \sim \Gamma$.

As it is can be seen in Fig. 2, 3 and 4, the system availability values for these cases are very close to each other. For all values of ρ the difference between analytical and simulation values does not exceed 1%. Increasing of the relative recovery rate of the system's components provides a rapid increase of probability $1-\pi_3$ (both analytical and simulation) for all examined distributions. This result shows the veracity of simulation tools and insensitivity of the system to the shape of its components' repair time distributions.

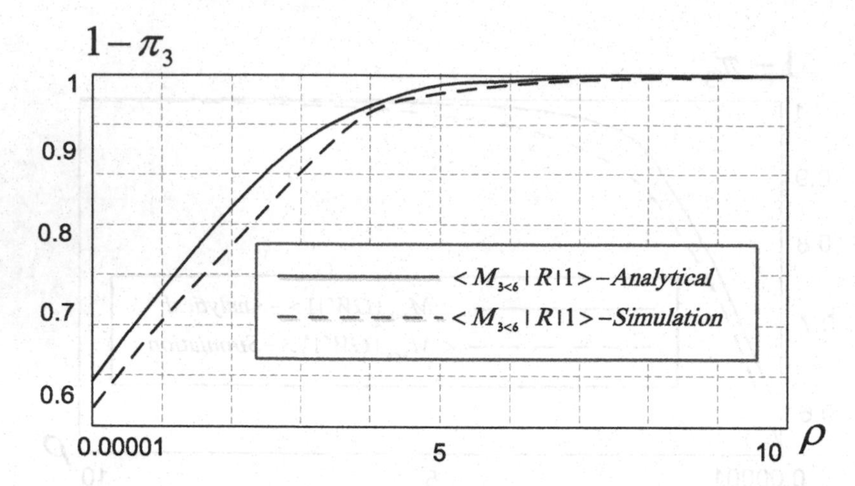

Fig. 4. Comparison of analytical and simulation results for the system availability $1 - \pi_3$, when $B(x) \sim R$.

The second experiment presents only simulation results (Fig. 5). The curves represent the system availability $1 - \pi_3$, when repair times for partial and full failures have Γ, GW and P distributions and when lifetime has Γ distribution.

In this case, the mean lifetime $\mathbf{E}A = 0.5$, the coefficient of variation $c = 10$ and ρ, as well as $\mathbf{E}B$, have the same values as in the previous example. The availability $1 - \pi_3$ tends rapidly to 1 and shows its asymptotic insensitivity to the shape of life- and repair time distribution of the system's components.

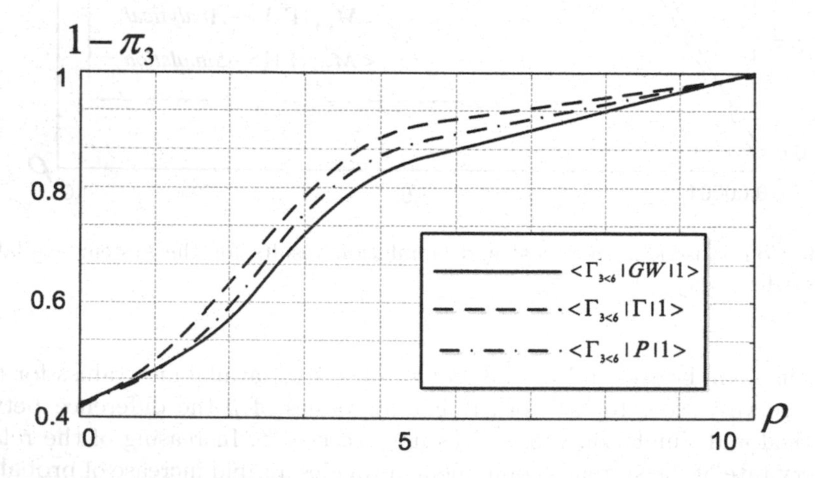

Fig. 5. The system availability $1 - \pi_3$ for the case of generally distributed life- and repair times (simulation results).

5 Conclusion

The analytical expressions for the s.s.p. of the 3-out-of-6 : F system with exponential lifetime and general repair time distributions have been found. For the same system the simulation model has been created in AnyLogic environment and applied for calculation of the system's reliability characteristics when both life- and repair times have general distributions. It was shown that as the relative recovery rate of the system's components increases, the system s.s.p. become asymptotically insensitive to the shapes of the life- and repair time distributions of the system's components. The subject of our further research is to find analytical expressions for the s.s.p. of a k-out-of-n : F system with exponential lifetime and general repair time distributions as well as investigation of more complex systems with dependent failures of their components.

References

1. Kala, Z.: Sensitivity analysis in probabilistic structural design: a comparison of selected techniques. Sustainability **12**(11), 19 (2020). https://doi.org/10.3390/su12114788
2. Kala, Z.: Quantile-oriented global sensitivity analysis of design resistance. J. Civ. Eng. Manage. **25**(4), 297–305 (2019). https://doi.org/10.3846/jcem.2019.9627. ISSN 1392-3730, E-ISSN 1822-3605
3. Kala, Z.: Estimating probability of fatigue failure of steel structures. Acta et Commentationes Universitatis Tartuensis de Mathematica **23**(2), 245–254 (2019). https://doi.org/10.12697/ACUTM.2019.23.21. ISSN 1406-2283, E-ISSN 2228-4699
4. Kala, Z.: Global sensitivity analysis of reliability of structural bridge system. Eng. Struct. **194**, 36–45 (2019). https://doi.org/10.1016/j.engstruct.2019.05.045. ISSN 1644–9665
5. Sevast'yanov, B.A.: An ergodic theorem for Markov processes and its application to telephone systems with refusals. Theory Probab. Appl. **2**(1), 104–112 (1957)
6. Kovalenko, I.N.: Investigations on analysis of complex systems reliability, p. 210. Naukova Dumka, Kiev (1976). (in Russian)
7. Rykov, V.: Multidimensional alternative processes reliability models. In: Dudin, A., Klimenok, V., Tsarenkov, G., Dudin, S. (eds.) BWWQT 2013. CCIS, vol. 356, pp. 147–156. Springer, Heidelberg (2013). https://doi.org/10.1007/978-3-642-35980-4_17
8. Gnedenko, B.V.: On cold double redundant system. Izv. AN SSSR. Texn. Cybern. **4**, 3–12 (1964). (in Russian)
9. Gnedenko, B.V.: On cold double redundant system with restoration. Izv. AN SSSR. Texn. Cybern. **5**, 111–118 (1964). (in Russian)
10. Solov'ev, A.D.: On reservation with quick restoration. Izv. AN SSSR. Texn. Cybern. **1**, 56–71 (1970). (in Russian)
11. Efrosinin, D., Rykov, V., Vishnevskiy, V.: Sensitivity of reliability models to the shape of life and repair time distributions. In: 9th International Conference on Availability, Reliability and Security (ARES 2014), pp. 430–437. IEEE (2014) https://doi.org/10.1109/ARES.2014.65

12. Rykov, V., Kozyrev, D.: On sensitivity of steady-state probabilities of a cold redundant system to the shapes of life and repair time distributions of its elements. In: Pilz, J., Rasch, D., Melas, V.B., Moder, K. (eds.) IWS 2015. SPMS, vol. 231, pp. 391–402. Springer, Cham (2018). https://doi.org/10.1007/978-3-319-76035-3_28
13. Rykov, V., Zaripova, E., Ivanova, N., Shorgin, S.: On sensitivity analysis of steady state probabilities of double redundant renewable system with Marshall-Olkin failure model. In: Vishnevskiy, V.M., Kozyrev, D.V. (eds.) DCCN 2018. CCIS, vol. 919, pp. 234–245. Springer, Cham (2018). https://doi.org/10.1007/978-3-319-99447-5_20
14. Trivedi, K.S.: Probability and Statistics with Reliability, Queuing and Computer Science Application. Wiley, New York (2002)
15. Vishnevsky, V.M., Kozyrev, D.V., Rykov, V.V., Nguyen, Z.F.: Reliability modeling of an unmanned high-altitude module of a tethered telecommunication platform. Inf. Technol. Comput. Syst. (4) (2020). (in Russian) (in print)
16. Kozyrev, D.V., Phuong, N.D., Houankpo, H.G.K., Sokolov, A.: Reliability evaluation of a hexacopter-based flight module of a tethered unmanned high-altitude platform. In: Vishnevskiy, V.M., Samouylov, K.E., Kozyrev, D.V. (eds.) DCCN 2019. CCIS, vol. 1141, pp. 646–656. Springer, Cham (2019). https://doi.org/10.1007/978-3-030-36625-4_52
17. Kuo, W., Zuo, M.J.: Optimal Reliability Modeling: Principles and Applications. Wiley, New York (2003)
18. She, J., Pecht, M.G.: Reliability of a k-out-of-n warm standby system. IEEE Trans. Reliab. **41**(1), 72–75 (1992)
19. Zhang, T., Xie, M., Horigome, M.: Availability and reliability of (k-out-of-($M+N$)): warm standby systems. Reliab. Eng. Syst. Saf. **91**, 381–387 (2006)
20. Yuge, T., Maruyama, M., Yanagi, S.: Reliability of a (k-out-of-n) system with common-cause failures using multivariate exponential distribution. Procedia Comput. Sci. **96**, 968–976 (2016)
21. El-Damcese, M., Shama, M.S.: Reliability analysis of a new k-out- of-n: G model. World J. Model. Simul. **16**(1), 3–17 (2020)
22. Rykov, V., Kozyrev, D., Filimonov, A., Ivanova, N.: On reliability function of a k-out-of-n system with general repair time distribution. Probab. Eng. Inf. Sci. 1–18 (2020). https://doi.org/10.1017/S0269964820000285
23. Cox, D.: The analysis of non-Markovian stochastic processes by the inclusion of supplementary variables. In: Mathematical Proceedings of the Cambridge Philosophical Society, vol. 51, no. 3, pp. 433–441 (1955). https://doi.org/10.1017/S0305004100030437
24. Rykov, V.V., Kozyrev, D.V.: Analysis of renewable reliability systems by Markovization method. In: Rykov, V.V., Singpurwalla, N.D., Zubkov, A.M. (eds.) ACMPT 2017. LNCS, vol. 10684, pp. 210–220. Springer, Cham (2017). https://doi.org/10.1007/978-3-319-71504-9_19

Prioritized Service of URLLC Traffic in Industrial Deployments of 5G NR Systems

Ekaterina Markova[1], Dmitri Moltchanov[1,2], Rustam Pirmagomedov[2], Daria Ivanova[1(✉)], Yevgeni Koucheryavy[2], and Konstantin Samouylov[1]

[1] Peoples' Friendship University of Russia (RUDN University), 6 Miklukho-Maklaya St., Moscow 117198, Russian Federation
{markova-ev,1042200068,samouylov-ke}@rudn.ru

[2] Tampere University, Kalevantie 4, 33100 Tampere, Finland
dmitri.moltchanov@tut.fi, rustam.pirmagomedov@tuni.fi, yk@cs.tut.fi

Abstract. The simultaneous support of enhanced mobile broadband (eMBB) and ultra-reliable low latency (URLLC) traffic types at the air interface in upcoming 5G New Radio systems is a challenging problem requiring new connection admission control and scheduling strategies. To enable this coexistence while still maintaining the prescribed quality-of-service guarantees the state of the art solutions utilize non-orthogonal multiple access and traffic isolation with explicit resource reservation. In this paper, we study an explicit prioritization of URLLC traffic over other services. Using the tools of queuing theory we mathematically characterize and investigate several techniques for priority-based resource allocation. Our results demonstrate that preemptive priority service is a viable option to fulfill strict delay and loss guarantees at the NR air interface. We also show that elasticity of lower priority eMBB service allows for additional capacity gains in terms of the eMBB session drop probability during the service.

Keywords: 5G · New Radio · Industrial NR · URLLC · Priority service

1 Introduction

In recent releases, 3GPP has specified three main types of services to be supported by 5G systems, enhanced mobile broadband (eMBB), massive machine-type communications (mMTC), and ultra-reliable low latency communication (URLLC). The former type is an enhancement of conventional mobility broadband service generated by existing and emerging applications requiring high rates

The publication has been prepared with the support of the "RUDN University Program 5-100" (recipient Samouylov K.). The reported study was funded by RFBR, project numbers 19-07-00933 and 20-37-70079 (recipients Markova E., Ivanova D.).

V. M. Vishnevskiy et al. (Eds.): DCCN 2020, LNCS 12563, pp. 497–509, 2020.
https://doi.org/10.1007/978-3-030-66471-8_38

at the access interface while the latter is a principally new one characterized by extreme reliability and latency requirements. While mMTC services addressed by technologies that are already available on the market (e.g., NB-IoT) eMBB and URLLC require innovations and expected to be supported by the developing NR technology.

Communication patterns in industrial environments can be characterized by two main attributes: periodicity and determinism [1]. Examples of a periodical transmission include updates of a position or the repeated monitoring of a characteristic parameter [2]. An aperiodic transmission is triggered instantaneously by an event, such events are defined by the control system (e.g., temperature or pressure threshold is exceeded) or by the user (e.g., remote diagnostic or maintenance events). Determinism refers to the latency of communication and jitter, which important for real-time services, such as closed-loop control or control-to-control communication. Actual manufacturing facilities commonly include all those communication patterns.

The resource demands of periodic communication sessions can be well addressed using static network planning methods, while aperiodic sessions may have high spatial and temporal variations, and thus less predictable. If the network planned based on the maximum demands possible in the covered area, the resource utilization of such network facilities would be low, while CAPEX and OPEX high. While reduced in the network capacity will compromise the reliability of communication. To reach a healthy balance between cost and reliability, a service model for different types of traffic is required for industrial deployments.

The high throughput requirements of URLLC traffic posse most sever demands for radio resource management subsystem [3–5]. While the reliability requirements can be satisfied using conventional repetition coding benefiting from the large bandwidth at NR radio interface the latency requirement of just 1 ms one-day delay is much harder to satisfy. So far researchers have envisioned two approaches for support of URLLC traffic. These are non-orthogonal multiple access (NOMA, [6,7]) and explicit resource reservation [8,9]. The NOMA approach is based on the use of special codes allowing for simultaneous reception of intentionally overlapped transmissions. One of the major NOMA advantages is the capability of satisfying one-way delay constrain of 1 ms at the radio interface as URLLC transmission can be scheduled for transmission in the frame they were received. However, this approach has not been included in NR standards [10].

A case when URLLC services require high throughput is considered as the most challenging from the perspective of radio resource management [3,4]. Resource reservation is a conventional approach for resource allocation for high and low priority traffic at the wireless interface. According to it, a certain fraction of resources can be exclusively assigned to URLLC traffic while the rest of the resources can be used for eMBB traffic. However, this approach is known to suffer from system severe resource under-utilization and/or potential loss of URLLC traffic as its arrival pattern may fluctuate in time [11]. The use of priority scheduling of resources efficiently alleviates these shortcomings by allowing URLLC and eMBB traffic to share the transmission resources [12]. However, once the aggregated load exceeds the number of available resources eMBB sessions might be dropped. Thus, assessing eMBB traffic performance in the presence of high priority URLLC sessions is one of the open research problems.

In this paper, we study the priority-based resource allocation for URLLC traffic in the presence of elastic eMBB traffic of lower priority with minimum rate requirements. To this aim, we formulate the set of queuing theoretic models with different types of priority preemptive discipline with two arriving flows of constant bit rate and elastic nature, respectively. The metrics of interest are related to eMBB traffic performance in the presence of high priority URLLC traffic as well as overall system throughput.

The paper is organized as follows. The system model is formulated in Sect. 2. The queuing models capturing the service process of URLLC and eMBB traffic at NR BS are formulated in Sect. 3. Numerical results are provided in Sect. 4. The conclusions are drawn in the last section.

2 System Model

In this section, we introduce the system model by specifying its components including deployment and service models at the NR air interface. Finally, we specify metrics of interest.

2.1 Deployment and Air Interface

We consider a single NR BS serving two types of traffic: deterministic URLLC sessions and elastic eMBB sessions. Such a scenario may reflect the service process of e.g., remote monitoring sessions and direct communications between operational entities in the industrial environment as illustrated in Fig. 1. We specifically concentrate on the shortcomings and advantages of service disciplines at the air interface and abstract the wireless specifics by directly formalizing the service process of URLLC and eMBB types of traffic.

2.2 Traffic Types and Service Models

URLLC sessions are assumed to generate streaming traffic (e.g., maintenance messaging service between operational entities), which is characterized by extremely small randomly distributed service duration and first come, first served (FCFS) service discipline. eMBB service (e.g., remote monitoring) is assumed to be elastic in nature, i.e., can adapt to the changing network conditions by regulating the encoding rate at the sending side. Recall that usually, elastic traffic is parameterized by a certain file size to be transmitted and characterized by a variable service duration and processor sharing (PS) service discipline. In contrast, in our study eMBB sessions are still elastic but characterized by randomly distributed service duration that is independent of the network conditions over the lifetime of a session.

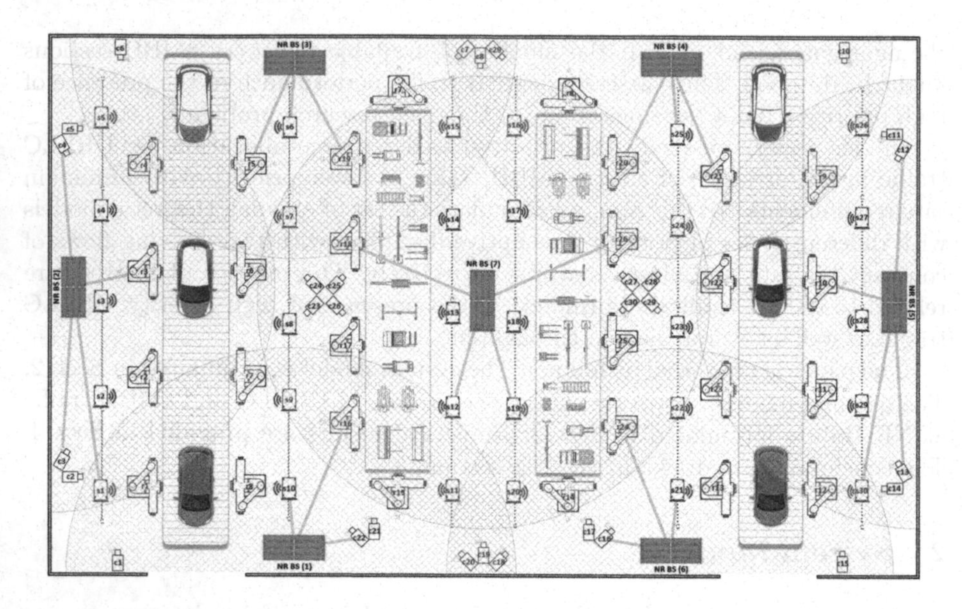

Fig. 1. Illustration of the service process.

We now introduce the main notation utilized in this work, see Fig. 2. We consider a system operating using a channel with a raw rate of C bandwidth units (b.u.). URLLC and eMBB sessions arrive according to the Poisson process with rates λ_1 and λ_2 sessions per second, respectively (Fig. 2). The session service times are exponentially distributed with means μ_1^{-1} and μ_2^{-1} seconds, respectively. Since URLLC sessions correspond to streaming traffic, their resource requirements are guaranteed and equal to b_1 b.u., $b_1 \geq 1$. We consider two types of minimum allocations [13] for eMBB traffic:

- **Strict strategy:** $b_2^{\min}$, $b_2^{\min} \geq 1$: in this case there is a single minimum threshold for eMBB sessions corresponding to the minimum monitoring session performance with satisfactory visual quality – if this threshold if violated the corresponding session is dropped during the service;
- **Flexible strategy:** $b_2^{\min_1}$ b.u., $b_2^{\min_1} \geq 1$: in this case there are two thresholds to protect eMBB sessions – if the amount of resources available for one or few eMBB sessions falls below the first threshold but remains above the second one, the session may still be served at reduced quality; the rationale is that URLLC sessions are extremely short-lived inducing just tiny load to the system implying that the resources available for eMBB sessions may recover.

Note that URLLC sessions are latency-sensitive, so we assume these are provided exclusive access to the transmission resources via preemptive-priority service discipline. This means that if there is a lack of system resources for a newly arriving URLLC session, the amount of resources allocated for the eMBB session can be reduced to the minimum requirement threshold $b_2^{\min_2}, b_2^{\min_1} > b_2^{\min_2} > 0$. If the amount of resources is still not sufficient to accept an arriving URLLC session one or more eMBB sessions can be interrupted. According to this service discipline, URLLC sessions can only be lost if their aggregated instantaneous load exceeds the system transmission resources.

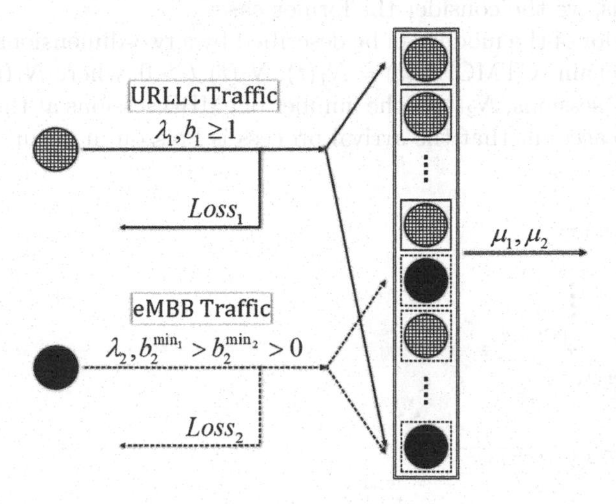

Fig. 2. Scheme model of the service process.

2.3 Parameterization and Metrics of Interest

To assess performance of the proposed service models in realistic environment one needs to provide the following metrics as a function of NR BS parameters, propagation characteristics and environment: (i) coverage of NR BS, (ii) mean resource request of URLLC and eMBB sessions. The former can be determined using the procedure offered in [14,15] while the latter can be found as discussed in [16]. Note that we also do not explicitly specify the operational frequency of NR technology. In case of millimeter wave band one also needs to account for blockage phenomena using the models proposed in [17,18].

In our study we concentrate on five main metrics of interest: (i) URLLC session drop probability, (ii) eMBB session drop probability, (iii) eMBB preemption probability (drop during the service), and (iv) resource utilization coefficient and (v) fraction of time eMBB session operates at reduced quality. These metrics are further utilized to make conclusions of the performance of the considered system.

3 Performance Evaluation Model

In this section we analyze the introduced system model for performance metrics of interest. We start with formalizing the model as a multi-dimensional Markov chain and then proceed characterizing its stationary state distribution and performance metrics of interest.

3.1 Model Formalization

The model with two resource requirements thresholds for eMBB sessions, $b_2^{\min_1}$ b.u., $b_2^{\min_1} \geq 1$, is a generalization of the one with a single threshold $b_2^{\min}$, $b_2^{\min} \geq 1$. Thus, we the consider the former case.

The behavior of the model can be described by a two-dimensional continuous-time Markov chain (CTMC) $\mathbf{X}(t) = N_1(t), N_2(t), t > 0$, where $N_1(t)$ is the number of URLLC sessions, $N_2(t)$ is the number of eMBB sessions at the time instant t. Taking into account that the arrival process is Poisson in nature while service

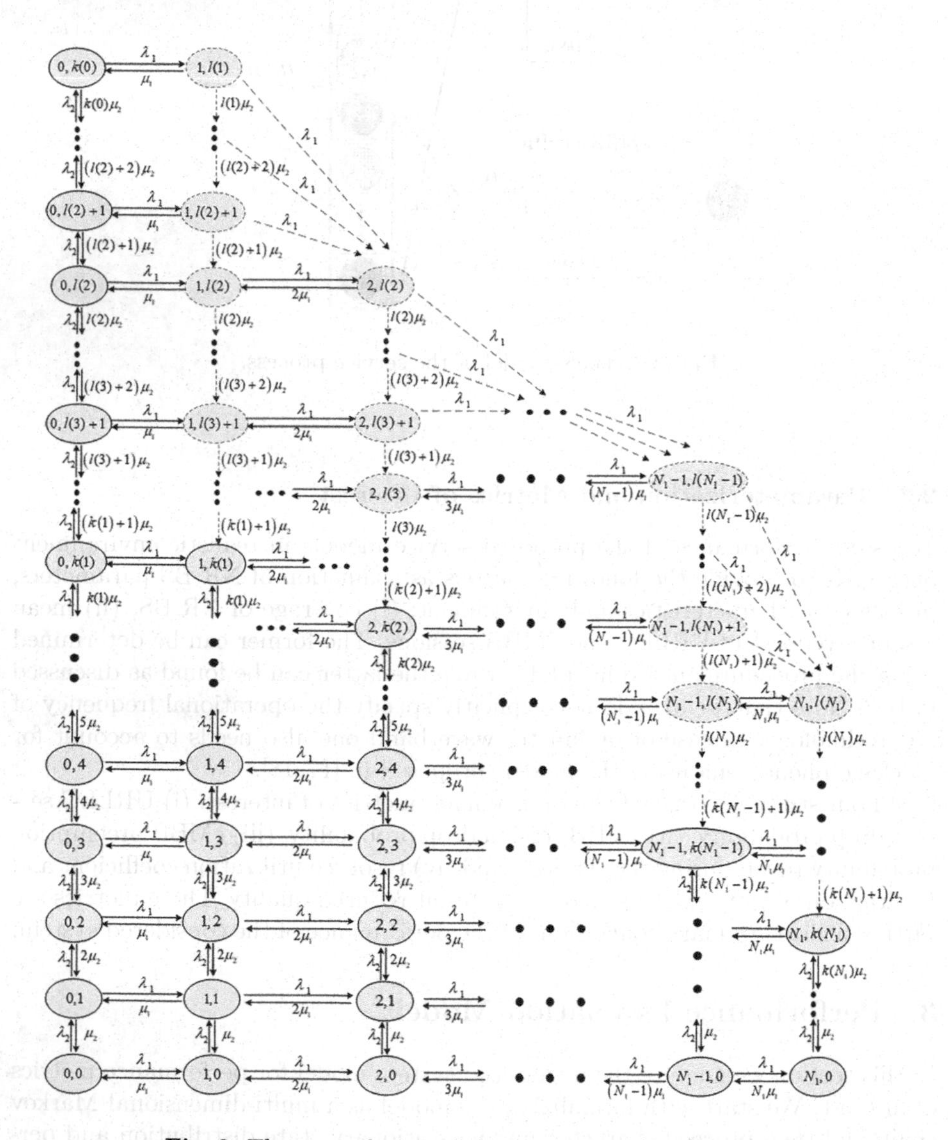

Fig. 3. The general view of the state transition diagram.

times are exponentially distributed, the state space of the Markov model $\mathbf{X}(t)$ is provided by

$$\mathbf{X} = \left\{ (n_1, n_2) : n_1 = 0, .., \left\lfloor \frac{C}{b_1} \right\rfloor, n_2 = 0, .., \left\lfloor \frac{C}{b_2^{\min_1}} \right\rfloor, n_1 b_1 + n_2 b_2^{\min_2} \leq C \right\}, \quad (1)$$

where n_1 and n_2 are the number of URLLC and eMBB sessions in the system.

Introduce the following notations: (i) $N_1 = \lfloor C/b_1 \rfloor$ – the maximum number of URLLC sessions that can be simultaneously served, (ii) $N_2 = \lfloor C/b_2^{\min_1} \rfloor$ – the maximum number of eMBB sessions that can be simultaneously served, (iii) $k(n_1) = \lfloor (C - n_1 b_1)/b_2^{\min_1} \rfloor$ – the maximum number of eMBB sessions, which could be simultaneously served, if the current number of current URLLC sessions in the system isn_1, (iv) $l(n_1) = \min(N_2, \lfloor (C - n_1 b_1)/b_2^{\min_2} \rfloor)$ – the maximum number of eMBB sessions, that can be served occupying the minimum of $b_2^{\min_2}$ resources, if the current number of URLLC sessions in the system is n_1. Further, we note that due to the elastic nature of eMBB sessions, characterized by the uniform distribution of all system resources between all the simultaneously served eMBB sessions, an achievable amount of occupied resources $b_2(n_1, n_2)$ for this type of sessions depends on the state $(n_1, n_2) \in \mathbf{X}$ of the system,

$$b_2(n_1, n_2) = \left\lfloor \frac{C - n_1 b_1}{n_2} \right\rfloor. \quad (2)$$

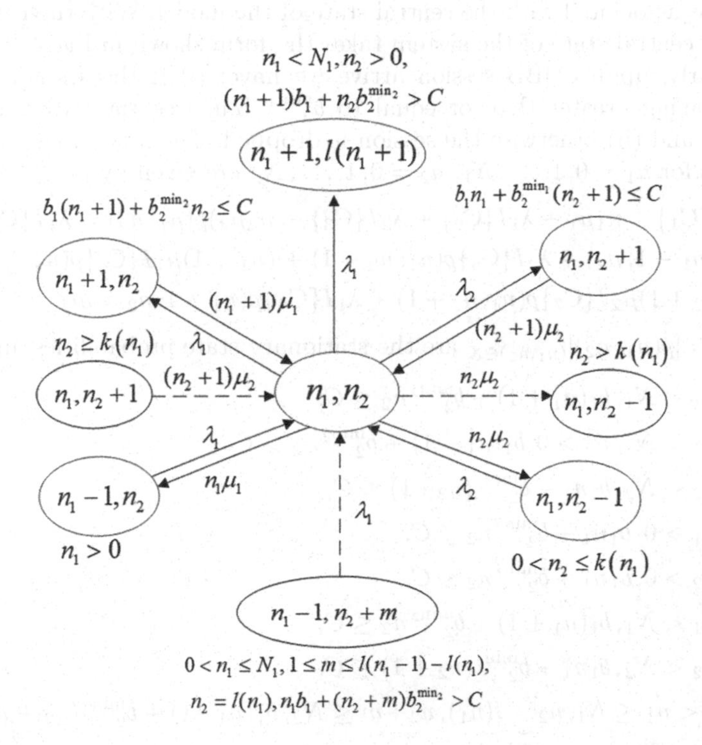

Fig. 4. Central state of the state transition diagram.

Further, we formalize the admission control mechanism, first, from the perspective of URLLC sessions, and, second, from the standpoint of the eMBB sessions. When a new URLLC session arrives, the following options are possible:

- if the system having free b.u. greater or equal to b_1, the URLLC session is accepted, the current number of eMBB sessions remains unchanged, and the amount occupied by eMBB sessions resources is higher than $b_2^{\min_1}$ b.u.;
- if the system having less than b_1 free b.u., the current number of served URLLC sessions is less than N_1, and the amount of occupied by eMBB sessions resources subject to the establishment of a new URLLC session is higher than $b_2^{\min_2}$ b.u., i.e. $b_2(n_1+1, n_2) \geq b_2^{\min_2}$, the URLLC session is accepted, and the amount of occupied resources by eMBB sessions is reduced to $\lfloor (C-(n_1+1)b_1)/n_2 \rfloor$ b.u., $b_2^{\min_2} \leq \lfloor (C-(n_1+1)b_1)/n_2 \rfloor < b_2^{\min_1}$;
- if the system having less than b_1 b.u. free, the current number of served URLLC sessions is less than N_1, the current number of served eMBB sessions are higher than 1, and the amount of occupied by eMBB sessions resources subject to the establishment of a new URLLC session is less than $b_2^{\min_2}$ b.u., i.e. $b_2(n_1+1, n_2) < b_2^{\min_2}$, the URLLC session is accepted and $\lfloor b_1/(b_2(n_1, n_2)) \rfloor$ served eMBB sessions will be preempted (dropped);
- otherwise the arrived URLLC session is dropped.

With the considered admission control rules, the overall structure of the state transition diagram of the CTMC is illustrated in Fig. 3, while Fig. 4 highlights the transitions associated with the central state of the model. With these admission rules, the central state of the system takes the form shown in Fig. 4.

Similarly, upon eMBB session arrive we have: (i) if the session finds the system having greater than or equal to $b_2^{\min_1}$ b.u. free the eMBB session is accepted, and (ii) otherwise the session is dropped. The associated equilibrium equations for $n_1 = 0, 1, \ldots, N_1$, $n_2 = 0, 1, \ldots, N_2$ are given by

$$
\begin{aligned}
&(\lambda_1 I\{\mathrm{C}_1\} + n_1\mu_1 + \lambda_1 I\{\mathrm{C}_2\} + \lambda_2 I\{\mathrm{C}_3\} + n_2\mu_2) p(n_1, n_2) = \lambda_1 I\{\mathrm{C}_4\} \\
&\times p(n_1-1, n_2) + \lambda_2 I\{\mathrm{C}_5\} p(n_1, n_2-1) + (n_1+1)\mu_1 I\{\mathrm{C}_6\} p(n_1+1, n_2) \\
&+ (n_2+1)\mu_2 I\{\mathrm{C}_7\} p(n_1, n_2+1) + \lambda_1 I\{\mathrm{C}_8\} p(n_1-1, n_2+m),
\end{aligned} \tag{3}
$$

where $\mathbf{p} = [p(n_1, n_2)]_{(n_1,n_2)\in\mathbf{X}}$ are the stationary state probabilities and

$$
\begin{aligned}
&\mathrm{C}_1 := n_1 < N_1, b_1(n_1+1) + b_2^{\min_2} n_2 \leq C, \\
&\mathrm{C}_2 := n_1 < N_1, n_2 > 0, b_1(n_1+1) + b_2^{\min_2} n_2 > C, \\
&\mathrm{C}_3 := n_2 < N_2, b_1 n_1 + b_2^{\min_1}(n_2+1) \leq C, \\
&\mathrm{C}_4 := n_1 > 0, b_1 n_1 + b_2^{\min_2} n_2 \leq C, \\
&\mathrm{C}_5 := n_2 > 0, b_1 n_1 + b_2^{\min_1} n_2 \leq C, \\
&\mathrm{C}_6 := n_1 < N_1, b_1(n_1+1) + b_2^{\min_2} n_2 \leq C, \\
&\mathrm{C}_7 := n_2 < N_2, b_1 n_1 + b_2^{\min_2}(n_2+1) \leq C, \\
&\mathrm{C}_8 := 0 < n_1 \leq N_1, n_2 = l(n_1), n_2+m \leq N_2, b_1(n_1-1) + b_2^{\min_2}(n_2+m) \leq C, \\
&\quad b_1 n_1 + b_2^{\min_2}(n_2+m) > C.
\end{aligned} \tag{4}
$$

3.2 Solution Methodology

As a result of the preemptive-priority service mechanism, the CTMC $\mathbf{X}(t)$ is not a reversible Markov chain, so the stationary state probability distribution $p(n_1, n_2), (n_1, n_2) \in \mathbf{X}$ does not have product form. However, one can determine it numerically. For this purpose, we rewrite the system of equilibrium equations (3) as follows

$$\mathbf{pA} = \mathbf{0}, \mathbf{p1}^T = 1, \tag{5}$$

where $\mathbf{A}$ is the infinitesimal generator.

Having the stationary state probability distribution $p(n_1, n_2), (n_1, n_2) \in \mathbf{X}$, one can compute the performance measures of the considered system as follows:

- URLLC session drop probability B_1 is given by

$$B_1 = \sum_{i=0}^{l(N_1)} p(N_1, i). \tag{6}$$

- eMBB session drop probability B_2 is given by

$$B_2 = \sum_{i=0}^{N_1} \sum_{j=k(i)}^{l(i)} p(i, j). \tag{7}$$

- eMBB session preemption probability Π is given by

$$\Pi = \sum_{i=0}^{N_1 - 1} \sum_{j=l(i+1)+1, l(i) \neq l(i+1)}^{l(i)} \frac{\lambda_1 p(i,j)}{\lambda_1 + \lambda_2 I\{j < k(i)\} + i\mu_1 + j\mu_2}. \tag{8}$$

- mean resource utilization, U, is given by

$$U = C \sum_{i=0}^{N_1} \sum_{j=1}^{l(i)} (i + j) p(i, j) + b_1 \sum_{i=1}^{N_1} i p(i, 0). \tag{9}$$

- fraction of time ω an arbitrary eMBB session is served with reduced quality is given by

$$\omega = \sum_{i=1}^{N_1} \sum_{j=k(i)+1, k(i)<l(i)}^{l(i)} j p(i, j) \Big/ \sum_{i=0}^{N_1} \sum_{j=0}^{l(i)} j p(i, j). \tag{10}$$

4 Numerical Results

In this section we elaborate on numerical results. We mainly concentrate on evaluating the use of preemptive-priority serving discipline for serving mixture of URLLC and eMBB traffic and comparing the proposed service schemes for eMBB traffic.

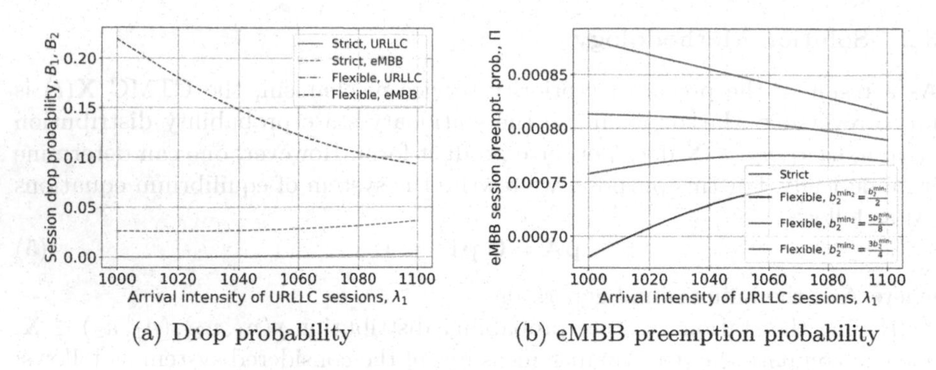

(a) Drop probability (b) eMBB preemption probability

Fig. 5. Session drop and preemption probabilities.

The default system parameters are shown in Table 1. The coverage of NR BS is determined using the procedure offered in [14,15] while the mean resource requests of URLLC and eMBB sessions are obtained found as discussed in [16]. To approximate URLLC traffic transmission latency, we set the associated mean service time to NR frame duration, 1 ms. To ensure reliable delivery of data within this deadline we assume repetition coding, i.e., the same replica of the URLLC message is repeated three times within the same NR frame.

Table 1. Parameters utilized for numerical assessment.

Parameter	Value
Carrier frequency	28 GHz
NR BS bandwidth	20 MHz
Transmit power	0.2 W
NR BS side antenna array	16×4
UE side antenna array	4×4
NR BS and UE heights	6 m, 1 m
Arrival intensity of URLLC sessions	$1000 \rightarrow 1100$
Arrival intensity of eMBB sessions	2
Mean service time of URLLC sessions	0.1 s
Mean service time of eMBB sessions	120 s
Rate of URLLC sessions	0.8 Mbps
Minimum rate of eMBB sessions	10 Mbps

We start analyzing the system with Fig. 5 showing the session drop and preemption probabilities as a function of URLLC session arrival intensity. As one may notice, the URLLC session drop probabilities remain unchanged for both considered strategies. However, the considered strategies not only lead to drastically different absolute values of eMBB session drop probabilities but characterized by principally different qualitative behavior. The absolute values of these strategies coincide only when the URLLC arrival intensity increases. It is important that the flexible strategy leads to much higher eMBB session drop probabilities by it is compensated by significantly lower preemption probabilities.

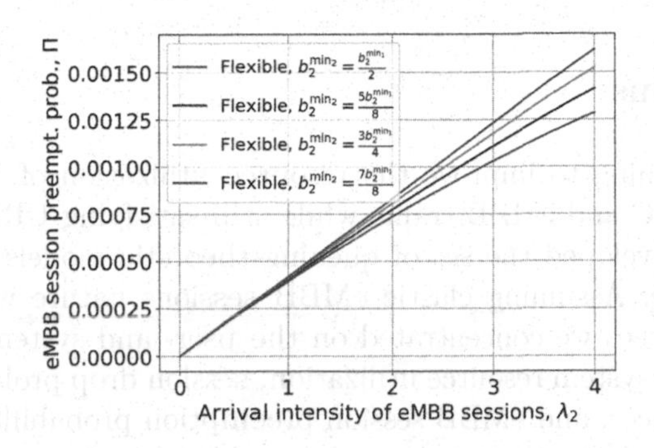

Fig. 6. Scheme model of the service process.

To complement the results illustrated in Fig. 5(b) we also demonstrate the dependence of eMBB session preemption probability on the arrival intensity of eMBB sessions in Fig. 6 for flexible strategy with different values of $b_2^{\min_2}$. As one may observe, contrarily to the effect of URLLC session arrival intensity, the difference induced by choosing $b_2^{\min_2}$ is negligible and the major effect is produced by the eMBB session arrival intensity.

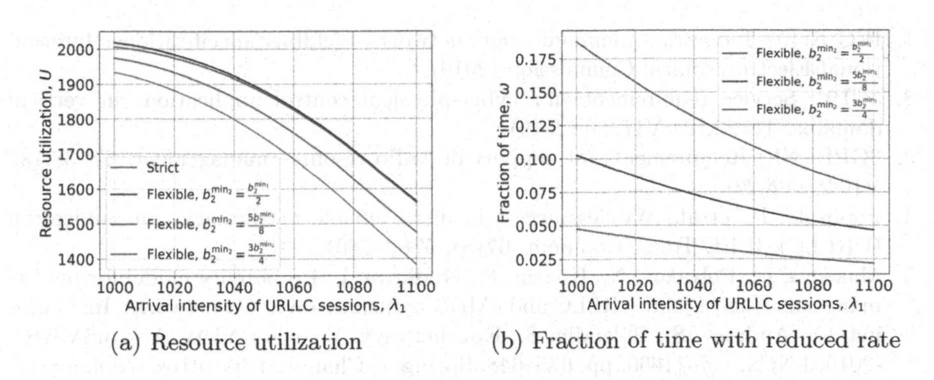

(a) Resource utilization (b) Fraction of time with reduced rate

Fig. 7. Resource utilization and fraction of time with reduced rate.

Assessing the effect of considered strategies on the resource utilization shown in Fig. 7, we note that the flexible strategy allowing eMBB sessions to spend some time with degraded quality leads to much better usage across the whole range of considered URLLC arrival intensities. The price one has to pay for this gain is shown in Fig. 7(b) illustrating the fraction of time eMBB sessions served with degraded quality. We note that the usage of two limits of acceptable rates for eMBB sessions provides a simple and efficient method for network operators to balance QoS and resource utilization while maintaining absolute service priority for URLLC traffic.

5 Conclusions

In this paper, aiming to improve the resource utilization of NR BS serving a mixture of URLLC and eMBB traffic while still satisfying URLLC traffic delay constraints we developed the set of queuing-theoretic models with preemptive priority discipline. Assuming elastic eMBB sessions nature with two types of minimum guarantees we concentrated on the user- and system-level metrics of interest including system resource utilization, session drop probability for eMBB and URLLC services, and eMBB session preemption probability.

Our results indicate that compared to static resource reservation mechanism, priority-based service allows to reach 80–90% of system resource utilization while still maintaining latency guarantees to URLLC sessions. On top of this, compared to dynamic resource reservation, the proposed resource allocation procedure does not require on-line resource reallocation procedures. The proposed service strategy can be utilized in those environments, where NR BS are expected to serve dynamically changing loads of URLLC and eMBB traffic, e.g., autonomous factories.

References

1. IEC 61158 Industrial communication networks-Fieldbus specifications. International Electrotechnical Commission (2019)
2. 3GPP: Service requirements for cyber-physical control applications in vertical domains. TS 22.104 V17.2.0 (2019)
3. 3GPP: NR; Requirements for support of radio resource management. TS 38.133 V16.2.0 (2020)
4. Popovski, P., et al.: Wireless access in ultra-reliable low-latency communication (URLLC). IEEE Trans. Commun. **67**(8), 5783–5801 (2019)
5. Makeeva, E., Polyakov, N., Kharin, P., Gudkova, I.: Probability model for performance analysis of joint URLLC and eMBB transmission in 5G networks. In: Galinina, O., Andreev, S., Balandin, S., Koucheryavy, Y. (eds.) NEW2AN/ruSMART -2019. LNCS, vol. 11660, pp. 635–648. Springer, Cham (2019). https://doi.org/10.1007/978-3-030-30859-9_55
6. Ding, Z., et al.: Application of non-orthogonal multiple access in LTE and 5G networks. IEEE Commun. Mag. **55**(2), 185–191 (2017)
7. Ding, Z., Lei, X., Karagiannidis, G.K., Schober, R., Yuan, J., Bhargava, V.K.: A survey on non-orthogonal multiple access for 5G networks: research challenges and future trends. IEEE J. Sel. Areas Commun. **35**(10), 2181–2195 (2017)

8. She, C., Yang, C., Quek, T.Q.S.: Radio resource management for ultra-reliable and low-latency communications. IEEE Commun. Mag. **55**(6), 72–78 (2017)
9. Pandey, S.R., Alsenwi, M., Tun, Y.K., Hong, C.S.: A downlink resource scheduling strategy for URLLC traffic. 2019 IEEE International Conference on Big Data and Smart Computing (BigComp), pp. 1–6. IEEE (2019)
10. Chen, Y., et al.: Toward the standardization of non-orthogonal multiple access for next generation wireless networks. IEEE Commun. Mag. **56**(3), 19–27 (2018)
11. Begishev, V., et al.: Resource allocation and sharing for heterogeneous data collection over conventional 3GPP LTE and emerging NB-IoT technologies. Comput. Commun. **120**, 93–101 (2018)
12. Korshykov, M.V., Daraseliya, A.V., Sopin, E.S.: Development of Analytical Framework for Evaluation of LTE-LAA Probabilistic Metrics. In: Galinina, O., Andreev, S., Balandin, S., Koucheryavy, Y. (eds.) NEW2AN/ruSMART -2019. LNCS, vol. 11660, pp. 318–328. Springer, Cham (2019). https://doi.org/10.1007/978-3-030-30859-9_27
13. Naumov, V., Beschastnyi, V., Ostrikova, D., Gaidamaka, Y.: 5G new radio system performance analysis using limited resource queuing systems with varying requirements. In: Vishnevskiy, V.M., Samouylov, K.E., Kozyrev, D.V. (eds.) DCCN 2019. LNCS, vol. 11965, pp. 3–14. Springer, Cham (2019). https://doi.org/10.1007/978-3-030-36614-8_1
14. Begishev, V., et al.: Quantifying the impact of guard capacity on session continuity in 3GPP new radio systems. IEEE Trans. Veh. Technol. **68**, 12345–12359 (2019)
15. Samuylov, A., et al.: Characterizing resource allocation trade-offs in 5G NR serving multicast and unicast traffic. IEEE Trans. Wirel. Commun. **19**, 3421–3434 (2020)
16. Kovalchukov, R., Moltchanov, D., Gaidamaka, Y., Bobrikova, E.: An accurate approximation of resource request distributions in millimeter wave 3GPP new radio systems. In: Galinina, O., Andreev, S., Balandin, S., Koucheryavy, Y. (eds.) NEW2AN/ruSMART -2019. LNCS, vol. 11660, pp. 572–585. Springer, Cham (2019). https://doi.org/10.1007/978-3-030-30859-9_50
17. Gapeyenko, M., et al.: Analysis of human-body blockage in urban millimeter-wave cellular communications. 2016 IEEE International Conference on Communications (ICC), pp. 1–7. IEEE (2016)
18. Kovalchukov, R., et al.: Evaluating SIR in 3D millimeter-wave deployments: direct modeling and feasible approximations. IEEE Trans. Wirel. Commun. **18**(2), 879–896 (2019)

Milestone Developments in Quantum Information and No-Go Theorems

K. K. Sharma[1], V. P. Gerdt[2,3](✉), and P. V. Gerdt[2]

[1] DY Patil International University, Sect-29, Nigdi Pradhikaran, Akurdi, Pune 411044, Maharashtra, India
iitbkapil@gmail.com

[2] Joint Institute for Nuclear Research, Dubna 141980, Russian Federation
gerdt@jinr.ru, gapon1970@yandex.ru

[3] Peoples' Friendship University of Russia, Moscow 117198, Russian Federation

Abstract. In this article we present milestone developments in the theory and application of quantum information from historical perspectives. The domain of quantum information is very promising to develop quantum computer, quantum communication and varieties of other applications of quantum technologies. We also give the light on experimental manifestations of major theoretical developments. In addition, we present important no-go theorems frequently used in quantum information along with ideas of their respective mathematical proofs.

Keywords: Quantum information · Quantum communication · Milestone development · No-go theorem

1 Introduction

A breakthrough in the classical information theory has been done by C. Shannon in 1948 [1]. This theory has its practical manifestations in communication systems, computing devices, gaming, imaging and with countless applications in real world. The theory of quantum information, which generalizes Shannon's theory stems from the postulates of quantum mechanics and based on such fundamental ingredients of quantum mechanics as superposition and entanglement. Based on quantum information theory, the efforts to develop quantum computer are on the way by using several physical techniques. Many companies in the market are eagerly applying efforts to push this area towards commercialization and to develop quantum computer with different physical approaches like super conducting approach, ion trap system, ultra cold atoms, Majorana fermions, etc. (see recent review [2] and its bibliography). A number of quantum computers with several dozens qubits have been created at the laboratory level, and in January 2019 IBM (https://www.ibm.com/quantum-computing/)

The publication has been prepared with the support of the "RUDN University Program 5-100".

V. M. Vishnevskiy et al. (Eds.): DCCN 2020, LNCS 12563, pp. 510–525, 2020.
https://doi.org/10.1007/978-3-030-66471-8_39

announced the first commercial quantum computer. We may fire a natural question: can we store and process the information in these physical systems more efficiently than classical one? And what are the physical constraints responsible to execute quantum information [3,4]? So, investigating situations which are not possible is also an important paradigm. These impossible physical conditions are expressed by no-go theorems [5,6]. Before applying the fruitful efforts to develop any quantum application it is always good to keep in view the structure of no-go theorems, which is always helpful to tackle the feasibility and non-feasibility of physical situations. On the other hand, towards development of quantum computing, there are always challenges to manipulate and control the qubits and to protect them from decoherence. The phenomenon of decoherence is the killer of superposition in quantum systems and a serious restriction on the way to perfect quantum computation. Recent developments in quantum computation are very progressive and more rapid than the past historical developments. So, in this direction it is very important to understand the gradual milestone developments and track them what may be useful for further progress in the future. In the following sections, we present some major developments with theoretical aspects and touching the experimental issues as well. We managed the time periods of developments in two slots (1970–2000) and (2001–2020).

2 Duration (1970–2000)

This period is most significant for the theoretical development of quantum information and known for producing the idea of reversible computation by C. Bennett [7] and also by T. Toffoli [8] who invented the first reversible n-bit quantum gates which are heavily used in the circuit model of quantum computation. Another milestone development is the foundation of Holevo bound. A. Holevo has established the upper bound for amount of information that can be contained in a quantum system [9]. Then one of the first attempts to create the quantum information theory was made by R. Ingarden, a Polish mathematical physicist, who published a seminal paper entitled "Quantum information theory" in 1976 [10]. This work generalizes Shannon's information theory [1] for quantum mechanics of open systems. With the progress of quantum information theory, the idea towards quantum computing was proposed by Yu. Manin [11] in 1980. Then R. Feynman argued [12] that for simulation of quantum systems one should use computational devices based on quantum physics, since such simulation tasks are very hard for classical computers. In 1982, a major result of no-cloning in quantum physics was discovered by W. Wootters and W. Zurek [14] and independently by D. Dieks [15]. The no-cloning theorem states that it is not possible to clone arbitrary quantum states. This theorem became the milestone for quantum information. We are inclined to present this theorem with its proof in Sect. 4.1. With this major development, P. Benioff proposed a first theoretical model for quantum computation based on quantum Hamiltonian [16]. He made the first attempt to quantize the Turing machine and the framework of quantum Turing machine has taken place. The concept of entanglement has already taken birth during 1935 and 1936 with the debate of A. Einstein and E. Schrödinger [17,18]. The advantage of entanglement and no-cloning theorem together captured the discovery of the first protocols for quantum cryptography (quantum key distribution) done by Ch. Bennet and G. Brassard in 1984 [19] and A. Ekert in 1991 [20]. The development of quantum cryptography opened the new field of secure quantum communication, which is very promising to this date.

The last decade of the XX century extensively contributed to the development of entanglement-based quantum algorithms. In 1992, D. Deutsch and R. Jozsa proposed a deterministic quantum algorithm to test weather a function is balanced or constant by using black box model in quantum computation [21]. With continuation of this work, the first milestone quantum algorithm was formulated by P. Shor at Bell Labs (New Jersey) in 1994 and published in 1997 [22]. The algorithm allows a quantum computer to factor an integer number very fast and runs in polynomial time. This algorithm can easily break the public-key cryptographic schemes as RSA scheme [24]. Meanwhile, as the development on quantum algorithms, P. Shor and A. Steane proposed the schemes for quantum error corrections in 1995 [23,25]. Quantum error corrections protocols are used to protect quantum information from decoherence and essentially needed in "noisy" quantum hardware. After design of Shor's algorithm, L. Grover in 1996 [26] invented the quantum search algorithm in an unsorted database. That algorithm is the fastest one and has become a landmark in quantum computation. Here we mention that the period of (1990–1997) has been recognized as the golden period for theoretical as well as experimental developments in quantum computation. Beside the quantum algorithms development, there also was discovered important protocol of quantum information processing called quantum teleportation. It was proposed by C. Bennett et al. in 1993 [27] and experimentally verified in 1997 [28]. During 1997 the scientific community was strongly focused on experimental manifestations of quantum information around the world. The first experimental approach to realize the quantum gates by using nuclear magnetic resonance (NMR) technique was performed by N. Gershenfeld and I. Chuang in 1997 [29]. NMR technique came out as a useful resource to produce fruitful experimental manifestations of quantum computation. In 1998, the first execution of Deutsch-Jozsa algorithm was performed by using just NMR technique. This has been done by J. Jones and M. Mosca at Oxford University and shortly later by I. Chuang at IBM's Almaden Research Center together with co-workers at Stanford University and MIT [30]. In the same year Grover's algorithm was also experimentally verified with NMR quantum computation [31]. This experimental development encouraged the further investigations. Beside the theoretical and experimental manifestations, it was also major interest to look into some physical situations which are not feasible like the non-cloning theorem. Towards this direction, there is one important quantum no-deletion theorem proved by A. Pati and S. Braunstein [32], which states that, given two copies of arbitrary qubit states, one cannot delete one of them if it is unknown. This theorem has its own important implications in quantum information [32]. We are inclined to discuss this theorem in Sect. 4.4.

3 Duration (2001–2020)

This period is well known for the role of quantum optics in quantum information, in parallel with another major developments towards the implementation of quantum networks. In 2001, the first experimental execution of Shor's algorithm at IBM's Almaden Research Center and Stanford University was implemented by using NMR technique [33]. The number 15 was factored by using 10^{18} identical molecules. In the same year, the scenario of optical quantum computing has been started. E. Knill, R. Laflamme and G. Milburn showed that optical quantum computing is possible with beam splitters, phase shifters, single photon sources and photo-detectors [34]. They also have shown that quantum teleportation can be performed with beam splitters by using photonic qubits. Their contribution opened the avenues of usage of optics in quantum information. The role of optics is very promising nowadays to establish long distance quantum communication. The implementation of quantum gates with optics is an essential requirement to perform quantum computation. In this direction, quantum Controlled-Not gates using linear optical elements has been developed by T. Pittman and collaborators at Applied Physics Laboratory, Johns Hopkins University in 2003 [35]. The similar results have been produced independently by J. O'Brien and collaborators at the University of Queensland [36]. Quantum optics not only had its applications in quantum cryptography but DARPA Quantum network also became operational by using optical fibers supporting the transmission of entangled photons [37]. Quantum networks use the protocol called quantum repeater for long distance quantum communication to overcome decoherence. These quantum repeaters transmit the quantum states to receiver with the help of quantum memories. The recognizable framework of quantum optics with atom-photon interaction proved to be a successful framework and assisted to develop quantum memories [38], which are essential to establish quantum Internet [39]. In 2005, at Harvard University and Georgia Institute of Technology researchers succeeded in transferring quantum information between "quantum memories" from atoms to photons and back again [40]. Along with the advancement of quantum networks, the concept of distributed quantum computing has taken place and a protocol called quantum telecloning was proposed by M. Murao et al. in 1999 [41]. This is the protocol in which the optical clones of an unknown quantum state are created and distributed over distant parties. Of course, the quantum no-cloning theorem implies that these copies cannot be perfect. S. Braunstein at the University of York together with his colleagues from the University of Tokyo and the Japan Science and Technology agency gave the first experimental demonstration of quantum telecloning in 2006 [42]. Quantum networks and quantum repeaters attracted much attention of quantum community, and along this line of research the concept of entanglement swapping [43] was developed by S. Pirandola et al. in 2006 which has its important application in quantum repeaters. Beside the developments on quantum memory by using the optical techniques, there was also interest to develop such memory by using the condensed matter approach. It was done in 2007 by using the Bose-Einstein condensation [44]. Till 2007, the experimental manifestation of two-qubit entanglement has been successfully performed, but the entanglement in hybrid systems also attracted attention of quantum community. Much progress was done in 2008 to perform photonic qubit-qutrit entanglement [45]. In the direction of implementation of quantum networks and towards the reality of quantum Internet, the logic gates have been implemented in optical fibers by P. Kumar, which became the foundation of quantum networks [46].

The continuity of past developments in quantum information and its experimental manifestations were maintained in this era with two major center of interest: how to develop efficient quantum processors and how to increase the coherence time in quantum systems. On the other hand, few past records also broken in this era. As a continuation, in 2011, the von Neumann's architecture was employed in quantum computing with superconducting approach [47]. This work contributes in developing quantum central processing unit that exchanges the data with a quantum random-access memory integrated on a chip. There was a breakthrough in 2014, as scientists transferred data by quantum teleportation over the distance of 10 feet with zero percent error rate, this was a vital step towards a feasible quantum internet [48]. In the same year, N. Dattani and N. Bryans broken the record for factoring the unbeatably largest number 56153 using NMR with 4 qubits only on a quantum device which outperformed the record established in 2012 with factoring the number 143 [49]. After a long journey for development in quantum computation, there are still many theoretical and experimental open problems inherited in the essence of quantum information. One of the major issues is controlling entanglement and its manipulation in many-body quantum systems with its protection from decoherence. There have been successful efforts to increase the coherence time in 2015 to six hours in nuclear spins [50]. With the advancement of quantum processors, there was breakthrough in 2017 by D-wave. This company developed commercially available quantum annealing based quantum processor, which is fully functional now and have been used for varieties of optimization problems [51] with applications in quantum machine learning. The recent progress in quantum computing hardware is described in [2]. With the connection of improving coherence time and deeper theoretical investigations on entanglement, here we mention that entanglement is a fragile phenomenon and very sensitive to quantum measurements and environmental interactions. It may die for a finite time in a quantum system and alive again as time advances. This phenomenon is called entanglement sudden death (ESD) which was observed earlier by Yu-Eberly [52,53] and investigated in [54–60]. The phenomenon of ESD is a threat to quantum applications, so overcoming from it is again an issue and needs fruitful solutions. During the period 2011–2020, there have been vast research on entanglement and related aspects such as distillable entanglement and bound entanglement in quantum information theory initiated in [61]. Quantum community has investigated various mathematical tools of entanglement detections and quantification, distillable protocols, monogamy of entanglement. However, these aspects are lacking for higher dimensional quantum systems. The efforts of quantum community is always to search the quantum systems which can sustain long coherence time, which is an important topic of research. For more recent advances in quantum information and communication we refer to the review [62]. We single practical achievements in quantum cryptography [83,84].

4 No-Go Theorems

In this section we consider important no-go theorems in theory of quantum information [63]. A no-go theorem implies the impossibility of a particular physical situation [5,6]. These theorems have the major impact on the experimental development of quantum information. All these theorems are developed by taking the linear property of quantum mechanics. Here in the following subsections we discuss important no-go theorems with their corresponding proofs.

4.1 No-Cloning Theorem

The no-cloning theorem states that one can not create an identical copy of an arbitrary pure quantum state. No-cloning theorem is provided by L. Park in 1970 [13], then further re-investigated in 1982 by W. Wootters and W. Zurek [14] and separately by D. Dieks [15]. The theorem of quantum cloning is easy to prove. Here we give a proof of this theorem based on unitarity of the underlying operation.

Proof. Consider two pure states as $|\psi\rangle$, $|\phi\rangle$ and a blank state $|b\rangle$. Mixing each pure state with blank state and perform the unitary operation which has the goal to copy the pure state into a blank state. So we get,

$$U(|\psi\rangle \otimes |b\rangle) = |\psi\rangle \otimes |\psi\rangle\,, \qquad U(|\phi\rangle \otimes |b\rangle) = |\phi\rangle \otimes |\phi\rangle. \tag{1}$$

Taking the complex conjugate of both the sides of both the above equations we obtain

$$(\langle\psi| \otimes \langle b|)U^{\dagger} = \langle\psi| \otimes \langle\psi|. \tag{2}$$

By multiplying the left and right sides of Eq. (1) and Eq. (2) we find

$$(\langle\psi| \otimes \langle b|)U^{\dagger}U(|\phi\rangle \otimes |b\rangle) = (\langle\psi| \otimes \langle\psi|)(|\phi\rangle \otimes |\phi\rangle).$$

We know that $U^{\dagger}U = I$ and therefore

$$\langle\psi|\phi\rangle = \langle\psi|\phi\rangle^2.$$

The equation is conflicting and it is true only if $\langle\psi|\phi\rangle = 0$, or $|\psi\rangle = |\phi\rangle$. These conditions reveal that there is no unitary operation which can be used to clone an arbitrary quantum state. □

4.2 No-Broadcast Theorem

This theorem generalizes the no-cloning theorem to mixed states. The first attempt to prove that non-commuting mixed states can no be broadcast was done by H. Barnum et al. in [68]. Further extensions are done by many authors. To look into broad view of no-broadcasting and different paradigms, we refer to [68–73]. Before formulation of the theorem we explain what is meant by broadcasting (cf. [63]). Let we have a mixed state ρ in H and a map E that maps ρ to a state ρ_{AB} on $H_A \otimes H_B$. Then E broadcasts ρ if

$$\mathrm{Tr}_A(\rho_{AB}) = \mathrm{Tr}_B(\rho_{AB}) = \rho\,, \tag{3}$$

where Tr_A and Tr_B denote partial trace over the subsystem A and subsystem B, respectively. Note that the final state ρ_{AB} is not necessarily a product state.

The no-broadcast theorem states that no such map E exists for an arbitrary quantum state, i.e., arbitrary quantum states cannot be broadcasted. Alternatively, given an arbitrary quantum state ρ, it is impossible to create a state ρ_{AB} such that the equality (3) holds. Moreover, the two mixed states ρ_1, ρ_2 in H can satisfy (3) if and only if they commute [68].

Proof. We outline here the main idea of the proof given in [72] that is the most general one among known proofs and relies on fundamental principles of information theory, especially on entropy-based arguments. This proof allows to establish a link of the quantum theorem to its classical analogue [74].

The relative entropy $S(\rho_1|\rho_2) = \mathrm{Tr}[\rho_1(\log\rho_1 - \log\rho_2)]$ of states ρ_1 and ρ_2 is conserved under unitary time evolution

$$S(\rho_1(t)|\rho_2(t)) = S(\rho_1(0) \mid \rho_2(0)) \tag{4}$$

and it is monotone, i.e.,

$$S(\rho_{1,AB}|\rho_{2,AB}) \geq S(\rho_{1,B}|\rho_{2,B})\,, \tag{5}$$

where $\rho_{1,AB}$, $\rho_{2,AB}$ are density operators of the composite system AB and $\rho_{1,B}$, $\rho_{2,B}$ are the corresponding density operators of the subsystem B. From Eqs. (4) it follows

$$S(\rho_1|\rho_2) = S(\rho_{1,AB}|\rho_{2,AB})\,,$$

whereas for non-commuting ρ_1 and ρ_2. On the other hand, Eq. (5) implies the strict inequality

$$S(\rho_1|\rho_2) < S(\rho_{1,AB}|\rho_{2,AB})\,,$$

a contradiction. □

4.3 No-Deletion Theorem

The no-deletion theorem states [32] that, given two copies of an arbitrary pure quantum state, it is impossible to delete one of them by a unitary interaction.

Proof. Let we define the quantum deleting machine as follows

$$(\exists U, A, A')\ (\forall |\psi\rangle)\ [U|\psi\rangle|\psi\rangle|A\rangle = |\psi\rangle|0\rangle|A'\rangle\,] \tag{6}$$

On the left-hand side of the equation, U a the unitary operator acting on the composited state $|\psi\rangle|\psi\rangle|A\rangle \in H \otimes H \otimes H_A$ where $|A\rangle$ is an ancilla state, which is independent of $|\psi\rangle$. In the right-hand side of Eq. (6) the state $|0\rangle$ is a blank one of the same dimension as $|\psi\rangle$ signifying deletion of the last state, and $|A'\rangle$ is the final state of the ancilla. If the equation in (6) holds, then

$$|\psi\rangle|\psi\rangle|A\rangle = U^{-1}|\psi\rangle|0\rangle|A'\rangle.$$

But this stands in contradiction to the no-cloning theorem. □

4.4 No-Teleportation Theorem

In quantum information, the no-teleportation theorem [75] states that neither an arbitrary quantum state can be converted into a sequence of classical bits nor the classical bits can create original quantum state. This theorem is the consequence of no-cloning theorem. If an arbitrary quantum states allow producing sequence of classical bits, then as we know the classical bits can always be copied and hence the quantum state also can be copied, which violate the no cloning theorem. So the conversion of an arbitrary quantum states in sequence of classical bits is not possible.

Proof. The similarity of two states is defined as follows. Two quantum states ρ_1 and ρ_2 are identical if the measurement results of any physical observable have the same expectation value for ρ_1 and ρ_2. Let prepare an arbitrary mixed quantum state ρ_{input}, then perform the measurement on the state and obtain the classical measurement results. Now by using these classical measurement results the original quantum states is recovered as ρ_{output}. Both the input and output states are not equal

$$\rho_{input} \neq \rho_{output}. \tag{7}$$

This result holds irrespective to the state preparation process and measurement outcome. Hence Eq. (7) proves that one can not convert an arbitrary quantum states into a sequence of classical bits. □

This theorem does not have any relation to teleportation of a quantum state based on the phenomenon of quantum entanglement as a means of transmission (see, for example, textbooks [18,63]).

4.5 No-Communication Theorem

No-communication theorem [76,77] is also known as the no-signaling principle. This theorem essentially states that is not possible to transmit classical bits of information by means of carefully prepared mixed or pure states, whether entangled or not. Therefore, the theorem disallows any communication, not just superluminal (i.e., faster than the speed of light in vacuum), by means of shared quantum states.

Proof. Let Alice and Bob share a quantum state. Our goal is to show that the measurement action performed at the end of Alice is not detectable by Bob at his end, and this is true in either case, weather the composite state of Alica and Bob is separable or entangled. We first consider the case when the composite state is separable. Denote this state by ρ, and let Alice perform a measurement on her end. Such measurement can be modeled by Kraus operators, which may not commute. Denote the Kraus operator(s) at the end of Alice by A_m. Now the probability of measurement outcome x can be written as

$$p_x = \sum_m Tr(A_{x,m}\,\rho\,A^{\dagger}_{x,m}) = Tr[\rho\,V_x\,], \quad V_x = \sum_m A^{\dagger}_{x,m}A_{x,m}\,, \quad \sum_x V_x = 1.$$

Now denote the Kraus operator(s) at the end of Bob by B_n. The probability of measurement outcome y at the end of Bob, irrespective to what Alice has found, is given by

$$p_y = \sum_x Tr\big(\sum_{m,n} B_{y,n}A_{x,m}\,\rho\,A^{\dagger}_{x,m}B^{\dagger}_{y,n}\big). \tag{8}$$

The order of measurements on te separable composite system does not matter, so the following commutation relation $[A_{x,m}, B_{y,n}] = 0$ holds, and Eq. (8) can be rewritten as

$$p_y = \sum_x Tr\big(\sum_{mn} A_{x,m}B_{y,n}\,\rho\,B^{\dagger}_{y,n}A^{\dagger}_{x,m}\big).$$

Using the cyclic property of trace operation and expanding the summation we obtain

$$p_y = Tr\big(\sum_n B_{y,n}\,\rho\,B^{\dagger}_{y,n}\big).$$

In this equation all the operators of Alice disappear, so Bob is not able to detect which Alice's measurements has been performed at her end. Hence, the statistics of measurements at the end of Bob has not been effected by Alice.

Suppose now that the composite state is entangled. Assume for simplicity that this state is the two-particle spin-singlet one $|\psi\rangle = \frac{1}{\sqrt{2}}(|01\rangle + |10\rangle)$, where $|0\rangle$ is the spin down state and $|1\rangle$ is the spin up state. Let Alice and Bob perform the measurement of the particles at their disposal by using the detectors D^A and D^B, respectively. Following the Bell's experiment (see, for example, [78], Sect. 2.5.), assume that the detectors are initially oriented along the z axis and rotated independently at the end of Alice and Bob in such a way that the difference between the angles of detectors becomes θ. Then it yields the conditional probabilities of measurements outcome (cf. [78], Eq. (2.182)):

$$\{A(0), B(0)\},\; P_{00} = \frac{1}{2}\sin^2\Big(\frac{\theta}{2}\Big), \qquad \{A(0), B(1)\},\; P_{01} = \frac{1}{2}\cos^2\Big(\frac{\theta}{2}\Big),$$
$$\{A(1), B(0)\},\; P_{10} = \frac{1}{2}\cos^2\Big(\frac{\theta}{2}\Big), \qquad \{A(1), B(1)\},\; P_{11} = \frac{1}{2}\sin^2\Big(\frac{\theta}{2}\Big),$$

with the normalization condition $P_{00} + P_{01} + P_{10} + P_{11} = 1$. Calculating the probabilities of the measurement outcome for spin up ($|1\rangle$) and spin down ($|0\rangle$)

at the Alice end gives $P_1^A = P_{11} + P_{10} = \frac{1}{2}$, $P_0^A = P_{01} + P_{11} = \frac{1}{2}$. Similarly, for the probabilities at the Bob end, we obtain $P_1^B = P_0^B = \frac{1}{2}$. Thus, we the probabilities of measurement outcomes are totally independent on the angle θ, and the actions performed by measurements at either end of Alice or Bob are not detected at another end and vice versa. This is the essence of no-communication theorem. □

4.6 No-Hiding Theorem

No-hiding theorem [79] is an important theorem in quantum information, which indicates the conservation of information. The idea of no-hiding theorem come out from the correlation of quantum information processing with the classical one in Vernam's cipher used in cryptography. This one-time pad cipher is based on addition of a random (secret) key to the original information to be transmitted. C. Shanon in [1] proved that the original information neither reside in encoded message nor in the key, so where the information is gone? In the one-time pad cipher this information is hidden in the correlations with the key. One can think of the same scenario in quantum mechanical sense. The teleportation can be assumed as a quantum analogue of one-time pad cipher. In teleportation, there are two parties Alice and Bob both share an entangled state. Alice applies few unitary operations at her end and sends the measurement results to Bob. Bob applies the corresponding measurements and recovers the relevant information from Alice. In this whole process, the decoherence is not considered. If one considers the decoherence in teleportation process, then one has to take into account interaction with environment. If a quantum system interacts with an environment in the form of decoherence, then it destroys the original information. So a natural question arises: where the lost information from the original system has gone? In the quantum mechanical case it does not reside in correlations. This idea leads to the "No-hiding theorem". The theorem states that the original information resides in the subspace of the environmental Hilbert space and not in the part of correlation of the system and environment.

Proof. From a subspace I we chose an arbitrary input mixed quantum state ρ_I and encode it into a larger Hilbert space. With respect to a hiding process, there exists an output mixtd state σ_O of a subspace O of the encoded Hilbert space. The remainder of the last space is regarded as an ancilla space A. The hiding process is characterized by the following mapping $\rho_I \longmapsto \sigma_O$, ($\sigma$ is fixed $\forall \rho$).

To be physical, the hiding process must be linear and unitary. Because of linearity, it is sufficient to consider a pure input state $\rho_I = |\psi\rangle_I {}_I\langle\psi|$. In this case the hiding process can be established as the Schmidt decomposition of the final state

$$|\psi\rangle_I \mapsto \sum_{i=1}^{N} \sqrt{p_i} |i\rangle_O \otimes |A_i(\psi)\rangle_A. \tag{9}$$

Here p_i are nonzero eigenvalues of state σ_O and $\{|i\rangle \mid 1 \leq i \leq N\}$ are their eigenvectors. The sets $\{|i\rangle\}$ and $\{|A_n\rangle\}$ are orthonormal. By linearity,

$$|A_i(\mu|\psi\rangle + \nu|\psi_\perp\rangle)\rangle = \mu|A_i(\psi)\rangle + \nu|A_i(\psi_\perp)\rangle\,,$$

where $|\psi_\perp\rangle$ denotes a state orthogonal to $|\psi\rangle$. Hence, the inner product of two such ancilla states gives

$$\mu^*\nu\langle A_m(\psi)|A_n(\psi_\perp)\rangle + \mu\nu^*\langle A_m(\psi_\perp)|A_n(\psi)\rangle = 0.$$

For arbitrary complex numbers of μ and ν, all cross-terms must vanish. Thus, an orthonomal basis $\{|\psi_j\rangle\}$ defines an orthonomal set of states $|A_{n,j}\rangle = |A_n(\psi_j)\rangle$ spanning a Hilbert space that completely describes the reduced state of the ancilla. Hence, $|A_{n,j}\rangle = |q_n\rangle \otimes |\psi_j\rangle \oplus 0$, where $\{|q_n\rangle\}$ is an orthonomal set and $\oplus 0$ means that we pad unused dimensions of the ancilla space by zero vectors. Thereby, Eq. (9) takes the form

$$|\psi_I\rangle = \sum_{i=1}^{n} \sqrt{p_i}|i\rangle_O \otimes (|q_n\rangle \otimes |\psi\rangle \oplus 0)_A.$$

Since we may swap the state $|\psi\rangle$ with any other state in the ancilla using purely ancilla-local operations, we conclude that any information about $|\psi\rangle$ that is encoded globally is in fact encoded entirely within the ancilla. Neither the information about $|\psi\rangle$ is encoded in system-ancilla correlations nor in the system-system correlations. □

It is significant that this theorem has been verified experimentally [80].

4.7 No-Programming Theorem

This theorem states [81] that no deterministic universal quantum processor, i.e. a processor which can be programmed to perform any unitary operation, can be realized.

Proof. Let distinct (up to a global phase) unitary operators $U_1, \ldots, U_N$ be implemented by some programmable quantum gate array. If so, the program register is at least $N-$dimensional, i.e., contains at least $\log_2 N$ qubits. It is straightforward) to show that the corresponding programs $|\mathcal{P}_1\rangle, \ldots, |\mathcal{P}_N\rangle$ are mutually orthogonal.

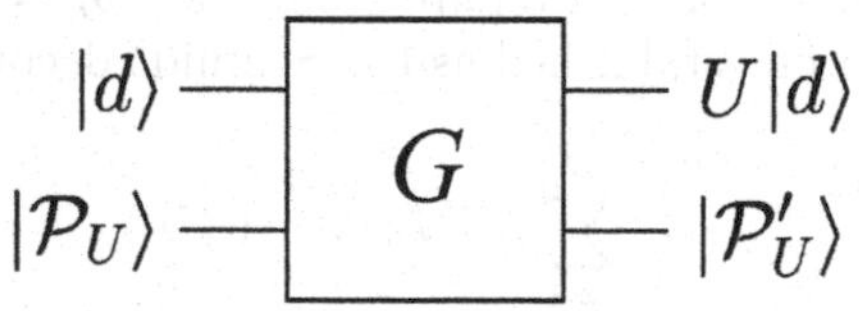

$|d\rangle$ — G — $U|d\rangle$
$|\mathcal{P}_U\rangle$ — G — $|\mathcal{P}'_U\rangle$

Fig. 1. Schematic circuit for a programmable quantum processor which implements the unitary operation U determined by the quantum program $|\mathcal{P}_U\rangle$.

Let $\mathcal{P}$ and $\mathcal{Q}$ be programs implementing unitary operators U_p and U_q, respectively (see Fig. 1). Then for arbitrary data $|d\rangle$ we have equalities

$$G(|d\rangle \otimes |\mathcal{P}\rangle) = (U_p|d\rangle) \otimes |\mathcal{P}'\rangle, \quad G(|d\rangle \otimes |\mathcal{Q}\rangle) = (U_q|d\rangle) \otimes |\mathcal{Q}'\rangle. \quad (10)$$

Taking the inner product of the equations in (10) we obtain

$$\langle \mathcal{Q}|\mathcal{P}\rangle = \langle \mathcal{Q}'|\mathcal{P}'\rangle \langle d|U_q^\dagger U_p|d\rangle.$$

Assume that $\langle \mathcal{Q}'|\mathcal{P}'\rangle \neq 0$. Then the left-hand side does not depend on $|d\rangle$, and hence $U_q^\dagger U_p = \gamma I$. It follows that if $\langle \mathcal{Q}'|\mathcal{P}'\rangle \neq 0$, then U_p and U_q are the same up to a global phase. This contradicts to our assumption on distinction of U_p and U_q. Thus, $\langle \mathcal{Q}'|\mathcal{P}'\rangle = 0$ what implies $\langle \mathcal{Q}|\mathcal{P}\rangle = 0$. This result demonstrate that no *deterministic* universal quantum processor exists. □

Although the deterministic universal quantum array cannot be realized, it is still possible to conceive it in an approximate (probabilistic) fashion [81]. In [82] some bounds for the minimal resources necessary for this aim are given.

5 Conclusions

In this article, we discussed milestone developments in quantum information. We captured the experimental manifestations of theoretical developments as well. In addition, we discussed physical situations, which are not possible in no-go theorems with their mathematical proof. These theorems have important consequences for quantum information processing and for quantum communication. Covering the broad aspects of milestone developments in this article may be useful for quantum information community.

References

1. Shannon, C.E.: A Mathematical theory of communication. Bell Syst. Tech. J. **27**, 379–423, 623–656 (1948)
2. Bishnoi, B.: Quantum-computation and applications. arXiv:2006.02799 (2020)
3. Landauer, R.: Information is physical. Phys. Today **44**, 23–29 (1991)
4. Landauer, R.: Information is a physical entity. Physica A **263**, 63–67 (1999)
5. Oldofredi, A.: No-go theorems and the foundations of quantum physics. J. Gen. Philos. Sci. **49**(3), 355–370 (2018). https://doi.org/10.1007/s10838-018-9404-5
6. Luo, M.-X., Li, H.-R., Lai, H., Wang, X.: Unified quantum no-go theorems and transforming of quantum pure states in a restricted set. Quantum Inf. Process. **16**(12), 1–32 (2017). https://doi.org/10.1007/s11128-017-1754-0
7. Bennett, C.H.: Logical reversibility of computation. IBM J. Res. Dev. **17**(6), 525–532 (1973)
8. Toffoli, T.: Reversible computing, Tech. Memo MIT/LCS/TM-151, MIT Lab for Computer Science (1980)
9. Holevo, A.S.: Bounds for the quantity of information transmitted by a quantum communication channel. Probl. Inf. Transm. **9**(3), 177–183 (1973)

10. Ingarden, R.S.: Quantum information theory. Rept. Math. Phys. **10**, 43–72 (1976)
11. Manin, Y.I.: Vychislimoe i nevychislimoe. Sov. Radio, Moskva (1980). (in Russian)
12. Feynman, R.P.: Simulating physics with computers. Int. J. Theoret. Phys. **21**(6), 467–478 (1982)
13. Park, L.: The concept of transition in quantum mechanics. Found. Phys. **1**, 23–33 (1970). https://doi.org/10.1007/BF00708652
14. Wootters, W.K., Zurek, W.H.: A single quantum cannot be cloned. Nature **299**, 802–803 (1982)
15. Dieks, D.: Communication by EPR devices. Phys. Lett. A. **92**, 271–272 (1982)
16. Benioff, P.: Quantum mechanical Hamiltonian models of Turing machines. J. Stat. Phys. **29**(3), 515–546 (1982). https://doi.org/10.1007/BF01342185
17. Einstein, A., Podolsky, B., Rosen, N.: Can quantum-mechanical description of physical reality be considered complete? Phys. Rev. **47**, 777–780 (1935)
18. Nielsen, M.A., Chuang, I.L.: Quantum Computation and Quantum Information. 10th Anniversary edition. Cambridge University Press, Cambridge (2010)
19. Bennet, C.H., Brassard, G.: Quantum cryptography: public key distribution and coin tossing. In: Proceedings of the IEEE International Conference on Computers, Systems and Signal Processing, Bangalore, December pp. 175–179 (1984)
20. Ekert, A.K.: Quantum cryptography based on Bell's theorem. Phys. Rev. Lett. **67**, 661–663 (1991)
21. Deutsch, D., Jozsa, R.: Rapid solutions of problems by quantum computation. Proc. R. Soc. London **439**, 553–558 (1992)
22. Shor, P.W.: Polynomial time algorithms for prime factorization and discrete logarithms on a quantum computer. SIAM J. Comput. **26**, 1484–1509 (1997)
23. Shor., P.W.: Scheme for reducing decoherence in quantum computer memory. Phys. Rev. A **52**, R2493–R2496 (1995)
24. Rivest, R., Shamir, A., Adleman, L.: A method for obtaining digital signatures and public-key cryptosystems. Commun. ACM **21**, 120–126 (1978)
25. Steane, A.M.: Error correcting codes in quantum theory. Phys. Rev. Lett. **77**, 793–797 (1996)
26. Grover, L.K.: A fast quantum mechanical algorithm for database search. In: Proceedings of the 28th Anual ACM Symposium on the Theory of Computing, Philadelphia, Pensylvania, pp. 212–219 (1996)
27. Bennett, C.H., et al.: Teleporting an unknown quantum state via dual classical and Einstein-Podolsky-Rosen channels. Phys. Rev. Lett. **70**, 1895–1899 (1993)
28. Bouwmeester, D., et al.: Experimental quantum teleportation. Nature **390**, 575–579 (1997)
29. Gershenfeld, N.F., et al.: Bulk spin-resonance quantum computation. Science **275**, 350–356 (1997)
30. Jones, J.A., Mosca, M.: Implementation of a quantum algorithm to solve Deutsch's problem on a nuclear magnetic resonance quantum computer. J. Chem. Phys. **109**, 1648 (1998)
31. Chuang, I.L., Gershenfeld, N., Kubinec, M.: Experimental implementation of fast quantum searching. Phys. Rev. Lett. **80**, 3408–3411 (1998)
32. Pati, A.K., Braunstein, S.L.: Impossibility of deleting an unknown quantum state. Nature **404**, 164–165 (2000)
33. Vandersypen, L.M.K., et al.: Experimental realization of Shor's quantum factoring algorithm using nuclear magnetic resonance. Nature **414**, 883–887 (2001)
34. Knill, E., Laflamme, L., Milburn, G.J.: A scheme for efficient quantum computation with linear optics. Nature **409**, 46–52 (2001)

35. Pittman, T.B., et al.: Experimental controlled-NOT logic gate for single photons in the coincidence basis. Phys. Rev. A **68**, 032316 (2003)
36. O'Brien, J.L., Branning, D., et al.: Demonstration of an all-optical quantum controlled-NOT gate. Nature **426**, 264–267 (2003)
37. Elliott, C.: The DARPA quantum network. arXiv:quant-ph/0412029v1 (2004)
38. Brennen, G., Giacobino, E., Simon, C.: Focus on Quantum Memory. New J. Phys. **17**, 050201 (2015)
39. Kimble, H.: The quantum internet. Nature **453**, 1023–1030 (2008)
40. Chaneliere, T., et al.: Storage and retrieval of single photons transmitted between remote quantum memories. Nature **438**, 833–836 (2005)
41. Murao, M., et al.: Quantum telecloning and multiparticle entanglement. Phys. Rev. A **59**, 156–161 (1999)
42. Koike, S., et al.: Demonstration of quantum telecloning of optical coherent states. Phys. Rev. Lett. **96**, 060504 (2006)
43. Pirandola, S., et al.: Macroscopic entanglement by entanglement swapping. Phys. Rev. Lett. **97**, 150403 (2006)
44. Brennecke, F., et al.: Cavity QED with a Bose Einstein condensate. Nature **450**, 268–271 (2007)
45. Lanyon, B.P., et al.: Manipulating biphotonic qutrits. Phys. Rev. Lett. **100**, 060504 (2008)
46. Chen, J., et al.: Demonstration of a quantum controlled-NOT gate in the telecommunications band. Phys. Rev. Lett. **100**, 133603 (2008)
47. Mariantoni, M., et al.: Implementing the quantum von Neumann architecture with superconducting circuits. Science **334**, 61–65 (2011)
48. Pfaff, W., et al.: Unconditional quantum teleportation between distant solid-state quantum bits. Science **345**, 532–535 (2014)
49. Dattani, N.S., Bryans, N.: Quantum factorization of 56153 with only 4 qubits. arXiv:1411.6758 (2014)
50. Zhong, M., et al.: Optically addressable nuclear spins in a solid with a six-hour coherence time. Nature **517**, 177–180 (2015)
51. Gibney, E.: D-Wave upgrade: how scientists are using the world's most controversial quantum computer. Nature **541**, 447–448 (2017)
52. Yu, T., Eberly, J.H.: Finite-time disentanglement via spontaneous emission. Phys. Rev. Lett. **93**, 140404 (2004)
53. Yu, T., Eberly, J.H.: Sudden death of entanglement. Science **30**, 598–601 (2009)
54. Sharma, K.K., Awasthi, S.K., Pandey, S.N.: Entanglement sudden death and birth in qubit-qutrit systems under Dzyaloshinskii-Moriya interaction. Quantum Inf. Process. **12**, 3437–3447 (2013). https://doi.org/10.1007/s11128-013-0607-8
55. Sharma, K.K., Pandey, S.N.: Entanglement Dynamics in two parameter qubit-qutrit states under Dzyaloshinskii-Moriya interaction. Quantum Inf. Process. **13**, 2017–2038 (2014). https://doi.org/10.1007/s11128-014-0794-y
56. Sharma, K.K., Pandey, S.N.: Influence of Dzyaloshinshkii-Moriya interaction on quantum correlations in two qubit Werner states and MEMS. Quantum Inf. Process. **14**, 1361–1375 (2015). https://doi.org/10.1007/s11128-015-0928-x
57. Sharma, K.K., Pandey, S.N.: Dzyaloshinshkii-Moriya interaction as an agent to free the bound entangled states. Quantum. Info. Process. **15**, 1539 (2016)
58. Sharma, K.K., Pandey, S.N.: Dynamics of entanglement in two parameter qubit-qutrit states with x-component of DM interaction. Commun. Theor. Phys. **65**, 278–284 (2016)

59. Sharma, K.K., Pandey, S.N.: Robustness of Greenberger-Horne-Zeilinger and W states against Dzyaloshinskii-Moriya interaction. Quantum Inf. Process **15**, 4995–5009 (2016)
60. Sharma, K.K., Gerdt, V.P.: Entanglement sudden death and birth effects in two qubits maximally entangled mixed states under quantum channels. Int. J. Theoret. Phys. **59**, 403–414 (2020). https://doi.org/10.1007/s10773-019-04332-z
61. Horodecki, R., et al.: Quantum entanglement. Rev. Mod. Phys. **81**, 865–942 (2009)
62. Acin, A., et al.: The quantum technologies roadmap: a European community view. New J. Phys. **20**, 080201 (2018)
63. Pathak, A.: Elements of Quantum Computation and Quantum Communication. Taylor & Francis Group, Boca Raton (2013)
64. Coffman, V., Kundu, J., Wootters, W.K.: Distributed entanglement. Phys. Rev. A **61**, 052306 (2000)
65. Terhal, B.M.: Is entanglement monogamous? IBM J. Res. Dev. **48**, 71–78 (2004)
66. Koashi, M., Winter, A.: Monogamy of quantum entanglement and other correlations. Phys. Rev. A **69**, 022309 (2004)
67. Osborne, T.J., Verstraete, F.: General Monogamy inequality for bipartite qubit entanglement. Phys. Rev. Lett. **96**, 220503 (2006)
68. Barnum, H., et al.: Noncommuting mixed states cannot be broadcast. Phys. Rev. **76**, 2818–2821 (1996)
69. D'Ariano, G.M., Macchiavello, C., Perinotti, P.: Superbroadcasting of mixed states. Phys. Rev. Lett. **95**, 060503 (2005)
70. Lindblad, G.: A general no-cloning theorem. Lett. Math. Phys. **47**, 189–196 (1999)
71. Barnum, H., et al.: Generalized no-broadcasting theorem. Phys. Rev. Lett. **99**, 240501 (2007)
72. Kalev, A., Hen, I.: No-broadcasting theorem and its classical counterpart. Phys. Rev. Lett. **100**, 210502 (2008)
73. Piani, M., Horodecki, P., Horodecki, R.: No-local-broadcasting theorem for multipartite quantum correlations. Phys. Rev. Lett. **100**, 090502 (2008)
74. Daffertshofer, A., Plastino, A.R., Plastino, A.: Classical no-cloning theorem. Phys. Rev. Lett. **88**, 210601 (2002)
75. Gruska, J., Imai, H.: Power, puzzles and properties of entanglement. In: Margenstern, M., Rogozhin, Y. (eds.) MCU 2001. LNCS, vol. 2055, pp. 25–68. Springer, Heidelberg (2001). https://doi.org/10.1007/3-540-45132-3_3
76. Popescu, S., Rohrlich, D.: Causality and nonlocality as axioms for quantum mechanics. In: Hunter, G., Jeffers, S., Vigier, J.P. (eds.) Causality and Locality in Modern Physics. Fundamental Theories of Physics (An International Book Series on The Fundamental Theories of Physics: Their Clarification, Development and Application), vol. 97, pp. 383–390. Springer, Dordrecht. https://doi.org/10.1007/978-94-017-0990-3_45
77. Peres, A., Terno, D.R.: Quantum information and relativity theory. Rev. Mod. Phys. **76**, 93–123 (2004)
78. Benenti, G., Casatti, G., Strini, G.: Principles of Quantum Computation and Information. Vol. I: Basic Concepts. World Scientific, Singapore (2004)
79. Braunstein, S.L., Pati, A.K.: Quantum information cannot be completely hidden in correlations: implications for black hole information paradox. Phys. Rev. Lett. **98**, 080502 (2007)
80. Pani, S.J., Pati, A.K.: Experimental test of the quantum no-hiding theorem. Phys. Rev. Lett. **106**, 080401 (2011)
81. Nielsen, M.A., Chuang, I.L.: Programmable quantum gate arrays. Phys. Rev. Lett. **79**, 321–324 (1997)

82. Kubicki, A.M., Palazuelos, C., Peréz-Garcia, D.: Resource quantification for the no-programing theorem. Phys. Rev. Lett. **122**, 080505 (2019)
83. Diamanti, E. et al.: Practical challenges in quantum key distribution. npj Quantum Inf. **2**, 16025 (2016). https://doi.org/10.1038/npjqi.2016.25
84. Bedington, R., Arrazola, J.M., Ling, A.: Progress in satellite quantum key distribution. npj Quantum Inf. **3**, 30 (2017)

Practical Application of the Multi-model Approach in the Study of Complex Systems

Anna V. Korolkova[1], Dmitry S. Kulyabov[1,2(✉)], and Michal Hnatič[2,3,4]

[1] Peoples' Friendship University of Russia (RUDN University), 6 Miklukho-Maklaya St, Moscow 117198, Russian Federation
{korolkova_av,kulyabov_ds}@rudn.university
[2] Joint Institute for Nuclear Research, 6 Joliot-Curie St., Dubna , Moscow Region 141980, Russian Federation
[3] Department of Theoretical Physics, SAS, Institute of Experimental Physics, Watsonova 47, 040 01 Košice, Slovak Republic
hnatic@saske.sk
[4] Pavol Jozef Šafárik University in Košice (UPJŠ), Šrobárova 2, 041 80 Košice, Slovak Republic

Abstract. Different kinds of models are used to study various natural and technical phenomena. Usually, the researcher is limited to using a certain kind of model approach, not using others (or even not realizing the existence of other model approaches). The authors believe that a complete study of a certain phenomenon should cover several model approaches. The paper describes several model approaches which we used in the study of the random early detection algorithm for active queue management. Both the model approaches themselves and their implementation and the results obtained are described.

Keywords: Active queue management · Mathematical modeling · Simulation · Surrogate modeling · Stochastic systems

1 Introduction

Scientific research is easy to start but difficult to complete. Our study of the Random Early Detection (RED) algorithm stood out from the study of approaches and mechanisms of traffic control in data transmission networks. But the further we went, the less satisfied we were with the results. The originally constructed mathematical model seemed to us somewhat artificial and non-extensible. To build a more natural mathematical model from first principles, we have developed a method of stochastization of one-step processes. To verify the mathematical model, we have built physical and simulation models. To conduct optimization studies, we began to build a surrogate model for the RED algorithm.

The publication has been prepared with the support of the "RUDN University Program 5-100" and funded by Russian Foundation for Basic Research (RFBR) according to the research project No 19-01-00645.

V. M. Vishnevskiy et al. (Eds.): DCCN 2020, LNCS 12563, pp. 526–537, 2020.
https://doi.org/10.1007/978-3-030-66471-8_40

At the last we came to an understanding that all of our models form some kind of emergent structure, with the help of which we can investigate various phenomena—stochastic and statistical systems in particular.

In this paper, we try to present our understanding of the multi-model approach to modeling.

2 Model Approaches

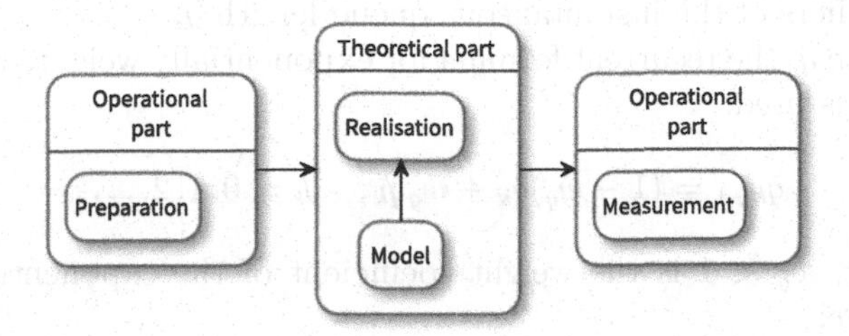

Fig. 1. Generic structure of the model approach

Modeling as a discipline encompasses different types of model approaches. From our point of view, these approaches can be schematically described in a unified manner (see Fig. 1). In this case, the research structure consists of operational and theoretical parts. The operational parts are represented by procedures of the system preparation and measurement. It is also common to describe the operational part as input and output data.

The theoretical part consists of two layers: a model layer and an implementation layer. The implementation layer describes the specific structure of the evolution of the system. Depending on the type of implementation, different types of models can be obtained: a mathematical model (implementation—mathematical expressions), a simulation model (implementation—an algorithm), a physical model (implementation—an analog system), a surrogate model (implementation—approximation of behavior). Each type of model has its area of applicability, its advantages and disadvantages. The use of the entire range of models allows the most in-depth and comprehensive study of the modeled system.

3 RED Active Queue Management Algorithm

Random Early Detection (RED) algorithm is at the heart of several mechanisms to prevent and control congestion in router queues. Its main purpose is to smooth out temporary bursts of traffic and prevent prolonged network congestion by notifying traffic sources about the need to reduce the intensity of information transmission.

The operation of a module implementing a RED-type algorithm can be schematically represented as follows.

When a packet of transmitted data enters the system, it enters the reset module. The decision to remove the package is made based on the value of the $p(\hat{q})$ function received from the control unit. The function $p(\hat{q})$ depends on the exponentially weighted moving average of the queue length $\hat{q}$, also calculated by the ontrol unit based on the current value of the queue length q.

The classic RED algorithm is discussed in detail in [3]. Only the formulas for calculating the reset function $p(\hat{q})$ and the exponentially weighted moving average queue length $\hat{q}$ are given here. The use of $\hat{q}$ is associated with the need to smooth outliers of the instantaneous queue length q.

To calculate $\hat{q}$, the recurrent formula for exponentially weighted moving average (EWMA) is used:

$$\hat{q}_{k+1} = (1 - w_q)\hat{q}_k + w_q q_k, \quad k = 0, 1, 2, \dots,$$

where w_q, $0 < w_q < 1$ is the weight coefficient of the exponentially weighted moving average:

$$w_q = 1 - e^{-1/C},$$

where C is the channel bandwidth (packets per second).

The $p(\hat{q})$ packet discard function linearly depends on the $\hat{q}$, the minimum q_{min} and maximum q_{max} thresholds and the maximum discard parameter p_{max}, which sets the maximum packet discard level when $\hat{q}$ reaches the value q_{max}, and the function $p(\hat{q})$ is calculated as follows:

$$p(\hat{q}) = \begin{cases} 0, & 0 < \hat{q} \leqslant q_{\mathrm{min}}, \\ \dfrac{\hat{q} - q_{\mathrm{min}}}{q_{\mathrm{max}} - q_{\mathrm{min}}} p_{\mathrm{max}}, & q_{\mathrm{min}} < \hat{q} \leqslant q_{\mathrm{max}}, \\ 1, & \hat{q} > q_{\mathrm{max}}. \end{cases} \tag{1}$$

The drop function (1) describes the classic RED algorithm. And the main effort in the design of new algorithms like RED is directed at various modifications of the type of the drop function.

Since the complete simulated system consists of interoperable TCP and RED algorithms, it is necessary to simulate the evolution of the TCP source as well. Since the original model was based on the TCP Reno protocol, we simulated this particular protocol.

TCP uses a sliding window mechanism to deal with congestion. The implementation of this mechanism depends on the specific TCP protocol standard.

In TCP Reno the congestion control mechanism consists of the following phases: slow start, congestion avoidance, fast transmission, and fast recovery. The dynamics of the congestion window (CWND) size depends on the specific phase.

TCP Reno monitors two types of packet loss:

- Triple Duplicate ACK (TD). Let the n-th packet not to be delivered, and the subsequent packets ($n+1$, $n+2$, etc.) are delivered. For each packet delivered out of order (for $n+1$, $n+2$, etc.), the receiver sends an ACK message for the last undelivered (n-th) packet. Upon receipt of three such packets, the source resends the n-th packet. Besides, the window size is reduced by 2 times $cwnd \to cwnd/2$.
- Timeout (TO). When a packet is sent, the timeout timer is started. Each time a confirmation is received, the timer is restarted. The window is then set to the initial value of the overload window. The first lost packet is resent. The protocol enters the slow start phase.

The general congestion control algorithm is of the AIMD type (Additive Increase, Multiplicative Decrease)—an additive increase of the window size and its multiplicative decrease.

4 Mathematical Model

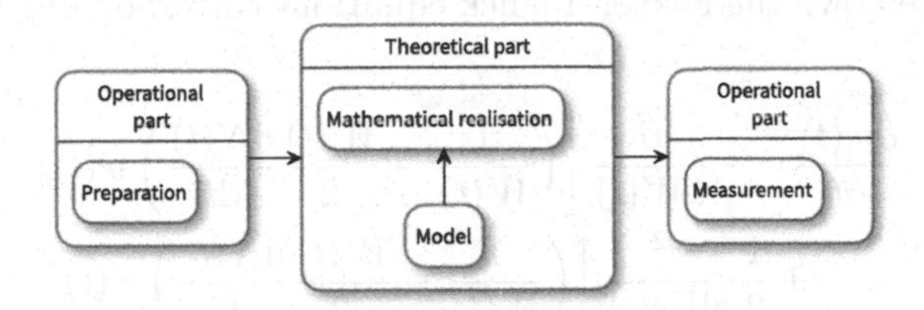

Fig. 2. Generic structure of the mathematical model

The most rigorous research is usually based on a mathematical model (see Fig. 2). In this case, the model layer is realized through mathematical expressions describing the evolution of the system.

There are several approaches to modeling RED-type algorithms. The most famous approach is modeling using the automatic control theory approach [8,14,15]. To us, this approach seems somewhat artificial and inconsistent. We prefer to do our modeling from first principles.

We have developed a method of stochastization of one-step processes, which allows us to obtain models from first principles. Moreover, the resulting model models are immanently stochastic [6,7,11]. Our model of interaction between the TCP source and the RED algorithm is based on these methods and is mathematically represented in the form of stochastic differential equations with Wiener and Poisson processes [12,22,23].

We will use the following notation. $W(t)$ is the TCP Reno window size, $Q(t)$ is the queue size, $T := T(Q(t))$ is the round-trip time (taking into account equipment delays), t is the time, C is the queue service intensity, $\hat{Q}(t)$ is the Exponentially Weighted Moving-Average (EWMA) [3]:

$$\hat{Q}(t) = (1 - w_q)\hat{Q}(t) + w_q Q(t),$$

where w_q, $0 < w_q < 1$ is the weight coefficient.

We used the method of stochastization of one-step processes to obtain the Fokker–Planck and Langevin equations for the stochastic processes $W(t)$ and $Q(t)$.

The kinetic equations will be:

$$\begin{cases} 0 \xrightarrow{\frac{1}{W(t)}} W(t), \\ W(t) \xrightarrow{\frac{1}{2}\frac{\mathrm{d}N(t)}{\mathrm{d}t}} 0, \\ 0 \xrightarrow{\frac{W(t)}{T}} Q(t), \\ 0 \xrightarrow{-C} Q(t), \end{cases} \tag{2}$$

where $\mathrm{d}N(t)$ is the Poisson process [15].

Let us write down the Fokker–Planck equations corresponding to the kinetic Eqs. (2):

$$\begin{aligned} \frac{\partial w(t)}{\partial t} &= -\frac{\partial}{\partial W(t)}\left[\left(\frac{1}{W(t)} - \frac{W(t)}{2}\frac{\mathrm{d}N(t)}{\mathrm{d}t}\right) w(t)\right] \\ &+ \frac{1}{2}\frac{\partial^2}{\partial W^2(t)}\left[\left(\frac{1}{W(t)} + \frac{W(t)}{2}\frac{\mathrm{d}N(t)}{\mathrm{d}t}\right) w(t)\right], \\ \frac{\partial q(t)}{\partial t} &= -\frac{\partial}{\partial Q(t)}\left[\left(\frac{W(t)}{T} - C\right) q(t)\right] \\ &+ \frac{1}{2}\frac{\partial^2}{\partial Q^2(t)}\left[\left(\frac{W(t)}{T} - C\right) q(t)\right], \end{aligned} \tag{3}$$

where $w(t)$ is the density of the random process $W(t)$, $q(t)$ is the density of the random process $Q(t)$.

The Langevin equations corresponding to the Eqs. (3) have the form:

$$\begin{cases} \mathrm{d}W(t) = \frac{1}{T}\mathrm{d}t - \frac{W(t)}{2}\mathrm{d}N(t) + \sqrt{\frac{1}{T} + \frac{W(t)}{2}\frac{\mathrm{d}N(t)}{\mathrm{d}t}}\mathrm{d}V^1(t), \\ \mathrm{d}Q(t) = \left(\frac{W(t)}{T} - C\right)\mathrm{d}Q(t) + \sqrt{\frac{W(t)}{T} - C}\mathrm{d}V^2(t), \end{cases} \tag{4}$$

where $\mathrm{d}V^1(t)$ is the Wiener process corresponding to the random process $W(t)$, $\mathrm{d}V^2(t)$ is the Wiener process corresponding to the random process $Q(t)$.

The Eqs. (4) are supplemented by the constraint equation (written in differential form for convenience):

$$\frac{\mathrm{d}\hat{Q}(t)}{\mathrm{d}t} = w_q C(Q(t) - \hat{Q}(t)).$$

This mathematical model can be investigated both in the form of stochastic differential equations and by writing them down in moments. The equation in moments is naturally easier to study.

5 Physical Model

The resulting mathematical model should be compared with experimental data and verified. Unfortunately, we do not have the resources to take data from a working network or build a full-scale test bench on real network equipment. Therefore, we tried to create a virtual experimental installation based on the virtual machines [20]. Virtual machines run images of real routers operating systems. This is what allows us to call this model *physical*.

To create the stand, the software package GNS3 (Graphical Network Simulator) [25] was chosen. This allows you to simulate a virtual network of routers and virtual machines. The software package GNS3 works on almost all platforms. It can be considered as a graphical interface for different virtual machines. To emulate Cisco devices the Dynamips emulator is used. Alternatively, emulators such as VirtualBox and Qemu can be used. The latter is especially useful when it used with a KVM system which allows a hardware processor implementation. GNS3 coordinates the operation of various virtual machines and also provides the researcher with a convenient interface for creation and customization of the required stand configuration. Also, the developed topology can be linked to an external network to manage and control data packets.

The stand consists of a Cisco router, a traffic generator, and a receiver. D-ITG (Distributed Internet Traffic Generator) is used as a traffic generator (see Fig. 3). D-ITG allows us to obtain estimates of the main indicators of the quality of service (average packet transmission delay, delay variation (jitter), packet loss rate, performance) with a high degree of confidence.

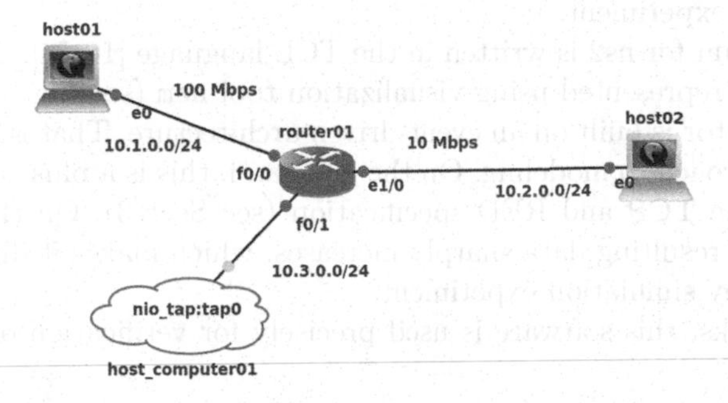

Fig. 3. Virtual stand for studying the functioning of the RED algorithm. `host01` is the packet source; `host02` is the recipient.

6 Simulation Model

With the development of computer technology it became possible to specify a model implementation not in the form of a mathematical description, but in the form of some algorithm (Fig. 4). This type of model is called simulation models and the approach itself is called simulation.

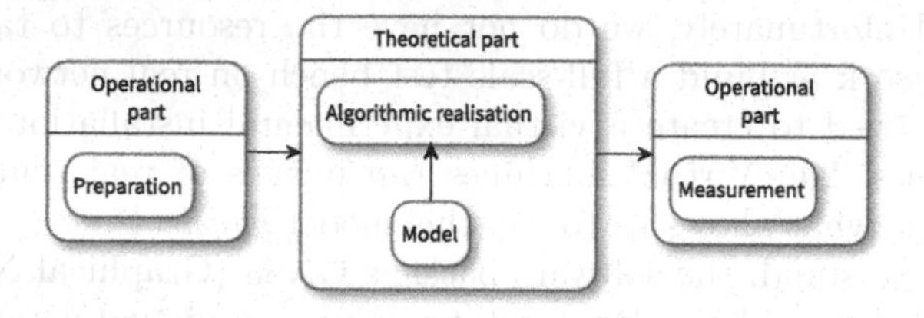

Fig. 4. Generic structure of the simulation model

The simulation model plays a dual role. The simulation model, debugged and tested on experimental data and a physical model, can itself serve the purposes of the mathematical model verification. On the other hand, the simulation model makes it possible to study the behavior of the modeled system more effectively than the mathematical model for different variants of the input data.

6.1 Simulation Model on NS-2

The `ns2` [1,9] package is the network protocol simulation tool. During its existence, the functionality has been repeatedly verified by data from field experiments. Therefore, this package itself has become a reference modeling tool. This is exactly the case when a simulation model is a replacement for a physical model and a natural experiment.

The program for ns2 is written in the TCL language [16,24]. The simulation results can be represented using visualization tool `nam` (see Fig. 5).

The simulator is built on an event-driven architecture. That is, it implements a discrete approach to modeling. On the one hand, this is a plus, since it directly implements the TCP and RED specification (see Sect. 3). On the other hand, the amount of resulting data sharply increases, which makes it difficult to carry out any lengthy simulation experiment.

In our works, this software is used precisely for verification of the obtained results [19,21].

6.2 Hybrid Model for RED Algorithm

To study the RED algorithm we developed the prototype of the simulation model. We wanted to avoid the resource intensiveness of discrete modeling approaches. However, it was necessary to take into account the discrete specifics

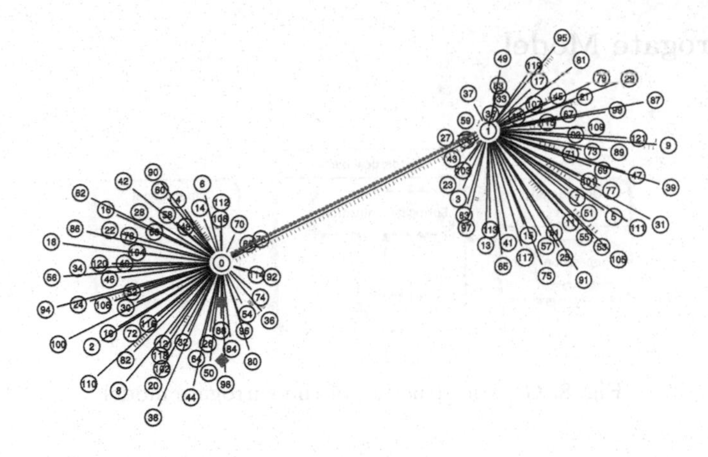

Fig. 5. Visualization of the simulation. Packets drop is shown

of TCP and RED (see Sect. 3). Therefore, we have chosen a hybrid (continuous–discrete) approach. The model was implemented in the hybrid modeling language Modelica [4,5].

Since we are building the hybrid continuous–discrete model, then to describe each phase of TCP functioning, we will turn to the model with continuous time. The transition between phases will be described by discrete states.

To build a hybrid model, we need:

- write a dynamic model for each state;
- replace systems with piecewise constant parameters with systems with variable initial conditions;
- write the state diagram of the model (Fig. 6 and 7).

The resulting diagrams are directly converted into a Modelica program [2, 13,18].

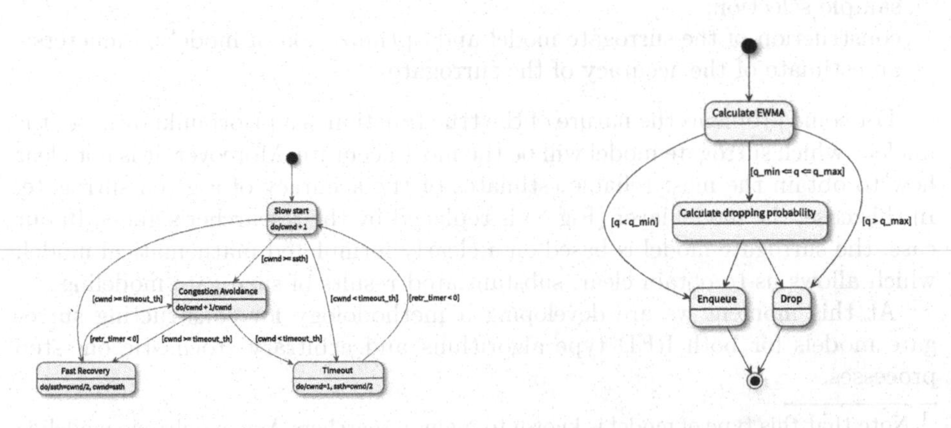

Fig. 6. TCP state diagram

Fig. 7. RED state diagram

7 Surrogate Model

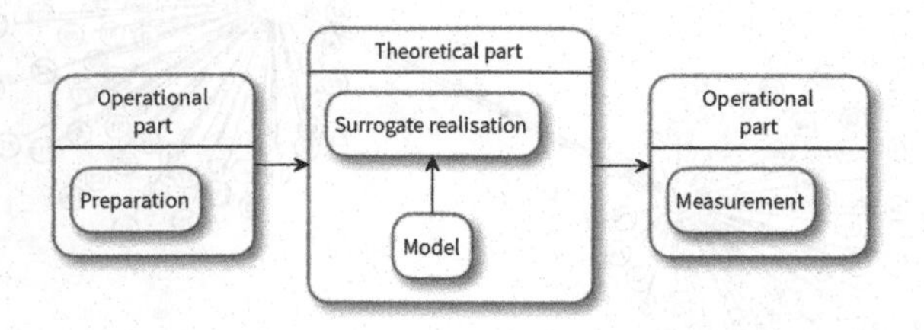

Fig. 8. Generic structure of the surrogate model

Most scientific and technical problems require experiments and simulations to obtain results, to determine the limitations imposed on the result. However, for many real-world problems simulation alone can take minutes, hours, days. As a result, routine tasks such as decision optimization, decision space exploration, sensitivity analysis, and what-if analysis become impossible as they require thousands or millions of modeling evaluations.

One way to simplify research is to build surrogate models (approximation models, response surface models, metamodels, black box models) (see Fig. 8) that mimic the behavior of the original model so closely as much as possible, while being computationally cheap [10]. Surrogate models are built using a data-driven approach. The exact inner workings of the simulation code are not supposed to be known (or even understood), only the input—output (preparation—measurement) behavior is important. The model is built based on modeling the response to a limited number (sometimes quite large) of selected data points [1].

The scientific challenge for surrogate modeling is to create a surrogate that is as accurate as possible using as few modeling estimates as possible. The process consists of the following main stages [17]:

- sample selection;
- construction of the surrogate model and optimization of model parameters;
- an estimate of the accuracy of the surrogate.

For some problems the nature of the true function is a priori unknown, so it is unclear which surrogate model will be the most accurate. Moreover, it is not clear how to obtain the most reliable estimates of the accuracy of a given surrogate. In this case, the model layer (Fig. 8) is replaced by the researcher's guess. In our case, the surrogate model is based on a clearly formulated mathematical model, which allows us to obtain clear, substantiated results of surrogate modeling.

At this moment we are developing a methodology for constructing surrogate models for both RED type algorithms and arbitrary stochastic one-step processes.

[1] Note that this type of model is known to many researchers. When only one modeling variable is involved, the process of building the surrogate model is called curve fitting.

8 Conclusion

The authors tried to outline the concept of a multi-model approach to the study of physical and technical systems using the example of the interaction between the TCP protocol and the RED-type active queue management algorithm. This research is in line with the research of stochastic models in science and technology.

The multi-model approach makes it possible to increase the efficiency of the study of the phenomenon, to consider it from different angles, and to create effective software systems.

References

1. Altman, E., Jiménez, T.: NS simulator for beginners. In: Synthesis Lectures on Communication Networks, vol. 5, no. 1, pp. 1–184, January 2012. https://doi.org/10.2200/S00397ED1V01Y201112CNT010
2. Apreutesey, A.M.Y., Korolkova, A.V., Kulyabov, D.S.: Modeling RED algorithm modifications in the OpenModelica. In: Kulyabov, D.S., Samouylov, K.E., Sevastianov, L.A. (eds.) Proceedings of the Selected Papers of the 9th International Conference Information and Telecommunication Technologies and Mathematical Modeling of High-Tech Systems (ITTMM-2019), Moscow, Russia, 15–19 April 2019, vol. 2407, pp. 5–14. CEUR Workshop Proceedings, Moscow, April 2019
3. Floyd, S., Jacobson, V.: Random early detection gateways for congestion avoidance. IEEE/ACM Trans. Netw. **1**(4), 397–413 (1993). https://doi.org/10.1109/90.251892
4. Fritzson, P.: Principles of Object-Oriented Modeling and Simulation with Modelica 2.1. Wiley-IEEE Press (2003)
5. Fritzson, P.: Introduction to Modeling and Simulation of Technical and Physical Systems with Modelica. Wiley, Hoboken (2011). https://doi.org/10.1002/9781118094259
6. Gevorkyan, M.N., Demidova, A.V., Velieva, T.R., Korol'kova, A.V., Kulyabov, D.S., Sevast'yanov, L.A.: Implementing a method for stochastization of one-step processes in a computer algebra system. Program. Comput. Softw. **44**(2), 86–93 (2018). https://doi.org/10.1134/S0361768818020044
7. Hnatič, M., Eferina, E.G., Korolkova, A.V., Kulyabov, D.S., Sevastyanov, L.A.: Operator approach to the master equation for the one-step process. In: EPJ Web of Conferences, vol. 108, p. 02027 (2016). https://doi.org/10.1051/epjconf/201610802027
8. Hollot, C.V.V., Misra, V., Towsley, D.: A control theoretic analysis of RED. In: Proceedings IEEE INFOCOM 2001. Conference on Computer Communications. Twentieth Annual Joint Conference of the IEEE Computer and Communications Society (Cat. No.01CH37213), vol. 3, pp. 1510–1519. IEEE (2001). https://doi.org/10.1109/INFCOM.2001.916647
9. Issariyakul, T., Hossain, E.: Introduction to Network Simulator NS2. Springer, Boston (2012). https://doi.org/10.1007/978-1-4614-1406-3
10. Jin, Y.: Surrogate-assisted evolutionary computation: recent advances and future challenges. Swarm Evol. Comput. **1**(2), 61–70 (2011). https://doi.org/10.1016/j.swevo.2011.05.001

11. Korolkova, A.V., Eferina, E.G., Laneev, E.B., Gudkova, I.A., Sevastianov, L.A., Kulyabov, D.S.: Stochastization of one-step processes in the occupations number representation. In: Proceedings 30th European Conference on Modelling and Simulation, pp. 698–704, June 2016. https://doi.org/10.7148/2016-0698
12. Korolkova, A.V., Kulyabov, D.S., Velieva, T.R., Zaryadov, I.S.: Essay on the study of the self-oscillating regime in the control system. In: Iacono, M., Palmieri, F., Gribaudo, M., Ficco, M. (eds.) 33 European Conference on Modelling and Simulation, ECMS 2019. Communications of the ECMS, vol. 33, pp. 473–480. European Council for Modelling and Simulation, Caserta, June 2019. https://doi.org/10.7148/2019-0473
13. Korolkova, A.V., Velieva, T.R., Abaev, P.A., Sevastianov, L.A., Kulyabov, D.S.: Hybrid simulation of active traffic management. In: Proceedings 30th European Conference on Modelling and Simulation, pp. 685–691, June 2016. https://doi.org/10.7148/2016-0685
14. Misra, V., Gong, W.B., Towsley, D.: Stochastic differential equation modeling and analysis of TCP-windowsize behavior. In: Proceedings of PERFORMANCE, vol. 99 (1999)
15. Misra, V., Gong, W.B., Towsley, D.: Fluid-based analysis of a network of AQM routers supporting TCP flows with an application to RED. ACM SIGCOMM Comput. Commun. Rev. **30**(4), 151–160 (2000). https://doi.org/10.1145/347057.347421
16. Nadkarni, A.P.: The TCL Programming Language: A Comprehensive Guide. CreateSpace Independent Publishing Platform (2017)
17. Sevastianov, L.A., Sevastianov, A.L., Ayrjan, E.A., Korolkova, A.V., Kulyabov, D.S., Pokorny, I.: Structural approach to the deep learning method. In: Korenkov, V., Strizh, T., Nechaevskiy, A., Zaikina, T. (eds.) Proceedings of the 27th Symposium on Nuclear Electronics and Computing (NEC-2019), vol. 2507, pp. 272–275. CEUR Workshop Proceedings, Budva, September 2019
18. Velieva, T.R., Eferina, E.G., Korolkova, A.V., Kulyabov, D.S., Sevastianov, L.A.: Modelica-based TCP simulation. J. Phys. Conf. Ser. **788**(100), 012036.1–012036.7 (2017). https://doi.org/10.1088/1742-6596/788/1/012036
19. Velieva, T.R., Korolkova, A.V., Demidova, A.V., Kulyabov, D.S.: Software package development for the active traffic management module self-oscillation regime investigation. In: Zamojski, W., Mazurkiewicz, J., Sugier, J., Walkowiak, T., Kacprzyk, J. (eds.) DepCoS-RELCOMEX 2018. AISC, vol. 761, pp. 515–525. Springer, Cham (2019). https://doi.org/10.1007/978-3-319-91446-6_48
20. Velieva, T.R., Korolkova, A.V., Kulyabov, D.S.: Designing installations for verification of the model of active queue management discipline RED in the GNS3. In: 6th International Congress on Ultra Modern Telecommunications and Control Systems and Workshops (ICUMT), pp. 570–577. IEEE Computer Society (2015). https://doi.org/10.1109/ICUMT.2014.7002164
21. Velieva, T.R., Korolkova, A.V., Kulyabov, D.S., Abramov, S.A.: Parametric study of the control system in the TCP network. In: 10th International Congress on Ultra Modern Telecommunications and Control Systems, Moscow, pp. 334–339 (2019). https://doi.org/10.1109/ICUMT.2018.8631267
22. Velieva, T.R., Korolkova, A.V., Kulyabov, D.S., Dos Santos, B.A.: Model Queue Management on Routers, vol. 2, 81–92. Bulletin of Peoples' Friendship University of Russia. Series Mathematics. Information Sciences, Physics (2014)

23. Velieva, T.R., Kulyabov, D.S., Korolkova, A.V., Zaryadov, I.S.: The approach to investigation of the the regions of self-oscillations. J. Phys. Conf. Ser. **937**, 012057.1–012057.8 (2017). https://doi.org/10.1088/1742-6596/937/1/012057
24. Welch, B., Jones, K.: Practical Programming in TCL and TK, 4th edn. Prentice Hall (2003)
25. Welsh, C.: GNS3 Network Simulation Guide. PACKT Publisher (2013)

Three Approaches in the Study of Recurrent Markovian and Semi-Markovian Processes

Boyan Dimitrov[1], Vladimir Rykov[2](✉), and Sahib Esa[3]

[1] Department of Mathematics, Kettering University, Flint, MI, USA
bdimitro@kettering.edu

[2] Peoples' Friendship University of Russia (RUDN University), 6 Miklukho-Maklaya St, Moscow 117198, Russian Federation
rykov-vv@rudn.ru

[3] University of Kurdistan, Erbil, Iraq
sahib_esa@yahoo.com

Abstract. We present three classical methods in the study of dynamic and stationary characteristic of processes of Markovian or Semi-Markovian type which possess points of regeneration. Our focus is on the stationary distributions and conditions of its existence and use.

The first approach is based on detailed probability analysis of time dependent passages between the states of the process at a given moment. We call this approach Kolmogorov approach.

The second approach uses the probability meaning of Laplace-Stieltjes transformation and of the probability generating functions/ Some additional arteficial excrement construction is used to show how derive direct relationships between these functions and how to find them explicitly.

The third approach obtains relationships between the stationary characteristics of the process by use of so called "equations of equilibrium". The input flow in each state must be equal to the respective output flow from that state. In such a way no accumulations should happen on each of that states when process gets its equilibrium.

In all the illustrations of the these approaches we analyze a dynamic Marshal-Olkin reliability model with dependent components functioning in parallel. Results on this example are new.

Keywords: Equilibrium balances · Kolmogorov equations · Probability interpretation of probability generating functions · Marshal-Olkin model in dynamics

1 Introduction

Most of process studies are usually focused on transient (in time) behaviour of process characteristics, and from there - how they behave after long time (limits at infinity). Especially, for processes with some exponential component, these

V. M. Vishnevskiy et al. (Eds.): DCCN 2020, LNCS 12563, pp. 538–554, 2020.
https://doi.org/10.1007/978-3-030-66471-8_41

are suitable Markovian and Semi-Markovian type of processes for which lots of theoretical results and particular applications have been obtained.

Following those ideas, we found out, that there are different approaches which produce at the end same, or similar results.

In this work we discuss three approaches in these studies on an example, which at the end of the day produces equivalent results about the possibly existing stationary characteristics of classes of processes similar to it. It is up to the researchers what they would prefer, but according to us, they should know about all of these opportunities.

In what follows, we discuss the nature of each of these three approaches, and demonstrate how they work on a Marshal-Olkin dynamic model not discussed yet in the reliability articles.

2 The Marshal-Olkin Dynamic Mode

In 1967 Marshall and Olkin [10] proposed a bivariate distribution, briefly (MO), with dependent components, defined via three independent Poisson processes, which represent three types of shocks: individual to each component and commons to both. This model possesses what is known as a Bivariate Lack of Memory Property -henceforth BLMP.

Consider a heterogeneous two-component redundant hot standby renewable system, wherein components work in parallel and fail according to the original MO model, but are repaired according to description below. For lifetimes T_1 and T_2, the MO model is specified by the representation

$$(T_1, T_2) = (\min(A_1, A_3), \min(A_2, A_3)), \qquad (1)$$

where non-negative continuous random variables A_1 and A_2 are the times to occurrence of independent "individual risk strikes" affecting individually each of the two devices working in parallel. The first risk strike affects only the first component, the second one affects only the second one, while the third type of risk strike represents the time to occurrence of the "common failure" A_3 that affects both components simultaneously, or just the working one, and leads to the failure of the entire system in any case. It is supposed that the risk strikes are governed by independent homogeneous Poisson processes, i.e., A_i's in (1) are exponentially distributed with parameters α_i $(i = 1, 2, 3)$.

In dealing with a renewable model, we need to consider the system's renovation after its partial and/or complete failure. Here it is assumed that after a partial failure (when only one component say i, fails) the repair time of type i, with random duration B_i $(i = 1, 2)$ begins. This means that the system continues to function with the one working component. After a complete system failure a repair time of the whole system (both components) begins, and lasts some random time, say B_3. It is assumed that the repair times B_k $(k = 1, 2, 3)$ have cumulative distribution functions (CDF) $B_k(x)$ $(k = 1, 2, 3)$ respectively. All repair times are assumed independent from the other random durations.

Situations like this could be found in practice as real. Imagine, two power station providing energy in certain region. each may fail individually for some internal reason. But common failures may be due to some weather on other environmental conditions. Common failures must be repaired simultaneously by common services, and be started simultaneously on same times for security reason,

The system state space can be represented by $E = \{E_0, E_1, E_2, E_3\}$, where E_0 means that both components are working; E_1 shows that the first component is being repaired, and the second one is working; E_2 indicates that the second component is being repaired, and the first one is working; E_3 says that both components are in down states, the system has failed and is being repaired. To describe the system's behavior we introduce a random process $\{J(t),\ t \geq 0\}$ which takes values in the phase space E, such that

$$J(t) = j, \quad \text{if at time } t \text{ the system is in state } E_j \ (j = 0, 1, 2, 3).$$

Further, for short, we will use the following notations:

- $\alpha = \alpha_1 + \alpha_2 + \alpha_3$ is the summary risk intensity of the system failure;
- $b_k = \int_0^\infty x\, dB_k(x))$, $(k = 1, 2, 3)$ is the mean repair time of a k-th component and of the whole system when $k = 3$;
- $\beta_k(s) = \int_0^\infty e^{-sx} dB_k(x)$ $(k = 1, 2, 3)$ is the LST of the repair time c.d.f. of a k-th component and the whole system when $k = 3$;
- $T = \inf\{t : J(t) = 3\}$ is the system lifetime. It is assumed to start with both components working, and ends with a failure of both components (either both hit by risk 3, or one fails by its own, or due to a strike from risk 3 while other one is in repair);
- W = the system life cycle which represents the portion of time interval when the system starts after a whole repair or both components being working, and ends with the complete repair of the whole system, i.e. $W == T + B_3$. Its LST is denoted by $\omega(s)$.
- $\tilde{b}_k(x) = (1 - B_k(x))^{-1} b_k(x)$ $(k = 1, 2)$ is the conditional repair intensity (hazard rate function) of a k-th component and the whole system (when $k = 3$) given that elapsed repair time is x;
- $F(t) = \mathbf{P}\{T \leq t\}$ and $f(s)$ its LST;
- also to shorter some formulas the following notation is used

$$\begin{aligned} \phi_1(s) &= \alpha_1(\beta_1(s + \alpha_2)\ \phi_2(s) \\ &= \alpha_2(\beta_2(s + \alpha_1)\ \psi(s) = \phi_1(s) + \phi_2(s), \end{aligned} \tag{2}$$

3 Kolmogorov Equations in Detailed Time Analysis

It is well known, that adding additional information about the situation at a given time moment almost every time dependent process can be treated as Markov, possibly for the exigences of dimension. In our system behavior study the method of *additional variable* or the so-called Markovization method will be used. It consists of introducing an additional variables in order to describe the system's behavior via a Markov processes. In the case considered here, we use as such additional variable the time, spent by the state component in its J-th state subject to its last entry in it (the so-called elapsed time). We thus consider a two-dimensional Markov process $Z = \{Z(t), t \geq 0\}$, with

$$Z(t) = (J(t), X(t)),$$

where $J(t)$ is the system state at time t, and $X(t)$ represents the elapsed time of the process in the $J(t)$-th state after its last entering in it. The process phase space is given by $\mathcal{E} = \{0, (1, x), (2, x), (3, x)\}$. Corresponding probabilities (densities with respect to additional variables) are denoted by $\pi_0(t)$, $\pi_1(t; x)$, $\pi_2(t; x)$, $\pi_3(t; x)$ and we will refer to them as to the process (and the system) *micro-state* probabilities. The probabilities $\pi_i(t) = \mathbf{P}\{J(t) = j\}$ $(j = 0, 1, 2, 3)$ are called as *macro-state* process (and system) probabilities.

To calculate the time dependent system state probabilities during its life cycle the Markov process Z with absorbing state 3 should be used. Under the above assumptions, the following statement is true.

Theorem 1. *The LT $\tilde{\pi}_i(s)$ of the time dependent system state probabilities $\pi_i(t)$, $(i = 0, 1, 2\}$ and LT $\tilde{R}(s)$ of the reliability function $R(t)$ for the considered system are*

$$\begin{aligned}
\tilde{\pi}_0(s) &= \frac{1}{s+\alpha_3+\psi(s)},\\
\tilde{\pi}_1(s) &= \frac{\phi_1(s)}{(s+\alpha_2)(s+\alpha_3+\psi(s))},\\
\tilde{\pi}_2(s) &= \frac{\phi_2(s)}{(s+\alpha_1)(s+\alpha_3+\psi(s))},\\
\tilde{\pi}_3(s) &= \frac{\alpha_1(s+\alpha_1)\phi_2(s)+\alpha_1(s+\alpha_2)\phi_1(s)+\alpha_3(s+\alpha_1)(s+\alpha_2)}{s(s+\alpha_1)(s+\alpha_2)(s+\alpha_3+\psi(s))},\\
\tilde{R}(s) &= \frac{(s+\alpha_1)(s+\alpha_2)+(s+\alpha_1)\phi_1(s)+(s+\alpha_1)\phi_2(s)}{(s+\alpha_2)(s+\alpha_1)(s+\alpha_3+\psi(s))},
\end{aligned}$$

where notations above are used.

Proof. By the usual method the system of Kolmogorov forward partial differential equations for the process time dependent state probabilities can be obtained:

$$\begin{aligned}
&\tfrac{d}{dt}\pi_0(t) = -\alpha\pi_0(t) + \textstyle\int_0^t \pi_1(t, x)\tilde{b}_1(x)dx\\
&+ \textstyle\int_0^t \pi_2(t, x)\tilde{b}_2(x)dx;\\
&\left(\tfrac{\partial}{\partial t} + \tfrac{\partial}{\partial x}\right)\pi_i(t; x) = -(\alpha_j + \alpha_3 + \tilde{b}_i(x))\pi_i(t; x)\ (i, j = 1, 2);\ i \neq j\\
&\tfrac{d}{dt}\pi_3(t) = \alpha_3\pi_0(t) + \alpha_2 \textstyle\int_0^t \pi_2(t; x)dx\\
&+ \alpha_1 \textstyle\int_0^t \pi_1(t; x)dx,
\end{aligned}$$

jointly with the initial $\pi_0(0) = 1$ and boundary conditions

$$\pi_i(t, 0) = \alpha_i\pi_0(t), \quad (i = 1, 2). \tag{3}$$

To solve this system the method of characteristics for solving the first-order partial differential equations (Petrovskiy [13]) is used. Accordingly to this method the equations for $\pi_i(t;x)$ $i=1,2$ in (3) is determined by the system of the system of ordinary differential equations, which in symmetric form are

$$dt = dx = \frac{-d\pi_i(\cdot)}{(\alpha_j + \bar{b}_i(x))}\pi_i(x) \;\; i,j=1,2, \;\; i \neq j.$$

The solution of this system along characteristics $t = x$ for $x \leq t$ is

$$\pi_1(t;x) = h_1(t-x)e^{-\alpha_2 x}(1 - B_1(x)),$$
$$\pi_2(t;x) = h_2(t-x)e^{-\alpha_1 x}(1 - B_2(x)).$$

where $h_1(\cdot)$ is some function, which is constant along characteristic and can be found from the boundary conditions.[1]

Further, from boundary conditions (3) it holds

$$\pi_i(t;0) = h_i(t) = \alpha_i \pi_0(t) \;\; (i=1,2). \tag{4}$$

Substitution of these solutions to the first of equation in (3) gives

$$\begin{aligned}\tfrac{d}{dt}\pi_0(t) &= -\alpha\pi_0(t)\\ &+ \textstyle\int_0^t h_1(t-x)e^{-(\alpha_2+\alpha_3)x}b_1(x)dx\\ &+ \textstyle\int_0^t h_2(t-x)e^{-(\alpha_1+\alpha_3)x}b_2(x)dx.\end{aligned}$$

In terms of LT with $\pi_0(0) = 1$ it holds that

$$(s+\alpha)\tilde{\pi}_0(s) - 1 = \beta_1(s+\alpha_1)+)\beta_2(s+\alpha_2).$$

By substituting into this equation the Laplace transform $\tilde{h}_i(s) = \alpha_i\tilde{\pi}_0(s)$ of functions $h_i(t)$ $(i=1,2)$ from (4) after some algebra one can obtain

$$\begin{aligned}&(s+\alpha)\tilde{\pi}_0(s) - \alpha_1\beta_1(s+\alpha_1)\tilde{\pi}_0(s)\\ &-\alpha_2\beta_2(s+\alpha_2)\tilde{\pi}_0(s) = 1.\end{aligned}$$

From this equality taking into account that $\alpha = \alpha_1+\alpha_2+\alpha_3$ and the notations (2) the following representation for $\tilde{\pi}_0(s)$ follows:

$$\tilde{\pi}_0(s) = [s + \alpha_3 + \psi(s)]^{-1}. \tag{5}$$

[1] The functions $h_i(\cdot)$ in Eq. (4) are the result of application of the characteristics method to the second one of Eqs. (3). However, these functions have a clear probabilistic interpretation. The states $(i,0)$ $(i=1,2)$ of the process Z can be considered as partially regenerative states (the state 0 is the state of full regeneration). Times of entering into these states are consequently the times of partial and full regeneration. Thus, the functions $h_i(\cdot)$ can be considered as renewal densities of the process Z for these partial regenerative times, while the other two multipliers in formula (4) show that during time x neither failure, nor repair occurs.

Applying LT to functions $\pi_i(t) = \int_0^t \pi_i(t;x)dx$ in Eqs. (4) with the help of expressions for $\tilde{h}_i(s)$ from (4) one can obtain

$$\begin{aligned}\tilde{\pi}_i(s) &= \int_0^\infty e^{-st} \int_0^t \pi_i(t;x)dxdt \\ &= \int_0^\infty e^{-st} \int_0^t h_i(t-x)e^{-\alpha_j x}(1-B_i(x))dxdt \\ &= \tilde{h}_i(s)\tfrac{1-\beta_i(s+\alpha_j)}{s+\alpha_j} \\ &= \tfrac{\phi_i(s)}{s+\alpha_j\tilde{\pi}_0(s)}\ (i,j=1,2;\ \ i\neq j)\end{aligned}$$

that coincides with the second expression in (3).

To find $\tilde{\pi}_3(s)$ we apply LT to the last equation of system (3). Usage of the expressions (4) for probabilities $\pi_i(t;x)$ $(i=1,2)$ leads as above to the following equality

$$\begin{aligned}s\tilde{\pi}_3(s) &= \alpha_3\tilde{\pi}_0(s) + \alpha_2\tilde{h}_2(s)\tfrac{1-\beta_2(s+\alpha_2)}{s+\alpha_2} \\ &+ \alpha_1\tilde{h}_1(s)\tfrac{1-\beta_1(s+\alpha_1)}{s+\alpha_1}.\end{aligned}$$

Substitution of representations for $\tilde{h}_i(s)$ from (4) in terms of $\tilde{\pi}_0(s)$ gives

$$\begin{aligned}s\tilde{\pi}_3(s) &= \tilde{\pi}_0(s)\left(\tfrac{\alpha_2\phi_2(s)}{s+\alpha_2} + \tfrac{\alpha_1\phi_1(s)}{s+\alpha_1} + \alpha_3\right) \\ &= \tfrac{\alpha_2(s+\alpha_1)\phi_2(s)+\alpha_1(s+\alpha_2)\phi_1(s)+\alpha_3(s+\alpha_2)(s+\alpha_1)}{(s+\alpha_2)(s+\alpha_1)(s+\alpha_3+\psi(s))},\end{aligned}$$

from which the expression for $\tilde{\pi}_3(s)$ in (3) follows.

Finally, taking into account that

$$R(t) = 1 - \mathbf{P}\{T \leq t\} = 1 - \pi_3(t)$$

after some cumbersome calculations one can find

$$\begin{aligned}\tilde{R}(s) &= \tfrac{1}{s} - \tilde{\pi}_3(s) \\ &= \tfrac{1}{s}\left[1 - \pi_0\left(\tfrac{\alpha_2\phi_2(s)}{s+\alpha_2} + \tfrac{\alpha_1\phi_1(s)}{s+\alpha_1} + \alpha_3\right)\right] \\ &= \tfrac{(s+\alpha_2)(s+\alpha_1)+(s+\alpha_1)\phi_1(s)+(s+\alpha_2)\phi_2(s)}{(s+\alpha_2)(s+\alpha_1)(s+\alpha_3+\psi(s))},\end{aligned} \tag{6}$$

which ends the proof.

As a corollary, by a substitution $s = 0$ one can find the mean time to the system failure.

Corollary 1. *The mean system life time with the help of notations (2) can be represented as follows:*

$$\begin{aligned}\mathbf{E}[T] &= \tilde{R}(0) \\ &= \frac{\alpha_1\alpha_2 + \alpha_1\phi_1(0) + \alpha_2\phi_2(0)}{\alpha_1\alpha_2(\alpha_3 + \psi(0))} \\ &= \frac{\alpha_1\alpha_2 + \alpha_1\alpha_1(1-\beta_1(\alpha_1)) + \alpha_2\alpha_1(1-\beta_2(\alpha_2))}{\alpha_1\alpha_2(\alpha_3 + \alpha_1(1-\beta_1(\alpha_1)) + \alpha_2(1-\beta_2(\alpha_2))}.\end{aligned}$$

The last expression coincides with corresponding result of corollary 4.3 in next section.

How these results can be used in sensitivity analysis of this process one may see in [14].

4 Probability Interpretations of Generating Functions

The system-level characteristics in terms of its Laplace-Stieltjes transform (LST) for this model are derived, by use of probability meaning of the LSTs and avoiding cumbersome analytic mathematical details. In the previous section the Laplace Transforms (LT) of state probabilities were found by use of somewhat direct probability analysis. In this section the *stationary* probabilities for such this MO renewable failure model are derived and investigated using common means of Markov chains. For this reason the time dependent probabilities are not required.

4.1 Life Cycle and System Life Time

Since every life cycle W consists of a system work portion of time T and ends with next system repair time B_3, a repair type 3, it is true that $W = T + B_3$, and T and B_3 are independent. Therefore it holds:

Lemma 1. *The LST $\omega(s)$ is solution of the equation*

$$\begin{aligned}
\omega(s) &= \tfrac{\alpha_3}{\alpha+s}\beta_3(s) \\
&+ \tfrac{\alpha_1}{\alpha+s}\beta_1(s+\alpha_2+\alpha_3)\omega(s) \\
&+ \tfrac{\alpha_2}{\alpha+s}\beta_2(s+\alpha_1+\alpha_3)\omega(s) \\
&+ \tfrac{\alpha_1}{\alpha+s}\tfrac{\alpha_2+\alpha_3}{\alpha_2+\alpha_3+s}[1-\beta_1(s+\alpha_2+\alpha_3)]\beta_3(s) \\
&+ \tfrac{\alpha_2}{\alpha+s}\tfrac{\alpha_1+\alpha_3}{\alpha_1+\alpha_3+s}[1-\beta_2(s+\alpha_1+\alpha_3)]\beta_3(s)
\end{aligned} \tag{7}$$

Proof. In this proof we will use the probability meaning of the LST, and the exponential distributions of the risks. The probability meaning of the LST was originally introduced by Kesten and Runnenburg [6]. It became public in the book of Klimov [7] and extensively used in the monograph of Gnedenko et al. [5].

In what follow next we explain the used meaning.

Introduce a complement process S_t of "catastrophes" - a Poisson process with parameter $s > 0$, and let S be the time to its first occurrence. Then the LST

$$\omega(s) = \int_0^\infty e^{-sx} dW(x) = P(S > W)$$

is the probability that during a time of duration W there will not happen any "catastrophes".

If we have two competing risks of parameters s and α, then probability that a risk of parameter α will happen first, and no risk of parameter s will happen meanwhile is

$$\frac{\alpha}{\alpha+s}[1-\omega(\alpha+s)] = \int_0^{\infty} \alpha e^{-(\alpha+s)x} dW(x).$$

Now, $\omega(s)$ is the probability that "no catastrophes" will happen during a cycle. The first line in the statement reflects the chance that one of the following sequence of independent events occurs:

(a1) The first comes a risk of type 3, no "catastrophes" until it, and then "no catastrophes" happen in the following repair time B_3 that will follow;

(a2) The first risk that comes is of type 1, and "no catastrophes" happen before it (probability of it is $\frac{\alpha_1}{s+\alpha}$, and then "no catastrophes" and no other risks of type 2 or 3 happen during the time B_1 (probability of this is $\beta_1(s+\alpha_2+\alpha_3)$, and then in the following new cycle "no catastrophes" happen (probability of what equals $\omega(s)$);

(a3) Analogously to the sequence described in (a2), the first risk that comes is of type 2, and "no catastrophes" happen before it, and then "no catastrophes" and no other risks of type 1 or 3 happen, and then during the following new cycle "no catastrophes" happen;

(a4) The first risk that comes is of type 1, and "no catastrophes" happen before it (probability of it is $\frac{\alpha_1}{s+\alpha}$), then "no catastrophes" but risks of type 2 or 3 happen during repair B_1 (probability is $\frac{\alpha_2+\alpha_3}{s+\alpha_2+\alpha_3][1-\beta_1(s+\alpha_2+\alpha_3]}$, and then in the following repair of type 3 "no catastrophes" happen (probability is $\beta_3(S)$);

(a5) Analogously to the sequence described in (a4), the *sequence starts with risk 2 occurring first* then the sequence ends with repair of type 3 during which "no catastrophes" happen;

These are the 5 particular cases in realizations of the event that during a time of duration W there will not happen any "catastrophes". By the total probability rule equals to the sum of probabilities of its particular cases.

These derived relations hold for $s > 0$ but are valid for any real and complex values of s according to the theory of continuation of the analytic functions.

Corollary 2. *The distribution of the system life cycle duration W is determined by its LST*

$$\omega(s) = \frac{\alpha_3 + \sum_{i,j=1,\, i\neq j}^{2} \alpha_i \frac{\alpha_j+\alpha_3}{\alpha_j+\alpha_3+s}[1-\beta_i(s+\alpha_j+\alpha_3)]\beta_3(s)}{\alpha+s-\alpha_1\beta_1(s+\alpha_2+\alpha_3)-\alpha_2\beta_2(s+\alpha_1+\alpha_3)}.$$

Proof. Solving the equation obtained in the above Lemma we get the statement in the corollary.

Corollary 3. *The life time of the system T is determined by its LST*

$$\tau(s) = \frac{\alpha_3 + \sum_{i,j=1,\, i\neq j}^{2} \alpha_i \frac{\alpha_j+\alpha_3}{\alpha_j+\alpha_3+s}[1-\beta_i(s+\alpha_j+\alpha_3)]}{\alpha+s-\alpha_1\beta_1(s+\alpha_2+\alpha_3)-\alpha_2\beta_2(s+\alpha_1+\alpha_3)}.$$

Proof. Use that

$$W = T + B_3$$

and T and B_3 are independent. Therefore $\omega(s) = \tau(s)\beta_3(s)$. Hence, $\tau(s) = \omega(s)/\beta_3(s)$. Substitute here $\omega(s)$ from Corollary refcoro2, and get the presentation in the statement.

Corollary 4. *The mean work time $E(T)$ of the system during a cycle is determined by the expression*

$$E(T) = \frac{1 + \frac{\alpha_1}{\alpha_2+\alpha_3}[1-\beta_1(\alpha_2+\alpha_3)] + \frac{\alpha_2}{\alpha_2+\alpha_3}[1-\beta_2(\alpha_1+\alpha_3]}{\alpha-\alpha_1\beta_1(\alpha_2+\alpha_3)-\alpha_2\beta_2(\alpha_1+\alpha_3)}.$$

Proof. Use that $E(T) = (-1)\frac{d\tau(s)}{ds}|_{s=0}$, and after some calculations get the statement. Calculations are significantly simplified if one does differentiation in s in the equation $\tau(s)denom(s) = num(s)$, where $denom(s)$ and $num(s)$ are notations for denominator and numerator in the expression for $\tau(s)$.

The mean work time $E(T)$ of the system during a cycle is finite, when the right hand side in last expression is finite.

Comment: If the repair times B_1, B_2 are instant, then $\beta_i(s) = 1$, $(i = 1, 2)$ and the only break is of type 3. Then $E(T) = 1/\alpha_3$. If $P(B_i > 0) > 0$, $(i = 1, 2)$, then $0 < \beta_i(s) < 1$, $(i = 1, 2)$, the numerator and denominator in $E(T)$ are finite. Therefore, $E(T)$ is always finite, hence, the life time of the system is always having finite expectation. Moreover, if $b_3 < \infty$, the cycle has a finite duration, and stationary regime is guaranteed.

Theorem 2. *If $\alpha_i > 0$, $(i = 1, 2, 3)$ and $0 < b_3 < \infty$, then the process is stable, and the macro state stationary probabilities*

$$\lim_{t\to\infty} P(E_0 \cup E_1 \cup E_2, t) = \frac{E(T)}{E(T) + b_3},$$

and

$$\lim_{t\to\infty} P(E_3, t) = \frac{b_3}{E(T) + b_3}$$

do exist for any distributions of the repair times B_i $(i = 1, 2)$.

Proof. In a long run the system process is an alternating renewal process, where a work time of duration T and a down time of duration B_3 alternatively change. By the renewal theory for alternating times of finite expectations the statement holds.

4.2 Number of Passages Between the States During a Cycle

We study the number of changes between the states during a cycle of the system. It uses another probability interpretation of the probability generating functions together with the LST when changes occur. Again, we use the probability meaning of the PGF's combined with the LST, as referred above to the monograph of Gnedenko et al. [5].

Introduce the random variables (Symbol # means "counts in the set")

$$N_i = \#(passages\ into\ E_i\ during\ a\ cycle).$$

Call a passage "green" with a probability $z_i \in [0, 1]$ independently on the color of other passages, and any other events. Then the function

$$\omega(\boldsymbol{z}, s) = E(z_0^{N_0} z_1^{N_1} z_2^{N_2} z_3^{N_3} e^{-sW})$$
$$= \int_0^\infty \sum_{k_i=0}^\infty P\left(\sum_{i=0}^3 N_i = k_i\right) z_0^{k_0} z_1^{k_1} z_2^{k_2} z_3^{k_3} e^{-sx} dW(x)$$

can be interpreted for $z_i \in [0, 1]$ and $s > 0$ as probability, that "during a cycle no catastrophes will happen, and all passages inside will be green".

Notice that

$$\omega(\boldsymbol{1}, s) = \omega(s) \text{ and } \omega(\boldsymbol{z}, 0) = \omega(z_0, z_1, z_2, z_3) \tag{8}$$

are the LST of the cycle duration, and the PGF of the number of passages in a cycle correspondingly. It is true:

Lemma 2. *The function $\omega(\boldsymbol{z}, s)$ is solution of the equation*

$$\begin{aligned}
\omega(\boldsymbol{z}, s) &= \frac{\alpha_3}{\alpha + s} z_3 \beta_3(s) \\
&+ \frac{\alpha_1}{\alpha + s} z_1 \beta_1(s + \alpha_2 + \alpha_3) z_0 \omega(\boldsymbol{z}, s) \\
&+ \frac{\alpha_2}{\alpha + s} z_2 \beta_2(s + \alpha_1 + \alpha_3) z_0 \omega(\boldsymbol{z}, s) \\
&+ \frac{\alpha_1}{\alpha + s} z_1 \frac{\alpha_2 z_2 + \alpha_3}{\alpha_2 + \alpha_3 + s} [1 - \beta_1(s + \alpha_2 + \alpha_3)] z_3 \beta_3(s) \\
&+ \frac{\alpha_2}{\alpha + s} z_2 \frac{\alpha_1 z_1 + \alpha_3}{\alpha_1 + \alpha_3 + s} [1 - \beta_2(s + \alpha_1 + \alpha_3)] z_3 \beta_3(s)
\end{aligned} \tag{9}$$

Proof. In this proof we use the probability meaning of the PGF $\omega(\boldsymbol{z}, s)$ combined with the LST when introduce a complimentary process S_t of "catastrophes" – a Poisson process with parameter $s > 0$, and "the green colors" of all the passages, as defined above. Then $\omega(\boldsymbol{z}, s)$ is the probability that during a time of duration W there will not happen any "catastrophes", and all the passages between the states are "green".

We have two independent competing risks of parameters s and α, then probability that a risk of parameter α will happen first, and no risk of parameter s will happen and the particular passage is "green" (probability equals z) is

$$\frac{\alpha}{\alpha + s} z[1 - \omega(\alpha + s)] = z \int_0^\infty \alpha e^{-(\alpha+s)x} dW(x),$$

since only one passage (one count) nay happen. Now, the probability $\omega(\boldsymbol{z}, s)$ that during a time of duration W there will not happen any "catastrophes", and all the passages between the states are "green" is probability of an event which has five particular cases. The first line in the statement reflects the chance that:

(a1) The first comes a risk of type 3, no "catastrophes" until it, this passage $E_0 \to E_3$ is green, and then "no catastrophes" happen in the following repair time B_2;

(a2) The first risk that comes is of type 1, and "no catastrophes" happen before it, this passage is "green" (with probability z_1), then "no catastrophes" and no other risks of type 2 or 3 happen, and then in the following new cycle passage is "green", "no catastrophes happen and all passages are green" probability of what is $z_0\omega(\boldsymbol{z}, s)$. This case explains line 2, presenting probability of the second particular case;

(a3) Same interpretation of this case as in (a2) when the first risk that comes is of type 2. No need to repeat details to explain line 3;

(a4) Line 4 presents the particular case, when the first risk that comes is of type 1, "no catastrophes" happen before it, the passage $E_0 \to E_1$ is green, then "no catastrophes" but risks of type 2 or 3 happen, the passage $E_1 \to E_2$ is green (probability z_2) and then during the following passage to repair of type 3 "no catastrophes" happen, and the last passage is also "green" with probability z_3;

(a5) The last line 5 reflects the probability of a particular case similar to that explained for line 4 when the first risk that comes is of type 2. We skip detailed explanation again.

These are the five particular cases in realizations of the event whose total probability equals to the sum of probabilities of its particular cases. These relations hold for $s > 0$ and $z_i \in (0, 1)$, but are valid for any real and complex values of s and z_i's according to the theory of analytic functions.

Corollary 5. *The PGF of the number of passages in a cycle $\omega(z_0, z_1, z_2, z_3)$ is determined by the equation*

$$\omega(\boldsymbol{z}) = \frac{\alpha_3 z_3 + \sum_{i,j=1,\, i\neq j}^{2} \alpha_i z_i \frac{\alpha_j z_j + \alpha_3}{\alpha_j + \alpha_3}[1 - \beta_i(\alpha_j + \alpha_3)] z_3}{\alpha - \alpha_1 z_1 \beta_1(\alpha_2 + \alpha_3) z_0 - \alpha_2 z_2 \beta_2(\alpha_1 + \alpha_3) z_0}.$$

Proof. Solving the Eq. (10) with respect to $\omega(\boldsymbol{z}, s)$ first, and let $s = 0$ in the obtained expression, and get the statement. By the way, if in the expression obtained for $\omega(\boldsymbol{z}, s)$ you put $\boldsymbol{z} = \boldsymbol{1}$, you find the result of Corollary 1. In this sense Lemma 2 presents a more detailed analysis of probabilities what happens within a cycle.

Corollary 6. *(a0) The average number of visits in state E_0 during a cycle equals*

$$E(N_0) = \frac{\alpha_1\beta_1(\alpha_2 + \alpha_3) + \alpha_2\beta_2(\alpha_1 + \alpha_3)}{\alpha - \alpha_1\beta_1(\alpha_2 + \alpha_3) - \alpha_2\beta_2(\alpha_1 + \alpha_3)};$$

(ai) The average number of visits in state E_i, $(i = 1, 2)$ during a cycle equals

$$E(N_i) = \frac{\alpha_i + \alpha_j \frac{\alpha_1}{\alpha_i+\alpha_3}\beta_j(\alpha_i+\alpha_3)}{\alpha - \alpha_i\beta_i(\alpha_j+\alpha_3) - \alpha_j\beta_j(\alpha_i+\alpha_3)}, \quad i,j = 1,2; \quad i \neq j'$$

(a3) The average number of visits in state E_3 during a cycle equals

$$E(N_3) = 1.$$

Proof. It is well known that

$$E(N_i) = \frac{\partial}{\partial z_i}\omega(z_0, z_1, z_2, z_3)|_{z=1}, \ i = 0, 1, 2, 3.$$

After taking the partial derivatives in expression for $\omega(z)$ and let all z_i's equal to one, by solving obtained equations with respect to $E(N_i)$ we get the stated expressions. Again, differentiation is simplified, if one multiplies by denominator both sides in result of Corollary 4.4, and then takes derivatives.

No wonder that $E(N_3) = 1$, since just once system may fail during a cycle, and this is the sure end of each cycle.

One might continue further, by finding the variances $Var(N_i)$, co-variances mixed moments $E(N_iN_j)$ and the correlation coefficients $\rho(N_iN_j)$ between these variables. By doing this one will get the results obtained in the work of Dimitrov and Rykov [14]. We do not attach it here.

Comment. Having the LST or the PGF of a distribution, one can investigate its asymptotic behaviour for small or large values of their arguments, applying Abelian or Tauberian theorems, as recommended in the work of Omey and Willenkens [11] approach can be found in Dimitrov [3]. In this way several useful approximations valid in large scope of situations could be found and practically used instead of detailed characteristics that follow from exact relationships.

Also it is important here to notice, that when some relationships are valid on an interval of continuity for the functions arguments the analytic contention theorem stays then these relations stay valid for all the regions where participating functions remain analytic.

4.3 Sojourn Times During a Life Cycle

To calculate the sojourn times G_i in each state E_i during a life cycle we will use the expressions relating numbers of visits in a state N_i and the individual sojourn times g_i $(i = 0, 1, 2, 3)$ at each visit. It holds

$$G_i = \sum_{k=0}^{N_i} g_i. \tag{10}$$

For our purposes, we are interested on the average sojourn times $E(G_i)$ in each state. Since we already know the distributions and mean times of the numbers N_i, we need the mean times $E(g_i)$ only. The use of the Wald identity

$$E(G_i) = E(N_i)E(g_i), \tag{11}$$

applied to (10) will give us the desired results.

Let look at the average sojourn times in each of the states.

Each stay in the state E_0 is the minimum of the three exponents of parameter α_i $(i = 1, 2, 3)$ each. Therefore

$$E(g_0) = \frac{1}{\alpha_1 + \alpha_2 + \alpha_3}.$$

Each stay g_i in the state E_i, $(i = 1, 2)$ is made either by a non-interrupted repair service B_i, or by an interrupted by failure of the other component, or by the risk of type 3. We use the probability meaning of the LST $g_i(s)$ to express these relationships. Our demo is on the case of $g_i(s)$ $i = 1, 2$. It holds

$$\begin{aligned} g_i(s) &= \beta_i(s + \alpha_j + \alpha_3) \\ &+ \tfrac{\alpha_j+\alpha_3}{s+\alpha_j+\alpha_3}[1 - \beta_i(s + \alpha_j + \alpha_3)], \quad i, j = 1, 2, i \neq j. \end{aligned} \tag{12}$$

This identity gives the probability $g_i(s)$ fo no "catastrophes" during the repair time B_i as a probability of the next particular cases: (1) None of the four risks (for catastrophes, interruptions of type j or 3 risks) will occur probability of what is $\beta_i(s+\alpha_j+\alpha_3)$, and (2) Some of the 3 competing risks occur, probability of what is $1 - \beta_i$, and first comes either risk j or risk 3, probability of what is $\dfrac{\alpha_j + \alpha_3}{s + \alpha_j + \alpha_3}$.

Now using that $E(g_i) = (-1)\frac{dg_i(s)}{ds}|_{s=0}$ we get

$$E(g_i) = \frac{1}{\alpha_j + \alpha_3}[1 - \beta_i(\alpha_j + \alpha_3)]. \tag{13}$$

Combine the ideas and results in this section above with the results of Corollary 5 we come to the following:

Theorem 3. *(A0) The average sojourn time in state E_0 during a cycle equals*

$$\begin{aligned} E(G_0) &= \tfrac{\alpha_1\beta_1(\alpha_2+\alpha_3)+\alpha_2\beta_2(\alpha_1+\alpha_3)1}{\alpha-\alpha_1\beta_1(\alpha_2+\alpha_3)-\alpha_2\beta_2(\alpha_1+\alpha_3)}\tfrac{1}{\alpha}, \\ i, j &= 1, 2; \; i \neq j; \end{aligned}$$

(Ai) The average sojourn time in state E_i during a cycle equals

$$\begin{aligned} E(G_i) = &\frac{\alpha_i + \alpha_j\frac{\alpha_i}{\alpha_i+\alpha_3}\beta_j(\alpha_i + \alpha_3)}{\alpha - \alpha_i\beta_i(\alpha_j + \alpha_3) - \alpha_j\beta_j(\alpha_i + \alpha_3)} \\ &\times\frac{1}{\alpha_j + \alpha_3}[1 - b_i(\alpha_j + \alpha_3)] \\ &i, j = 1, 2; \; i \neq j; \end{aligned}$$

(A3) The average sojourn time in state E_3 during a cycle equals

$$E(G_3) = E(B_3) = b_3.$$

An interesting dissection could be found if you compare the result of Corollary 4.3 and the last theorem. It must be true

$$E(T) = E(G_0) + E(G_1) + E(G_2),$$

since both expressions represent the work time on average during a life cycle.

4.4 Stationary Probabilities

The transitions between the macro states E_i in the considered process form a Markov chain with finite number of states. According the theory (Feller, [4]). Such chains always have stationary state and the stationary probabilities do exist. Namely, if $\pi_i(t)$ are the probabilities at the instant t the process

$$\pi_i = \lim_{t\to\infty} \pi_i(t), \quad (i = 0, 1, 2, 3)$$

are the stationary ones. We do not focus on the time dependent probabilities $\pi_i(t)$, but use the meaning of the stationary probabilities π_i. These are the portions of time in one unit of time, when the process spends in the state E_i, no matter how many times the process changes its states. Hence

Theorem 4. *(P0) The Stationary probability to find the process in state E_0 when both components are functioning is*

$$\pi_0 = \frac{E(G_0)}{E(T) + E(B_3)}$$

(Pi) The Stationary probability to find the process in state E_i $i = 1, 2$ when only component i is functioning is

$$\pi_i = \frac{E(G_i)}{E(T) + E(B_3)}, \quad i = 1, 2;$$

(P3) The Stationary probability to find the process in state E_3 when both components 1 and 2 are not functioning, and the whole system is under repair is

$$\pi_3 = \frac{E(B_3)}{E(T) + E(B_3)}$$

where $E(G_i)$ and $E(T)$ are determined by the expressions in Theorems 3 and Corollary 3.

Proof. The proof is a simple consequence of the rule

$$\pi_i = \frac{E(G_i)}{\sum_{j=0}^{3} E(G_j)}, \quad (i = 0, 1, 2, 3)$$

which is a consequence of the meaning of the stationary probabilities of a finite Markov chain.

In our opinion, this approach can be successfully applied in studying n-component systems with various modifications of the Marshal-Olkin type f maintenance models with renewals, as well as in modeling of k-out-of-n reliability systems under similar to ours assumptions.

5 Stationary Equilibrium Equations Approach

Ergodicity is a classical and broad topic. It has many applications in many fields [12].

The equilibrium equation has the following interpretation. On the left hand side, π_j is the long run proportion of time that the process is in state j, while λ_j is that rate of leaving state j when the process is in state j. Thus, the product $\pi_j\lambda_j$ is interpreted as the long run rate of leaving state j. On the right hand side, μ_{ij} is the rate of going to state j when the process is in state i, so the product $\pi_i\mu_{ij}$ is interpreted as the long run rate of going from state i to state j. Summing over all $i \neq j$ then gives the long run rate of going to state j. That is, the equation

$$\pi_j\lambda_j = \sum_{i\neq j} \pi_i\mu_{ij}),\ i,j \in E$$

is interpreted as "the long run rate out of state j" = "the long run rate into state j". These equations called the Global Balance Equations, or just Balance Equations. They express the fact that when the process is made stationary, there must be equality, or balance, in the long run rates into and out of any state. In addition, the sum of probabilities that the process is in the either state equals 1. A good use of this approach in the study of finite sources queuing systems can be found in [2].

Theorem 5. *The system of equilibrium equations*

$$\begin{aligned}
&\pi_0\alpha = \pi_1\tfrac{\beta_1(\alpha_2+\alpha_3)}{b_1} + \pi_2\tfrac{\beta_2(\alpha_1+\alpha_3)}{b_2} + \pi_3\tfrac{1}{b_3};\\
&\pi_i\left(\tfrac{\beta_i(\alpha_j+\alpha_3)}{b_i} + (\alpha_j+\alpha_3)\tfrac{1-\beta_i(\alpha_j+\alpha_3)}{b_i}\right)\\
&= \tfrac{\alpha_i}{\alpha}\pi_0;,\ i,j=1,2,\ i\neq j;\\
&\pi_3\left\{(\alpha_2+\alpha_3)\left(1-\beta_1(\alpha_2+\alpha_3)\right)\right.\\
&\left.+\left((\alpha_1+\alpha_3)(1-\beta_2(\alpha_1+\alpha_3))\right)+\tfrac{\alpha_3}{\alpha}\right\}\\
&= \tfrac{1}{b_3}\pi_0,
\end{aligned}$$

and the normalizing equation

$$\pi_0+\pi_1+\pi_2+\pi_3 = 1$$

uniquely determine the stationary probabilities of the embedded Markov chain in the MO dynamic renewable model.

Proof. Using the input-output intensities for each of the four states, for our MO dynamic process we get the equilibrium equations in the first block. The normalizing equation is natural.

Solutions of these equations coincide with the results obtained in Theorems 4.3 and 4.2 with the expression of Corollary 4.3 are used respectively. We omit details.

6 Conclusion

The three discussed approaches produce equivalent results in regard the stationary probabilities. However, each approach offers different details about the behaviour of the process progress and allows the use of these details for studying various process characteristics. If one is interested just in the steady state relationships, maybe then third approach is sufficient.

When non-stationary behaviour is important, especially in prose's control, we would recommend first approach.

If one likes to find extra details within a process cycle (between two points of regeneration), then probably the Second approach should be preferred.

Our detailed discussion on MO dynamic model are new and we are glad to have the opportunity to present it here and share with you.

In our opinion, these approaches can be successfully applied in studying n-component systems with various modifications of the Marshal-Olkin type of maintenance models with renewals, as well as in modeling of k-out-of-n reliability systems under similar to ours assumptions.

Acknowledgements. The publication has been prepared with the support of the "RUDN University Program 5–100" and funded by RFBR according to the research project No. 20-01-00575.

References

1. Barlow, R.E., Proshan, F.: Statistical theory of Reliability and Life Testing: Probability Models (To Begin With). Silver Spring, Maryland (1981)
2. Bocharov, P.P., D'Apice, C., Pechinkin, A.V.: Queueing Theory (Modern Probability and Statistics). De Gruiter, Berlin (2001)
3. Dimitrov, B.: Asymptotic expansions of characteristics for queuing systems of the type M/G/1. Bulletin de l'Inst. de Math. Acad. Bulg. Sci. **XV**, 237–263 (1984). (in Bulgarian)
4. An, F.W.: Introduction to Probability Theory and its Applications, vol. 2. Wiley, London (1966)
5. Gnedenko, B., Danielyan, E., Klimov, G., Matveev, V., Dimitrov, B.: Prioritetnye Sistemy Obslujivania. Moscow State University (1973). (in Russian)
6. Kesten, H., Runnenburg, J.T.: Priority in Waiting Line Problems. Koninklijke Netherlands Akademie van Wetenschappen, vol. 60. pp. 312–336 (1957)
7. Klimov, G.P.: Stochastic Queuing Systems. Nauka, Moscow (1966). (in Russian)
8. Li, X., Pellerey, F.: Generalized Marshall-Olkin distributions and related bivariate aging properties. J. Multivariate Anal. **102**, 1399–1409 (2011)
9. Lin, J., Li, X.: Multivariate generalized Marshall-Olkin distributions and copulas. Method. Comput. Appl. Probab. **16**, 53–78 (2014). https://doi.org/10.1007/s11009-012-9297-4
10. Marshall, A., Olkin, I.: A multivariate exponential distribution. J. Am. Stat. Assoc. **62**, 30–44 (1967)
11. Omey, E., Willenkens, E.: Abelian and Tauberian theorems for the Laplace transform of functions in several variables. J. Multivariate Anal. **30**, 292–306 (1989)

12. Pakes, A.G.: Some conditions for ergodicity and recurrence of Markov chains. Oper. Res. **17**, 1048–1061 (1969)
13. Petrovsky, I.G.: Lectures on the theory of ordinary differential equations (1951). M.-L.: GITTL 1952, 232 p. (in Russian)
14. Rykov, V., Dimitrov, B.: Renewal redundant systems under the Marshall-Olkin failure model. sensitivity analysis. In: Vishnevskiy, V.M., Samouylov, K.E., Kozyrev, D.V. (eds.) DCCN 2019. LNCS, vol. 11965, pp. 234–248. Springer, Cham (2019). https://doi.org/10.1007/978-3-030-36614-8_18

The Remaining Busy Time in a Retrial System with Unreliable Servers

Evsey Morozov[1,2,3(✉)] and Taisia Morozova[1]

[1] Petrozavodsk State University, Lenin str. 33, Petrozavodsk 185910, Russia
[2] Institute of Applied Mathematical Research of the Karelian Research Centre of RAS, Pushkinskaya str. 11, Petrozavodsk 185910, Russia
emorozov@karelia.ru, tiamorozova@gmail.com
[3] Moscow Center for Fundamental and Applied Mathematics, Moscow State University, Moscow 119991, Russia

Abstract. In this paper, we consider a multiserver retrial queuing system with unreliable servers class-dependent retrial rates and N classes of customers following Poisson input processes. We analyze the distribution of the stationary generalized remaining service time which includes all unavailable periods (setup times) occurring during service of the customer. During service of a class-i customer, the interruptions occur according to the i-dependent Poisson process and the following i-dependent random *setup time* of the server. We consider two following disciplines caused by the *service interruptions*: *preemptive repeat different* and *preemptive resume*. Using coupling method and regenerative approach, we derive the stationary distribution of the generalized remaining service time in an arbitrary server. For each class i, this distribution is expressed as a convolution of the corresponding original service times and setup times, and in general is available in the terms of the Laplace-Stieltjes transform allowing to calculate the moments of the target distribution. Some numerical examples are included as well.

Keywords: Retrial system · Generalized service time · Unreliable servers · Distribution of the remaining busy time

1 Introduction

This paper complements the research presented in the work [16] and is devoted to analysis of the stationary remaining service time in the multiclass retrial queueing systems with constant class-dependent retrial rates and *non-reliable servers*. In the paper [16], we study this problem for the classical buffered system with *reliable servers*. A common idea connecting these two papers is to apply the *regenerative approach* to deduce the stationary distribution of the remaining

The research is supported by Russian Foundation for Basic Research, projects No. 19-07-00303, 18-07-00156, 18-07-00147.

V. M. Vishnevskiy et al. (Eds.): DCCN 2020, LNCS 12563, pp. 555–566, 2020.
https://doi.org/10.1007/978-3-030-66471-8_42

service time using the renewal process generated by the sequence of the actual service times realized in the given server.

In this paper we consider this problem in continuous-time setting to find the limiting (stationary) distribution of the remaining service time of a customer being in the server at instant t, as $t \to \infty$. Another approach suggests that the remaining service time is estimated at the arrival instants of the customers and it leads to the discrete-time setting of the problem. It is worth mentioning that, because the input flow is Poisson, then both approaches give the same stationary distribution of the remaining service time due to the property PASTA [1].

In general, the analysis of the remaining service time is a challenging problem. On the other hand, it is an important Quality of Service parameter, especially in the retrial systems. The main reason is that, unlike the classic systems, in the retrial system after each departure of a customer, the server *remains unused* for a time, until the next exogenous customer arrives or a customer from an orbit occupies the server. These idle periods of the server occur after each departure, and thus the estimated remaining service time can be used by the orbital customers, for instance, to speed-up the attempts, in the appropriate time interval, to increase the chance to enter server. In this regard, the applications of this analysis in the framework of the game theory seem to be quite promising, see for instance, [7,8]. It is well-known that the remaining service time is important in the regenerative stability analysis [10,11]. However, in this case the *tightness* of the remaining service time plays a key role. To the best of our knowledge, a few papers only study the remaining service time as the main object of the research, see [4,10,18].

Now we describe the main underlying idea of the approach used in this paper to analyze the remaining service time. This approach is also developed in the paper [14] for the classic buffered queueing systems with *reliable servers*. However, as we show in this research, the method is extended to a wide class of the multiclass queueing systems in which the service times of a given class are independent identically distributed (iid) random variables; and we call them *class-dependent* service times. As we mention above, the systems we study possess *regeneration property*, and this is a critical requirement. We first assume that the basic regenerative queueing processes are *positive recurrent*, that is the mean regeneration period length is finite. In fact it implies *stability* of the system meaning that the stationary distribution of the basic process exists [1]. As to the stability conditions of the retrial systems, we refer to the papers [3,17] in which such conditions, for a wide class of the systems with the *constant retrial rate policy*, are established. The servers are assumed to be unreliable, and the interruptions imply a service delay. In this work, we consider the following two service disciplines caused by the *service interruptions*: *preemptive repeat different*, and *preemptive resume*. To reflect the delay caused by the interruptions, we consider the so-called *generalized service time* of a customer which equals the full time that customer spends in the server, see [5]. This time includes in particular, all setup times following interruptions which occurring during service of this customer. These generalized service times, for each class i, are iid and thus the method we develop in this research is applicable to study the remaining generalized service time by the analogy with the system with reliable servers [14]. Using the same approach as in [14], we construct a simpler system in which the stationary distribution of the remaining generalized service time remains unchanged. Then, using the iid generalized service times of class-i customers, we construct a class-i renewal process and show that the target distribution coincides with the distribution of the *stationary overshoot* in this process, multiplied by the stationary probability that the server is busy by a class-i customer. The type of service interruptions caused by the failure of the server plays a key role in our analysis, while the retrial policy is important for the motivation of the research. The interrupted customer blocks the server for other customers until he finishes service.

This model is a generalization of the model considered in previous works [12,15,16], where reliable servers are only considered. We stress that it is not the purpose of this work to discuss the systems with service interruptions in details, instead we refer to the following survey papers on this topic [6,9].

The main contribution of this paper is intuitive and states that the stationary distribution of the remaining generalized service time of a class-i customer coincides with the *stationary overshoot* in the renewal process generated by the class-i service times, multiplied by the stationary busy probability of the server by class-i customer. Because the distribution of the generalized service time is a convolution of the original service times and setup times, we apply the Laplace-Stieltjes transform (LST) representation of this distribution obtained earlier in the paper [5]. In particular this approach allows to calculate the moments of the target variable, and we give an example of this calculation in the last Sect. 5. Note that, in general, it is a challenging problem to obtain the entire distribution function by means of the inverting of the corresponding LST.

The paper is organized as follows. In Sect. 2 we describe the basic retrial system and its regenerative structure. Then, in Sect. 3, we introduce the generalized service time and give the representation of its the density for both preemptive repeat and for preemptive resume policies. Then the corresponding LST representations, adapted from the paper [5], are given. In Sect. 4 we outline the regenerative proof of the explicit form of stationary distribution of the generalized service time. In particular, we consider a simpler system, which is easier to be investigated, in which the limiting distribution of the remaining generalized service time is the same as in the original system. Then we deduce an explicit expression for the target distribution and the explicit expression for the stationary busy probability. To use the limiting distribution of the remaining generalized service time in practice, we need to be able to invert the corresponding LSTs. On the other hand, this expression allows in general to calculate numerically the moments of the target distribution, and in the last Sect. 5 we give some numerical examples which illustrate this point.

2 Description of the Model

In general, our analysis is applied to a wider class of the queueing systems with unreliable servers. Nevertheless, we describe in brief the following basic retrial system which, as we mentioned in the Introduction, motivates the present analysis of the remaining service time. It is assumed that the system has m indentical servers and is fed by N classes of customers following Poisson independent inputs. We denote by λ_i the input rate of class-i customers, and by $\lambda = \sum_{i=1}^{N} \lambda_i$, the total input rate. It is clear that $p_i := \lambda_i/\lambda$, is the probability that an arbitrary arrival belongs to class i; $i = 1, \ldots, N$. We assume that the service times of class-i customers, $\{S_n^{(i)}, n \geq 1\}$, are iid with rate $\mu_i = 1/\mathsf{E}S^{(i)}$ (where $S^{(i)}$ denotes the generic service time). We further denote by $F_i(x)$ the distribution function of $S^{(i)}$, by $f_i(x)$ its density and by $\mathcal{F}_i(z)$ the corresponding LST, that is

$$\mathcal{F}_i(z) = \mathsf{E}e^{-zS^{(i)}} = \int_0^\infty e^{-zx} f_i(x)dx, \ z \geq 0; \ i = 1, \ldots, N. \tag{1}$$

Each server becomes unavailable time to time, interrupting current service if any. It is assumed that the interruptions of each server occur according to a Poisson process with rate ν_i if a class-i customer is being served, $i = 1, \ldots, N$. After any interruption the server remains unavailable for a random *setup time* R_i, assuming that a class-i customer is interrupted. It is important to note that the interrupted customer continues to occupy the service area to return for service after the setup period. Thus this server is blocked for other customers until current customer finishes service and leaves the system. We consider preemptive repeat different, and preemptive resume service rules. In the former case, the interrupted service starts again as a new independently sampled service time, while in the latter case, the service continues after each interruption, when the server becomes available again after setup time, until it is exhausted. We denote by $r_i(x)$ the density of the setup time R_i, and by $\mathcal{R}_i(z)$ its LST,

$$\mathcal{R}_i(z) = \int_0^\infty e^{-zx} r_i(x)dx, \ z \geq 0; \ i = 1, \ldots, N. \tag{2}$$

An arriving class-i customer meeting all servers occupied, joins a class-i (virtual) orbit and then attempts to occupy servers again following a Poisson process with i-dependent rate. (The exact values of the retrial rates are not used in the subsequent analysis.)

Now we define the *regenerations* of the system as the arrival instants when the new customer meets *fully idle system*. Denote by T the generic *regeneration period length*. Then stability means that the mean length is finite, $\mathsf{E}T < \infty$, in which case the basic regenerative processes describing the dynamics of the system are called *positive recurrent* [1]. More detailed description of the regenerative structure of the retrial systems can be found in [5,12], while the stability analysis of the constant retrial rate systems has been developed in particular in the papers [2,3,17].

3 Generalized Service Time

In the systems with server interruptions, the total time a class-i customer spends in the server becomes bigger than the "pure" service time $S^{(i)}$ because it includes setup periods which occur during his service. By this reason, the distribution of this total time includes a convolution of the setup time distributions and in general is not easy to be calculated. We will call *generalized service time* the total time that a customer spends in the server which is precisely the time from the start of the service until the customer leaves the system. Denote by $C_n^{(i)}$ the generalized service time of the nth arriving class-i customer. For each i, the sequence $\{C_n^{(i)}, n \geq 1\}$ is iid with generic element $C^{(i)}$, $i = 1, \dots, N$.

3.1 Preemptive Repeat Service Discipline

We first consider the *preemptive repeat* interruption rule, which works as follows. If an interruption occurs while a class-i customer is being served then the attained service time is lost and a new independent service time, sampled from the common distribution F_i, starts anew when server becomes available after the setup time. For a distribution F, denote by $\bar{F} = 1 - F$ its *tail distribution.* Denote by G_i the distribution function of the generalized service time of class-i customer, and let g_i be its density. It is shown in [5] that the density g_i satisfies the following functional equation:

$$g_i(x) = e^{-\nu_i x} f_i(x) + \nu_i e^{-\nu_i x} \bar{F}_i(x) * r_i(x) * g_i(x), \ x \geq 0, \tag{3}$$

where $*$ means convolution. (A detailed discussion of this formula can be found in [5].) To solve this equation, we apply the LST. Define the LST of the generalized service time

$$\mathcal{G}_i(z) = \int_0^\infty e^{-zx} g_i(x) dx, \ z > 0.$$

Then, using (1)–(3) and the property of the LST of the convolution, we obtain, after a simple algebra (also see [5]), the following expression:

$$\mathcal{G}_i(z) = \frac{(\nu_i + z)\, \mathcal{F}_i(\nu_i + z)}{\nu_i + z - \nu_i \big(1 - \mathcal{F}_i(\nu_i + z)\big)\, \mathcal{R}_i(z)}, \ i = 1, \dots, N. \tag{4}$$

In particular, using (4) and the property of the LST, we derive the mean generalized service time of the class-i customer,

$$\mathsf{E}C^{(i)} = -\frac{d}{ds} \mathcal{G}_i(z)|_{z=0} = \frac{1 - \mathcal{F}_i(\nu_i)}{\mathcal{F}_i(\nu_i)} \left(\frac{1}{\nu_i} + \mathsf{E}R_i \right). \tag{5}$$

This in turn allows to calculate the traffic intensity of class-i customers,

$$\rho_i = \lambda_i \mathsf{E}C^{(i)}, \ \ i = 1, \dots, N, \tag{6}$$

which is used in the subsequent analysis.

Remark 1. Traffic intensity is also used in the stability analysis of the retrial systems. For instance, as it is expected, condition $\sum_i \rho_i < m$ is necessary for the positive recurrence of this retrial system. However the sufficient stability (positive recurrence) conditions are more complicated, see for instance the recent papers [12,17].

3.2 Preemptive-Resume Service Discipline

In this case, each customer spends in the server his initial service time and all setup periods caused by the interruptions which occur during his service. As it is shown in [5], in this case the density of the generalized service time satisfies

$$g_i(x) = e^{-\nu_i x} f_i(x) + f_i(x) \sum_{n=1}^{\infty} e^{-\nu_i x} \frac{(\nu_i x)^n}{n!} * r_i^{(n)}(x), \; x \geq 0, \tag{7}$$

where $r_i^{(n)}(x)$ denotes the nth convolution of the density $r_i(x)$ with itself. Then, as above, it follows that,

$$\mathcal{G}_i(z) = \mathcal{F}_i(\nu_i(1 - \mathcal{R}_i(z)) + z), \tag{8}$$

(also see [5]) implying in particular that

$$\mathsf{E}C^{(i)} = \mathsf{E}S^{(i)}(1 + \nu_i \mathsf{E}R_i), \; i = 1, \ldots, N. \tag{9}$$

4 Performance Analysis

Now we fix an arbitrary server and suppress server index in the following analysis. Denote by $C_i(t)$ the remaining generalized service time at instant t^-, and let $C_i(t) = 0$ if server is idle or serves a class-k customer with $k \neq i$. Then the process

$$B_i(t) = \int_0^t 1(C_i(u) > 0)du, \; i = 1, \ldots, N, \tag{10}$$

where $1(\cdot)$ denotes indicator function, describes the aggregated time, in the interval $[0,\, t]$, when the server is busy by class-i customers. Denote the weak limit, if exists,

$$C_i(t) \Rightarrow \mathbb{C}_i, \quad t \to \infty,$$

which is the *stationary remaining generalized service time.* When the system is positive recurrent, then it follows from the regenerative theory (and because the aggregated input process is Poisson) that the latter limit exists, and moreover,

$$\lim_{t\to\infty} \frac{B_i(t)}{t} =: \mathsf{P}_B^{(i)} = \mathsf{P}(\mathbb{C}_i > 0), \tag{11}$$

where $\mathsf{P}_B^{(i)}$ is the stationary probability that the server is occupied by class-i customer. Our main result is as follows.

Theorem 1. *The stationary distribution of the remaining generalized service time of class-i customer has the following form:*

$$\mathsf{P}(\mathbb{C}_i \le x) = 1 - \frac{\mathsf{P}_B^{(i)}}{\mathsf{E}C^{(i)}} \int_x^\infty (1 - G_i(u))du, \; x \ge 0; \; i = 1, \ldots, N. \tag{12}$$

Proof. The proof of this statement is similar to that is given in the paper [14] for the multiclass system with *reliable* servers. For the completeness we outline below the main steps of the proof. Denote by T_n, $n \ge 1$, the regeneration instants of the system, define the iid increments

$$B_n^{(i)} = \int_{T_{n-1}}^{T_n} 1(C_i(t) > 0)dt$$

of the process (10) over regeneration cycles, and let $B^{(i)}$ represent a generic increment of this busy time process over a generic regeneration period T. Now we consider a modified system, denoted by $\tilde{\Sigma}$, in which we couple together all class-i service times within a regeneration cycle, and then shift the obtained busy period, *distributed as* $B^{(i)}$, to the beginning of the cycle. We then denote by $\tilde{C}_i(t)$ the remaining (generalized) service time of a class-i customer in the given server in system $\tilde{\Sigma}$ at the instant t. By definition, $\tilde{C}_i(t) = 0$ if the server is free or serves a class-k customer, $k \ne i$. Using regenerative arguments, it is easy to show that the limiting distributions of $C_i(t)$ and $\tilde{C}_i(t)$ coincide, and in particular,

$$\lim_{t\to\infty} \mathsf{P}(C_i(t) > 0) = \lim_{t\to\infty} \mathsf{P}(\tilde{C}_i(t) > 0) = \mathsf{P}_B^{(i)}. \tag{13}$$

Because $\tilde{C}_i(t) = 0$ when $t \in [B^{(i)}, T)$, then we obtain, after some algebra, that

$$\lim_{t\to\infty} \frac{1}{t} \int_0^t 1(C_i(u) > x)du = \frac{1}{\mathsf{E}T} \mathsf{E} \int_0^{B^{(i)}} 1(\tilde{C}_i(u) > x)du. \tag{14}$$

Then we construct a *renewal process* $\hat{\mathbb{Z}}_i$ generated by the iid generalized service times $\{C_n^{(i)}\}$ realized in the given server, and denote by $\{\hat{C}_i(t)\}$ the remaining renewal time at instant t in this process. Denote the weak limit

$$\hat{C}_i(t) \Rightarrow \hat{\mathbb{C}}_i, \; t \to \infty, \tag{15}$$

which is the stationary *overshoot* in the renewal process $\hat{\mathbb{Z}}_i$. The following evident representation, which holds for each $x \ge 0$, is important for the further analysis:

$$\mathsf{P}(C_i(t) > x) = \mathsf{P}(C_i(t) > x | C_i(t) > 0)\mathsf{P}(C_i(t) > 0), \; t \ge 0.$$

Our purpose is to show that the conditional probability satisfies

$$\lim_{t\to\infty} \mathsf{P}(C_i(t) > x | C_i(t) > 0) = \mathsf{P}(\hat{\mathbb{C}}_i > x), \tag{16}$$

that is, coincides with the (tail) distribution of the overshoot (14) in the renewal process $\hat{\mathbb{Z}}_i$. (Note that, unlike (13), this convergence is not evident.) A key observation for our analysis is that each renewal point can be considered as a regeneration point of the process $\{\hat{C}_i(t)\}$. On the other hand, we can also split the renewal process $\hat{\mathbb{Z}}_i$ onto iid blocks distributed as a generic busy period $B^{(i)}$, and then treat these blocks as the regeneration periods "embedded" in the renewal process $\hat{\mathbb{Z}}_i$. By assumption, both these "regeneration periods" have finite means, $\mathsf{E}C^{(i)} < \infty$, $\mathsf{E}B^{(i)} < \infty$, and we obtain, using the well-known regenerative arguments [1], the following twofold representation, which holds for each $x \geq 0$:

$$\lim_{t\to\infty} \frac{1}{t}\int_0^t 1(\hat{C}_i(u) > x)du = \frac{1}{\mathsf{E}B^{(i)}}\mathsf{E}\int_0^{B^{(i)}} 1(\hat{C}_i(u) > x)du$$
$$= \mu_i \int_x^\infty (1 - G_i(u))du. \tag{17}$$

Combining (13)–(17) and after some algebra, we arrive to (12). □

Denote by $C(t)$ the *unconditional generalized remaining service time* at instant t, and then $C(t) = 0$ only if the server is free. Because, by construction, for $i = 1, \ldots, N$, only one indicator $1(C_i(t) > 0)$ maybe positive at each instant t, then we obtain the following representation

$$\mathsf{P}(C(t) > x) = \sum_{i=1}^N \mathsf{P}(C_i(t) > x | C_i(t) > 0)\mathsf{P}(C_i(t) > 0). \tag{18}$$

Denote the weak limit $C(t) \Rightarrow \mathbb{C}$. Then relation (18) immediately implies the following generalization of the statement (12). (Some details of the proof can be found in [14].)

Theorem 2. *The stationary distribution of the remaining generalized service time (in an arbitrary server) has the following form:*

$$\mathsf{P}(\mathbb{C} \leq x) = 1 - \sum_{i=1}^N \frac{\mathsf{P}_B^{(i)}}{\mathsf{E}C^{(i)}} \int_x^\infty (1 - G_i(u))du, \; x \geq 0. \tag{19}$$

It remains to find the probability $\mathsf{P}_B^{(i)}$. It can be done exactly as in [14] using the following balance equation connecting the work $V_i(t)$, generated by class-i arrivals in the interval $[0, t]$, the busy time $B_i(t)$ and the remaining work (workload) needed to serve class-i customers present at instant t,

$$V_i(t) = W_i(t) + B_i(t), \; i = 1, \ldots, N, \tag{20}$$

where, by the positive recurrence, $W_i(t) = o(t)$, $t \to \infty$ with probability 1 [19]. Recall that the servers are assumed to be equivalent and then, in the classic buffered systems, the realised *class-i work* $B_i(t)$ in the limit is equally divided among all m servers. This is because, by the FCFS discipline, each server accepts class-i customers with the same probability $p_i = \lambda_i/\lambda$ as in the input flow. In the retrial systems however the FCFS rule no longer holds. Nevertheless, because of the equivalence of the servers and by the positive recurrence implying the equality of the arrived and departed customers at the regeneration instants, we accept a plausible *assumption* that the mentioned uniform division of the work $B_i(t)$ holds true in the limit in this case as well. Then we easily find from (20) and (6) that

$$\lim_{t\to\infty}\frac{B_i(t)}{t} = m\mathsf{P}_B^{(i)} = \lim_{t\to\infty}\frac{V_i(t)}{t} = \lambda_i\mathsf{E}C^{(i)} = \rho_i,$$

implying

$$\mathsf{P}_B^{(i)} = \frac{\rho_i}{m},\ i = 1,\ldots,N.$$

Thus in particular the distribution function (19) can be rewritten as

$$D_i(x) := \mathsf{P}(\mathbb{C} \le x) = 1 - \sum_{i=1}^{N}\frac{\lambda_i}{m}\int_x^{\infty}(1 - G_i(u))du. \tag{21}$$

Note that, unlike the buffered system [14], we can not extend this analysis to the non-stationary (non-positive recurrent) retrial system because in this case the distribution of class-i customers among servers becomes unknown.

We stress that the distribution G_i of the generalized service time which is contained in (21) is obtained in the terms of the LST only. Thus, the extracting of the distributions G_i and D_i in an explicit form requires inversion of the LST, and as a rule is hardly available. Nevertheless in some cases the corresponding moments can be found. Indeed, it follows from (12), that the density $d_i(x)$ of the distribution $D_i(x)$ equals

$$d_i(x) = \frac{dD_i(x)}{dx} =: D_i^{(1)}(x) = \frac{\lambda_i}{m}(1 - G_i(x)),\ i = 1,\ldots,N;\ x \ge 0. \tag{22}$$

This gives the following expression for the nth moment of the target variable $\mathbb{C}_i$:

$$\int_0^{\infty} x^n d_i(x)dx = \mathsf{E}[\mathbb{C}_i]^n = \frac{\lambda_i}{m}\frac{\mathsf{E}[C^{(i)}]^{n+1}}{n+1},\ n > 0, \tag{23}$$

provided the moment $\mathsf{E}[C^{(i)}]^{n+1}$ exists. Thus if we can calculate $\mathsf{E}[C^{(i)}]^{n+1}$ using the LST representation (4) or (8) then we obtain the moment $\mathsf{E}[\mathbb{C}_i]^n$ in an explicit form.

It is worth mentioning that, by the property PASTA, expression (23) gives also the moments of the stationary remaining generalized service time observed by an arrival customer.

5 Examples

In this section we calculate some moments of the generalized service time for the preemptive-resume service interruption policy and class-i customers, using representation (8):

$$\mathcal{G}_i(z) = \mathcal{F}_i(\nu_i - \nu_i \mathcal{R}_i(z) + z). \tag{24}$$

The second derivation of $\mathcal{G}_i(z)$, denoted $\mathcal{G}_i^{(2)}(z)$, equals

$$\begin{aligned}\mathcal{G}_i^{(2)}(z) &= -\nu_i \mathcal{R}_i^{(2)}(z)\mathcal{F}_i^{(1)}(\nu_i - \nu_i \mathcal{R}_i(z) + z) \\ &+ (1 - \nu_i \mathcal{R}_i^{(1)}(z))^2 \mathcal{F}_i^{(2)}(\nu_i - \nu_i \mathcal{R}_i(z) + z).\end{aligned} \tag{25}$$

Because

$$\begin{aligned}&\mathcal{R}_i(0) = 1,\ \mathcal{R}_i^{(1)}(0) = -\mathsf{E}R_i,\ \mathcal{R}_i^{(2)}(0) = \mathsf{E}R_i^2,\ \mathcal{R}_i^{(3)}(0) = -\mathsf{E}R_i^3, \\ &\mathcal{F}_i^{(1)}(0) = -\mathsf{E}S^{(i)},\ \mathcal{F}_i^{(2)}(0) = \mathsf{E}[S^{(i)}]^2,\ \mathcal{F}_i^{(3)}(0) = -\mathsf{E}[S^{(i)}]^3,\end{aligned} \tag{26}$$

then, substituting $z = 0$ in both sides of (25), we obtain

$$\mathcal{G}_i^{(2)}(0) = \mathsf{E}C_i^2 = \nu_i\, \mathsf{E}R_i^2\, \mathsf{E}S^{(i)} + (1 + \nu_i \mathsf{E}R_i)^2\, \mathsf{E}[S^{(i)}]^2. \tag{27}$$

Now (23) gives the mean stationary generalized remaining service time of class-i customer:

$$\mathsf{E}\mathbb{C}_i = \frac{\lambda_i}{2m}\Big[\nu_i\, \mathsf{E}R_i^2\, \mathsf{E}S^{(i)} + (1 + \nu_i \mathsf{E}R_i)^2\, \mathsf{E}[S^{(i)}]^2\Big],\quad i = 1, \dots, N. \tag{28}$$

To calculate the second moment $\mathsf{E}\mathbb{C}_i^2$, we find the third derivative of $\mathcal{G}_i(z)$:

$$\begin{aligned}\mathcal{G}_i^{(3)}(z) = &- \nu_i \mathcal{R}_i^{(3)}(z)\mathcal{F}_i^{(1)}(\nu_i - \nu_i R(z) + z)(-\nu_i \mathcal{R}_i^{(1)}(z) + 1) \\ &- \nu_i \mathcal{R}_i^{(2)}(z)\mathcal{F}_i^{(2)}(\nu_i - \nu_i R(z) + z)(-\nu_i \mathcal{R}_i^{(1)}(z) + 1) \\ &+ 2(1 - \nu_i \mathcal{R}_i^{(1)}(z))(-\nu_i \mathcal{R}_i^{(2)}(z))\mathcal{F}_i^{(2)}(\nu_i - \nu_i R(z) + z) \\ &+ (1 - \nu_i \mathcal{R}_i^{(1)}(z))^3 \mathcal{F}_i^{(3)}(\nu_i - \nu_i R(z) + z),\end{aligned}$$

which, after substituting $z = 0$, gives

$$\begin{aligned}\mathsf{E}\mathbb{C}_i^2 &= -\frac{\lambda_i}{3m}\mathcal{G}_i^{(3)}(0) \\ &= \frac{\lambda_i}{3m}\Big[\mathsf{E}[S^{(i)}]^3(1 + \nu_i \mathsf{E}R_i)^3 + 3\nu_i \mathsf{E}R_i^2(1 + \nu_i \mathsf{E}R_i)\mathsf{E}[S^{(i)}]^2 + \nu_i \mathsf{E}R_i^3 \mathsf{E}S^{(i)}\Big].\end{aligned}$$

Now we present two numerical examples which illustrate formula (28) and show the impact of the preemptive-resume service interruptions on the mean generalized service time. In both examples we allow arbitrary number of servers m and arbitrary input rate λ_i.

Pareto Service Time. First we consider Pareto service time distribution

$$F_i(x) = 1 - \left(\frac{1}{x}\right)^3, \; x \geq 1 \; (F_i(x) = 0, \; x < 1), \tag{29}$$

and the exponential setup period R_i with rate 4. Moreover we take the interruption rate $\nu_i = 2$. As a result, we obtain the following values of the required moments:

$$\mathsf{E}R_i = 0.25, \; \mathsf{E}R_i^2 = 0.125, \; \mathsf{E}S^{(i)} = 1.5, \; \mathsf{E}[S^{(i)}]^2 = 3. \tag{30}$$

Now, substituting these values in formula (28), we obtain the mean stationary remaining generalized service time:

$$\mathsf{E}\mathbb{C}_i = \frac{\lambda_i}{2m} 7.125. \tag{31}$$

Weibull Service Time. In this example we choose the Weibull service time distribution

$$F_i(x) = 1 - e^{-x^2}, \; x \geq 0, \tag{32}$$

leaving all other parameters unchanged. This gives

$$\mathsf{E}S^{(i)} = \Gamma\left(\frac{3}{2}\right) = \frac{\sqrt{\pi}}{2} \approx 0.89, \; \mathsf{E}[S^{(i)}]^2 = \Gamma(2) = 1,$$

where Γ denotes Gamma-function, and by (28), we obtain

$$\mathsf{E}\mathbb{C}_i = \frac{\lambda_i}{2m} \frac{\sqrt{\pi} + 18}{8} \approx \frac{\lambda_i}{2m} 2.47.$$

It is seen that in both examples the influence of interruptions on the mean remaining generalized service time turns out to be not very significant. Also note that the Weibull service time with the shape parameter 2 as in (32) has the New-Better-Than-Used property [1], and it may cause a decreasing of the remaining generalized service time.

6 Conclusion

In this work we consider a multisever multiclass retrial system with unreliable servers. We investigate the remaining generalized service time of a customer observed at a random instant as time goes to infinity. The generalized service time includes possible unavailable periods (setup times) caused by the interruptions of the server occurring during service. We consider both preemptive repeat different service and preemptive resume service disciplines. Using the regenerative arguments, we obtain the stationary distribution of the remaining generalized service time, which is available in the terms of the LST. In general, this allows to calculate the moments of the remaining generalized service time when the moments of the generalized service time can be found. Some numerical examples are included as well.

References

1. Asmussen, S.: Applied Probability and Queues, 2nd edn. Springer-Verlag, New York (2003). https://doi.org/10.1007/b97236
2. Avrachenkov, K., Morozov, E.: Stability analysis of GI/G/c/K retrial queue with constant retrial rate. Math. Meth. Oper. Res. **79**(79), 273–291 (2014). https://doi.org/10.1007/s00186-014-0463-z
3. Avrachenkov, K., Morozov, E., Steyaert, B.: Sufficient stability conditions for multi-class constant retrial rate systems. Queueing Syst. 149–171 (2015). https://doi.org/10.1007/s11134-015-9463-9
4. Boer, P.T.D., Nicola, V.F., van Ommeren, J.K.C.: The remaining service time upon reaching a high level in $M/G/1$ queues. Queueing Syst. **39**, 55–78 (2001)
5. Dimitriou, I., Morozov, E., Morozova, T.: Multiclass retrial system with coupled orbits and service interruptions: Verification of stability conditions. In: Conference of Open Innovations 133 Association, FRUCT, 2019, vol. 24, pp. 75–81 (2019). ISSN: 2305-7254, eISSN: 2343-0737
6. Federgruen, A., Green, L.: Queueing systems with service interruptions. Oper. Res. **34**(5), 752–768 (1986). https://doi.org/10.1287/opre.34.5.752
7. Kerner, Y.: The conditional distribution of the residual service time in the M n/G/1 queue. Stoch. Models **24**(3), 364–375 (2008)
8. Kerner, Y.: Equilibrium joining probabilities for an M/G/1 queue. Games Econ. Behav. **71**(2), 521–526 (2011)
9. Krishnamoorthy, A., Pramod, P.K., Chakravarthy, S.R.: Queues with interruptions: a survey. TOP **22**(1), 290–320 (2012). https://doi.org/10.1007/s11750-012-0256-6
10. Morozov, E.: The tightness in the ergodic analysis of regenerative queueing processes. Queueing Syst. **27**, 179–203 (1997)
11. Morozov, E., Delgado, R.: Stability analysis of regenerative queues. Autom. Remote Control **70**(12), 1977–1991 (2009)
12. Morozov, E., Morozova, T.: Analysis of a generalized system with coupled orbits. In: FRUCT23: Proceedings of the 23rd Conference of Open Innovations Association FRUCT 2018, pp. 253–260, November (2018)
13. Morozov, E., Phung-Duc, T.: Stability analysis of a multiclass retrial system with classical retrial policy. Perform. Eval. **112**, 15–26 (2017)
14. Morozov, E., Morozova, T.: On stationary remaining service time in queuieng systems (submitted for Smarty 2020)
15. Morozov, E., Dimitriou, I.: Stability analysis of a multiclass retrial system with coupled orbit queues. In: Reinecke, P., Di Marco, A. (eds.) EPEW 2017. LNCS, vol. 10497, pp. 85–98. Springer, Cham (2017). https://doi.org/10.1007/978-3-319-66583-2_6
16. Morozov, E., Morozova, T., Dimitriou, I.: Simulation of multiclass retrial system with coupled orbits. In: Proceedings of SMARTY 2018: First International Conference Stochastic Modeling and Applied Research of Technology Petrozavodsk, 21–25 September 2018, Russia, pp. 6–16 (2018)
17. Morozov, E., Rumyantsev, A., Dey, S., Deepak, T.G.: Performanace analysis and stability of multiclass orbit queue with constant retrial rates and balking. Perform. Eval. **102005** (2019). https://doi.org/10.1016/j.peva.2019.102005
18. Ross, S.M., Seshadri, S.: Hitting time in an $M/G/1$ queue. J. Appl. Prob. **36**, 934–940 (1999)
19. Smith, W., L.: Regenerative stochastic processes. Proc. Roy. Soc., ser. A **232**, 6–31 (1955)

Diffusion Approximation for Multiserver Retrial Queue with Two-Way Communication

Anatoly Nazarov[1], Tuan Phung-Duc[2], Svetlana Paul[1](✉), and Olga Lizyura[1]

[1] Institute of Applied Mathematics and Computer Science, National Research Tomsk State University, 36 Lenina ave., Tomsk 634050, Russia
nazarov.tsu@gmail.com, paulsv82@mail.ru, oliztsu@mail.ru

[2] Faculty of Engineering Information and Systems, University of Tsukuba, 1-1-1 Tennodai, Tsukuba, Ibaraki 305-8573, Japan
tuan@sk.tsukuba.ac.jp

Abstract. In this paper, we consider a multiserver retrial queue with two-way communication. Incoming calls arrive according to the stationary Poisson process and occupy the servers. Durations of incoming calls have an exponential distribution. If all the servers are busy upon arrival, the incoming call joins the orbit. The time spent by the call in the orbit is an exponential random variable. Idle servers also make outgoing calls whose durations follow an exponential distribution. We derive the diffusion limits of the number of calls in the orbit and the approximation of its stationary probability distribution.

Keywords: Multiserver retrial queue · Two-way communication · Incoming call · Outgoing call · Asymptotic-diffusion analysis · Diffusion approximation

1 Introduction

Currently, more and more services are partially or fully working in a call-center mode. Banks use call centers to advise customers and to advertise their services; online stores utilize call centers to refine and confirm orders, and call centers also exist independently to conduct social surveys.

A lot of research papers are devoted to modeling call centers. Studies [1,4,5,8] are devoted to the quality of customer service in telephone services. In [3], the authors present a study of incoming processes in real call-centers. A statistical analysis of the work of call-centers taking into account various aspects of the functioning of the service is presented in the study [6].

Recently, retrial queues with two-way communication have been used as a model of a blended call-center. The most detailed studies on retrial queues can

The publication has been prepared with the support of RFBR according to the research project No.18-01-00277.

V. M. Vishnevskiy et al. (Eds.): DCCN 2020, LNCS 12563, pp. 567–578, 2020.
https://doi.org/10.1007/978-3-030-66471-8_43

be found in [2,7]. The main feature of model with two-way communication is that during idle time, the server makes outgoing calls and serves them along with incoming calls. As an outgoing call we refer to a call of an operator to a client, or any other type of alternative activities of the operators rather than serving incoming calls. Thus, retrial queueing models with two-way communication are flexible and allow us to simulate most modern telephone services. In [10] the model of multiserver retrial queue with two-way communication was proposed and numerical analysis was presented.

In this paper, we consider the same model as in [10] but rather than numerical analysis of the stationary distribution, our focus is to obtain the diffusion limit of the underlying time-dependent Markov process [9]. As a byproduct, the limiting results are then used to approximate the probability distribution of the number of customers in the orbit in the stationary state.

The rest of our paper is organized as follows. Section 2 is devoted to the presentation of the model and preliminary analysis of the underlying Markov chain. Our main results are in Sect. 3 where we present the asymptotic-diffussion analysis to obtain the diffusion limit of the underlying Markov chain. In Sect. 4 we describe an algorithm for constructing a probability distribution approximation of the system state. Section 5 shows the diffusion approximation accuracy for several values of system parameters. Finally, Sect. 6 is devoted to some concluding remarks.

2 Mathematical Model

We consider a multiserver retrial queue with two-way communication. The input process is the stationary Poisson process with rate λ. Service times of incoming calls are exponentially distributed with rate μ_1. Calls that find all the servers fully occupied join the orbit and reattempt to access the server after an exponentially distributed delay with rate σ. When the server is idle it makes outgoing calls with rate α and provides the service for an exponentially distributed time with rate μ_2. We denote N as the number of servers in the system.

Let $n_1(t), n_2(t)$ denote the numbers of servers busy serving incoming and outgoing calls at the time t, respectively. Also denote $i(t)$ as the number of calls in the orbit. Thus, we can see that three-dimensional random process $\{n_1(t),\ n_2(t),\ i(t)\}$ is a continuous time Markov chain.

Let $P(n_1, n_2, i, t) = P\{n_1(t) = n_1, n_2(t) = n_2, i(t) = i\}$ denote the probability distribution of the process $\{n_1(t),\ n_2(t),\ i(t)\}$, which is the solution of Kolmogorov's system of equations

$$\frac{\partial P(0,0,i,t)}{\partial t} = -(\lambda + i\sigma + N\alpha)P(0,0,i,t) + \mu_1 P(1,0,i,t) + \mu_2 P(0,1,i,t),$$

$$\frac{\partial P(n_1,n_2,i,t)}{\partial t} = -(\lambda + i\sigma + (N - n_1 - n_2)\alpha + n_1\mu_1 + n_2\mu_2)P(n_1,n_2,i,t)+$$
$$+\lambda P(n_1-1,n_2,i,t) + (i+1)\sigma P(n_1-1,n_2,i+1,t)+$$

$$+(N-n_1-n_2+1)\alpha P(n_1,n_2-1,i,t)+(n_1+1)\mu_1 P(n_1+1,n_2,i,t)+$$
$$+(n_2+1)\mu_2 P(n_1,n_2+1,i,t),\ 0<n_1+n_2<N,$$
$$\frac{\partial P(n_1,n_2,i,t)}{\partial t}=-(\lambda+n_1\mu_1+n_2\mu_2)P(n_1,n_2,i,t)+\lambda P(n_1-1,n_2,i,t)+$$
$$+\lambda P(n_1,n_2,i-1,t)+(i+1)\sigma P(n_1-1,n_2,i+1,t)+$$
$$+\alpha P(n_1,n_2-1,i,t),\ n_1+n_2=N. \tag{1}$$

Then we transform the system (1) into system for partial characteristic functions $H(n_1,n_2,u,t)=\sum\limits_{i=0}^{\infty}e^{jui}P(n_1,n_2,i,t)$, where $j=\sqrt{-1}$

$$\frac{\partial H(n_1,n_2,u,t)}{\partial t}=-(\lambda+N\alpha)H(n_1,n_2,u,t)+j\sigma\frac{\partial H(n_1,n_2,u,t)}{\partial u}+$$
$$+\mu_1 H(n_1+1,n_2,u,t)+\mu_2 H(n_1,n_2+1,u,t),\ n_1+n_2=0,$$
$$\frac{\partial H(n_1,n_2,u,t)}{\partial t}=-(\lambda+(N-n_1-n_2)\alpha+n_1\mu_1+n_2\mu_2)H(n_1,n_2,u,t)+$$
$$+j\sigma\frac{\partial H(n_1,n_2,u,t)}{\partial u}+\lambda H(n_1-1,n_2,u,t)-j\sigma e^{-ju}\frac{\partial H(n_1-1,n_2,u,t)}{\partial u}+$$
$$+(N-n_1-n_2+1)\alpha H(n_1,n_2-1,u,t)+(n_1+1)\mu_1 H(n_1+1,n_2,u,t)+$$
$$+(n_2+1)\mu_2 H(n_1,n_2+1,u,t),0<n_1+n_2<N,$$
$$\frac{\partial H(n_1,n_2,u,t)}{\partial t}=-(\lambda+n_1\mu_1+n_2\mu_2)H(n_1,n_2,u,t)+\lambda H(n_1-1,n_2,u,t)+$$
$$+\lambda e^{ju}H(n_1,n_2,u,t)-j\sigma e^{-ju}\frac{\partial H(n_1-1,n_2,u,t)}{\partial u}+$$
$$+\alpha H(n_1,n_2-1,u,t),\ n_1+n_2=N. \tag{2}$$

We denote $\mathbf{H}(u,t)$ as a matrix of functions $H(n_1,n_2,u,t)$ and rewrite the system (2) in the form of

$$\frac{\partial \mathbf{H}(u,t)}{\partial t}=(\mathbf{A}+\lambda e^{ju}\mathbf{B})\mathbf{H}(u,t)+j\sigma(\mathbf{I}_0-e^{-ju}\mathbf{I}_1)\frac{\partial \mathbf{H}(u,t)}{\partial u}, \tag{3}$$

where $\mathbf{A}$, $\mathbf{B}$, $\mathbf{I}_0$, $\mathbf{I}_1$ are operators, which set in the following form

$$\mathbf{A}\mathbf{H}(u,t)=\begin{cases}-(\lambda+N\alpha)H(n_1,n_2,u,t)+\mu_1 H(n_1+1,n_2,u,t)+\\ +\mu_2 H(n_1,n_2+1,u,t),\ n_1+n_2=0,\\ \\ -(\lambda+(N-n_1-n_2)\alpha+n_1\mu_1+n_2\mu_2)H(n_1,n_2,u,t)+\\ +(N-n_1-(n_2-1))\alpha H(n_1,n_2-1,u,t)+\\ +\lambda H(n_1-1,n_2,u,t)+(n_1+1)\mu_1 H(n_1+1,n_2,u,t)+\\ +(n_2+1)\mu_2 H(n_1,n_2+1,u,t),\ 0<n_1+n_2<N,\\ \\ -(\lambda+n_1\mu_1+n_2\mu_2)H(n_1,n_2,u,t)+\\ +\lambda H(n_1-1,n_2,u,t)+\alpha H(n_1,n_2-1,u,t),\\ n_1+n_2=N,\end{cases} \tag{4}$$

$$\mathbf{B}\mathbf{H}(u,t) = \begin{cases} 0, & n_1 + n_2 < N, \\ H(n_1, n_2, u, t), & n_1 + n_2 = N, \end{cases} \tag{5}$$

$$\mathbf{I}_0\mathbf{H}(u,t) = \begin{cases} H(n_1, n_2, u, t), & n_1 + n_2 < N, \\ 0, & n_1 + n_2 = N, \end{cases} \tag{6}$$

$$\mathbf{I}_1\mathbf{H}(u,t) = \begin{cases} 0, & n_1 + n_2 = 0, \\ H(n_1 - 1, n_2, u, t), & n_1 + n_2 > 0. \end{cases} \tag{7}$$

We denote $\mathbf{E}$ as an operator that sums up over all available values of n_1, n_2 and present the following additional equation

$$\mathbf{E}\frac{\partial \mathbf{H}(u,t)}{\partial t} = \mathbf{E}(\mathbf{A} + \lambda e^{ju}\mathbf{B})\mathbf{H}(u,t) + j\sigma\mathbf{E}(\mathbf{I}_0 - e^{-ju}\mathbf{I}_1)\frac{\partial \mathbf{H}(u,t)}{\partial u}. \tag{8}$$

We also note that

$$\mathbf{E}(\mathbf{A} + \lambda\mathbf{B}) = 0, \; \mathbf{E}(\mathbf{I}_0 - \mathbf{I}_1) = 0. \tag{9}$$

3 Asymptotic-Diffusion Analysis

In operator Eqs. (3) and (8), we introduce the following notations

$$\sigma = \varepsilon, \; u = \varepsilon w, \; \tau = \varepsilon t, \; \mathbf{H}(u,t) = \mathbf{F}(w, \tau, \varepsilon),$$

to obtain the equations

$$\varepsilon\frac{\partial \mathbf{F}(w,\tau,\varepsilon)}{\partial \tau} = (\mathbf{A} + \lambda e^{jw\varepsilon}\mathbf{B})\mathbf{F}(w,\tau,\varepsilon) + j(\mathbf{I}_0 - e^{-jw\varepsilon}\mathbf{I}_1)\frac{\partial \mathbf{F}(w,\tau,\varepsilon)}{\partial w}, \tag{10}$$

$$\varepsilon\mathbf{E}\frac{\partial \mathbf{F}(w,\tau,\varepsilon)}{\partial \tau} = \mathbf{E}(\mathbf{A} + \lambda e^{jw\varepsilon}\mathbf{B})\mathbf{F}(w,\tau,\varepsilon) + j\mathbf{E}(\mathbf{I}_0 - e^{-jw\varepsilon}\mathbf{I}_1)\frac{\partial \mathbf{F}(w,\tau,\varepsilon)}{\partial w}. \tag{11}$$

We solve the system (11) taking the limit as $\varepsilon \to 0$ and present the result in the following theorem.

Theorem 1. *In considered retrial queue, the limiting stationary probability distribution $R(n_1, n_2)$ of the two-dimensional process $\{n_1(t),\; n_2(t)\}$ is a solution of the system of operator equations*

$$(\mathbf{A} + \lambda\mathbf{B} - x(\tau)(\mathbf{I}_0 - \mathbf{I}_1))\mathbf{R} = 0,$$

$$\mathbf{E}\mathbf{R} = 1, \tag{12}$$

where $\mathbf{R}$ is a matrix of probabilitites $R(n_1, n_2)$ and function $x(\tau)$ is a solution of differential equation

$$x'(\tau) = \mathbf{E}\left[\lambda\mathbf{B} - x(\tau)\mathbf{I}_1\right]\mathbf{R}.$$

Proof. We consider the Eq. (10) in the limit by $\varepsilon \to 0$, denoting $\lim_{\varepsilon\to 0} \mathbf{F}(w,\tau,\varepsilon) = \mathbf{F}(w,\tau)$

$$(\mathbf{A}+\lambda\mathbf{B})\mathbf{F}(w,\tau)+j(\mathbf{I}_0-\mathbf{I}_1)\frac{\partial \mathbf{F}(w,\tau)}{\partial w}=0. \tag{13}$$

We find the solution of the operator Eq. (13) in the following form

$$\mathbf{F}(w,\tau)=\mathbf{R}e^{jwx(\tau)}, \tag{14}$$

where $\mathbf{R}$ is the matrix with elements $R(n_1,n_2)$, which are the joint stationary distribution of two-dimentional process $\{n_1(t),n_2(t)\}$, then we obtain the operator equation

$$(\mathbf{A}+\lambda\mathbf{B}-x(\tau)(\mathbf{I}_0-\mathbf{I}_1))\mathbf{R}=0. \tag{15}$$

For the probability distribution $\mathbf{R}$ we also have the normalization condition

$$\mathbf{E}\mathbf{R}=1, \tag{16}$$

hence, the probability distribution $\mathbf{R}$ is the solution of operator system of Eqs. (12).

From the Eq. (11) in the limit by $\varepsilon \to 0$ we obtain

$$\mathbf{E}\frac{\mathbf{F}(w,\tau)}{\partial \tau}=jw\left(\lambda\mathbf{E}\mathbf{B}\mathbf{F}(w,\tau)+j\mathbf{E}\mathbf{I}_1\frac{\partial \mathbf{F}(w,\tau)}{\partial w}\right). \tag{17}$$

Using (14) we derive the differential equation for the function $x(\tau)$

$$x'(\tau)=\mathbf{E}\left[\lambda\mathbf{B}-x(\tau)\mathbf{I}_1\right]\mathbf{R}. \tag{18}$$

Denoting

$$a(x)=\mathbf{E}\left[\lambda\mathbf{B}-x\mathbf{I}_1\right]\mathbf{R} \tag{19}$$

and making the following replacements in the operator Eqs. (3) and (8)

$$\mathbf{H}(u,t)=e^{j\frac{u}{\sigma}x(\sigma t)}\mathbf{H}^{(2)}(u,t), \tag{20}$$

we have

$$\frac{\partial \mathbf{H}^{(2)}(u,t)}{\partial t}+jux'(\sigma t)\mathbf{H}^{(2)}(u,t)=$$

$$=(\mathbf{A}+\lambda e^{ju}\mathbf{B}-x(\sigma t)(\mathbf{I}_0-e^{-ju}\mathbf{I}_1))\mathbf{H}^{(2)}(u,t)+j\sigma(\mathbf{I}_0-e^{-ju}\mathbf{I}_1)\frac{\partial \mathbf{H}^{(2)}(u,t)}{\partial u}, \tag{21}$$

$$\mathbf{E}\frac{\partial \mathbf{H}^{(2)}(u,t)}{\partial t}+jux'(\sigma t)\mathbf{E}\mathbf{H}^{(2)}(u,t)=$$

$$=\mathbf{E}(\mathbf{A}+\lambda e^{ju}\mathbf{B}-x(\sigma t)(\mathbf{I}_0-e^{-ju}\mathbf{I}_1))\mathbf{H}^{(2)}(u,t)+$$

$$+j\sigma\mathbf{E}(\mathbf{I}_0-e^{-ju}\mathbf{I}_1)\frac{\partial \mathbf{H}^{(2)}(u,t)}{\partial u}. \tag{22}$$

In operator Eqs. (21) and (22) we introduce the following notations

$$\sigma=\varepsilon^2,\ \tau=\varepsilon^2 t,\ u=\varepsilon w,\ \mathbf{H}^{(2)}(u,t)=\mathbf{F}^{(2)}(w,\tau,\varepsilon), \tag{23}$$

and taking (19) into account we obtain the equations

$$\varepsilon^2 \frac{\partial \mathbf{F}^{(2)}(w,\tau,\varepsilon)}{\partial \tau} + jw\varepsilon a(x)\mathbf{F}^{(2)}(w,\tau,\varepsilon) =$$

$$= (\mathbf{A} + \lambda e^{jw\varepsilon}\mathbf{B} - x(\mathbf{I}_0 - e^{-jw\varepsilon}\mathbf{I}_1))\mathbf{F}^{(2)}(w,\tau,\varepsilon) +$$

$$+ j\varepsilon(\mathbf{I}_0 - e^{-jw\varepsilon}\mathbf{I}_1)\frac{\partial \mathbf{F}^{(2)}(w,\tau,\varepsilon)}{\partial w}, \tag{24}$$

$$\varepsilon^2 \mathbf{E}\frac{\partial \mathbf{F}^{(2)}(w,\tau,\varepsilon)}{\partial \tau} + jw\varepsilon a(x)\mathbf{E}\mathbf{F}^{(2)}(w,\tau,\varepsilon) =$$

$$= \mathbf{E}(\mathbf{A} + \lambda e^{jw\varepsilon}\mathbf{B} - x(\mathbf{I}_0 - e^{-jw\varepsilon}\mathbf{I}_1))\mathbf{F}^{(2)}(w,\tau,\varepsilon) +$$

$$+ j\varepsilon\mathbf{E}(\mathbf{I}_0 - e^{-jw\varepsilon}\mathbf{I}_1)\frac{\partial \mathbf{F}^{(2)}(w,\tau,\varepsilon)}{\partial w}. \tag{25}$$

Solving the Eqs. (24) and (25) taking the limit as $\varepsilon \to 0$ we present the following theorem.

Theorem 2. *Probability density of diffusion limit $z(\tau)$ of the number of calls in the orbit given as follows*

$$\Pi(z) = \frac{C}{b(z)} \exp\left\{ \frac{2}{\sigma} \int_0^z \frac{a(x)}{b(x)} dx \right\}, \tag{26}$$

where C is a normalization factor, function $a(x)$ is defined by (19), function $b(x)$ has the following form

$$b(x) = a(x) + 2[\mathbf{E}(\lambda\mathbf{B} - x\mathbf{I}_1)\mathbf{g} + x\mathbf{E}\mathbf{I}_1\mathbf{R}]. \tag{27}$$

Here $\mathbf{g}$ is the matrix of additional values. It has the same dimension as $\mathbf{R}$ and appears as the solution of the system of operator equations

$$(\mathbf{A} + \lambda\mathbf{B} - x(\tau)(\mathbf{I}_0 - \mathbf{I}_1))\mathbf{g} = a(x)\mathbf{R} - (\lambda\mathbf{B} - x(\tau)\mathbf{I}_1)\mathbf{R},$$

$$\mathbf{E}\mathbf{g} = 0. \tag{28}$$

Proof. We rewrite the Eq. (24) up to $O(\varepsilon^2)$

$$jw\varepsilon a(x)\mathbf{F}^{(2)}(w,\tau,\varepsilon) =$$

$$= (\mathbf{A} + \lambda\mathbf{B} + jw\varepsilon\lambda\mathbf{B} - x(\mathbf{I}_0 - \mathbf{I}_1 + jw\varepsilon\mathbf{I}_1))\mathbf{F}^{(2)}(w,\tau,\varepsilon) +$$

$$+ j\varepsilon(\mathbf{I}_0 - \mathbf{I}_1)\frac{\partial \mathbf{F}^{(2)}(w,\tau,\varepsilon)}{\partial w} + O(\varepsilon^2). \tag{29}$$

We assume that the solution of the operator Eq. (29) has the following form

$$\mathbf{F}^{(2)}(w,\tau,\varepsilon) = \Phi(w,\tau)\{\mathbf{R} + jw\varepsilon\mathbf{f}\} + O(\varepsilon^2), \tag{30}$$

where $\mathbf{R}$ is a matrix obtained in theorem 1; $\mathbf{f}$ is a matrix of additional values, which has the same structure as $\mathbf{R}$ and $\Phi(w,\tau)$ is a scalar function.

$$jw\varepsilon a(x)\Phi(w,\tau)\mathbf{R} =$$
$$= \Phi(w,\tau)(\mathbf{A} + \lambda\mathbf{B} + jw\varepsilon\lambda\mathbf{B} - x(\mathbf{I}_0 - \mathbf{I}_1 + jw\varepsilon\mathbf{I}_1))\{\mathbf{R} + jw\varepsilon\mathbf{f}\} +$$
$$+ j\varepsilon\tfrac{\partial\Phi(w,\tau)}{\partial w}(\mathbf{I}_0 - \mathbf{I}_1)\mathbf{R} + O(\varepsilon^2), \tag{31}$$

Taking (15) into account we obtain

$$jw\varepsilon a(x)\Phi(w,\tau)\mathbf{R} =$$
$$= jw\varepsilon\Phi(w,\tau)\left[(\mathbf{A} + \lambda\mathbf{B} - x(\mathbf{I}_0 - \mathbf{I}_1))\mathbf{f} + (\lambda\mathbf{B} - x\mathbf{I}_1)\mathbf{R}\right] +$$
$$+ j\varepsilon\frac{\partial\Phi(w,\tau)}{\partial w}(\mathbf{I}_0 - \mathbf{I}_1)\mathbf{R} + O(\varepsilon^2). \tag{32}$$

Dividing the operator Eq. (32) by $j\varepsilon\Phi(w,\tau)$, we obtain an equation for $\mathbf{f}$

$$(\mathbf{A} + \lambda\mathbf{B} - x(\mathbf{I}_0 - \mathbf{I}_1))\mathbf{f} =$$
$$= a(x)\mathbf{R} - (\lambda\mathbf{B} - x\mathbf{I}_1)\mathbf{R} - \frac{\partial\Phi(w,\tau)/\partial w}{w\Phi(w,\tau)}(\mathbf{I}_0 - \mathbf{I}_1)\mathbf{R}. \tag{33}$$

We represent the additional values $\mathbf{f}$ in the following form

$$\mathbf{f} = C\mathbf{R} + \mathbf{g} - \varphi\frac{\partial\Phi(w,\tau)/\partial w}{w\Phi(w,\tau)}, \tag{34}$$

where $\mathbf{g}$ and φ are matrices the same structure as $\mathbf{f}$ and $\mathbf{R}$. Then we obtain two operator equations

$$(\mathbf{A} + \lambda\mathbf{B} - x(\mathbf{I}_0 - \mathbf{I}_1))\mathbf{g} = a(x)\mathbf{R} - (\lambda\mathbf{B} - x\mathbf{I}_1)\mathbf{R}, \tag{35}$$

$$(\mathbf{A} + \lambda\mathbf{B} - x(\mathbf{I}_0 - \mathbf{I}_1))\varphi = (\mathbf{I}_0 - \mathbf{I}_1)\mathbf{R}. \tag{36}$$

From (36) we can see, that

$$\varphi = \frac{d\mathbf{R}}{dx}, \tag{37}$$

which is true since the matrix $\mathbf{R}$ is a solution of the system (12) depending on parameter x and the Eq. (36) is a derivative of Eq. (12) by x. Thus, we can write the additional condition

$$\mathbf{E}\varphi = 0.$$

As for the operator Eq. (35) we can see that both left and right parts of (35) will become zero if we apply the summing operator $\mathbf{E}$. Therefore, we add an additional condition

$$\mathbf{E}\mathbf{g} = 0.$$

Let us consider the scalar Eq. (25) in the following form

$$\varepsilon^2 \mathbf{E}\frac{\partial \mathbf{F}^{(2)}(w,\tau,\varepsilon)}{\partial \tau} + jw\varepsilon a(x)\mathbf{E}\mathbf{F}^{(2)}(w,\tau,\varepsilon) =$$

$$= \mathbf{E}\left(\mathbf{A} + \lambda\mathbf{B} + jw\varepsilon\lambda\mathbf{B} + \frac{(jw\varepsilon)^2}{2}\lambda\mathbf{B} - \right.$$

$$\left. -x\left(\mathbf{I}_0 - \mathbf{I}_1 + jw\varepsilon\mathbf{I}_1 - \frac{(jw\varepsilon)^2}{2}\mathbf{I}_1\right)\right)\mathbf{F}^{(2)}(w,\tau,\varepsilon) +$$

$$+ j\varepsilon\mathbf{E}(\mathbf{I}_0 - \mathbf{I}_1 + jw\varepsilon\mathbf{I}_1)\frac{\partial \mathbf{F}^{(2)}(w,\tau,\varepsilon)}{\partial w} + O(\varepsilon^3). \quad (38)$$

We rewrite the equation using decomposition (30) and properties (9)

$$\varepsilon^2 \frac{\partial \Phi(w,\tau)}{\partial \tau}\mathbf{E}\mathbf{R} + (jw\varepsilon)^2 a(x)\Phi(w,\tau)\mathbf{E}\mathbf{f} =$$

$$= (jw\varepsilon)^2\Phi(w,\tau)\mathbf{E}(\lambda\mathbf{B} - x\mathbf{I}_1)\mathbf{f} + \frac{(jw\varepsilon)^2}{2}\Phi(w,\tau)\mathbf{E}(\lambda\mathbf{B} + x\mathbf{I}_1)\mathbf{R} +$$

$$+ j\varepsilon(jw\varepsilon)\frac{\partial \Phi(w,\tau)}{\partial w}\mathbf{E}\mathbf{I}_1\mathbf{R} + O(\varepsilon^3). \quad (39)$$

Dividing both parts of the last equation by ε^2 and taking the limit as $\varepsilon \to 0$ we obtain

$$\frac{\partial \Phi(w,\tau)}{\partial \tau} = w^2 a(x)\Phi(w,\tau)\mathbf{E}\mathbf{f} + (jw)^2\Phi(w,\tau)\mathbf{E}(\lambda\mathbf{B} - x\mathbf{I}_1)\mathbf{f} +$$

$$+ \frac{(jw)^2}{2}\Phi(w,\tau)\mathbf{E}(\lambda\mathbf{B} + x\mathbf{I}_1)\mathbf{R} - w\frac{\partial \Phi(w,\tau)}{\partial w}\mathbf{E}\mathbf{I}_1\mathbf{R}.$$

We substitute the equality (34) into the equation

$$\frac{\partial \Phi(w,\tau)}{\partial \tau} = w\frac{\partial \Phi(w,\tau)}{\partial w}(\mathbf{E}(\lambda\mathbf{B} - x\mathbf{I}_1)\varphi - \mathbf{E}\mathbf{I}_1\mathbf{R}) +$$

$$+ \frac{(jw)^2}{2}\Phi(w,\tau)(\mathbf{E}(\lambda\mathbf{B} + x\mathbf{I}_1)\mathbf{R} + 2\mathbf{E}(\lambda\mathbf{B} - x\mathbf{I}_1)\mathbf{g}). \quad (40)$$

We note that

$$\mathbf{E}(\lambda\mathbf{B} - x\mathbf{I}_1)\varphi - \mathbf{E}\mathbf{I}_1\mathbf{R} = a'(x). \quad (41)$$

Denoting

$$b(x) = a(x) + 2[\mathbf{E}(\lambda\mathbf{B} - x\mathbf{I}_1)\mathbf{g} + x\mathbf{E}\mathbf{I}_1\mathbf{R}], \quad (42)$$

we rewrite the Eq. (40) in the following form

$$\frac{\partial \Phi(w,\tau)}{\partial \tau} = a'(x)w\frac{\partial \Phi(w,\tau)}{\partial w} + b(x)\frac{(jw)^2}{2}\Phi(w,\tau). \quad (43)$$

This equation is a Fourier transform of the Fokker-Planck equation for the probability density $P(y,\tau)$ of the centered and normalized number of calls in the orbit. We apply an inverse Fourier transform to the Eq. (43) to have

$$\frac{\partial P(y,\tau)}{\partial \tau} = -\frac{\partial}{\partial y}\{a'(x)yP(y,\tau)\} + \frac{1}{2}\frac{\partial^2}{\partial y^2}\{b(x)P(y,\tau)\}. \tag{44}$$

The Eq. (44) is a Fokker-Planck equation for $P(y,\tau)$. Hence, the function $P(y,\tau)$ is a probability density of some diffusion process, which we denote as $y(\tau)$ with drift coefficient $a'(x)y$ and diffusion coefficient $b(x)$. This process is a solution of the stochastic differential equation

$$dy(\tau) = a'(x)yd\tau + \sqrt{b(x)}dw(\tau). \tag{45}$$

We denote $z(\tau)$ as diffusion limit of the number of calls in the orbit and build this process as follows

$$z(\tau) = x(\tau) + \varepsilon y(\tau), \tag{46}$$

where $\varepsilon = \sqrt{\sigma}$. Using $dx(\tau) = a(x)d\tau$ we obtain

$$dz(\tau) = d(x(\tau) + \varepsilon y(\tau)) = (a(x) + \varepsilon ya'(x))d\tau + \varepsilon\sqrt{b(x)}dw(\tau). \tag{47}$$

We consider decompositions

$$a(z) = a(x + \varepsilon y) = a(x) + \varepsilon ya'(x) + o(\varepsilon^2),$$

$$\varepsilon\sqrt{b(z)} = \varepsilon\sqrt{b(x+\varepsilon y)} = \varepsilon\sqrt{b(x) + o(\varepsilon)} = \varepsilon\sqrt{b(x)} + o(\varepsilon^2).$$

Then the Eq. (47) we rewrite in the following form up to $o(\varepsilon^2)$

$$dz(\tau) = a(z)d\tau + \sqrt{\sigma b(z)}dw(\tau). \tag{48}$$

The probability density for the process $z(\tau)$ we denote as

$$\Pi(z,\tau) = \frac{\partial P\{z(\tau) < z\}}{\partial z}.$$

As the process $z(\tau)$ is a solution of the stochastic differential Eq. (48), then it is a diffusion process and for its probability density $\Pi(z,\tau)$ we can write a Fokker-Planck equation

$$\frac{\partial \Pi(z,\tau)}{\partial \tau} = -\frac{\partial}{\partial z}\{a(z)\Pi(z,\tau)\} + \frac{1}{2}\frac{\partial^2}{\partial z^2}\{\sigma b(z)\Pi(z,\tau)\}.$$

We assume that the retrial queue is functioning in stationary regime, then

$$\Pi(z,\tau) = \Pi(z).$$

Thus, the Fokker-Planck equation for the stationary probability distribution $\Pi(z)$ has following form

$$(-a(z)\Pi(z))' + \frac{\sigma}{2}(b(z)\Pi(z))'' = 0,$$

$$-a(z)\Pi(z) + \frac{\sigma}{2}(b(z)\Pi(z))' = 0.$$

Solving this differential equation we obtain the probability density $\Pi(z)$ of the diffusion limit of the number of calls in the orbit

$$\Pi(z) = \frac{C}{b(z)} \exp\left\{ \frac{2}{\sigma} \int_0^z \frac{a(x)}{b(x)} dx \right\}. \tag{49}$$

From the obtained probability density $\Pi(z)$ we build the approximation of the probability distribution of the number of calls in the orbit using expression

$$PD(i) = \frac{\Pi(i\sigma)}{\sum\limits_{n=0}^{\infty} \Pi(n\sigma)}. \tag{50}$$

4 Algorithm of Calculating Drift and Diffusion Coefficients

To obtain the function $a(x)$ we need to calculate the elements of the matrix $\mathbf{R}$. We rewrite the system of operator Eqs. (12) in scalar form

$$-(\lambda + N\alpha + x(\tau))R(n_1, n_2) + \mu_1 R(n_1+1, n_2) +$$
$$+\mu_2 R(n_1, n_2+1) = 0, \ n_1 + n_2 = 0,$$
$$-(\lambda + (N - n_1 - n_2)\alpha + n_1\mu_1 + n_2\mu_2 + x(\tau))R(n_1, n_2) +$$
$$+(N - n_1 - (n_2 - 1))\alpha R(n_1, n_2 - 1) + (\lambda + x(\tau))R(n_1 - 1, n_2) +$$
$$+(n_1+1)\mu_1 R(n_1+1, n_2) + (n_2+1)\mu_2 R(n_1, n_2+1) = 0, \ 0 < n_1 + n_2 < N,$$
$$-(n_1\mu_1 + n_2\mu_2)R(n_1, n_2) + (\lambda + x(\tau))R(n_1 - 1, n_2) +$$
$$+\alpha R(n_1, n_2 - 1) = 0, \ n_1 + n_2 = N.$$
$$\sum_{n_1=0}^{N} \sum_{n_2=0}^{N-n_1} R(n_1, n_2) = 1. \tag{51}$$

We transform the system of Eqs. (51) to the system of linear algebraic equations renumbering the elements of the matrix $\mathbf{R}$ in the following way

$$(n_1, n_2) \to 2n_1 + n_2 + \frac{(n_1+n_2)^2 - (n_1+n_2)}{2}. \tag{52}$$

Then we obtain the system of equations

$$\widetilde{\mathbf{R}}\mathbf{D}(x) = 0, \ \widetilde{\mathbf{R}}\mathbf{e} = 1, \tag{53}$$

where $\widetilde{\mathbf{R}}$ is a vector of probabilities $R(n_1, n_2)$, $\mathbf{e}$ is a unit vector, the matrix $\mathbf{D}(x)$ is a matrix of the system with renumbered elements.

Solving the system of Eqs. (53) we can express the function $a(x)$ as follows

$$a(x) = (\lambda + x)\tilde{\mathbf{R}}\mathbf{e}_1 - x, \tag{54}$$

where $\mathbf{e}_1$ is a vector, in which the last $N+1$ elements are ones and the other elements are zeroes.

To derive the function $b(x)$ we need to calculate the elements of the matrix $\mathbf{g}$. We rewrite the system of operator Eqs. (28) in scalar form

$$\begin{gathered}
-(\lambda + N\alpha + x(\tau))g(n_1,n_2) + \mu_1 g(n_1+1,n_2) + \mu_2 g(n_1,n_2+1) = \\
= a(x)R(n_1,n_2),\ n_1+n_2=0, \\
-(\lambda + (N-n_1-n_2)\alpha + n_1\mu_1 + n_2\mu_2 + x(\tau))g(n_1,n_2) + \\
+(N-n_1-(n_2-1))\alpha g(n_1,n_2-1) + (\lambda + x(\tau))g(n_1-1,n_2) + \\
+(n_1+1)\mu_1 g(n_1+1,n_2) + (n_2+1)\mu_2 g(n_1,n_2+1) = \\
= a(x)R(n_1,n_2) + x(\tau)R(n_1-1,n_2),\ 0 < n_1+n_2 < N, \\
-(n_1\mu_1 + n_2\mu_2)g(n_1,n_2) + (\lambda + x(\tau))g(n_1-1,n_2) + \alpha g(n_1,n_2-1) = \\
= (a(x)-\lambda)R(n_1,n_2) + x(\tau)R(n_1-1,n_2),\ n_1+n_2=N. \\
\sum_{n_1=0}^{N}\sum_{n_2=0}^{N-n_1} g(n_1,n_2) = 0.
\end{gathered} \tag{55}$$

We transform the system of the operator Eqs. (28) to the system of linear algebraic equations renumbering the elements of the matrix $\mathbf{g}$ using (52). Thus, we obtain the system of equations

$$\tilde{\mathbf{g}}\mathbf{D}(x) = \mathbf{d}(x),\ \tilde{\mathbf{g}}\mathbf{e} = 0, \tag{56}$$

where $\tilde{\mathbf{g}}$ is a vector of elements of the matrix $\mathbf{g}$ with renumbered elements, vector $\mathbf{d}(x)$ is a vector of right parts of the system (55) with renumbered elements.

Solving the system of Eqs. (56) we can express the function $b(x)$ as follows

$$b(x) = a(x) + 2[(\lambda + x)\tilde{\mathbf{g}}\mathbf{e}_1 + x(1 - \tilde{\mathbf{R}}\mathbf{e}_1)]. \tag{57}$$

5 Numerical Examples

We fix the number of servers in the system $N = 5$. We assume that the rates of service times are $\mu_1 = 1$, $\mu_2 = 2$. Outgoing calls rate is $\alpha = 1$.

The accuracy of approximation we will determine using Kolmogorov range

$$\Delta = \max_{0 \leqslant i \leqslant \infty} \left| \sum_{n=0}^{i} (P(n) - PD(n)) \right|, \tag{58}$$

where $P(n)$ is the probability distribution of the number of calls in the orbit, obtained with simulation, $PD(n)$ is a diffusion approximation defined by (50).

We assume that the approximation is acceptable when its accuracy $\Delta < 0.05$. Table 1 depicts the accuracy of the diffusion approximation $PD(n)$ depending on parameters σ and ρ, where ρ characterizes the system load $\rho = \lambda / N\mu_1$.

Table 1. Kolmogorov range

Δ	$\sigma = 5$	$\sigma = 2$	$\sigma = 1$	$\sigma = 0.5$	$\sigma = 0.2$	$\sigma = 0.1$
$\rho = 0.6$	**0,04**	**0,039**	**0,039**	**0,022**	**0,015**	**0,023**
$\rho = 0.7$	0,07	0,063	0,051	**0,036**	**0,031**	**0,032**
$\rho = 0.8$	0,09	0,078	0,064	**0,049**	**0,039**	**0,041**
$\rho = 0.9$	0,081	0,069	0,065	0,053	**0,044**	**0,05**

6 Conclusion

We have considered multiserver retrial queue with two-way communication. Using asymptotic-diffusion analysis method we have built the diffusion process the distribution density using which we construct the approximation of the number of calls in the orbit.

We have presented numerical experiments where we have estimated the accuracy of the approximation by comparing to simulation.

References

1. Aguir, S., Karaesmen, F., Akşin, O.Z., Chauvet, F.: The impact of retrials on call center performance. OR Spect. **26**(3), 353–376 (2004)
2. Artalejo, J.R., Gómez-Corral, A.: Retrial Queueing Systems: A Computational Approach. Springer-Verlag, Heidelberg (2008)
3. Avramidis, A., Deslauriers, A., L'Ecuyer, P.: Modeling daily arrivals to a telephone call center. Manag. Sci. **50**(7), 896–908 (2004)
4. Bernett, H.G., Fischer, M.J., Masi, D.M.B.: Blended call center performance analysis. IT Prof. **4**(2), 33–38 (2002)
5. Bhulai, S., Koole, G.: A queueing model for call blending in call centers. IEEE Trans. Autom. Control **48**(8), 1434–1438 (2003)
6. Brown, L., et al.: Statistical analysis of a telephone call center: a queueing-science perspective. J. Am. Stat. Assoc. **100**(469), 36–50 (2005)
7. Falin, G., Templeton, J.G.C.: Retrial Queues, vol. 75. CRC Press, Boca Raton (1997)
8. Gilmore, A., Moreland, L.: Call centres: how can service quality be managed? Irish Mark. Rev. **13**(1), 3 (2000)
9. Moiseev, A., Nazarov, A., Paul, S.: Asymptotic diffusion analysis of multi-server retrial queue with hyper-exponential service. Mathematics **8**(4), 531 (2020)
10. Phung-Duc, T., Kawanishi, K.: An efficient method for performance analysis of blended call centers with redial. Asia-Pac. J. Oper. Res. **31**(02), 1–35 (2014)

On a Single Server Queueing Inventory System

K. A. K. AL Maqbali(✉), V. C. Joshua, and A. Krishnamoorthy

Department of Mathematics, CMS College Kottayam, Kerala, India
{khamis,vcjoshua,krishnamoorthy}@cmscollege.ac.in

Abstract. In this paper, we consider a queueing inventory model. Customers arriving to a single server queueing system in which, server uses some commodities from the inventory to fulfill the service. Service of a customer is initiated only when the server is free and the commodities for the service are available in the inventory. Otherwise, the arriving customer joins the queue. The inventory level is managed with (s,S) policy. Lead time is assumed to be positive. Steady state analysis of the model is performed. Various performance measures are estimated. A numerical example is provided.

Keywords: Queueing inventory · Lead time · Phase type distribution · Matrix analytic method

1 Introduction

In many real life phenomena, customers waiting in the queue need inventoried items to complete the service. The study of such models fall under the category of Queueing Inventory Models. $M/M/1$ Queues with attached inventories were studied by Krishnamoorthy et al. [2,3], Neethu and Dhanya [4] and Saffari et al. [5]. Sometimes, customer may go through different phases of service to get the items. Krishnamoorthy et al. [1] studied a $PH/PH/1$ inventory system under (s, S) policy when the lead time is zero. Another paper by Sigman and Simchi-Levi [6] studied an $M/G/1$ queue with limited inventory.

In this paper, we consider an $M/PH/1$ queueing system with inventory level under (s, S) policy. When the inventory level reaches s, an order is placed. We assume that the customer immediately takes the item when he starts to go through different phases of service. Also, we assume that the lead time is positive and follows exponential distribution with rate θ. The shortage is permitted.

Customers arrive to a single server according to Poisson Process with arrival rate λ. An arriving customer can immediately obtain the service if the server is idle and at least one item is available in the inventory. Otherwise, the customer

* Supported by the Indian Council for Cultural Relations (ICCR) and Ministry of Higher Education in Sultanate of Oman.

V. M. Vishnevskiy et al. (Eds.): DCCN 2020, LNCS 12563, pp. 579–588, 2020.
https://doi.org/10.1007/978-3-030-66471-8_44

must wait in the queue. We assume that the service time follows a phase-type distribution (PH-distribution) with representation (β, T) of order m. The customer takes one item of inventory at the beginning of the service.

In other words, this model can be described as follows: customers occur according to a Poisson Process with arrival rate λ and the service time, which is attached with inventory level, follows PH-distribution with representation (β, T) of order m. Under (s, S) policy, the lead time follows exponential distribution with rate θ. The order is placed when the inventory level reaches s. Hence, replenishment of $s, s-1, s-2, ..., 0$ to become S, occurs according to the rate of lead time. The arriving customer can not go through different phases of the service when inventory level is zero and he must wait in queue until inventory level becomes S.

The first arriving customer can immediately go through different phases of the service when at least one item is available. At the same time, he directly takes one item of inventory at the beginning of his service. Instantaneously, if inventory level reaches s, an order decision is made based on the reorder point s and replenishment depends on the lead time, which follows exponential distribution with rate θ. The next arriving customer can directly go through different phases of service when the previous customer completed his service. Otherwise, he has to wait in the queue until the previous customer completes his service. This keeps going.

The motivation for the model comes from the hospital inventory management. For example, heart patients are waiting for angioplasty stent placement for cardio vascular treatment. In general, hospitals keep coronary artery stents in store. When coronary artery stents are available in store and the operation theatre (service station) is available, the heart patient may go through different phases in hospital and the coronary artery stent is immediately taken from store to the patient at the beginning of his service. If the coronary artery stent is not available in store, the heart patient has to wait until the availability of the stents.

Another motivating example is the vehicle service repair station, which deals only with one type of parts. When parts are available in store and service station is free, the vehicle may be taken to service station and one inventory is directly taken from store at the beginning of its service. If these parts are not available in store of the service station, the vehicle has to wait until the availability of the parts.

2 Mathematical Description of the Model

For the analysis of the model, we introduce the following notations.

Let

$N(t)$ be the number of customers in the system at time t.

$I(t)$ be the number of items, which are the same type, in the inventory at time t.

$M(t)$ be the phase of service.

$X(t) = \{(N(t), I(t), M(t)); t \geq 0\}$ is a continuous time Markov Chain on the state space.

Therefore, this model can be studied as a level Independent Qusi-Birth-Death (LIQBD) process with state space is given by

$\Omega = \{(0,i); 0 \leq i \leq S\} \cup \{(n,0); n \geq 1\} \cup \{(n,i,m); n \geq 1; 0 \leq i \leq S; m = 1, 2, \ldots\}$.

The terms of transitions of the states are given in the Table 1.

Table 1. Intensities of Transitions

From	To		Transition Rate
$(0,i)$	$(1, i-1, m)$	$1 \leq i \leq S$	$\lambda\beta_m$
(n,i,m)	$(n+1, i, m)$	$n \geq 1;\ 0 \leq i \leq S$	λ
$(n,0)$	$(n+1, 0)$	$n \geq 0$	λ
$(0,i)$	$(0,S)$	$0 \leq i \leq s$	θ
$(n,0)$	$(n, S-1, m)$	$n \geq 1$	θ_{β_m}
(n,i,m)	(n,S,m)	$i \leq s; n \geq 1$	θ
$(1,i,m)$	$(0,i)$	$0 \leq i \leq S$	τ_m^0
(n,i,m)	$(n-1, i-1, m')$	$0 < i \leq S\ ; n > 1$	$\tau_m^0\beta_{m'}$
(n,i,m)	(n,i,m')	$1 \leq n; 0 \leq i \leq S; m \neq m'$	$\tau_{mm'}$
$(n,0,m)$	$(n-1,0)$	$n \geq 2$	τ_m^0

The infinitesimal generator Q of the level Independent Qusi-Birth-Death (LIQBD) process with state space is of the form

$$Q = \begin{pmatrix} B_{00} & B_{01} & & & \\ B_{10} & A_1 & A_0 & & \\ & A_2 & A_1 & A_0 & \\ & & A_2 & A_1 & A_0 \\ & & & \ddots & \ddots & \ddots \end{pmatrix};$$

where

$$B_{00} = \begin{pmatrix} -(\lambda+\theta)I_{(s+1)} & \Gamma \\ O_{(S-(s+1))\times(s+1)} & -\lambda I_{(S-s)} \end{pmatrix}_{(S+1)\times(S+1)};$$

where $\Gamma = \begin{pmatrix} O_{(s+1)\times(S-(s+1))} & \theta e_{(s+1)\times 1} \end{pmatrix}$;

$$B_{01} = \begin{pmatrix} \lambda & & & O \\ & \lambda\beta_{1\times m} & & \\ & & \ddots & \\ & & \lambda\beta_{1\times m} & O \end{pmatrix}_{(S+1)\times(1+m\times(S+1))};$$

$$B_{10} = \begin{pmatrix} 0 & & & & \\ T^0_{m\times 1} & & & & \\ & T^0_{m\times 1} & & & \\ & & \ddots & & \\ & & & T^0_{m\times 1} & \end{pmatrix}_{(1+m\times(S+1))\times(S+1)};$$

$$A_2 = \begin{pmatrix} 0 & & & & \dots & 0 \\ T^0_{m\times 1} & & & & \dots & O \\ & & & & & \vdots \\ & \wedge & & & & \\ & & \wedge & & & \\ & & & \ddots & & \\ & & & & \wedge & O \\ & & & & \wedge & O \end{pmatrix}_{(1+m\times(S+1))\times(1+m\times(S+1))};$$

where $\wedge = T^0_{m\times 1} \otimes \beta_{1\times m}$;

$$A_0 = \begin{pmatrix} \lambda & & & & \\ & \lambda I_m & & & \\ & & \lambda I_m & & \\ & & & \ddots & \\ & & & & \lambda I_m \end{pmatrix}_{(1+m\times(S+1))\times(1+m\times(S+1))};$$

$$A_1 = \begin{pmatrix} -(\lambda+\theta) & & & & & O\ \phi\ O \\ & \varphi\ O & & & & O\ \theta I_m \\ & O\ \varphi\ O & & & & \vdots\ \ \vdots \\ & & O\ \ddots\ O & & & \vdots \\ & & & O\ \varphi\ O \dots & & O\ \theta I_m \\ & & & O\ \omega\ O & & O \\ & & & & O\ \ddots\ O & \vdots \\ & & & & & O\ \omega\ O \\ & & & & & O\ \omega \end{pmatrix}_{(1+m\times(S+1))\times(1+m\times(S+1))}$$

where $\varphi = T_{m\times m} - (\lambda+\theta)I_m$; $\omega = T_{m\times m} - \lambda I_m$ and $\phi = \theta \otimes \beta_m$.

3 Steady-State Analysis

3.1 Stability Condition

Theorem 1. *The Markov chain with the infinitesimal generator Q of the level Independent Qusi-Birth-Death (LIQBD) is stable if and only if*

$$\lambda < \mu$$

Where $\mu = \boldsymbol{\pi_1}\, T^0_{m\times 1} + [(\sum_{i=2}^{S+1} \boldsymbol{\pi}_i)\ (T^0_{m\times 1} \otimes \beta_{1\times m})]e_{m\times 1}$ *; where* $\boldsymbol{\pi_1}$ *and* $\boldsymbol{\pi}_i$ *are sub vectors of order* $1 \times m$.

Proof. Let $A = A_2 + A_1 + A_0$. We can notice that A is an irreducible matrix. Thus, the stationary vector $\boldsymbol{\pi}$ of A exists such that

$$\boldsymbol{\pi} A = 0$$

$$\boldsymbol{\pi} e = 1.$$

The Markov chain with the infinitesimal generator Q of the level Independent Qusi-Birth-Death (LIQBD) is stable if and only if

$$\boldsymbol{\pi} A_0 e < \boldsymbol{\pi} A_2 e.$$

Recall, A_2 is a square matrix of order $(1 + m \times (S + 1))$ and

$$A_2 = \begin{pmatrix} 0 & & \dots 0 \\ T^0_{m\times 1} & & \dots O \\ & \wedge & \vdots \\ & & \ddots & \\ & & \wedge \; O \end{pmatrix}; \text{ where } \wedge = T^0_{m\times 1} \otimes \beta_{1\times m}.$$

$$\boldsymbol{\pi} A_2 e = (\pi_0, \boldsymbol{\pi}_1, \boldsymbol{\pi}_2, \dots, \boldsymbol{\pi}_{S+1}) A_2 e;$$

where π_0 is a single value and $\boldsymbol{\pi}_1, \boldsymbol{\pi}_2, \dots, \boldsymbol{\pi}_{S+1}$ are sub vectors of order $1 \times m$.

$$\begin{aligned} \boldsymbol{\pi} A_2 e &= [\boldsymbol{\pi}_1 T^0_{m\times 1} + \boldsymbol{\pi}_2 (T^0_{m\times 1} \otimes \beta_{1\times m}) + \dots + \boldsymbol{\pi}_{S+1} (T^0_{m\times 1} \otimes \beta_{1\times m})] e; \\ &= \boldsymbol{\pi}_1 T^0_{m\times 1} + [\boldsymbol{\pi}_2 (T^0_{m\times 1} \otimes \beta_{1\times m}) + \dots + \boldsymbol{\pi}_{S+1} (T^0_{m\times 1} \otimes \beta_{1\times m})] e_{m\times 1}; \\ &= \boldsymbol{\pi}_1 T^0_{m\times 1} + [\boldsymbol{\pi}_2 + \dots + \boldsymbol{\pi}_{S+1}] (T^0_{m\times 1} \otimes \beta_{1\times m}) e_{m\times 1}; \\ &= \boldsymbol{\pi}_1 T^0_{m\times 1} + (\sum_{i=2}^{S+1} \boldsymbol{\pi}_i)(T^0_{m\times 1} \otimes \beta_{1\times m}) e_{m\times 1}. \end{aligned}$$

Then, $\boldsymbol{\pi} A_2 e = \boldsymbol{\pi}_1 T^0_{m\times 1} + [(\sum_{i=2}^{S+1} \boldsymbol{\pi}_i)(T^0_{m\times 1} \otimes \beta_{1\times m})] e_{m\times 1} = \mu$.

Since $\pi A_0 e = \lambda$ and $\pi A_2 e = \mu$, then the queueing inventory system is stable if and only if

$$\lambda < \mu$$

3.2 Stationary Distribution

The stationary distribution of the Markov chain under consideration can obtained by solving the set of Eqs. 1 and 2 [7].

$$\mathbf{X} Q = 0; \tag{1}$$

$$\mathbf{X} e = 1. \tag{2}$$

Let $\mathbf{X}$ be decomposed with Q as following :

$\mathbf{X} = (\mathbf{X}_0, \mathbf{X}_1, \ldots)$ where $\mathbf{X}_0 = (x_{00}, x_{01}, x_{02}, \ldots, x_{0S})$;
$\mathbf{X}_i = (\mathbf{X}_{i0}, \mathbf{X}_{i1}, \ldots, \mathbf{X}_{iS})$; $\mathbf{X}_{i0} = (x_{i0}, x_{i01}, x_{i02}, \ldots, x_{i0m})$;
$\mathbf{X}_{ij} = (x_{ij_1}, x_{ij_2}, \ldots x_{ij_m})$ for $j = 1, \ldots, S$ and $i = 1, 2, \ldots$.
From Eq. 1, we obtain the following equations.

$$\mathbf{X}_0 B_{00} + \mathbf{X}_1 B_{10} = 0; \tag{3}$$

$$\mathbf{X}_0 B_{01} + \mathbf{X}_1 A_1 + \mathbf{X}_2 A_2 = 0; \tag{4}$$

$$\vdots$$

$$\mathbf{X}_{i-1} A_0 + \mathbf{X}_i A_1 + \mathbf{X}_{i+1} A_2 = 0 \text{ for } i = 2, 3, \ldots. \tag{5}$$

It may be shown that there exists a constant matrix R such that

$$\mathbf{X}_i = \mathbf{X}_{i-1} R \text{ for } i = 2, 3, \ldots.$$

The sub-vectors $\mathbf{X}_i$ are geometrically related by the equation

$$\mathbf{X}_i = \mathbf{X}_1 R^{i-1} \text{ for } i = 2, 3, \ldots. \tag{6}$$

We can obtain the constant matrix R by using the matrix quadratic Eq. 7

$$R^2 A_2 + R A_1 + A_0 = 0. \tag{7}$$

By successive substitution procedure [7], we can compute the constant matrix R from

$$R_{k+1} = -V - R_k^2 W \text{ and } R_0 = 0;$$

where $V = A_0 A_1^{-1}$ and $W = A_2 A_1^{-1}$.

Knowing the matrix R, $\mathbf{X_0}$ and $\mathbf{X_1}$ can be obtained by solving the Eqs. 3 and 4 by substituting $\mathbf{X}_2 = \mathbf{X}_1 R$ into Eq. 4.

Since there is no unique solution, $\mathbf{X_0}$ and $\mathbf{X_1}$ must be normalized by using the normalizing condition 8

$$\mathbf{X_0} + \mathbf{X_1}(I - R)^{-1} e = 1. \tag{8}$$

Then, we use the formula 6 to compute remainder probabilities $\mathbf{X_2}$, $\mathbf{X_3}$, ...

4 Performance Measures

Then, we obtain some performance measures of the system as following:

1. Expected number of customers in the system

$$E[N] = \sum_{i=0}^{\infty} i \mathbf{X}_i e.$$

2. Expected number of items in inventory.

$$E[I] = \sum_{i=0}^{\infty}\sum_{j=0}^{S} j\mathbf{X}_{ij}e.$$

3. Probability that the server is idle

$$b_0 = \sum_{j=0}^{S} x_{0j} + \sum_{i=1}^{\infty} x_{i0}.$$

4. Probability that a customer waits for service due to shortage of inventory

$$b_1 = \sum_{i=1}^{\infty} x_{i0} + \sum_{i=2}^{\infty} \mathbf{X}_{i0}e.$$

5. Expected number of customers waiting in the system due to lack of inventory

$$E[N_0] = \sum_{i=2}^{\infty} (i-1)\mathbf{X}_{i0}e.$$

5 Numerical Example

In order to illustrate the performance measures of the system numerically and the effect of S on the performance measures of this system, we fix arrival rate $\lambda = 2$, the rate of lead time $\theta = 0.6$ and $s = 3$.

For PH-representation(β, T) of order m, we fix $m = 3$, $\beta = \left(0.5, 0.4, 0.1\right)$,

$$T = \begin{pmatrix} -6 & 3 & 2 \\ 1 & -8 & 2 \\ 2.5 & 1.5 & -8 \end{pmatrix} \text{ and } T^0 = -Te = \begin{pmatrix} 1 \\ 5 \\ 4 \end{pmatrix}.$$

We find the performance measures of the system with respect to S corresponding to above parameters in the Table 2.

5.1 The Effect of S on the Performance Measures of the System

In this section, we show the effect of S on the performance measures of the system as following:

1. From Fig. 1, we can see that the expected number of customers in the system $E[N]$ decreases as S increases and the expected number of customers waiting in the system due to lack of inventory $E[N_0]$ decreases when S increases, too.
2. Also, we realize from Fig. 2 that the expected number of items in inventory $E[I]$ increases when S increases.

3. In addition, from Fig. 3 we note that the probability that a customer waits for service due to shortage of inventory b_1 decreases when S increases and probability that the server is idle b_0 has almost no change as S increases. In this system, the server is idle when the inventory level is zero or there is no customer in the system or both.

Table 2. Effect of S on various performance measures

S	E[N]	E[I]	b_0	b_1	$E[N_0]$
7	13.0391	28.0000	0.3570	0.2853	4.3938
8	8.4679	36.0000	0.3570	0.2334	2.5157
9	6.5122	45.0000	0.3570	0.1972	1.7319
10	5.4328	55.0000	0.3570	0.1705	1.3089
11	4.7512	66.0000	0.3570	0.1502	1.0469
12	4.2827	78.0000	0.3570	0.1341	0.8700
13	3.9415	91.0000	0.3570	0.1212	0.7431
14	3.6823	105.0000	0.3570	0.1105	0.6479
15	3.4787	120.0000	0.3570	0.1016	0.5740
16	3.3147	136.0000	0.3570	0.0940	0.5151

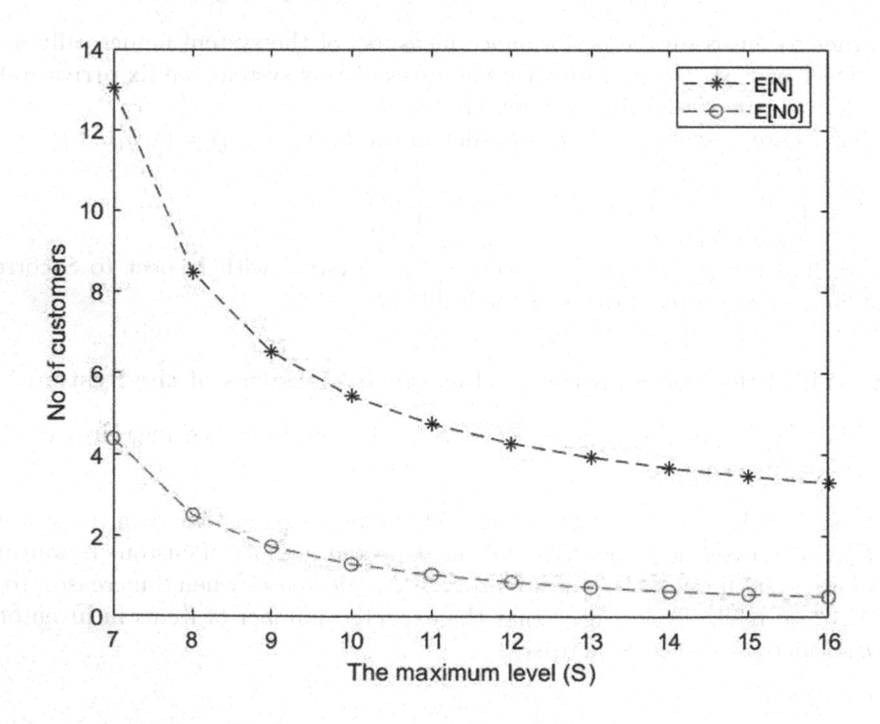

Fig. 1. Effect of S on expected number of customers

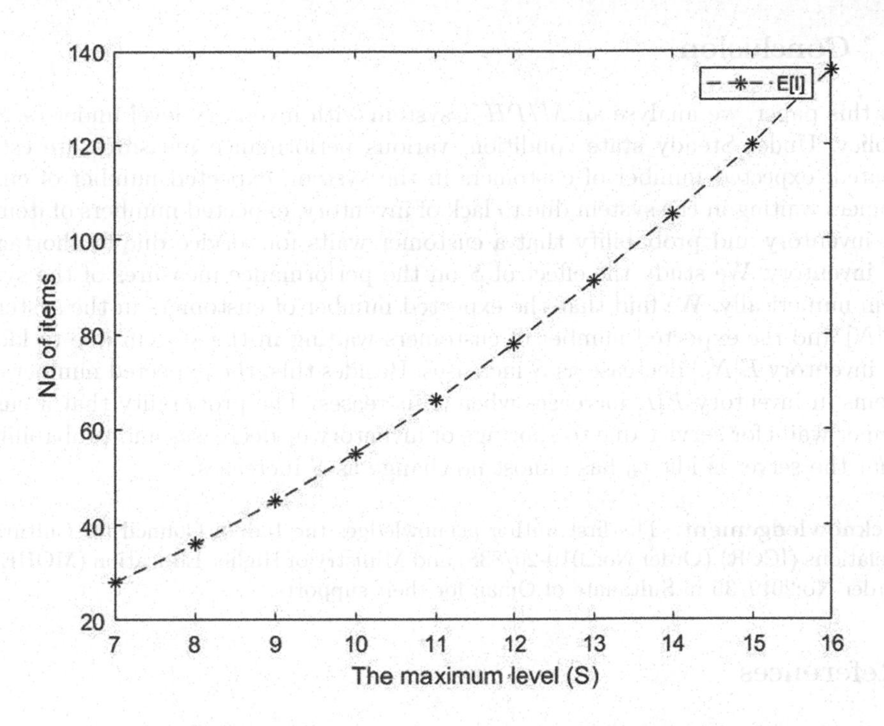

Fig. 2. Effect of S on expected number of items in inventory

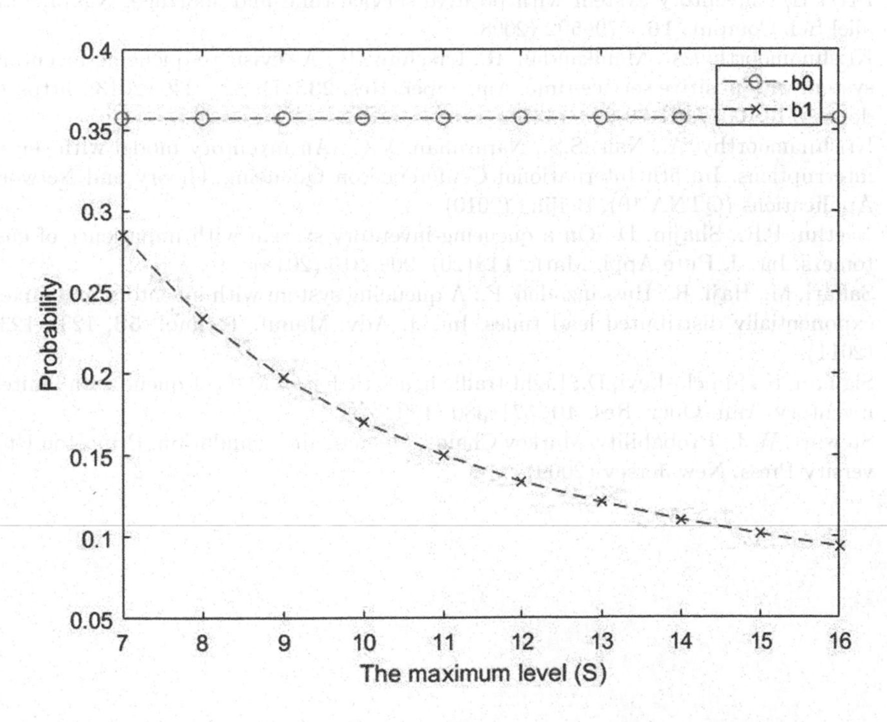

Fig. 3. Effect of S on probability that the server is idle and on probability that a customer waits for service due to shortage of inventory

6 Conclusion

In this paper, we analyse an $M/PH/1$ system with inventory level under (s, S) policy. Under Steady state condition, various performance measures are estimated: expected number of customers in the system, expected number of customers waiting in the system due to lack of inventory, expected numbers of items in inventory and probability that a customer waits for service due to shortage of inventory. We study the effect of S on the performance measures of the system numerically. We find that the expected number of customers in the system $E[N]$ and the expected number of customers waiting in the system due to lack of inventory $E[N_0]$ decrease as S increases. Besides this, the expected number of items in inventory $E[I]$ increases when S increases. The probability that a customer waits for service due to shortage of inventory b_1 decreases and probability that the server is idle b_0 has almost no change as S increases.

Acknowledgement. The first author acknowledges the Indian Council for Cultural Relations (ICCR) (Order No:2019-20/838) and Ministry of Higher Education (MOHE), Order No:2019/35 in Sultanate of Oman for their supports.

References

1. Krishnamoorthy, A., Jose, K.P., Narayanan, V.C.: Numerical investigation of a PH/PH/1 inventory system with positive service time and shortage. Neural Parallel Sci. Comput. **16**, 579–592 (2008)
2. Krishnamoorthy, A., Manikandan, R., Lakshmy, B.: A revisit to queueing-inventory system with positive service time. Ann. Oper. Res. **233**(1), 221–236 (2013). https://doi.org/10.1007/s10479-013-1437-x
3. Krishnamoorthy, A., Nair, S.S., Narayanan, V.C.: An inventory model with server interruptions. In: 5th International Conference on Queueing Theory and Network Applications (QTNA'10), Beijing (2010)
4. Neethu, P.R., Shajin, D.: On a queueing-inventory system with impatience of customers. Int. J. Pure Appl. Math. **118**(20), 909–916 (2018)
5. Saffari, M., Haji, R., Hassanzadeh, F.: A queueing system with inventory and mixed exponentially distributed lead times. Int. J. Adv. Manuf. Technol. **53**, 1231–1237 (2011)
6. Sigman, K., Simchi-Levi, D.: Light traffic heuristic for an M/G/1 queue with limited inventory. Ann. Oper. Res. **40**, 371–380 (1992)
7. Stewart, W.J.: Probability, Markov Chains, Queues, and Simulation. Princeton University Press, New Jersey (2009)

Rare-Event Simulation for the Hitting Time of Gaussian Processes

Oleg Lukashenko[1,2](✉) and Michele Pagano[3]

[1] Institute of Applied Mathematical Research of the Karelian Research Centre of RAS, Petrozavodsk, Russia
lukashenko@krc.karelia.ru
[2] Petrozavodsk State University, Petrozavodsk, Russia
[3] University of Pisa, Pisa, Italy
michele.pagano@iet.unipi.it

Abstract. In reliability theory and network performance analysis a relevant role is played by the time needed to reach a given threshold, known in probability theory as hitting time. Although such issue has been widely investigated, closed-form results are available only for independent increments of the input process. Hence, in this paper we focus on the estimation of the upper tail of the hitting time distribution for general Gaussian processes by means of discrete-event simulation. Indeed, Gaussian processes often arise as a powerful modelling tool in many real-life systems and suitable *ad-hoc* techniques have developed for their analysis and simulation. Since the event of interest becomes rare as the threshold increases, a variant of Conditional Monte Carlo, based on the bridge process, is introduced and the explicit expression of the estimator is derived. Finally, simulation results highlight the unbiasedness and effectiveness (in terms of relative error) of the proposed approach.

Keywords: Hitting time · Gaussian processes · Conditional Monte Carlo · Bridge process · Random threshold

1 Introduction

In several ICT scenarios the time needed to reach a given level, known in reliability theory as failure time and in probability theory as hitting time or first passage time, plays a relevant role in determining the overall system performance. For instance, the time to buffer overflow in a backbone router (compared to the *typical* round trip time) has a significant impact on the *optimal* choice

* The study was carried out under state order to the Karelian Research Centre of the Russian Academy of Sciences (Institute of Applied Mathematical Research KarRC RAS) and supported by the Russian Foundation for Basic Research, projects 18-07-00147, 18-07-00187, 19-07-00303 as well as by the University of Pisa under the PRA 2018–2019 Research Project "CONCEPT – COmmunication and Networking for vehicular CybEr-Physical sysTems".

V. M. Vishnevskiy et al. (Eds.): DCCN 2020, LNCS 12563, pp. 589–603, 2020.
https://doi.org/10.1007/978-3-030-66471-8_45

of the RED (Random Early Detection) parameters [6], including buffer size, thresholds and smoothing factor in the calculation of the average queue size. Another interesting application is related to the estimation of the runtime of a Desktop Grid application, where non-dedicated hosts over the Internet process loosely coupled computational tasks and the target level represents the required amount of work [14]. In both cases, invoking central limit theorem arguments, the input process can be modelled as a Gaussian process with stationary (and, in general, correlated) increments. A similar problem may arise in reliability theory, where the failure time can be interpreted as the time at which the value of the degradation (aging) index reaches a given critical level. In this framework the Wiener process is one of the most popular degradation models (see [4] and references therein) thanks to its analytic tractability.

In this paper we focus on the general case of correlated increments (for which analytic results are not available) and, starting from an abstract formulation of the problem, propose a simulation approach based on the bridge process, able to cope with the estimation of the tail of the distribution (i.e., with rare events). Then the method is extended to the case of random threshold, which naturally arises in the above-mentioned Desktop Grid scenario, due to an incomplete a-priori knowledge of the overall workload that might depend on intermediate tasks. It is worth mentioning that the application of the bridge process for the efficient simulation of Gaussian processes is not new. Indeed, in [15] we described its use for the estimation of the overflow probability and busy-period length in a queue, but in our case the definition of the estimator is slightly different (as pointed out in Sect. 4) and, to the best of our knowledge, in previous works the doubly stochastic case (with a random parameter in the probability under investigation) was not considered.

The rest of the paper is organized as follows. At first in Sect. 2 we provide the precise mathematical definition of the addressed problem and point out the available analytic results, since they will permit to verify the goodness of the proposed estimator. Then Sect. 3 recalls the basic concepts related to Monte Carlo and Conditional Monte Carlo approaches, focusing on the simulation of rare events, while our approach is detailed in Sect. 4. Then simulation results are reported in Sect. 5, where at first the Wiener case is considered (to verify the goodness of the estimator and to check the influence of its parameters on the estimation), and then some results for the Fractional Brownian Motion are presented to highlight the applicability of our approach to the general case of correlated processes as well as the strong influence of such correlation on performance indexes. Finally, Sect. 6 concludes the paper, summarizing the main contributions of the paper and highlighting open issues that require further investigation.

2 Model Description and Performance Measures

The issue addressed in the paper consists in the probabilistic characterization of the *(first) hitting time* τ_D of a random process $\{A(t)\}$

$$\tau_D := \min\{t: \ A(t) \geq D\}, \quad (1)$$

i.e., the first time the process $\{A(t)\}$ hits the threshold D.

In the following we will suppose that

$$A(t) = mt + X(t) \tag{2}$$

i.e., $A(t)$ consists of a linear term and a centered Gaussian process $\{X(t)\}$ with stationary increments, which describes random fluctuations around the linearly increasing mean. Such model is motivated by functional limit theorems [18] and is widely used in the ICT framework for different modelling purposes, ranging from the cumulative processed work in Desktop Grid applications [14] to the aggregated traffic in broadband computer networks (see, for instance, [17] for more details). For sake of completeness, it is worth mentioning that in reliability theory more complex degradation models (characterized by the nonlinearity of the degradation paths or by changing values of the linear drift) are also considered (see, for instance, [5,7] and the references therein).

Moreover, we will assume that the variance

$$v(t) := \mathbb{V}\text{ar}X(t)$$

is a strictly increasing function since, as discussed in the following, some analytic results are available only under such hypothesis and it provides a sufficient condition for the applicability of the bridge process. However, such assumption on $v(t)$ does not represent a significant limitation since it holds for a wide range of processes, including the ones typically used in ICT modelling: Brownian Motion (BM), Fractional Brownian Motion (FBM) [18], superposition of independent FBMs [18] and Integrated Ornstein-Uhlenbeck process (IOU) [12]. Note that, since $A(t)$ is a Gaussian process, it is unambiguously characterized by the covariance function $\Gamma(t,s)$ of the process X, which in turn is fully defined by the variance $v(t)$:

$$\Gamma(t,s) \;:=\; \mathbb{E}\left[X(t)X(s)\right] \;= \frac{1}{2}\left(v(t) + v(s) - v(|t-s|)\right). \tag{3}$$

2.1 Available Analytic Results

Although the *hitting time* of Gaussian processes is well investigated in the literature, exact analytic results are available only when the increments are not only stationary, but also independent, i.e. in the case of BM. In more detail, if X is a BM (i.e., $v(t) = \sigma^2 t$), the probability density function of τ_D is available in explicit form [2]:

$$f_\tau(t|D) = \frac{D}{\sqrt{2\pi}\sigma t^{3/2}} \exp\left(-\frac{(D-mt)^2}{2\sigma^2 t}\right) \tag{4}$$

and the corresponding cumulative distribution function can be expressed through the standard normal distribution function Ψ, while the corresponding expected value $\mathbb{E}[\tau_D]$ is simply

$$\mathbb{E}[\tau_D] = \frac{D}{m}. \tag{5}$$

In the FBM case (i.e., $v(t) = \sigma^2 t^{2H}$ with $H \neq 0.5$), only asymptotic results (for large values of D) and some bounds (quite inaccurate when the Hurst parameter H is close to one) for the distribution of τ_D are available. For instance, the following asymptotic was derived in [16] for the large values of D:

$$\lim_{D\to\infty} \frac{\mathbb{E}[\tau_D^n]}{D^n} = m^{-n}, \tag{6}$$

for all $n \geq 1$, $m > 0$, from which it is quite straightforward to show that for all $n \geq 1$

$$\frac{\tau_D}{D} \xrightarrow{L_n} \frac{1}{m} \quad \text{as} \quad D \to \infty, \tag{7}$$

where $\xrightarrow{L_n}$ means convergence in L_n space.

In [1] a perturbative expression (in $\epsilon = H - 1/2$) of the first-passage-time density was derived in a more general case with linear and non-linear (of the form νt^{2H}) drifts, which appears as a consequence of non-linear variable transformations. Moreover, the theoretical approximation is verified by simulation, using a recently proposed adaptive bisection method [19], which was specifically developed for the estimation of hitting times and is characterized by a high grid resolution only near the target.

Finally, for a general Gaussian process with stationary increments and strictly monotonically increasing and convex variance such that $\lim_{t\to 0} v(t)/t = 0$, the following asymptotic holds [3]:

$$\mathbb{P}(\tau_D \leq T) \sim \Phi\left(\frac{D - mT}{\sqrt{v(T)}}\right) \quad \text{as } D \to \infty, \tag{8}$$

where Φ denotes the tail distribution of the standard normal random variable $N(0,1)$.

To the best of our knowledge, "better" analytic results are not available, and the "goodness" of the previous asymptotic results is a relevant issue both from a theoretical point of view as well as for their practical applicability.

3 Estimation via Monte Carlo

A more flexible alternative to analytic methods is provided by Monte Carlo (MC) simulation, that in our case can be used to estimate the tail of the distribution of τ_D:

$$\pi(T) := \mathbb{P}(\tau_D \geq T) = \mathbb{E}I(\tau_D \geq T), \tag{9}$$

where I denotes the indicator function and the value of the probability depends on the random process X in Eq. (2). The MC estimation requires the generation of N independent trajectories $X_1, ..., X_N$ of the process X and the calculation of the sample mean

$$\widehat{\pi}_N := \frac{1}{N} \sum_{n=1}^{N} I_n, \tag{10}$$

which provides an unbiased estimation of the target probability. An important parameter in defining the accuracy of a simulation algorithm as a function of the number of replications N is given by the relative error (RE), defined as

$$\mathrm{RE}\left[\widehat{\pi}_N\right] := \frac{\sqrt{\mathbb{V}\mathrm{ar}\left[\widehat{\pi}_N\right]}}{\mathbb{E}\widehat{\pi}_N}. \tag{11}$$

3.1 Rare Event Simulation via Conditional MC

For large values of T, the probability (9) can be extremely small and its accurate estimation requires to generate a huge number of sample paths of the process X. Indeed, in traditional (crude) MC simulation the RE diverges for small values of the target probability π, i.e.

$$\mathrm{RE}\left[\widehat{\pi}_N\right] \sim \frac{1}{\sqrt{\pi N}} \quad \text{as} \quad \pi \to 0 \tag{12}$$

and rare event simulation techniques must be used. For instance, Importance Sampling (IS), when applied properly, can lead to an enormous variance reduction (several orders of magnitude with respect to crude MC [11]), but the optimal *change of measure* is known only in very simple cases and under an improper choice the RE may even grow infinitely [10].

An interesting alternative, which always leads to variance reduction, is the well-known Conditional Monte Carlo (CMC) method. In a nutshell, denoting by Z the indicator function of the target event in (9), the basic idea consists in finding an auxiliary random variable Y correlated with Z such that $\mathbb{E}[Z|Y]$ is available in explicit form. Then,

$$\pi(T) = \mathbb{E}Z = \mathbb{E}\left[\mathbb{E}[Z|Y]\right] \tag{13}$$

and the unbiased CMC estimator becomes

$$\widehat{\pi}_N = \frac{1}{N}\sum_{n=1}^{N} \mathbb{E}[Z|Y_n], \tag{14}$$

where $Y_1, ..., Y_N$ are N samples of Y. Note that, as stated above, the variance of (14) is always less than the variance of (10) since

$$\mathbb{V}\mathrm{ar}[Z] = \mathbb{E}[\mathbb{V}\mathrm{ar}[Z|Y]] + \mathbb{V}\mathrm{ar}[\mathbb{E}[Z|Y]]. \tag{15}$$

Unfortunately, in general, it is not easy to identify a suitable random variable Y for which $\mathbb{E}[Z|Y]$ is known in explicit form. However, in the Gaussian case, a general solution is provided the *bridge* process that is at the basis of the so-called Bridge Monte Carlo (BMC) method, originally proposed by some of the authors in [8] for the estimation of the overflow probability in single server queues and briefly described in the following section.

4 Bridge Monte Carlo Approach

The *bridge* $Y := \{Y(t)\}$ of a Gaussian process X is obtained by conditioning X to reach a certain level at some prefixed time instant s:

$$Y(t) = X(t) - \psi(t)X(s), \tag{16}$$

where ψ can be easily expressed in terms of the covariance function $\Gamma(t,s)$ of the process X:

$$\psi(t) := \frac{\Gamma(t,s)}{\Gamma(s,s)}.$$

Since $v(t)$ is an increasing function of t by hypothesis, it is easy to verify that $\psi(t) > 0$ for all $t > 0$. Moreover, for any t, $Y(t)$ is independent of $X(s)$ since

$$\mathbb{E}[X(s)Y(t)] = \Gamma(s,t) - \frac{\Gamma(t,s)}{\Gamma(s,s)}\Gamma(s,s) = 0$$

and $(X(s), Y(t))$ has bivariate normal distribution.

The use of the bridge was originally applied to the estimation of the overflow probability in [8] (see also [9,13] for a more detailed analysis of the estimator) and then to the distribution of the busy period lengths [15]. The interest for the BMC approach is motivated by the general variance reduction property of CMC and by [9], where it was shown that BMC has higher asymptotic efficiency than IS with a change of measure based on the most likely path (which is the optimal approach at least in the BM case). In the next subsection, we derive the expression of the estimator for the hitting probability.

4.1 Bridge Monte Carlo Estimation of the Hitting Time Distribution

Denoting by $\mathbb{T}$ the interval $[0, T]$, the target probability (9) can be rewritten in terms of the corresponding bridge process as follows:

$$\begin{aligned}
\pi(T) &= \mathbb{P}(\tau_D \geq T) = \mathbb{P}\left(\sup_{t\in\mathbb{T}} A(t) \leq D\right) \\
&= \mathbb{P}\left(\forall t \in \mathbb{T}:\ mt + X(t) \leq D\right) \\
&= \mathbb{P}\left(\forall t \in \mathbb{T}:\ X(s) \leq \frac{D - Y(t) - mt}{\psi(t)}\right) \\
&= \mathbb{P}\left(X(s) \leq \inf_{t\in\mathbb{T}} \frac{D - Y(t) - mt}{\psi(t)}\right) = \mathbb{P}\left(X(s) \leq \overline{Y}\right),
\end{aligned}$$

where

$$\overline{Y} := \inf_{t\in\mathbb{T}} \frac{D - Y(t) - mt}{\psi(t)}, \tag{17}$$

and, thanks to the independence between $\overline{Y}$ and $X(s)$:

$$\pi(T) = \mathbb{P}\left(X(s) \leq \overline{Y}\right) = \mathbb{E}\left[\Psi\left(\frac{\overline{Y}}{\sqrt{v(s)}}\right)\right],$$

where Ψ denotes the distribution function of a standard normal variable.

Hence, given N independent samples $\{\overline{Y}^{(n)}, n = 1, ..., N\}$ of $\overline{Y}$, the BMC estimator of $\pi(T)$ is

$$\widehat{\pi}_N^{\mathrm{BMC}} := \frac{1}{N}\sum_{n=1}^{N} \Psi\left(\frac{\overline{Y}^{(n)}}{\sqrt{v(s)}}\right). \tag{18}$$

Note that, although the derivation of the conditional probability is similar to the one in [13] for the overflow probability, $\overline{Y}$ is now calculated over the finite interval $\mathbb{T}$ (and not $\mathbb{R}^+$) so that the systematic error due to finite-size samples is not present in our case. Moreover, it is worth highlighting that

$$\Psi\left(\frac{\overline{Y}}{\sqrt{v(s)}}\right) = \mathbb{E}\left[I(X(s) \leq \overline{Y})|\overline{Y}\right], \tag{19}$$

and therefore the BMC approach is actually a special case of the CMC method (and hence leads to a variance reduction with respect to crude MC).

Finally, it is quite natural to extend the analysis to the case in which the threshold D is itself a random variable, which is independent of the input process X. Indeed, in practice the exact value of D may be only partially known and this incomplete information can be reflected by introducing the probability density function f_D of D. Then, the density of the RV τ_D can be rewritten as

$$f_\tau(x) = \int_{y=0}^{\infty} f_\tau(x|y) f_D(y) dy. \tag{20}$$

In general, this density can be evaluated only by numerical methods or, in some special cases, in terms of special functions (see [14] for further details). With our approach this can be performed without any additional complexity, just sampling the value of D from the considered distribution. In more detail, formula (18) still holds, with the only difference that the value of D in (17) is not a constant, but it is sampled according to f_D.

5 Simulation Results

In this section, we present the simulation analysis of the proposed BMC estimator considering BM and FBM as input. Unless otherwise stated, we used $N = 10000$ replications and the following values of the system parameters: $m = 1$; $D = 20$; $\sigma = 1$.

5.1 Brownian Motion

In the first set of simulation experiments we focused on the Wiener case (i.e., FBM with $H = 0.5$) in order to check the unbiasedness of the estimator. Indeed, as highlighted in Sect. 2.1, only in this special case it is possible to calculate the target probability by (4) and verify the impact of different simulation parameters on the goodness of the estimator.

At first we evaluated the influence of the conditioning point s. To this aim, in Fig. 1 the estimations of the hitting probability for different values of s are compared with the analytic values (continuous line). In general, $s = 120$ seems to be a good choice and permits to estimate even probabilities of the order of 10^{-20} in spite of the relatively small value of replications ($N = 10^4$).

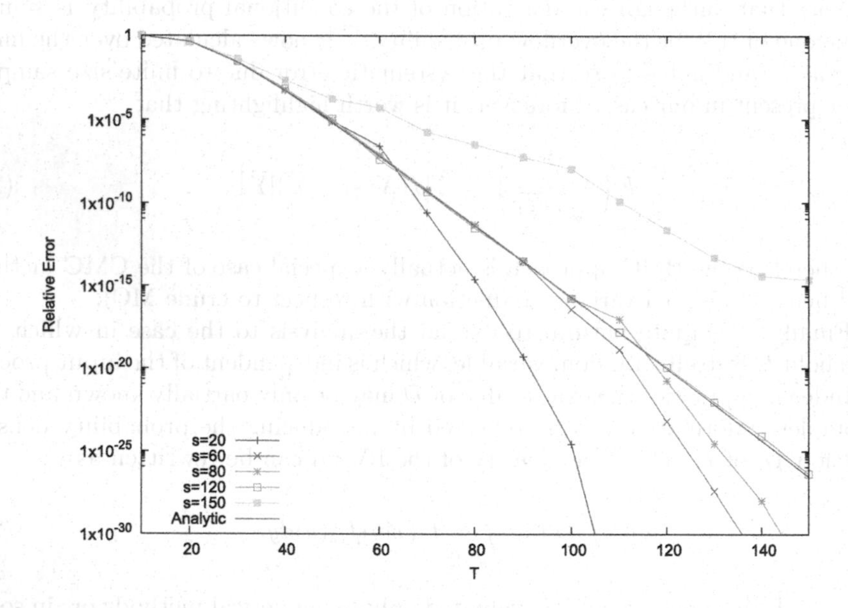

Fig. 1. Hitting probability as a function of the conditioning point in the Wiener case

It is important to highlight that, although BM and FBM are continuous time processes, in the simulation a discrete lattice must be used. To better understand the effect of the time discretization, let us denote by h the number of samples drawn per time unit (all the above results were obtained for $h = 10$) and check the dependence of the hitting probability on h (for $s = 120$). From Fig. 2 we can conclude that, at least in the Wiener case, the process discretization does not significantly affect the estimation.

Finally, to analytically verify the applicability of our BMC estimator to the case of random thresholds, at first we considered a simple case in which D can take just two values, $D_1 = 20$ with probability p and $D_2 = 40$ with probability $1 - p$. In this case, the distribution of τ_D is just the linear combination (with

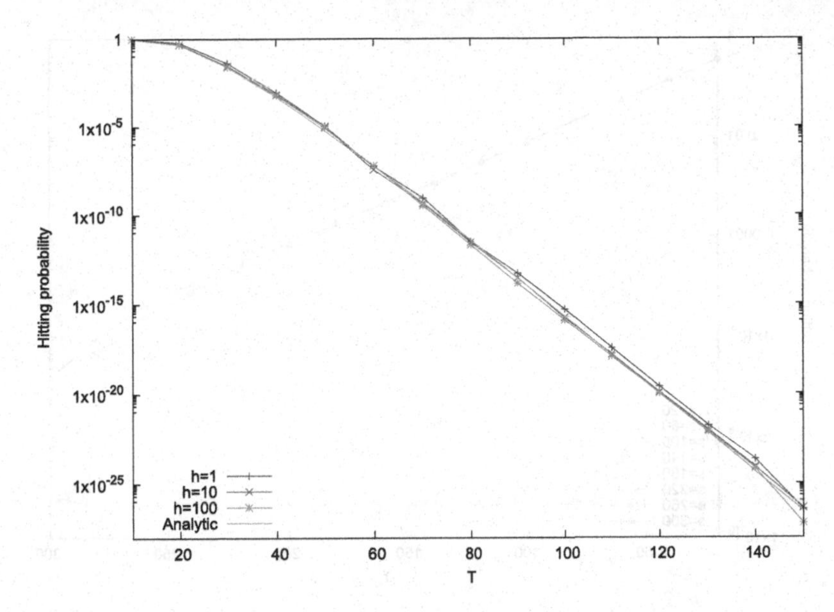

Fig. 2. Hitting probability as a function of the discretization factor in the Wiener case

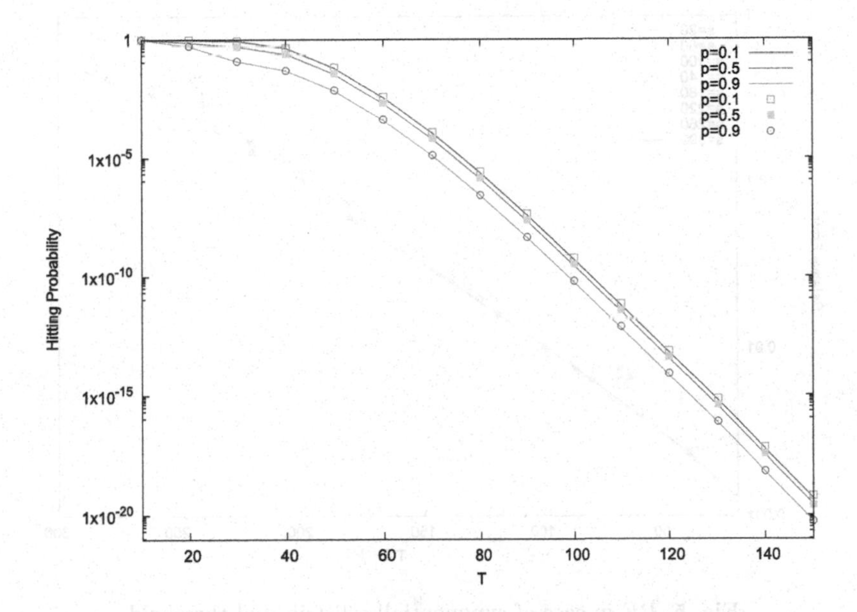

Fig. 3. Analytic (lines) vs. simulation (dots) results in case of random threshold

weights p and $1 - p$, respectively) of the distributions for the two values D_1 and D_2 of the threshold.

In Fig. 3 the simulation estimates (using $N = 10^6$ trajectories) are compared with the analytic values for three different values of p (using $h = 10$ and $s = 120$

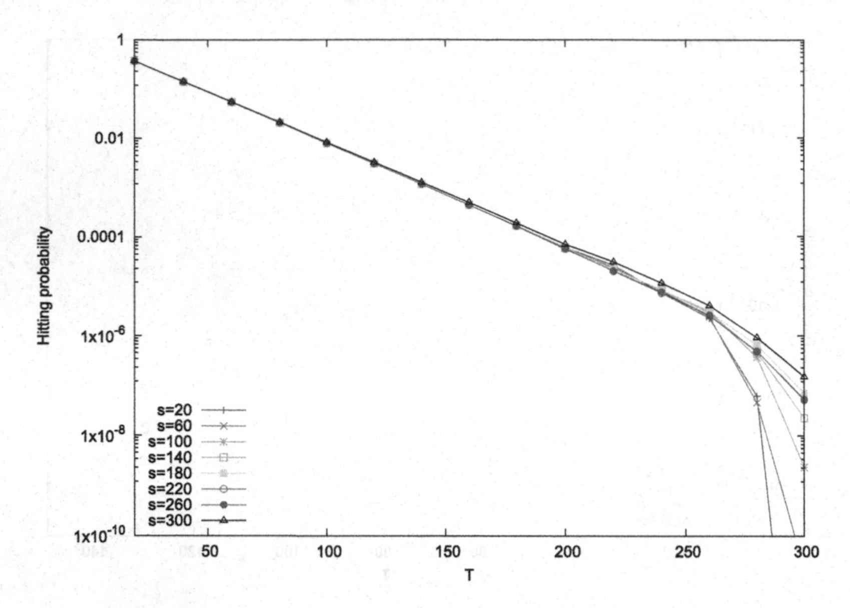

Fig. 4. Hitting probability in case of exponentially distributed threshold

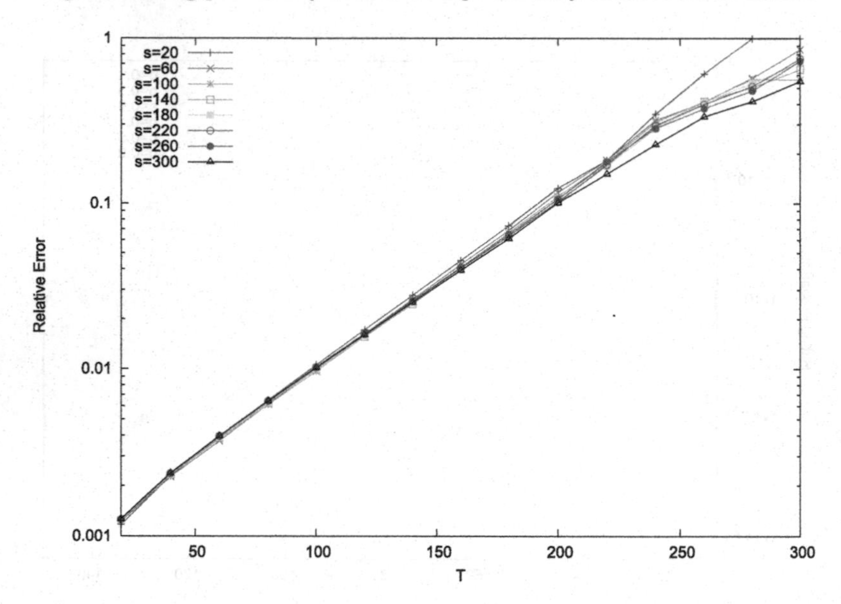

Fig. 5. RE in case of exponentially distributed threshold

in agreement with the previous considerations). As expected, the increment of N led to a reduction of the RE and over the entire range of probabilities the simulation results are very close to the theoretical ones.

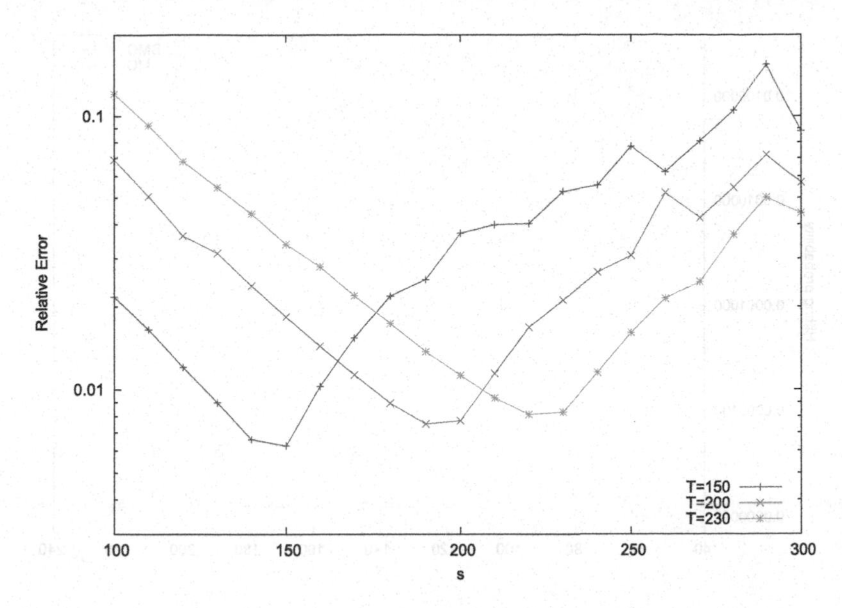

Fig. 6. RE as a function of the conditioning point in the FBM case ($H = 0.7$) for different values of T

The performance are slightly worse in case of continuous random variables with infinite support, as highlighted by Figs. 4 and 5, in which the hitting probability and the RE, estimated according to formula (11), are shown for an exponential distribution (with mean 20) of the threshold. Note that the conditioning point (which is fixed for all the realizations) has a very limited effect, probably due to the high variability of the value of D in different simulation runs.

5.2 Fractional Brownian Motion

The next set of experiments is devoted to the more general FBM case. For sake of brevity we present only the results for $H = 0.7$, i.e. in presence of long range dependence ($0.5 < H < 1$), which typically arises in ICT models.

At first we focused on the *optimal* choice of the conditioning point s: as highlighted by Fig. 6, in all considered cases the choice $s \approx T$ corresponds to a local minimum of the RE. Hence all the following simulations have been carried out for $s = T$.

Then, in order to verify the effectiveness of the proposed estimator, we considered the dependence of the RE on the rarity parameter T in comparison with the standard MC estimator. The results, presented in Figs. 7 and 8, point out the superiority of BMC: for probabilities of the order of 10^{-7} the RE is still less than 1%. Moreover, as expected, the tail distribution decays much slower than in the Wiener case (compare Fig. 7 with Fig. 2), confirming the relevant impact of long range dependence on system performance.

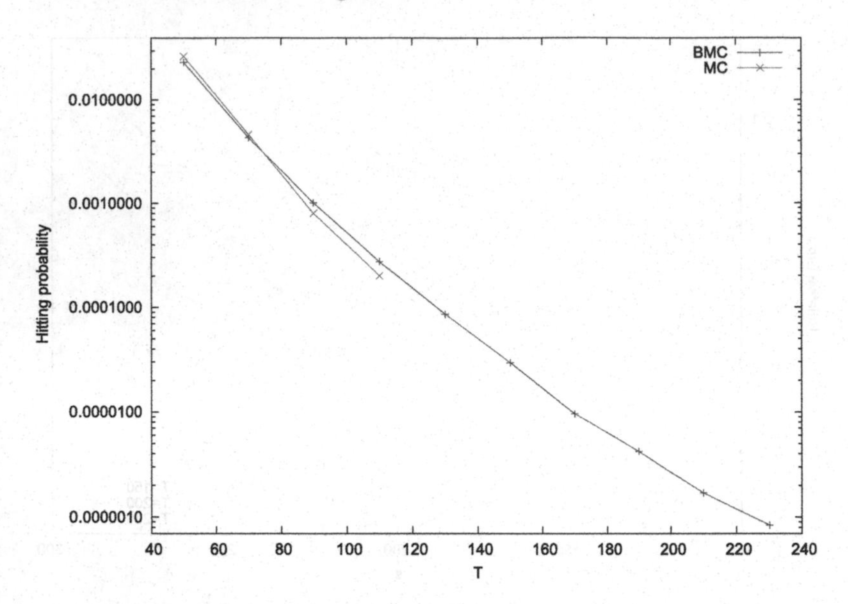

Fig. 7. Hitting probability: MC vs. BMC for FBM with $H = 0.7$

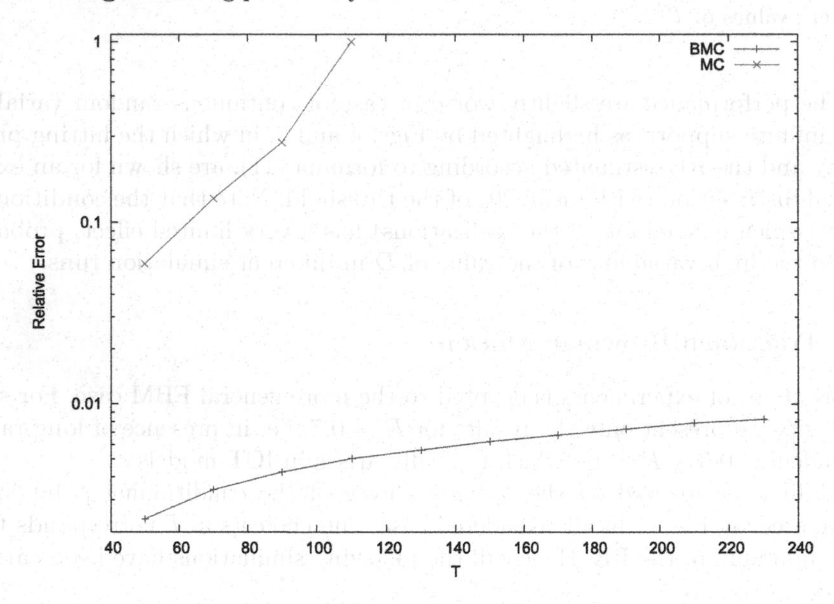

Fig. 8. RE : MC vs. BMC for FBM with $H = 0.7$

Furthermore, Fig. 9 compares the simulation results with the asymptotic (8) for $T = 150$; it is worth highlighting that the RE is less than 5% already for $D > 80$.

Finally, the effect of random thresholds is shown in Figs. 10 and 11, focusing on uniform (in the interval $[10, 30]$) and exponential (with mean 20) distributions.

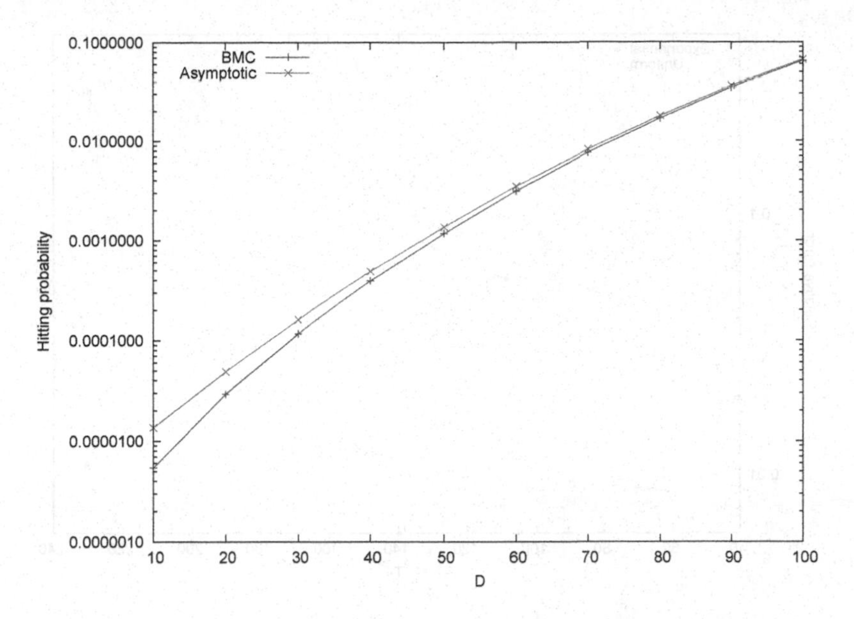

Fig. 9. Hitting probability for FBM with $H = 0.7$ for $T = 150$

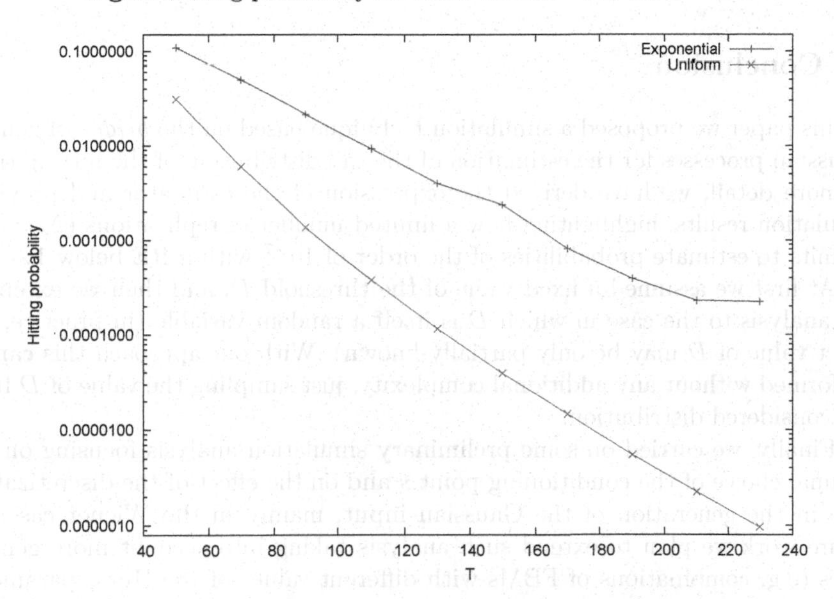

Fig. 10. Hitting probability in case of random threshold for FBM with $H = 0.7$

Also in case of FBM, the higher variability of the threshold has a significant impact on the hitting probability as well as on the RE of the estimator.

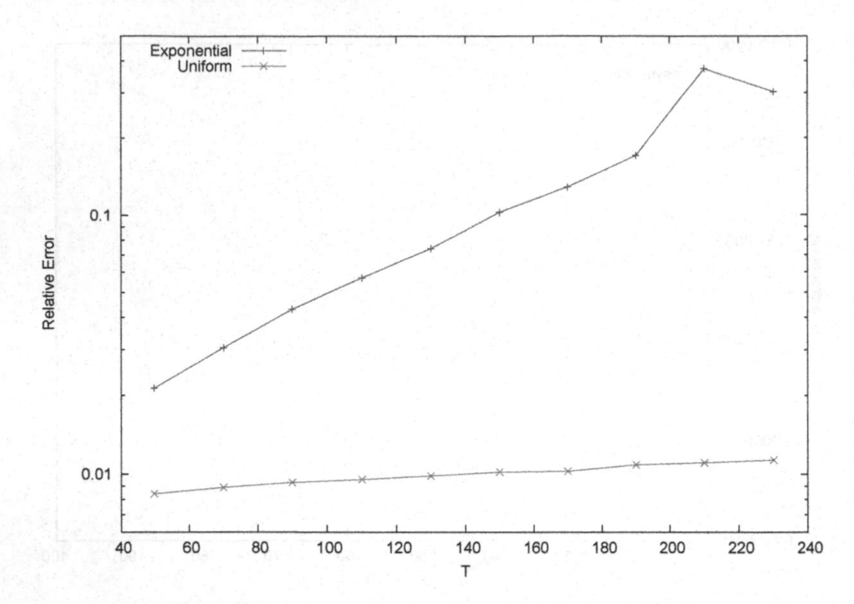

Fig. 11. RE in case of random threshold for FBM with $H = 0.7$

6 Conclusion

In this paper we proposed a simulation technique based on the *bridge* of general Gaussian processes for the estimation of the tail distribution of the hitting time. In more detail, we have derived the expression of the estimator and provided simulation results, highlighting how a limited number of replications ($N = 10^4$) permits to estimate probabilities of the order of 10^{-7} with a RE below 1%.

At first we assumed a fixed value of the threshold D, and then we extended the analysis to the case in which D is itself a random variable (in practice, the exact value of D may be only partially known). With our approach this can be performed without any additional complexity, just sampling the value of D from the considered distribution.

Finally, we carried on some preliminary simulation analysis focusing on the optimal choice of the conditioning point s and on the effect of the discretization step in the generation of the Gaussian input, mainly in the Wiener case. As future work we plan to extend such analysis taking into account more general cases (e.g. combinations of FBMs with different values of the Hurst parameter H) as well as the relation between the Hölder regularity of the sample paths (which depends on H) and the proper choice of the discretization step.

References

1. Arutkin, M., Walter, B., Wiese, K.J.: Extreme events for fractional brownian motion with drift: theory and numerical validation. Phys. Rev. E. **102**, 022102 (2020). https://doi.org/10.1103/PhysRevE.102.022102

2. Borodin, A.N., Salminen, P.: Handbook of Brownian Motion - Facts and Formulae. Birkhauser, Basel (2002)
3. Caglar, M., Vardar, C.: Distribution of maximum loss of fractional brownian motion with drift. Stat. Prob. Lett. **83**, 2729–2734 (2013)
4. Chetvertakova, E.S., Chimitova, E.V.: The Wiener degradation model in reliability analysis. In: 2016 11th International Forum on Strategic Technology (IFOST), pp. 488–490 (2016)
5. Dong, Q., Cui, L.: A study on stochastic degradation process models under different types of failure thresholds. Reliab. Eng. Syst. Saf. **181**, 202–212 (2019). https://doi.org/10.1016/j.ress.2018.10.002
6. Floyd, S., Jacobson, V.: Random early detection gateways for congestion avoidance. IEEE/ACM Trans. Netw. **1**(4), 397–413 (1993)
7. Gao, H., Cui, L., Kong, D.: Reliability analysis for a Wiener degradation process model under changing failure thresholds. Reliab. Eng. Syst. Saf. **171**, 1–8 (2018). https://doi.org/10.1016/j.ress.2017.11.006
8. Giordano, S., Gubinelli, M., Pagano, M.: Bridge Monte-Carlo: a novel approach to rare events of Gaussian processes. In: Proceedings of the 5th St.Petersburg Workshop on Simulation, pp. 281–286. St. Petersburg, Russia (2005)
9. Giordano, S., Gubinelli, M., Pagano, M.: Rare events of gaussian processes: a performance comparison between bridge monte-carlo and importance sampling. In: Koucheryavy, Y., Harju, J., Sayenko, A. (eds.) NEW2AN 2007. LNCS, vol. 4712, pp. 269–280. Springer, Heidelberg (2007). https://doi.org/10.1007/978-3-540-74833-5_23
10. Glasserman, P., Wang, Y.: Counterexamples in importance sampling for large deviations probabilities. Ann. Appl. Prob. **7**(3), 731–746 (1997)
11. Heidelberger, P.: Fast simulation of rare events in queueing and reliability models. ACM Trans. Model. Comput. Simul. **5**(1), 43–85 (1995)
12. Kulkarni, V., Rolski, T.: Fluid model driven by an Ornstein-Uhlenbeck process. Prob. Eng. Inform. Sci. **8**, 403–417 (1994)
13. Lukashenko, O.V., Morozov, E.V., Pagano, M.: On the efficiency of Bridge Monte-Carlo estimator. Inform. Appl. **11**(2), 16–24 (2017)
14. Lukashenko, O.V., Morozov, E.V., Pagano, M.: A Gaussian approximation of the distributed computing process. Inform. Appl. **13**(2), 109–116 (2019)
15. Lukashenko, O., Morozov, E., Pagano, M.: On the use of a bridge process in a conditional monte carlo simulation of gaussian queues. In: Dudin, A., Gortsev, A., Nazarov, A., Yakupov, R. (eds.) Information Technologies and Mathematical Modelling - Queueing Theory and Applications, pp. 207–220. Springer International Publishing, Cham (2016)
16. Michna, Z.: On tail probabilities and first passage times for fractional brownian motion. Math. Methods Oper. Res. **49**, 335–354 (1999)
17. Norros, I.: A storage model with self-similar input. Queueing Syst. **16**, 387–396 (1994)
18. Taqqu, M.S., Willinger, W., Sherman, R.: Proof of a fundamental result in self-similar traffic modeling. Comput. Commun. Rev. **27**, 5–23 (1997)
19. Walter, B., Wiese, K.J.: Sampling first-passage times of fractional brownian motion using adaptive bisections. Phys. Rev. E. **101**, 043312 (2020). https://doi.org/10.1103/PhysRevE.101.043312

A Queueing Inventory System with Two Channels of Service

Nisha Mathew[1(✉)], V. C. Joshua[2], and A. Krishnamoorthy[2]

[1] Department of Mathematics, B.K College Amalagiri, Kottayam, India
nishamathew@cmscollege.ac.in

[2] Department of Mathematics, CMS College, Kottayam, India
{vcjoshua,krishnamoorthy}@cmscollege.ac.in

Abstract. We consider a queueing inventory model with positive service time and two service channels. Channel I is a single server facility and channel II is a bulk service facility. There are two types of customers, type-I and type-II. The same type of commodity is served to both types of customers. Channel I provides service to type-I customers and channel II provides service to type-II customers. Service is initiated only if inventory is available. Bulk service is initiated at the end of a random clock or by the accumulation of N type-II customers. The inventory replenishment follows the (s, S) policy with positive leadtime. The service time follows phase type distribution. Steady state analysis of the model is performed. Some performance measures are evaluated.

Keywords: Queueing-inventory · Lead time · Positive service time

1 Introduction

In most of the real life problems, it takes some time to serve the item to a customer. Such models are called queueing inventory models or inventory with positive service time. Inventory models with positive service time (queueing inventory model) were introduced by Sigman and Simchi-Levi in [1] . A survey of inventory with positive service time is given by Krishnamoorty et al. in [2]. Chakravarthy et al. in [3] studied a single server queueing model in which the customers are served in batches of varying size. Stochastic inventory system with two types of services that follows (s, S) replenishment policy is given by Anbazhagan et al. in [4]. A single server retrial queueing model with two types of customers with service time distribution following phase type is studied by Krishnamoorty et al. in [5]. Queueing inventory model with exponential lead time are found in [6–8].

The modern day retail business is driven by multiple platforms which are either offline or online. Long gone are the days where customers had to be physically present for shopping. Sometimes customers give orders online through various virtual platforms and on a few ocassions go out for shopping. The sellers move into all the possibilities of customer interaction so that the sales are always boosted. This requires algorithms which could determine the stock pile required

V. M. Vishnevskiy et al. (Eds.): DCCN 2020, LNCS 12563, pp. 604–616, 2020.
https://doi.org/10.1007/978-3-030-66471-8_46

by the sellers well in advance, so that the customer demands are met without delay or failure.

This model is motivated by two types of demands that arrive at supermarkets: one is physically arriving customers and the other is online customers. The physically present customers are attended by the system on a FIFO basis. Online customer demands are attended only when the accumulated number of such demands reach a threshhold N or a random clock realizes, whichever occurs first.

2 Model Description

Our model includes one product which is been sold through two different platforms - a physical shop and an online platform. We consider a single server queueing inventory system with two types of customers, namely physical customers (type-I) and online customers (type-II). We assume that both the arrivals of type-I and type-II customers follow independent Poisson process. Let λ_1 be the rate of arrival of type-I customers and λ_2 be that of type-II customers. We also assume that the service time is positive. Type-I customers can form an infinite queue. Type-II customers are served in batches with maximum batch size N. A clock is also set, which starts ticking with the first arrival of type-II customer in every cycle. An (s, S) inventory policy is used. Service of a customer requires an inventory item. Each customer (both type-I and type-II) demands one unit of item, having a random duration of service time. Service time distribution of both type-I and type-II customers are assumed to be of phase type with irreducible representations $\mathrm{PH}(\alpha, T)$ with m_1 phases and $\mathrm{PH}(\beta, U)$ with m_2 phases respectively. The vectors T^0 and U^0 are given by $T^0 = -Te$ and $U^0 = -Ue$.

We use the (s, S) replenishment policy here. Lead time is assumed to be exponential with parameter γ. If the server is idle, type-I customer enters into the service. A clock is set for type-II customers. Let θ be the rate of realization (parameter) of the exponential clock. The clock starts at the arrival of first type-II customer. The type-II customers are served only when the order becomes N or the clock time expires, whichever occurs first. They are served in bulk, provided two or more type-II customers have joined before the expiry/realization of the clock. It may happen that no such demand/exactly one demand arrived before clock expiry.

When the service in channel I begins, the inventory level drops by one unit. But when the service in channel II begins, the inventory level drops by n_2 , where n_2 is the number of type-II customers present at that time. When the clock expires or when number of type-II customers reaches N, the service of all these type-II customers is provided in a batch, provided the server was idle at that time. On the other hand, if the server is busy at that time, type-II customers are served immediately on completion of service of the current type-I customer. No type II customer is allowed to join the system once the clock expires/ N type II are in the system.

3 Mathematical Formulation

Let
- $N_1(t)$ be the number of type-I customers in the queue at time t
- $N_2(t)$ be the number of type-II customers in the finite buffer at time t
- $B(t)$ be the server status at time t;

$$B(t) = \begin{cases} 0, & \text{if the server is idle} \\ 1, & \text{if the server is busy with a type-I customer} \\ 2, & \text{if the server is busy with a type-II customer} \end{cases}$$

- $C(t)$ be the clock status at time t

$$C(t) = \begin{cases} 0, \text{if the clock is off} \\ 1, \text{if the clock is on} \end{cases}$$

- $I(t)$ be the number of items in the inventory at time t
- $J(t)$ be the phase of the service process at time t

Then $\{(N_1(t), N_2(t), B(t), C(t), I(t), J(t)); t \geq 0\}$ is a continuous time Markov chain on the state space to be described below. This model can be considered as a Level Independent Quasi-Birth-Death (LIQBD) process and a solution is obtained by Matrix Analytic Method. We define the state space of the QBD under consideration and analyze the structure of its infinitesimal generator.

The state space $\Omega = \Omega_1 \bigcup \Omega_2 \bigcup \Omega_3 \bigcup \Omega_4 \bigcup \Omega_5$ where
$\Omega_1 = \{(0,0,0,0,i)/0 \leq i \leq S\}$,
$\Omega_2 = \{(n_1,0,0,0,0) \bigcup (n_1,n_2,0,1,0)/n_1 \geq 1; 1 \leq n_2 \leq N-1;\}$,
$\Omega_3 = \{(n_1,n_2,0,0,i)/n_1 \geq 0; 1 \leq n_2 \leq N; 0 \leq i \leq n_2 - 1\}$
$\Omega_4 = \{(n_1,n_2,1,c,i,j)/n_1 \geq 0; 0 \leq n_2 \leq N; c = 0,1; 0 \leq i \leq S; j = 1,2,\ldots,m_1\}$,
$\Omega_5 = \{(n_1,n_2,2,1,i,j) \bigcup (n_1,0,2,0,i,j) \bigcup (n_1,N,2,0,i,j)/n_1 \geq 0; 0 \leq n_2 \leq N-1; 0 \leq i \leq S; j = 1,2,\ldots,m_2\}$,

The transitions are given in the table below (Tables 1, 2).

Table 1. Transition table

From	T0	Rate	Description
$(0,0,0,0,i)$	$(0,0,1,0,i-1,j)$	$\lambda_1\alpha_j$	$i \geq 1, j = 1,2,\cdots m_1$
$(0,n_2,0,1,i)$	$(0,n_2,1,1,i-1,j)$	$\lambda_1\alpha_j$	$i \geq 1, j = 1,2,\cdots m_1, 1 \leq n_2 \leq N-1$
$(n_1,n_2,1,1,i,j)$	$(n_1+1,n_2,1,1,i,j)$	λ_1	$n_1 \geq 0, j = 1,2,\cdots m_1, 1 \leq n_2 \leq N-1$
$(n_1,0,1,0,i,j)$	$(n_1+1,0,1,0,i,j)$	λ_1	$n_1 \geq 0, j = 1,2,\cdots m_1$
$(n_1,n_2,1,0,i,j)$	$(n_1+1,n_2,1,0,i,j)$	λ_1	$n_1 \geq 0, j = 1,2,\cdots m_1, 1 \leq n_2 \leq N$
$(n_1,0,1,0,i,j)$	$(n_1,1,1,1,i,j)$	λ_2	$n_1 \geq 0, j = 1,2,\cdots m_1$
$(0,0,0,0,i)$	$(0,1,0,1,i)$	λ_2	$i \geq 0$

Table 2. Transition table

From	T0	Rate	Description
$(n_1, n_2, 1, 1, i, j)$	$(n_1, n_2+1, 1, 1, i, j)$	λ_2	$n_1 \geq 0, j=1,2,\cdots m_1$, $1 \leq n_2 \leq N-2$
$(0, n_2, 0, 1, i)$	$(0, n_2+1, 0, 1, i)$	λ_2	$1 \leq n_2 \leq N-2$
$(n_1, N-1, 1, 1, i, j)$	$(n_1, N, 1, 0, i, j)$	λ_2	$n_1 \geq 0, j=1,2,\cdots m_1$
$(n_1, 0, 2, 0, i, j)$	$(n_1, 1, 2, 1, i, j)$	λ_2	$n_1 \geq 0, j=1,2,\cdots m_2$
$(n_1, n_2, 2, 1, i, j)$	$(n_1, n_2+1, 2, 1, i, j)$	λ_2	$n_1 \geq 0, 1 \leq n_2 \leq N-2$, $j=1,2,\cdots m_2$
$(n_1, N-1, 2, 1, i, j)$	$(n_1, N, 2, 0, i, j)$	λ_2	$n_1 \geq 0, j=1,2,\cdots m_2$
$(0, N-1, 0, 1, i)$	$(0, 0, 2, 0, i-N, j)$	$\lambda_2\beta_j$	$i \geq N, j=1,2,\cdots m_2$
$(n_1, n_2, 1, 1, i, j)$	$(n_1, n_2, 1, 0, i, j)$	θ	$n_1 \geq 0, j=1,2,\cdots m_1$ $1 \leq n_2 \leq N-1$
$(0, n_2, 0, 1, i)$	$(0, 0, 2, 0, i-n_2, j)$	$\theta\beta_j$	$i \geq n_2, j=1,2,\cdots m_2$
(n_1, n_2, b, c, i, j)	(n_1, n_2, b, c, S, j)	γ	$n_1 \geq 0, n_2 \geq 0$, $b=0,1,2, c=0,1, 0 \leq i \leq s$
$(n_1, n_2, 0, 1, 0)$	$(n_1-1, n_2, 1, 1, S-1, j)$	$\gamma\alpha_j$	$n_1 \geq 1, j=1,2,\cdots m_1$
$(n_1, n_2, 0, 0, i)$	$(n_1, 0, 2, 0, S-n_2, j)$	$\gamma\beta_j$	$n_1 \geq 0, 1 \leq n_2 \leq N$, $i < n_2, j=1,2,\cdots m_2$
$(0, n_2, 1, 1, i, j)$	$(0, n_2, 0, 1, i)$	T_j^0	$n_2 \geq 1, j=1,2,\cdots m_1$
$(n_1, n_2, 1, 1, 0, j)$	$(n_1, n_2, 0, 1, 0)$	T_j^0	$n_2 \geq 1, j=1,2,\cdots m_1$
$(n_1, n_2, 1, 1, i, j)$	$(n_1-1, n_2, 1, 1, i-1, k)$	$T_j^0\alpha_k$	$n_1 \geq 1, n_2 \geq 1$, $i \geq 1, j, k=1,2,\cdots m_1$
$(n_1, n_2, 1, 0, i, j)$	$(n_1, 0, 2, 0, i-n_2, k)$	$T_j^0\beta_k$	$i \geq n_2$, $j=1,2,\cdots m_1, k=1,2,\cdots m_2$
$(0, 0, 1, 0, i, j)$	$(0, 0, 1, 0, i, k)$	T_{jk}	$i \geq 0, j, k=1,2,\cdots m_1$
$(n_1, n_2, 1, 1, i, j)$	$(n_1, n_2, 1, 1, i, k)$	T_{jk}	$n_1 \geq 0, j, k=1,2,\cdots m_1$
$(0, 0, 2, 0, i, j)$	$(0, 0, 0, 0, i)$	U_j^0	$j=1,2,\cdots m_2$
$(n_1, 0, 2, 0, 0, j)$	$(n_1, 0, 0, 0, 0)$	U_j^0	$j=1,2,\cdots m_2$
$(n_1, n_2, 2, 1, 0, j)$	$(n_1, n_2, 0, 1, 0)$	U_j^0	$n_1 \geq 1, 1 \leq n_2 \leq N-1$, $j=1,2,\cdots m_2$
$(n_1, N, 2, 0, i, j)$	$(n_1, N, 0, 0, i)$	U_j^0	$n_1 \geq 0, i \leq N-1$, $j=1,2,\cdots m_1$
$(n_1, 0, 2, 0, i, j)$	$(n_1-1, 0, 1, 0, i-1, k)$	$U_j^0\alpha_k$	$n_1 \geq 1, i \geq 1$, $j=1,2,\cdots m_2, k=1,2,\cdots m_1$
$(n_1, n_2, 2, 1, i, j)$	$(n_1-1, n_2, 1, 1, i-1, k)$	$U_j^0\alpha_k$	$n_1 \geq 1, 1 \leq n_2 \leq N-1, i \geq 1$, $j=1,2,\cdots m_2, k=1,2,\cdots m_1$
$(n_1, N, 2, 0, i, j)$	$(n_1, 0, 2, 0, i-N, k)$	$U_j^0\beta_k$	$n_1 \geq 0, i \geq N$, $j, k=1,2,\cdots m_1$
$(n_1, 0, 2, 0, i, j)$	$(n_1, 0, 2, 0, i, k)$	U_{jk}	$n_1 \geq 0, j, k=1,2,\cdots m_2$
$(n_1, n_2, 2, 1, i, j)$	$(n_1, n_2, 2, 1, i, k)$	U_{jk}	$n_1 \geq 0, j, k=1,2,\cdots m_2$

The infinitesimal generator Q of the LIQBD describing the above single server queueing inventory system is of the form

$$\begin{pmatrix} B_{00} & B_{01} & O & \dots\dots\dots \\ B_{10} & A_1 & A_0 & O \dots\dots\dots \\ O & A_2 & A_1 & A_0 & O \dots\dots \\ O & O & A_2 & A_1 & A_0 & O \dots \\ & \ddots & \ddots & \ddots & \ddots & \ddots \\ & & \ddots & \ddots & \ddots & \ddots & \ddots \end{pmatrix} \tag{1}$$

where B_{00}, A_0, A_1, A_2 are all square matrices of appropriate order whose entries are block matrices. A_0 represents the arrival of a customer to the system; that is transition from level $n_1 \rightarrow n_1 + 1$. A_2 represents transition from level: $n_1 \rightarrow n_1 - 1$, A_1 describes all transitions in which the level does not change (transitions within levels). The structure of the matrices $B_{00}, B_{01}, B_{10}, A_0, A_1, A_2$ are as follows:

$$A_0 = \lambda_1 I_K \tag{2}$$

where $K = (S+1)(2m_1N + m_2(N+1)) + (N/2)(N+3)$

$$A_2 = \begin{pmatrix} H_1^0 & & & & \\ & H_1^1 & & & \\ & & \ddots & & \\ & & & H_1^{N-1} & \\ & & & & H_1^N \end{pmatrix} \tag{3}$$

where

$$H_1^0 = \begin{pmatrix} O & Y & O \\ O & Z & O \end{pmatrix},$$

H_1^0 is a square matrix of order $1 + (S+1)(m_1 + m_2)$

$$H_1^j = \begin{pmatrix} O & O & O \\ O & Y & O \\ O & O & O \\ O & Z & O \end{pmatrix}, \text{ for } j = 1 \text{ to } N-1,$$

and H_1^j are square matrices of order $(j+1) + (S+1)(2m_1 + m_2)$,
H_1^N is a zero square matrix of order $N + (S+1)(m_1 + m_2)$,
$Y = \begin{pmatrix} O & \gamma\alpha & O \end{pmatrix}$, is a matrix of order $1 \times [(S+1)(m_1)]$,

$$Z = \begin{pmatrix} O & O \\ I_S \otimes T^0 \otimes \alpha & O \\ O & O \\ I_S \otimes U^0 \otimes \alpha & O \end{pmatrix} \text{ is a matrix of order } [(S+1)(m_1+m_2)] \times [(S+1)(m_1)]$$

$$B_{10} = \begin{pmatrix} E_1^0 & & & & \\ & E_1^1 & & & \\ & & \ddots & & \\ & & & E_1^{N-1} & \\ & & & & E_1^N \end{pmatrix} \quad (4)$$

$E_1^0 = \begin{pmatrix} O & Y & O \\ O & Z & O \end{pmatrix}$,
E_1^0 is a matrix of order $[1 + (S+1)(m_1 + m_2)] \times [(S+1)(m_1 + m_2 + 1)]$

$E_1^j = \begin{pmatrix} O & O & O \\ O & Y & O \\ O & O & O \\ O & Z & O \end{pmatrix}$ for $j = 1$ to $N - 1$, and

E_1^j are matrices of order $[j+1+(S+1)(2m_1+m_2)] \times [j+(S+1)(2m_1+m_2+1)]$,
E_1^N is a zero square matrix of order $N + (S+1)(m_1 + m_2)$,

$$B_{01} = \begin{pmatrix} W_1^0 & & & & \\ & W_1^1 & & & \\ & & \ddots & & \\ & & & W_1^{N-1} & \\ & & & & W_1^N \end{pmatrix} \quad (5)$$

where
$W_1^0 = \begin{pmatrix} V & 0 \\ 0 & I_{S+1} \otimes \lambda_1 I_{m_1} \end{pmatrix}$,
W_1^0 is a matrix of order $[(S+1)(m_1 + m_2 + 1)] \times [1 + (S+1)(m_1 + m_2)]$

$W_1^j = \begin{pmatrix} \lambda_1 I_j & 0 & 0 \\ 0 & V & 0 \\ 0 & 0 & I_{S+1} \otimes \lambda_1 I_{m_1+m_2} \end{pmatrix}$, for $j = 1$ to $N - 1$, and

W_1^j are matrices of order $[j+(S+1)(2m_1+m_2+1)] \times [(j+1)+(S+1)(2m_1+m_2)]$,
$W_1^N = \lambda_1 I_{K_1}$, $K_1 = N + (S+1)(m_1 + m_2)$,

$V = \begin{pmatrix} \lambda_1 & 0 & 0 \\ 0 & 0 & 0 \\ 0 & I_S \otimes \lambda_1 I_{m_1} & 0 \\ 0 & 0 & \lambda_1 I_{m_1} \end{pmatrix}$,

V is a matrix of order $[(S+1)(m_1+1)] \times [1+(S+1)m_1]$

$$B_{00} = \begin{pmatrix} K_0^1 & K_0^2 & 0 & 0 & \dots & \dots \\ K_1^0 & K_1^1 & K_1^2 & 0 & \dots & \dots \\ K_2^0 & 0 & K_2^1 & K_2^2 & 0 & \dots \\ K_3^0 & 0 & 0 & K_3^1 & K_3^2 & \dots \\ \vdots & \vdots & \vdots & \ddots & \ddots & \\ K_{N-1}^0 & \dots & & & K_{N-1}^1 & K_{N-1}^2 \\ K_N^0 & \dots & \dots & & 0 & K_N^1 \end{pmatrix} \tag{6}$$

where

$$K_0^1 = \begin{pmatrix} F_0 & F_6 & 0 \\ I_{s+1} \otimes T^0 & F_1 & 0 \\ I_{s+1} \otimes U^0 & 0 & F_2 \end{pmatrix},$$

K_0^1 is a square matrix of order $[(S+1)(m_1+m_2+1)]$

$$K_0^2 = \begin{pmatrix} 0 & \lambda_2 I_{S+1} & 0 & 0 \\ 0 & 0 & 0 & I_{S+1} \otimes \lambda_2 I_{m_1+m_2} \end{pmatrix},$$

K_0^2 is a matrix of order $[(S+1)(m_1+m_2+1)] \times [1+(S+1)(2m_1+m_2+1)]$

$$K_j^0 = \begin{pmatrix} 0 & Y_j \\ 0 & F^j \\ 0 & C_j \\ 0 & 0 \end{pmatrix}, \text{ for } j = 1, 2, ..., (N-2)$$

K_j^0 are matrices of order $[j+(S+1)(2m_1+m_2+1)] \times [(S+1)(m_1+m_2+1)]$

$$K_j^1 = \begin{pmatrix} -(\lambda_1+\gamma)I_j & 0 & 0 & 0 & 0 \\ \Theta^j & F_7 & 0 & F_6 & 0 \\ T^j & 0 & F_3 & 0 & 0 \\ 0 & I_{s+1} \otimes T^0 & I_{S+1} \otimes \theta I_{m_1} & F_4 & 0 \\ 0 & I_{s+1} \otimes U^0 & 0 & 0 & F_2 \end{pmatrix}, \text{ for } j = 1, 2, ..., (N-1)$$

K_j^1 are square matrices of order $j+(S+1)(2m_1+m_2+1)$

$$K_j^2 = \begin{pmatrix} 0 & 0 & 0 & 0 \\ 0 & \lambda_2 I_{s+1} & 0 & 0 \\ 0 & 0 & 0 & 0 \\ 0 & 0 & 0 & I_{S+1} \otimes \lambda_2 I_{m_1+m_2} \end{pmatrix}, \text{ for } j = 1, 2, ..., (N-2)$$

K_j^2 are matrices of order $[j+(S+1)(2m_1+m_2+1)] \times [j+1+(S+1)(2m_1+m_2+1)]$

$$K_{N-1}^0 = \begin{pmatrix} 0 & Y_{N-1} \\ 0 & F^{N-1} \\ 0 & C_{N-1} \\ 0 & 0 \end{pmatrix},$$

K_{N-1}^0 is a matrix of order $[N-1+(S+1)(2m_1+m_2+1)] \times [(S+1)(m_1+m_2+1)]$

$$K_{N-1}^2 = \begin{pmatrix} 0 & 0 \\ \Lambda & 0 \\ 0 & 0 \\ 0 & I_{S+1} \otimes \lambda_2 I_{m_1+m_2} \end{pmatrix},$$

K_{N-1}^2 is a matrix of order $[N-1+(S+1)(2m_1+m_2+1)] \times [N+(S+1)(m_1+m_2)]$

$$K_N^0 = \begin{pmatrix} 0 & Y_N \\ 0 & C_N \\ 0 & C \end{pmatrix},$$

K_N^0 is a matrix of order $[N + (S+1)(m_1+m_2)] \times [(S+1)(m_1+m_2+1)]$

$$K_N^1 = \begin{pmatrix} -(\lambda_1+\gamma)I_N & 0 & 0 \\ T_N & F_3 & 0 \\ U_N & 0 & F_5 \end{pmatrix},$$

K_N^1 is a square matrix of order $N + (S+1)(m_1+m_2)$

$\lambda = \lambda_1 + \lambda_2$,

$$C_j = \begin{pmatrix} 0 & 0 \\ I_{S-j+1} \otimes T^0 \otimes \beta & 0 \end{pmatrix}, \text{ for } j = 1, 2, ..., N$$

$$C = \begin{pmatrix} 0 & 0 \\ I_{S-N+1} \otimes U^0 \otimes \beta & 0 \end{pmatrix},$$

$$T^j = \begin{pmatrix} I_j \otimes T^0 \\ 0 \end{pmatrix}, \text{ for } j = 1, 2, ..., N$$

$$U^N = \begin{pmatrix} I_N \otimes U^0 \\ 0 \end{pmatrix},$$

$$\Theta^j = \begin{pmatrix} \theta I_j \\ 0 \end{pmatrix}, \text{ for } j = 1, 2, ..., N-1$$

$Y_j = e_j \otimes Y_j^1$ where $Y_j^1 = \begin{pmatrix} O & \gamma\beta & O \end{pmatrix}$, for $j = 1, 2, ..., N$

$$F_0 = \begin{pmatrix} I_{s+1}F_{01} & 0 & \Gamma \\ 0 & I_{S-s-1}F_{02} & 0 \\ 0 & 0 & F_{02} \end{pmatrix},$$

$$F_1 = \begin{pmatrix} I_{s+1} \otimes F_{11} & 0 & \Gamma_1 \\ 0 & I_{S-s-1} \otimes F_{12} & 0 \\ 0 & 0 & F_{12} \end{pmatrix},$$

$$F_2 = \begin{pmatrix} I_{s+1} \otimes F_{21} & 0 & \Gamma_2 \\ 0 & I_{S-s-1} \otimes F_{22} & 0 \\ 0 & 0 & F_{22} \end{pmatrix},$$

$$F_3 = \begin{pmatrix} I_{s+1} \otimes F_{31} & 0 & \Gamma_1 \\ 0 & I_{S-s-1} \otimes F_{32} & 0 \\ 0 & 0 & F_{32} \end{pmatrix},$$

$$F_4 = \begin{pmatrix} I_{s+1} \otimes F_{41} & 0 & \Gamma_1 \\ 0 & I_{S-s-1} \otimes F_{42} & 0 \\ 0 & 0 & F_{42} \end{pmatrix},$$

$$F_5 = \begin{pmatrix} I_{s+1} \otimes F_{51} & 0 & \Gamma_2 \\ 0 & I_{S-s-1} \otimes F_{52} & 0 \\ 0 & 0 & F_{52} \end{pmatrix},$$

$$F_6 = \begin{pmatrix} 0 & 0 \\ I_S \otimes \lambda_1\alpha & 0 \end{pmatrix},$$

$$F_7 = \begin{pmatrix} I_{s+1}F_{71} & 0 & \Gamma \\ 0 & I_{S-s-1}F_{72} & 0 \\ 0 & 0 & F_{72} \end{pmatrix},$$

$$F^j = \begin{pmatrix} 0 & 0 \\ I_{S-j-1} \otimes \theta\beta & 0 \end{pmatrix}, \text{ for } j = 1, 2, ..., (N-2)$$

$$F^{N-1} = \begin{pmatrix} 0 & 0 \\ F_{12}^{N-1} & 0 \end{pmatrix},$$

$$\Lambda = \begin{pmatrix} \lambda_2 I_N \\ 0 \end{pmatrix},$$

$$\Gamma = e_{s+1} \otimes \gamma$$

$$\Gamma_j = e_{s+1} \otimes \gamma I_{m_j} \text{ , for } j = 1, 2 \text{ ,}$$

$$F_{01} = -(\lambda + \gamma),\ F_{02} = -\lambda$$

$$F_{11} = T - (\lambda + \gamma)I_{m_1},\ F_{12} = T - \lambda I_{m_1},$$

$$F_{21} = U - (\lambda + \gamma)I_{m_2},\ F_{22} = U - \lambda I_{m_2},$$

$$F_{31} = T - (\lambda_1 + \gamma)I_{m_1},\ F_{32} = T - \lambda_1 I_{m_1},$$

$$F_{41} = T - (\lambda + \gamma + \theta)I_{m_1},\ F_{42} = T - (\lambda + \theta)I_{m_1},$$

$$F_{51} = U - (\lambda_1) + \gamma)I_{m_2},\ F_{52} = U - \lambda_1 I_{m_2}$$

$$F_{71} = -(\lambda + \gamma + \theta),\ F_{72} = T - (\lambda + \theta),$$

$$F_{12}^{N-1} = I_{S-N+2} \otimes \theta\beta + \begin{pmatrix} 0 & 0 \\ I_{S-N+1} \otimes \lambda_2\beta & 0 \end{pmatrix}$$

$$A_1 = \begin{pmatrix} L_0^1 & L_0^2 & 0 & 0 & \dots & \dots \\ L_1^0 & L_1^1 & L_1^2 & 0 & \dots & \dots \\ L_2^0 & 0 & L_2^1 & L_2^2 & 0 & \dots \\ L_3^0 & 0 & 0 & L_3^1 & L_3^2 & \dots \\ \vdots & \vdots & \vdots & \ddots & \ddots & \\ L_{N-1}^0 & \dots & & & L_{N-1}^1 & L_{N-1}^2 \\ L_N^0 & \dots\dots & & & 0 & L_N^1 \end{pmatrix} \quad (7)$$

where

$$L_0^1 = \begin{pmatrix} -(\lambda + \gamma) & 0 & 0 \\ T^1 & F_1 & 0 \\ U^1 & 0 & F_2 \end{pmatrix},$$

L_0^1 is a square matrix of order $1 + (S+1)(m_1 + m_2)$

$$L_0^2 = \begin{pmatrix} 0 & \lambda_2 & 0 & 0 \\ 0 & 0 & 0 & I_{S+1} \otimes \lambda_2 I_{m_1+m_2} \end{pmatrix},$$

L_0^2 is a matrix of order $[1 + (S+1)(m_1 + m_2)] \times [2 + (S+1)(2m_1 + m_2)]$

$$L_j^0 = \begin{pmatrix} 0 & Y_j \\ 0 & 0 \\ 0 & C_j \\ 0 & 0 \end{pmatrix}, \text{ for } j = 1, 2, ..., (N-1)$$

L_j^0 are matrices of order $[j + 1 + (S+1)(2m_1 + m_2)] \times [1 + (S+1)(m_1 + m_2)]$

$$L_j^1 = \begin{pmatrix} -(\lambda_1 + \gamma)I_j & 0 & 0 & 0 & 0 \\ \theta_j & -(\lambda + \gamma + \theta) & 0 & 0 & 0 \\ T^j & 0 & F_3 & 0 & 0 \\ 0 & T^1 & I_{S+1} \otimes \theta I_{m_1} & F_4 & 0 \\ 0 & U^1 & 0 & 0 & F_2 \end{pmatrix}, \text{ for } j = 1, 2, ..., (N-1)$$

L_j^1 are square matrices of order $j+1+(S+1)(2m_1+m_2)$

$$L_j^2 = \begin{pmatrix} 0 & 0 & 0 & 0 \\ 0 & \lambda_2 & 0 & 0 \\ 0 & 0 & 0 & 0 \\ 0 & 0 & 0 & I_{S+1} \otimes \lambda_2 I_{m_1+m_2} \end{pmatrix}, \text{ for } j = 1, 2, ..., (N-2)$$

L_j^2 are matrices of order $[j+1+(S+1)(2m_1+m_2)] \times [j+2+(S+1)(2m_1+m_2)]$

$$L_{N-1}^2 = \begin{pmatrix} 0 & 0 \\ \Lambda_N & 0 \\ 0 & 0 \\ 0 & I_{S+1} \otimes \lambda_2 I_{m_1+m_2} \end{pmatrix},$$

L_{N-1}^2 is a matrix of order $[N+(S+1)(2m_1+m_2)] \times [N+(S+1)(m_1+m_2)]$

$$L_N^0 = \begin{pmatrix} 0 & Y_N \\ 0 & C_N \\ 0 & C \end{pmatrix},$$

L_N^0 is a matrix of order $[N+(S+1)(m_1+m_2)] \times [1+(S+1)(m_1+m_2)]$

$$L_N^1 = \begin{pmatrix} -(\lambda_1+\gamma)I_N & 0 & 0 \\ T_N & F_3 & 0 \\ U_N & 0 & F_5 \end{pmatrix},$$

L_N^1 is a square matrix of order $[N+(S+1)(m_1+m_2)]$

$$T^1 = \begin{pmatrix} T^0 \\ 0 \\ \vdots \\ 0 \end{pmatrix}, U^1 = \begin{pmatrix} U^0 \\ 0 \\ \vdots \\ 0 \end{pmatrix},$$

$\theta_j = (\theta\, 0 \cdots 0)$, for $j = 1, 2, ..., N-1$ and θ_j is a matrix of order $1 \times j$

$\Lambda_N = \lambda_2[e_N(1)]'$, where $'$ denotes transpose

e_r denotes column vector of dimension r consisting of all 1's

$e_r j$ denotes column vector of dimension r with 1 in the j th position and 0 elsewhere

3.1 Stability Condition

The Markov chain with generator Q is positive recurrent if and only if

$$\lambda_1 < \sum_{j=0}^{N} \pi_j H_1^j \mathbf{e} \tag{8}$$

where the stationary vector π of A is obtained by solving

$$\pi A = 0; \pi \mathbf{e} = 1. \tag{9}$$

where the matrix A be defined as $A = A_0 + A_1 + A_2$.

3.2 Stationary Distribution

The stationary distribution of the Markov process under consideration is obtained by solving the set of equations

$$\mathbf{x}\ Q = 0; \mathbf{xe} = 1. \tag{10}$$

Let $\mathbf{x}$ be decomposed in conformity with Q. Then
$\mathbf{x} = (\mathbf{x_0}, \mathbf{x_1}, \mathbf{x_2}, \dots)$ where $\mathbf{x_i} = (\mathbf{x_{i0}}, \mathbf{x_{i1}}, \dots\dots \mathbf{x_{iN}})$

$$\mathbf{x_{ij}} = (\mathbf{x_{ij0}}, \mathbf{x_{ij1}}, \mathbf{x_{ij2}})$$

for $j = 1, 2, \dots, N$ whereas for $k = 0, 1, 2$,the vectors

$$\mathbf{x_{ijk}} = (\mathbf{x_{ijk0}}, \mathbf{x_{ijk1}})$$

$\mathbf{x_{ijkl}} = (\mathbf{x_{ijkl1}}, \mathbf{x_{ijkl2}}, \dots\dots \mathbf{x_{ijklS}})$ for $l = 0, 1$

$$\mathbf{x_{ijklr}} = (x_{ijklr1}, x_{ijklr2}, \dots\dots x_{ijklrt})$$

where $t = m_k$. x_{ijklru} is the probability of being in state (i, j, k, l, r, u) for $i \geq 0 : j = 1, 2, \dots, N; k = 1, 2; l = 0, 1; 0 \leq r \leq S; u = 1, 2, \dots, m_k$ and x_{ij0lr} is the probability of being in state $(i, j, 0, l, r)$. From $\mathbf{x}Q = 0$, we get the following equations:

$$\mathbf{x_0}B_{00} + \mathbf{x_1}B_{10} = 0 \tag{11}$$

$$\mathbf{x_0}B_{01} + \mathbf{x_1}A_1 + \mathbf{x_2}A_2 = 0 \tag{12}$$

$$\mathbf{x_1}A_0 + \mathbf{x_2}A_1 + \mathbf{x_3}A_2 = 0 \tag{13}$$

$$\mathbf{x_{i-1}}A_0 + \mathbf{x_i}A_1 + \mathbf{x_{i+1}}A_2 = 0, i = 2, 3, .. \tag{14}$$

It may be shown that there exists a constant matrix R such that

$$\mathbf{x_i} = \mathbf{x_{i-1}}R, i = 2, 3, ... \tag{15}$$

The sub vectors $\mathbf{x_i}$ are geometrically related by the equation

$$\mathbf{x_i} = \mathbf{x_1}R^{i-1}, i = 2, 3, ... \tag{16}$$

R can be obtained from the matrix quadratic equation

$$R^2A_2 + RA_1 + A_0 = O \tag{17}$$

4 Performance Measures

In this section we evaluate some performance measures of the system.

1. Expected number of type-I customers in the system

$$E[N_1] = \sum_{i=0}^{\infty} i\mathbf{x_i}\mathbf{e} \tag{18}$$

2. Expected number of type-II customers in the system

$$E[N_2] = \sum_{i=0}^{\infty}\sum_{j=0}^{N} j\mathbf{x_{ij}e} \tag{19}$$

3. Expected number of items in the inventory

$$E[I] = \sum_{i=0}^{\infty}\sum_{j=0}^{N}\sum_{k=0}^{2}\sum_{l=0}^{1}\sum_{r=0}^{S} r\mathbf{x_{ijklr}e} \tag{20}$$

4. Expected number of customers waiting in the system due to lack of inventory

$$E[W] = \sum_{i=0}^{\infty}\sum_{j=0}^{N}\sum_{l=1}^{2} i\mathbf{x_{ij0lo}e} + \sum_{i=0}^{\infty}\sum_{j=0}^{N}\sum_{l=1}^{2} j\mathbf{x_{ij0o}e} \tag{21}$$

5. Probability that the server is idle

$$b_0 = \sum_{i=0}^{\infty}\sum_{j=0}^{N}\mathbf{x_{ij0}e} \tag{22}$$

6. Probability that the server is busy with type-I customer

$$b_1 = \sum_{i=0}^{\infty}\sum_{j=0}^{N}\mathbf{x_{ij1}e} \tag{23}$$

7. Probability that the server is busy with type-II customer

$$b_2 = \sum_{i=0}^{\infty}\sum_{j=0}^{N}\mathbf{x_{ij2}e} \tag{24}$$

8. Probability that the clock is on

$$c_1 = \sum_{i=0}^{\infty}\sum_{j=0}^{N}\sum_{k=0}^{2}\mathbf{x_{ijk1}e} \tag{25}$$

9. Expected rate at which replenishment of inventory occurs

$$E_R = \sum_{i=0}^{\infty}\sum_{j=0}^{N}\sum_{k=0}^{2}\sum_{l=0}^{1}\sum_{r=0}^{s} r\mathbf{x_{ijklr}e} \tag{26}$$

10. Probability that the type-II customer is blocked from entering the service

$$p_b = \sum_{i=0}^{\infty}\sum_{j=1}^{N-1}\sum_{k=1}^{2}\sum_{t=0}^{S}\mathbf{x_{ijkot}e} + \sum_{i=0}^{\infty}\mathbf{x}_{iN}\mathbf{e} \tag{27}$$

5 Conclusion

In this paper, we considered a single server queueing inventory model with two channels of service. Service to both channels is provided by a single server. Various performance measures are evaluated at steady state conditions. We plan to analyse the problem for cost effectiveness.

References

1. Sigman, K., Simchi-Levi, D.: Light traffic heutrestic for an M/G/1 queue with limited inventory. Ann. Oper. Res. **40**, 371–380 (1992)
2. Krishnamoorthy, A., Shajin, D., Narayanan, V.C.: Inventory with positive service time: a survey. Advanced Trends in Queueing Theory: Series of Books "Mathematics and Statistics", Sciences. ISTE & Wiley, London (2019). Opsearch **48**(2), 153–169 (2011)
3. Chakravarthy, S.R., Maity, A., Gupta, U.C.: An (s,S) inventory in a queueing system with batch service facility. Ann. Oper. Res. **258**(2), 263–283 (2015). https://doi.org/10.1007/s10479-015-2041-z
4. Anbazhagan, N., Vigneshwaran, B., Jeganathan, K.: Stochastic inventory system with two types of services. Int. J. Adv. Appl. Math. Mech. **2**(1), 120–127 (2014)
5. Krishnamoorthy, A., Joshua, V.C., Mathew, A.P.: A retrial queueing system with abandonment and search for priority customers. In: Vishnevskiy, V.M., Samouylov, K.E., Kozyrev, D.V. (eds.) DCCN 2017. CCIS, vol. 700, pp. 98–107. Springer, Cham (2017). https://doi.org/10.1007/978-3-319-66836-9_9
6. Krishnamoorthy, A., Benny, B., Shajin, D.: A revisit to queueing-inventory system with reservation, cancellation and common life time. Opsearch **54**(2), 336–350 (2016). https://doi.org/10.1007/s12597-016-0278-1
7. Shajin, D., Lakshmy, B., Manikandan, R.: On a two stage queueing-inventory system with rejection of customers. Neural Parallel Sci. Comput. **23**, 111–128 (2015)
8. Krishnamoorthy, A., Manikandan, R., Lakshmy, B.: A revisit to queueing inventory system with positive service time. Ann. Oper. Res. **233**(1), 221–236 (2013)

The Analytical Model of Six-Dimensional Linear Dynamic Systems with Arbitrary Piecewise-Constant Parameters

K. A. Vytovtov[1,2(✉)], E. A. Barabanova[1,2], V. M. Vishnevsky[1], and I.Yu. Kvyatkovskaya[2]

[1] V.A. Trapeznikov Institute of Control Sciences RAS, Profsoyuznaya 65 street, Moscow, Russia
vytovtov_konstan@mail.ru

[2] Astrakhan State Technical University, Tatischva 16 street, Astrakhan, Russia

Abstract. In this paper the 6×6 fundamental matrix of a linear six - dimensional system with arbitrary piecewise-constant parameters is obtained in the analytical form in elementary functions for the first time. To write this matrix the new sign-function $F_q^{(i,j)}$ is introduced and described here. This function determines the order of interaction of eigen-oscillations of intervals with constant parameters in a resulting equivalent oscillation. The fundamentally new concept of equivalent oscillations for six-dimensional systems is introduced for the first time too. The stability problem of these systems is considered and the stability conditions are obtained in the analytical form in original parameters of a linear six-dimensional system. The results obtained allow solving a number of problems in mechanics, electrical engineering, communication systems, optoelectronics, automatic control theory and information theory. It can be, for example, the problems of determining the stability of dynamical systems, the problems of synthesizing dynamical systems of various nature, etc.

Keywords: Fundamental matrix · Six-dimensional system · Dynamical system

1 Introduction

Investigation of dynamic systems with three degrees of freedom is a very important problem for applied mechanics, microwave technique, quantum physics, communication theory [1–10]. Three coupled mechanical oscillators with varying masses or spring stiffnesses can be considered as the simplest example of a six-dimensional linear dynamic system. The system of three connected oscillatory circuits are analogous electrical system. In microwave theory and optics a stratified anisotropic structure with all non-zero elements in permittivity (permeability) dyadic can also be considered as a wave analogue of a considering system.

The reported study was funded by RFBR, project number 19-29-06043.

V. M. Vishnevskiy et al. (Eds.): DCCN 2020, LNCS 12563, pp. 617–630, 2020.
https://doi.org/10.1007/978-3-030-66471-8_47

In literature these systems can also be called as six-dimensional or sixth-order ones.

Such systems with constant parameters have been written in detail, for example, in [1–10]. However, linear parametric systems are least studied for today. Indeed, numerical methods only have been used to study their behaviour [7–9]. And it is well-known that there are no an analytical method of investigating six-dimensional dynamic systems with variable parameters. At the same time the accurate analytical methods have been widely applied to systems with one- and two-degrees of freedom [11–16]. In [11] the two-dimensional periodic electromagnetic system has been considered, the fundamental matrix is obtained in the analytical form, and the sign-function $f_{q,i}$ has been introduced for the first time. In [12] the results of [11] have been generalized for systems of any nature. The analytical method for a linear four-dimensional system can be found, for example, in [13]. The application of fundamental matrix to investigation electromagnetic structures has been demonstrated in [14–16]. Moreover, in [14] the sign-function $F_{p,i}$ has been introduced for description of considered systems.

In this paper the authors present the 6×6 fundamental matrix of a linear six-dimensional dynamic system with arbitrary piecewise-constant parameters for the first time. Note, that in the considered problem, any parameter of a system can jump at arbitrary moment of time. It also is very important to note that the solution is obtained in the analytical form in elementary functions, and the resulting matrix is expressed in original system parameters. To obtain the general analytical form of this matrix the authors introduce the new sign-function describing the interaction law of eigenmodes of intervals with constant parameters. The stability problem of a six-dimensional system with piecewise parameters is solved here too, and the stability conditions are obtained in analytical form in original system parameters [17,18].

2 Statement of the Problem

In this paper we consider a linear six-dimensional dynamic system described by the six order differential equation

$$\frac{d^6U(t)}{dt^6} + a_4(t)\frac{d^4U(t)}{dt^4} + a_2(t)\frac{d^2U(t)}{dt^2} + a_0(t)U = 0 \tag{1}$$

with the piecewise constant coefficients

$$a_s(t) = \begin{cases} a_{s1} & 0 < t < t_1^{(s)} \\ a_{s2} & t_1^{(s)} < t < t_2^{(s)} \\ \ldots & \ldots \\ a_{sN} & t_{N-1}^{(s)} < t < t_N^{(s)} \end{cases} \tag{2}$$

The coefficients a_0, a_2, a_4 can jump at any t by any value, and these coefficients are independent each other (Fig. 1).

Our main purpose is finding the fundamental matrix of (1) with the arbitrary piecewise-constant coefficients (2) in an analytical form in elementary functions.

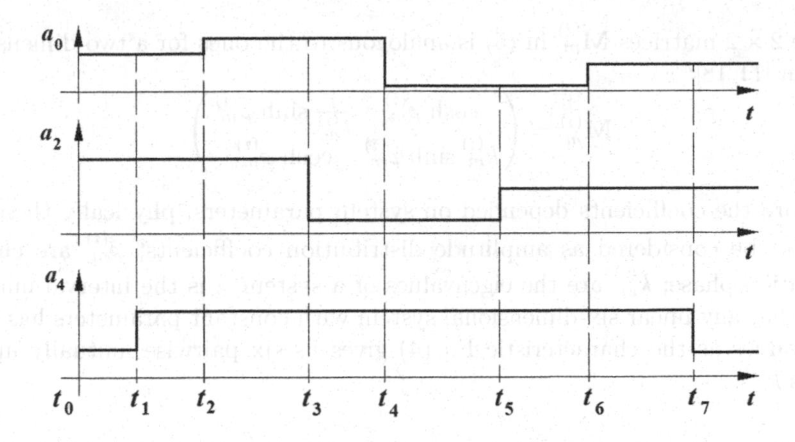

Fig. 1. The example of system parameters

Here we must take into account the well-known fact that a fundamental matrix of a linear system with piecewise-constant parameters is the product of matrices of intervals with constant parameters. This multiplication makes it possible to satisfy the boundary conditions for the function and its derivatives (from the first to the fifth) [17,18]. Moreover it is important to study the stability problem for periodic linear six-dimensional systems.

3 The System with Constant Parameters

The equation describing a linear six-dimensional system on an interval with constant parameters is

$$\frac{d^6U(t)}{dt^6} + a_4^{(i)}\frac{d^4U(t)}{dt^4} + a_2^{(i)}\frac{d^2U(t)}{dt^2} + a_0^{(i)}U = 0 \tag{3}$$

and it can be solved analytically. Here $a_0 = const$, $a_2 = const$, $a_4 = const$. Indeed, the characteristic equation

$$[k^{(i)}]^6 + a_4^{(i)}[k^{(i)}]^4 + a_2^{(i)}[k^{(i)}]^2 + a_0^{(i)} = 0 \tag{4}$$

can be solved analytically by using the Cardano method [19]. Then in accordance to the method [17] we can write the fundamental matrix of the linear homogeneous differential equation with constant coefficients (3) as

$$\mathbf{L}^{(i)} = \begin{pmatrix} \sum_{m=1}^{3} \gamma_{11m}^{(i)}\mathbf{M}_m^{(i)} & \sum_{m=1}^{3} \gamma_{12m}^{(i)}\mathbf{M}_m^{(i)} & \sum_{m=1}^{3} \gamma_{13m}^{(i)}\mathbf{M}_m^{(i)} \\ \sum_{m=1}^{3} \gamma_{21m}^{(i)}\mathbf{M}_m^{(i)} & \sum_{m=1}^{3} \gamma_{22m}^{(i)}\mathbf{M}_m^{(i)} & \sum_{m=1}^{3} \gamma_{23m}^{(i)}\mathbf{M}_m^{(i)} \\ \sum_{m=1}^{3} \gamma_{31m}^{(i)}\mathbf{M}_m^{(i)} & \sum_{m=1}^{3} \gamma_{32m}^{(i)}\mathbf{M}_m^{(i)} & \sum_{m=1}^{3} \gamma_{33m}^{(i)}\mathbf{M}_m^{(i)} \end{pmatrix} \tag{5}$$

where 2×2 matrices $\mathbf{M}_m^{(i)}$ in (5) is analogous to the ones for a two-dimensional system [11,18]:

$$\mathbf{M}_m^{(i)} = \begin{pmatrix} \cosh\varphi_m^{(i)} & \frac{1}{k_m^{(i)}}\sinh\varphi_m^{(i)} \\ k_m^{(i)}\sinh\varphi_m^{(i)} & \cosh\varphi_m^{(i)} \end{pmatrix} \tag{6}$$

$\gamma_{klm}^{(i)}$ are the coefficients depended on system parameters, physically these values can be considered as amplitude distribution coefficients; $\varphi_m^{(i)}$ are current oscillation phase; $k_m^{(i)}$ are the eigenvalues of a system, i is the interval number. Note that any linear six-dimensional system with constant parameters has three eigenvalues as the characteristic Eq. (4) gives us six pairwise mutually inverse values $k^{(i)}$.

4 The System with Piecewise-Constant Parameters

Now let us consider a six-dimensional linear system with arbitrary piecewise-constant parameters (for example, Fig. 1). The behaviour of such a system is described by the differential Eq. (1) with the coefficients (2). It is very important that any of the coefficients (1) can change its value abruptly at any time by any value.

In accordance to the method the fundamental matrix for the considered case must be found as the product of matrices (5) of intervals with constant parameters [17,18]

$$\mathbf{L}_N(T) = \prod_{i=1}^{N} \mathbf{L}_i \tag{7}$$

Here i is the interval number, N is the number of the intervals, $\mathbf{L}_i$ is the fundamental matrix of the i-th interval, T is the resulting interval containing all intervals with constant parameters. However, finding the resulting matrix in accordance with (7) gives a sum of products of hyperbolic functions for its elements. As the result, further analytical study of the fundamental matrix $\mathbf{L}_N(T)$ is impossible. Thus, the analytical study of six-dimensional systems requires additional transformations of (7).

Theorem 1. *The matrix of fundamental solutions of a linear ordinary differential equation of the sixth order (1) with arbitrary piecewise-constant coefficients (2) on the interval T is the finite sum of $3 \cdot 6^{N-1}$ unimodular hyperbolic 6×6 matrices*

$$\mathbf{L}_N(T) = \sum_{p=1}^{3^N} \eta_p \sum_{q=1}^{2^{N-1}} \xi_{pq} \mathbf{L}_{pq} \tag{8}$$

with the contribution coefficients

$$\eta_p = \sqrt{\det|\zeta_{m,n}^{(N)}|_1^3} \tag{9}$$

and

$$\xi_{pq} = \frac{1}{2^{N-1}}\sqrt{\frac{\sum_{j=1}^{3} k_j^{(1)} f_q^{(1)} F_p^{(1,j)}}{\sum_{j=1}^{3} k_j^{(N)} f_q^{(N)} F_p^{(N,j)}}} \times \prod_{i=1}^{N}\left(1+\frac{\sum_{j=1}^{3} k_j^{(i+1)} f_q^{(i+1)} F_p^{(i+1,j)}}{\sum_{j=1}^{3} k_j^{(i)} f_q^{(i)} F_p^{(i,j)}}\right) \tag{10}$$

where N is the number of intervals with constant parameters on the interval T. The unimodular 6×6 *matrix in (8)* $\mathbf{L}_{pq}$ *is*

$$\mathbf{L}_{pq} = \begin{pmatrix} \frac{\zeta_p^{1,1}}{\sqrt{\det|\zeta_{mn}^{(N)}|_1^3}}\mathbf{M}_{pq} & \frac{\zeta_p^{1,2}}{\sqrt{\det|\zeta_{mn}^{(N)}|_1^3}}\mathbf{M}_{pq} & \frac{\zeta_p^{1,3}}{\sqrt{\det|\zeta_{mn}^{(N)}|_1^3}}\mathbf{M}_{pq} \\ \frac{\zeta_p^{2,1}}{\sqrt{\det|\zeta_{mn}^{(N)}|_1^3}}\mathbf{M}_{pq} & \frac{\zeta_p^{2,2}}{\sqrt{\det|\zeta_{mn}^{(N)}|_1^3}}\mathbf{M}_{pq} & \frac{\zeta_p^{2,3}}{\sqrt{\det|\zeta_{mn}^{(N)}|_1^3}}\mathbf{M}_{pq} \\ \frac{\zeta_p^{3,1}}{\sqrt{\det|\zeta_{mn}^{(N)}|_1^3}}\mathbf{M}_{pq} & \frac{\zeta_p^{3,2}}{\sqrt{\det|\zeta_{mn}^{(N)}|_1^3}}\mathbf{M}_{pq} & \frac{\zeta_p^{3,3}}{\sqrt{\det|\zeta_{mn}^{(N)}|_1^3}}\mathbf{M}_{pq} \end{pmatrix} \tag{11}$$

the elements of the 2×2 *matrices* $\mathbf{M}_{pq}$ *in (11) are*

$$\begin{aligned} M_{pq}^{1,1} &= \sqrt{\frac{\sum_{j=1}^{3} k_j^{(N)} f_q^{(N)} F_p^{(N,j)}}{\sum_{j=1}^{3} k_j^{(1)} f_q^{(1)} F_p^{(1,j)}}}\cosh\psi_{pq} \\ M_{pq}^{1,2} &= \frac{f_q^{(N)}}{\sqrt{\sum_{j=1}^{3} k_j^{(1)} f_q^{(1)} F_p^{(1,j)} \sum_{j=1}^{3} k_j^{(N)} f_q^{(N)} F_p^{(N,j)}}}\sinh\psi_{pq} \\ M_{pq}^{2,1} &= \sqrt{\sum_{j=1}^{3} k_j^{(1)} f_q^{(1)} F_p^{(1,j)} \sum_{j=1}^{3} k_j^{(N)} f_q^{(N)} F_p^{(N,j)}}\sinh\psi_{pq} \\ M_{pq}^{2,2} &= f_q^{(N)}\sqrt{\frac{\sum_{j=1}^{3} k_j^{(1)} f_q^{(1)} F_p^{(1,j)}}{\sum_{j=1}^{3} k_j^{(N)} f_q^{(N)} F_p^{(N,j)}}}\cosh\psi_{pq} \end{aligned} \tag{12}$$

$$|\zeta_{m,n}^{(N)}|_1^3 = \prod_{i=1}^{N}\left[\sum_{j=1}^{3}\left(|\zeta_{m,n}^{(i)}|_1^3 F_p^{(i,j)}\right)\right] \tag{13}$$

$$\psi_{pq} = \sum_{i=1}^{N}\left[\sum_{j=1}^{3}\left(k_j^i f_q^i F_p^{(i,j)} t_i\right)\right] \tag{14}$$

f_q^i *and* $F_p^{(i,j)}$ *are the sign-functions:*

$$f_{q,i} = \operatorname{sign}\{\sin[\frac{\pi}{2^{N+1-i}}(2q-1)]\} \tag{15}$$

$$F_p^{(i,j)} = \frac{1}{2}\langle 1 + \mathrm{sign}\{\sin[3^{i-1-N}\pi\left(p - \frac{1}{2}\right) + \frac{\pi}{3}(2-j)] \times \tan[\frac{3^{i-N}}{2}\pi\left(p - \frac{1}{2}\right) - \frac{\pi}{2}(j-1)]\}\rangle \tag{16}$$

Proof. To prove the theorem, we use here the method of mathematical induction. First, let us find the fundamental matrix for two intervals with constant parameters as the product of the interval matrix (5):

$$\mathbf{L}^{(2)} = \begin{pmatrix} \mathbf{L}_{11}^{(2)} & \mathbf{L}_{12}^{(2)} & \mathbf{L}_{13}^{(2)} \\ \mathbf{L}_{21}^{(2)} & \mathbf{L}_{22}^{(2)} & \mathbf{L}_{23}^{(2)} \\ \mathbf{L}_{31}^{(2)} & \mathbf{L}_{32}^{(2)} & \mathbf{L}_{33}^{(2)} \end{pmatrix} \tag{17}$$

where 2×2 blocks in (17) are

$$\begin{aligned}
\mathbf{L}_{11}^{(2)} &= \sum_{n=1}^{3}\sum_{m=1}^{3}\left(\gamma_{11n}^{(1)}\gamma_{11m}^{(2)} + \gamma_{12n}^{(1)}\gamma_{21m}^{(2)} + \gamma_{13n}^{(1)}\gamma_{31m}^{(2)}\right)\mathbf{M}_n^{(1)}\mathbf{M}_m^{(2)} \\
&= \zeta_{11(nm)}^{(2)}\mathbf{M}_n^{(1)}\mathbf{M}_m^{(2)} \\
\mathbf{L}_{12}^{(2)} &= \sum_{n=1}^{3}\sum_{m=1}^{3}\left(\gamma_{11n}^{(1)}\gamma_{12m}^{(2)} + \gamma_{12n}^{(1)}\gamma_{22m}^{(2)} + \gamma_{13n}^{(1)}\gamma_{32m}^{(2)}\right)\mathbf{M}_n^{(1)}\mathbf{M}_m^{(2)} \\
&= \zeta_{12(nm)}^{(2)}\mathbf{M}_n^{(1)}\mathbf{M}_m^{(2)} \\
\mathbf{L}_{13}^{(2)} &= \sum_{n=1}^{3}\sum_{m=1}^{3}\left(\gamma_{11n}^{(1)}\gamma_{13m}^{(2)} + \gamma_{12n}^{(1)}\gamma_{23m}^{(2)} + \gamma_{13n}^{(1)}\gamma_{33m}^{(2)}\right)\mathbf{M}_n^{(1)}\mathbf{M}_m^{(2)} \\
&= \zeta_{13(nm)}^{(2)}\mathbf{M}_n^{(1)}\mathbf{M}_m^{(2)}
\end{aligned}$$

$$\begin{aligned}
\mathbf{L}_{21}^{(2)} &= \sum_{n=1}^{3}\sum_{m=1}^{3}\left(\gamma_{21n}^{(1)}\gamma_{11m}^{(2)} + \gamma_{22n}^{(1)}\gamma_{21m}^{(2)} + \gamma_{23n}^{(1)}\gamma_{31m}^{(2)}\right)\mathbf{M}_n^{(1)}\mathbf{M}_m^{(2)} \\
&= \zeta_{21(nm)}^{(2)}\mathbf{M}_n^{(1)}\mathbf{M}_m^{(2)} \\
\mathbf{L}_{22}^{(2)} &= \sum_{n=1}^{3}\sum_{m=1}^{3}\left(\gamma_{21n}^{(1)}\gamma_{12m}^{(2)} + \gamma_{22n}^{(1)}\gamma_{22m}^{(2)} + \gamma_{23n}^{(1)}\gamma_{32m}^{(2)}\right)\mathbf{M}_n^{(1)}\mathbf{M}_m^{(2)} \\
&= \zeta_{22(nm)}^{(2)}\mathbf{M}_n^{(1)}\mathbf{M}_m^{(2)} \\
\mathbf{L}_{23}^{(2)} &= \sum_{n=1}^{3}\sum_{m=1}^{3}\left(\gamma_{21n}^{(1)}\gamma_{13m}^{(2)} + \gamma_{22n}^{(1)}\gamma_{23m}^{(2)} + \gamma_{23n}^{(1)}\gamma_{33m}^{(2)}\right)\mathbf{M}_n^{(1)}\mathbf{M}_m^{(2)} \\
&= \zeta_{23(nm)}^{(2)}\mathbf{M}_n^{(1)}\mathbf{M}_m^{(2)}
\end{aligned} \tag{18}$$

$$\begin{aligned}
\mathbf{L}_{31}^{(2)} &= \sum_{n=1}^{3}\sum_{m=1}^{3}\left(\gamma_{31n}^{(1)}\gamma_{11m}^{(2)} + \gamma_{32n}^{(1)}\gamma_{21m}^{(2)} + \gamma_{33n}^{(1)}\gamma_{31m}^{(2)}\right)\mathbf{M}_n^{(1)}\mathbf{M}_m^{(2)} \\
&= \zeta_{31(nm)}^{(2)}\mathbf{M}_n^{(1)}\mathbf{M}_m^{(2)} \\
\mathbf{L}_{32}^{(2)} &= \sum_{n=1}^{3}\sum_{m=1}^{3}\left(\gamma_{31n}^{(1)}\gamma_{12m}^{(2)} + \gamma_{32n}^{(1)}\gamma_{22m}^{(2)} + \gamma_{33n}^{(1)}\gamma_{32m}^{(2)}\right)\mathbf{M}_n^{(1)}\mathbf{M}_m^{(2)} \\
&= \zeta_{32(nm)}^{(2)}\mathbf{M}_n^{(1)}\mathbf{M}_m^{(2)} \\
\mathbf{L}_{33}^{(2)} &= \sum_{n=1}^{3}\sum_{m=1}^{3}\left(\gamma_{31n}^{(1)}\gamma_{13m}^{(2)} + \gamma_{32n}^{(1)}\gamma_{23m}^{(2)} + \gamma_{33n}^{(1)}\gamma_{33m}^{(2)}\right)\mathbf{M}_n^{(1)}\mathbf{M}_m^{(2)} \\
&= \zeta_{33(nm)}^{(2)}\mathbf{M}_n^{(1)}\mathbf{M}_m^{(2)}
\end{aligned}$$

Here the elements with the indexes (1) and (2) describe the parameters of first and second intervals correspondingly. Using the results presented in [11], the product of matrices $\mathbf{M}_n^{(1)}\mathbf{M}_m^{(2)}$ can be represented as the sum

$$\mathbf{M}_n^{(1)}\mathbf{M}_m^{(2)} = \sum_{q=1}^{4} \xi_{q(m,n)}\mathbf{L}_{q(m,n)} \tag{19}$$

where

$$\mathbf{L}_{q(m,n)} = \begin{pmatrix} \sqrt{\frac{k_{m,n}^{(2)}}{k_{m,n}^{(1)}}}\cosh\psi_{q,(m,n)} & \frac{f_{q,2}}{\sqrt{k_{m,n}^{(1)}k_{m,n}^{(2)}}}\sinh\psi_{q,(m,n)} \\ \sqrt{k_{m,n}^{(1)}k_{m,n}^{(2)}}\sinh\psi_{q,(m,n)} & \sqrt{\frac{k_{m,n}^{(1)}}{k_{m,n}^{(2)}}}\cosh\psi_{q,(m,n)} \end{pmatrix} \tag{20}$$

Note that the matrces (20) are unimodular. $k_{(m,n)}^{(1)}$, $k_{(m,n)}^{(2)}$ are the eigenvalues for the first and second layers. Analogously to [11] the value $\xi_{q(m,n)}$ is called as the contribution coefficient and it is defined as

$$\xi_{q(m,n)} = \frac{1}{2}\sqrt{\frac{k_m^{(1)}}{k_n^{(2)}}}\left(1+\frac{k_n^{(2)}}{k_m^{(1)}}\right) \tag{21}$$

$$\psi_{qm,n} = \varphi_m^{(1)} f_{q,1} + \varphi_n^{(2)} f_{q,2} \tag{22}$$

The coefficients $\zeta_{ab(mn)}^{(2)}$ in (18) are defined as the elements of the matrix

$$|\zeta_{mn}^{(2)}|_1^2 = \begin{pmatrix} \gamma_{11m}^{(1)} & \gamma_{12m}^{(1)} & \gamma_{13m}^{(1)} \\ \gamma_{21m}^{(1)} & \gamma_{22m}^{(1)} & \gamma_{23m}^{(1)} \\ \gamma_{31m}^{(1)} & \gamma_{32m}^{(1)} & \gamma_{33m}^{(1)} \end{pmatrix} \begin{pmatrix} \gamma_{11n}^{(2)} & \gamma_{12n}^{(2)} & \gamma_{13n}^{(2)} \\ \gamma_{21n}^{(2)} & \gamma_{22n}^{(2)} & \gamma_{23n}^{(2)} \\ \gamma_{31n}^{(2)} & \gamma_{32n}^{(2)} & \gamma_{33n}^{(2)} \end{pmatrix} \tag{23}$$

The sign-function

$$f_{q,i} = \text{sign}\{\sin[\frac{\pi}{2^{N+1-i}}(2q-1)]\} \tag{24}$$

has been described in detail in [11,12].

Now let us consider a system containing tree intervals with constant parameters

$$\mathbf{L}^{(3)} = \begin{pmatrix} \mathbf{L}_{11}^{(3)} & \mathbf{L}_{12}^{(3)} & \mathbf{L}_{13}^{(3)} \\ \mathbf{L}_{21}^{(3)} & \mathbf{L}_{22}^{(3)} & \mathbf{L}_{23}^{(3)} \\ \mathbf{L}_{31}^{(3)} & \mathbf{L}_{32}^{(3)} & \mathbf{L}_{33}^{(3)} \end{pmatrix} \tag{25}$$

For this system with three intervals, the 2×2 block $\mathbf{L}_{11}^{3}$ is

$$
\begin{aligned}
\mathbf{L}_{11}^{(3)} &= \sum_{l=1}^{3}\sum_{n=1}^{3}\sum_{m=1}^{3}\Big[\left(\gamma_{11n}^{(1)}\gamma_{11m}^{(2)}+\gamma_{12n}^{(1)}\gamma_{21m}^{(2)}+\gamma_{13n}^{(1)}\gamma_{31m}^{(2)}\right)\gamma_{11l}^{(3)} \\
&\quad+\left(\gamma_{11n}^{(1)}\gamma_{12m}^{(2)}+\gamma_{12n}^{(1)}\gamma_{22m}^{(2)}+\gamma_{13n}^{(1)}\gamma_{32m}^{(2)}\right)\gamma_{21l}^{(3)} \\
&\quad+\left(\gamma_{11n}^{(1)}\gamma_{13m}^{(2)}+\gamma_{12n}^{(1)}\gamma_{23m}^{(2)}+\gamma_{13n}^{(1)}\gamma_{33m}^{(2)}\right)\gamma_{31l}^{(3)}\Big]\mathbf{M}_{n}^{(1)}\mathbf{M}_{m}^{(2)}\mathbf{M}_{l}^{(3)} \\
&= \sum_{l=1}^{3}\sum_{n=1}^{3}\sum_{m=1}^{3}\zeta_{11(mnl)}^{(3)}\mathbf{M}_{n}^{(1)}\mathbf{M}_{m}^{(2)}\mathbf{M}_{l}^{(3)}
\end{aligned}
\tag{26}
$$

Other elements of the fundamental solution matrix can be found in a similar way. Now, we carry out mathematical transformations of (26). As the result the product $\mathbf{M}_{n}^{(1)}\mathbf{M}_{m}^{(2)}\mathbf{M}_{l}^{(3)}$ in (26) is following:

$$
\mathbf{M}_{n}^{(1)}\mathbf{M}_{m}^{(2)}\mathbf{M}_{l}^{(3)} = \sum_{q=1}^{4}\xi_{p}\mathbf{L}_{q(mnl)} \tag{27}
$$

where

$$
\mathbf{L}_{q(mnl)} = \begin{pmatrix} \sqrt{\dfrac{k_{mnl}^{(3)}}{k_{mnl}^{(1)}}}\cosh\psi_{q(mnl)} & \dfrac{f_{q,3}}{\sqrt{k_{mnl}^{(1)}k_{mnl}^{(3)}}}\sinh\psi_{q(mnl)} \\ \sqrt{k_{mnl}^{(1)}k_{mnl}^{(3)}}\sinh\psi_{q(mnl)} & f_{q,3}\sqrt{\dfrac{k_{mnl}^{(1)}}{k_{mnl}^{(3)}}}\cosh\psi_{q(mnl)} \end{pmatrix} \tag{28}
$$

$$
\psi_{qm,n} = \varphi_{m}^{(1)}f_{q,1} + \varphi_{n}^{(2)}f_{q,2} + \varphi_{3}^{(3)}f_{q,3} \tag{29}
$$

$$
\xi_{q(mn)} = \frac{1}{4}\sqrt{\frac{k_{m}^{(1)}}{k_{n}^{(3)}}}\left(1+\frac{k_{n}^{(2)}f_{q,2}}{k_{m}^{(1)}f_{q,1}}\right)\left(1+\frac{k_{n}^{(3)}f_{q,3}}{k_{m}^{(2)}f_{q,2}}\right) \tag{30}
$$

The coefficient $\zeta_{ab(mnl)}^{(3)}$ in (26) is equal to the corresponding element of the matrix

$$
\begin{aligned}
|\zeta_{mn}^{(3)}|_{1}^{3} &= \begin{pmatrix} \gamma_{11m}^{(1)} & \gamma_{12m}^{(1)} & \gamma_{13m}^{(1)} \\ \gamma_{21m}^{(1)} & \gamma_{22m}^{(1)} & \gamma_{23m}^{(1)} \\ \gamma_{31m}^{(1)} & \gamma_{32m}^{(1)} & \gamma_{33m}^{(1)} \end{pmatrix}\begin{pmatrix} \gamma_{11n}^{(2)} & \gamma_{12n}^{(2)} & \gamma_{13n}^{(2)} \\ \gamma_{21n}^{(2)} & \gamma_{22n}^{(2)} & \gamma_{23n}^{(2)} \\ \gamma_{31n}^{(2)} & \gamma_{32n}^{(2)} & \gamma_{33n}^{(2)} \end{pmatrix} \\
&\quad\times\begin{pmatrix} \gamma_{11l}^{(3)} & \gamma_{12l}^{(3)} & \gamma_{13l}^{(3)} \\ \gamma_{21l}^{(3)} & \gamma_{22l}^{(3)} & \gamma_{23l}^{(3)} \\ \gamma_{31l}^{(3)} & \gamma_{32l}^{(3)} & \gamma_{33l}^{(3)} \end{pmatrix}
\end{aligned}
\tag{31}
$$

Thus, considering a system containing N intervals with constant parameters we can write the expressions (8)–(14). Then we can write the analogous expressions of the fundamental matrix for a system containing $N+1$ intervals with constant parameters. Then it can be verified that the expressions for $N+1$ intervals take the form (8) after replacing $M=N+1$. Here we don't present all these transformations since they are very unwieldy.

5 The Equivalent Oscillations of a Six-Dimensional System

Taking into account the kind of the fundamental matrix (8) and the well-known fact that an unimodular matrix describes a harmonic oscillation, we can assume that the resulting oscillation of this system on the interval T is the superposition of $3 \cdot 6^{N-1}$ oscillations. In accordance with these points, let us introduce the new concepts.

Definition 1. The oscillation described by the matrix $\mathbf{L}_{pq}$ in (8) is an equivalent oscillation of a linear homogeneous six-dimensional system.

Definition 2. The oscillations with the same index p is the oscillation group.

Thus we have 3^N oscillation groups with indexes p including 2^{N-1} oscillations in each.

Definition 3. The coefficient ξ_{pq} is called as the contribution coefficient of q-th oscillation into the p-th group.

Definition 4. The coefficient η_p is called as the contribution coefficient of the p-th oscillation group into a resulting oscillation.

Note that in Lyapunov's theory the so-called characteristic exponents have been introduced in solving the stability problems of linear systems. In our problem we also introduce the analogous values to describe system behaviour. Thus let us introduce the next new concepts.

Definition 5. The value

$$\alpha_p = \frac{1}{T}\mathrm{Ln}\eta_p \tag{32}$$

is the characteristic exponent of the oscillation group of a six-dimensional system.

Definition 5. The value

$$\alpha_{pq} = \frac{1}{T}\mathrm{Ln}\xi_{pq} \tag{33}$$

is the characteristic exponent of the equivalent oscillation of a six-dimensional system. Here the complex logarithm is defined as [17]

$$\mathrm{Ln}a = \ln|a| + j(\beta + 2\pi k) \tag{34}$$

where $|a|$ is the module of the complex number, β is the argument of the complex number a, j is the imaginary unite, $k = 0, 1, 2, ...$ Taking into account the above definitions we can state the following theorem:

Theorem 2. *The resulting oscillation $\mathbf{U}_N(T)$ of a six-dimensional dynamic system with arbitrary piecewise-constant parameters on the interval T is the superposition of the $3 \times 6^{N-1}$ equivalent oscillations $\mathbf{U}_{pq}$ with the given contribution coefficients*

$$\mathbf{U}_N(T) = \sum_{p=1}^{3^N} e^{\alpha_p t} \sum_{q=1}^{2^{N-1}} e^{\alpha_{pq} t} \mathbf{U}_{pq} \tag{35}$$

Proof. The proof of this theorem is given above in Theorem 1 proof. Here additionally it is necessary take into account the Definition 1–Definition 5.

The equivalent wave approach is very important when it is solved the problem with approximating continuous parameters by step functions. Indeed, in this case, there are no real interfaces between intervals and the corresponding boundary phenomena.

6 The Sing-Functions

To develop the analytical expression of the fundamental matrix (8) the new sign-function $F_p^{i,j}$ is introduced in this treatment. It also is used the sign-function $f_q^{(i)}$ that have been described in detail by the authors in [11,12].

To describe the sign-function $F_p^{(i,j)}$ let us consider a system with three intervals with constant parameters. For the given case the phases in (17) are following

$$\begin{aligned}
\psi_{p=1,q=1} &= k_1^{(1)} t^{(1)} + k_1^{(2)} t^{(2)} + k_1^{(3)} t^{(3)} \\
\psi_{p=1,q=2} &= k_1^{(1)} t^{(1)} + k_1^{(2)} t^{(2)} - k_1^{(3)} t^{(3)} \\
\psi_{p=1,q=3} &= k_1^{(1)} t^{(1)} - k_1^{(2)} t^{(2)} + k_1^{(3)} t^{(3)} \\
\psi_{p=1,q=4} &= k_1^{(1)} t^{(1)} - k_1^{(2)} t^{(2)} - k_1^{(3)} t^{(3)} \\
&\cdots\cdots\cdots\cdots \\
\psi_{p=2,q=1} &= k_1^{(1)} t^{(1)} + k_1^{(2)} t^{(2)} + k_2^{(3)} t^{(3)} \\
\psi_{p=2,q=2} &= k_1^{(1)} t^{(1)} + k_1^{(2)} t^{(2)} - k_2^{(3)} t^{(3)} \\
\psi_{p=2,q=3} &= k_1^{(1)} t^{(1)} - k_1^{(2)} t^{(2)} + k_2^{(3)} t^{(3)} \\
\psi_{p=2,q=4} &= k_1^{(1)} t^{(1)} - k_1^{(2)} t^{(2)} - k_2^{(3)} t^{(3)} \\
&\cdots\cdots\cdots\cdots \\
\psi_{p=27,q=1} &= k_3^{(1)} t^{(1)} + k_3^{(2)} t^{(2)} + k_3^{(3)} t^{(3)} \\
\psi_{p=27,q=2} &= k_3^{(1)} t^{(1)} + k_3^{(2)} t^{(2)} - k_3^{(3)} t^{(3)} \\
\psi_{p=27,q=3} &= k_3^{(1)} t^{(1)} - k_3^{(2)} t^{(2)} + k_3^{(3)} t^{(3)} \\
\psi_{p=27,q=4} &= k_3^{(1)} t^{(1)} - k_3^{(2)} t^{(2)} - k_3^{(3)} t^{(3)}
\end{aligned} \tag{36}$$

To obtain the general form of (36) we can write the following expression

$$\begin{aligned}\psi_{pq} = {} & k_1^{(1)} t^{(1)} f_q^{(1)} F_p^{(1,1)} + k_2^{(1)} t^{(1)} f_q^{(1)} F_p^{(1,2)} + k_3^{(1)} t^{(1)} f_q^{(1)} F_p^{(1,3)} \\ & + k_1^{(2)} t^{(2)} f_q^{(2)} F_p^{(2,1)} + k_2^{(2)} t^{(2)} f_q^{(2)} F_p^{(2,2)} + k_3^{(2)} t^{(2)} f_q^{(2)} F_p^{(2,3)} \\ & + k_1^{(3)} t^{(3)} f_q^{(3)} F_p^{(3,1)} + k_2^{(3)} t^{(3)} f_q^{(3)} F_p^{(3,2)} + k_3^{(3)} t^{(3)} f_q^{(3)} F_p^{(3,3)}\end{aligned} \qquad (37)$$

The signs of the eigenmodes on the intervals ("+" or "–") are taken into account by the function $f_q^{(i)}$ [11,12,14]. To select the necessary eigenmodes in (37), the function $F_q^{(i,j)}$ must have the values presented in Table 1 and Table 2.

The choice of the sign in accordance with the Tables 1 and 2 is made by the function

$$F^{(i,j)}(x) = \mathrm{sign}\{\sin[3^{i-2}2x + \frac{\pi}{3}(2-j)] \tan[3^{i-1}x - \frac{\pi}{2}(j-1)]\} \qquad (38)$$

within the interval $0 < x < 3\pi/2$ (Fig.1). In accordance to Table 1 and Table 2 the number of the function values is $p = 3^N = 27$ and the resulting sign-function is

$$\begin{aligned}F_p^{(i,j)} = {} & \frac{1}{2}\langle 1 + \mathrm{sign}\{\sin[3^{i-1-N}\pi\left(p - \frac{1}{2}\right) + \frac{\pi}{3}(2-j)] \\ & \times \tan[\frac{3^{i-N}}{2}\pi\left(p - \frac{1}{2}\right) - \frac{\pi}{2}(j-1)]\}\rangle\end{aligned} \qquad (39)$$

Table 1. The values of the sign-function.

p	01	02	03	04	05	06	07	08	09	10	11	12	13	14	15
$\phi_1^{(1)}$	1	1	1	1	1	1	1	1	1	0	0	0	0	0	0
$\phi_2^{(1)}$	0	0	0	1	0	0	0	0	0	1	1	1	1	1	1
$\phi_3^{(1)}$	0	0	0	0	0	0	0	0	0	0	0	0	0	0	0
$\phi_1^{(2)}$	1	1	1	0	0	0	0	0	0	1	1	1	0	0	0
$\phi_2^{(2)}$	0	0	0	1	1	1	0	0	0	0	0	0	1	1	1
$\phi_3^{(2)}$	0	0	0	0	0	0	1	1	1	0	0	0	0	0	0
$\phi_1^{(3)}$	1	0	0	1	0	0	1	0	0	1	0	0	1	0	0
$\phi_2^{(3)}$	0	1	0	0	1	0	0	1	0	0	1	0	0	1	0
$\phi_3^{(3)}$	0	0	1	0	0	1	0	0	1	0	0	1	0	0	1

The function $F_p^{(i,j)}$ differs from the function F_{pq} presented in [13–16] when describing four-dimensional systems, since the law of sign change for six-dimensional systems is different (Fig. 2).

Table 2. The values of the sign-function.

p	16	17	18	19	20	21	22	23	24	25	26	27
$\phi_1^{(1)}$	0	0	0	0	0	0	0	0	0	0	0	0
$\phi_2^{(1)}$	1	1	1	0	0	0	0	0	0	0	0	0
$\phi_3^{(1)}$	0	0	0	1	1	1	1	1	1	1	1	1
$\phi_1^{(2)}$	0	0	0	1	1	1	0	0	0	0	0	0
$\phi_2^{(2)}$	0	0	0	0	0	0	1	1	1	0	0	0
$\phi_3^{(2)}$	1	1	1	0	0	0	0	0	0	1	1	1
$\phi_1^{(3)}$	1	0	0	1	0	0	1	0	0	1	0	0
$\phi_2^{(3)}$	0	1	0	0	1	0	0	1	0	0	1	0
$\phi_3^{(3)}$	0	0	1	0	0	1	0	0	1	0	0	1

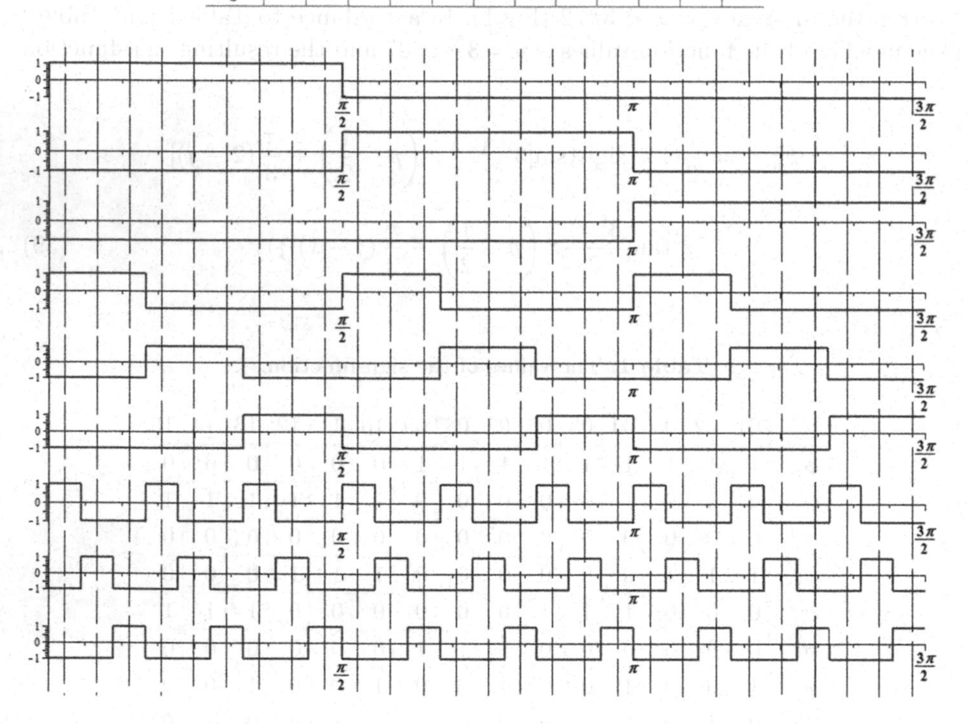

Fig. 2. The sigh-function $F^{(i,j)}(s)$ for the case of three intervals with constant parameters

7 Stability of a Linear Six-Dimensional System

The stability problem of a linear six-dimensional system can be solved by using Liapunov's theory [17]. In accordance to this theory a system is stable if all eigennumbers of the fundamental matrix (multipliers) are less than or equal to one. These numbers λ_i are the roots of the characteristic equation

$$\lambda^6 + b \cdot \lambda^4 + c \cdot \lambda^2 + d = 0 \tag{40}$$

of the matrix (8). Here it is necessary to take into account the fact that the matrix $\mathbf{L}_N(T)$ is unimodular. Moreover the determinant of this matrix is equal to one ($\det \mathbf{L}_N(T) = b = 1$) if the parameters (2) of the system are real. The equation (40) can be solved analytically by using Cardano method. Therefore the eigennumbers can be expressed as a function of the system parameters (2). Taking into account the method we can say that the system is stable if the conditions

$$\begin{cases} |\alpha + \beta| \le 1 \\ |-\dfrac{\alpha+\beta}{2} + j\dfrac{\alpha-\beta}{2}\sqrt{3}| \le 1 \\ |-\dfrac{\alpha+\beta}{2} - j\dfrac{\alpha-\beta}{2}\sqrt{3}| \le 1 \end{cases} \tag{41}$$

are satisfied. Here

$$\begin{aligned} \alpha &= \frac{1}{18}\Big[972bc - 216b^3 - 2916d + 72\left(-5b^6 + 45b^4c + 108b^3d \right.\\ &\qquad \left. -162b^2c^2 - 486bcd + 243c^3 + 729d^2\right)^{1/2}\Big]^{1/3} \\ \beta &= \frac{1}{18}\Big[972bc - 216b^3 - 2916d - 72\left(-5b^6 + 45b^4c + 108b^3d \right.\\ &\qquad \left. -162b^2c^2 - 486bcd + 243c^3 + 729d^2\right)^{1/2}\Big]^{1/3} \end{aligned} \tag{42}$$

The conditions (41) are expressed in the parameters of a linear six dimensional system with arbitrary picewise-constant coefficients as the matrix (8) is written in the system parameters.

8 Conclusion

In this paper the 6×6 fundamental matrix of a linear six-dimensional system with arbitrary piecewise-constant parameters is obtained in the analytical form in elementary functions for the first time. For this the new sign-function $F_q^{(i,j)}$ is introduced and described here. The kind of this matrix is analogous to the matrices [11–14] presented early for two- and four-dimensional systems. The function $F_q^{(i,j)}$ is similar in meaning to similar functions described in [14–16], but its mathematical form is significantly different from the previous expressions. Moreover the concept of the so-call equivalent oscillations for linear six-dimensional dynamic systems is introduced for the first time for six-dimensional systems in our treatment. These oscillations are also similar to those used to describe two-dimensional and four-dimensional dynamical systems. Therefore we can say about new general approach to study linear dynamic systems with arbitrary piecewise-constant parameters.

References

1. Sadettin, M.R., Orhan O.: Self-excited vibration of the three-degrees of freedom system. In: The 25th Annual International Conference on Mechanical Engineering ISME2017. Tehran, Iran, pp. 1–2 (2017)
2. Atkins, K.M., Hutson, J.M.: Phase space structures in 3 and 4 degrees of freedom: application to chemical reactions. In: Sim, C. (ed.) Hamiltonian Systems with Three or More Degrees of Freedom. NATO ASI Series (Series C: Mathematical and Physical Sciences), vol. 533, pp. 295–299. Springer, Dordrecht (1994). https://doi.org/10.1007/978-94-011-4673-9_26
3. Arnol'd, V.I.: Instability of dynamical systems with many degrees of freedom. Dokl. Akad. Nauk SSSR **156**(1), 9–12 (1964)
4. Segal, I.E.: Foundations of the theory of dynamical systems of infinitely many degrees of freedom. II. Can. J. Math. **131961**, 1–18 (1961)
5. Ladzinski, R.: Dynamic systems with a finite degrees of freedom number. Arch. Control Sci. **24**(LX)(4), 207–234 (2014)
6. Enaibe, A.E., Osafile, A.E., Omosode, E.: Theoretical mechanics of 6-dimensional (6d) central-force motion with tangential oscillation. J. Adv. Nat. Sci. **1**(1), 22–31 (2014)
7. Wazwaz, A.M.: The numerical solution of sixth order boundary value problems by the modified decomposition method. Appl. Math. Comput. **118**, 311–325 (2001)
8. El-Gamel, M., Cannon, J.R., Zayed, A.I.: Sinc-Galerkin method for solving linear sixth-order boundary-value problems. Math. Comput. **73**(247), 1325–1343 (2002)
9. Viswanadham, K.N., Kasi, S., Ch, S., Kiranmayi, V.: Nnumerical solution of sixth order boundary value problems by Petrov-Galerkin method with quartic B-splines as basic functions and quintic B-splines as weight functions. ARPN J. Eng. Appl. Sci. **12**(22), 6301–6308 (2017)
10. Sado, D., Krzysztof, G.: Analysis of vibration of three-degree-of-freedom dynamical system with double pendulum. J. Theor. Appl. Mech. **46**(1), 141–156 (2008)
11. Vytovtov, K.: Analytical investigation of stratified isotropic media. J. Opt. Soc. Am. A **22**(4), 689–696 (2005)
12. Vytovtov, K., Barabanova, E., Vishnevskiy, V.: Accurate mathematical model of two-dimensional parametric systems based on 2 × 2 matrix. In: Communications in Computer and Information Science, pp. 199–211 (2019)
13. Vytovtov, K., Barabanova, E.: mathematical model of four-dimensional parametric systems based on block diagonal matrix with 2 × 2 blocks. In: Communications in Computer and Information Science, pp. 139–151 (2019)
14. Vytovtov, K.A., Bulgakov, A.A.: Investigation of photonic crystals containing bian-isotropic layers. In: 35th European Microwave Conference, vol. 2, pp. 1359–1362 (2005)
15. Vytovtov, K.A., Barabanova, E.A., Zouhdi, S.: Optical switching cell based on metamaterials and ferrite films. In: 12th International Congress on Artificial Materials for Novel Wave Phenomena - Metamaterials, Espoo, Finland, pp. 424–426 (2018)
16. Vytovtov, K., Barabanova, E., Zouhdi, S.: Penetration effect in uniaxial anisotropic metamaterials. Appl. Phys. A Mater. Sci. Process. **124**(2), 137 (2018)
17. Gantmacher, F.R.: The Theory of Matrices, vol. 2. Chelsea Public. Comp., New York (1959)
18. Arnold, V.I.: Ordinary Differential Equations. Springer, Heidelberg (1992)
19. Kurosh, A.G.: Higher Algebra Course. Nauka, Moscow (1965). (in Russian)

Evaluation of the End-to-End Delay of a Multiphase Queuing System Using Artificial Neural Networks

A. V. Gorbunova(✉), V. M. Vishnevsky, and A. A. Larionov

V.A. Trapeznikov Institute of Control Sciences of Russian Academy of Sciences, 65, Profsoyuznaya Street, Moscow 117997, Russia
avgorbunova@list.ru, vishn@inbox.ru, larioandr@gmail.com

Abstract. The article discusses a multiphase queuing system with a recurrent incoming flow, storage units of unlimited capacity, and recurrent service times at its nodes. To analyze and evaluate the main parameters of the performance of this network, a new approach is presented, which consists in combining simulation with data mining methods. This approach is universal in nature, and its area of application is not limited to studies of tandem networks only, but, on the contrary, extends to any other networks of arbitrary topology, as well as to queuing systems with various combinations of its main components. To assess the quality of the characteristics obtained, namely the average end-to-end delay, a comparative analysis is carried out with the known approximate results obtained in earlier works.

Keywords: Queuing network · Tandem (multiphase) queuing system · Average response time · Multilayer perceptron · Artificial neural networks · Machine learning methods

1 Introduction

Queuing networks (QNs) are traditionally used for modeling and, accordingly, performance evaluation and optimization of many complex systems. For example, the analysis of computer systems together with their subsystems, communication networks, and many other various kinds of technical, economic, industrial, transport, medical, military, etc. [1] systems may be of concern.

This paper is more focused on the so-called multiphase or tandem queuing systems, the structure of which will be described in more detail in the next section. As for real physical systems, which are adequately modeled by such networks, one of the most striking examples is the operation of broadband wireless communication networks with a linear topology [2–4].

Moreover, it is noted that in this context, open systems, in which requests come from the external environment and, after sequential servicing at all its

The reported study was funded by RFBR, project number 19-29-06043.

V. M. Vishnevskiy et al. (Eds.): DCCN 2020, LNCS 12563, pp. 631–642, 2020.
https://doi.org/10.1007/978-3-030-66471-8_48

phases or nodes, leave the network, are specifically considered. It is also assumed that each node on the network has an unlimited storage capacity.

Let us consider the existing methods of research and, as a consequence, performance evaluation of the systems described above. Accurate analytical results were obtained for a very small class of QNs [5–9]. In particular, open Jackson networks with Poisson incoming flow and exponential service times are suggested. In this case, the network nodes are queuing systems of the $M|M|1$ type, and the joint stationary distribution of the number of claims at the network nodes has a rather simple multiplicative form. Therefore, finding all possible indicators of the quality of functioning of such systems is not difficult. It is also noted that the form of the solution in the form of a product is also valid for some other cases indicated in the BCMP theorem [10].

For most other types of networks, exact closed-form solutions do not exist. Therefore, various approximation methods and, accordingly, their combinations are used to analyze them. This is especially true for QNs, the number of nodes in which exceeds two. As a result, an approximate analysis of general networks actually becomes the only possible solution. Although, of course, it should be noted that in some individual cases, accurate analytical research methods are still possible, but due to the serious dimension of the state space of the studied networks, even in the conditions of further use of the essential capabilities of modern computing systems, they are unreasonable [1,11,12]. Thus, one of the most common methods for analyzing a general queuing network is the decomposition method, which implies a transition to the analysis of each network node separately, followed by studying their interaction with each other and further compiling the results obtained to assess the main performance indicators of the network as a whole [13]. In this case, there is a need to obtain estimates of the first and second moments of the output flows of each of the network nodes. Therefore, this approach is also called the second-order approximate theory for QNs [14].

In the context of this approach, one of the methods for studying a general queuing network is called the diffusion approximation method, which is actually used in the analysis of isolated fragments of a $GI|G|1$ type network. Perhaps one of the significant disadvantages of this type of approximation is the relatively low accuracy under conditions of low and medium loads on the network nodes, as well as its strong dependence on the choice of a specific distribution in the case of analyzing a one-dimensional diffusion process. That, however, can be partially leveled out due to, for example, the introduction of a similar, but two-dimensional process, which approximately describes the process of forming a queue [15,16]. In this case, the relative error in the approximation of the indicators of an isolated node can be significantly reduced as shown by numerical experiments.

In addition, the work [18] described for tandem networks the so-called bottleneck effect, which is the last phase of maintenance. Even improved diffusion approximation procedures, one of which is described in [19], do not always show satisfactory results here, although later algorithms were proposed to improve the accuracy of the analysis in [20, 21]. Nevertheless, despite some disadvantages that are sometimes quite significant in terms of the accuracy of approximations, the described method allows determining the necessary parameters of the output flows from the nodal subsystems, which is essential in the analysis of networks, and not queuing systems (QS).

Finally, the third approach to the study of QNs is simulation. This is probably one of the most realistic approaches to the analysis of physically existing QNs. However, it does not always imply a kind of universality inherent in queuing theory, since modeling can be performed for a single real physical system and, accordingly, it is possible to exclude the stage of constructing a mathematical model as such.

It should be said that there are many works devoted to the study of two-node tandem systems with the Markovian Arrival Process (MAP) and its generalization—the Batch Markovian Arrival Process (BMAP), which, generally speaking, are not recurrent, but nevertheless make it possible to take into account the complex, including correlated, nature of flows in modern telecommunication networks, such as, for example, 4G broadband networks or next-generation 5G networks. However, despite all the complexity and generality of such models, they cannot be used to analyze networks with non-Markovian flows, which are described, for example, by distribution laws such as gamma distribution, Weibull, Pareto, lognormal, uniform, deterministic, etc. [17].

This paper proposes a new approach to the analysis of QNs, based on the methods of data mining and, in particular, artificial neural networks (ANNs). The few publications dealing with the application of neural networks not only to the analysis of queuing models appeared relatively long ago, nevertheless, they were rather scattered and did not have a clearly formulated concept regarding their applicability to solving problems of the queuing theory.

One of the first articles of this kind mentions the use of machine learning methods, namely the ID3 (Iterative Dichotomiser 3) decision tree algorithm, in [22]. It describes in general the potential of machine learning methods, as well as the possibility of their application in simulating the behavior of complex systems and the advantages of their use in the implementation of expert systems. Therefore, the much later work [23], in which non-Markovian QSs with "warm up" of the $H_2|M|M|3$ and $M|H_2|M|3$ type are investigated, can be referred to a more detailed and more rigorous in the mathematical sense application of ANNs to the analysis of queuing systems (not networks).

The article is organized as follows. Section 2 considers a mathematical model of a tandem QN, discusses existing methods for assessing its main characteristics, namely end-to-end delay, and also describes a new approach—the use of ANNs. Finally, in Sect. 3, a numerical experiment is presented.

2 Mathematical Model of a Multiphase QN. Average End-to-End Delay

Let us consider a queuing network with a linear topology, which consists of K nodes, $K \geq 2$ (Fig. 1). Moreover, the first network node is a queuing system of the $GI|G|1$, type, and the rest are $\cdot|G|1$. Such networks actually refer to the so-called multiphase or tandem queuing systems, in which a request, after servicing in one phase or node, sequentially switches to servicing in the next phase, and this happens until it leaves the system after the end of servicing at the last K-th node.

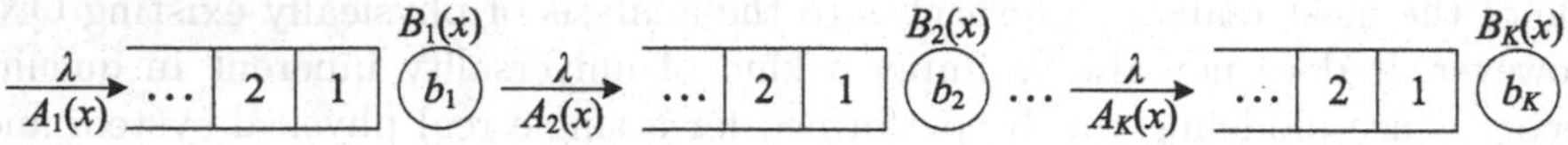

Fig. 1. Scheme of functioning of the tandem QN

One of the most important performance characteristics of any queuing system or network is its response time. When it comes to networks, this characteristic is often referred to as end-to-end latency. In fact, this value allows estimating the time spent by a network user waiting for a response to his request.

2.1 Assessment of the Characteristics of a QN Based on the Decomposition Method

As known, the exact expression for the average value of the end-to-end delay, when it comes to more than two consecutive subsystems, was obtained only in isolated cases, in particular, when the network nodes are queuing systems of the $M|M|1$ type and for some extensions of this QN [8,10]. In other situations, when the incoming flow is not Poisson, various approximations of the expected response time have been proposed.

Since a tandem system is considered, the average end-to-end delay is the sum of the average sojourn times of requests at each node

$$v = v_1 + v_2 + ... + v_K.$$

Thus, in order to obtain an approximation for the desired value, it is necessary to separately consider each i-th node of the network, which is characterized by the incoming flow from the distribution functions of time between the arrivals of requests $A_i(x)$ and the distribution function of the service time of requests $B_i(x)$, $i = \overline{1, K}$. Moreover, among all the distribution functions describing the flows entering the nodes, it is assumed that the distribution function is known only for the flow entering the network, that is, $A_1(x)$.

The main issue is the need to estimate the parameters of the incoming flow for the second and subsequent nodes of the network, in addition, of course, to the fact that until now there are no methods for finding an exact expression for the average sojourn time of requests in the QS of the $GI|G|1$ type, but only its approximation.

Nevertheless, due to the unlimited storage of nodes in the network under consideration, the intensity of the incoming flow of requests to the i node coincides with the intensity of the outgoing flow from it. This parameter is actually invariant regarding the $A_i(x)$ and $B_i(x)$ functions. Therefore, if the intensity of the incoming flow to the first phase is denoted by λ, then the intensity of the incoming flows to all other nodes will also be equal to λ due to the linearity of the QN.

The basic formula used to approximate the average response time of the i-th phase was proposed in [24] and is:

$$\widehat{v}_i = \frac{b_i \rho_i}{2(1-\rho_i)}(C_{A_i}^2 + C_{B_i}^2) g(\rho_i, C_{A_i}, C_{B_i}) + b_i, \tag{1}$$

where

$$g(\rho, C_A, C_B) = \begin{cases} \exp\left\{ -\frac{2(1-\rho)}{3\rho} \cdot \frac{(1-C_A^2)^2}{C_A^2+C_B^2} \right\}, & \text{if} \quad C_A \leq 1, \\ \exp\left\{ -(1-\rho) \cdot \frac{(1-C_A^2)^2}{C_A^2+C_B^2} \right\}, & \text{if} \quad C_A > 1, \end{cases} \tag{2}$$

$\rho_i = \lambda b_i$—the load of the node i, C_{A_i}—the coefficient of variation for incoming flow to the node i,, i.e., for $A_i(x)$ distribution, C_{B_i} is the variation coefficient for a random value of the service time on the device at the i node, i.e., for the service time distribution function $B_i(x)$, and b_i is mathematical expectation of service time on the device in the same network node, $i = \overline{1, K}$.

This expression was obtained empirically in the study of the $GI|G|1$ QS and for $g(\rho, C_A, C_B) = 1$, it completely coincides with the formula for the first time instant of the request stay at $M|G|1$ QS.

As a result, in order to calculate the average residence time of a request, for example, at the $(i+1)$ node, it is also necessary to estimate the coefficient of variation of the flow entering this phase. For this purpose, the following formulas exist:

$$C_{A_{i+1}} = C_{B_i}, \tag{3}$$

$$C_{A_{i+1}} = \rho_i(1-\rho_i) + \rho_i^2 C_{B_i}^2 + (1-\rho_i) C_{A_i}^2, \tag{4}$$

$$C_{A_{i+1}} = C_{A_i}^2 + 2\rho_i C_{B_i}^2 - \rho_i (C_{A_i}^2 + C_{B_i}^2) g(\rho_i, C_{A_i}, C_{B_i}), \tag{5}$$

$$C_{A_{i+1}} = \rho_i^2 C_{B_i}^2 + (1-\rho_i^2) C_{A_i}^2, \tag{6}$$

where $g(\rho_i, C_{A_i}, C_{B_i})$ is defined in (2). The first of them was proposed in the work [25] and is based on the assumption that under heavy network load, the probability that no maintenance is performed on phase i is close to zero, which implies that leaving this node, the flow tends to recurrent and its distribution coincides with the distribution of the service time here.

The formula (4) was obtained in [26]. The expression numbered (5) was presented by Kuhn in his work [28]; it is a consequence of the formula for the coefficient of variation of the outgoing flow from [27] for the $GI|G|1$ QS:

$$C_{A_{i+1}} = C_{A_i}^2 + 2\rho_i C_{B_i}^2 - \frac{2\rho_i(1-\rho_i)(\widehat{v}_i - b_i)}{b_i},$$

Finally, the relation (6) is a simplification of (5) under the assumption that $g(\rho_i, C_{A_i}, C_{B_i}) = 1$ [28].

Despite the fact that all of the above results were obtained relatively long ago, unfortunately, no significant shifts in research in this direction have taken place. More accurate approximations were obtained only for systems that are limited, as a rule, by a set of conditions. One of the most common is the duality of tandem systems, that is, that there are only two consecutive maintenance phases [1].

2.2 Evaluation of QN Characteristics Using an ANN

An approach based on a combination of simulation modeling with data mining methods and with training an artificial neural network, in particular, is proposed as an alternative to the above-described method of isolated analysis of network nodes. The idea is to use simulation to obtain a set of values of the network characteristic of interest and then use this set to train the ANN and use it to predict the desired value with a sufficiently high degree of accuracy in the event of a change in the values of the parameters affecting the operation of the network under study. For example, for other (intermediate) values of the network load, which is essentially a continuous value.

As of today, there are many different simulation software, a list of which can be found, for example, in [29]. The most popular systems are probably the GPSS World, AnyLogic, and Arena environments. Nevertheless, simulation modeling can turn out to be quite an expensive process in terms of the time required for simulation, which directly depends on the number of simulated parameters. Besides, two of the software tools listed above are commercial products. Therefore, a combination of two approaches, simulation modeling and machine learning methods, which include ANNs, turns out to be more economical from the point of view of resources spent. The training time for the neural network is relatively short, and the forecast itself is practically time-free. To demonstrate the effectiveness of the described approach, the results of estimating the average end-to-end delay obtained using the decomposition method and using the ANN in a numerical experiment will be compared.

3 Numerical Experiment

To estimate the response time, a tandem QN with a different number of phases and several types of probability distributions for the incoming stream and service times is considered. The decomposition method in the context of formulas for the approximate calculation of the mathematical expectation of the response time gives completely different accuracies depending on the load of nodes and types of distribution for the time intervals between the arrivals of requests and the times of their servicing. The relative calculation error can be relatively small and take a value within a fraction of a percent for an individual node. Along with this, in some cases, its value can reach several tens of percent.

Two variants of distributions, which show good results in the case of an isolated analysis of network nodes with linear topology, namely, uniform distribution and Pareto distribution under certain constraints on the shape parameter (see below) are considered for comparative analysis. Since in a situation with a large relative error exceeding tens of percent, the application of the second-order approximate theory will not be that rational.

The root mean square error (MSE), mean absolute error (MAE), and mean absolute relative error (MAPE) will be used to estimate the approximation error. Those are determined by the following expressions

$$MSE = \frac{1}{N}\sum_{j=1}^{N}(v_j - \widehat{v}_j)^2, \quad MAE = \frac{1}{N}\sum_{j=1}^{N}|v_j - \widehat{v}_j|,$$

$$MAPE = \frac{1}{N}\sum_{j=1}^{N}\left|\frac{(v_j - \widehat{v}_j)}{v_j}\right| \cdot 100\%,$$

where $\widehat{v}_j$ is the estimate of the studied characteristic, obtained either using an ANN on a test sample, or using one of the formulas (3)–(6) on the same set of input parameters, v_j is the real value returned by the system, i.e., in this case, it is the result of network simulation, N is the number of estimated elements (the number of elements in the test sample), $j = \overline{1, N}$. To carry out a numerical experiment, a simulation model was developed in the Phyton software environment, in which the ANN was trained using the backpropagation method.

So, in the first case, consider a tandem QN with a uniform distribution of the flow entering it on the $[1; d]$ segment, and also with a uniform service time distribution function on a server with parameters 1 and 29, that is, $U[1; 29]$ same for each of the network nodes, as well as the load $\rho_i = 30/(1 + d)$, $i = \overline{1, K}$. The values of the d parameter are selected in such a way that the load factor changes from 0.1 to 0.9 with a step of 0.1. In this case, the number of service phases K will vary from 2 to 200, inclusive. As the structure of the trained neural network, let us choose a three-layer perceptron with two input parameters—the number of service phases K and the load factor ρ, as well as the only output parameter – the average end-to-end delay, and the logistic activation function on each neuron $\varphi(x) = 1/(1 + e^{-x})$ (Fig. 2).

In order to improve the quality of estimation using ANN, the data of both the training and test samples were subjected to preliminary processing, namely, standardization and normalization. By standardization, the authors mean bringing a set of input data for each of the parameters influencing the system to such a form that its mathematical expectation becomes equal to zero, and the standard deviation—to one. Data normalization here means scaling the values of the input parameters in the range from 0 to 1.

Let us calculate the error of the approximation estimate for a particular value of the parameter $d = 99$ and, accordingly, $\rho = 0,3$. As can be seen from Table 1, in the case of a uniform distribution, the formulas show a relatively low approximation error—no more than 6%. Nevertheless, in this case, it is possible

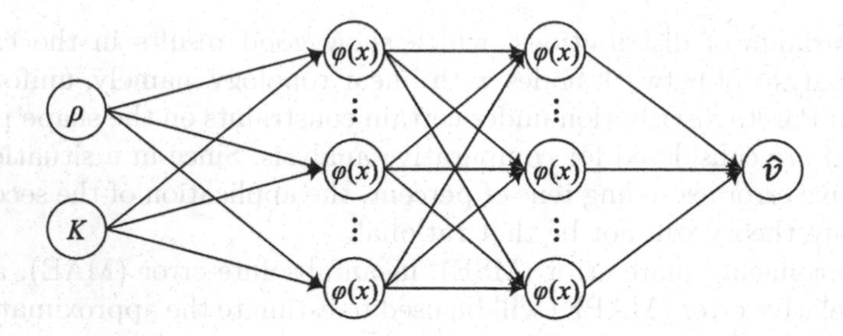

Fig. 2. Diagram of a three-layer perceptron used in training a neural network

Table 1. Approximation errors when calculating the average end-to-end delay using the formulas (3)–(6) and ANN in the case of a uniform distribution for $\rho = 0,3$.

Type of error	MSE	MAE	MAPE, %
Formula (3)	15435,175	108,865	5,984
Formula (4)	12386,602	94,106	4,753
Formula (5)	12330,974	94,128	4,755
Formula (6)	12393,388	94,140	4,756
Neural network	4,703	1,829	0,137

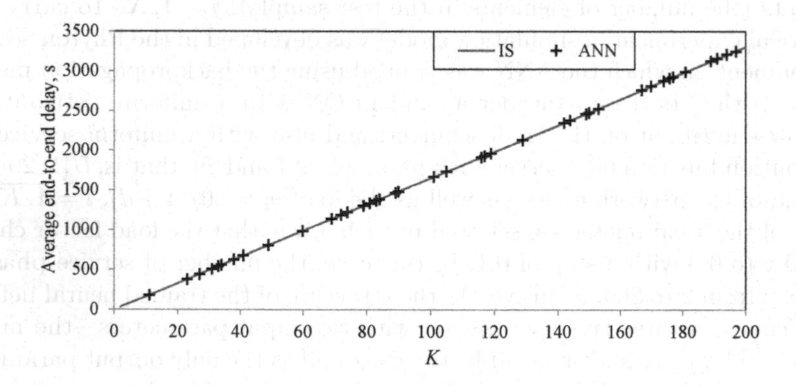

Fig. 3. The result of ANN training in the case of a uniform distribution for the flow entering the network and service times at the network nodes, $\rho = 0,3$

to train the neural network in such a way that the result predicted with its help turns out to be much better.

As for the Pareto distribution, it is worth recalling that it is two-parameter. The first parameter, the form factor, will be denoted by α. For the second parameter, the scale factor m, which in fact determines the minimum value accepted by a random variable with the specified distribution, let us choose the value one $(\alpha, m > 0)$.

It should be noted that in practice, the case is interesting when the shape parameter for a random variable with a Pareto distribution varies from 1 to 2, inclusive, then its average value is finite, and the variance is infinite. Thus, for example, fractal traffic, characterized by a high level of ripple, can be modeled within a certain physical system. However, due to the fact that the two approaches are compared, the option with infinite variance does not make sense, so let us consider only the case with $\alpha > 2$.

So, for the service time, let us select the value of the form parameter 25, then the values of the analogous parameter for the incoming stream will be in the range from $\approx 2,119$ to $15,625$. Therefore, this will allow getting the interval $[0,55;0,975]$ for the values of the load factor with a step of $0,025$, taking the above limit on the α parameter into account.

Table 2. Approximation errors when calculating the average end-to-end delay using the formulas (3)–(6) and ANN in the case of the Pareto distribution.

Type of error	MSE	MAE	MAPE, %
Formula (3)	1,080	0,655	2,321
Formula (4)	1,648	1,106	6,387
Formula (5)	1,595	1,088	6,224
Formula (6)	1,604	1,091	6,209
Neural network	0,335	0,319	0,927

As an ANN structure, let us also use a three-layer perceptron with two input parameters, namely, the number of network nodes K and the load factor ρ (Fig. 2). The results of training using the ANN are presented in Table 2. The approximation error here is calculated on the basis of a test sample containing 248 elements, while the number of elements in the training sample is 994. The formulas (4)–(6), in contrast to (3), in this situation show lower accuracy, which was expected, since the output of the formula (3) is just based on the assumption of a high system load. However, the result of training ANN is still better.

Note that when training ANN, the data was divided into training and test samples in a ratio of 80% and 20% randomly in both examples. For example, Fig. 3 shows the dependence of the end-to-end delay on the number of network nodes K for a fixed load $\rho = 0,3$. The solid line shows the result of imitation simulation (IS), and the markers mark the values obtained by training the neural network.

Figure 4 similarly shows the dependence of the end-to-end delay on the load of nodes for $K = 17$ by the solid line. The markers mark the values obtained using the neural network for the intermediate load values of the nodes, which were not used as input parameters when training the network. The value of $MAPE$ calculated for this data is approximately 0,7% when the value of the same error for the formulas (3)–(6) varies from 1,8% to 6%. In this case, the

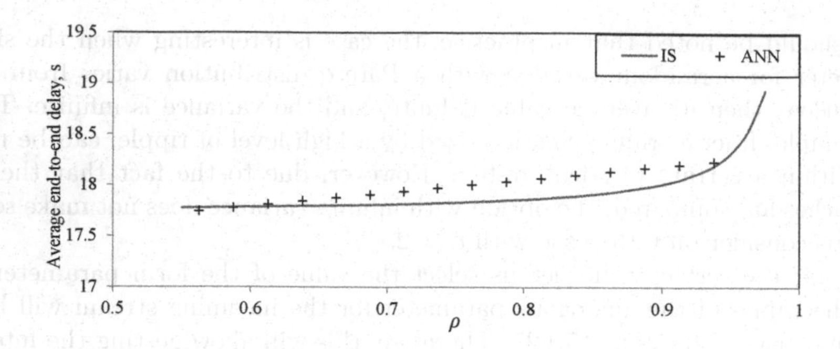

Fig. 4. The result of ANN training in the case of the Pareto distribution for the flow entering the network and the service times at the network nodes for $K = 17$

maximum value of the relative error for the neural network is only 1,4%, while for the analytical formula (3), showing the best approximation, the maximum is already 6,3%.

4 Conclusion

The article presents a new approach to the study of performance characteristics of multiphase queuing systems. The novelty of this approach lies in the use of ANNs for predicting the desired characteristics with an acceptable level of accuracy, while other machine learning methods can also be used for the same purposes.

The advantage of this technique is the ability to analyze non-Markovian models of systems or QNs not only of linear but also of arbitrary topology.

At the same time, as shown by a numerical experiment, even if the chosen ANN structure is not the most optimal in terms of the number of hidden layers and the number of neurons in them, it is possible to obtain a much better quality and, what is important, an admissible approximation with a relatively small time spent on selection of the main components of the neural network architecture.

References

1. Vishnevsky, V.M., Dudin, A.N., Klimenok, V.I.: The Theory of Queuing Systems with Correlated Flows, 1st edn. Springer, Heidelberg (2019). https://doi.org/10.1007/978-3-030-32072-0
2. Vishnevski, V.M., Larionov, A.A., Ivanov, R.E.: An open queueing network with a correlated input arrival process for broadband wireless network performance evaluation. In: Dudin, A., Gortsev, A., Nazarov, A., Yakupov, R. (eds.) ITMM 2016. CCIS, vol. 638, pp. 354–365. Springer, Cham (2016). https://doi.org/10.1007/978-3-319-44615-8_31

3. Vishnevsky, V.M., Dudin, A.N., Kozyrev, D.V., Larionov, A.A.: Methods of performance evaluation of broadband wireless networks along the long transport routes. In: Vishnevsky, V., Kozyrev, D. (eds.) DCCN 2015. CCIS, vol. 601, pp. 72–85. Springer, Cham (2016). https://doi.org/10.1007/978-3-319-30843-2_8
4. Vishnevsky, V.M., Krishnamoorthy, A., Kozyrev, D.V., Larionov, A.A.: Review of methodology and design of broadband wireless networks with linear topology. Indian J. Pure Appl. Math. **47**(2), 329–342 (2016). https://doi.org/10.1007/s13226-016-0190-7
5. Jackson, R.R.P.: Queueing systems with phase type service. J. Oper. Res. Soc. **5**(4), 109–120 (1954)
6. Jackson, J.R.: Networks of waiting lines. Operations Research. **5**(4), 518–521 (1957)
7. Jackson, J.R.: Jobshop-like queueing systems. Manag. Sci. **10**(1), 131–142 (1963)
8. Kelly, F.P.: Networks of queues with customers of different types. J. Appl. Probab. **12**(3), 542–554 (1975)
9. Kelly, F.P.: Networks of queues. Adv. Appl. Probab. **8**(2), 416–432 (1976)
10. Vishnevsky, V.M.: Teoreticheskie osnovy proektirovaniya komp'yuternyh setej [Theoretical foundations of computer network design], 1st edn. Technosphera, Moscow (2003). (in Russian)
11. Baldwin, R., Davis IV, N., Midkiff, S., Kobza, J.: Queueing network analysis: concepts, terminology, and methods. J. Syst. Softw. **66**(2), 99–117 (2003)
12. Klimenok, V., Dudin, A., Vishnevsky, V.: On the stationary distribution of tandem queue consisting of a finite number of stations. In: Kwiecień, A., Gaj, P., Stera, P. (eds.) CN 2012. CCIS, vol. 291, pp. 383–392. Springer, Heidelberg (2012). https://doi.org/10.1007/978-3-642-31217-5_40
13. Rabta, B.: A review of decomposition methods for open queueing networks. In: Reiner, G. (ed.) Rapid Modelling for Increasing Competitiveness, pp. 25–42. Springer, London (2009). https://doi.org/10.1007/978-1-84882-748-6_3
14. Basharin, G.P., Bocharov, P.P., Kogan, Ya.A.: Analiz ocheredej v vychislitel'nyh setyah. Teoriya i metody rascheta [Analysis of queues in computer networks. Theory and calculation methods]. Nauka, Moscow (1989). (in Russian)
15. Tarasov, V.N., Bahareva, N.F., Konnov, A.L.: Network serving system decomposition without queue length limitation. St. Petersburg Polytech. Univ. J. **2**, 31–36 (2008)
16. Bakhareva, N.F., Tarasov, V.N., Ushakov, U.A.: The generalized two-dimensional diffusion waiting line model of type $GI|G|1$. Telecommunications **7**, 2–8 (2009)
17. Klimenok, V.I., Taramin, O.S.: A two-phase $GI|PH|1 \rightarrow \cdot|PH|1|0$ system with losses. Autom. Remote Control **72**(5), 1004–1016 (2011)
18. Suresh, S., Whitt, W.: The heavy-traffic bottleneck phenomenon in open queueing networks. Oper. Res. Lett. **9**(6), 355–362 (1990)
19. Whitt, W.: The queueing network analyzer. Bell Syst. Tech. J. **62**(9), 2779–2815 (1983)
20. Whitt, W.: Variability functions for parametric-decomposition approximations of queueing networks. Manag. Sci. **41**(10), 1704–1715 (1995)
21. Dai, J.G., Nguyen, V., Reiman, M.I.: Sequential bottleneck decomposition: an approximation method for generalized Jackson networks. Oper. Res. **42**(1), 119–136 (1994)
22. Khoshnevis, B., Parisay, S.: Machine learning and simulation: application in queuing systems. Simulation **61**(5), 294–302 (1993)
23. Gindin, S.I., Khomonenko, A.D., Adadurov, S.E.: Numerical calculations of multichannel queuing system with recurrent input and "warm up". Proc. Petersburg Transp. Univ. **37**(4), 92–101 (2013)

24. Krämer, W., Langenbach-Belz, M.: Approximate formulae for the delay in the queueing system $GI|G|1$. In: Proceedings of the 8th International Teletraffic Congress, Melbourne, vol. 235, pp. 1–8. (1976)
25. Reiser, M., Kobayashi, H.: Accuracy of the diffusion approximation for some queuing systems. IBM J. Res. Dev. **18**(2), 110–124 (1974)
26. Gelenbe, E., Pujolle, G.: The behaviour of a single queue in a general queueing network. Acta Informatica **7**(2), 123–136 (1976)
27. Marshall, K.T.: Some inequalities in queuing. Oper. Res. **16**(3), 651–668 (1968)
28. Kühn, P.: Analysis of complex queuing networks by decomposition. In: Proceedings of the 8th International Teletraffic Congress, Melbourne, vol. 236, pp. 1–8 (1976)
29. Dias, L.M.S., Vieira, A.A.C., Pereira, G.A.B., Oliveira, J.A.: Discrete simulation software ranking - a top list of the worldwide most popular and used tools. In: Proceedings of the 2016 Winter Simulation Conference (WSC), Arlington Virginia, pp. 1060–1071. IEEE Press (2016)

Model of Navigation and Control System of an Airborne Mobile Station

V. M. Vishnevsky[1], K. A. Vytovtov[1,2(✉)], and E. A. Barabanova[1,2]

[1] V.A. Trapeznikov Institute of Control Sciences of RAS, Profsoyuznaya 65 Street, Moscow, Russia
vishn@inbox.ru, vytovtov_konstan@mail.ru
[2] Astrakhan State Technical University, Tatischva 16 Street, Astrakhan, Russia

Abstract. In this paper the theoretical foundation of the navigation system for airborne mobile communication objects is proposed. This system is served for detection of a positioning location, a space shift, speed of an air mobile station based on a drone. The working frequency of the radio-frequency navigation system is chosen to be 2.4 GHz. First of all the new approach of positioning location control and the method of space coordinate calculation are proposed by authors. To design such the system the helix antenna is proposed. The main parameters of it are calculated and the numerical simulation is carried out. The analytical method of velocity and acceleration determination for airborne mobile communication objects is developed too.

Keywords: Airborne mobile communication station · Positioning location · Automatic tracking

1 Introduction

Today, research in region of practical use of airborne mobile communication stations [1–4] is very relevant point for application of modern 5G communication systems [5]. One of the most important problem in this direction is determining a positioning location of a station and its automatic tracking. This is an essential part of the functioning of airborne mobile stations, in particular, in the case of absence of satellite navigation signals.

Many various approaches to solving this problem have been described in scientific literature [6–13]. The solution to this problem has many obstacles for today. Indeed, the well-known triangulation method [14] would be used for this purpose as this method gives us best measurement accuracy. However, it does require the deployment of radar stations over long distances. The simplest method of determining location using one transmitting and one receiving antennas [6–14] is also unacceptable due to low accuracy of angular coordinate determination. Moreover, the well-known Doppler effect [14] cannot be used to

The reported study was funded by RFBR, project number 19-29-06043.

V. M. Vishnevskiy et al. (Eds.): DCCN 2020, LNCS 12563, pp. 643–657, 2020.
https://doi.org/10.1007/978-3-030-66471-8_49

determine the station speed, since the mobile station in the normal mode must be stationary and, therefore, the radial velocity increment with gusts of wind will be too small. The use of optical navigation means [15] requires a lot of energy since the efficiency of modern optical radars is very small. Additionally, characteristics of an optical signal in the visible and infra-red domains are highly dependent on weather conditions.

In this paper, it is proposed to use both active and semi-active radar systems [14,16,17] in the microwave domain (2.4 GHz) simultaneously for control and automatic tracking of an airborne mobile station. In accordance to offered approach the transmitting antenna with a narrow radiation pattern is installed on the aircraft, the ground receiving station has the three identical antennas located at vertices of a regular triangle (Fig. 1). To reduce the measurement errors the so-called difference method is used in the receiving part of the system. In accordance with the method, the difference between the levels of the received signals in the channels is calculated in the signal subtractor of the receiving device and the drone control signal is generated by using the calculation results. Thus, the error in measuring the drone coordinates is determined only by the width of the transmitting antenna pattern. Note that the use of three antennas in the difference method is pioneering, from our point of view. Additionally, it also is offered the new methods of space shift and speed determination of airborne mobile communication stations.

2 Statement of the Problem

In this paper, we solve the problem of determining a positioning location of an airborne mobile station located on a drone, and the problem of its automatic tracking and stabilization. The flight altitude of the station is assumed from 50 to 100 m. The station is powered by an electric cable from a ground-based generator [4].

Our main aim is developing a microwave navigation system of such an airborne mobile station. First of all it is necessary to build the theoretical bases of designing the positioning location subsystem, the speed detection subsystem, the range measurement subsystem. Moreover, it is very important to choose antenna systems for solving the considered problems.

3 Determination of Angular Coordinates

3.1 The Difference Strength Method and Block Scheme

To determine and control a positioning location of the airborne mobile communication stations we use the so-call difference strength method described in numerous scientific works in detail, for example, in [14,16,17]. In the literature this method has been demonstrated, as rule, for the simplest case of positioning location determination in a plane.

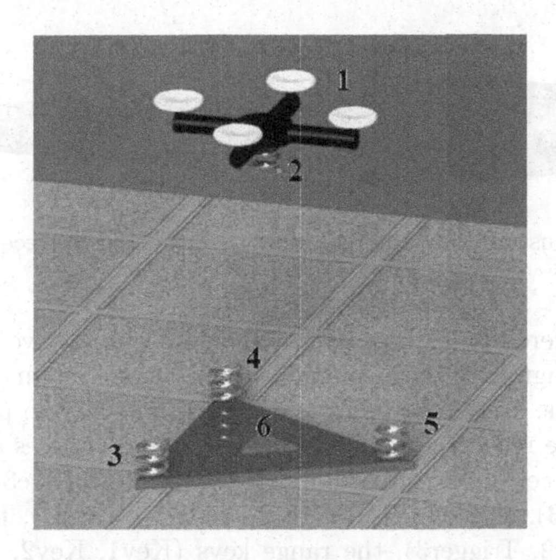

Fig. 1. The illustration of the navigation system

Let's briefly describe the control principle of the airborn communication station located on a drone in accordance with the method. A signal radiated by a transmitting antenna located on the drone are received by two identical antennas of a ground station and they go to identical received channels. After its processing, the difference of these signals is calculated in a subtractor of the ground station. This difference between received signal levels must be equal to zero if a drone is symmetrical about both antennas. And this difference is not equal to zero if a drone is offset from the symmetry axis of ground antennas. A voltage proportional to the difference between received signal levels goes to a driver of an executing signal for spatial displacement of the drone.

Considering a space location in accordance with the literature [14,16,17], we must use the analogous method in two perpendicular planes. Thus now we should have had four channels for positioning location determination (Fig. 2). Such the way gives us the simplest calculation method and high accuracy. However it requests expensive equipment, and the complexity of the system will be very large.

To solve this problem we propose to use one transmit and three receiver antennas (Fig. 3) for the first time. It is important that a distance between the receiving antennas to be small in comparison with triangular method [14].

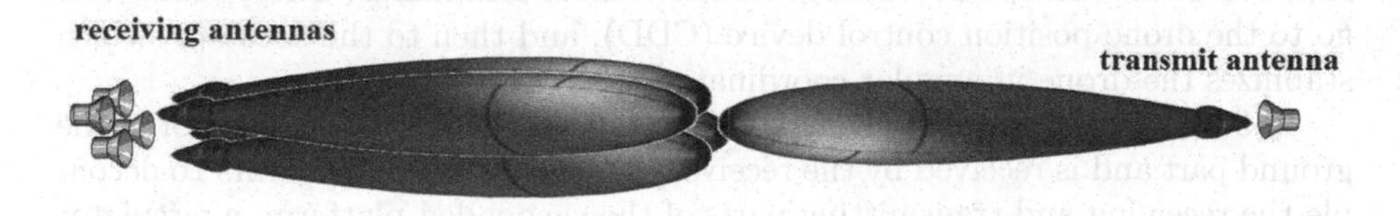

Fig. 2. The illustration of the radiation system with four received antennas

Fig. 3. The illustration of the radiation system with tree received antennas

The system offered by us uses both active and semi-active radar systems. In Fig. 4 the block diagram of the automatic stabilization system of the mobile platform is shown. The system includes the ground and mobile parts. The ground part includes three receiving antennas located at the vertices of the right triangle (Fig. 1), the receivers of three channels (Rec1, Rec2, Rec3), the subtractors (Diff1, Diff2, Diff3), the analog-to-digital converter (ADC), the range triggers (Trigger1, Trigger2, Trigger3), the range keys (Key1, Key2, Key3), the range counters (Counter1, Counter2, Counter3), the clock generator (CPG1), the controller (Controller). Moreover, in accordance with the functional purpose, the ground part can be divided into

1) The automatic stabilization subsystem containing (Fig. 4) the receivers (Rec1, Rec2, Rec3), the subtractors (Diff1, Diff2, Diff3), the analog-to-digital converter (ADC), the controller (Controller)

2) The subsystem for calculating the spatial displacement of the suspended mobile station containing (Fig. 4) the receivers (Rec1, Rec2, Rec3), the triggers (Trigger1, Trigger2, Trigger3), the keys (Key1, Key2, Key3), the counters (Counter1, Counter2, Counter3), the clock pulse generator (CPG1), the controller (Controller).

To increase the effective reflecting surface, the half-wave reflector is placed in the center of the regular triangle (Fig. 1).

The mobile part contains the receiving-transmitting antenna, the pulse transmitter (Trans), the receiver (Rec4), the circulator (Circulator), the range trigger (Trigger), the key (Key), the clock pulse generator (CPG), the range counter (Counter), the drone position control device (CDD), actuator.

The receiving-transmitting antenna of the mobile platform radiates the range pulses, which are received by all three receiving antennas of the ground part. Then after processing in the receivers, these pusses go to the subtractors. If the drone is located at the same distance from all three antennas (to be at the system axis), then all differential signals are zero. Otherwise, one or more differential signals are not equal to zero, and these signals are converted to digital form and supplied to the controller, which generates control commands. These commands go to the drone position control device (CDD), and then to the actuator, which stabilizes the drone in angular coordinates.

In addition, the radiated signal is reflected from the half-wave reflector of the ground part and is received by the receiver of the suspended platform. To decouple the receiving and transmitting parts of the suspended platform, a circulator is installed in the circuit of the mobile part. The range trigger is triggered by a

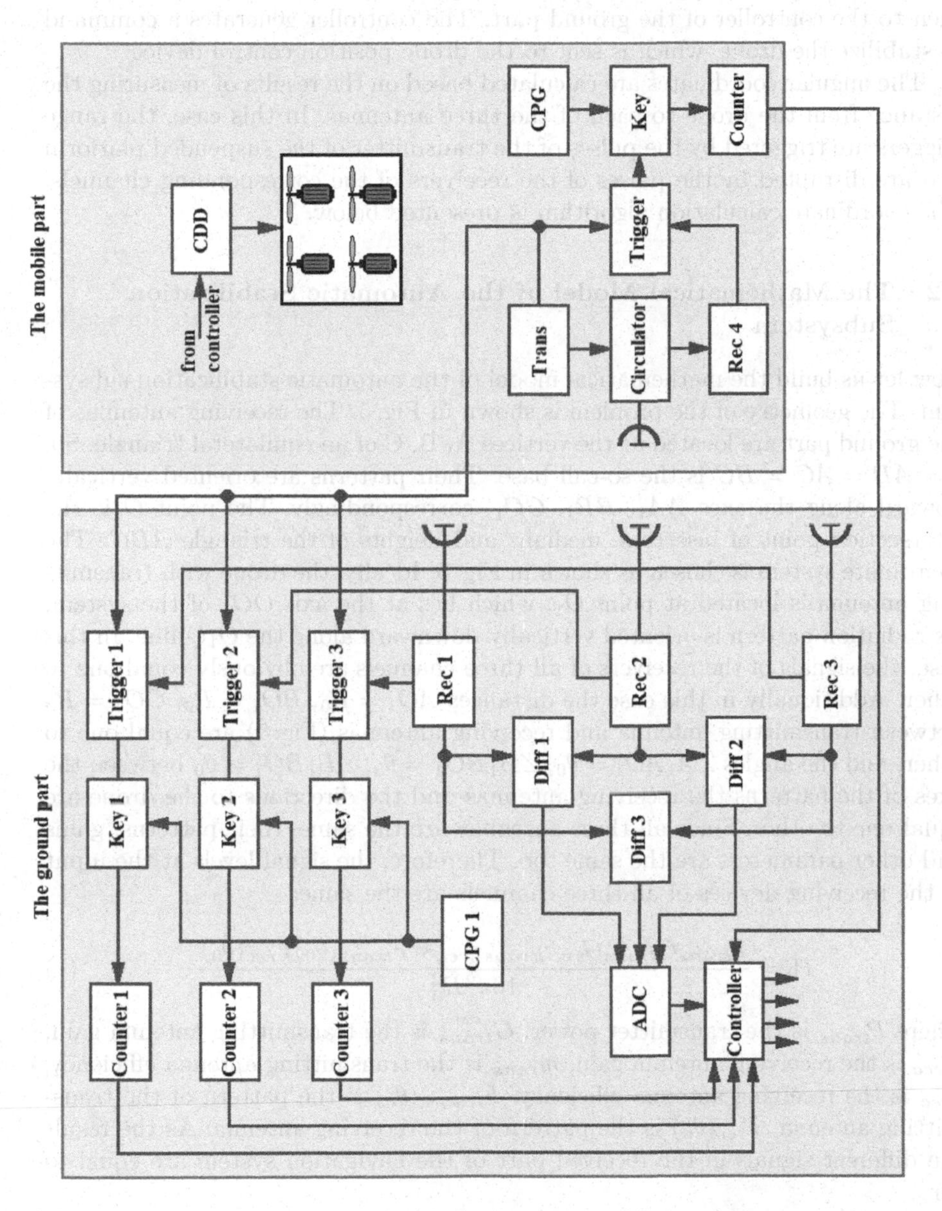

Fig. 4. The block diagram of the system

transmitter signal and cancelled by the receiver signal. Thus it forming a rectangular pulse, the duration of which is proportional to the distance between the drone and the ground equipment. Then the key must be opened by this pulse, and this key passes the pulses of the clock pulse generator to the counter, and then to the controller of the ground part. The controller generates a command to stabilize the drone, which is sent to the drone position control device.

The angular coordinates are calculated based on the results of measuring the distance from the drone to each of the three antennas. In this case, the range triggers are triggered by the pulses of the transmitter of the suspended platform and are disrupted by the pulses of the receivers of the corresponding channels. The coordinate calculation algorithm is presented below.

3.2 The Mathematical Model of the Automatic Stabilization Subsystem

Now let us build the mathematical model of the automatic stabilization subsystem. The geometry of the problem is shown in Fig. 5. The receiving antennas of the ground part are located at the vertices A, B, C of an equilateral triangle. So, $b = AB = AC = BC$ is the so-call base. Their patterns are oriented vertically upward along the axes AA_1, BB_1, CC_1, correspondingly. The point O is the intersection point of bisectors, medians and heights of the triangle ABC. The coordinate system is chosen as shown in Fig. 5. Ideally, the drone with transmitting antenna is located at point O_1, which lies at the axis OO_1 of the system. Its radiation pattern is oriented vertically downward along the O_1O line. In this case, the signals of the receivers of all three channels are obviously equal one to other. Additionally in this case the distances $AO_1 = R_0$, $BO_1 = R_0$, $CO_1 = R_0$ between transmitting antenna and receiving antennas (Fig. 5) are equal one to other, and the angles $\angle A_1AO_1 = \theta_0$, $\angle B_1BO_1 = \theta_0$, $\angle B_1BO_1 = \theta_0$ between the axes of the patterns the receiving antennas and the directions to the drone are equal one to other. Since all three antennas are the same, their patterns, gains and other parameters are the same too. Therefore, the signal levels at the input of the receiving devices of all three channels are the same:

$$P_0 = \frac{P_{trans}G_{trans}G_{rec}\eta_{trans}\eta_{rec}\lambda^2 F_{trans}(\theta_0)F_{rec}(\theta_0)}{16\pi^2 R_0^2} \tag{1}$$

where P_{trans} is the transmitter power, G_{trans} is the transmitting antenna gain, G_{rec} is the receiving antenna gain, η_{trans} is the transmitting antenna efficiency, η_{rec} is the receiving antenna efficiency, $F_{trance}(\theta_0)$ is the pattern of the transmitting antenna, $F_{rec}(\theta_0)$ is the pattern of the receiving antenna. As the result the different signals in the received part of the navigation system are equal to zero.

Now let us assume that the drone is shifted to point O_3 as a result of external influence (Fig. 5 and Fig. 6). In this case the distances from the transmitting antenna to the receiving antennas $d_{A1} = AO_3$, $d_{B1} = BO_3$, $d_{C1} = CO_3$ and the angles $\angle A_1AO_3 = \theta_{A1}$, $\angle B_1BO_3 = \theta_{B1}$, $\angle C_1CO_3 = \theta_{C1}$ between the pattern

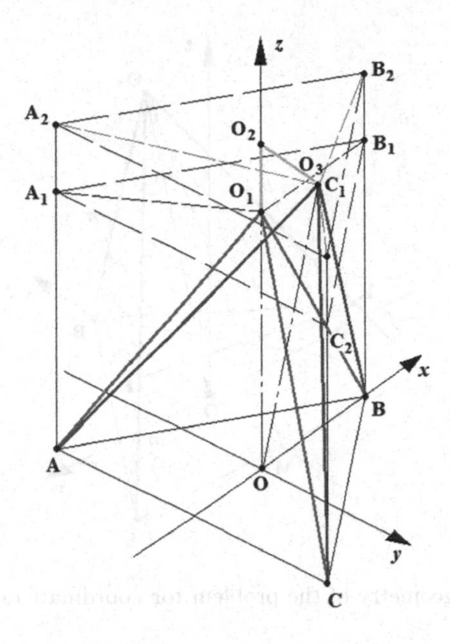

Fig. 5. The geometry of the problem for coordinate determination

axes and the directions to the drone are different. Thus, at small distances, when losses in atmosphere can be neglected, the powers at the receiver inputs of each of the channels are different and equal

$$
\begin{aligned}
P_{A1} &= \frac{P_{trans}G_{trans}G_{rec}\eta_{trans}\eta_{rec}\lambda^2 F_{trans}(\theta_{A1})F_{rec}(\theta_{A1})}{16\pi^2 d_{A1}^2} \\
P_{B1} &= \frac{P_{trans}G_{trans}G_{rec}\eta_{trans}\eta_{rec}\lambda^2 F_{trans}(\theta_{B1})F_{rec}(\theta_{B1})}{16\pi^2 d_{B1}^2} \\
P_{C1} &= \frac{P_{trans}G_{trans}G_{rec}\eta_{trans}\eta_{rec}\lambda^2 F_{trans}(\theta_{C1})F_{rec}(\theta_{C1})}{16\pi^2 d_{C1}^2}
\end{aligned}
\tag{2}
$$

Here P_{A1}, P_{B1}, P_{C1} are the received powers of A, B, C receivers, correspondingly. Then the signals proportional to the difference of the received powers are generated in the ground part. Thus we have the voltages

$$
\begin{aligned}
U_{A1B1} &= g\left(f(P_{A1}) - f(P_{B1})\right) \\
U_{A1C1} &= g\left(f(P_{A1}) - f(P_{C1})\right) \\
U_{B1C1} &= g\left(f(P_{B1}) - f(P_{C1})\right)
\end{aligned}
\tag{3}
$$

connected with the drone space shift. Here f is a function proportional to the received signal power, g is the function proportional to the signal difference. The specific kind of these functions is determined by the equipment of the system ground part. These signals go to the drone's actuator to restore its original position.

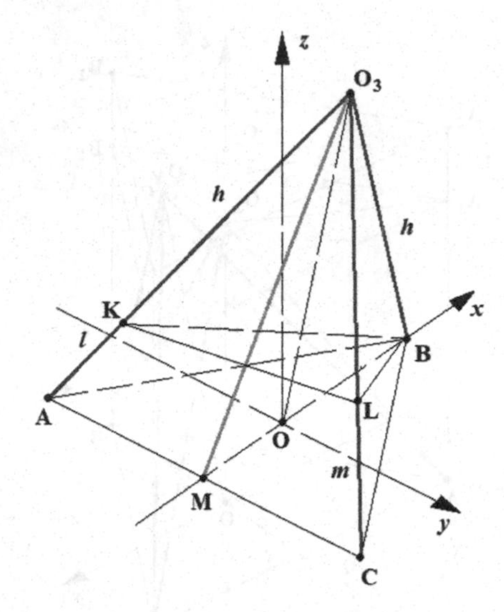

Fig. 6. The geometry of the problem for coordinate determination

Note that in accordance to method the accuracy of the automatic stabilization system depends to a greater extent on the steepness of the receiving antenna patterns at small angles, and, to a lesser extent, on the pattern width.

3.3 The Subsystem for Calculating the Spatial Displacement of the Mobile Station

Documenting the current position of the drone is the very important problem of navigation. Therefore, we must not only stabilize the drone, but also calculate its current spatial coordinates. This procedure is also very important for extending the attitude of the drone.

The geometry of the problem is shown in Fig. 5 and Fig. 6. Now, taking into account the geometry of the problem, the coordinates of the point O_3 can be calculated as

$$\begin{gathered} x_{O3} = \frac{d_{A1}^2 - 2d_{B1}^2 + d_{C1}^2}{2\sqrt{3}b} \\ y_{O3} = \frac{d_{A1}^2 - d_{C1}^2}{2b} \\ z_{O3} = \left[b^4 - (d_{A1}^2 + d_{B1}^2 + d_{C1}^2)b^2 + d_{A1}^4 + d_{B1}^4 + d_{C1}^4 \right. \\ \left. - d_{A1}^2 d_{B1}^2 - d_{A1}^2 d_{C1}^2 - d_{B1}^2 d_{C1}^2\right]^{1/2} / (\sqrt{3}b) \end{gathered} \tag{4}$$

However the determination of d_{A1}^2, d_{B1}^2, d_{C1}^2 cannot be accurate when the above method (measurement of received power) is used. Indeed this method is well applicable for the spatial stabilization of the drone, but it requires high accuracy of antenna alignment and taking into account all external factors affecting the patterns. Therefore, the authors propose the new technique based on determining the signal delay time during its passage from the transmitting antenna to each of the receiving antennas. However, to determine this time, it is necessary to accurately synchronize the operation of all units of the system, that is impossible in practical conditions. To overcome this obstacle the relative distances m and l are measured (Fig. 6). First of all let us calculate the height $h = BO_3 = KO_3 = LO_3$ taking into account the geometry of the problem. Thus, we have

$$h^2 + \frac{2}{3}(m+l)h + \frac{1}{3}(l^2 + m^2 + 3d^2 - b^2) = 0 \tag{5}$$

Here $d = OO_3$, $l = AK$, $m = CL$. Thus

$$h = -\frac{1}{3}(m+l) + \sqrt{\frac{1}{9}(m+l)^2 - \frac{1}{3}(l^2 + m^2 - 3d^2 - b^2)} \tag{6}$$

Then

$$\begin{aligned} d_A &= \frac{1}{3}(2l-m) + \sqrt{\frac{1}{9}(m+l)^2 - \frac{1}{3}(l^2 + m^2 - 3d^2 - b^2)} \\ d_B &= \frac{1}{3}(l+m) + \sqrt{\frac{1}{9}(m+l)^2 - \frac{1}{3}(l^2 + m^2 - 3d^2 - b^2)} \\ d_C &= \frac{1}{3}(2m-l) + \sqrt{\frac{1}{9}(m+l)^2 - \frac{1}{3}(l^2 + m^2 - 3d^2 - b^2)} \end{aligned} \tag{7}$$

Thus the expressions (4)–(7) allows us to calculate the space coordinates of the drone. The distance $d = OO_3$ between drone and the center of the ground station can be found as $d = c\tau/2$, where c is the wave speed in free space, τ is the delay time of a signal radiated and received by the mobile station. The values m and n can be found as the relative delay times. On other words, the system does not measure the delay times on the signal passing from the transmitting antenna to each of the receiving antennas, but the difference between these times. However, numerical calculations show us that for an accurate calculation of the drone coordinates (displacement about a few meters), the system must distinguish between a pulse delay of the order of units and tenths of a nanosecond. For example, the time delays for the channels A and B must be $3 \cdot 10^{-10}$ s and $2.5 \cdot 10^{-10}$ s, correspondingly, in the case of $\Delta x = 4.76$ m, $\Delta y = 0.75$ m, $\Delta z = 0.12$ m. Measuring this time can provide the frequency of 5 GHz and higher. Additionally it is required broadband properties of the devices in the system. Moreover, the measurement requires an increase in the pulse power to ensure reliable operation of the transmission and reception channels [18].

4 The Antenna Calculations

To design the system we propose to use helical antenna. The main advantage of such antennas is their high quality factor. However a pattern width of these ones cannot be too narrow. Indeed, the resolution in angular coordinates is determined by the width of the pattern of a transmitting antenna in accordance with the applied method. Therefore, it is proposed to use an array of four helical antennas in the transmitting part of the system. The electrical field components of this antennas can be calculated as [19]

$$\begin{aligned} E_\theta = -E_0 \sum_{m=-\infty}^{\infty} J_m(t)\{(-1)^{(n+1)n+m} \cos\vartheta \\ \times 2(\nu+m)\frac{\sin(\nu n\pi)}{(\nu+m)^2-1} + (-1)^{m(n+1)+1} \\ \times 2\tan\alpha\sin\vartheta\frac{2\sin(\nu n\pi)}{\nu+m} \end{aligned} \tag{8}$$

$$E_\varphi = jE_0 \sum_{m=-\infty}^{\infty} (-1)^{(m+1)n+m} J_m(t)\frac{2\sin(\nu n\pi)}{(\nu+m)^2-1}$$

here

$$E_0 = \frac{j\omega\mu_0 a}{4\pi} I_{01} \frac{\exp(-jkr)}{r} \tag{9}$$

$$t = ka\sin\vartheta \tag{10}$$

$$\nu = ka\tan\alpha\cos\vartheta - \frac{\beta_1 a}{\cos\alpha} \tag{11}$$

I_{10} is the current amplitude of the wave T_1, $J_m(t)$ is the Bessel function of the m-th order, a is the spiral radius, α is the helix angle, n is the number of the turns.

In the engineering calculations the pattern of such an antenna can be found as [19]

$$\begin{aligned} F_\theta(\theta) = F_{1\theta} F_C(\theta) \\ F_\varphi(\theta) = F_{1\varphi} F_C(\theta) \end{aligned} \tag{12}$$

where

$$\begin{aligned} F_{1\theta}(\theta) = \cos\theta \cdot J_0(ka\sin\theta) \\ F_{1\varphi}(\theta) = \nu \cdot J_0(ka\sin\theta) \end{aligned} \tag{13}$$

$$\nu = 1 + ka(1-\cos\theta)\tan\alpha \tag{14}$$

$$F_C(\theta) = \frac{2}{\pi n} \cdot \frac{\sin(\pi n\nu)}{(\nu^2-1)} \tag{15}$$

k is a wavenumber in free space

The pattern of the offered helical antenna is presented in Fig. 7 and Fig. 8. Here the frequency is $f = 2.4$ GHz, the number of the loops is $n = 10$, the loop length is $L = 2\lambda$, the loop angle is $\alpha = \pi/12$.

The results of the numerical simulation obtained by the finite difference method are presented too in Fig. 9. The parameters of the antenna are the following: the frequency is 2.4 GHz, the helix loop length is 0.75λ, the helix pitch is

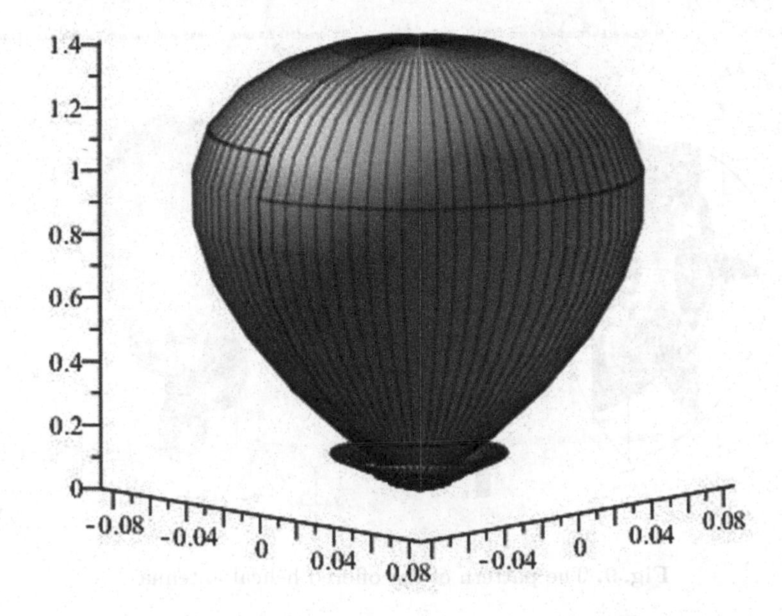

Fig. 7. The pattern of the offered helical antenna

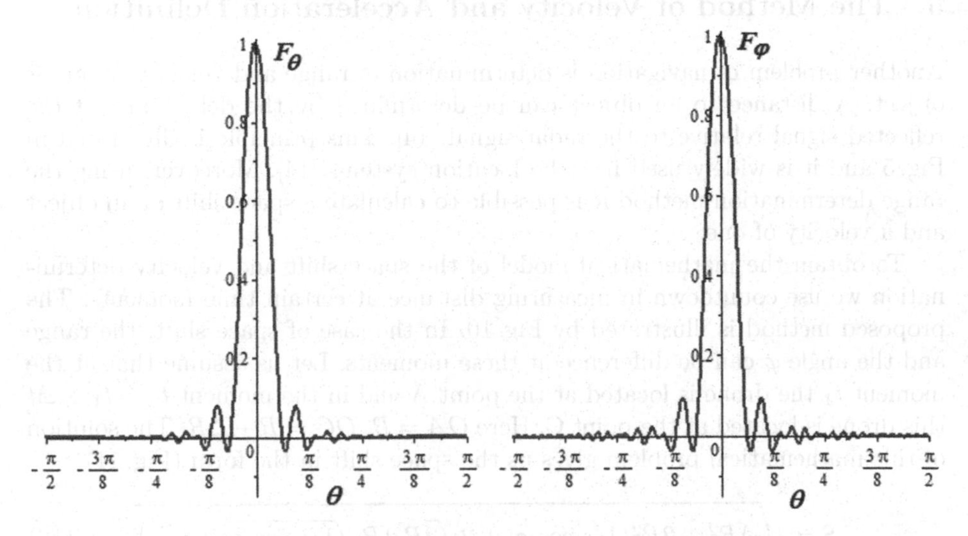

Fig. 8. The pattern of the offered helical antenna

0.24λ, the wire radius is 0.035λ, the screen radius is 0.4λ, the screen thickness is 5 mm, the number of the spiral turns is 5, the antenna input impedance 50O m. The distribution of the electric field in the near zone is shown in the left figure of Fig. 9 The result of calculation of the 3D pattern is present in the right figure of Fig. 9. The directional factor of this antenna is 9.72 dB, $\theta = 4.09°$, $\varphi = 320°$.

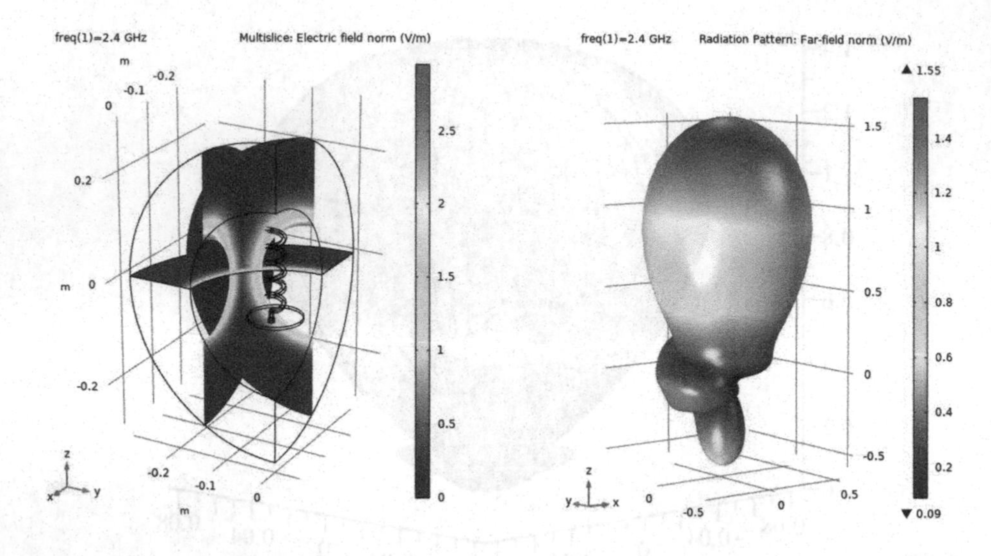

Fig. 9. The pattern of the offered helical antenna

5 The Method of Velocity and Acceleration Definition

Another problem of navigation is determination of range and velocity of an air object. A distance to an object can be determined by the delay time of the reflected signal relative to the radar signal [14]. This principle is illustrated in Fig. 5 and it is widely used in radio location systems [14]. Moreover, using the range determination method it is possible to calculate a space shift of an object and a velocity of one.

To obtain the mathematical model of the space shift and velocity determination we use countdown in measuring distance at certain time moments. The proposed method is illustrated by Fig. 10. In the case of space shift, the range and the angle φ can be difference at these moments. Let us assume that at the moment t_1 the drone is located at the point A and in the moment $t_2 = t_1 + \Delta t$ this drone is located at the point C. Here $OA = R$, $OC = R + \Delta R$. The solution of this mathematical problem gives us the space shift in the form (Fig. 11)

$$S = \sqrt{\Delta R^2 + 2R^2(1 - \cos\varphi) + 2\sqrt{2}R\Delta R\sqrt{1 - \cos\varphi}\sin\left(\frac{\varphi}{2}\right)} \qquad (16)$$

In the case when the distance is not change ($R = \text{const}$) then the space shift is

$$S = R\sqrt{2(1 - \cos\varphi)} \qquad (17)$$

Now the velocity of the drone shift along S is

$$v = \frac{1}{\Delta t}\sqrt{\Delta R^2 + 2R^2(1 - \cos\varphi) + 2\sqrt{2}R\Delta R\sqrt{1 - \cos\varphi}\sin\left(\frac{\varphi}{2}\right)} \qquad (18)$$

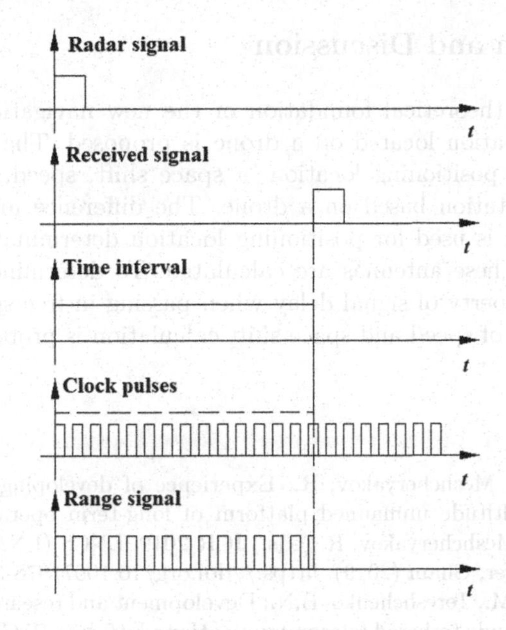

Fig. 10. The pattern of the offered helical antenna

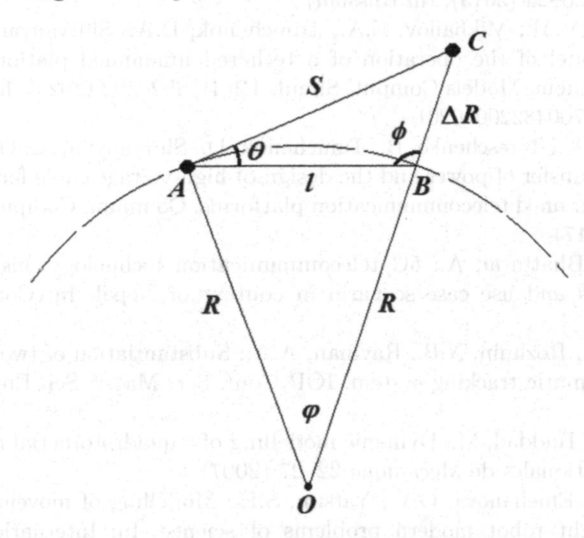

Fig. 11. The pattern of the offered helical antenna

If $R = \text{const}$ then the velocity of the drone is

$$v = \frac{R}{\Delta t}\sqrt{2(1-\cos\varphi)} \tag{19}$$

6 Conclusion and Discussion

In this paper the theoretical foundation of the new navigation system for an airborne mobile station located on a drone is proposed. The system is served for detection of a positioning location, a space shift, speed, and acceleration of an air mobile station based on a drone. The difference method with three receiving antennas is used for positioning location determination. The general characteristics of these antennas are calculated. To determine the distance to the drone, the property of signal delay when passing in free space is used. The analytical method of speed and space shift calculation is proposed too.

References

1. Vishnevsky, V., Meshcheryakov, R.: Experience of developing a multifunctional tethered high-altitude unmanned platform of long-term operation. In: Ronzhin, A., Rigoll, G., Meshcheryakov, R. (eds.) ICR 2019. LNCS (LNAI), vol. 11659, pp. 236–244. Springer, Cham (2019). https://doi.org/10.1007/978-3-030-26118-4_23
2. Vishnevsky, V. M., Tereshchenko, B.N.: Development and research of a new generation of high-altitude tethered telecommunication platforms. T-Comm Telecommun. Transp. (7) 20–24 (3013). (in Russian)
3. Vishnevsky, V.M., Mikhailov, E.A., Tumchenok, D.A., Shirvanyan, A.M.: Mathematical model of the operation of a tethered unmanned platform under wind loading. Mathem. Models Comput. Simul. **12**(4), 492–502 (2020). https://doi.org/10.1134/S2070048220040201
4. Vishnevsky, V., Tereschenko, B., Tumchenok, D., Shirvanyan, A.: Optimal method for uplink transfer of power and the design of high-voltage cable for tethered high-altitude unmanned telecommunication platforms. Commun. Comput. Inf. Sci. **700**, 240–247 (2017)
5. Paudel, P., Bhattarai, A.: 5G telecommunication technology: history, overview, requirements and use case scenario in context of Nepal. In: Conference: IT4D (2018)
6. Sheval, V.V., Rozhnin, N.B., Rayman, A.A.: Substantiation of two-channel structure of automatic tracking system. IOP Conf. Ser. Mater. Sci. Eng. **537**, 032108 (2019)
7. Chettibi, T., Haddad, M.: Dynamic modelling of a quadrotoraerial robot. Journees D'etudes Nationales de Mecanique 22–27 (2007)
8. Popov, N.I., Emelianova, O.V., Yatsun, S.F.: Modelling of movement of quadrotoraerial flight robot modern problems of science. In: International Scientific-Practical Conference (Moscow: Sputnik+), pp. 6–8 (2013)
9. Gur'yanov, A.E.: Modelling of quadcopter control Youth Scientific-Technical Journal. Moscow: Bauman Moscow State Technical University 522–534 (2014)
10. Gong, X., Hou, Z.-C., Zhao, C.-J., et al.: Adaptive backstepping mode trajectory tracking control for a quad-rotor. Int. J. Autom. Comput. **9**(5), 555–560 (2012). https://doi.org/10.1007/s11633-012-0679-4
11. Krasovskiy, A.N., Suslova, O.A.: Optimal control of drone-quadcopter movement according to energy consumption criterion Ural Technical Institute of Communication and Information. Success Mod. Sci. Educ. **4**(3), 193–197 (2017)

12. Sheval, V.V., Ogoltsov, I.I., Terskov, V.G.: Aggregation Methods of Associations of Onboard Aircraft Complexes Feedback Actuators. BIBLIO-GLOBUS Publishing House, Moscow (2016)
13. Jami, I., Ali, M., Ormondroyd, R.: Comparison of methods of locating and tracking cellular mobiles. In: IEE Colloquium on Novel Methods of Location and Tracking of Cellular Mobiles and Their System Applications, 1/1-1 (1999)
14. Rahman, H.: Fundamental Principles of Radar. Taylor & Francis Group (2019)
15. Paluszek, M., Littman, M., Mueller, J.: Optical navigation system. In: AIAA Infotech@Aerospace, pp. 1–24 (2010)
16. Sheval, V.V., Rozhnin, N.B., Rayman, A.A.: Substantiation of two-channel structure of automatic tracking system. IOP Conf. Ser. Mater. Sci. Eng. **537**, 032108 (2019)
17. Fred, E., Nathanson, J., Reilly, P., Marvin, N.: Cohen Radar Design Principles. McGmw-Hill, Inc. (1969)
18. Mesiats, G.F., Yalandin, M.I.: High power picosecond electronics. Usp. Phys. Nauk. **175**(3), 225–246 (2005). (in Russian)
19. Drapkin, A.L., Zuenko, V.L., Kislov, A.G.: Antenna-Feeder Devices. Soviet Radio, Moscow (1974). (in Russian)

Modeling D2D-Enhanced IoT Connectivity: An Approach Through the Simplified Analytical Framework

Tatiana Milovanova[1] and Dmitry Kozyrev[1,2](✉)

[1] Peoples' Friendship University of Russia (RUDN University), 6 Miklukho-Maklaya Street, Moscow 117198, Russian Federation
milovanova-ta@rudn.ru, kozyrev-dv@rudn.ru

[2] V.A. Trapeznikov Institute of Control Sciences of Russian Academy of Sciences, 65 Profsoyuznaya Street, Moscow 117997, Russian Federation

Abstract. In the light of the proliferation of the Internet of Things (IoT), the device-to-device (D2D) communication is becoming a promising technology and a key enabler for enhancing the energy efficiency of the wireless network environment and reducing the traffic latency between user equipments (UEs) within their communication range. This papers considers one simplified analytical framework for modeling of communication offloading scenario within the D2D communications underlying cellular network, in which a UE generates a session toward another UE in the same cell. The model aims to improve the system capacity and energy-efficiency by offloading cellular traffic onto D2D communications, when the source and destination UEs are proximate enough to satisfy their QoS requirements.

Keywords: D2D communications · Energy-efficient communications · Green IoT · Quasi-birth-death process

1 Introduction

The development of information and communication technologies (ICT) and increasing interest to their applications go hand in hand. The potential and capabilities of advanced ICT systems are still growing exponentially fueled by the progress in hardware, micro-systems, networking, data processing and human machine interfaces. This has led to an increased need for reliable connectivity of various electronic devices. This massive interconnection of proliferating heterogeneous physical objects together with services has formed a new ecosystem

The publication has been prepared with the support of the "RUDN University Program 5-100" and funded by RFBR according to the research project number 20-07-00804 (recipient T.A. Milovanova, mathematical model development, numerical analysis) and project number 19-29-06043 (recipient D.V. Kozyrev, formal analysis, validation). Numerical experiment was carried out using the infrastructure of the shared research facilities CKP "Informatics" of the FRC CSC RAS [14].

V. M. Vishnevskiy et al. (Eds.): DCCN 2020, LNCS 12563, pp. 658–665, 2020.
https://doi.org/10.1007/978-3-030-66471-8_50

which is technically termed as the Internet of Things [1]. However, this increase in connectivity creates many serious challenges. The rapid growth in the number of interconnected IoT devices creates a significant burden on existing and emerging wireless networks [2]. One of the ways to stem the tide of the number of newly deployed base stations is to take advantage of the device-to-device connectivity. D2D communication is a promising technology that utilizes the proximity of communicating devices [3], and is a potential enabler of green communication for it allows reducing power consumption [4,5]. Although the idea of proximal communications is not new [6], the industrial standardization of D2D technology has only recently been started [7,8]. D2D Communications underlying cellular networks provide mobile broadband operators with supplementary transport and increase network capacity through spatial reuse of radio resources for cellular and D2D communications [9].

In a series of previous studies the attention was focused on the mobility-centric analysis of communication offloading for interconnected IoT devices. In [10] an architecture of (beyond-) 5G systems was proposed, in which a mobile user is equipped with wearable devices that are continuously exchanging data with the surrounding nodes as well as the cloud via a gateway node. In that scenario, an aerial access point is used together with D2D communication to achieve the ultimate goal of offloading the existing cellular network. In [10] the authors proposed a novel mathematical framework that enabled assessing the impact of network offloading on the probabilistic characteristics related to the quality of service (e.g. connection unavailability and connection loss probabilities).

In [11] the authors elaborated on a simulation model of an enclosed area (a warehouse) with three types of wireless communication links: D2D links, drone-assisted and infrastructure-based links. As a result of the conducted research there was proposes the software tool, which provides the graphical representation of mobility of the objects inside the warehouse with given parameters. It allows one to assess the network coverage area, study the IoT connectivity issues and study the impact of different mobility models on the system-level performance of the considered data transmission system in terms of its network connectivity, defined as the ratio of connected devices.

In the current paper we propose and numerically study a simple mathematical model of the wireless communication between mobile UEs inside an IoT cell. The area of interest has a base station providing infrastructure coverage and a constant number of UE devices. All UEs in the cell can initiate sessions that can be serviced by utilizing both the D2D links and the infrastructure links. We describe an analytical framework for modeling the considered communications system, which turns out to be a closed multi-server queueing system. Numerical solutions of the equations describing the stationary distribution of the system's content allows one to gain some insight into the system's performance.

2 Mathematical Model

Assume that the total number of UEs is fixed and equal to N. The total number of infrastructure links is also fixed and equal to c, whereas the maximum total number of D2D links is limited by the integer nearest to $N/2$. Each UE initiates a session[1] according to a Poisson flow and occupies either an infrastructure or a D2D link. The session initiation rate via an infrastructure link is equal to α and the one via a D2D link is equal to β. Duration of a UEs session through both the infrastructure link and the D2D link (does not depend on the UE type and cannot be interrupted) has an exponential distribution with parameters μ_I and μ_D respectively. Due to the adopted assumptions the pair[2] $(D(t), I(t))$ is the two-dimensional Markov chain in continuous time t with the discrete state space

$$\{(i, j) : 0 \leq i \leq \lfloor N/2 \rfloor,\ 0 \leq j \leq c\}.$$

The Markov chain $\{(D(t), I(t)), t \geq 0\}$ is in fact the level-dependent QBD process. Following the terminology of [12] the level of this QBD process is the value of $D(t)$ and the phase is the value of $I(t)$. Denoting the infinitesimal generator of $\{(D(t), I(t)), t \geq 0\}$ by $\mathbb{A}$, the joint stationary distribution (which is unique and always exists whenever the μ_I and μ_D are not equal to zero simultaneously)

$$p_{ij} = \lim_{t \to \infty} \mathbf{P}\{D(t) = i, I(t) = j\},$$

can be found by solving (using one of the many methods available in the literature) the system of linear algebraic equations $\boldsymbol{p}\mathbb{A} = \mathbf{0}$, $\boldsymbol{p}\mathbf{1} = 1$, where $\boldsymbol{p} = (p_{00}, \ldots, p_{0c}, p_{10}, \ldots \ldots, p_{1c}, \ldots, p_{\lfloor \frac{N}{2} \rfloor, c})$.

The Kolmogorov–Chapman equations for the considered system have the following form:

$$p_{0,0} = p_{1,0} \cdot \frac{\mu_D}{\mu_D + 2(n-1)(\alpha+\beta)} + p_{0,1} \cdot \frac{\mu_I}{\mu_I + 2(n-1)(\alpha+\beta)}, \tag{1}$$

$$\begin{aligned} p_{i,j} = p_{i-1,j} &\cdot \frac{2(n-i-j+1)\beta u(i-1)}{(i-1)\mu_D + j\mu_I + 2(n-i-j+1)(\alpha+\beta)} \\ &+ p_{i,j-1} \cdot \frac{2(n-i-j+1)\alpha u(j-1)}{i\mu_D + (j-1)\mu_I + 2(n-i-j+1)(\alpha+\beta)} \\ &+ p_{i+1,j} \cdot \frac{(i+1)\mu_D u(n-i-j-1)}{(i+1)\mu_D + j\mu_I + 2(n-i-j-1)\gamma(j)} \\ &+ p_{i,j+1} \cdot \frac{(j+1)\mu_I u(n-i-j-1)u(c-j-1)}{i\mu_D + (j+1)\mu_I + 2(n-i-j-1)\gamma(j)}, \\ &0 \leq i \leq n,\ 0 \leq j \leq \min(c, n-i), \end{aligned} \tag{2}$$

where $u(x) = 0$ if $x \geq 0$ and $u(x) = 1$ otherwise, and $\gamma(j) = \alpha u(c-j) + \beta$.

[1] With the other UE, i.e. whenever an infrastructure or a D2D link is busy, it is busy simultaneously by 2 UEs.

[2] $D(t)$ is the total number of UE pairs, connected through the D2D links; $I(t)$ is the total number of UE pairs, connected through the infrastructure links.

Note that in the above equations, instead of the intensities α and β, one can introduce various call intensities α_n and β_n, depending on the number n of vacant pairs. This introduction allows to model a more general case of initiation of communication sessions in the cell.

Another way to obtain the joint distribution is to build the uniformized two-dimensional discrete-time Markov chain $\{(D_t, I_t), t = 0, 1, 2, \dots\}$ from $\{(D(t), I(t)), t \geq 0\}$. For the new, uniformized Markov chain the only possible transitions are as follows:

$$(D_t, I_t) = \begin{cases} (D_{t-1} + 1, I_{t-1}), & \text{w.p.} \frac{2\beta(N/2 - D_{t-1} - I_{t-1})}{\Delta_{t-1}}, \\ (D_{t-1}, I_{t-1} + 1), & \text{w.p.} \frac{2\alpha(N/2 - D_{t-1} - I_{t-1})}{\Delta_{t-1}}, \\ (D_{t-1} - 1, I_{t-1}), & \text{w.p.} \frac{\mu_D D_{t-1}}{\Delta_{t-1}}, \\ (D_{t-1}, I_{t-1} - 1), & \text{w.p.} \frac{\mu_I I_{t-1}}{\Delta_{t-1}}, \end{cases}$$

where $\Delta_{t-1} = \mu_D D_{t-1} + \mu_I I_{t-1} + 2(\alpha + \beta)(N/2 - D_{t-1} - I_{t-1})$.

From this relation it can be directly seen that within the considered model the session initiation rates and the session duration rates can be state-dependent (i.e. α, β, μ_D and μ_I can depend on the total number of idle UEs). This fact will be used in the next section to demonstrate the performance of the model under various assumptions on UEs behaviour.

3 Numerical Example

Since under the adopted assumptions a requested UE session can always be established, the most natural QoS measure is the distribution of the number of busy D2D and infrastructure links. Below we present the numerical results for the two qualitatively different cases: (i) constant session initiation rate with $\alpha_n = \beta_n = 2$, $0 \leq n \leq N/2$ and (ii) pulsing session initiation rate with

$$\alpha_n = \beta_n = \left(\frac{N + 2 - 2n}{2}\right) \left|\sin\left(\frac{N + 2 - 2n}{2}\right)\right|. \tag{3}$$

Both in (i) and (ii) cases the mean duration of a UEs session was held fixed and equal to 1 i.e. $\mu_I = \mu_D = 1$. The distributions of the number of busy infrastructure links and busy D2D links, depending on the total number of users N and total number of infrastructure links c, are plotted in Fig. 1 and 2. The numbers above the graphs indicate the respective mean value of the distributions. From the figures it can be seen that the distributions behave quite regularly, showing tendency towards normal-shaped curves.

In case (ii), with quite a different, pulsing session initiation rates given by (3) (see Fig. 3), the QoS performance remains qualitatively the same as can be seen from Fig. 4 and 5.

Note that it is very appealing to conjecture that the distribution of busy D2D links are normal. But as can be seen from Fig. 5, such conjecture is too optimistic since it can happen that the distributions are skewed.

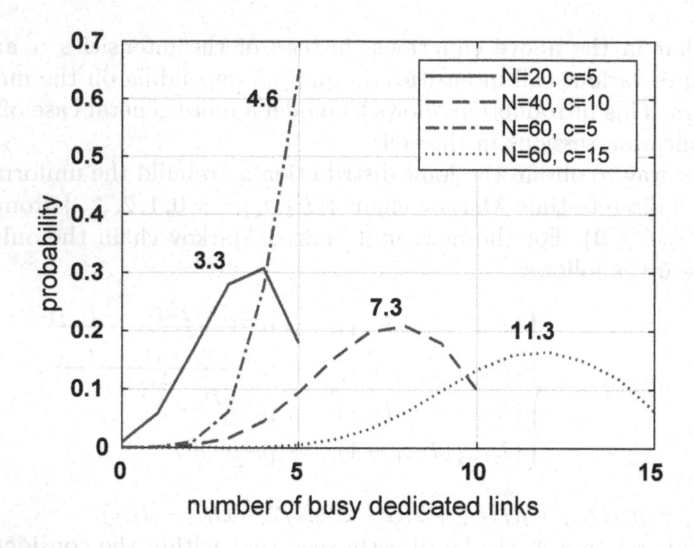

Fig. 1. Distribution of the number of busy infrastructure links for various values of N and c for case (i).

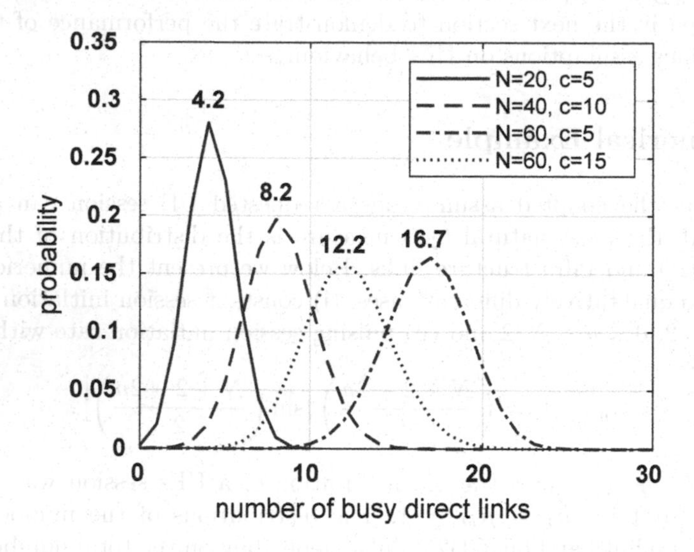

Fig. 2. Distribution of the number of busy D2D links for various values of N and c for case (i).

If the model assumptions hold, then, as the numerical experiments show, the system never behaves irregularly in the sense that the joint distribution of the number of busy D2D and infrastructure links is visually smooth. For example, there never appear two (or more) spikes i.e. the distribution is always unimodal (see Fig. 6 for the joint distribution in the case $N = 60$, $c = 15$ and the arrival rates as given by (3)).

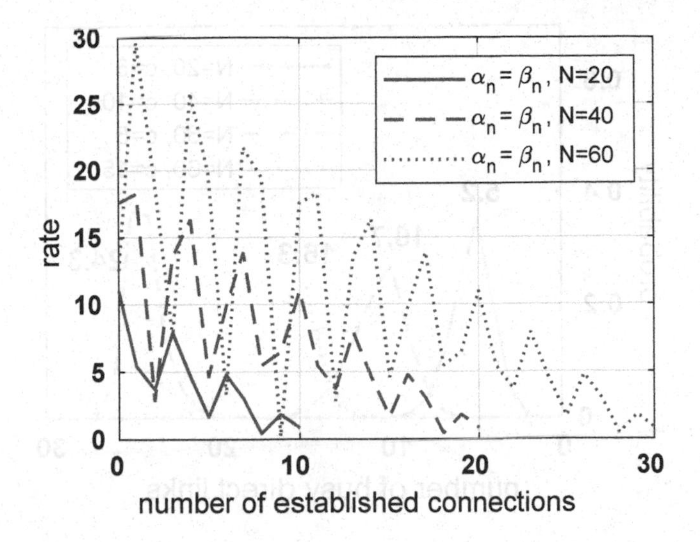

Fig. 3. Pulsing session initiation rates for various values of N.

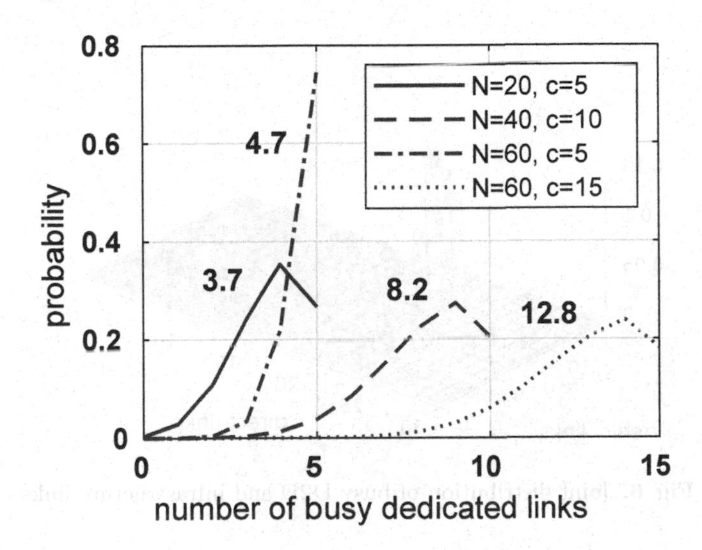

Fig. 4. Distribution of the number of busy infrastructure links for various values of N and c for case (ii).

Therefore, within the model assumptions, the shape of the empirical joint distribution (estimated from the available data) may serve as the indication of the irregularity of the UEs' behaviour.

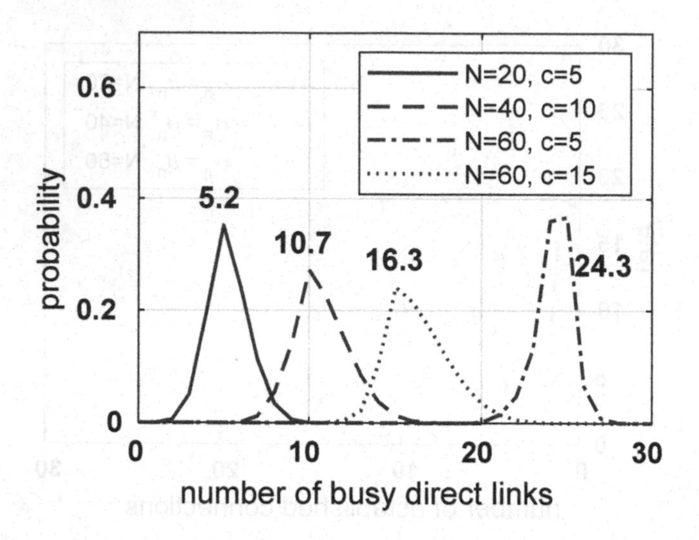

Fig. 5. Distribution of the number of busy D2D links for various values of N and c for case (ii).

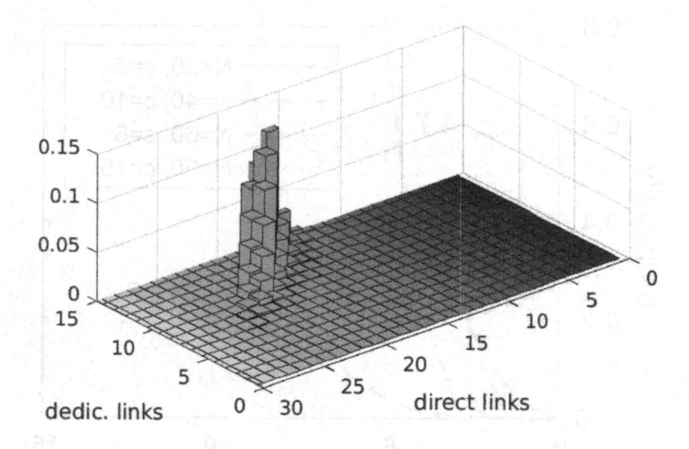

Fig. 6. Joint distribution of busy D2D and infrastructure links.

4 Conclusion

The performance evaluation study presented above is for probably the simplest model, which remains tractable but ignores such features as the mobility of the UEs, possible impatience of the UEs, not 100% reliability of the D2D and infrastructure links etc. As our numerical experiments show such necessary additional features do have impact on the quality of service within the proposed model. Yet, as long as we assume memoryless distribution for the ongoing processes within the model, the general picture (as in Fig. 3 and 5) remains the same (even if all the rates depend on the current state of the process). Incorporation of other distributions in the model requires further study and effort to put into the mathematical formulation. With this respect it is worth mentioning that (assuming memoryless distributions) there is a resemblance between the presented model and the 3-urn Ehrenfest model [13]. Indeed the evolution of the Markov chain $\{(D_t, I_t), t = 0, 1, 2, \dots\}$ can be imagined as the interchange of balls between the three labelled urns, under the restriction that the total number of balls remains fixed and equal to N. The only crucial difference is that in the classical multi-urn Ehrenfest model the rearrangement probabilities (although may depend on the total number of balls in some urns) are the same for any of the balls drawn. Yet in the presented model the rearrangement probabilities depend on the urn from which the ball is drawn. This fact makes the available analytical results for the multi-urn Ehrenfest model inapplicable and may motivate further studies.

References

1. Kalla, A., Prombage, P., Liyanage, M.: Introduction to IoT. In: IoT Security (2020)
2. Cisco. Global Mobile Data Traffic Forecast 2016–2021: White Paper (2017)
3. Fodor, G., et al.: An overview of device-to-device communications technology components in METIS. IEEE Access **4**, 3288–3299 (2016)
4. An, J., Yang, K., Wu, J., Ye, N., Guo, S., Liao, Z.: Achieving sustainable ultra-dense heterogeneous networks for 5G. IEEE Commun. Mag. **55**(12), 84–90 (2017)
5. Sakr, A.H., Tabassum, H., Hossain, E., Kim, D.I.: Cognitive spectrum access in device-to-device-enabled cellular networks. IEEE Commun. Mag. **53**(7), 126–133 (2015)
6. Fitzek, F.H.P.: Cellular controlled short range communication for cooperative P2P networking. Wirel. Pers. Commun. **48**(1), 141–155 (2009)
7. 3GPP TS 23.303 V15.1.0: Technical specification group services and system aspects. Proximity-based services (ProSe), Stage 2. Rel-15 (2018)
8. 3GPP TS 36.746 V15.1.1: Study on further enhancements to LTE Device to Device (D2D). UE to network relays for Internet of Things (IoT) and wearables. Rel-15 (2018)
9. Andreev, S., Hosek, J., Olsson, T., et al.: A unifying perspective on proximity-based cellular-assisted mobile social networking. IEEE Commun. Mag. **54**(4), 108–116 (2016)
10. Kozyrev, D., et al.: Mobility-centric analysis of communication offloading for heterogeneous Internet of Things devices. Wirel. Commun. Mob. Comput. **2018**, 3761075 (2018)
11. Rykov, V.V., Kimenchezhi, V., Kozyrev, D.V.: Mobility simulation of connected objects in a heterogeneous wireless data transmission system. CEUR Workshop Proc. **2177**, 11–18 (2018)
12. Neuts, M.F.: Matrix-Geometric Solutions in Stochastic Models: An Algorithmic Approach. The Johns Hopkins University Press, Baltimore (1981)
13. Karlin, S., McGregor, J.: Ehrenfest urn models. J. Appl. Probab. **2**(2), 352–376 (1965)
14. Regulations of CKP "Informatics". http://www.frccsc.ru/ckp. Accessed 29 Oct 2020

Optimization of SPTA Acquisition for a Distributed Communication Network of Weather Stations

Evgeny Golovinov[1], Dmitry Aminev[2], Sergey Tatunov[3], Sergey Polesskiy[3], and Dmitry Kozyrev[2,4](✉)

[1] Federal State Budgetary Scientific Institution All-Russian Research Institute for Hydraulic Engineering and Land Reclamation (VNIIGiM), Moscow, Russia
evgeny@golovinov.info

[2] V. A. Trapeznikov Institute of Control Sciences of Russian Academy of Sciences, 65 Profsoyuznaya Street, Moscow 117997, Russia
aminev.d.a@ya.ru, kozyrev-dv@rudn.ru

[3] National Research University "Higher School of Economics", 20 Myasnitskaya Ulitsa, Moscow 101000, Russia
bestdk2@gmail.com, spolessky@hse.ru

[4] Peoples' Friendship University of (Russia RUDN University), 6 Miklukho-Maklaya St, Moscow 117198, Russian Federation

Abstract. We consider a topology of a distributed communication network of weather stations and its hardware configuration. The operability criteria for the considered network are formulated. The theoretical foundations for optimizing the acquisition of spare parts, tools and accessories (SPTA) for distributed systems are considered. An algorithm for SPTA optimization for distributed systems has been developed. Optimization of the SPTA acquisition for the hardware components of the distributed communication network of 24 weather stations was carried out. The acquisition options are proposed, taking into account various criteria.

Keywords: Reliability · Agrometeorological parameters · Weather station · Redundancy · Failure rate · Operability · Spare parts

1 Introduction

While implementing agricultural digitalization programs, an important task is to equip agricultural fields with automatic means of monitoring agrometeorological parameters, such as temperature, humidity of the surface layer and atmospheric layer, precipitation, soil moisture at various levels, etc. Meteorological stations register such agrometeorological parameters and transmit them to monitoring servers for further analysis. Since these agrometeorological parameters are used

The publication has been prepared with the support of the "RUDN University Program 5-100" and funded by RFBR according to the research project number 19-29-06043.

V. M. Vishnevskiy et al. (Eds.): DCCN 2020, LNCS 12563, pp. 666–679, 2020.
https://doi.org/10.1007/978-3-030-66471-8_51

to determine the water consumption, quality, efficiency and speed of growing crops, and, consequently, the yield in general, their high importance poses stringent requirements for reliability and recovery time of weather stations in case of failure. Earlier, a comprehensive study for determining the reliability of a distributed network of weather stations was carried out [1–3]. In [4] a general methodology has been proposed for the reliability study of the considered network, and the apparatus of the multidimensional alternating stochastic processes has been applied. In addition to the analytical approach, the general topology cases were studied with the use of the developed discrete-event simulation model. Generally, another suitable complement to probabilistic reliability analysis of complex systems is structural sensitivity analysis [5]. Despite a significant amount of scientific work in this area, the problem of optimization of acquisition of spare parts for this class of distributed systems is not sufficiently developed.

A distributed automated weather stations network (AWSN), which is a mesh topology network (Fig. 1), distributed on the ground, consists of access points and weather stations (WS), remote from each other at distances of several kilometers and connected to the nearest access points via wireless communication channels. The equipment of each WS includes a microcontroller, memory card, Wi-Fi, GPS modules, GSM modem, antennas, agrometeorological sensors.

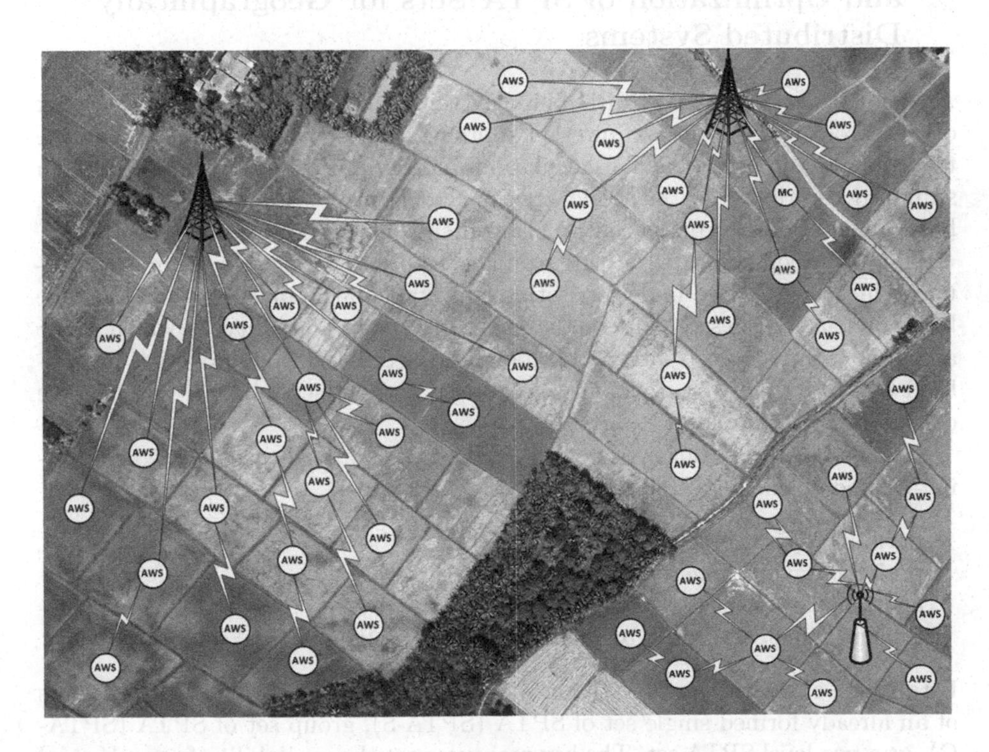

Fig. 1. Distributed communication network of weather stations

From the practice of agrometeorological measurements for the restoration of one weather station in AWSN, a recovery time of up to 1 day is permissible. This weather station can then be repaired within a week.

One of the ways to ensure the uninterrupted reliable operation of AWSN are spare parts, tools and accessories (SPTA) sets. However, the main methodological basis for calculating the sufficiency of these sets is based on the assessment of single or group spare parts without taking into account the possibility of combining all spare parts into a single warehouse, which will reduce maintenance costs compared to the standard scheme, when each facility has its own separate spare parts, but this increases the time required to satisfy the replacement request. The main methodological sources for calculating the SPTA indicators are the GOST 27.507 standard [6], the textbook by V.V. Rykov [7], the textbook by I.A. Ushakov [8], as well as the American standard MIL-HDBK-472 [9]. However, their use is not optimal for calculating SPTA for geographically distributed products. Therefore, there is a need to create a methodology for assessing spare parts sets for geographically distributed systems, such as AWSN.

2 Analysis of Existing Approaches to the Assessment and Optimization of SPTA Sets for Geographically Distributed Systems

Assessment of a territorially distributed system, which is the considered network of meteorological stations, requires an approach that is slightly different from the traditional approach to SPTA calculation and optimization. The main Russian standard describing the methodology for calculating SPTA, GOST 27.507 [6], has a number of drawbacks that appear when considering such systems.

The methods described in this standard apply to restorable technical products which are serviced under operating conditions (at application sites, storage bases or repair bodies), to ensure the reliability (maintainability) of which single and (or) group sets of spare parts, tools and accessories are provided, or two-level SPTA systems, or stocks in warehouses, formed according to the average consumption rates of spare parts.

In this standard, four types of replenishment strategies are used when evaluating and optimizing SPTA sets:

- periodic replenishment;
- periodic replenishment with emergency deliveries;
- continuous replenishment;
- replenishment according to the level of the irreducible stock.

Estimation of reserves in this standard refers to the calculation of all parameters of an already formed single set of SPTA (SPTA-S), group set of SPTA (SPTA-G), or a two-level SPTA set. The key measures are the availability factor K_r and the average delay time in satisfying requests for a spare part Δt (total number of spare parts is N):

$$K_r = e^{-\sum_{i=1}^{N} R_i} \quad (1)$$

$$\Delta t = \frac{\sum_{i=1}^{N} R_i}{\sum_{i=1}^{N} m_i \lambda_i} \quad (2)$$

where R_i is the inventory insufficiency indicator of type i and its calculation method depends on the strategy, m_i is the quantity of stock of type i, λ_i – replacement rate of an i-th spare part.

The differences between the estimates of SPTA-S and SPTA-G are very insignificant. They consist in calculating the consumption rates a_i. When evaluating SPTA-G, to calculate this parameter, the number of devices supported by this set, is taken into account.

Assessment of a SPTA set is performed in two stages. At the first stage, the recovery time is adjusted for all single sets of SPTA included in the system. At the second stage, the SPTA-G set is assessed, and then the STPA-S is assessed. The calculated indicator of sufficiency in the assessment of SPTA-S will be the indicator of sufficiency for the estimated SPTA set.

Another approach described in the American standard MIL-HDBK-472 introduces a warehouse system for organizing spare parts. The approach to problem solving used in this standard implies a good knowledge of the details of the project and how to maintain the system being assessed. MIL-HDBK-472's methodology is predictive, similar to recovery tasks. Basically, this technique is empirical, since it is used only if it is consistent with the known values of the recovery time for specific systems. Samples for recovery times are selected based on failure rates. It is assumed that the time spent on diagnosing and correcting the failure of a given component remains unchanged for the remaining components of a given type, which is not always true, as shown by operational data.

If recovery means replacement of the entire large blocks at once, then the sample is taken from the failure rate of these blocks. The estimated recovery time for each considered system is obtained from the maintainability properties checklist.

In MIL-HDBK-472 standard [9], the maintenance concept differs sharply from the conventional system of SPTA sets. The standard considers the use of a warehouse system. In total, three types of warehouses are considered. A central warehouse is a two-level SPTA system, a regional warehouse is a group SPTA set and an individual warehouse is a single SPTA set. At the same time, there is no division into separate types or different systems of stock formation within the standard. It is understood that there is always a central, a regional and an individual warehouse.

Fast deliveries to individual warehouses are made only from regional warehouses, while the central warehouse is used as a buffer storage for remanufactured spare parts that come from the repair base. Thus, a kind of smoothing filter appears, since part of the spare parts at the time of the necessary replenishment of regional warehouses are in the repair department. Additional stock is available in the central warehouse to compensate for delays due to repair times.

In addition, problems are solved using repair bodies and replenishment from them, as well as the restoration of failed parts of technical products with subsequent return to the warehouse.

The considered MIL-HDBK-472 standard uses optimal redundancy instead of optimal SPTA sets. In total, there are two types of optimal redundancy problems under consideration:

- by means of redundancy, achieve the maximum possible value of the selected reliability indicator of the system under the given restrictions on the total costs associated with the introduction of backup elements;
- by means of redundancy, achieve the required value of the system reliability indicator at the lowest possible cost for backup elements.

These optimal redundancy problems are solved in the following ways:

- using the Pareto method for optimal solutions;
- using the method of Lagrange multipliers;
- using fast coordinate descent methods;
- using the dynamic programming method.

Both standards, despite the detailed description of the mathematical apparatus and typical schemes, have a number of shortcomings: the American one is tightly bound to a specific structure of warehouses, the Russian one does not have stock reservation methods and support for identical SPTA-S for SPTA-G. Therefore, to solve the problem of SPTA calculation and optimization for distributed systems, more highly specialized methodologies are often developed. For example, in [10] and [11], the problem of optimal distribution of spare parts between an inexhaustible supply, local warehouses (SPTA-G) and operational facilities (SPTA-S) is considered. The article describes the application of strategies of periodic replenishment, continuous replenishment and replenishment at the level of an irreducible stock for organizing SPTA for various systems with a distributed structure. However, these papers do not deal with the periodic replenishment strategy with emergency deliveries, the application of which is well described in [12] and [13], however, these papers do not consider the continuous replenishment strategy, which is one of the most commonly used in real systems.

Among the many works on reliability and SPTA, there are papers that consider the network of weather stations: [14] and [15]. These studies focus on the SPTA calculation without estimating the cost of repairing the stations replaced in case of a breakdown, which does not allow making the most optimal estimate of the cost of SPTA-G.

Of all the above considered approaches, none can be fully applied to the selected network structure. This is due to the clear dependence of the type of spare parts, their replenishment strategy and the number of such parts. At that, in a geographically distributed system, there can be several spare parts of the same type, and each part can be replenished according to different strategies. Therefore, the optimal solution would be to develop a new technique for SPTA optimization for such systems.

3 Methodology for SPTA Calculation and Optimization for AWSN

The approach, according to which a single stock (warehouse) is used for various geographically distributed objects, allows to reduce the number of spare parts compared to the approach using a separate SPTA set for each operational object. A similar approach is used in geographically distributed systems (Fig. 2) – systems in which the reliability and operability of several operation facilities widely spaced across the terrain are supported by a single SPTA set. In addition, the same type of stock can be delivered using different strategies: periodic replenishment, emergency deliveries, continuous. Despite the fact that inside the warehouse the stock must be taken into account for each object, we must not forget that for each object there may be identical parts of the stock and for calculating the optimal set, this condition must be taken into account [14].

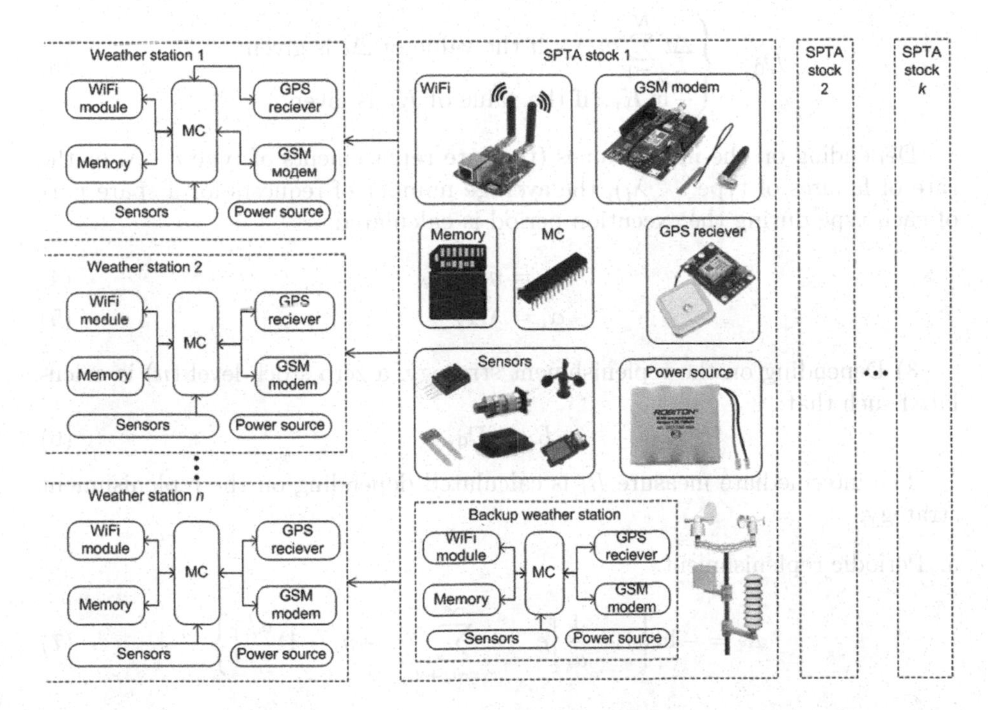

Fig. 2. Scheme of a geographically distributed SPTA for AWSN

When applying any technique, it is necessary to determine what input parameters are required for the calculation: the required value of the sufficiency indicator for optimization, the type of inventory costs and their unit of measurement, the number of inventory types, the description of each type of inventory, the calculation accuracy, the number of operational objects. In addition, stocks have their own parameters: the number of parts of a certain type in the product (m), the replacement rate of a part, the cost of a spare part (c), the average number of requests arriving for the stock of each type during the replenishment period (α), parameters describing the replenishment strategy (T, β), initial stock level for each type of parts (n). All indicators for the sake of formalization and ease of access must be recorded in a table, separately for each object.

After accounting for all operational objects, it is necessary to divide the types of stock into those that are unique for each object and into the general ones. After preparing the inpit data, it is necessary to form the parameters of reserves and their types in the form of tables of eight-position forms for each operational facility.

Common parts can be calculated according to the well-known method from GOST standard [1], but to calculate the parts that are unique for the object, a new method should be applied:

1) Calculate the indicator of the direct optimization problem using the following formula:

$$D_0 = \begin{cases} \Delta t \sum\limits_{i=1}^{N} m_i \lambda_i, \text{ if the value of } \Delta t \text{ is given;} \\ -\ln K_r, \text{ if the value of } K_r \text{ is given.} \end{cases} \tag{3}$$

Depending on the initial values (the rate replacements of type i – λ_i or the rate of failures of type i – λ_i), the average number of requests for a spare part of each type during the execution period is calculated:

$$\alpha_i = m_i \lambda_i T_i \tag{4}$$

$$\alpha_i = \lambda_i T_i \tag{5}$$

3) Depending on the replenishment strategy, a zero stock level (n) is calculated such that:

$$R_i \leq D_0. \tag{6}$$

The intermediate measure R_i is calculated depending on the replenishment strategy:

a. Periodic replenishment:

$$R_i = -\ln\left\{1 - \frac{1}{a_i}\left[e^{-a_i} \sum_{\gamma=n_i+2}^{\infty} (\gamma - n_i - 1)\frac{a_i^{\gamma}}{\gamma!}\right]\right\}. \tag{7}$$

b. Periodic replenishment with emergency deliveries (e.d.);

$$R_i = -\ln\left\{1 - \frac{T_{e.d.,i}}{T_i}\left[e^{-a_i} \sum_{l=1}^{\infty}\sum_{\gamma=l}^{\infty} (n_i + 1)\frac{a_i^{\gamma}}{\gamma!}\right]\right\}. \tag{8}$$

c. Continuous replenishment:

$$R_i = -\ln\left[1 - \frac{a_i^{n+1}}{(n_i+1)!\sum_{\gamma=0}^{n+1}\frac{a_i^\gamma}{\gamma!}}\right]. \tag{9}$$

4) The value for each type is determined:

$$R_i(n_i^0 + 1). \tag{10}$$

5) The value for each type is determined:

$$\Delta_i = \frac{R_i(n_i^0) - R_i(n_i^0 + 1)}{c_i}. \tag{11}$$

6) The cost-optimal stock is calculated:

$$R_\Sigma^0 = \sum_{i=1}^{N} R_i. \tag{12}$$

If the condition $R_\Sigma^0 \leq D_0$ is satisfied, the calculated values of n are optimal, if not, then a spare part of the type on which the condition is not satisfied is added and the calculation is repeated anew.

7) Availability factor for unique parts:

$$K_\mathrm{r} = e^{-\sum_{i=1}^{N} R_i(n_i)}. \tag{13}$$

Next, we carry out the calculation for groups of spare parts common to any operational objects. The basis for calculating these groups is the general probability formula or Bayes' theorem. Since the availability factor is the probability that the required spare part is available in the SPTA set at the time of the request, and the same spare part can be used for several operational objects, then the total probability is the sum of the probabilities that the required spare part is available for at least one of supported operational objects. Let's consider the case when there is a type of spare parts for two operational objects at once. Assume there is a separate stock for one and the other object. Then the total probability that, at the time of the request, the required spare part is available in stock is the sum of the probability that it is in both stocks, the probability that it is in stock 1 and not in stock 2 and the probability that it is in stock 2, but not in stock 1. Formula for the case of two operational objects is as follows:

$$K_{\mathrm{r\ overall}} = K_{r1}K_{r2} + K_{r1}(1 - K_{r2}) + K_{r2}(1 - K_{r1}), \tag{14}$$

where $K_{r1}K_{r2}$ is the probability that both stocks are ready, $K_{r1}(1 - K_{r2})$ is the probability that one is ready and the second is not, and $K_{r2}(1 - K_{r1})$ is the probability that the second stock is ready and the first is not. However, if there are more than two operational objects, then the formula for the total probability becomes bulky. At that, it becomes possible to calculate the availability factors for the first and second objects, taking into account different delivery times and replenishment strategies. Accordingly, optimization is carried out in such a way that the final overall availability factor is equal or close to the specified one. That is, a spare part is added iteratively, the availability factor is calculated using the algorithm given above, the overall availability factor is calculated using the general probability formula, and a comparison is made with the specified factor. A spare part is added for the type for which the value of Δ_i is the highest. This ensures a cost-effective addition of spare parts to the set. In order to obtain a cost-effective set, it is necessary to achieve the required product of the availability factors of an already optimized set of unique spare parts and general spare parts. If it is equal to the specified one, then the optimized set for the geographically distributed system is obtained.

The overall system availability factor is the product of the availability factors of unique parts and the availability factors of common groups. The resulting coefficient must be close to the specified one in the problem with a given accuracy, and then the resulting stock set is the required one.

The algorithm for optimizing the acquisition of SPTA for distributed systems is presented in Fig. 3.

The IDEF0 diagram of the optimization methodology for the SPTA acquisition for AWSN is shown in Fig. 4. This method allows the researcher to evaluate and optimize the SPTA according to the specified sufficiency indicators, technical requirements and conditions.

The initial data here are the technical task (TT) and technical specifications (TS) for SPTA sets or the AWSN as a whole.

The technique includes 5 stages:

- Assembling the SPTA schemes for AWSN parts and the product as a whole;
- SPTA assessment;
- designing the SPTA sets;
- SPTA sets optimization;
- analysis and development of recommendations for the SPTA calculation.

At the first stage, the formation of SPTA schemes for AWSN parts and the product as a whole is carried out taking into account the topology and structure of the equipment. In the case of a positive completion of the first stage, an assessment of SPTA is carried out, in compliance with the rules for determining the sufficiency indicators.

At the stage of designing SPTA sets, a scheme is developed according to the SPTA calculating standards and calculation projects are created in the ASONIKA-K-SPTA software [16]. When optimizing the SPTA sets using the proposed algorithm, the requirements for the cost and quantity of SPTA, and operating conditions are considered, and calculations are made in the ASONIKA-K-SPTA software. As a result of the technique, recommendations are developed and reports on the calculation of SPTA sets are released.

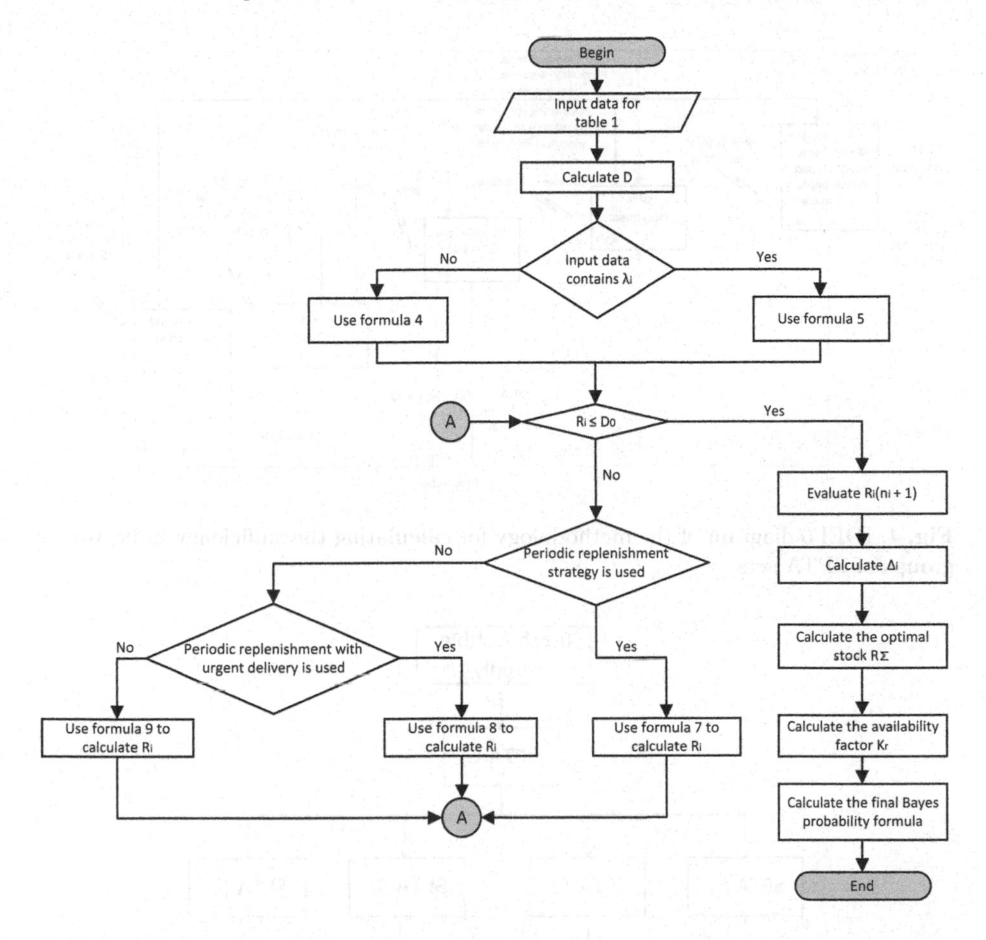

Fig. 3. Algorithm for optimizing the SPTA acquisition for distributed systems

4 Optimization of SPTA Sets for a Distributed Communication Network of 24 Weather Stations

To assess and optimize the SPTA set, a part of the network will be used, including one mobile communication tower, 6 weather stations connected via GSM modules and 18 weather stations connected via WiFi modules. The final SPTA scheme (see Fig. 5) consists of two types of SPTA-S: weather stations with WiFi connection and weather stations with GSM connection, and one type of SPTA-G, which includes the constituent parts of the weather stations.

Applying the above-described assessment and optimization methodology, we determine the types and parameters of stocks for a single weather station (Table 1) and for a set of SPTA-S (Table 2).

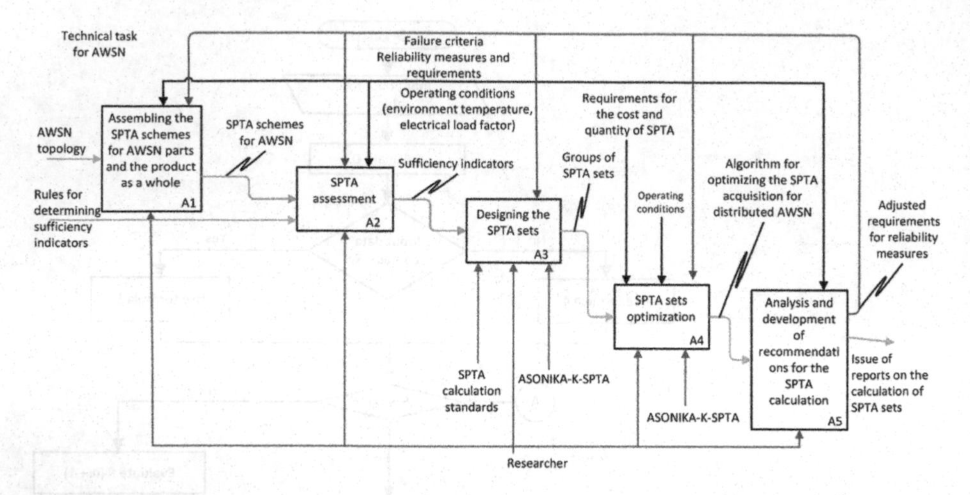

Fig. 4. IDEF0 diagram of the methodology for calculating the sufficiency indicators of groups of SPTA sets

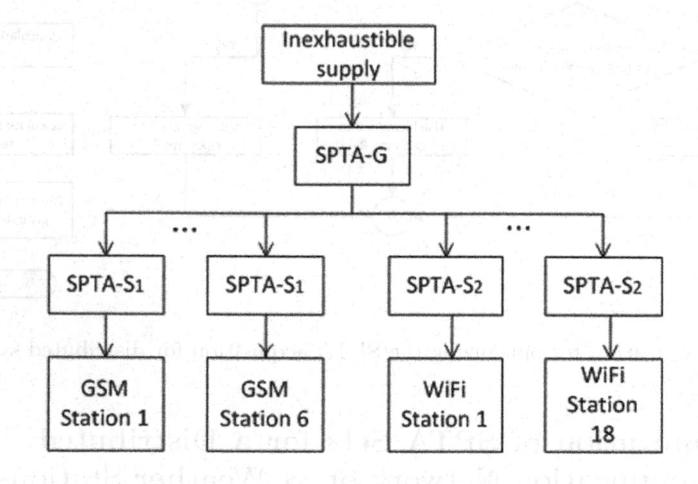

Fig. 5. SPTA diagram for the network of weather stations

Table 1. Types and parameters of weather station stocks

Spare part name	i	m, pcs	$\lambda \cdot 10^{-6}$, 1/hour	c, cost units	α, replenishment strategy	T, hours
Microcontroller	1	1	0.02	2000	3	168
RAM memory	2	1	0.03	1000	3	168
Sensors	3	1	0.03	1200	3	168
GPS module	4	1	5	800	3	168
GSM module	5	1	10	800	3	168
WiFi module	6	1	10	800	3	168
Power source	7	1	1	50	3	168

Table 2. Types and parameters of SPTA-S set stocks

Spare part name	i	m, pcs	$\lambda \cdot 10^{-6}$, 1/hour	c, cost units	α, replenishment strategy	T, hours
GSM weather station	1	6	26.08	6650	3	24
WiFi weather station	2	18	16.08	5850	3	24

Table 3. Sufficiency indicators for a set of SPTA-G for AWSN

Sufficiency indicators	Estimated values for the continuous replenishment strategy (7 days)	Required values
Average delay time in satisfying requests for a spare part, Δt, hours	0.2984	3.83455
Availability factor, K_r, rel. units	0.99999	0.9999
Total shortage level for n, $\sum R(n)$, rel. units	7.78226×10^{-6}	0.0001
Total shortage level for $n+1$, $\sum R(n+1)$, rel. units	4.2996×10^{-10}	Not specified
Total cost per set	2450	Not specified
Total number of spare parts in the set, pcs	4	Not specified

Table 4. Sufficiency indicators for a set of SPTA-S for AWSN

Sufficiency indicators	Estimated values for the continuous replenishment strategy (24 h)	Required values
Average delay time in satisfying requests for a spare part, Δt, hours	0.02323	2.04165
Availability factor, K_r, rel. units	0.99999	0.99909
Total shortage level for n, $\sum R(n)$, rel. units	0.00001	0.00091
Total shortage level for $n+1$, $\sum R(n+1)$, rel. units	1.61107×10^{-8}	Not specified
Total cost per set	12500	Not specified
Total number of spare parts in the set, pcs	2	Not specified

It is necessary to achieve the availability factor $K_r = 0.999$. Costs are presented as cost in rubles. The initial number of spare parts is zero, since there are no unique parts in the proposed stations, the SPTA-G set is combined without changes.

As a result of calculating separately SPTA-S (Table 3) and SPTA-G (Table 4) using the Bayes' formula, the final value of the availability factor is obtained:

$$K_{\text{r overall}} = K_{\text{SPTA-S}} \cdot K_{\text{SPTA-G}} = 0.99998121008569$$

The obtained value satisfies the posed requirements, but while using the chosen methodology, one can find out that the requirements can be met with less cost for spare parts and less waiting time of the recovery (Fig. 6).

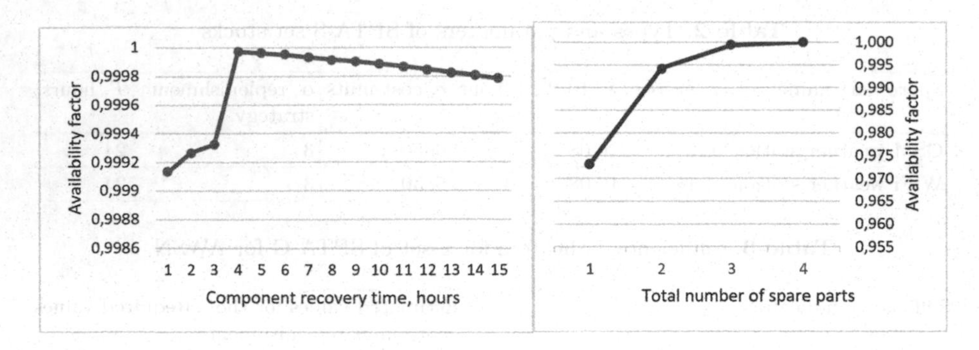

Fig. 6. Dependence of the availability factor on the recovery time (a) and on the number of spare parts (b)

Applying the proposed methodology for the SPTA evaluation and optimization, it was possible to calculate the optimal set of spare parts for a section of the network of weather stations under the condition of 24-h (1 day) recovery.

5 Conlusions

The proposed methodology for assessing and optimizing SPTA sets, taking into account the peculiarities of geographically distributed systems, made it possible to obtain an optimal set for a distributed communication network of 24 weather stations. The overall availability factor is 0.99998121, which meets the reliability requirements for this class of systems without strong loss of the recovery waiting time and at a reduced economic cost. The methodology has shown its effectiveness in optimizing SPTA sets for larger network sections without increasing the calculation time, based on which it can be concluded that the methodology is highly scalable and can be utilized in the future, when optimizing SPTA sets for large-scale AWSNs and similar systems.

Thus, it is advisable to modify the existing software for assessing and optimizing SPTA sets for geographically distributed systems, such as ASONIKA-K-SPTA, taking into account the provisions of the proposed methodology.

References

1. Aminev, D., Zhurkov, A., Polesskiy, S., Kulygin, V., Kozyrev, D.: Comparative analysis of reliability prediction models for a distributed radio direction finding telecommunication system. In: Vishnevskiy, V.M., Samouylov, K.E., Kozyrev, D.V. (eds.) DCCN 2016. CCIS, vol. 678, pp. 194–209. Springer, Cham (2016). https://doi.org/10.1007/978-3-319-51917-3_18
2. Rykov, V.V., Kozyrev, D.V.: Reliability model for hierarchical systems: regenerative approach. Autom. Remote Control **71**(7), 1325–1336 (2010). https://doi.org/10.1134/S0005117910070064

3. Kozyrev, D., Rykov, V., Kolev, N.: Reliability function of renewable system under Marshall-Olkin failure model. Reliab. Theory Appl. **13**(1), 39–46 (2018)
4. Aminev, D., Golovinov, E., Kozyrev, D., Larionov, A., Sokolov, A.: Reliability evaluation of a distributed communication network of weather stations. In: Vishnevskiy, V.M., Samouylov, K.E., Kozyrev, D.V. (eds.) DCCN 2019. LNCS, vol. 11965, pp. 591–606. Springer, Cham (2019). https://doi.org/10.1007/978-3-030-36614-8_45
5. Kala, Z.: Estimating probability of fatigue failure of steel structures. Acta et Commentationes Universitatis Tartuensis de Mathematica **23**(2), 245–254 (2019). https://doi.org/10.12697/ACUTM.2019.23.21. ISSN 1406-2283. E-ISSN 2228-4699
6. GOST 27.507-2015 Reliability engineering. Spare parts, tools and accessories. Evaluation of reserves (2015)
7. Rykov, V.: Reliability of Engineering Systems and Technological Risks. Wiley, Hoboken (2016). https://doi.org/10.1002/9781119347194
8. Ushakov, I.A.: Handbook of Reliability Engineering. Wiley, Hoboken (1994). 704 pages
9. MIL-HDBK-472 Maintainability Prediction. SESS - Systems Engineering Standards and Specifications
10. Tapia-Ubeda, F.J., Miranda, P.A., Roda, I., Macchi, M., Durán, O.: An inventory-location modeling structure for spare parts supply chain network design problems in industrial end-user sites. https://www.sciencedirect.com/science/article/pii/S2405896318316100
11. González-Varona, J.M., Poza, D., Acebes, F., Villafáñez, F., Pajares, J., López-Paredes, A.: New business models for sustainable spare parts logistics: a case study. Sustainability **12**, 3071 (2020)
12. Girgin, N.: Allocation of stocks in a multi-echelon spare parts distribution system (2011)
13. Durán, O., Afonso, P.S., Durán, P.A.: Spare parts cost management for long-term economic sustainability: using fuzzy activity based LCC. Sustain. (Switzerland) **11**(7) (2019). https://doi.org/10.3390/su11071835
14. Bagiorgas, H., Assimakopoulos, M., Konofaos, N.: The design, installation and operation of a fully computerized, automatic weather station for high quality meteorological measurements. Fresenius Environ. Bull. **16**(8), 948–962 (2007)
15. Zahumensky, I.: Guidelines on quality control procedures for data from automatic weather stations. In: Proc of the WMO, Instruments and Observing Methods, TECO-2005, Bucharest, Romania, Annex I, Report No. 3(15) (2005)
16. http://asonika.com/

Distributed Systems Applications

Crystal-Ball and Magic Wand Combined: Predicting Situations and Making Them Happen

Arkady Zaslavsky[1(✉)], Ali Hassani[1], Pari Delir Haghighi[2], Antonio Robles-Kelly[1], and Panos K. Chrysanthis[3]

[1] Deakin University, Geelong, VIC 3217, Australia
arkady.zaslavsky@deakin.edu.au
[2] Monash University, Melbourne, VIC 3168, Australia
[3] University of Pittsburgh, Pittsburgh, PA 15260, USA

Abstract. The Internet of Things (IoT) envisions an ecosystem in which everyday objects are enhanced with sensing, computation, and communication capabilities. These 'smart' devices (i.e., IoT devices) can sense and collect considerable amounts of data and share it with each other via the Internet. This paper proposes an IoT middleware platform enhanced with context- and situation-prediction capability, called Context-Prediction-as-a-Service (CPaaS). CPaaS offers real-time context prediction capabilities to a variety of IoT applications as a service and enables more effective decision support using relevant validated dependable information. A number of use cases where CPaaS can be deployed are also discussed.

Keywords: Context · Context prediction · IoT · Situational awareness · Distributed context management platform

1 Introduction and Background

The Internet of Things (IoT) envisions an ecosystem in which everyday objects (e.g., refrigerator, air conditioner, smartphones, weather stations, cars, industrial robots, just to name a few) are enhanced with sensing, computation, and communication capabilities. These 'smart' devices (i.e., IoT devices) can sense and collect very large amounts of data and share it with each other via the Internet. Due to proliferation of IoT devices, their numbers are expected to reach 20 to 30 billion in 2021 [1]. It is then possible to build services that can share rich, useful and relevant information with users about an 'entity' and situation of interest. We define data external to such an entity and interpreted by the IoT application as context. Sharing context enables a wide range of context-aware and smart applications that can adapt their behavior according to the current context of one or several entities.

The need for contextual intelligence is a fundamental and critical factor for delivering IoT intelligence, efficiency, effectiveness, performance, and sustainability. Contextual intelligence enables intelligent interactions between IoT devices,

V. M. Vishnevskiy et al. (Eds.): DCCN 2020, LNCS 12563, pp. 683–697, 2020.
https://doi.org/10.1007/978-3-030-66471-8_52

such as sensors/actuators, mobile smart phones, smart vehicles to name a few. Context management platforms (CMP) for IoT applications are emerging as the ETSI (European Telecommunications Standards Institute) efforts on standardisation [2] prove. Existing CMPs only take the current context of IoT devices and entities into account. However, in many IoT applications, it is essential to predict the future context of IoT entities with acceptable confidence above specified threshold and to provide important relevant dependable real-time information and valuable recommendations for better decision support and actuation. For example, with context prediction and proactive adaptation, it would be possible to predict direction, speed and scope of fire spread to proactively deploy mobile hardware assets like robots and/or drones to set up virtual fences, herd, direct and save wildlife in case of bushfires, which are common to Australia.

To address this shortcoming of existing CMPs, in this paper, we propose a novel framework, coined Context-Prediction-as-a-Service (CPaaS). CPaaS can create new capabilities for context management platforms (CMPs) that enrich IoT applications with proactive and preventive behaviour. Such applications can predict the future context and complex situations and take pre-emptive actions, and continuously re-evaluate the impact of the actions and update/extend the existing knowledge. CPaaS will be extremely beneficial to building intelligent systems as it will provide the following novel components:

- A context prediction selector that can match the requirements of IoT applications to the prediction techniques and determine the most appropriate prediction technique.
- An evolutionary learning approach that constantly re-evaluates context prediction and updates the existing model with new knowledge.
- An actuation mechanism that enhances context prediction with the capability to support preventive and mitigating actions.
- A standard, formal and flexible context prediction model that will extend an existing context query language (CDQL) [3] developed by authors.

In the rest of this paper, we will first discuss the vision of CPaaS and highlight its importances and also the main challenges that need to be addressed to develop such a framework. Then, we will describe our proposed architecture for CPaaS and explain its underlying components.

2 Motivational Use-Case

An important application domain in dire need of context and situation prediction is wildlife conservation. Australia is home to distinctive wildlife and a number of extant species. While a great deal of effort is spent on wildlife conservation, bushfires pose a significant danger to already endangered species [4]. In 2009, Black Saturday bushfires in Victoria burnt over 450,000 hectares, killing about one million wild and domesticated animals, reported by RSPCA [5]. Recent bushfires in New South Wales, Victoria, Queensland, South Australia and Kangaroo Island destroyed millions of hectares, and left more than one billion animals dead [6].

In Kangaroo Island bushfires, it was estimated almost 30,000 koalas perished [6]. Figure 1 shows map of recent Australian bushfires during Summer of 2019–2020. Accurate prediction of fire behaviour, size and spread (including its shape, area and speed) can significantly help with mitigating and reducing the catastrophic effects of bushfires on wildlife and allow managing and sustaining fire-prone and safe ecosystems for them. A promising solution that can mitigate the effects of bushfires on wildlife is to create a virtual fence that translocates animals to safe fire-free areas. Virtual fence devices have been already tested in another significant threat to Australian wildlife, which is roadkill [7,8]. A trial of virtual fences in Tasmania over three years showed a reduction rate of 50% [7]. Virtual fencing is also used as an animal-friendly system to move or confine livestock. Context prediction can be used to predict future fire threats, and animals can be moved to a safe area (where no fire is predicted) by creating a virtual fence.

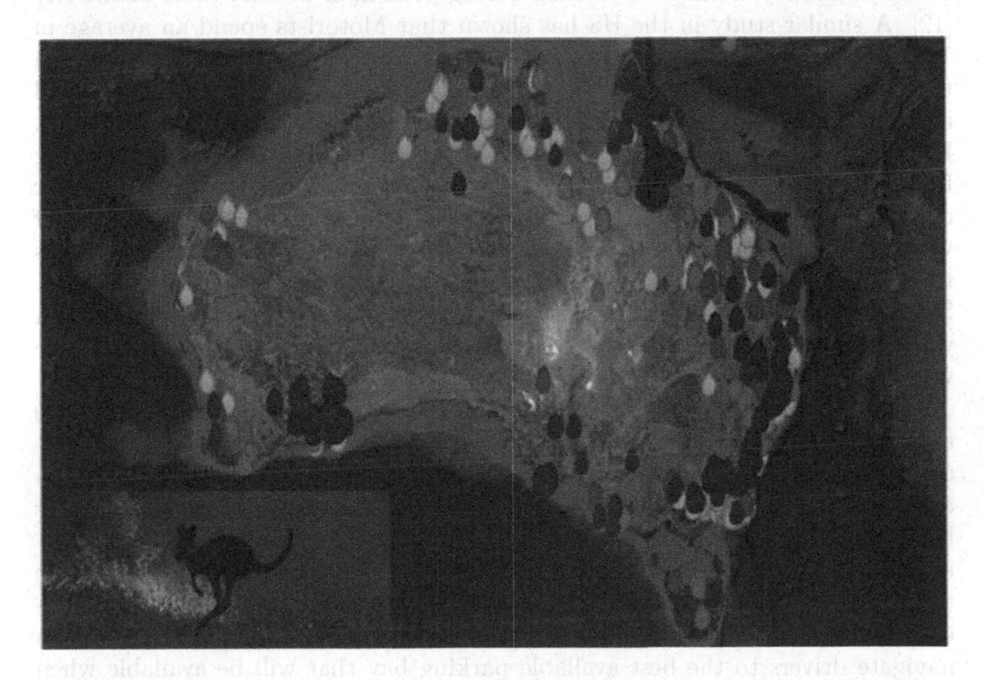

Fig. 1. Map of Australia with recent bishfires (https://sydneynews.sydney/sydney-news/1-billion-animals-perish-in-australian-bushfires/5762/, accessed on 12 September, 2020)

More practical and hotly needed motivational scenario is related to context-aware car parking availability prediction. Australia's capital cities have been transforming at a staggering pace. In 2011 the total population of the top five largest cities of Australia were around 13.5 million people [9]. Today, that figure is more than 16 million people, which means around 20% growth of population [10]. By 2055, the expected population of Australia's capital cities is predicted to reach more than 26 million people [11]. This growth in the population of the urban areas of Australia will undoubtedly put significant strain on the environment and infrastructures of the capital cities. For instance, without having a sophisticated plan to deal with the population explosion, in the near future, Australians will face several major problems such as traffic, air pollution, and water problems.

Therefore, to mitigate the possible negative impact of population growth, it is vital to design and develop effective and efficient solutions for better management of urban cities that contribute to environmental and urban sustainability and resilience.

One of the new raising challenges due to the population growth in large cities is searching for parking. As cities become more congested, the direct and indirect costs of parking are growing quickly. A survey completed in Melbourne and Sydney in 2014 showed that Australian drivers spend on average around 20 min a journey looking for parking during peak hour at busy areas of the city [12]. A similar study in the US has shown that Motorists spend an average of 17 h a year searching for carparks on-streets or in parking facilities [13]. Based on this report, the amount of wasted time, fuel, and emissions for each driver will add up to around $345 per year. This problem becomes worsen in large cities. For instance, In New York City drivers on average spend 107 h a year looking for parking spots, which about $2,243 in wasted time, fuel, and emissions per driver. To deal with the aforementioned problem and minimise the amount of wasted time, fuel, and emissions during parking search, a promising solution is to design, and implement a smart parking application by utulising IoT data. Such a solution can work with existing infrastructure and provide benefits to a wide range of stakeholders - from drivers to car manufacturers, parking space vendors and government.

However, most of the existing research in this domain take only real-time availability of parking facilities into account during the decision-making procedure. For example, if a driver is planning for a trip to a location that is 30 min away from its current location, and the IoT application suggests parking options based on their current availability, that parking might not be available anymore by the time the smart vehicle reaches the destination. Hence, to maximise the potential of such an application, it is vital to predict the future availability/capacity of parking facilities. Therefore, the smart parking application can navigate drivers to the best available parking bay that will be available when the vehicle reaches its destination.

3 Related Work and State-of-the-Art

3.1 Context- and Situation-Awareness

Context is a key characteristic of modern IoT-enabled systems. According to the widely acknowledged definition given by Dey and Abowd [14], context is "any in-formation that can be used to characterize situation of an entity". In plain words, any piece of information that the system has is a part of the system's context. The aspects of context include, but are not limited to, location, identity, activity, time. The system is context aware "if it uses context to provide relevant information and/or services to the user, where relevancy depends on the user's task". In simple words, the definition means that the system is context aware if it can use the context information to improve its performance, efficiency, effectiveness and utility. Although recognised as an interdisciplinary area, context-awareness is often associated with pervasive computing, and more recently with the Internet of Things (IoT). Context awareness is a core functionality in IoT, and any pervasive computing system is context aware to some extent.

IoT devices have sensing, actuation, computational and storage capabilities. These devices directly measure the environment characteristics (like temperature, light, humidity). Observation can be also considered as direct user input using keyboards, touchscreens, and voice recognition. Sensor information and user inputs are often processed in a similar manner. After highly heterogeneous input data is delivered, the first processing step is the data fusion and low-level validation of sensor information. Sometimes raw sensor data, collected in a single vector of values, are already viewed as low-level context. The distinction between different levels of context depends on the amount of pre-processing performed upon the collected sensor information. Usually raw or minimally pre-processed sensor data is referred to as low-level context, while the generalized and evaluated information is referred to as high-level context [15].

The situation awareness in pervasive computing and IoT can be viewed as the highest level of context generalisation [16]. Situation awareness aims to formalise and infer real-life situations out of context data. From the perspective of a context aware IoT system, the situation can be identified as "external semantic interpretation of sensor data", where the interpretation means "situation assigns meaning to sensor data" and external means "from the perspective of applications, rather than from sensors" [15]. Therefore, the concept of a situation generalises the context data and elicits the most important information from it. Properly designed situation awareness extracts the most relevant information from the context data and provides it in a clear manner.

3.2 Prediction Techniques

Context prediction aims to predict future context information. It can be done on any level of context processing, starting from low-level context prediction and ending with situation prediction. The existing prediction techniques which can be adapted to context prediction include [17]:

Sequence Prediction Approach. This approach to context prediction is based on the sequence prediction task from theoretical computer science and can be applied if the context can be decomposed into some kind of event flow.

Markov Chains Approach. Context prediction techniques based on Markov chains are quite widespread. Markov chains provide an easily understandable view of the system and can be applied if the context can be decomposed into a finite set of non-overlapping states.

Bayesian Network Approach. This can be viewed as the generalisation of the Markov models. It provides more flexibility but requires more training data in turn.

Neural Networks Approach. Neural networks are biologically inspired formal models that imitate the activity of an interconnected set of neurons. Neural networks are quite popular in machine learning. Context prediction approaches based on neural networks are extensively used and perform well.

Branch Prediction Approach. This approach initially comes from the task of predicting the instruction flow in a microprocessor after the branching command. Some context prediction systems use similar algorithms.

Trajectory Prolongation Approach. Some context prediction approaches treat the vector of context data as a point in multidimensional space. Then the context predictor approximates or interpolates those points with some function, and that function is extrapolated to predict future values.

Expert Systems Approach. Based on expert systems and rule-based engines, the expert systems approach appears in some works on context prediction. The goal of the approach is to construct the rules for prediction. It provides a clear view of the system.

One of the context prediction research challenges is the development of a general approach to context prediction. Many context prediction approaches were designed to fit a particular task and most of them were not designed to be generaliseable (although some of them have generalisation capability). The context prediction process consists of several steps [18]:

Sensor Data Acquisition. This step takes data received from multiple sensors and arranges them into the vector of values. Feature extraction. This step transforms raw sensor data for further usage. From vector of sensor data, vector of features is formed.

Classification. Performs searches for recurring patterns in context feature space. Growing neural gas algorithm was considered to be the best choice. The result of the classification step is a vector of values that represents degrees of membership of a current vector to a certain class.

Labelling. This is the only step that involves direct user interaction. The frequency of involvement depends on a quality of clustering step if classes are often overwritten and replaced that will result in more frequent user involvements.

Prediction. This step takes the history of class vectors and estimates a future expected class membership vector. Context prediction is a relatively new problem for computer science research. The area of context prediction is just being developed and still there are numerous challenges yet to be addressed. Those challenges include [17]:

Lack of General Approaches to the Context Prediction Problem. Most current solutions predict context for particular situations. There have been only a few attempts to define and solve the context prediction task in general.

Lack of Automated Decision-Making Approaches. Most context prediction-related works focused the efforts on prediction itself, but proper acting on prediction results usually was not considered. Most context prediction systems employed an expert system with pre-defined rules to define the actions based on prediction results. With one notable exception of Markov decision processes, almost no systems considered a problem like "learning to act".

Mutual Dependency Between System Actions and Prediction Results Is Not Resolved. This challenge is somewhat related to the previous one. Many context prediction systems considered the tasks of predicting the context and acting on predicted context in sequence: predict and then act on prediction results. That approach can handle only simplified use cases when actions do not affect prediction results. For example, in a smart home the system can employ any policy for switching the light or opening the door in advance, depending on user movement prediction results. But whatever the system does, it will not affect user intentions to go to a particular room. However, in a general case system, actions do affect prediction results. For example, consider a system which is capable of automatic purchases to some degree and which needs to plan the expenses, or in a more serious use case, consider a pervasive system that is capable of calling the ambulance and that needs to decide whether to do it or not depending on observed user conditions. In those and many more use cases, prediction results clearly will depend on what the system does. However, there are almost no work that considered the problem of mutual dependency between system actions and prediction results. So far, the only works that did address that problem were the ones on the Markov decision processes as discussed above. The task of resolving that dependency is actually a special case of a reinforcement learning task. In our opinion, although comparing to most reinforcement learning task pervasive computing systems have their own specifics (e.g., relatively obscure cost and reward functions, high cost of errors and therefore very limited exploration capabilities), recent advancement in the reinforcement learning area can help to over-come that problem.

If all those context prediction challenges are resolved, it will IoT systems handle more sophisticated use cases, enhance the applicability and the effectiveness of context prediction techniques and therefore enhance overall usability of IoT-enabled context-aware systems.

The next section will present and discuss the proposed Context Prediction-as-a-Service (CPaaS) engine.

4 Context Prediction as a Service—Vision and Open Challenges

As a step towards operationalising context-awareness in the realm of IoT, IoT middleware platforms, also known as Context Management Platform (CMP), have become a significant research challenge. CMPs manage interactions with sources of context (context providers (CP)) and offer contextual information to context-aware applications (context consumers (CC)) as a service. In our earlier research, we have developed a novel CMP called Context-as-a-Service (CoaaS) [19]. As it is illustrated in Fig. 2, CoaaS acts as a middleware that facilitates communication between CC and CP. One of the main features of CoaaS that distinguishes it from other existing CMPs is its generic and flexible query language that allows developers of context-aware IoT applications to query and monitor context of the entities of interest in real-time [3]. More importantly, Context Definition and Query Language (CDQL) supports queries about multiple entities and their situations (i.e., high-level, inferred context), which can be defined as part of the query at run time [3]. Another unique distinguishing feature of CoaaS is continuous reasoning and applying AI over IoT data streams and situations monitoring. Our experimental results show that CoaaS platform

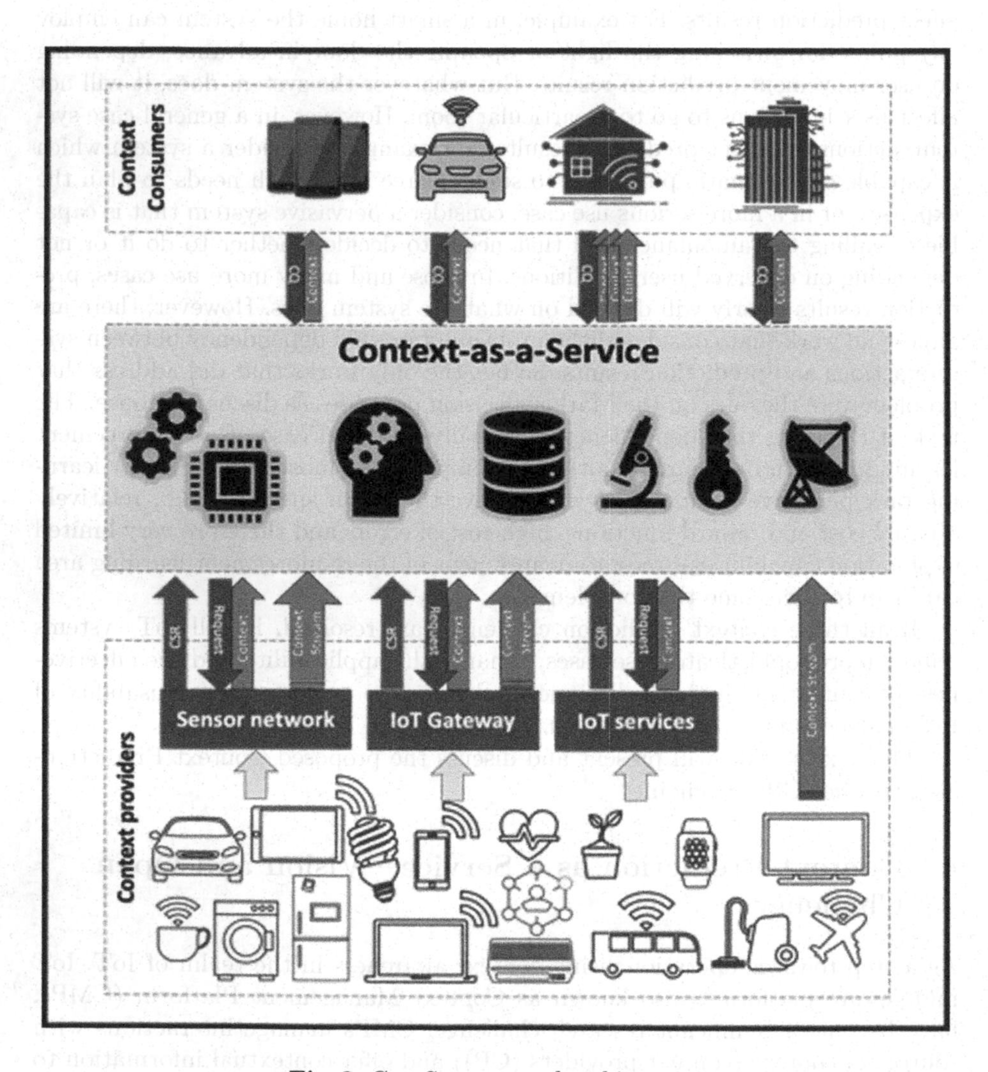

Fig. 2. CoaaS conceptual architecture

has a better overall performance in the execution of complex context queries compared to its main rival Fiware Orion [20] developed as part of EU research projects.

While current version of CoaaS platform can be utilised by context-aware IoT applications to seamlessly acquire and monitor the current context/situation of IoT entities, it does not support predicting future context/situations. IoT applications generally operate in unstable and unpredictable environments, where the context of entities constantly changes. It is insufficient to perform querying and adaptation solely based on the current context. To fully realise the potential benefits of context-awareness, it is essential to provide IoT applications with the capability to predict the future context and situations so they can adapt proactively to future changes, and take preemptive smart actions to mitigate any undesired or negative consequences. Context prediction will also enable applications to make better decisions and manage resources more efficiently.

A CMP that supports context prediction can offer distinct advantages to IoT ecosystems. It will enable a range of intelligent and proactive IoT applications and services in multiple domains. For example, it can enable smart parking with near real-time accurate prediction of parking availability and navigate drivers to available parking spaces hence saving fuel, time, and reducing emissions' impact on the environment.

The existing context prediction research is rather limited in meeting IoT applications' prediction demands. First, they do not support predicting multiple contexts and real-world complex situations that are highly important for operating in dynamic IoT ecosystems. Second, they mostly lack the ability to update their prediction model and suffer from the 'stale' model problem that can degrade accuracy. Third, they generally do not incorporate any actuation mechanism for mitigation purposes. Finally, the major shortcoming in existing context prediction approaches is that they are application-specific, and not applicable and accessible to IoT applications as a general service. The key challenges of context prediction and learning over predicted context in IoT ecosystems are listed below:

- Developing theoretical underpinnings and enhancing the prediction techniques for context/situation prediction, and incorporate them into a library where the best technique can be selected at run time according to the requirements of IoT applications in order to increase the accuracy and efficiency of prediction.
- Developing an advanced learning approach, using the state-of-the-art AI and machine learning techniques, to monitor system evolution and adaptation based on predicted context, and continuously update and extend the existing knowledge and heuristics to increase the prediction accuracy and efficiency.
- Developing a proactive actuation mechanism that can automatically re-evaluate and react to predicted context/situations by recommending actuations and mitigating actions to context consumers (entities and/or applications) that might be affected by the predicted context/situations.

- Developing new query language constructs for context and situation prediction to allow interactions between the CMPs and context consumers in a uniform way.

To address these challenges, in this paper, we propose to extend CoaaS by introducing a new component, called Context Prediction Engine (CPE). CPE will provide a generic mechanism for real-time context prediction, i.e., Context-Prediction-as-a-Service (CPaaS). Accordingly, we will extend the CDQL language with new constructs for context prediction, which can be used to satisfy the needs of context consumers.

5 Context Prediction Engine (CPE)

Significant challenges have to be addressed in researching, advancing, integrating and validating a generic real-time context prediction mechanism to enhance CoaaS and supporting real-time context prediction over IoT data. In [17], we conducted a comprehensive investigation into the context prediction techniques and challenges and identified the essential requirements of such a system while also identifying knowledge gaps in the current literature. Seven prediction approaches have been identified that can be used to predict context of IoT entities. These prediction approaches are Sequence prediction approach, Markov chains approach, Bayesian network approach, Neural networks/deep learning approach, Branch prediction approach, Trajectory Prolongation/Approximation approach, and Expert systems approach. Each of these approaches works best for a certain type of context data and would also depend on other corresponding meta-data, such as what amount of data is available, frequency of observations, and seasonality of the data to name a few. Hence, development of a generic context prediction mechanism, implies a dynamic context prediction mechanism that contains the library of the above-mentioned prediction techniques integrated, and based on the type of context data (e.g., location), the characteristics of data (e.g., observation frequency), and requirements of context provide (e.g., accuracy), applies the matching prediction techniques (e.g., Trajectory prolongation/approximation).

5.1 Problem Definition

To formulate the context prediction algorithm selection problem, consider a library of prediction techniques with n registered prediction algorithms and a given prediction task. The set of prediction techniques is denoted by $P = \{p_1, p_2, \ldots, p_n\}$, and the prediction task is denoted by:

$$pr_i = \langle metaContext_i, metaData_i, requirments_i, contextValues_i \rangle \tag{1}$$

where

$$metaContext_i = \langle entitytype, contextattributetype, fressness, ontology, \ldots \rangle \tag{2}$$

$$requrments = \langle ccuracy, responsetime, predictioninterval, \ldots\rangle$$
$$contextValues = \{cv_t, cvv_{t-1}, cv_{t-2}, \ldots\}$$

The goal of a context prediction selection is choosing the most appropriate prediction technique for a given prediction task in a way that the prediction accuracy is maximised. Hence, the problem of context prediction selection can be formulated as an optimization problem:

$$min_j(ca_{t+h}^{entity_i} - p_j(pr_i)), wherep_j \in \{p_1, p_2, \ldots, p_n\} \quad (3)$$

5.2 Context Prediction Engine Framework

Figure 3 presents the architecture of the proposed CPaaS framework, with a specific focus on the context prediction engine (CPE). CPaaS has five main components, namely Communication & Security Manager (CASM), Context Reasoning Engine (CRE), Context Storage Management System (CSMS), Context

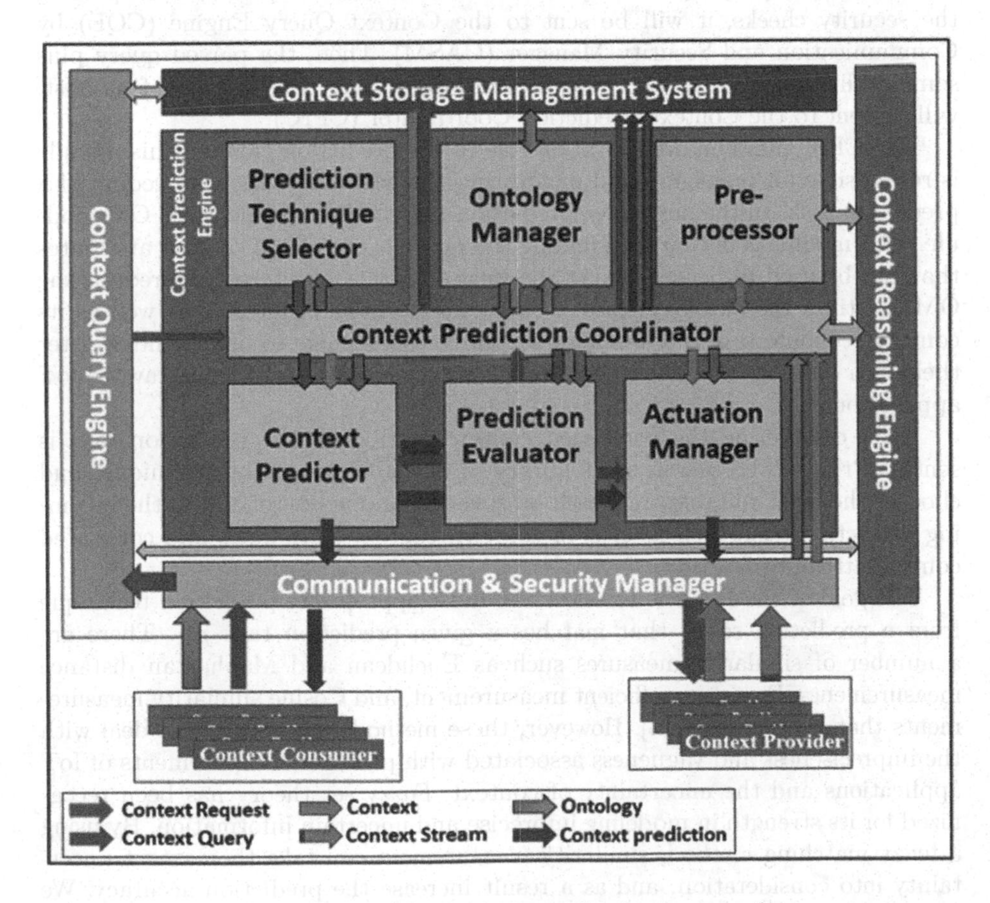

Fig. 3. Context-Prediction-as-a-Service framework

Query Engine (CQE), and Context Prediction Engine (CPE). CASM acts as a proxy and distributes all the incoming messages from CPs and CCs to the corresponding components after performing authentication and authorisation. CQE is responsible for parsing the incom-ing queries, generating and orchestrating the query execution plan, and producing the final query result. CSMS is in charge of storing descriptions of context services and facilitates service discovery, caching contextual information, and storing and analysing the historical context. The main task of CRE is to infer situations from raw sensory data or existing primitive low-level context. The details of these components are available in our earlier publications [21,22].

Lastly, CPE, which is the main focus of this paper is responsible for predicting the future context and complex situations and taking preemptive actions to mitigate any undesired or negative consequences. CPE consists of six main subcomponents, namely, Context Prediction Coordinator (CPrC), Ontology Manager (OM), Prediction Technique Selector (PrTS), Context Predictor (CPr), Prediction Evaluator (PrE), and Actuation Manager (AM).

When a CDQL query with a prediction task is issued to CPaaS, after passing the security checks, it will be sent to the Context Query Engine (CQE) by Communication and Security Manager (CASM). Then, the parsed query plus some additional information, such as meta-data about the context of interest, will be sent to the Context Prediction Coordinator (CPrC).

The CPrC plays an orchestration role in the prediction engine. This module is responsible for managing and monitoring the whole execution procedure of a prediction task. In the next step, prediction tasks will be pushed into OM module. This module is in charge of finding the possible correlated context attributes that can be used to better predict the future context of interest. Moreover, the OM identifies the ontology class of the context entity of interest as well as its context attribute type. CPaaS allows Context Consumers to define and register their own ontology to offer more flexibility to developers of context-aware IoT applications.

After discovering the correlated context attributes, the prediction task is sent to PrTS. PrTS searches the library of available prediction techniques and chooses the most suitable one based on several parameters related to the incoming prediction task, such as type of context, context meta-data, and correlated context-attributes.

The goal of the PrTS is to select the most appropriate prediction technique from a prediction set P that matches a given prediction task pr_i. There are a number of similarity measures such as Euclidean and Manhattan distance measurements, Pearson coefficient measurement, and Cosine similarity measurements that can be used [23]. However, these methods are not able to deal with the impreciseness and vagueness associated with prediction requirements of IoT applications and the uncertainty of context. Fuzzy set theory has been recognized for its strength in modeling imprecise and uncertain information. By using a fuzzy matching method, similarity measurement can take the context uncertainty into consideration, and as a result increase the prediction accuracy. We can represent the fuzzy selection of a prediction algorithm as below:

$$fuzzyselection_m = \{(pr_m, p_m, \mu_{(pr_m, p_m)}), m = 1, \ldots, n\} \quad (4)$$

The membershiop degree of p_m for pr_m is represented by $\mu_{(pr_m,p_m)}$. The fuzzy selection aim is to find a prediction algorithm p_m for the application which has a fuzzy matching with the prediction task pr_m, using a fuzzy rule that implies $pr_m \rightarrow p_m$.

In parallel with the previous step, a request is sent to the Context Storage Management System (CSMS) to fetch the related historical context of the attribute of interest and other correlated attributes. Lastly, all the retrieved information in previous steps is passed to the CPr module.

CPr builds a prediction model based on the provided information and performs the requested context prediction. The outcome of the prediction is sent back to the Context Consumer though the CASM. CPr also caches the prediction result and model for future usage. Moreover, the predicted context/situation is sent to AM.

AM is responsible for identifying the possible actions that need to be taken proactively based on the outcome of the prediction and notifying corresponding context consumers. To achieve this goal, AM uses OM to discover the severity of the predicted situation and possible actions. Then, AM query for all the context consumers that subscribed for the predicted situation and push an actuation signal to them.

After the prediction task is completed, the CPE keeps monitoring the real-time value of the context attribute of interest for a certain amount of time. The motivation behind this procedure is to evaluate the generated prediction model against the new knowledge and update it if needed to improve the prediction accuracy at the next cycle. PrE is in charge of this task. To do so, the real-time value of the context attribute is sent to the CPrC. Then, after performing a pre-processing on the value, it is passed into the PrE. In addition to the cleaned, real-time value of the context-attribute, other relevant information, such as the predicted value(s), generated prediction model, the prediction task, and historical values is also shared with the PrE. Then, PrE measures the accuracy of the prediction engine and tries to enhance the prediction accuracy by applying machine learning techniques, in particular, evolutionary learning.

Here, one of the major drawbacks of discriminative machine learning techniques, such as deep networks, are the lack of explainability and interpretability. More even so since deep networks do not have an inherent representation of causality. Moreover, with the rapid development of autonomous sensing platforms and decision support using deep learning there is an urgent need to add an automated interpretation and identification of the underlying processes and parameter states that govern the predictive behaviour of the network. This is even more important for legal and ethical evaluation and compliance particularly with mission-critical applications. Expressing the knowledge implicit in the network using hierarchical models so as to represent the knowledge in the network in an easily interpretable manner has analogies with mixtures of learners in the literature. This is an important observation since these have an explicit syntactic richness to support the extraction of declarative rules, a property that has been used in syntactical and structural pattern recognition for SVMs and shallow networks. Further, these mixtures of learners have been used in formulations based upon mixture models, Markov Logic Networks, decision trees and Bayesian Networks, all of which provide the ability to extract complex probabilistic relationships and impose constraints on the inference process.

6 Conclusion and Future Work

In this paper a novel Context Prediction Engine was proposed. It supports real-time context prediction and machine learning using deep networks, such as GANNs. Future work is concerned with prototyping, integrating and testing the software components, as well as collaborating with ETSI in further developing and improving standards related to context management platforms.

References

1. van der Meulen, R.: Gartner says 8.4 billion connected ‘Things’ will be in use in 2017 up 31 percent from 2016. Gartner. Letzte Aktual. **7**, 2017 (2017)
2. ETSI - ETSI ISG CIM group releases first specification for context exchange in smart cities. https://www.etsi.org/newsroom/news/1300-2018-04-news-etsi-isg-cim-group-releases-first-specification-for-context-exchange-in-smart-cities. Accessed 18 Feb 2019
3. Hassani, A., Medvedev, A., Delir Haghighi, P., Ling, S., Zaslavsky, A., Prakash Jayaraman, P.: Context definition and query language: conceptual specification, implementation, and evaluation. Sensors **196** (2019). https://doi.org/10.3390/s19061478
4. Australia bushfires: Which animals typically fare best and worst? - BBC News. https://www.bbc.com/news/world-australia-50511963. Accessed 10 Feb 2020
5. Counting the terrible cost of a state burning. https://www.smh.com.au/national/counting-the-terrible-cost-of-a-state-burning-20090208-811f.html. Accessed 10 Feb 2020
6. Bushfire Emergency. https://www.wwf.org.au/get-involved/bushfire-emergency#gs.wj4zec. Accessed 10 Feb 2020
7. Fox, S., Potts, J.M., Pemberton, D., Crosswell, D.: Roadkill mitigation: trialing virtual fence devices on the west coast of Tasmania. Aust. Mammal. (2019). https://doi.org/10.1071/AM18012
8. Englefield, B., Candy, S.G., Starling, M., McGreevy, P.D.: A trial of a solar-powered, cooperative sensor/actuator, opto-acoustical, virtual road-fence to mitigate roadkill in Tasmania, Australia. Animals **9**(10), 752 (2019)
9. 3218.0 - Regional Population Growth, Australia (2011). https://www.abs.gov.au/AUSSTATS/abs@.nsf/DetailsPage/3218.02011?OpenDocument. Accessed 10 Feb 2020
10. 3218.0 - Regional Population Growth, Australia, 2017–18. https://www.abs.gov.au/AUSSTATS/abs@.nsf/Lookup/3218.0Main+Features12017-18?OpenDocument. Accessed 10 Feb 2020
11. Australia's capital cities - McCrindle. https://mccrindle.com.au/insights/blog/australias-capital-cities/. Accessed 10 Feb 2020

12. Australian drivers spend over 3,000 hours looking for parking in their lifetime - Parkhound. https://www.parkhound.com.au/blog/australian-drivers-spend-over-3000-hours-looking-for-parking-in-their-lifetime/. Accessed 10 Feb 2020
13. Drivers spend an average of 17 hours a year searching for parking spots. https://www.usatoday.com/story/money/2017/07/12/parking-pain-causes-financial-and-personal-strain/467637001/. Accessed 10 Feb 2020
14. Dey, A.K.: Understanding and using context. Pers. Ubiquitous Comput. **5**(1), 4–7 (2001). https://doi.org/10.1007/s007790170019
15. Ye, J., Dobson, S., McKeever, S.: Situation identification techniques in pervasive computing: a review. Pervasive Mob. Comput. **8**(1), 36–66 (2012). https://doi.org/10.1016/j.pmcj.2011.01.004
16. Endsley, M.R.: Toward a theory of situation awareness in dynamic systems. Hum. Factors **37**(1), 32–64 (1995)
17. Boytsov, A., Zaslavsky, A.: Context prediction in pervasive computing systems: achievements and challenges (2011)
18. Mayrhofer, R.: An architecture for context prediction. Adv. Pervasive Comput. (2004)
19. Moore, P., Xhafa, F., Barolli, L.: Context-as-a-service: a service model for cloud-based systems. In: Proceedings of the 2014 8th International Conference on Complex, Intelligent and Software Intensive Systems, CISIS 2014, pp. 379–385 (2014). https://doi.org/10.1109/CISIS.2014.53
20. Fiware-Orion. https://fiware-orion.readthedocs.io/en/master/. Accessed 18 Feb 2019
21. Hassani, A., Medvedev, A., Zaslavsky, A., Haghighi, P. D., Jayaraman, P.P., Ling, S.: Efficient execution of complex context queries to enable near real-time smart IoT applications. Sensors (2019). https://doi.org/10.3390/s19245457
22. Hassani, A., Medvedev, A., Zaslavsy, A., Delir Haghighi, P., Ling, S., Indrawan-Santiago, M.: Context-as-a-service platform: exchange and share context in an IoT ecosystem. In: Percom 2018 (2018)
23. Siddiquee, M.M.R., Haider, N., Rahman, R.M.: A fuzzy based recommendation system with collaborative filtering. In: SKIMA 2014–8th International Conference on Software, Knowledge, Information Management and Applications (2014). https://doi.org/10.1109/SKIMA.2014.7083524

Architectural ML Framework for IoT Services Delivery Based on Microservices

Kristina Dineva and Tatiana Atanasova(✉)

Institute of Information and Communication Technologies - BAS, Acad. G. Bonchev, Bl.2, 1113 Sofia, Bulgaria
{k.dineva,atanasova}@iit.bas.bg

Abstract. The Internet of Things (IoT) is the interconnection of devices and services that allows free data flow. Managing and analyzing this data is the actual added value that IoT is beneficial for. Machine learning plays an increasingly important role in performing data analysis in IoT solutions. This paper presents an architectural framework with machine learning solutions implemented as a service in the microservice group. This architectural framework for IoT services delivery is designed following the Agile methodology. The requirements for the software architecture and expected functionalities of the system are defined. The microservices collection is explained by providing a separate description for every service. Machine learning (ML) analytics on IoT (as the processing paradigm for intelligently handling the IoT data) is represented as a part of the microservice platform. Several strategic advantages of the proposed microservice-based IoT architecture over others are discussed together with implementation issues.

Keywords: Microservice architecture (MSA) · Internet of things (IoT) · Machine learning (ML) · Prediction · Scalability · MongoDB

1 Introduction

Internet of Things (IoT) is steadily invading all areas of our life. With the better network connectivity more and more "things" are getting available covering varieties of applications. Wide range of IoT solutions is available nowadays. The Internet of Things can now be considered a mature technology with many options both in terms of connectivity and hardware. IoT Analytics Research talks about over 600 IoT Platforms in 2019 [1]. A whole number of practical use cases are already available [2].

Increased number of implementations of IoT solutions leads to enormously increased data collection and transmission of all types of data. Managing and analyzing this data represents the real added value that IoT can bring. Machine learning (ML) plays an increasingly important role in data analyzing in IoT applications. Knowing how to analyze and understand the collected data from IoT devices is a crucial element in the successful system realizations [3]. It is

V. M. Vishnevskiy et al. (Eds.): DCCN 2020, LNCS 12563, pp. 698–711, 2020.
https://doi.org/10.1007/978-3-030-66471-8_53

a challenging task to apply analytics over IoT data streams to discover new information. Analytics on IoT-enabled devices require a platform consisting of machine learning frameworks. Beside this the integrating software with internet connectivity is needed to implement an effective IoT system.

Among the most significant impacts of IoT is the demonstration of the concept of shifting from computing with centralized servers to distributed ones deployed on modern technical infrastructures. IoT devices can interact by providing new services across different domains and environments, generating added value for different users of domain-independent services. This poses a number of challenges for IoT, namely: the technical (connectivity, compatibility, intelligent analysis and security), societal, business and legal. One of the main challenges IoT technologies face is related to the big heterogeneity of the IoT world. This heterogeneity is associated with the variety of use cases where these technologies can be applied, the many different types of devices, the diverse connectivity mechanisms supported and the number of protocols and IoT platforms. Interoperability and integration are key requirements for these existing solutions [4].

Recently, Software as a Service (SaaS) is gradually becoming the norm in the IoT world. Research in IoT is directed to solutions for the development of IoT applications from the perspective of the Service Oriented Architecture (SOA) style, which aims to provide loosely coupled systems to control the use and reuse of IoT services at the middle-ware layer, to minimize system integration problems. However, in SOA applications some challenges of integrating, scaling and ensuring resilience in IoT systems persist. Major key reasons for poor integration in IoT systems is the lack of an intelligent, connection-aware framework to support interaction in IoT systems [5]. IoT systems tend to expand, and over time the appropriate SOA framework becomes too rigid to cope with system extensibility. Thus the importance of analyzing the adoption of Microservices in the development of IoT applications is arising. Microservice aims to fragment different IoT systems based on the System of Systems (SOS) paradigm to adequately provide for system evolution and extensibility [5].

In contrast to traditional architectures for designing IoT applications, the Microservices architecture offers advantages such as those discussed in [6]: Technology Heterogeneity; Resilience; Scalability; Ease of Deployment; Organizational Alignment, which will enable the development of large-scale IoT applications. Microservice practitioners give some common characteristics of microservices as tools which aim is "to build, manage, and design architectures of small autonomous units" [7]. In the research [8] the applications based on the Microservices architecture defined as "polyglot, that is, they are constructed from the programming language of the choice of the developer and the deployments are carried out continuously".

Microservice architecture (MSA) is a service-oriented pattern architecture. Microservices are described using simple API which is thinly layered (lightweighted compared to SOA) [5]. Microservices are similar to SOA, they are both service-based architecture with an emphasis on service usability and reusability. They differ in terms of their architectural style, architectural characteristics, service characteristics and capabilities. Service atomity and taxonomy present a great distinctive feature between SOA and microservices. Interactions between different services are appropriately handled in microservices architecture and as such will be advantageous in inherently asynchronous IoT system [9].

Machine learning is the processing paradigm for intelligently handling the IoT data. However, the resource of IoT devices, render them a challenging platform for deployment of desired data analytics [10]. This research is focused on providing independent service deployment and ML service atomicity on the base of microservices architecture (share-as-little-as-possible). With microservices, the functionality of the IoT system is broken down into components to its lowest reasonable level. The microservice-based architecture for designing efficient, resilient, and scalable machine learning frameworks for IoT applications is provided in the paper. It is shown that microservices enable efficient and faster development by breaking down IoT functionalities into small, modular and independent units that work in isolation without affecting the overall performance of the IoT ecosystem. The used technologies and software libraries for implementation of the architectural ML framework are specified.

2 Problem Statement and Defining System Requirements

The task is to design an innovative system built of IoT devices with a cloud server part and a web user interface that interact and exchange data with each other. The developed architecture of the IoT devices has to be of modular type, which will bring several significant advantages to the system, as follows:

1. Effective work with different types of hardware components and devices, easy replacement of existing ones or adding new ones when there is a need to change the scale or functionality of the system.
2. The support of different communication channels by IoT devices allows work with different hardware components depending on the specific needs of the user, the specifics of the terrain and the number of devices used.
3. Much of the logic performed by the devices can be controlled and changed remotely through the software application.

The *main requirements* for the development of server architecture can be defined as:

- Robust functionalities - clearly defined functionalities are essential.
- Services - clearly defined services, each service to perform a specific task without interfering with the tasks of other services.
- Solid APIs – all system entry points should be concrete, consistent, secured and well described.
- Flexibility - Easily add or remove elements from the system without compromising its integrity.
- Scalable - the server should scalable both performance-wise and data-wise.

- Maintenance - easy troubleshooting by isolating faults, this process does not interfere with the operation of the entire system.

The expected *supported functionalities* of the system are:

- User support - creation, delete and edit users and all data related to them.
- Data handling - implementation and control of sustainable processes for acquiring heterogeneous data from remote devices; transformation of new data and the existing one.
- Machine Learning - implementation, usage and monitoring of a trained machine learning model against prepared data.
- IoT devices - workflow management of the IoT devices with sending eligible commands to them by authorized users.
- Logging - recording events in the system that show when and how they occurred and the results after their occurrence.

Based on the described functionalities, the technological stack is defined, with which the required implementation will be achieved. The technology stack is the aggregation of all infrastructure and software tools and technologies that are used during the software process. Choosing the right technology stack is of great importance for obtaining a successful overall result.

3 Software Architectural Framework

Design refers to the process of defining the software architecture. It, in turn, determines the overall structure of the software system, software components, properties and relationships between them.

There are different types of software server architectures - Single tier, Single layer, Traditional n-layer, Microservice architecture, DDD (Domain Driven Design), EDA (Event-Driven Architecture). After determining the functional requirements and the technological stack for system development, it is decided that the most appropriate architecture is based on the use of microservices (Microservice architecture - MSA) [https://microservices.io]. It allows the creation and maintenance of distributed software systems, usage of various databases and decoupled user interfaces that can communicate with the backend server. The microservice architecture is an advanced one with a high complexity level. Specific knowledge and capabilities are required to properly implement microservice-based solutions in a real environment. In essence, it is a set of several or many stand-alone software applications, each of which performs its specific tasks. They can communicate with each other and exchange information.

Microservices are small, modular, independently deployable and loosely coupled stand-alone application, which help to reduce the integration complexity faced while using monolithic architecture. The research [11] further specifies the main aim of microservices in the way "to coordinate distributed applications as a collective implementation of services running autonomously and independently based on its described process context". This MSA framework promotes independent service deployment, service atomicity, prevention to single-point failure, enhance security transaction management and service orchestration.

Unlike monolithic and other server architectures, the microservice architecture has several features that give it a strategic advantage over others:

- *Reducing complexity* - in essence, the business logic of the software system is distributed among the required number of microservices, which leads to a lack of dependencies between the individual elements in the system. Each one is an independent application and is responsible for some specific features in the system. Thus, all the complexity in the implementation of the business logic of the system is divided into small, specifically defined tasks, which are delegated to a specific microservice.
- Each microservice *is deployed independently.* This is important because the requirement for 24/7 availability of the system implies that it is not stopped during an upgrade or maintenance process.
- Each microservice *is scaled individually.* This is a mandatory feature for distributed software applications.

The proposed software architectural framework (Fig. 1) consists on a group of microservices - Identity Service, Data Acquisition, Data Transformation, Machine Learning, Device Management, Logging Service.

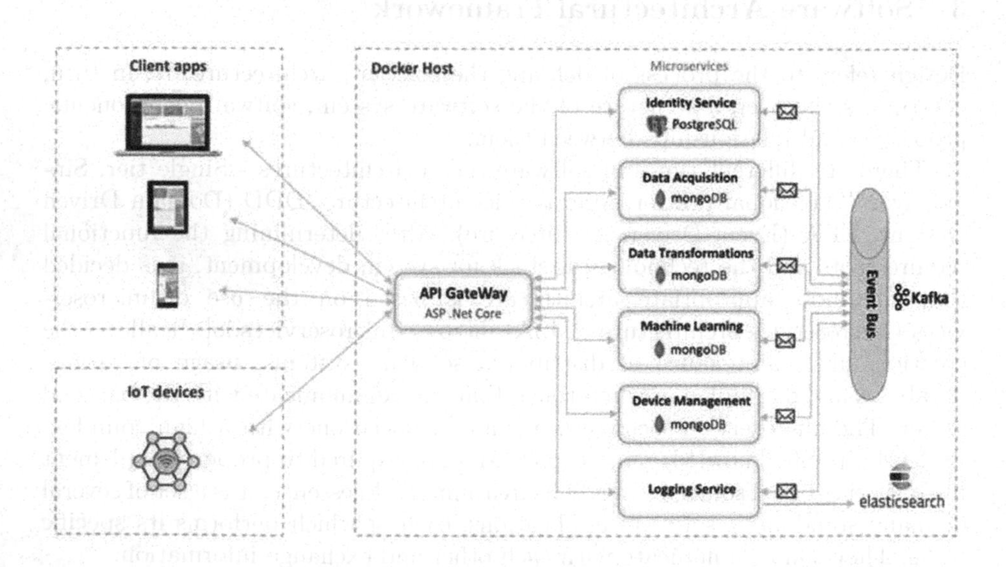

Fig. 1. Microservice software architecture.

3.1 Microservices Group

Identity Service is responsible for all operations related to the user, such as creating new ones, editing existing ones, deleting users, changing users' pass-words, recovering a forgotten password, setting user roles and others. All these actions, which are available for each user, allow having uniquely identified users with properly assigned needed access rights to the system and to provide the appropriate level of protection of the user's data. The technology used is Asp.NET Core 3, which is the latest stable version at the moment. The database used is Postgres, which runs in a Linux environment. The connection between the two technologies is the Entity Framework (EF) Core, which is an ORM (Object-relational mapping) for working with databases and applications using Asp.NET Core. EF Core allows the creation of mapping, where software objects correspond to specific tables in the database. This provides the required level of abstraction and makes the process of developing and maintaining a software application much easier and understandable for developers.

Figure 2 shows a general diagram of the relational database for the Identity Service. It contains 12 tables with relationships between them of type 1:1 and 1:n.

Data Acquisition is responsible for all data pipelines (a set of all steps taken in the process of sustainable data transfer between remote physically different locations) through which the system collects data from remote IoT devices. The system has the ability to initiate a process for self-retrieval of data via SFTP (Secure File Transfer Protocol), allows data to be pushed to the system using REST (Representational state transfer) and/or Streams and two-way data exchange using Binary over TCP protocol for messaging.

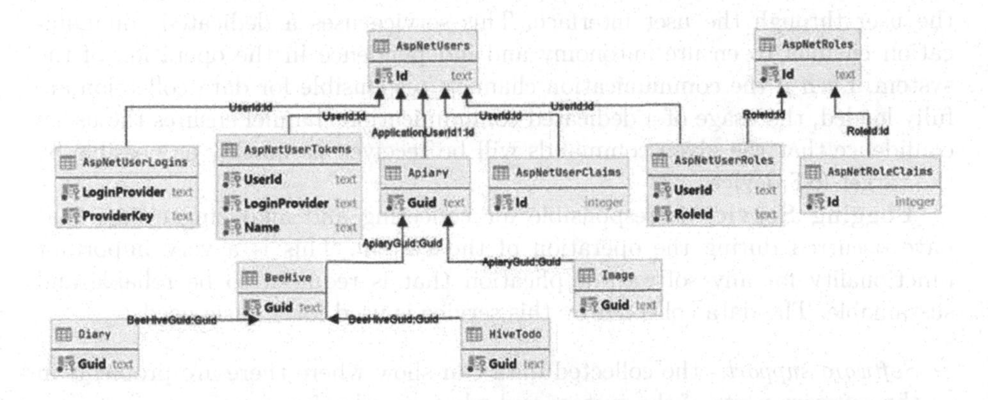

Fig. 2. Schematic of a relational database for Identity Service.

Once the data has been successfully stored in the system and validated as a correct transaction, another microservice is notified that new data is available and it can start working with it. The programming language is Python 3.8, and the technologies used are Numpy, Pandas, Scikit Learn, Kafka [12], FaspAPi, Sql Alchemy. All logged data are stored in a MongoDB database. It was chosen because it is a database that can be easily adapted if there is a change in the data scheme - something that is common when working with unstructured heterogeneous data.

Data Transformation is responsible for all transformations on the newly received raw data until they acquire the required type, suitable for analysis and application of machine learning models on them. To achieve sustainable results in the machine learning model prediction process, all transformations on the data are consistent with those undertaken during the training of the machine learning model. This is an important requirement for achieving maximum reliability of the predicted results. For this reason, a microlayer has been created in the structure of this microservice, which can be easily modified if changes are required in the process of transforming the input data.

Machine Learning Service is responsible for applying ready-made, trained machine learning models to predict values [13] and possible events. Only new and successfully transformed data is fed into the model. This is a condition of high importance to achieve near real-time prediction speed. The microservice has a modular structure, which makes it easy to exchange an existing model with a new one, as well as the integration of a new model in this service. The microservice does not store information about the status of the data. It works on a submitted data, provides a result without recording in a database neither the number of predictions made nor the results of the prediction itself. The programming language is Python, the technology stack includes the Scikit-Learn library and the Kafka streaming platform.

Device Management Service is responsible for sending commands to IoT devices by an authorized and eligible user. These commands are initiated by the user through the user interface. This service uses a dedicated communication channel to ensure autonomy and independence in the operation of the system. Even if the communication channels responsible for data collection are fully loaded, the usage of a dedicated communication channel ensures the user's confidence that the given commands will be received as quickly as possible by the target IoT devices.

Logging Service is responsible for collecting and analyzing events that have occurred during the operation of the system. This is a very important functionality for any software application that is required to be reliable and sustainable. The data collected by this service is used for 4 basic needs:

- *Software support* - the collected data can show where there are problems in the software parts of the system and what exactly they are.
- *System support* - through the analysis of the collected data it is clear where the system performs slowly or unsustainably, and the conclusions made provide important guidelines for improving the overall architecture of the system.
- *Security* - with a centralized collection of data on the operation of the system, it can be filtered and records that are directly related to the security of data in the system can be quickly analyzed to take timely action. It is important

that this information can be traced through the timeline of the records. This, in turn, provides important consequences about the attack vectors that were used in the process of attacking the system.
- *Prediction* - This is necessary for intelligent IoT data handling. Achieving speed in the process of prediction by machine learning models is an important feature of the ML-based IoT system. The analysis of the records clearly shows which models are used most often and which models work with the largest volumes of data. The conclusions made are important in the optimization that is made during the life cycle of the ML microservice responsible for the analytics.

The **Gateway Service API** is used for decoupling the user interfaces from the system. In an architecture built through microservices, a single client request usually needs information requested from several microservices. This request should not be made directly to all services, because the client must maintain several communication channels at the same time, and this may lead to significant complications in its operation. Also, the client must authenticate against each service, which can make any system unstable and insecure. The Gateway API takes on all these responsibilities - the customer is authenticated only in once in the system and all data requests are aggregated into a single one.

Event Bus takes care of the transmission of messages to the targeted recipients. When a microservice needs to communicate with one or more services, it pushes a message to the Event Bus. All messages in the system that use Event Bus are an implementation of a command pattern, which is a type of behavioural pattern. It encapsulates in one object both the command itself and all the necessary information needed to execute the command to its end. The information includes the command initiator, the name of the method to be executed, and the arguments that this method needs. A callback method can also be included with the command, which can be called when the command is executed. This method can initiate a new command, or it can only notify part of the system for a change in the state of the data or service.

The types of communications in the system supported by the Event Bus are:

- *Publish – Subscribe.* Services can subscribe to certain communication channels. Each time a service publishes a message to Event Bus, it is delivered to all services that have subscribed to it. This type of communication is used in the Data Acquisition service when data is successfully received; two messages are sent - for successful work and for the availability of new data. The first message is sent to the Logging Service, the second is sent to the Data Transformation Service. Both messages are sent through the Event Bus, the services receive only the message that is sent through the communication channel to which they are subscribed.
- *Broadcast.* The message is delivered to all services. This type of communication is used upon successful login of a registered user. When this event occurs, all microservices are notified and the user's Id is sent to them along with his WebSocket's Id.

Each of the described modules and microservices has explicit functional boundaries, which are defined before the development process. The set of microservices form the unified business logic that meets the functional needs of the system. Changing one of the microservices will not lead to system instability. This is achieved by providing backward compatibility between all upgraded and unmodified microservices.

3.2 Microservice Maintenance

In the established microservice architecture it is important to maintain each microservice. They are different in their functionality and significance for the system. To achieve maximum flexibility in the process of installing new versions without compromising the continuity of the system, two approaches are used to update the services:

- *Blue-Green deployment* is used in the system to install updated microservices that are critical to the operation of the system. During this process, both the updated service and its previous version exist in the work environment for a certain period of time. Both the new and old versions are fully functional and working. Routing based versioning is used to easily recognize the current version of each microservice in the system in the URL (Uniform Resource Locator) of each request that this service can process, configure and update its version. The "/v1/users" segment indicates that the current version of the service is "1". During the process of updating the services, both the new version and the old one are maintained for a certain time, so the changes can be easily reversed (roll back) if there is such a need. The traffic that goes to them is initially directed mostly towards the old version. In time, when there is full confidence that the new version fully meets the requirements for functionality and security, the traffic is redirected entirely to it, and the old version is removed from the system.

  This method of updating needs to be applied to the Identity Service, Machine Learning Service, Data Management Service and Logging Service.
- *Canary deployment* is used to update a system of microservices, which have a heavy load on themselves. In this approach, the changes initially affect a small part of the functionality of the service, then more, until the moment when the whole microservice is completely updated. This update method needs to be applied to the Data Acquisition Service and the Data Transformation Service.

4 Context, Communication and Data Aspects

The process of developing microservices begins by identifying three main aspects: Context Constraint, Communication Types and Data Consistency.

Bounded Context (Boundaries) - each microservice must have certain logical boundaries to achieve a flexible architecture with no direct dependencies between services. A bounded context is a key approach in developing distributed applications. With this approach, large and complex business models are divided into smaller but logically meaningful and grouped ones, which are called "Bounds". Each service in the system deals with only one "bounded context", and they are defined and implemented in a way that there are no dependencies between them. Changing each of them does not lead to a direct reason for the change in any of the other service.

Types of Communications in the System - the creation of a continuous and stable communication channel for microservices communications in the system is primary when choosing its type.

There are two types of communication channels - synchronous and asynchronous.

Fig. 3. Synchronous communication client - API gateway – service.

- *Synchronous: Client - API Gateway - Service*
  Communications are performed via HTTPS request-response (Fig. 3). Started tasks can be completed after the response to the client has happened. This type of communication is used when the user's request can quickly obtain the necessary for the response information by accessing only one microservice. The system mainly avoids the use of synchronous communication.
- *Asynchronous communication: Service - Client*
  Communications use WebSocket (Fig. 4), which is extension of HTTP and creates a direct connection between the client and the server. Through this channel, the information in the client (browser) is updated in real-time.
- *Asynchronous communication: Service - Service*
  Communications use asynchronous messaging via Binary over TCP protocol, and the initiator of the communication does not receive a response. The system communicates with one or many recipients depending on the specific need (Fig. 5). Special attention is paid to the stability of the connection and the duration (durability) of the transmitted messages.

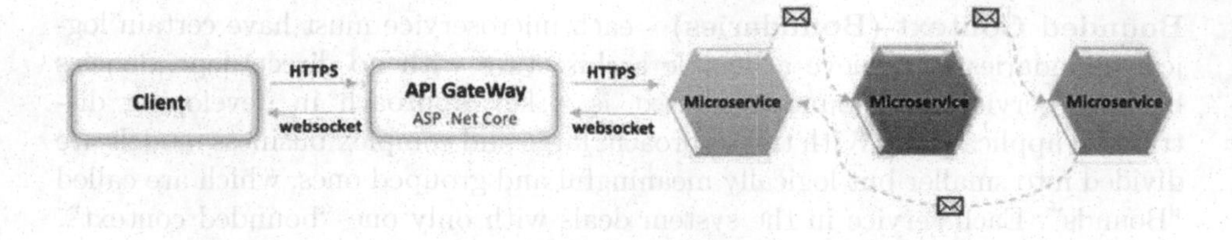

Fig. 4. Asynchronous communication: Service - Client communication.

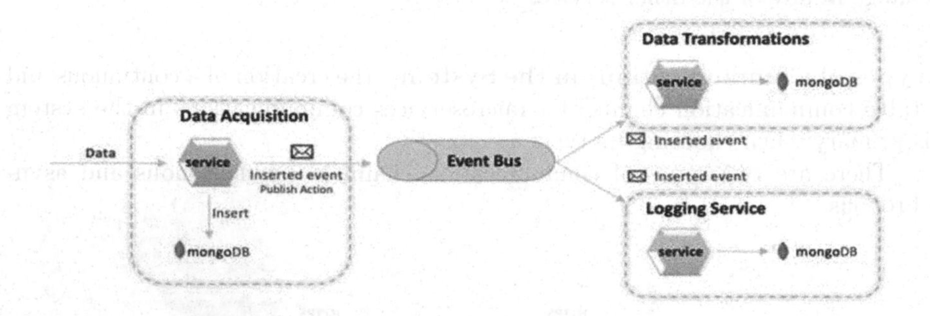

Fig. 5. Asynchronous communication: Service - Service communication.

Data Consistency - Requests needing data from more than one microservice is common in the work of the system. To avoid too frequent communication between services (which is a bad practice), some of the most frequently requested data are duplicated in the Cold Database for quick access by the client. This approach is used only for static data because it is not suitable for working with real-time data.

- Command Query Responsibility Segregation (CQRS) is an architectural model that divides reading and writing into two different models.This means that each model must be a command that executes an action or a query that returns data. The command cannot return data and the request cannot change the data. This model is used in the read-only advance data generation system. This approach significantly improves the performance of the system, considering that it is possible to work with inconsistent data [14] for a short period of time.

- The proposed architectural framework permits analysis of the requests made to it. This analysis allows for identifying cases when there is a need for too frequent aggregation of data from different microservices. This may be a sign that the system architecture can be optimized.

5 Technologies and Software for Implementation

The following technologies have been used to achieve the set goals and objectives:

- Infrastructure - the microservice-based IoT system uses cloud-based virtual servers (VPS - Virtual Private Server) with Debian 9 operating system.

Debian is an open-source Linux operating system that is popular in the development of various types of software systems. It provides a high level of security and reliability (with proper system settings). The connection to the remote virtual servers is done using SSH (Secure Shell) protocol.
- Programming languages - the system performs tasks of various nature. The different programming languages are suitable for specific tasks. The selected programming languages are:
  - Python - a high-level general-purpose programming scripting language.
  - C# - object-oriented high-level general-purpose programming language created by Microsoft, part of the .NET platform.
  - Typescript - an open-source scripting language created by Microsoft. It is built on JavaScript, adding several improvements that make the development process much easier. The language is used in the creation of user web interfaces (UI).
- Technologies and libraries - various (free) technologies and libraries are used in the system: Numpy, Pandas, Scikit-Learn, FastAPI, SqlAlchemy, Asp.NET Core, SignalR, Ocelot, Angular, Kafka, MongoDB, Postgres and others.
- Development Tools (IDE) - JetBrains products are of the highest level and allow solving all the tasks set before the system.
  - PyCharm - an intelligent environment for developing Python-based web and ML applications.
  - Raider - an intelligent environment for developing C# based applications.
  - DataGrid - intelligent environment for working with databases.
  - Visual Studio Code - a smart environment for software application development, developed by Microsoft. It is used to develop the system's user interface.

6 Conclusion

As more devices are connected to the internet, there comes the need for a scalable, extensible and fault-tolerant integration framework [15]. This article discusses the issues of creating a microservice-based architectural ML framework that enhances the reliability of an IoT system and also provides machine learning capabilities.

In the established microservice architecture (MSA) it is important to maintain each microservice. These microservices are different in their functionality and significance to the system.

The software architecture of MSA type gives the following advantages for IoT system implementation:

1. Focus on functionality, not technology. Microservices have no restrictions on the use of programming languages and databases for their creation and successful operation.
2. Improved performance and speed. Microservices can run simultaneously without having to wait.

3. Easy maintenance, flexibility and scalability of the whole system. Each microservice can be managed independently. If necessary and future development of the system, new independent microservices can be easily added. When changing one microservice or adding a new one, the functioning of the other participants in the system is not disturbed.
4. Autonomy and multi-functionality. The microservice architecture provides independence and automatic operation of microservices.

The integration issues still pose a great challenge. Nowadays, there is still a lack of standards for naming convention of IoT devices and related services, data information and the relevant description. There are poor context-awareness for services and poor device service classification, as well poor information visualization and analysis. In the proposed framework the machine learning microservice manages data analysis and provides tools to predict future values and to identify fluctuations in real-time.

The proposed investigation shows that the adoption of microservice architecture for IoT systems offers a way to remodel integration frameworks. Implementation issues are described accordingly.

References

1. IoT Architectures for Digital Twin with Apache Kafka. https://www.kaiwaehner.de/blog/2020/03/25/architectures-digital-twin-digital-thread-apache-kafka-iotplatforms-machine-learning/, Accessed 5 Jun 2020
2. Dineva, K., Atanasova, T.: Model of modular IoT-based bee-keeping system. In: 31st European Simulation and Modelling Conference ESM 2017, pp. 404–406. EUROSIS-ETI, Lisbon (2017)
3. Dineva, K., Atanasova, T.: Methodology for data processing in modular IoT system. In: Vishnevskiy, V.M., Samouylov, K.E., Kozyrev, D.V. (eds.) DCCN 2019. LNCS, vol. 11965, pp. 457–468. Springer, Cham (2019). https://doi.org/10.1007/978-3-030-36614-8_35
4. Fernandez, J.M., Vidal, I., Valera, F.: Enabling the orchestration of IoT slices through edge and cloud microservice platforms. Sensors **19**(13), 2980 (2019). https://doi.org/10.3390/s19132980
5. Uviase, O., Kotonya, G.: IoT architectural framework: connection and integration framework for IoT systems. In: Pianini, D., Salvaneschi, G. (eds.) First workshop on Architectures, Languages and Paradigms for IoT EPTCS 264, pp. 1–17 (2018). https://doi.org/10.4204/EPTCS.264.1
6. Newman, S.: Building Microservices: Designing Fine-Grained Systems. O'Reilly Media, Inc., Newton (2015)
7. Fowlerand M., Lewis, J.: Microservices (2014). http://martinfowler.com/articles/microservices.html
8. Humble, J., Farley, D.: Continuous Delivery: Reliable Software Releases Through Build, Test, and Deployment Automation. Addison-Wesley, Boston (2011)
9. Nemer J.: Blog/Cloud Adoption: Advantages and Disadvantages of Microservices Architecture. Accessed 13 Nov 2019, https://cloudacademy.com/blog/microservices-architecturechallenge-advantage-drawback/

10. Calabrese, M., et al.: SOPHIA: an event-based IoT and machine learning architecture for predictive maintenance in industry 4.0. Information **11**, 202 (2020)
11. Richards, M.: Microservices vs. Service-Oriented Architecture. O'Reilly Media, Newton (2016)
12. Moon, J., Shine, Y.: A study of distributed SDN controller based on apache kafka. In: Proceedings of EEE International Conference on Big Data and Smart Computing, BigComp, Busan, Korea. IEEE (2020)
13. Sajjad, A., Jarwar, M.A., Chong, I.: Design methodology of microservices to support predictive analytics for IoT applications. Sensors **18**, 4226 (2018). https://doi.org/10.3390/s18124226
14. Balabanov, T., Zankinski, I., Barova, M.: Distributed evolutionary computing migration strategy by incident node participation. In: Lirkov, I., Margenov, S.D., Waśniewski, J. (eds.) LSSC 2015. LNCS, vol. 9374, pp. 203–209. Springer, Cham (2015). https://doi.org/10.1007/978-3-319-26520-9_21
15. Macías, A., Navarro, E., González, P.: A microservice-based framework for developing internet of things and people applications. In: Proceedings, vol. 31, p. 85 (2018). https://doi.org/10.3390/proceedings2019031085

Applying Machine Learning to Data from a Structured Database in a Research Institute to Support Decision Making

Nina Bakanova[1], Tatiana Atanasova[2](✉), and Arsenii Bakanov[3]

[1] Keldysh Institute of Applied Mathematics - RAS, Miusskaya pl., 4, 125047 Moscow, Russia
nina@keldysh.ru

[2] Institute of Information and Communication Technologies - BAS, Acad. G. Bonchev, Bl.2, 1113 Sofia, Bulgaria
atanasova@iit.bas.bgm

[3] Institute of Psychology - RAS, 13 Yaroslavskaya Street, Moscow, Russia
arsb2000@pochta.ru

Abstract. The article deals with the problems of creating an information base for accumulating data on the scientific activities of employees in the research organization, to study these data as the basis of scientometrics indicators and to provide tools for making strategical decisions at research institutions. The study is based on Weka – open source machine learning environment that gives several mechanisms for finding consistent patterns in the data.

Keywords: Scientometrics · Research performance · Machine learning algorithms · Weka · Decision making

"The road reaches every place, the short cut only one"
— James Richardson [1]

1 Introduction

1.1 ICT Transformation

Information and Communication Technologies (ICT) have transformed the academic field like any other area of our lives. Scientific research has been transformed by information technology due to the rapid, widespread diffusion of electronic papers, digitalization of libraries and journals, web access to information and repositories among other facilities [2]. Today's capability for remote exchange of research results and worldwide communication, gives innovative advances on use of data (technical use) and elaboration and presentation of projects (academic use).

In this paper, it is proposed to use the progress in the field of machine learning as part of the ITC transformation for the processing of data contained in the

V. M. Vishnevskiy et al. (Eds.): DCCN 2020, LNCS 12563, pp. 712–722, 2020.
https://doi.org/10.1007/978-3-030-66471-8_54

information system of a research organization. This data is aimed to follow and mark various types of activities, conducted at research centers and institutions. Proper presentation of valuable information [3], whether it is systematically or manually collected, should be analyzed to provide support in the decision making to research leaders [4]. Strategic planning models and information provision for decision-making in complex strategic situations are frequent subjects for scientific research [5]. Decisions made at the strategic level of research institutions affect policies, strategies, and actions that the institutions make.

At the strategic level, decisions are less structured and related to the positioning of the organization when they face changes in their environment and are related to the planning of the internal consequences of this positioning [6]. Therefore, the usage of efficient information datastore and computational algorithms is vital to enhance this process. To choose the most suitable learning algorithm, a clear objective is required, and an analysis of previous data must be performed [7]. There are many ways to reach strategic goals as roads that can reach every place. The shortest one is the hardest to find.

This paper deals with the problems of creating an information base for accumulating data on the publication activity of employees of the organization, to study these data as the basis of scientometrics indicators and to provide tools for making strategical decisions at research institutions. The research is based on Weka – open source machine learning environment that gives several mechanisms for finding some consistent patterns in sets of data.

1.2 Scientometrics

The active introduction of information technologies opens new opportunities for the development of various areas of management, including the management of scientific activities. Numerous studies and discussions are being conducted, bibliometric databases are being created, and citation indexes are being developed. All this gives an impetus to the development of one of the branches of science – scientometrics [8].

Scientometrics is an integral part of science studies, an interdisciplinary scientific field that includes the Sociology of science, Economics of science, Psychology of scientific creativity, and other research areas of scientific knowledge and scientific activity. Scientometrics is the application of statistical methods to analyze data (Fig. 1) about the scientific activities of an individual, a team of scientists, or a whole organization. At this stage, the following databases in Scientometrics are most widely recognized and used:

- WoS - Web of Science (WoS) - an authoritative polythematic abstract-bibliographic and scientometric database. The database includes a collection of diverse sections operating on the ISI Web of Knowledge platform, developed by the Institute of Scientific Information (USA);
- Scopus - bibliographic and abstract database and tool for tracking the citation of articles published in scientific journals. The database indexes scientific journals, conference proceedings and serial books, as well as trade journals.

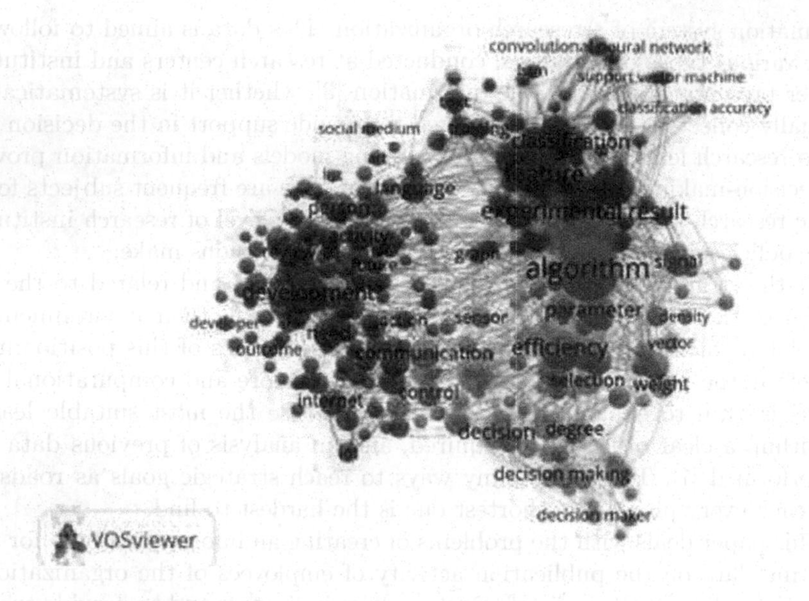

Fig. 1. Visualization of taxonomy for articles in Web of Science (WoS) core collection with "decision making" as keywords - VOSviewer [9].

The development of the methodological base of this scientific direction will allow:

- usage of Scientometrics indicators to evaluate the results of scientific activities,
- exploration the dynamics and interaction of scientific research with various social institutions,
- to plan funding for research projects based on Scientometrics indicators.

The research results in the field of Scientometrics and agreed Scientometrics indicators will allow scientific organizations to prepare reporting documents, indicating quantitative references to the publication activity of employees in the implementation of scientific projects, and administrators at various levels to use this data in the planning and management of scientific activities.

To effectively use data on the publication activity of employees in the work of a scientific organization, with the possibility of forming Scientometrics indicators, it is necessary to develop a database that provides the following features of working with publications:

- the system should take into account various types of publications of the organization's employees (depending on where the publication is issued; whether the publication house belongs to academic sphere or not; whether the publication source indexed in world recognition scientific databases; to which quartile is associated; whether there a link to publication, etc.);
- the system should provide for various sources of information about publications and record authors of data entry;

- the system should provide for remote access and teamwork of employees of the organization;
- the system should provide for a hierarchy of access to information, including heads of scientific departments of various levels, researchers, system administrators, controllers for checking publication indexes, etc.

The objectives are related to the creation of application systems and databases for storing consistent data on the publications of the organization. The creation of universal databases for such purposes is always complicated by the specifics of research areas [8], the predominant types of publication activity in this organization, and other tasks and indicators of the organization's activity that can be included in this system.

2 Problem Statement

The "Scientific activity of the organization" system should provide not only direct data entry by employees of the organization, but also data entry from specialized databases. In this case, there should be a mode of "recognizing" the publication by the author and adding it to the list of their own publications. In general, the mode of joining the publication to the author's personal list should be provided even if the publication has already been entered into the system by one of the co-authors. The implementation of this mode will ensure the formation of the "publication activity" feature, which characterizes the work of an employee.

When creating retrospective arrays of the publication database, it is necessary to check the data on bibliometric databases at the level of characteristics of publications in which the work of the organization's research staff is published.

Besides publication data, all other aspects of scientific activities have to be marked as:

- participation in applied projects,
- providing expertise for different public and government organizations,
- fulfillment of administrative and organizational obligations, and
- teaching hours and disciplines.

The workflow of accumulating and analyzing data is shown on Fig. 2.

2.1 Features of Structural Database Construction

The accumulation of retrospective arrays is performed in a database. The structural construction of the database has certain features. The objective complexity of the system is the need to resolve multiple "many – to – many" links. The following examples complicate the development process:

- variety of spelling of authors' names (in different languages) in publications,
- tracking co-authors from different departments or other organizations,
- numerous types of publications, the structure of which may differ significantly,

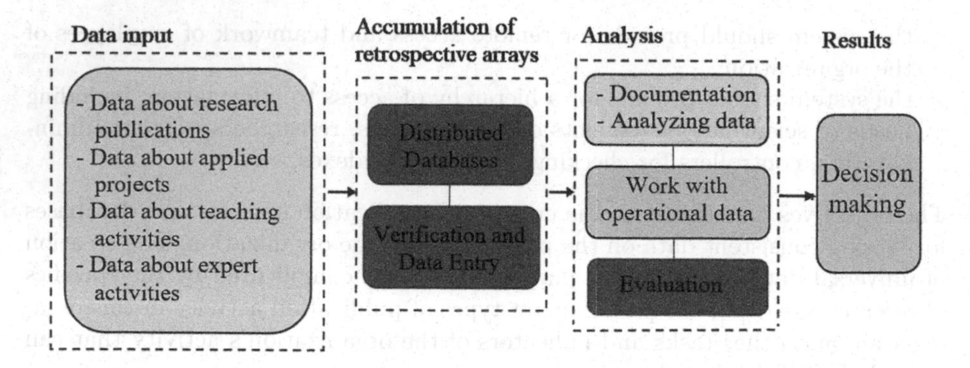

Fig. 2. Workflow of accumulating and analyzing data.

– an abundance of unstructured information on conferences, seminars, and symposiums that is presented differently by different participants.

The focus of the proposed system is on distributing the activities of all researches into the following categories: scientific research, applied projects, teaching activities, organizational activity and expert impacts. The collected data may give tools for analyzing different groups of researchers and their impact, thus allowing quality measurement and decision-making process.

3 Experimental Results

The accumulated data contains information about the work of scientific staff on different position. An excerpt from the data of 72 scientists in different administrative positions and with different scientific degrees were extracted from the specialized system. The data are grouped in 5 sections according to a certain type of activity. Each type of activity receives weighting factors, determined empirically according to the strategic goals of the organization. Part of the experimental data is shown in the Table 1. The data set is private, it belongs to a research institute. The whole data set is normalized after collecting and arranging raw data.

The data is being analysing, processing, classified and clustered by machine learning algorithms in WEKA [10] environment. Various classification, cluster and regression analysis algorithms built-in WEKA can be used to see all aspect of the collected data. Thus, all aspects of researchers' activities can be analyzed. The distribution of the data and clusterisation of normalized data is shown on Fig. 3 according to the employments' achievements in five directions: research and publication, participated in applied projects, teaching, organizational and expert activities.

Table 1. Part of the experimental data

Research	Applied	Teaching	Organizational	Expert
3056,47	37,37	15,5	263,18	151
2588,8	36	33,25	195,7	54
1836,18	20	28,8	213	688
1308,21	534,31	20,51	410,35	147,32
966,56	740,19	13	229,48	282
963,71	789,05	9	71	62
1444,1	87,78	7,98	125,02	94,43
949,81	29,38	3,6	247	460
1133,37	6	47,1	245,5	178
1031,82	242,71	16,75	159,75	96
1346,87	0	10,5	7,5	116
995,37	0	39	119	262
1078,35	0	11	32	20
503,8	107,3	11,6	135,5	272
534	90,5	88,25	74	223
425	127,3	58,5	151,12	112
562,3	27,5	41	107	68,4
1142,65	654,72	7	103	10
1154,67	11,5	14	5	32
663,05	174	20	169,5	179,8
731,19	227	23,25	144,36	32
883,07	20	64	43	72
827,19	157,23	20,5	10	17
862,48	30	8	42	79
612,79	63,6	22,69	157,94	85,85
232,87	53,59	33,49	252,86	178
466,12	251,75	3,5	0	18
372,42	86,25	0,5	64	94
460,5	0	58	38	59
476	0	30,1	0	40
417,87	0	0,5	18,25	79
421	36,16	15,5	5	2
486,02	0	0	0	0
180,25	0	129,75	60,18	102
299,55	41	4	71	34
366,75	0	4,75	15	62
324	0	15	0	12
323,5	0	0,5	9	0
8	50,1	0	6	36
812,76	110,73	0	35,25	2
404	40,07	289	15	12
620,99	0,00	0,00	20,00	4,00
592,80	3,50	11,25	0,00	16,00
329,43	243,00	0,00	5,00	42,00
457,98	64,17	0,00	12,00	14,00
517,90	0,00	0,00	21,00	2,00

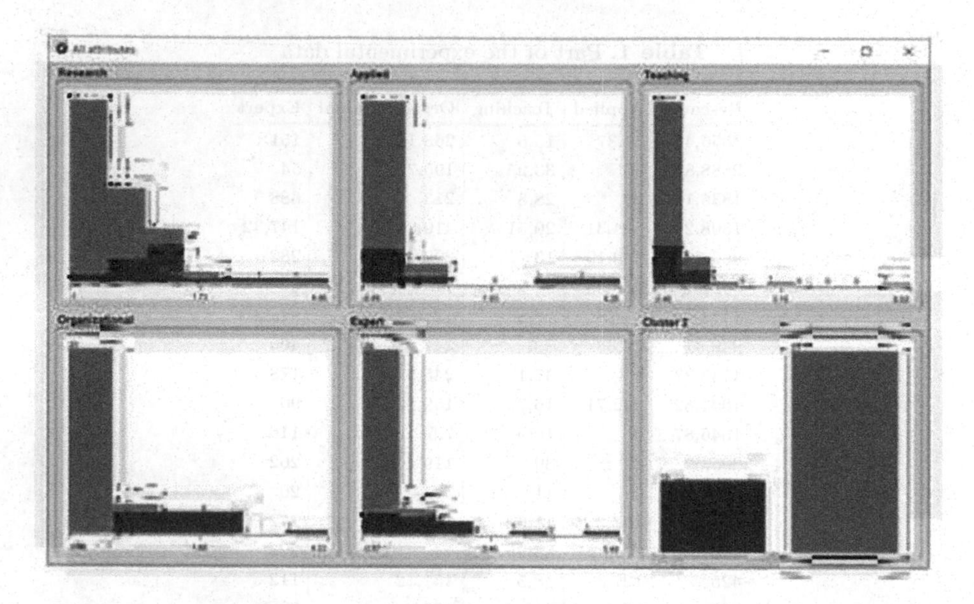

Fig. 3. The distribution of the data according to the employment's achievements and clusterisation of normalized data.

Different machine learning algorithms were tried.

For example, the Linear Regression Model gives the following result:

```
Organizational = 0.3516 * Research + 0.2482 * Applied + 0.4721 *
Expert
```

Correlation coefficient: 0.6823
Kendall's tau: 0.6157
Spearman's rho: 0.4602
Root mean squared error: 0.7591
Relative absolute error: 57.964 %
Root relative squared error: 74.2403 %
Total Number of Instances: 72.

And another model constructed relative to an alternative variable by linear regression:

```
Expert = -0.1487 * Applied + 0.7316 * Organizational
```

Correlation coefficient: 0.5055
Kendall's tau: 0.4856
Spearman's rho: 0.6489
Mean absolute error: 0.5032.

This model shows that expert activities have a higher weighting factor in the performance of organizational duties and, surprisingly, participation in applied projects has a negative impact on their expert work.

Another WEKA classifier Scheme (ElasticNet) gives the following model:

```
Expert = 0,145 * Research - 0,041 * Applied + 0,048 * Teaching
+0,422 * Organizational
```

The model shows a serious impact of research and publications together with activities in the organizational direction in the expert activity of researchers.

The SMOreg model gives another weights to the research activities in the expert work: `0.0162 * Research + 0.0133 * Applied + 0.1398 * Teaching + 0.3832 * Organizational`.

The results of applying different ML algorithms to the dataset are summarized in the Table 2.

Table 2. Different ML models and their performance evaluation metrics.

Performance criteria	1	2	3	4	5	6	7
Correlation coefficient	0.4786	0.6823	0.5677	0.5804	0.5618	0.6009	0.5185
Kendall's tau	0.5767	0.6157	0.5890	0.5564	0.3742	0.3391	0.5909
Spearman's rho	0.7453	0.7983	0.7724	0.7244	0.5159	0.5035	0.7655
Mean absolute error	0.4869	0.4602	0.3923	0.4268	0.4509	0.4591	0.4281
Root mean squared error	0.8839	0.7591	0.8389	0.8198	0.8329	0.8092	0.8508
Relative absolute error	76.41%	57.96%	59.57%	64.81%	64.48%	69.71%	65.01%
Root relative squared error	86.59%	74.24%	82.18%	80.32%	81.59%	79.28%	83.35%

Note: 1 - Elastic Net; 2 - Linear Regression Model; 3 - SMO Improved; 4 - Simple Linear Regression; 5 - Additive Regression; 6 - Regression by Discretization; 7 - REPTree.

It is possible to organize the data in the predefined number of classes. The clusterisation into two classes can be evaluated by the confusion matrix (Table 3). The Simple K-Means clusters all 72 instances into 2 classes. A confusion matrix is computed to evaluate the accuracy of the classification (see Table 4).

Table 3. A confusion matrix.

a	b	classified as
TP = 17	FP = 2	a = cluster 1/2
FN = 4	TN = 49	b = cluster 2/2

Table 4. Detailed accuracy by class

TP rate	FP rate	Precision	Recall	F-measure	MCC	ROC area	PRC area class
0.895	0.75	0.810	0.850	0.794	0.931	0.883	Cluster 1 of 2
0.925	0.105	0.925	0.942	0.794	0.931	0.956	Cluster 2 of 2
0.917	0.097	0.917	0.918	0.794	0.931	0.937	Weighted Avg

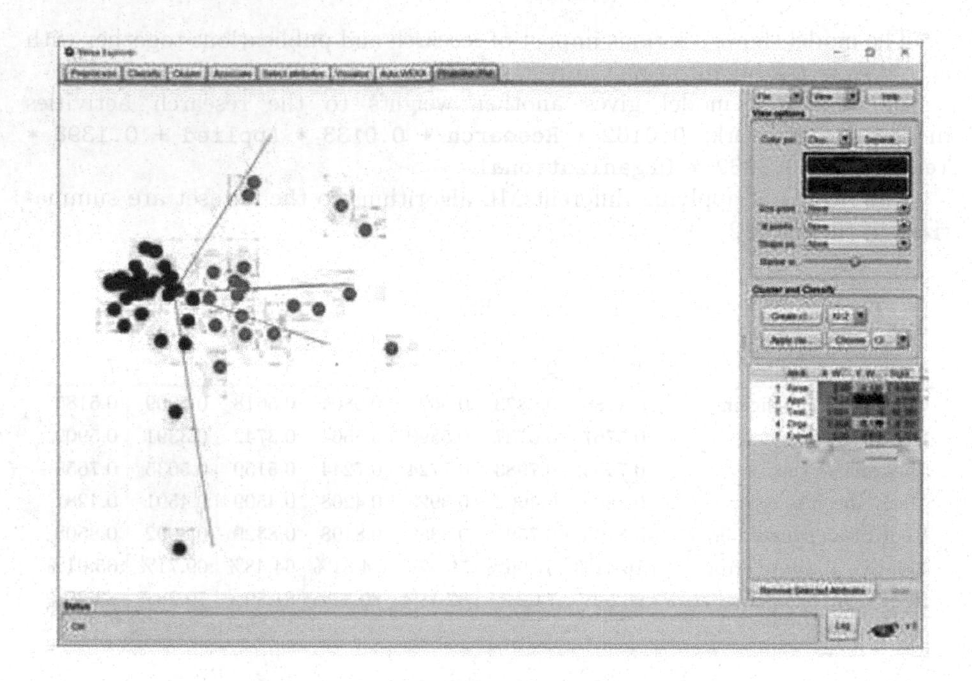

Fig. 4. Clustering in 2 classes

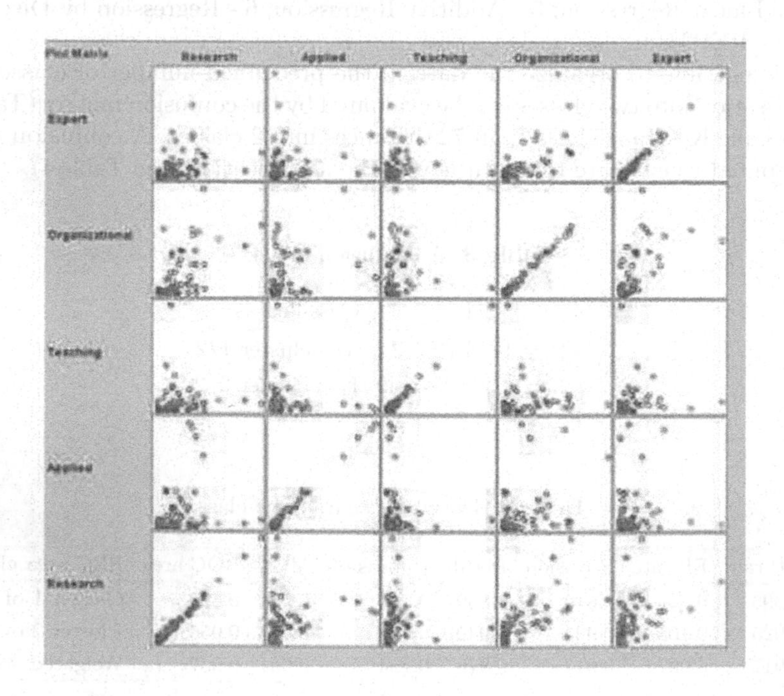

Fig. 5. Visualization

where:

TP = True Positive; FP = False Positive (Type 1 Error); FN = False Negative (Type 2 Error); TN = True Negative. MCC-Matthews correlation coefficient, and F-Measure is a measure of a test's accuracy. The ROC Area in the interval 0.9 to 1 estimates the classification as excellent.

The WEKA environment provides visualization tools for explanation of the received results. The classification can be visualized as it is shown on Fig. 4. On Fig. 5 the distribution of the data and outliers can be seen. Beside that data can be visualized both as a tree structure and from the point of view of the Cost-Benefit analysis.

4 Conclusion

The ICT transformation in research organisations has its impact that depends on the infrastructure of information and communication technology, its availability and intensity of use. More efficient and user-friendly applications provided by the ICT are needed for decision making in scientometrics. Machine learning arises with different algorithms that learn from data to support different tasks in research centres and organizations.

The volume, variety and heterogeneity of the data collected in the "Scientific activity of the organization" system makes it challenging for the stakeholders to take strategical decisions. Using of efficient ML algorithms may lead to faster decision making process and to design a vision of the "whole picture" with all necessary data. This article discusses the issues of data usage on the base of created database of information system "Scientific activity at research organization". The data can be used for analyzing the publication activity of employees of a scientific organization in order to use this data both in internal activities (seminars, employee certification, etc.) of the institution, and in preparing reports to higher organizations. The experimental results show that there are models which can discover hidden dependencies in the collected data.

The article also highlights the main difficulties of building a system for recording various scientific activities of employees at research center and features of the structural construction of the database of the information system.

The development of the system allows to analyze the accumulated data to build various summaries. The variety of indicators of scientific activity will allow using multi-criteria analysis to form a scientometrics assessment of the work of teams of scientists.

Acknowledgement. This article was prepared with the support of the RFBR grant 18-07-00833 A and the RFBR grant 18-29-03085 mk.

References

1. Richardson, J.: Vectors: Aphorisms and Ten-Second Essays. Ausable Press, Keene (2001)
2. Nieto, Y., Gacía-Díaz, V., Montenegro, C., González, C.C., Crespo, R.G.: Usage of machine learning for strategic decision making at higher educational institutions. IEEE Access **7**, 75007–75017 (2019)
3. Bakanov, A., Atanasova, T., Bakanova, N.: Cognitive approach to modeling human-computer interaction with a distributed intellectual information environment. In: BdKSCE 2019, 1–4. IEEE Xplore, Sofia (2019)
4. Ibrahim, O., Sundgren, D., Larsson, A.: An integrated decision support system framework for strategic planning in higher education institutions. In: Zaraté, P., Kersten, G.E., Hernández, J.E. (eds.) Conference: Group Decision and Negotiation. A Process-Oriented View, Proceedings Series: Lecture Notes in Business Information Processing, Toulouse, France, 180 (2014)
5. Peixoto, L., Golgher, A.B., Cyrino, Á.B.: Using Information Systems to strategic decision: an analysis of the values added under executive's perspective. Braz. J. Inf. Stud. Res. Trends **11**(1), 54–71 (2017)
6. Pincus, K.V., Stout, D.E., Sorensen, J.E., Stocks, K.D., Lawson, R.A.: Forces for change in higher education and implications for the accounting academy. J. Accounting Educ. **40**(9), 1–18 (2016)
7. Bakanova, N.B., Atanasova, T.V.: Use of information resources of organizational systems to support managerial decisions. In: International Conference on Big Data, Knowledge and Control Systems Engineering BdKCSE 2018, pp. 29–36, Sofia, Bulgaria (2018)
8. Scientometrics, An International Journal for all Quantitative Aspects of the Science of Science, Communication in Science and Science Policy. https://www.springer.com/journal/11192. Accessed 08 May 2020
9. VOSviewer Homepage. https://www.vosviewer.com/. Accessed 08 May 2020
10. Weka 3 - Data Mining with Open Source Machine Learning. https://www.cs.waikato.ac.nz/ml/weka/

Generation of Metadata for Network Control

Alexander Grusho(✉), Nick Grusho, Michael Zabezhailo, and Elena Timonina

Federal Research Center "Computer Science and Control" of the Russian Academy of Sciences, Vavilova 44-2, 119333 Moscow, Russia
grusho@yandex.ru, info@itake.ru, zabezhailo@yandex.ru, eltimon@yandex.ru

Abstract. In this paper we propose a method of ensuring network information security by controlling network connections using metadata. The metadata contains information about admissible task interactions and application locations in a corporate network.

Metadata is formed based on a mathematical model of an information technology. Information technology models are represented in the form of directed acyclic graphs. This paper describes hierarchical decompositions of directed acyclic graphs. Hierarchical decomposition makes it possible to optimally form blocks of information transformations for their placement on hosts of distributed information and computing system (DICS). Protocols of metadata usage are briefly discussed. The problem of transforming directed acyclic graphs into metadata is solved. Security risks introduced by user interaction with information technology process are considered.

Keywords: Information system · Metadata · Directed acyclic graph · Information security

1 Introduction

This paper considers two classes of information technology (IT) security threats that can be implemented in enterprise distributed information and computing systems (DICS). The first class includes threats posed by malicious code. The second class is related to hostile user actions and will be considered later.

As a rule, malicious code has the ability to carry out hostile actions anonymously and with impunity. In fact, an embedded malicious code can sleep indefinitely or perform hostile actions regularly.

However, finding malicious code or identifying the person who has embedded that code is complicated. Malicious code is effective if it has access to critical area of DICS. To achieve that, the adversary must solve several complicated tasks. Initially, he has to break into a device that is a part of a DICS. This can be

Partially supported by Russian Foundsation for Basic Research (projects 18-29-03081, 18-07-00274).

V. M. Vishnevskiy et al. (Eds.): DCCN 2020, LNCS 12563, pp. 723–735, 2020.
https://doi.org/10.1007/978-3-030-66471-8_55

achieved by use of vulnerabilities of computer systems and network equipment. The malicious code is able to achieve its objectives if it is able to transfer itself to the appropriate damage-critical area of DICS. For this purpose, it is necessary to perform transfer of malicious code and supplementary data through the corporate network and computer systems.

To ensure the information security of network interactions [1,2], it has been proposed to manage network connections using metadata. The metadata contains information about permitted task interactions and locations of applications that are required to solve these tasks in a distributed network.

Typically, the transmission of the malicious code from its initial penetration point to the critical area is possible if it is known which host contains targeted sensitive data and its vulnerabilities through which it can be penetrated.

The host information can be hidden on a corporate network. Then the malicious code will attempt to communicate with all hosts it can access. It is possible if multiple hosts on a network allow free interactions, and in this case the chances increase that the malicious code will cause damage. Note that some of the interactions are necessary to perform legal IT. Therefore, all interactions in the corporate network can be divided into legal and illegal and we should aim to prevent illegal interactions while allowing legal ones.

We'll call this approach "safe interactions" on corporate network.

IT requires control over host interactions on the network, which consists of monitoring interactions and managing connections. Managing network host interactions allows to reduce the threat of malicious code being introduced and distributed through network equipment and communication channels. Papers [1,2] show how network host interactions can be managed using metadata.

Interactions control by metadata uses the following logic. Legal interactions are based on business processes of an organization. Recall (see [3]) that a business process is defined as a logically complete system of interactions between recurring types of activity. A business process is modeled by its formalized description. A functional approach to describing business processes presents it as a diagram of interacting functions. The information support of the functional presentation of the business process contains the description of information flows (IF) that provide information to functions of business processes and related information transformations. The information support for the functional presentation of business processes is called information technology.

IT consists of information transformations (tasks solving) implemented by applications. The solution of tasks consists of three processes:

- collection of information required to solve the task (input data);
- information processing on computers by means of the software (applications);
- distribution of results of information processing (output data).

IT can be thought of as a set of interacting tasks.

Different tasks can be solved on different hosts of the network. The network allows collection of input data for tasks, and distribute the result of information processing.

Let's assume that a mathematical model has been created for IT that defines all system actions that are required to perform the requested computations or tasks. Complete information about performing IT can be represented by PERT (Project Evaluation and Review Technique) [4] diagrams, which are based on directed acyclic graphs (DAG).

Models in the form of DAG are often used in Computer Sciences. We list some examples related to this publication. Papers [4–6] address the problem of scheduling work to maximize the acceleration of IT execution.

In [7] DAG is used in fault search. In [8] DAG serves as the model for the analysis of cause-effect relationships. The paper [9] is devoted to research of hierarchy of interactions.

Metadata contains information about the queue of IT tasks to be solved and about the rules of data exchange between them. We call one or more interconnected tasks the "block of tasks". The concept of a block is necessary to build a hierarchical decomposition of the IT model and describe tasks that are solved on a single host. If different hosts contain different blocks, the metadata determines the sequence of legal connections between such hosts during IT execution.

In this paper, methods of transformation of DAG to metadata are constructed to manage legal connections between hosts containing IT execution blocks.

The important task of ensuring information security is to control the participation of users in the implementation of IT. User participation needs to be reflected in the DAG of IT.

2 Information Technology Description with DAG

Consider models of IT [3] presented as DAG. Capital Latin letters $A, B, \ldots$, denote data (objects) serving as input or output data of information transformations in IT. This data will be presented in figures as circles. The transformations are called blocks, denoted with lowercase Latin letters and represented by boxes in the figures. The set of input data, transformations and output data are denoted with lowercase Greek letters.

Each block corresponds to the transformation of information and corresponds to the solution of one or more tasks required for IT implementation. The DAG arcs correspond to the data transmission to blocks from previous blocks, that is, the arc exits the vertex corresponding to the output of the performed transformation and enters the vertex corresponding to the input data for the next transformation. Repeated transformations are valid if they have different inputs.

IT may contain directed cycles of transformations. Obviously, the input and output data of each cycle is different. Consequently, there are no directed cycles in the graph of IT. If the transformation output completely defines the input data of one or more of following transformations, multiple arcs exit from the transformation output. However, some output may be unused in the next block.

Most information technologies involve users. Their participation in IT comes down to two scenarios that can be reflected in DAG of IT.

Scenario 1. Users are considered as transformations of information and then considered as fragments of DAG of IT.

Scenario 2. Users are external objects of DAG and can receive information from solved tasks or use this data and alter the input data of other blocks. In this case, the presence of users may be temporary.

The simplest DAG describing the transformation of information is shown in Fig. 1, where A is transformation input, B is transformation output, f is the transformation.

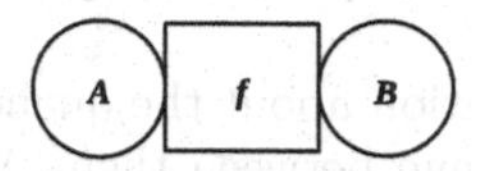

Fig. 1. The simplest DAG

If the output data A of some transformation and the output data B of some other transformation are used for the transformation f, then the input data for the transformation f is the vector (A, B) (see Fig. 2).

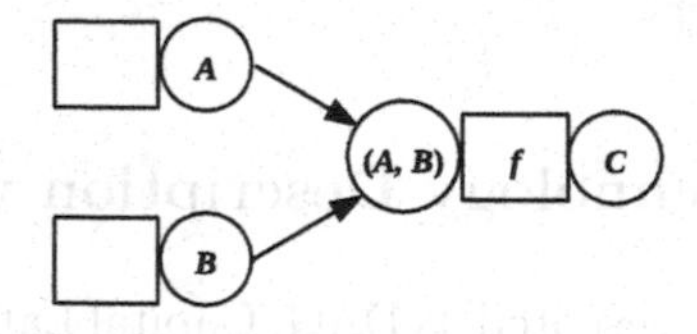

Fig. 2. Vector of input data

Let's construct a hierarchical decomposition of DAG. The hierarchical decomposition is determined iteratively by applying two operations.

1. *Block division operation.* Let the data transformation f in the current DAG be as shown in Fig. 1. Suppose that f can be represented as the superposition of two transformations $B = f(A) = f_2(f_1(A))$ and $f_1(A) = C$. The result of block division operation on DAG fragment in Fig. 1 can be graphically represented as shown in Fig. 3.

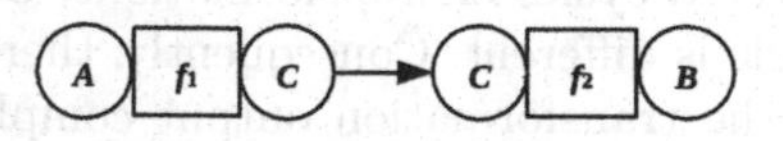

Fig. 3. Block divide operation

This transformation preserves the acyclicity of the source graph.

2. *Block detailing operation.* Let the transformation f depend on the input data (A, B) and can be described as $C = f(A, B)$. Suppose there are functions f_1 and f_2 such that $f(A, B) = f_2(f_1(A), B)$. The result of the block detailing operation is then as shown in Fig. 4.

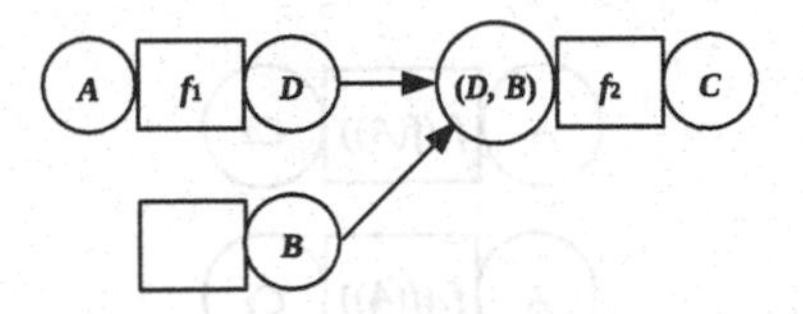

Fig. 4. Block detailing operation

Obviously, the block detail operation does not produce directed cycles in the generated graph.

Definition. DAG $\mathcal{G}$ is called an adequate model of the given IT if it can be proved that any source data of IT transformed by blocks with the required functionality produces the desired result of IT.

In practice, the test on adequacy involves implementation of the IT model, repeated execution of IT and reproduction of the results obtained for a representative sample of source data.

Theorem 1. *Each DAG $\mathcal{G}'$ obtained by applying a sequence of block detailing and block division operations on DAG $\mathcal{G}$, which is an adequate model of IT, produce a model of IT which is also adequate.*

Proof. Each step that uses a detailing or division operation does not change the functionality and source data of the model. Hence, $\mathcal{G}'$ is the model with the same functionality and on the same source data. Theorem 1 is proved.

The inverse problem, i.e. the integration of DAG $\mathcal{G}$, which is the adequate model of this IT, has the following solution.

a) Let DAG $\mathcal{G}$ be an adequate model of IT. We highlight fragments of graph $\mathcal{G}$ corresponding to Fig. 3, i.e. fragments having transformations of the form $f(A) = f_2(f_1(A)) = B$. For this case, we define the transformation f as a single block $B = f(A)$. We then replace the model of $\mathcal{G}$ that contains two blocks of transformations $f_1(A) = C$ and $f_2(C) = B$ by the model with a single block. It is clear that the resulting graph $\mathcal{G}'$ is an adequate model of IT. We apply this operation repeatedly wherever possible and get a graph that is the adequate model of IT, but has fewer blocks.

b) Consider the fragment of graph $\mathcal{G}$ shown in Fig. 5.

This fragment can be transformed into DAG $\mathcal{G}'$ as follows, retaining the adequacy property of the model of IT (Fig. 6).

Transforming the fragment in Fig. 5 to the fragment in Fig. 6 reduces the number of blocks in graph $\mathcal{G}'$ as compared to graph $\mathcal{G}$ and preserves the adequacy of IT model.

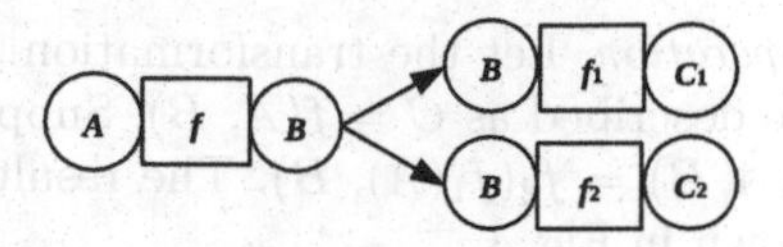

Fig. 5. Data transfer to different chains

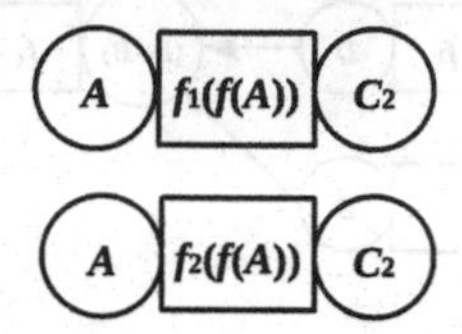

Fig. 6. Reduce the number of blocks by using parallel chains

c) The detailing fragment in Fig. 2 in the adequate model of IT can be reversed using vector data and vector transformations if the vector (A, B) is the input data of vector transformation (f_1, f_2). Namely, the next part of the adequate model of IT maintains the adequacy of the model of IT (Fig. 7).

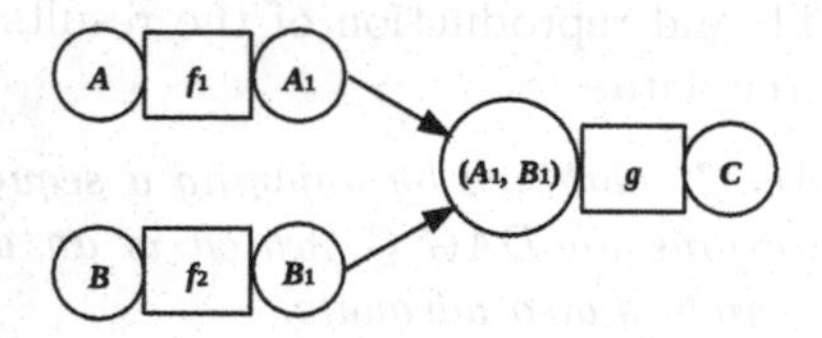

Fig. 7. Transformation using input from several other transformations

The fragment in Fig. 7 can be transformed into the fragment in Fig. 8 while maintaining the adequacy of the model of IT and reducing the number of blocks as compared to the source graph.

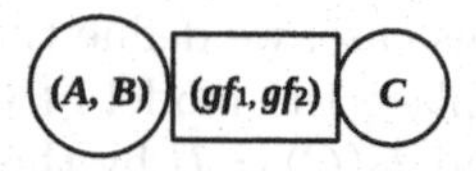

Fig. 8. Use of vector transformation to reduce the number of blocks

d) If the fragment (Fig. 9) uses the data from the previous block and at least partially uses the source data, the fragment can be reduced in the number of blocks while maintaining the adequacy of the model of IT.

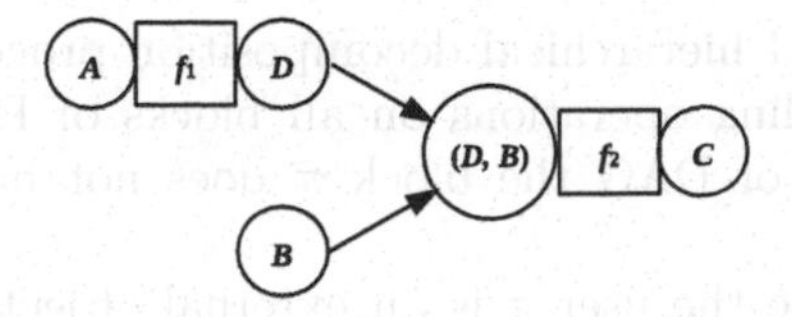

Fig. 9. Use of part of the source data while reducing the number of blocks

Figure 10 shows a diagram of the reduction of the number of blocks when some of the source data is added in the operation of IT at this stage.

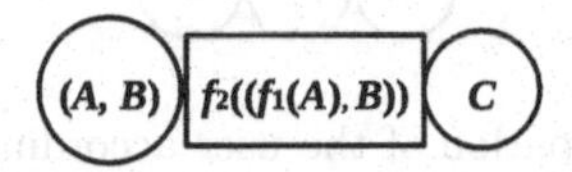

Fig. 10. Reduced block interaction diagram using additional source data

There can be no other fragments in the DAG (except for an increase in the multiplicity of input and output arcs). Indeed, each DAG vertex is described by input and output arcs. If one input arc and one output arc are associated with the given vertex, the case refers to a), c) or d). If the number of input arcs is greater than or equal to 2, then the case refers to c) or d). If the number of output arcs is greater than or equal to 2, then the case refers to b).

Thus, an iterative algorithm can be constructed that reduces the number of blocks in the next iteration and maintains the model of IT. This implies Theorem 2.

Theorem 2. *Each adequate model of IT in the form of DAG can be derived from the adequate Fig. 1 model using a sequence of block detailing and block division operations. Conversely, each adequate model of IT in the form of DAG can be converted into the adequate model of IT presented in Fig. 1.*

Let's consider division and integration of DAG while introducing the users into the model of an IT (Fig. 1). Assume there is an isolated fragment of DAG in Fig. 1 in which the user π participates. This block is then transformed into three blocks by the block division operation as it is shown in Fig. 11.

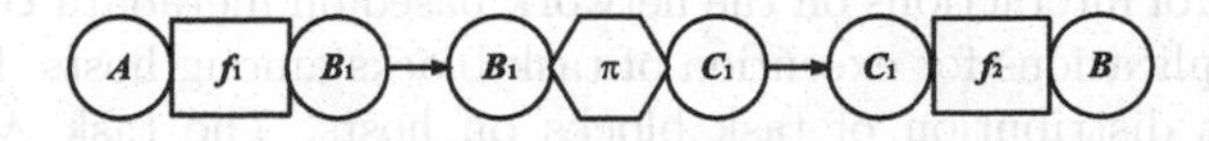

Fig. 11. Participation of the user according to scenario 1

The case when the user π has to obtain data from other tasks comes down to transformation of detailing in Fig. 9. The integration operation of DAG is constructed as an inverse operation to the divide and detail operations.

Obviously, the DAG hierarchical decomposition procedure does not require the division and detailing operations on all blocks of DAG. I.e. upon further division and detailing of DAG the block π does not have to be divided and detailed.

In scenario 2, where the user π is an external object of DAG and data for the block is described in the definition of detailing operation, the participation of the user can be represented by the following chart (Fig. 12).

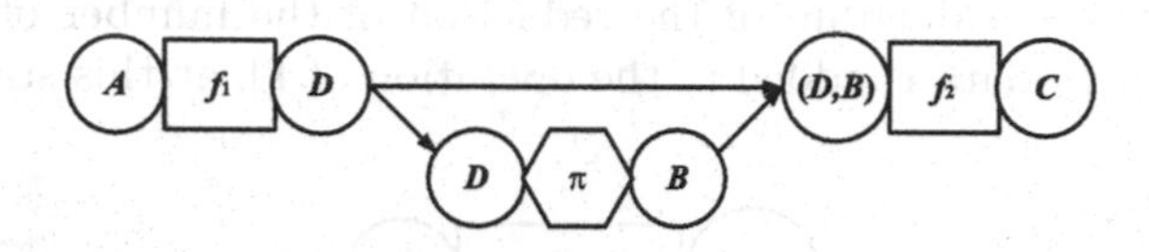

Fig. 12. Participation of the user according to scenario 2

The integration operation inverts the block detail operation that leads to removal of the fragment where the user obtains the intermediate data D.

3 Transformation DAG of Adequate Models of IT to Metadata

Usage of metadata for connection management in distributed information computing systems is described in detail in [10,11]. However, in those papers, metadata that describes the order of solving the tasks is obtained from graphs of reduction [11] and are based on a complicated tree traversal algorithm.

In this paper, the source models of IT are DAG and the DAG hierarchy, which allows aggregation of tasks into blocks on separate computers. Therefore, it is necessary to construct the transformation of DAG into connection management metadata.

For the sake of simplicity, we build a rigid order of execution of IT task blocks represented by DAG [11]. This order must be supplemented by the scheme of input and output data exchange between blocks. All this information is called metadata $\mathcal{B}$ in IT. The network management algorithm uses the metadata to uniquely determine the order of accesses between blocks on the network.

For blocks included in sequence $\mathcal{B}$, we define three additional tasks: $\mathcal{M}$, $\mathcal{N}$, $\mathcal{R}$, which control interactions on the network based on metadata $\mathcal{B}$. The task $\mathcal{M}$ distributes applications for execution of task blocks among hosts. For simplicity we will call it distribution of task blocks on hosts. The task $\mathcal{M}$ defines the binary relation $H(\alpha)$ meaning that the block α can be executed on host H. Note that the task $\mathcal{M}$ can use the hierarchical decomposition of DAG to form blocks and to distribute these blocks at network hosts while taking into account the information security requirements. In addition, task $\mathcal{M}$ allows for efficient block redundancy.

Results produced by the task $\mathcal{M}$ are used by the task $\mathcal{N}$. The task $\mathcal{N}$ keeps in contact with each host, and is responsible for permission checking and providing information to hosts on request for interaction of blocks at different hosts. Permission checking is based on metadata $\mathcal{B}$. The task $\mathcal{R}$ builds primary and backup routes for task $\mathcal{N}$.

Let the block α is legally running at host $H(\alpha)$. The block α_1 uses block output data α (denoted as (α, α_1)) and is located on another host. Each host H has an agent with cryptographic facilities and the key $k(H)$ for communication with the host $H(\mathcal{N})$. For each H, the connection to $H(\mathcal{N})$ does not produce an unacceptable delay.

In order to access block α_1 the block α contacts the task $\mathcal{N}$ through the agent of its host which checks whether (α, α_1) exists. Then, the information about whether it is necessary to connect to $H(\alpha)$, the connection protection key $k((\alpha, \alpha_1))$, the identifier, the port, and the time stamp are sent to the host $H(\alpha_1)$ through the agent of this host. Similar information is sent to the host $H(\alpha)$. After the data has been transmitted to the block α_1, the connection between hosts $H(\alpha)$ and $H(\alpha_1)$ is terminated.

Obviously, DAG defines the strict order [4] and the set of vertices of the graph $\mathcal{G}$ forms the partially ordered set. In order to construct metadata, first it is necessary to define the order of solving blocks of tasks $\{\alpha_1, ... \alpha_m\}$ so that this order (the permutation of blocks $(\alpha_{i_1}, ... \alpha_{i_m})$) has the property of consistency with the graph $\mathcal{G}$. That is, when metadata allows the transition from the block α_i to the block α_j, then the strict order of blocks defined by DAG $\mathcal{G}$ implies that α_i is less than α_j. In [4] it is proved that at least one such order exists, and algorithms for building permutations of task blocks under various additional restrictions are proposed. However, data exchange and the use of the built permutation are complicated.

Recall that some transformation may appear repeatedly in the set $\{\alpha_1, ... \alpha_m\}$, but as their input data mut differ for each repetition, we consider them as different blocks.

The built permutation of blocks does not conflict with the sequence of task blocks execution in the DAG, meaning that if the block β needs data from the block α, the block α must wait for the block β to be executed in queue determined by the permutation. If there are several blocks with data for the block β, they should all "remember" the data to be presented to the block β before that block appears and wait for their data transmission queue. Data collection is not provided in block execution order. Since there may be a lot of such cases and blocks may be repeated, each block in the permutation must remember which data it should transmit to which other block.

The second problem is that each particular block in the network needs to establish its own connection between hosts containing the respective blocks to transmit data. The host that is to transmit the data must initiate the connection, but does not know when to do it. In addition, queues to this particular application may appear due to transformation repetitions and other parallel IT.

It is possible to solve these problems by means of the matrix Γ, which is formed at the host $H(\mathcal{N})$ of the task $\mathcal{N}$. The square matrix $\Gamma = \|\gamma_{ij}\|$ of size $m \times m$, where m is the number of blocks, Γ is the matrix in which $\gamma_{ij} = 1$ if the block α_i is to transmit data to the block α_j, and $\gamma_{ij} = 0$ if the block α_i does not have to transmit data to the block α_j, or has already transmitted that data. When it's time for block α_j to execute, the j-th column of the matrix Γ defines blocks that wait for its queue to transmit their information or gain access to the block α_j.

The usage of the matrix Γ implies that the task $\mathcal{N}$ in turn establishes the connection to hosts containing blocks having nonzero values in the corresponding column of next block and indicates that data has to be transmitted to it.

If a user can be associated with regular transformation in IT (scenario 1), he can be incorporated by the standard protocol of building the queue of blocks and providing him with data. In this case, the priorities may be altered, the examples of such an alteration will be given for the scenario 2 of user integration into IT.

When a user is integrated into IT outside of a DAG, the following schema describes his impact on the metadata construction. Let blocks $\{\alpha_1, ..., \alpha_m\}$ be described in metadata by the permutation $(\alpha_{i_1}, ..., \alpha_{i_m})$ that is consistent with DAG of IT. Then the user π can be considered to be a new task (see Fig. 12) and can be inserted into any position in the queue after blocks which must produce the data required to theuser, but in front of other blocks to which he must provide the processed data. For example, such a position is determined in Figs. 11 and 12.

Note that the user cannot send his data to blocks that precede the position in the permutation from where he has received the data. Otherwise, the data will become misaligned. If it is known in advance where from does the user take and where does he send the data, then the user π can be temporarily incorporated into the matrix Γ. In fact, his position in the permutation following the sequence of blocks from where he has to obtain the data. In order to achieve that, a column corresponding to π must be added to the matrix Γ. In that row, nonzero elements indicate where the task $\mathcal{N}$ should to obtain data for π. The line in the matrix Γ corresponding to π is filled with nonzero elements for those blocks to which π has to transfer the data.

4 Information Security

Information security is governed by security policies [12]. Confidentiality and integrity are aimed by traditional discretionary access control (DAC) security policies and their modifications, such as Role Based Access Control (RBAC). Access control in the DAC is determined by the Access Control List (ACL) matrix of the permissions granted to the users and subjects on their behalf to access objects. The ACL must consider the value of the input data that is used or appears in IT during its execution.

In addition to the mentioned security policies, confidentiality protection is also achieved through the use of Multi-Level Security (MLS) policy, which prevents information flows from objects with valuable information to subjects and objects that are not allowed to access valuable information.

Obviously, information flow control supports the aforementioned classes of security policies. However, while providing information security in IT, the static picture of information flows as defined by DAG, does not take into account the possibilities of using valuable and not valuable source data simultaneously or generation of valuable information during IT execution. The valuable information may become exposed due to erroneous actions of users participating in IT. Thus, in many stages of IT, it is necessary to consider the possibility of access rules alteration, i.e. analyze input and output data and make decisions on access rights.

When information flow control policy is used, an access denial means that IT is stopped, which is equivalent to the failure of that IT. Recall that $\mathcal{M}$ and $\mathcal{R}$ tasks provide backup of IT blocks and backup network routes. Stopping IT execution due to failure or security policy violation engages these backup capabilities. For this purpose, the task $\mathcal{N}$ must obtain the information on a possible failure or an appearance of valuable information, which is determined by the information classification. This classification is performed either at the input of the IT or in the executed block α_i according to the specified criterion.

The criterion of the value of information is based on the characteristics of data value, which in turn are specified by the classification table. For example, the database field ID indicates that the identified data is not permitted to use in the executed IT. However, the user who has temporarily accessed executable IT in scenario 2 can make a mistake and input the valuable data into that IT. Since the detection of such cases results in a delay in the execution of IT, the access to the valuable information should occur only after the user identifies his rights to interfere with the execution of IT, which does not allow the usage of the valuable information.

In this case all non-zero elements in all columns of the i-th row of the matrix Γ are replaced with the symbol "v", which allows to engage the $\mathcal{N}$ task in order to create and use the specially protected IT continuation. Specifically, this requires network routes to be rebuilt in order to securely continue the IT and re-route non-valuable data. In fact, the system must be reconfigured and an additional secure IT perimeter must be created. That means that DAG and metadata must be changed, and the system itself becomes at least two-level (MLS).

Two tools are used to protect integrity:

- control of integrity;
- making backups of data and implementation paths of IT in the system.

The permissions to alter data are defined by the DAC. Therefore, an attempt to violate integrity also results in a failure. This case has been already discussed.

The control of integrity during data transmission over the network is implemented by using MAC (Message Authentication Code), but this method can not be used to control the integrity of transformation results in the blocks. It is therefore necessary to use backup paths of tasks execution in critical IT fragments, as suggested above for processing valuable information.

5 Conclusion

In this paper the hierarchical decomposition models of IT in the form of DAG has been constructed. It is devoted to the problem of control of interactions of hosts on a corporate network. Hierarchical decomposition makes it possible to optimally form blocks of information transformations for their distribution across hosts of distributed information and computing system.

Transformation of DAG to metadata consists of two tasks:

- ordering block execution in a way that does not contradict the DAG;
- distribution of data produced by executed blocks for use by other blocks of IT.

In order to distribute data, it was necessary to add data in the form of the matrix Γ to the task $\mathcal{N}$ field. The matrix Γ controls the queueing of already executed blocks to transmit its results to the next blocks.

The task of information access control in IT implementation according to information security requirements was addressed. Although access control must be refined during the business process modeling phase, it may come into conflict with security policy requirements. In order to overcome failures caused by such conflicts, it is proposed to use backup of IT fragments, which are developed on the basis of metadata.

The main concern of specialists opposing metadata usage is the client-server architecture which could cause time delays. An attempt has been made to exclude such delays by constructing an optimised protocol. There are several ways to achieve that. For example, several hosts can be implemented with the tasks $\mathcal{N}$ for different IT (with prioritization among them).

In future we plan to investigate the complexity of this task and support our research with experimental data. However, it should be noted that before the development of metadata concept we evaluated the speed characteristics of this class of architectures. Measurements were made in virtual networks. We got satisfactory results which were published in [13].

References

1. Grusho, A.A., Timonina, E.E., Shorgin, S.Ya.: Modelling for ensuring information security of the distributed information systems. In: 31th European Conference on Modelling and Simulation Proceedings. Digitaldruck Pirrot GmbHP Dudweiler, Germany, pp. 656–660 (2017)
2. Grusho, A., Grusho, N., Zabezhailo, M., Zatsarinny, A., Timonina, E.: Information security of SDN on the basis of meta data. In: Rak, J., Bay, J., Kotenko, I., Popyack, L., Skormin, V., Szczypiorski, K. (eds.) MMM-ACNS 2017. LNCS, vol. 10446, pp. 339–347. Springer, Cham (2017). https://doi.org/10.1007/978-3-319-65127-9_27

3. Samuylov, K.E., Chukarin, A.V., Yarkina, N.V.: Business Processes and Information Technologies in Management of the Telecommunication Companies. Alpina Publishers, Moscow (2009)
4. Tanayev, V.S., Shkurba, V.V.: Introduction to the Scheduling Theory. Science, Moscow (1975). (in Russian)
5. Fei, Y., Du, X., Jiang, C., Deng, R.: Directed acyclic task graph scheduling for heterogeneous computing systems by dynamic critical path duplication algorithm. J. Algorithms Comput. Technol. **3**(2), 247–270 (2009)
6. Mao, Y., Zhong, H., Wang, L., Li, X.: Delay-bounded associated tasks scheduling based on hierarchical graph model in the cloud. Int. J. Hybrid Inf. Technol. **9**, 367–386 (2016)
7. Behravan, A., Obermaisser, R., Basavegowda, D.H., Meckel, S.: Automatic model-based fault detection and diagnosis using diagnostic directed acyclic graph for a demand-controlled ventilation and heating system in Simulink. In: 2018 Annual IEEE International Systems Conference (SysCon), pp. 1–7 (2018)
8. Williams, T.C., Bach, C.C., Matthiesen, N.B., Henriksen, T.B., Gagliardi, L.: Directed acyclic graphs: a tool for causal studies in pediatrics. Pediatric Res. **84**, 487–493 (2018)
9. Balmas, F.: Displaying dependence graphs: a hierarchical approach. In: Proceedings Eighth Working Conference on Reverse Engineering, pp. 261–270 (2001)
10. Grusho, A., Grusho, N., Timonina, E.: Information flow control on the basis of meta data. In: Vishnevskiy, V.M., Samouylov, K.E., Kozyrev, D.V. (eds.) DCCN 2019. LNCS, vol. 11965, pp. 548–562. Springer, Cham (2019). https://doi.org/10.1007/978-3-030-36614-8_42
11. Grusho, A.A., Timonina, E.E., Shorgin, S.Ya.: Hierarchical method of meta data generation for control of network connections. J. Inform. Primen. **12**(2), 44–49 (2018). (Russian)
12. TCSEC. Department of Defense Trusted Computer System Evaluation Criteria. DoD (1985)
13. Grusho, N.A., Senchilo, V.V.: Modeling of secure architecture of distributed information systems on the basis of integrated virtualization. J. Syst. Means Inf. **28**(1), 110–122 (2018). (Russian)

Author Index

Printed in the United States
By Bookmasters

Printed in the United States
By Bookmasters